Grassmann Algebra

Volume 1: Foundations

Grassmann algebra extends vector algebra by introducing the exterior product to algebraicize the notion of linear dependence. With it, vectors may be extended to higher-grade entities: bivectors, trivectors, ... multivectors. The extensive exterior product also has a regressive dual: the regressive product. The pair behaves a little like the Boolean duals of union and intersection. By interpreting one of the elements of the vector space as an origin point, points can be defined, and the exterior product can extend points into higher-grade located entities from which lines, planes and multiplanes can be defined. Theorems of Projective Geometry are simply formulae involving these entities and the dual products. By introducing the (orthogonal) complement operation, the scalar product of vectors may be extended to the interior product of multivectors, which in this more general case may no longer result in a scalar. The notion of the magnitude of vectors is extended to the magnitude of multivectors: for example, the magnitude of the exterior product of two vectors (a bivector) is the area of the parallelogram formed by them. To develop these foundational concepts, we need only consider entities which are the sums of elements of the same grade. This is the focus of this volume.

But the entities of Grassmann algebra need not be of the same grade, and the possible product types need not be constricted to just the exterior, regressive and interior products. For example quaternion algebra is simply the Grassmann algebra of scalars and bivectors under a new product operation. Clifford, geometric and higher order hypercomplex algebras, for example the octonions, may be defined similarly. If to these we introduce Clifford's invention of a scalar which squares to zero, we can define entities (for example dual quaternions) with which we can perform elaborate transformations. Exploration of these entities, operations and algebras will be the focus of the volume to follow this.

There is something fascinating about the beauty with which the mathematical structures that Hermann Grassmann discovered describe the physical world, and something also fascinating about how these beautiful structures have been largely lost to the mainstreams of mathematics and science. He wrote his seminal *Ausdehnungslehre (Die Ausdehnungslehre. Vollständig und in strenger Form*) in 1862. But it was not until the latter part of his life that he received any significant recognition for it, most notably by Gibbs and Clifford. In recent times David Hestenes' Geometric Algebra must be given the credit for much of the emerging awareness of Grassmann's innovation.

In the hope that the book be accessible to scientists and engineers, students and professionals alike, the text attempts to avoid any terminology which does not make an essential contribution to an understanding of the basic concepts. Some familiarity with basic linear algebra may however be useful.

I hope you enjoy exploring this beautiful mathematical system.

Grassmann Algebra

Volume 1: Foundations

Exploring extended vector algebra with *Mathematica*

John Browne

Barnard Publishing

To discuss anything in this book or Grassmann algebra contact the author
John Browne
grassmannalgebra@gmail.com

For more information about the book or the *GrassmannAlgebra* software visit
http://sites.google.com/site/grassmannalgebra

Keywords: Grassmann, Grassmann algebra, *GrassmannAlgebra*, *Ausdehnungslehre*, exterior algebra, exterior product, regressive product, interior product, inner product, scalar product, dot product, cross product, complement operation, duality, projective geometry, geometric algebra, vector algebra, linear space, vector space, vector, point, bivector, bound vector, weighted point, grade, determinant, multilinear, common factor, multiplanes, orthogonality, metric, invariance.

Published by Barnard Publishing
180 Laughing Waters Road
Eltham VIC 3095 Australia

Printed by the print-on-demand process
and distributed world-wide by local distributors

ISBN: 978-1479197637

First printing October 2012
First edition

Contents

5 The Complement 289

6 The Interior Product 372

Preface

This book is the first of two volumes on the work of Hermann Grassmann. There is something fascinating about the beauty with which the mathematical structures Grassmann discovered (invented, if you will) describe the physical world, and something also fascinating about how these beautiful structures have been largely lost to the mainstreams of mathematics and science. *Volume 1: Foundations* develops the algebraic foundations of the structures, while *Volume 2: Applications* will explore some of their applications. The books have grown out of an interest in Grassmann's work over the past four decades.

Hermann Günther Grassmann was born in 1809 in Stettin, near the border of Germany and Poland. He was only 23 when he discovered the method of adding and multiplying points and vectors which was to become the foundation of his *Ausdehnungslehre*. In 1839 he composed a work on the study of tides entitled *Theorie der Ebbe und Flut*, which was the first work ever to use vectorial methods. In 1844 Grassmann published his first *Ausdehnungslehre* (*Die lineale Ausdehnungslehre ein neuer Zweig der Mathematik*) and in the same year won a prize for an essay which expounded a system satisfying an earlier search by Leibniz for an 'algebra of geometry'. Despite these achievements, Grassmann received virtually no recognition.

In 1862 Grassmann re-expounded his ideas from a different viewpoint in a second *Ausdehnungslehre (Die Ausdehnungslehre. Vollständig und in strenger Form)*. Again the work was met with resounding silence from the mathematical community, and it was not until the latter part of his life that he received any significant recognition from his contemporaries. Of these, most significant were J. Willard Gibbs who discovered his works in 1877 (the year of Grassmann's death), and William Kingdon Clifford who discovered them in depth about the same time. Both became quite enthusiastic about this new mathematics.

More details on the biography of Grassmann may be found at the end of the book, but for the most comprehensive biography see Petsche [2009].

The term '*Ausdehnungslehre*' is variously translated as 'extension theory', 'theory of extension', or 'calculus of extension'. In this book we will use these terms to refer to Grassmann's original work and to other early work in the same notational and conceptual tradition (particularly that of Edward Wyllys Hyde, Henry James Forder and Alfred North Whitehead).

The term 'exterior calculus' will be reserved for the calculus of *exterior differential forms*, originally developed by Elie Cartan from the *Ausdehnungslehre*. This is an area in which there are many excellent texts, and which is outside the scope of these books.

The term 'Grassmann algebra' will be used to describe that body of algebraic theory and results based on the *Ausdehnungslehre*, but extended to include more recent results and viewpoints. This will be the basic focus of this book.

Finally, the term '*GrassmannAlgebra*' will be used to refer to the *Mathematica* based software package which accompanies the books.

The intrinsic power of Grassmann algebra arises from its fundamental product operation, the exterior product. The exterior product codifies the property of linear dependence, so essential for modern applied mathematics, directly into the algebra. Simple non-zero elements of the algebra may be viewed as representing constructs of linearly independent elements. For example, a simple bivector is the exterior product of two vectors; a line is represented by the exterior product of two points; a plane is represented by the exterior product of three points.

The focus of these books is to provide a readable account in modern notation of Grassmann's major algebraic contributions to mathematics and science. I would like them to be accessible to scientists and engineers, students and professionals alike. Consequently I have tried to avoid all mathematical terminology which does not make an essential contribution to understanding the basic concepts. The only assumptions I have made as to the reader's background is that they have some familiarity with basic linear algebra.

The focus is also to provide an environment for exploring applications of Grassmann algebra. For general applications in higher dimensional spaces, computations by hand in any algebra become tedious, indeed limiting, thus restricting the hypotheses that can be explored. For this reason the book is integrated with a package for exploring Grassmann algebra, called *GrassmannAlgebra*. *GrassmannAlgebra* has been developed in *Mathematica*. You can read the book without using the package, or you can use the package to extend the examples in the text, experiment with hypotheses, or explore your own interests.

Mathematica is a powerful system for doing mathematics on a computer. It has an inbuilt programming language ideal for extending its capabilities to other mathematical systems like Grassmann algebra. It also has a sophisticated mathematical typesetting capability. This book, cover and text, was written entirely in *Mathematica*. The chapters are written in its standard notebook format, making the book interactive in its electronic version with the *GrassmannAlgebra* package. For the printed text, *Mathematica* also managed the style sheet, the graphics, specially constructed symbols, automatic section and equation numbering and referencing, generation of the table of contents, spelling checking, and analysis of the text for all pertinent index terms. Pdf generation, including font embedding and file stitching was done with Apple's Preview application.

Computational sections within the text use *GrassmannAlgebra* or *Mathematica*. Wherever possible, explanation is provided so that the results are still intelligible without a knowledge of *Mathematica*. *Mathematica* input/output dialogue is in indented Courier font, with the input bold and the output plain. For example:

```
GrassmannSimplify[1 + x + x ∧ x]

1 + x
```

Volume 1: Foundations comprises six chapters. *Chapter 1* provides a brief preparatory overview, introducing the seminal concepts of each chapter, and solidifying them with simple examples. This chapter is designed to give you a global appreciation with which better to understand the detail of the chapters which follow. However, it is independent of those chapters, and may be read as far as your interest takes you. *Chapter 2* discusses the exterior product - the fundamental product operation of the algebra - and shows how it creates the suite of linear spaces which form the algebra. *Chapter 3* discovers that the symmetry in this suite leads to another product, the regressive product, elegantly 'dual' to the exterior product. Equipped with these dual products, *Chapter 4* shows how the algebra can be interpreted geometrically to easily recreate projective geometry. *Chapter 5* lays the foundation for introducing a metric by defining for each element, a partner - its complement. Then *Chapter 6* shows how to combine all of these to define the interior product, a much more general product than the scalar product. These chapters form the essential core for a working knowledge of Grassmann algebra and for the explorations in the second volume. They are most profitably read (or at least scanned) sequentially. The rest is exploration!

Volume 2: Applications is currently a work in progress. When it is published you will find explorations using the fundamental theory developed in this volume of applied topics, for example screw algebra and mechanics, and of other branches of mathematics, for example the hypercomplex and Clifford algebras.

The *GrassmannAlgebra* website may be found at
http://sites.google.com/site/grassmannalgebra

From this site you will be able to: email me, download the *GrassmannAlgebra* application and earlier draft copies of the books, learn of any updates to them, check for known errata or bugs, let me know of any errata or bugs you find, get hints and FAQs for using the package, and link to where you can find the books in other formats. The *GrassmannAlgebra* application can turn an exploration requiring days by hand into one requiring just minutes of computing time.

Acknowledgements are many. First I would like to acknowledge Cecil Pengilley who originally supported me in applying Grassmann's work to engineering; the University of Melbourne, Xerox Corporation, the University of Rochester and Swinburne University of Technology which sheltered me while I pursued this interest; Janet Blagg who peerlessly edited the early draft text; my family who had to put up with my writing; Stephen Wolfram who created *Mathematica* and provided me with a visiting scholar's grant to work at Wolfram Research Institute where I began developing the *GrassmannAlgebra* package; Devendra Kapadia, André Kuzniarek, Lou D'Andria, Chris Carlson, Chris Hill, Rob Raguet-Schofield and John Fultz of Wolfram Research who tirelessly fielded my many *Mathematica* questions; Hans-Joachim Petsche for making the Graßmann Bicentennial Conference so fruitful and memorable; and Eckhard Hitzer, Alvin Swimmer, Jacques Riche, Allan Cortzen, Diane Demers, Gary Harper, and Gary Miller for their highly valued conversations and correspondences.

Writing this sort of book also involves a lot of *Mathematica* coding. In this I would like to especially acknowledge my great appreciation of David Park's interest in Grassmann algebra and the thought-provoking conversations which have ensued. And particularly, I would like to acknowledge his many contributions to the *GrassmannAlgebra* software. As well as contributing some seminal capabilities to the software, he is the creator of its documentation system - not a trivial task, given its several hundreds of functions [Park 2010].

Above all however, I must acknowledge Hermann Grassmann. His contribution to mathematics and science puts him among the great thinkers of the nineteenth century.

I hope you enjoy exploring this beautiful mathematical system.

John Browne
October, 2012

"For I have every confidence that the effort I have applied to the science reported upon here, which has occupied a considerable span of my lifetime and demanded the most intense exertions of my powers, is not to be lost. … a time will come when it will be drawn forth from the dust of oblivion and the ideas laid down here will bear fruit. … some day these ideas, even if in an altered form, will reappear and with the passage of time will participate in a lively intellectual exchange. For truth is eternal, it is divine; and no phase in the development of truth, however small the domain it embraces, can pass away without a trace. It remains even if the garments in which feeble men clothe it fall into dust."

Hermann Grassmann
in the foreword to the Ausdehnungslehre of 1862
translated by Lloyd Kannenberg

1 Introduction

1.1 Background

The mathematical representation of physical entities

Three of the more important mathematical systems for representing the entities of contemporary engineering and physical science are the (three-dimensional) vector algebra, the more general tensor algebra, and geometric algebra. Grassmann algebra is more general than vector algebra, overlaps aspects of the tensor algebra, and underpins geometric algebra. It predates all three. In this book we will show that it is only via Grassmann algebra that many of the geometric and physical entities commonly used in the engineering and physical sciences may be represented mathematically in a way which correctly models their pertinent properties and leads straightforwardly to principal results.

As a case in point we may take the concept of *force*. It is well known that a force is not satisfactorily represented by a (free) vector, yet contemporary practice is still to use a (free) vector calculus for this task. The deficiency may be made up for by verbal appendages to the mathematical statements: for example 'where the force `f` acts along the line through the point `P`'. Such verbal appendages, being necessary, and yet not part of the calculus being used, indicate that the calculus itself is not adequate to model force satisfactorily. In practice this inadequacy is coped with in terms of a (free) vector calculus by the introduction of the concept of *moment*. The conditions of equilibrium of a rigid body include a condition on the sum of the moments of the forces about any point. The justification for this condition is not well treated in contemporary texts. It will be shown later however that by representing a force correctly in terms of an element of the Grassmann algebra, both force-vector and moment conditions for the equilibrium of a rigid body may be united in one condition, a natural consequence of the algebraic processes alone.

Since the application of Grassmann algebra to mechanics was known during the nineteenth century one might wonder why, with the 'progress of science', it is not currently used. Indeed the same question might be asked with respect to its application in many other fields. To attempt to answer these questions, a brief biography of Grassmann is included as an appendix. In brief, the scientific world was probably not ready in the nineteenth century for the new ideas that Grassmann proposed, and now, in the twenty-first century, seems only just becoming aware of their potential.

The central concept of the *Ausdehnungslehre*

Grassmann's principal contribution to the physical sciences was his discovery of a natural language of geometry from which he derived a geometric calculus of significant power. For a mathematical representation of a physical phenomenon to be 'correct' it must be of a tensorial nature and since many 'physical' tensors have direct geometric counterparts, a calculus applicable to geometry may be expected to find application in the physical sciences.

The word '*Ausdehnungslehre*' is most commonly translated as '*theory of extension*', the fundamental product operation of the theory then becoming known as the *exterior product*. The notion of *extension* has its roots in the interpretation of the algebra in geometric terms: an element of the algebra may be 'extended' to form a higher order element by its (exterior) product with another, in the way that a point may be extended to a line, or a line to a plane, by a point exterior to it. The notion of *exteriorness* is equivalent algebraically to that of linear independence. If the exterior product of elements of grade 1 (for example, points or vectors) is non-zero, then they are independent.

A line may be defined by the exterior product of *any* two distinct points on it. Similarly, a plane may be defined by the exterior product of *any* three distinct points in it, and so on for higher dimensions. This independence with respect to the specific points chosen is an important and fundamental property of the exterior product. Each time a higher dimensional object is required it is simply created out of a lower dimensional one by multiplying by a new element in a new dimension. Intersections of elements are also obtainable as products.

Simple elements of the Grassmann algebra may be interpreted as defining subspaces of a linear space. The exterior product then becomes the operation for building higher dimensional subspaces (higher order elements) from a set of lower dimensional independent subspaces. A second product operation called the *regressive product* may then be defined for determining the common lower dimensional subspaces of a set of higher dimensional non-independent subspaces.

Comparison with the vector and tensor algebras

The Grassmann algebra is a tensorial algebra, that is, it concerns itself with the types of mathematical entities and operations necessary to describe physical quantities in an invariant manner. In fact, it has much in common with the algebra of anti-symmetric tensors – the exterior product being equivalent to the anti-symmetric tensor product. Nevertheless, there are conceptual and notational differences which make the Grassmann algebra richer and easier to use.

Rather than a sub-algebra of the tensor algebra, it is perhaps more meaningful to view the Grassmann algebra as a super-algebra of the three-dimensional vector algebra since both commonly use invariant (coordinate-free) notations. The principal differences are that the Grassmann algebra has a dual axiomatic structure, can treat higher order elements than vectors, can differentiate between points and vectors, generalizes the notion of 'cross product', is independent of dimension, and possesses the structure of a true algebra.

Algebraicizing the notion of linear dependence

Another way of viewing Grassmann algebra is as linear or vector algebra onto which has been introduced a product operation which algebraicizes the notion of linear dependence. This product operation is called the *exterior product* and is symbolized with a wedge $\wedge$.

If vectors $x_1, x_2, x_3, \ldots$ are linearly dependent, then it turns out that their exterior product is zero: $x_1 \wedge x_2 \wedge x_3 \wedge \ldots = 0$. If they are independent, their exterior product is non-zero.

Conversely, if the exterior product of vectors $x_1, x_2, x_3, \ldots$ is zero, then the vectors are linearly dependent. Thus the exterior product brings the critical notion of linear dependence into the realm of direct algebraic manipulation.

Although this might appear to be a relatively minor addition to linear algebra, we expect to demonstrate in this book that nothing could be further from the truth: the consequences of being able to model linear dependence with a product operation are far reaching, both in facilitating an understanding of current results, and in the generation of new results for many of the algebras and their entities used in science and engineering today. These include of course linear and multilinear algebra, but also vector and tensor algebra, screw algebra, hypercomplex algebras, and Clifford algebras.

Grassmann algebra as a geometric calculus

Most importantly however, Grassmann's contribution has enabled the operations and entities of all of these algebras *to be interpretable geometrically*, thus enabling us to bring to bear the power of geometric visualization and intuition into our algebraic manipulations.

It is well known that a vector x_1 may be interpreted geometrically as representing a *direction* in space. If the space has a metric, then the *magnitude* of x_1 is interpreted as its *length*. The introduction of the exterior product enables us to *extend* the entities of the space to higher dimensions. The exterior product of two vectors $x_1 \wedge x_2$, called a *bivector*, may be visualized as the two-dimensional analogue of a direction, that is, a *planar direction*. Neither vectors nor bivectors are interpreted as being located anywhere since they do not possess sufficient information to specify independently both a direction *and* a position. If the space has a metric, then the *magnitude* of $x_1 \wedge x_2$ is interpreted as its *area*, and similarly for higher order products.

We *depict* a simple bivector by its vector factors arranged head-to-tail linked by the ghost of a parallelogram.

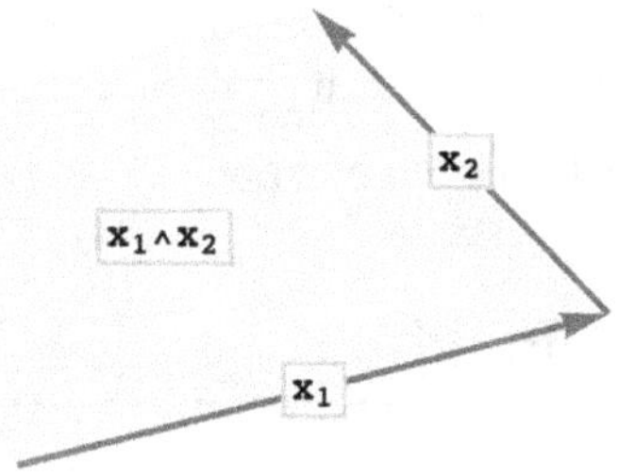

Depicting a simple bivector. This bivector is located nowhere.

For applications to the physical world, however, the Grassmann algebra possesses a critical capability that no other algebra possesses so directly: it can distinguish between *points* and *vectors* and treat them as separate entities. Lines and planes are examples of higher order constructs from points and vectors, which have both position and direction. A *line* can be represented by the exterior product of any two points on it, or by any point on it and a vector parallel to it.

Point∧vector depiction
Point∧point depiction

Two different depictions of a bound vector in its line.

A *plane* can be represented by the exterior product of any point on it and a bivector parallel to it, any two points on it and a vector parallel to it, or any three points on it.

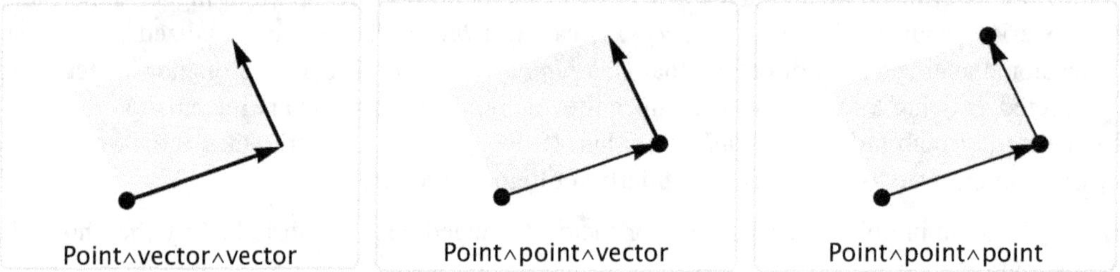
Point∧vector∧vector
Point∧point∧vector
Point∧point∧point

Three different depictions of a bound bivector.

Finally, it should be noted that the Grassmann algebra subsumes all of real algebra, the exterior product reducing in this case to the usual product operation among real numbers.

Here then is a geometric calculus *par excellence*.

1.2 The Exterior Product

The anti-symmetry of the exterior product

The exterior product of two vectors x and y of a linear space yields the bivector $x \wedge y$. The bivector is *not* a vector, and so does not belong to the original linear space. In fact the bivectors form their own linear space.

The fundamental defining characteristic of the exterior product is its *anti-symmetry*. That is, the product changes sign if the order of the factors is reversed.

$$x \wedge y == - y \wedge x \qquad 1.1$$

From this we can easily show the equivalent relation, that the exterior product of a vector with itself is zero.

$$x \wedge x == 0 \qquad 1.2$$

This is as expected because x is linearly dependent on itself.

The exterior product is associative, distributive, and behaves linearly as expected with scalars.

Exterior products of vectors in a three-dimensional space

By way of example, suppose we are working in a three-dimensional space, with basis e_1, e_2, and e_3. Then we can express vectors x and y as linear combinations of these basis vectors:

$$x == a_1 e_1 + a_2 e_2 + a_3 e_3$$

$$y == b_1 e_1 + b_2 e_2 + b_3 e_3$$

Here, the a_i and b_i are of course scalars. Taking the exterior product of x and y and multiplying out the product allows us to express the *bivector* $x \wedge y$ as a linear combination of *basis bivectors*.

$$x \wedge y == (a_1 e_1 + a_2 e_2 + a_3 e_3) \wedge (b_1 e_1 + b_2 e_2 + b_3 e_3)$$

$$x \wedge y == (a_1 b_1)\, e_1 \wedge e_1 + (a_1 b_2)\, e_1 \wedge e_2 + (a_1 b_3)\, e_1 \wedge e_3 + (a_2 b_1)\, e_2 \wedge e_1 + (a_2 b_2)\, e_2 \wedge e_2 + (a_2 b_3)\, e_2 \wedge e_3 + (a_3 b_1)\, e_3 \wedge e_1 + (a_3 b_2)\, e_3 \wedge e_2 + (a_3 b_3)\, e_3 \wedge e_3$$

The first simplification we can make is to put all basis bivectors of the form $e_i \wedge e_i$ to zero [1.2] . The second simplification is to use the anti-symmetry of the product [1.1] and collect the terms of the bivectors which are not *essentially* different (that is, those that may differ only in the order of their factors, and hence differ only by a sign). The product $x \wedge y$ can then be written:

$$x \wedge y == (a_1 b_2 - a_2 b_1)\, e_1 \wedge e_2 + (a_2 b_3 - a_3 b_2)\, e_2 \wedge e_3 + (a_3 b_1 - a_1 b_3)\, e_3 \wedge e_1$$

The scalar factors appearing here are just those which would have appeared in the usual vector *cross product* of x and y. However, there is an important difference. The exterior product expression does not require the vector space to have a metric, while the usual definition of the cross product, because it generates a vector *orthogonal* to x and y, necessarily assumes a metric. Furthermore, the exterior product is associative and valid for any number of vectors in spaces of arbitrary dimension, while the cross product is not associative and is necessarily confined to products of vectors in a space of three dimensions.

For example, we may continue the product by multiplying $x \wedge y$ by a third vector z.

$$z == c_1 e_1 + c_2 e_2 + c_3 e_3$$

$$x \wedge y \wedge z == ((a_1 b_2 - a_2 b_1)\, e_1 \wedge e_2 + (a_2 b_3 - a_3 b_2)\, e_2 \wedge e_3 + (a_3 b_1 - a_1 b_3)\, e_3 \wedge e_1) \wedge (c_1 e_1 + c_2 e_2 + c_3 e_3)$$

Adopting the same simplification procedures as before we obtain the *trivector* $x \wedge y \wedge z$ expressed in basis form.

$$x \wedge y \wedge z == (a_1 b_2 c_3 - a_3 b_2 c_1 + a_2 b_3 c_1 + a_3 b_1 c_2 - a_1 b_3 c_2 - a_2 b_1 c_3)\, e_1 \wedge e_2 \wedge e_3$$

We *depict* a simple trivector by its vector factors arranged head-to-tail linked by the ghost of a parallelepiped.

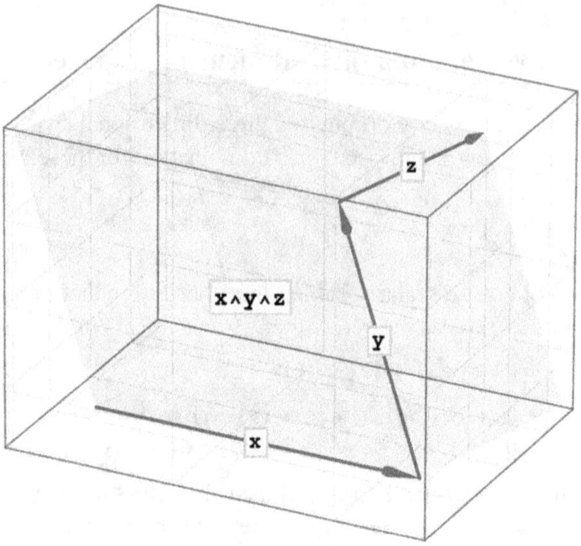

A trivector is the exterior product of a three vectors

A trivector in a space of three dimensions has just one component. Its coefficient is the *determinant* of the coefficients of the original three vectors. Clearly, if these three vectors had been linearly dependent, this determinant would have been zero. In a metric space, this coefficient would be proportional to the *volume* of the parallelepiped formed by the vectors x, y, and z. Hence the geometric interpretation of the algebraic result: if x, y, and z are lying in a planar direction, that is, they are dependent, then the volume of the parallelepiped defined is zero.

We see here also that the exterior product begins to give geometric meaning to the often inscrutable operations of the algebra of determinants. In fact we shall see that *all* the operations of determinants are straightforward consequences of the properties of the exterior product.

In three-dimensional *metric* vector algebra, the vanishing of the scalar triple product of three vectors is often used as a criterion of their linear dependence, whereas in fact the vanishing of their exterior product (valid also in a *non-metric* space) would suffice. It is interesting to note that the notation for the scalar triple product, or 'box' product, is Grassmann's original notation for the exterior product, *viz* [x y z].

Finally, we can see that the exterior product of more than three vectors in a three-dimensional space will always be zero, since they must be dependent.

Terminology: elements and entities

To this point we have been referring to the elements of the space under discussion as *vectors*, and their higher order constructs in three dimensions as *bivectors* and *trivectors*. In the general case we will refer to the exterior product of an unspecified number of vectors as a *multivector*, and the exterior product of *m* vectors as an *m-vector*.

The word 'vector' however, is in current practice used in two distinct ways. The first and traditional use endows the vector with its well-known geometric properties of direction, sense, and (possibly) magnitude. In the second and more recent, the term vector may refer to any element of a linear space, even though the space is devoid of geometric context.

In this book, we adopt the traditional practice and use the term vector only when we intend it to have its traditional geometric interpretation of a (free) arrow-like entity. When referring to an element of a linear space which we are not specifically interpreting geometrically, we simply use the term *element*. The exterior product of m 1-elements of a linear space will thus be referred to as an *m-element*. (In the *GrassmannAlgebra* package however, we have had to depart somewhat from this convention in the interests of common usage: symbols representing 1-elements are called VectorSymbols).

The reason this distinction is important is because it allows us to introduce points into the algebra. By adding a new element called the *Origin* into the basis of a vector space which has the interpretation of a *point*, we are able to distinguish two types of 1-element: vectors and points. This simple distinction leads to a sophisticated and powerful tableau of free (involving only vectors) and bound (involving points) entities with which to model geometric and physical systems.

Science and engineering make use of mathematics by endowing its constructs with geometric or physical interpretations. We will use the term *entity* to refer to such a construct of elements which we specifically wish to endow with a geometric or physical interpretation. For example we would say that (geometric) *points* and *vectors* and (physical) *positions* and *directions* are 1-entities, while (geometric) *bound vectors*, *bivectors* and *screws*, and (physical) *forces* and *angular momenta* are 2-entities. Points, vectors, bound vectors and bivectors and screws are examples of *geometric* entities. Positions, directions, forces and momenta are examples of *physical* entities.

Points, lines, planes and multiplanes may also be conveniently considered for computational purposes *as* geometric entities, or we may also define them in the more common way as a set of points: the set of all points *in* the entity. (A 1-element is in an m-element if their exterior product is zero.) We call such a set of points a *geometric object*. We *interpret* elements of a linear space geometrically or physically, while we *represent* geometric or physical entities by elements of a linear space.

An entity need not have a unique grade. For example, we will see in Volume 2 that hypercomplex entities like complex numbers, quaternions, and Clifford numbers are usually multigraded, a typical case being the sum of a scalar (of grade 0) and a bivector (of grade 2), with a typical interpretation being that of a *rotation*.

A more complete summary of terminology is given at the end of the book.

The grade of an element

The exterior product of m 1-elements is called an m-element. The value m is called the *grade* of the m-element. For example the element $u \wedge v \wedge x \wedge y$ is of grade 4.

An m-element may be *denoted* by a symbol underscripted with the value m. For example:

$$\underset{4}{\alpha} == u \wedge v \wedge x \wedge y$$

For simplicity, however, we do not generally denote 1-elements with an underscripted '1'.

The grade of a scalar is 0. We shall see that this is a natural consequence of the exterior product axioms formulated for elements of general grade.

The *dimension* of the underlying linear space of 1-elements is denoted by n. Elements of grade greater than n are zero.

The *complementary grade* of an m-element in an n-space is $n-m$.

GrassmannAlgebra recognizes the symbol `Dimension` as the numerical dimension of the current underlying linear space; and the symbol `★n` as the symbolic dimension of any underlying linear space under theoretical consideration.

Interchanging the order of the factors in an exterior product

The exterior product is defined to be associative. Hence we can isolate any two adjacent 1-element factors. Interchanging the order of these factors will change the sign of the product:

$$\ldots \wedge x \wedge y \wedge \ldots \;=\!=\; \ldots \wedge (x \wedge y) \wedge \ldots \;=\!=\; \ldots \wedge (-(y \wedge x)) \wedge \ldots \;=\!=\; -(\ldots \wedge y \wedge x \wedge \ldots)$$

In fact, interchanging the order of *any* two 1-element factors will also change the sign of the product.

$$\ldots \wedge x \wedge \ldots \wedge y \wedge \ldots \;=\!=\; -(\ldots \wedge y \wedge \ldots \wedge x \wedge \ldots)$$

To see why this is so, suppose the number of factors between `x` and `y` is m. First move `y` to the immediate left of `x`. This will cause $m+1$ changes of sign. Then move `x` to the position that `y` vacated. This will cause m changes of sign. In all there will be $2m+1$ changes of sign, equivalent to just one sign change.

Note that it is only elements of odd grade that anti-commute. If, in a product of two elements, at least one of them is of even grade, then the elements commute. For example, 2-elements commute with all other elements.

$$(x \wedge y) \wedge z \;=\!=\; z \wedge (x \wedge y)$$

A brief summary of the properties of the exterior product

In this section we summarize a few of the more important properties of the exterior product some of which we have already introduced informally. In Chapter 2: The Exterior Product, the complete set of axioms is discussed.

- The exterior product of an m-element and a k-element is an $(m+k)$-element.

- The exterior product is associative.

$$\left(\underset{m}{\alpha} \wedge \underset{k}{\beta}\right) \wedge \underset{r}{\gamma} \;=\!=\; \underset{m}{\alpha} \wedge \left(\underset{k}{\beta} \wedge \underset{r}{\gamma}\right) \qquad 1.3$$

- The unit scalar acts as an identity under the exterior product.

$$\underset{m}{\alpha} \;=\!=\; 1 \wedge \underset{m}{\alpha} \;=\!=\; \underset{m}{\alpha} \wedge 1 \qquad 1.4$$

- Scalars factor out of products.

$$\left(a\,\underset{m}{\alpha}\right)\wedge\underset{k}{\beta} = \underset{m}{\alpha}\wedge\left(a\,\underset{k}{\beta}\right) = a\left(\underset{m}{\alpha}\wedge\underset{k}{\beta}\right) \qquad 1.5$$

- An exterior product is anti-commutative whenever the grades of the factors are both odd.

$$\underset{m}{\alpha}\wedge\underset{k}{\beta} = (-1)^{m\,k}\,\underset{k}{\beta}\wedge\underset{m}{\alpha} \qquad 1.6$$

- The exterior product is both left and right distributive under addition.

$$\left(\underset{m}{\alpha}+\underset{m}{\beta}\right)\wedge\underset{r}{\gamma} = \underset{m}{\alpha}\wedge\underset{r}{\gamma}+\underset{m}{\beta}\wedge\underset{r}{\gamma} \qquad \underset{m}{\alpha}\wedge\left(\underset{r}{\beta}+\underset{r}{\gamma}\right) = \underset{m}{\alpha}\wedge\underset{r}{\beta}+\underset{m}{\alpha}\wedge\underset{r}{\gamma} \qquad 1.7$$

1.3 The Regressive Product

The regressive product as a dual product to the exterior product

One of Grassmann's major contributions, which appears to be all but lost to current mathematics, is the *regressive product*. The regressive product is the real foundation for the theory of the inner and scalar products (and their generalization, the interior product). Yet the regressive product is often ignored and the inner product defined as a new construct independent of the regressive product. This approach not only has potential for inconsistencies, but also fails to capitalize on the wealth of results available from the natural duality between the exterior and regressive products. The approach adopted in this book follows Grassmann's original concept. The regressive product is a simple dual operation to the exterior product and an enticing and powerful symmetry is lost by ignoring it, particularly in the development of metric results involving complements and interior products.

The underlying beauty of the *Ausdehnungslehre* is due to this symmetry, which in turn is due to the fact that linear spaces of m-elements and linear spaces of $(n-m)$-elements have the same dimension. This too is the key to the duality of the exterior and regressive products. For example, the exterior product of m 1-elements is an m-element. The dual to this is that the regressive product of m $(n-1)$-elements is an $(n-m)$-element. This duality has the same form as that in a Boolean algebra: if the exterior product corresponds to a type of 'union' then the regressive product corresponds to a type of 'intersection'.

It is this duality that permits the definition of complement in Chapter 5, and hence to the definition of the interior, inner and scalar products in Chapter 6. To underscore this duality it is proposed to adopt here the $\vee$ ('vee') for the regressive product operation. Unfortunately the now almost universal adoption of the 'wedge' for the exterior product (and hence the 'vee' for the regressive product) yields the reverse symbolic connotation to the notions of 'union' and 'intersection' in a Boolean algebra.

Unions and intersections of spaces

Consider a (non-zero) 2-element $x \wedge y$. We can test to see if any given 1-element z is in the subspace spanned by x and y by taking the exterior product of $x \wedge y$ with z and seeing if the result is zero. From this point of view, $x \wedge y$ is an element which can be used to *define* the subspace instead of the individual 1-elements x and y.

Thus we can define the *space* of $x \wedge y$ as the space of all 1-elements z such that $x \wedge y \wedge z = 0$. We extend this to more general elements by defining the space of a simple *m*-element A as the space of all 1-elements z such that $A \wedge z = 0$. (We discuss the notion of space in more detail in the section on terminology at the end of the book).

We will also need the notion of congruence. We will say that two elements (of any grade) are *congruent* if one is a scalar multiple of the other. For example x and $2\,x$ are congruent; $x \wedge y$ and $-x \wedge y$ are congruent. Congruent elements define the same subspace. We denote congruence by the symbol $\equiv$. The following concepts of union and intersection only make sense up to congruence.

A *union of elements* is an element defining the subspace they *together* span.

The dual concept to union of elements is *intersection of elements*. An *intersection of elements* is an element defining the subspace they span *in common*.

Suppose we have three independent 1-elements: x, y, and z. A union of $x \wedge y$ and $y \wedge z$ is any element congruent to $x \wedge y \wedge z$. An intersection of $x \wedge y$ and $y \wedge z$ is any element congruent to y.

The computation of unions and intersections by exterior and regressive products alone is limited to some special (but important) cases as we shall see below and in Chapter 3. The computation of unions and intersections in general is best done using the notion of *span*. See section 2.13: Unions and Intersections.

A brief summary of the properties of the regressive product

In this section we summarize a few of the more important properties of the regressive product. In Chapter 3: The Regressive Product, we develop the complete set of axioms from those of the exterior product. By comparing the axioms below with those for the exterior product in the previous section, we see that they are effectively generated by replacing $\wedge$ with $\vee$, and *m* by *n*–*m*. The unit element 1 in its form $\underset{0}{1}$ becomes $\underset{n}{1}$.

- The regressive product of an *m*-element and a *k*-element in an *n*-space is an (*m*+*k*–*n*)-element.

- The regressive product is associative.

$$\left(\underset{m}{\alpha} \vee \underset{k}{\beta}\right) \vee \underset{r}{\gamma} == \underset{m}{\alpha} \vee \left(\underset{k}{\beta} \vee \underset{r}{\gamma}\right) \qquad 1.8$$

- The unit *n*-element $\underset{n}{1}$ acts as an identity under the regressive product.

$$\underset{m}{\alpha} == \underset{n}{1} \vee \underset{m}{\alpha} == \underset{m}{\alpha} \vee \underset{n}{1} \qquad 1.9$$

- Scalars factor out of products.

$$\left(a \underset{m}{\alpha}\right) \vee \underset{k}{\beta} == \underset{m}{\alpha} \vee \left(a \underset{k}{\beta}\right) == a \left(\underset{m}{\alpha} \vee \underset{k}{\beta}\right) \qquad 1.10$$

- A regressive product is anti-commutative whenever the complementary grades of the factors are both odd.

$$\underset{m}{\alpha} \vee \underset{k}{\beta} == (-1)^{(n-m)\,(n-k)}\, \underset{k}{\beta} \vee \underset{m}{\alpha} \qquad 1.11$$

- The regressive product is both left and right distributive under addition.

$$\left(\underset{m}{\alpha} + \underset{m}{\beta}\right) \vee \underset{r}{\gamma} == \underset{m}{\alpha} \vee \underset{r}{\gamma} + \underset{m}{\beta} \vee \underset{r}{\gamma} \qquad \underset{m}{\alpha} \vee \left(\underset{r}{\beta} + \underset{r}{\gamma}\right) == \underset{m}{\alpha} \vee \underset{r}{\beta} + \underset{m}{\alpha} \vee \underset{r}{\gamma} \qquad 1.12$$

Note that when using the *GrassmannAlgebra* application, the unit *n*-element should be denoted by $\underset{\star n}{1}$ to ensure a correct interpretation of the symbol.

The Common Factor Axiom

Up to this point we have no way of connecting the dual axiom structures of the exterior and regressive products. That is, given a regressive product of an *m*-element and a *k*-element, how do we find the (*m*+*k*–*n*)-element to which it is equivalent, expressed only in terms of exterior products?

To make this connection we need to introduce a further axiom which we call the Common Factor Axiom. The form of the Common Factor Axiom may seem somewhat arbitrary, but it is in fact one of the simplest forms which enable intersections to be calculated. This can be seen in the following application of the axiom to a vector 3-space.

Suppose x, y, and z are three independent vectors in a vector 3-space. The Common Factor Axiom says that the regressive product of the two bivectors $x \wedge z$ and $y \wedge z$ may also be expressed as the regressive product of the trivector $x \wedge y \wedge z$ with their common factor z.

$$(x \wedge z) \vee (y \wedge z) == (x \wedge y \wedge z) \vee z$$

Since the space is 3-dimensional, we can write any trivector such as $x \wedge y \wedge z$ as a scalar factor (a, say) times the unit trivector (introduced in axiom 1.9).

$$(x \wedge y \wedge z) \vee z == \left(a \underset{3}{1}\right) \vee z == a\, z$$

This then gives us the axiomatic structure to say that the regressive product of two such elements possessing an element in common is congruent to that element.

$$(x \wedge z) \vee (y \wedge z) \equiv z$$

We *depict* this relation by showing the common factor (intersection) of the bivectors docked in a convenient position relative to each other. Remember, the bivectors are not actually located anywhere!

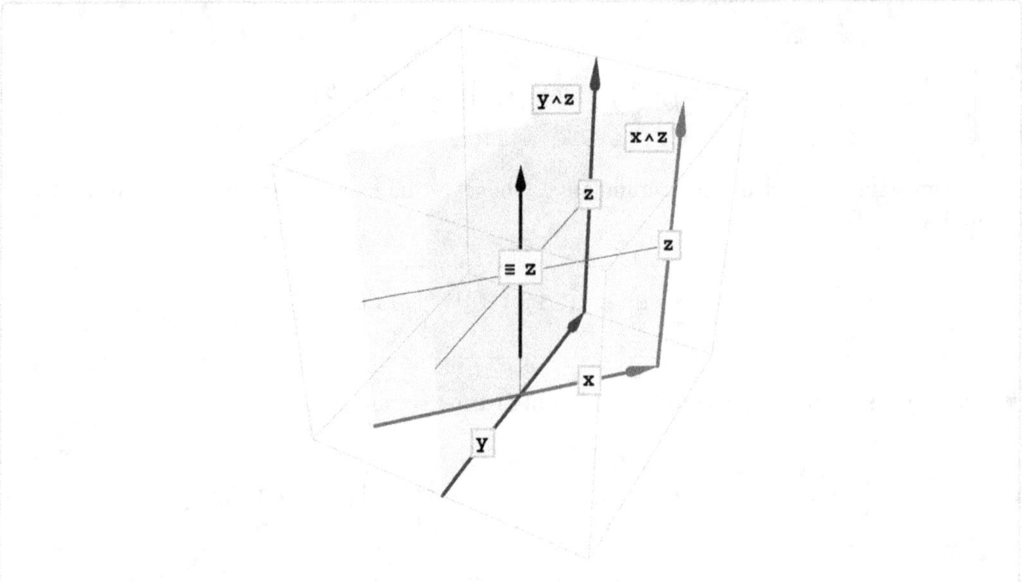

The intersection of two bivectors in a three-dimensional space

Of course this is just a simple case. More generally, let $\underset{m}{\alpha}$, $\underset{k}{\beta}$, and $\underset{s}{\mu}$ be *simple* elements with $m+k+s = n$, where n is the dimension of the space. Then the *Common Factor Axiom* states that

$$\left(\underset{m}{\alpha} \wedge \underset{s}{\mu}\right) \vee \left(\underset{k}{\beta} \wedge \underset{s}{\mu}\right) == \left(\underset{m}{\alpha} \wedge \underset{k}{\beta} \wedge \underset{s}{\mu}\right) \vee \underset{s}{\mu} \qquad m + k + s == n \qquad 1.13$$

There are many rearrangements and special cases of this formula which we will encounter in later chapters. For example, when s is zero, the Common Factor Axiom shows that the regressive product of an m-element with an $(n-m)$-element is a scalar which can be expressed in the alternative form of a regressive product with the unit 1.

$$\underset{m}{\alpha} \vee \underset{n-m}{\beta} == \left(\underset{m}{\alpha} \wedge \underset{n-m}{\beta}\right) \vee 1$$

The Common Factor Axiom allows us to prove a particularly useful result: the Common Factor Theorem. The Common Factor Theorem expresses *any* regressive product in terms of exterior products alone. This of course enables us to calculate intersections of more general elements.

Most importantly however we will see later that the Common Factor Theorem has a counterpart expressed in terms of exterior and interior products, called the Interior Common Factor Theorem. This forms the principal expansion theorem for interior products and from which we can derive many of the most important theorems relating exterior and interior products.

The Interior Common Factor Theorem, and the Common Factor Theorem upon which it is based, are possibly the most important theorems in the Grassmann algebra.

In the next section we informally apply the Common Factor Theorem to obtain the intersection of two bivectors in a three-dimensional space.

The intersection of two bivectors in a three-dimensional space

Suppose that $x \wedge y$ and $u \wedge v$ are non-congruent bivectors in a three dimensional space. Since the space has only three dimensions, the bivectors must have an intersection. We denote the regressive product of $x \wedge y$ and $u \wedge v$ by z:

$$z == (x \wedge y) \vee (u \wedge v)$$

We will see in Chapter 3: The Regressive Product that this can be expanded by the *Common Factor Theorem* to give

$$(x \wedge y) \vee (u \wedge v) == (x \wedge y \wedge v) \vee u - (x \wedge y \wedge u) \vee v \qquad 1.14$$

But we have already seen in section 1.2 that in a 3-space, the exterior product of three vectors will, in any given basis, give the basis trivector, multiplied by the determinant of the components of the vectors making up the trivector.

Additionally, we note that the regressive product (intersection) of a vector with an element like the basis trivector completely containing the vector, will just give an element congruent to itself. Thus the regressive product leads us to an explicit expression congruent to the intersection of the two bivectors.

$$z \equiv \text{Det}[x, y, v]\, u - \text{Det}[x, y, u]\, v$$

Here `Det[x,y,v]` is the determinant of the components of x, y, and v in the chosen basis. We could also have obtained an equivalent formula expressing z in terms of x and y instead of u and v by simply interchanging the order of the bivector factors in the original regressive product.

Note carefully however, that *this formula only finds the common factor up to congruence*, because until we determine an explicit expression for the unit *n*-element in terms of basis elements (which we do by introducing the complement operation in Chapter 5), we cannot usefully use axiom 1.9 above. Nevertheless, this is not to be seen as a restriction. Rather, as we shall see in the next section it leads to interesting insights as to what can be accomplished when we work in spaces without a metric, such as projective spaces.

1.4 Geometric Interpretations

Points and vectors

In this section we introduce *two* different types of *geometrically interpreted* elements which can be represented by the elements of a linear space: *vectors* and *points*. Then we look at the interpretations of the various higher grade elements that we can generate from them by the exterior product. Finally we see how the regressive product can be used to calculate intersections of these higher order elements.

As discussed in section 1.2, the term 'vector' is often used to refer to an element of a linear space with no intention of implying an interpretation. In this book however, we reserve the term for a particular type of geometric interpretation: that associated with representing *direction*. Exterior products of vectors then represent higher-dimensional analogues to the notion of direction.

But an element of a linear space may also be interpreted as a *point*. Of course vectors may also be used to represent points, but only *relative to another given point*. Hence they cannot represent absolute position. These vectors are properly called *position vectors*. Common practice often omits explicit reference to this other given point, or perhaps may refer to it verbally. Points can be represented satisfactorily in many cases by position vectors alone, but when *both position and direction* are required in the same element we must distinguish mathematically between the two.

To describe true position in a three-dimensional physical space, a linear space of *four* dimensions is required, one for an origin point, and the other three for the three spatial directions. Since the exterior product is independent of the dimension of the underlying space, it can deal satisfactorily with points and vectors together. The usual three-dimensional vector algebra however cannot.

Suppose x, y, and z are elements of a linear space interpreted as vectors. Vectors always have a direction. But only when the linear space has a metric do they also have a *magnitude*. Since to this stage we have not yet introduced the notion of metric, we will only be discussing interpretations and applications which do not require elements (other than congruent elements) to be commensurable.

Of course, vectors may be summed in a space with no metric, the standard geometric interpretation of this operation being either the 'triangle rule' or the 'parallelogram rule'.

Sums and differences of points

A *point* is defined as the sum of the *origin point* and a *vector*. If $\star\mathbb{O}$ is the origin, and x is a vector, then $\star\mathbb{O} + x$ is a point.

$$P == \star\mathbb{O} + x \qquad 1.15$$

The vector x is called the *position vector* of the point P.

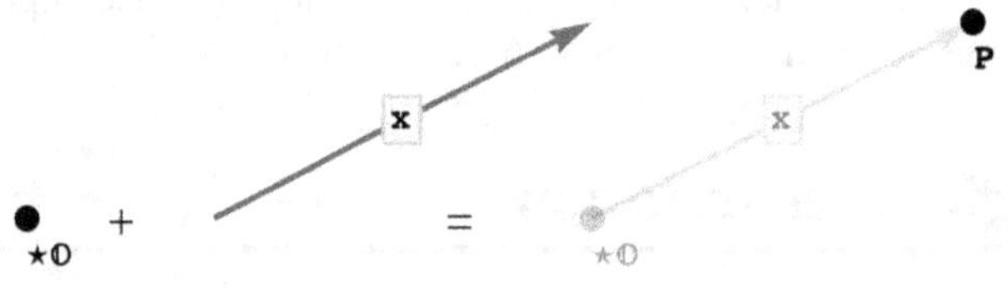

The sum of the origin and a vector is a point

The *sum of a point and a vector* is another point.

$$Q == P + y == (\star\mathbb{O} + x) + y == \star\mathbb{O} + (x + y)$$

We depict this by conveniently docking the tails of the vectors at the points.

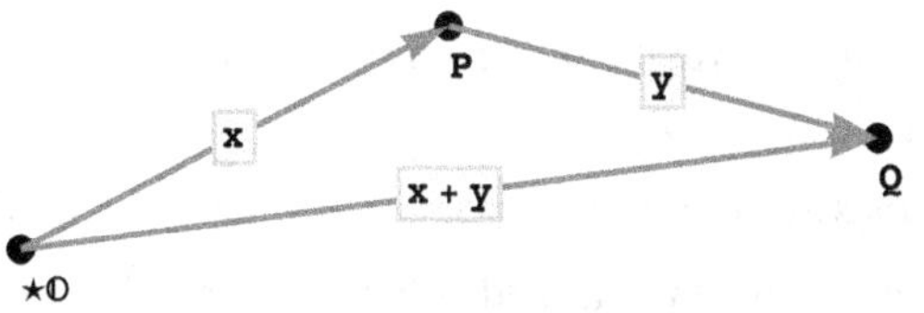

The sum of a point and a vector is another point

The *difference of two points* is a vector since the origins cancel.

$$\mathbf{P} - \mathbf{Q} == (\star\mathbb{O} + \mathbf{x}) - (\star\mathbb{O} + \mathbf{x} + \mathbf{y}) == \mathbf{y}$$

This simple result is actually the seminal underpinning of the relationship between points and vectors.

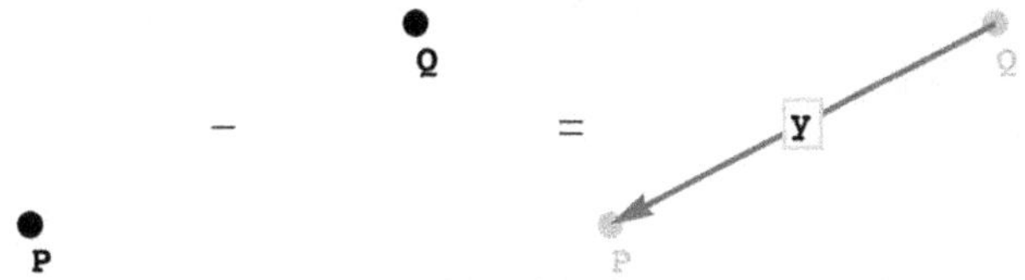

The difference of two points is a vector

A scalar multiple of a point is called a *weighted point*. For example, if m is a scalar, mP is a weighted point with weight m.

The *sum* of two points gives the point halfway between them with a weight of 2.

$$\mathbf{P}_1 + \mathbf{P}_2 == (\star\mathbb{O} + \mathbf{x}_1) + (\star\mathbb{O} + \mathbf{x}_2) == 2\left(\star\mathbb{O} + \frac{\mathbf{x}_1 + \mathbf{x}_2}{2}\right) == 2\,\mathbf{P}_c$$

Thus the sum of two points yields a result which is of quite a different nature to their difference. The sum of two points is a weighted point, while their difference is a vector - the first a located entity; the second un-located.

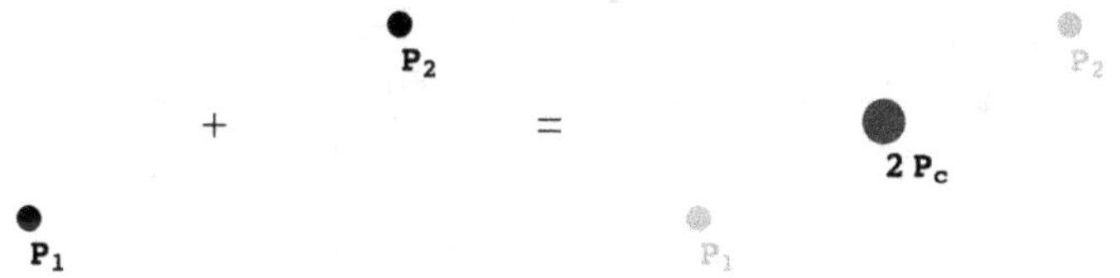

The sum of two points is a point of weight 2 located mid-way between them

◆ *Historical Note*

The point was originally considered the fundamental geometric entity of interest. However the difference of points was clearly no longer a point, since reference to the origin had been lost. Sir William Rowan Hamilton coined the term 'vector' for this new entity since adding a vector to a point 'carried' the point to a new point.

Determining a mass-centre

A classic application of a sum of weighted points is to the determination of a centre of mass.

Consider a collection of points P_i weighted with masses m_i. The sum of the weighted points gives the point P_G at the mass-centre (centre of gravity) weighted with the total mass M.

To show this, first add the weighted points and collect the terms involving the origin.

$$M\,P_G == m_1\,(\star\mathbb{O} + x_1) + m_2\,(\star\mathbb{O} + x_2) + m_3\,(\star\mathbb{O} + x_3) + \ldots$$
$$== (m_1 + m_2 + m_3 + \ldots)\,\star\mathbb{O} + (m_1\,x_1 + m_2\,x_2 + m_3\,x_3 + \ldots)$$

Dividing through by the total mass M gives the centre of mass.

$$P_G == \star\mathbb{O} + \frac{m_1\,x_1 + m_2\,x_2 + m_3\,x_3 + \ldots}{m_1 + m_2 + m_3 + \ldots}$$

- To fix ideas, we take a simple example demonstrating that centres of mass can be accumulated in any order. Suppose we have three points P, Q, and R with masses p, q, and r. The centres of mass taken two at a time are given by

$$(p + q)\,G_{PQ} == p\,P + q\,Q$$
$$(q + r)\,G_{QR} == q\,Q + r\,R$$
$$(p + r)\,G_{PR} == p\,P + r\,R$$

Now take the centre of mass of each of these with the other weighted point. Clearly, the three sums will be equal.

$$(p + q)\,G_{PQ} + r\,R == (q + r)\,G_{QR} + p\,P == (p + r)\,G_{PR} + q\,Q$$
$$== p\,P + q\,Q + r\,R == (p + q + r)\,G_{PQR}$$

It is straightforward to depict these relationships. In the diagram below we have depicted the mass of a point by its area.

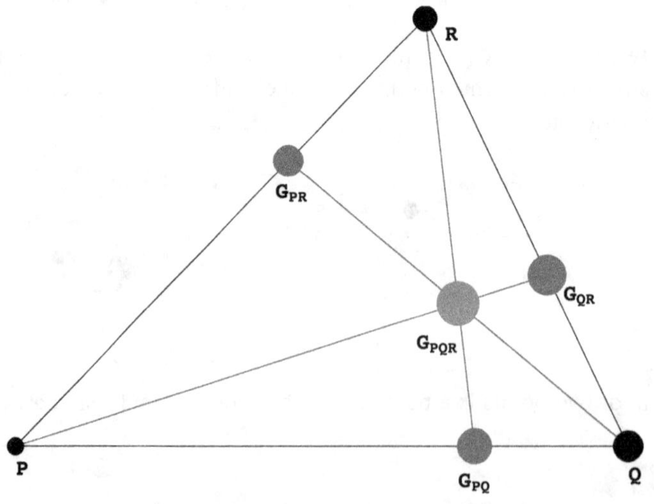

Centres of mass can be accumulated in any order

Lines and planes

The exterior product of a point and a vector gives a *bound vector*. Bound vectors are the entities we need for mathematically representing lines. A line is the set of points *in* the bound vector. That is, it consists of all the points whose exterior product with the bound vector is zero.

In practice, we usually *compute* with lines by computing with their bound vectors. For example, to get the intersection of two lines in the plane, we take the regressive product of their bound vectors. By abuse of terminology we may therefore often refer to a bound vector *as* a line, or to a line as a bound vector.

A bound vector can be defined by the exterior product of a point and a vector, or of two points. In the first case we represent the line L *through* the point P in the direction of x by any entity *congruent* to the exterior product of P and x. In the second case we can introduce Q as $P + x$ to get the same result.

Two different depictions of a bound vector in its line.

$$L \equiv P \wedge x \equiv P \wedge (Q - P) \equiv P \wedge Q$$

A line is independent of the specific point used to define it. To see this, consider any other point R on the line. Since R is on the line it can be represented by the sum of P with an arbitrary scalar multiple of the vector x:

$$L \equiv R \wedge x == (P + a\,x) \wedge x == P \wedge x$$

A line may also be represented by the exterior product of *any two points* on it.

$$L \equiv P \wedge R == P \wedge (P + a\,x) == a\,P \wedge x$$

Note that the bound vectors $P \wedge x$ and $P \wedge R$ are (in general) different, but congruent. They therefore define the same line.

$$L \equiv P \wedge x \equiv R \wedge x \equiv P \wedge R \qquad 1.16$$

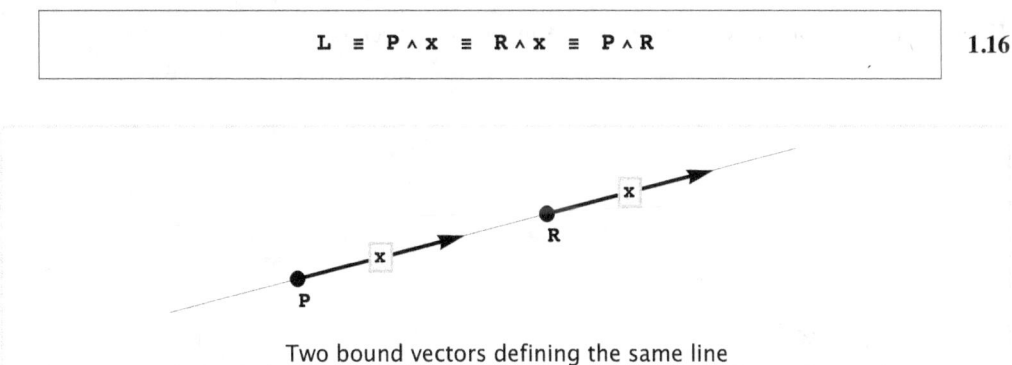

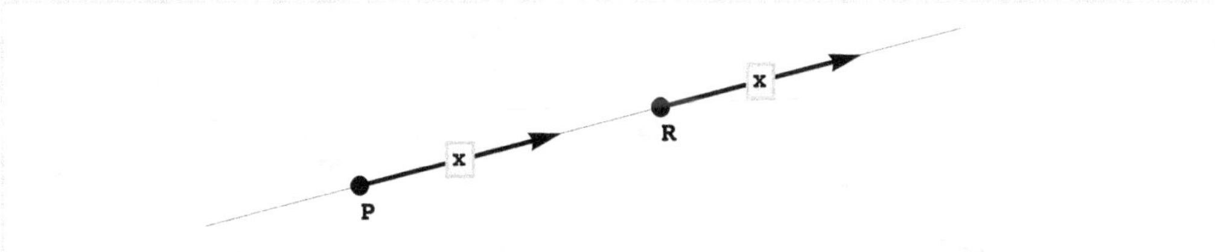

Two bound vectors defining the same line

These concepts extend naturally to higher dimensional constructs. For example a plane Π may be represented by the exterior product of single point on it together with a bivector in the direction of the plane, any two points on it together with a vector in it (not parallel to the line joining the points), or any three points on it (not in the same line).

$$\Pi \equiv P \wedge x \wedge y \equiv P \wedge Q \wedge y \equiv P \wedge Q \wedge R \tag{1.17}$$

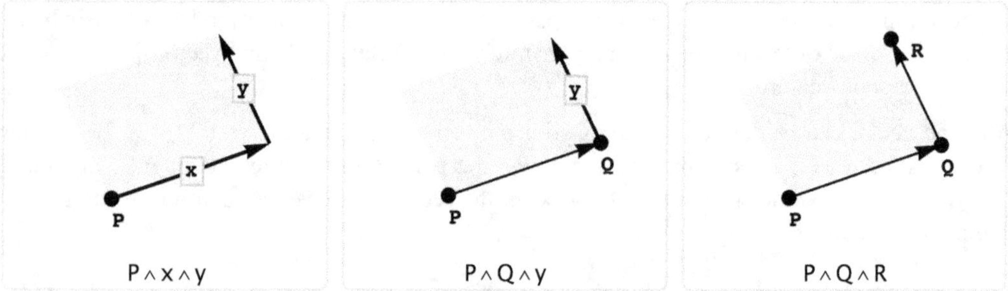

Three different depictions of a bound bivector in its plane.

To build higher dimensional geometric entities from lower dimensional ones, we simply take their exterior product. For example we can build a line by taking the exterior product of a point with any point or vector *exterior* to it. Or we can build a plane by taking the exterior product of a line with any point or vector exterior to it.

The intersection of two lines

We can use the regressive product to find the intersection of two geometric entities if together the entities span the whole space. For example, suppose we have two lines in a plane and we want to find the point of intersection P. As we have seen we can represent the lines in a number of ways. For example:

$$L_1 \equiv P_1 \wedge x_1 == (\star\mathbb{O} + v_1) \wedge x_1 == \star\mathbb{O} \wedge x_1 + v_1 \wedge x_1$$

$$L_2 \equiv P_2 \wedge x_2 == (\star\mathbb{O} + v_2) \wedge x_2 == \star\mathbb{O} \wedge x_2 + v_2 \wedge x_2$$

The point of intersection of L_1 and L_2 is the point P given by (congruent to) the regressive product of the lines L_1 and L_2.

$$P \equiv L_1 \vee L_2 == (\star\mathbb{O} \wedge x_1 + v_1 \wedge x_1) \vee (\star\mathbb{O} \wedge x_2 + v_2 \wedge x_2)$$

Here we depict the lines overlaid by their defining bound vectors.

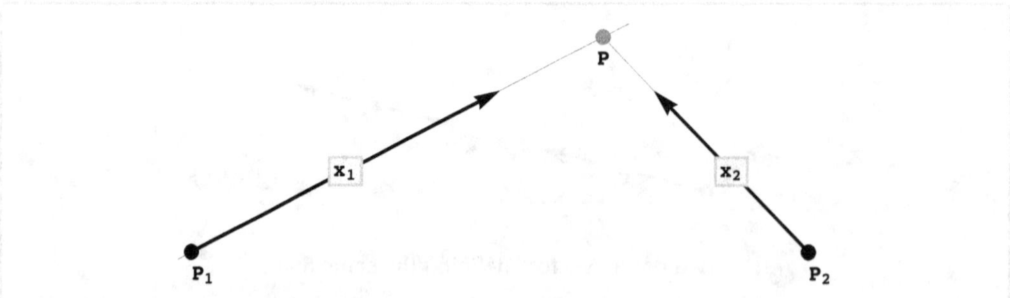

The intersection of lines as the common factor of bound vectors

Expanding the formula for P gives four terms.

$$P \equiv (\star O \wedge x_1) \vee (\star O \wedge x_2) + (v_1 \wedge x_1) \vee (\star O \wedge x_2) + (\star O \wedge x_1) \vee (v_2 \wedge x_2) + (v_1 \wedge x_1) \vee (v_2 \wedge x_2)$$

The Common Factor Theorem for the regressive product of elements of the form $(x \wedge y) \vee (u \wedge v)$ in a linear space of three dimensions was introduced as formula [1.14] in section 1.3 as

$$(x \wedge y) \vee (u \wedge v) == (x \wedge y \wedge v) \vee u - (x \wedge y \wedge u) \vee v$$

Since a bound 2-space is three dimensional (its basis contains three elements - the origin and two vectors), we can use this formula to expand each of the terms in P.

$$(\star O \wedge x_1) \vee (\star O \wedge x_2) == (\star O \wedge x_1 \wedge x_2) \vee \star O - (\star O \wedge x_1 \wedge \star O) \vee x_2$$

$$(v_1 \wedge x_1) \vee (\star O \wedge x_2) == (v_1 \wedge x_1 \wedge x_2) \vee \star O - (v_1 \wedge x_1 \wedge \star O) \vee x_2$$

$$(\star O \wedge x_1) \vee (v_2 \wedge x_2) == -(v_2 \wedge x_2) \vee (\star O \wedge x_1)$$
$$== -(v_2 \wedge x_2 \wedge x_1) \vee \star O + (v_2 \wedge x_2 \wedge \star O) \vee x_1$$

$$(v_1 \wedge x_1) \vee (v_2 \wedge x_2) == (v_1 \wedge x_1 \wedge x_2) \vee v_2 - (v_1 \wedge x_1 \wedge v_2) \vee x_2$$

The term $\star O \wedge x_1 \wedge \star O$ is zero because of the exterior product of repeated factors. The four terms involving the exterior product of three vectors, for example $v_1 \wedge x_1 \wedge x_2$, are also zero since any three vectors in a two-dimensional vector space must be dependent (The vector space is 2-dimensional since it is the vector sub-space of a bound 2-space). Hence we can express the point of intersection P as congruent to a weighted point.

$$P \equiv (\star O \wedge x_1 \wedge x_2) \vee \star O + (\star O \wedge v_2 \wedge x_2) \vee x_1 - (\star O \wedge v_1 \wedge x_1) \vee x_2$$

If we express the vectors in terms of a basis, e_1 and e_2 say, we can reduce this formula (after some manipulation) to:

$$P \equiv \star O + \frac{\mathrm{Det}[v_2, x_2]}{\mathrm{Det}[x_1, x_2]} x_1 - \frac{\mathrm{Det}[v_1, x_1]}{\mathrm{Det}[x_1, x_2]} x_2$$

Here, the determinants are the determinants of the coefficients of the vectors in the given basis.

To verify that P does indeed lie on both the lines L_1 and L_2, we only need to carry out the straightforward verification that the products $P \wedge L_1$ and $P \wedge L_2$ are both zero.

Although this approach *in this simple case* is certainly more complex than the standard algebraic approach in the plane, its interest lies in the facts that it is immediately generalizable to intersections of any geometric objects in spaces of any number of dimensions, and that it leads to easily computable solutions.

1.5 The Complement

The complement as a correspondence between spaces

The Grassmann algebra has a duality in its structure which not only gives it a certain elegance, but is also the basis of its power. We have already introduced the regressive product as the dual *product operation* to the exterior product. In this section we extend the notion of duality to define the *complement of an element*. The notions of metric, orthogonality, and interior, inner and scalar products are all based on the complement.

Consider a linear space of dimension n with basis $e_1, e_2, \ldots, e_n$. The set of all the essentially different m-element products of these basis elements forms the basis of another linear space, but this time of dimension $\binom{n}{m}$. For example, when n is 3, the linear space of 2-elements has three elements in its basis: $e_1 \wedge e_2, e_1 \wedge e_3, e_2 \wedge e_3$.

The anti-symmetric nature of the exterior product means that there are just as many basis elements in the linear space of (n–m)-elements as there are in the linear space of m-elements. Because these linear spaces have the same dimension, we can set up a *correspondence* between m-elements and (n–m)-elements. That is, given any m-element, we can define its corresponding (n–m)-element. The (n–m)-element is called the *complement* of the m-element. Normally this correspondence is set up between basis elements and extended to all other elements by linearity.

The Euclidean complement

Suppose we have a three-dimensional linear space with basis e_1, e_2, e_3. We *define* the *Euclidean complement* of each of the basis elements as the basis 2-element whose exterior product with the basis element gives the basis 3-element $e_1 \wedge e_2 \wedge e_3$. We denote the complement of an element by placing a 'bar' over it. Thus:

$$\overline{e_1} == e_2 \wedge e_3 \quad \Longrightarrow \quad e_1 \wedge \overline{e_1} == e_1 \wedge e_2 \wedge e_3$$

$$\overline{e_2} == e_3 \wedge e_1 \quad \Longrightarrow \quad e_2 \wedge \overline{e_2} == e_1 \wedge e_2 \wedge e_3$$

$$\overline{e_3} == e_1 \wedge e_2 \quad \Longrightarrow \quad e_3 \wedge \overline{e_3} == e_1 \wedge e_2 \wedge e_3$$

The Euclidean complement is the simplest type of complement and defines a Euclidean metric, that is, where the basis elements are mutually orthonormal. This was the only type of complement considered by Grassmann. In Chapter 5: The Complement, we will show however, that Grassmann's concept of complement is easily extended to more general metrics. Note carefully that we will be using the notion of complement to *define* the notions of orthogonality and metric, and until we do this, we will not be relying on their existence in Chapters 2, 3, and 4.

With the definitions above, we can now proceed to define the Euclidean complement of a general 1-element x in a three-dimensional space.

$$x == a\, e_1 + b\, e_2 + c\, e_3$$

To do this we need to endow the complement operation with the property of linearity so that it has meaning for linear combinations of basis elements.

$$\overline{x} == \overline{a\, e_1 + b\, e_2 + c\, e_3} == a\, \overline{e_1} + b\, \overline{e_2} + c\, \overline{e_3} == a\, e_2 \wedge e_3 + b\, e_3 \wedge e_1 + c\, e_1 \wedge e_2$$

In a vector 3-space, the complement of the vector x is the bivector $\overline{x}$, and the complement of the bivector $\overline{x}$ is the vector x. Hence they are mutual complements.

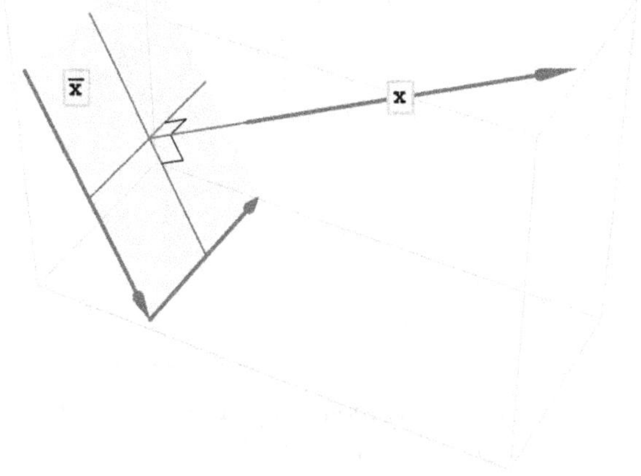

A vector and its bivector complement in a three-dimensional vector space

In section 1.6 we will see that the complement of an element is orthogonal to the element, because we will define the interior product (and hence inner and scalar products) using the complement. We can start to see how the scalar product of a 1-element with itself might arise by expanding the exterior product of x with its complement to exhibit the expected scalar as the coefficient of the basis *n*-element.

$$x \wedge \overline{x} == (a\, e_1 + b\, e_2 + c\, e_3) \wedge (a\, e_2 \wedge e_3 + b\, e_3 \wedge e_1 + c\, e_1 \wedge e_2)$$
$$== \left(a^2 + b^2 + c^2\right) e_1 \wedge e_2 \wedge e_3$$

The Euclidean complement of a basis 2-element can be defined in a manner analogous to that for 1-elements, that is, such that the exterior product of a basis 2-element with its complement is equal to the basis 3-element. The complement of a 2-element in 3-space is therefore a 1-element.

$$\overline{e_2 \wedge e_3} == e_1 \quad \Longrightarrow \quad e_2 \wedge e_3 \wedge \overline{e_2 \wedge e_3} == e_2 \wedge e_3 \wedge e_1 == e_1 \wedge e_2 \wedge e_3$$
$$\overline{e_3 \wedge e_1} == e_2 \quad \Longrightarrow \quad e_3 \wedge e_1 \wedge \overline{e_3 \wedge e_1} == e_3 \wedge e_1 \wedge e_2 == e_1 \wedge e_2 \wedge e_3$$
$$\overline{e_1 \wedge e_2} == e_3 \quad \Longrightarrow \quad e_1 \wedge e_2 \wedge \overline{e_1 \wedge e_2} == e_1 \wedge e_2 \wedge e_3$$

To complete the definition of Euclidean complement in a 3-space we note that

$$\overline{1} == e_1 \wedge e_2 \wedge e_3 \qquad \overline{e_1 \wedge e_2 \wedge e_3} == 1$$

Summarizing these results for the Euclidean complement of the basis elements of a Grassmann algebra in three dimensions shows the essential symmetry of the complement operation.

Complement Palette

Basis	Complement
1	$e_1 \wedge e_2 \wedge e_3$
e_1	$e_2 \wedge e_3$
e_2	$-(e_1 \wedge e_3)$
e_3	$e_1 \wedge e_2$
$e_1 \wedge e_2$	e_3
$e_1 \wedge e_3$	$-e_2$
$e_2 \wedge e_3$	e_1
$e_1 \wedge e_2 \wedge e_3$	1

The complement of a complement

Applying the Euclidean complement operation twice to a 1-element x shows that the complement of the complement of x in a 3-space is just x itself.

$$x == a\, e_1 + b\, e_2 + c\, e_3$$

$$\overline{x} == \overline{a\, e_1 + b\, e_2 + c\, e_3} == a\, \overline{e_1} + b\, \overline{e_2} + c\, \overline{e_3} == a\, e_2 \wedge e_3 + b\, e_3 \wedge e_1 + c\, e_1 \wedge e_2$$

$$\overline{\overline{x}} == \overline{a\, e_2 \wedge e_3 + b\, e_3 \wedge e_1 + c\, e_1 \wedge e_2} == a\, \overline{e_2 \wedge e_3} + b\, \overline{e_3 \wedge e_1} + c\, \overline{e_1 \wedge e_2}$$
$$== a\, e_1 + b\, e_2 + c\, e_3$$

$$\Longrightarrow \quad \overline{\overline{x}} == x$$

More generally, as we shall see in Chapter 5: The Complement, we can show that the complement of the complement of *any* element is the element itself, apart from a possible sign.

$$\overline{\overline{\underset{m}{\alpha}}} == (-1)^{m\,(n-m)}\, \underset{m}{\alpha} \qquad 1.18$$

This result is independent of the correspondence that we set up between the *m*-elements and (*n*–*m*)-elements of the space, *except that the correspondence must be symmetric*. This is equivalent to the requirement that the metric tensor (and inner product) be symmetric.

Whereas in a 3-space, the complement of the complement of a 1-element is the element itself, in a 2-space it turns out to be the negative of the element. Here is a palette of basis elements and their complements for a 2-space.

Complement Palette

Basis	Complement
1	$e_1 \wedge e_2$
e_1	e_2
e_2	$-e_1$
$e_1 \wedge e_2$	1

Although this sign dependence on the dimension of the space and grade of the element might appear arbitrary, it turns out to capture some essential properties of the elements in their spaces to which we have become accustomed. For example in a 2-space, taking the complement of a vector x once rotates it anticlockwise by $\frac{\pi}{2}$. Taking the complement twice rotates it anticlockwise by a further right angle into $-x$.

$$x == a\, e_1 + b\, e_2$$

$$\overline{x} == a\, \overline{e_1} + b\, \overline{e_2} == a\, e_2 - b\, e_1$$

$$\overline{\overline{x}} == a\, \overline{e_2} - b\, \overline{e_1} == -a\, e_1 - b\, e_2$$

Hence any vector and its complement can form an orthogonal basis for a vector 2-space.

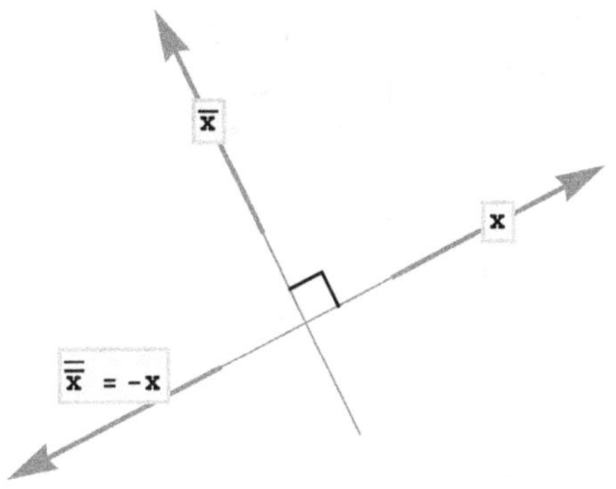

The complement operation rotating vectors in a vector 2-space

The Complement Axiom

From the Common Factor Axiom we can derive a powerful relationship between the Euclidean complements of elements and their exterior and regressive products. The Euclidean complement of the exterior product of two elements is equal to the regressive product of their complements.

$$\overline{A \wedge B} == \overline{A} \vee \overline{B} \qquad 1.19$$

However, although this may be *derived* in this simple case, to develop the Grassmann algebra for general metrics, we will *assume* this relationship holds independent of the metric. It thus takes on the mantle of an axiom.

This axiom, which we call the Complement Axiom, is the quintessential formula of the Grassmann algebra. It expresses the duality of its two fundamental operations on elements and their complements. We note the formal similarity to de Morgan's law in Boolean algebra.

We will also see that adopting this formula for general complements will enable us to compute the complement of any element of a space once we have defined the complements of its basis 1-elements.

- As an example consider two vectors x and y in 3-space, and their exterior product. The Complement Axiom becomes

$$\overline{x \wedge y} == \overline{x} \vee \overline{y}$$

The complement of the bivector $x \wedge y$ is a vector. The complements of x and y are bivectors.

The regressive product of these two bivectors is a vector. The following graphic depicts this relationship, and the orthogonality of the elements. We discuss orthogonality in the next section.

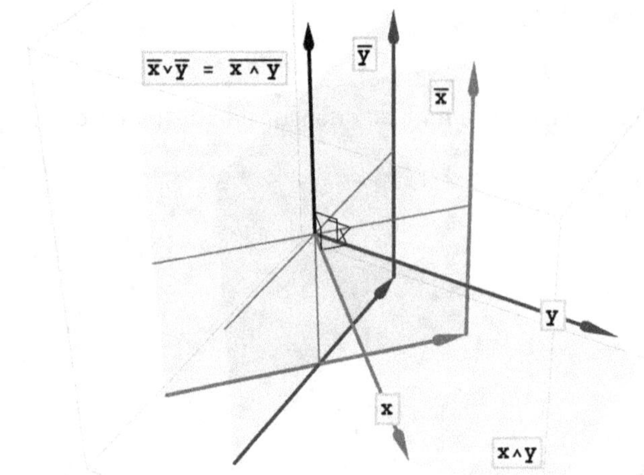

Visualizing the complement axiom in vector three-space

1.6 The Interior Product

The definition of the interior product

The interior product is a generalization of the inner or scalar product to elements of arbitrary grade. First we will define the interior product and then show how the inner and scalar products are special cases.

The *interior product* of an element $\underset{m}{\alpha}$ with an element $\underset{k}{\beta}$ is denoted $\underset{m}{\alpha} \ominus \underset{k}{\beta}$ and *defined* to be the *regressive* product of $\underset{m}{\alpha}$ with the complement of $\underset{k}{\beta}$.

$$\underset{m}{\alpha} \ominus \underset{k}{\beta} == \underset{m}{\alpha} \vee \overline{\underset{k}{\beta}} \qquad 1.20$$

The grade of an interior product $\underset{m}{\alpha} \ominus \underset{k}{\beta}$ may be seen from the definition to be $m+(n-k)-n = m-k$.

Note that while the grade of a regressive product depends on the dimension of the underlying linear space, the grade of an interior product is *independent* of the dimension of the underlying space. This independence underpins the important role the interior product plays in the Grassmann algebra - the exterior product sums grades while the interior product differences them. However, grades may be arbitrarily summed, but not arbitrarily differenced, since there are no elements of negative grade.

Thus the order of factors in an interior product is important. When the grade of the first element is *less* than that of the second element, the result is necessarily zero.

Inner products and scalar products

The interior product of two elements $\underset{m}{\alpha}$ and $\underset{m}{\beta}$ *of the same grade* is (also) called their *inner product*. Since the grade of an interior product is the difference of the grades of its factors, an inner product is always of grade zero, hence scalar.

In the case that the two factors of the product are of grade 1, the product is called a *scalar product*. This conforms to common usage.

In Chapter 6 we will show that *the inner product is symmetric*, that is, the order of the factors is immaterial.

$$\underset{m}{\alpha} \ominus \underset{m}{\beta} == \underset{m}{\beta} \ominus \underset{m}{\alpha} \qquad 1.21$$

When the inner product is between *simple* elements it can be expressed as the determinant of the array of scalar products according to the following formula:

$$(\alpha_1 \wedge \alpha_2 \wedge \cdots \wedge \alpha_m) \ominus (\beta_1 \wedge \beta_2 \wedge \cdots \wedge \beta_m) == \mathbf{Det}\left[\alpha_i \ominus \beta_j\right] \qquad 1.22$$

For example, the inner product of two 2-elements $\alpha_1 \wedge \alpha_2$ and $\beta_1 \wedge \beta_2$ may be written

$$(\alpha_1 \wedge \alpha_2) \ominus (\beta_1 \wedge \beta_2) == (\alpha_1 \ominus \beta_1)\ (\alpha_2 \ominus \beta_2) - (\alpha_1 \ominus \beta_2)\ (\alpha_2 \ominus \beta_1) \qquad 1.23$$

Sequential interior products

Definition [1.20] for the interior product leads to an immediate and powerful formula relating exterior and interior products by grace of the associativity of the regressive product and the Complement Axiom [1.19].

$$\begin{aligned} \mathbf{A} \ominus (\beta_1 \wedge \beta_2 \wedge \ldots \wedge \beta_k) &== \mathbf{A} \vee \overline{(\beta_1 \wedge \beta_2 \wedge \ldots \wedge \beta_k)} \\ == \mathbf{A} \vee \overline{\beta_1} \vee \overline{\beta_2} \vee \ldots \vee \overline{\beta_k} &== (((\mathbf{A} \ominus \beta_1) \ominus \beta_2) \ominus \ldots) \ominus \beta_k \end{aligned}$$

$$\begin{aligned} &\mathbf{A} \ominus (\beta_1 \wedge \beta_2 \wedge \ldots \wedge \beta_k) \\ &== (\mathbf{A} \ominus \beta_1) \ominus (\beta_2 \wedge \ldots \wedge \beta_k) \\ &== (((\mathbf{A} \ominus \beta_1) \ominus \beta_2) \ominus \ldots) \ominus \beta_k \end{aligned} \qquad 1.24$$

By reordering the β_i factors it becomes apparent that there are many different forms in which these formulae can be expressed. For example, the inner product of two bivectors can be rewritten to display the interior product of the first bivector with either of the vectors

$$(\mathbf{x} \wedge \mathbf{y}) \ominus (\mathbf{u} \wedge \mathbf{v}) == ((\mathbf{x} \wedge \mathbf{y}) \ominus \mathbf{u}) \ominus \mathbf{v} == -((\mathbf{x} \wedge \mathbf{y}) \ominus \mathbf{v}) \ominus \mathbf{u}$$

It is the straightforward and consistent derivation of formulae like [1.24] from definition [1.20] using only the fundamental exterior, regressive and complement operations, that shows how powerful Grassmann's approach is. The alternative approach of simply *introducing* an inner product onto a space cannot bring such power to bear.

Orthogonality

As is well known, two 1-elements are said to be *orthogonal* if their scalar product is zero.

More generally, a 1-element x is orthogonal to a simple element A if and only if their interior product $A \ominus x$ is zero.

However, for $A \ominus (x_1 \wedge x_2 \wedge \cdots \wedge x_k)$ to be zero it is only necessary that *one* of the x_i be orthogonal to A. To show this, suppose it to be (without loss of generality) x_1. Then by formula [1.24] we can write

$$A \ominus (x_1 \wedge x_2 \wedge \cdots \wedge x_k) == (A \ominus x_1) \ominus (x_2 \wedge \cdots \wedge x_k)$$

Hence it becomes immediately clear that if $A \ominus x_1$ is zero then so is the whole product.

$$A \ominus x_i == 0 \quad \Longrightarrow \quad A \ominus (\ldots \wedge x_i \wedge \ldots) == 0 \qquad 1.25$$

Measure and magnitude

The *measure* of a simple element A is denoted $|A|$, and is defined to be the positive square root of the interior product of the element with itself. Suppose A is expressed as the exterior product of 1-elements, then we can use formula [1.22] as the basis for computing its *measure*.

$$A == \alpha_1 \wedge \alpha_2 \wedge \cdots \wedge \alpha_m$$

$$|A|^2 == A \ominus A == (\alpha_1 \wedge \alpha_2 \wedge \cdots \wedge \alpha_m) \ominus (\alpha_1 \wedge \alpha_2 \wedge \cdots \wedge \alpha_m) == \text{Det}\left[\alpha_i \ominus \alpha_j\right] \qquad 1.26$$

Under a geometric interpretation of the space in which 1-elements are interpreted as vectors representing displacements, the concept of measure corresponds to the concept of *magnitude*. The magnitude of a vector is its length, the magnitude of a bivector is the area of the parallelogram formed by its two vectors, and the magnitude of a trivector is the volume of the parallelepiped formed by its three vectors. The magnitude of a scalar is the scalar itself.

The magnitude of a vector x is, as expected, given by the standard formula.

$$|x| == \sqrt{x \ominus x} \qquad 1.27$$

The magnitude of a bivector $x \wedge y$ is given by formula [1.26] as

$$|x \wedge y| == \sqrt{(x \wedge y) \ominus (x \wedge y)} == \sqrt{\text{Det}\begin{bmatrix} x \ominus x & x \ominus y \\ x \ominus y & y \ominus y \end{bmatrix}} \qquad 1.28$$

Of course, a bivector may be expressed in an infinity of ways as the exterior product of two vectors, since adding a scalar multiple of the first vector to the second does not change the bivector. For example

$$B == x \wedge y == x \wedge (y + a\,x)$$

From this, the square of its area may be written in either of two ways:

$$\mathbf{B} \ominus \mathbf{B} == (\mathbf{x} \wedge \mathbf{y}) \ominus (\mathbf{x} \wedge \mathbf{y})$$
$$\mathbf{B} \ominus \mathbf{B} == (\mathbf{x} \wedge (\mathbf{y} + a\,\mathbf{x})) \ominus (\mathbf{x} \wedge (\mathbf{y} + a\,\mathbf{x}))$$

However, multiplying out these expressions using formula [1.26] shows that terms cancel in the second expression, thus reducing them both to the same expression, and demonstrating the invariance of the definition.

$$\mathbf{B} \ominus \mathbf{B} == (\mathbf{x} \ominus \mathbf{x})\,(\mathbf{y} \ominus \mathbf{y}) - (\mathbf{x} \ominus \mathbf{y})^2$$

Thus the measure of a bivector is independent of the actual vectors used to express it. Geometrically interpreted, this confirms the elementary result that the area of the corresponding parallelogram (with sides corresponding to the displacements represented by the vectors) is independent of its shape. These results extend straightforwardly to simple elements of any grade.

The measure of an element is equal to the measure of its complement. By the definition of the interior product [1.20], and formulae [1.18] and [1.19] we have

$$\underset{m}{\alpha} \ominus \underset{m}{\alpha} == \underset{m}{\alpha} \vee \overline{\underset{m}{\alpha}} == \overline{\overline{\underset{m}{\alpha} \vee \overline{\underset{m}{\alpha}}}} == (-1)^{m\,(n-m)}\, \overline{\overline{\underset{m}{\alpha}}} \vee \overline{\underset{m}{\alpha}} == \overline{\underset{m}{\alpha}} \vee \overline{\overline{\underset{m}{\alpha}}} == \overline{\underset{m}{\alpha}} \ominus \overline{\underset{m}{\alpha}}$$

$$\mathbf{A} \ominus \mathbf{A} == \overline{\mathbf{A}} \ominus \overline{\mathbf{A}} \qquad 1.29$$

A *unit element* $\hat{A}$ can be defined by the ratio of the element to its measure.

$$\hat{\mathbf{A}} == \frac{\mathbf{A}}{|\mathbf{A}|} \qquad 1.30$$

Calculating interior products from their definition

We can use the interior product definition [1.20], the definitions of the Euclidean complement in section 1.5, and the regressive unit axiom [1.9] with $\underset{n}{1} = \overline{1} = e_1 \wedge e_2 \wedge e_3$ to calculate interior products directly from their definition. In what follows we calculate the interior products of representative basis elements of a 3-space with Euclidean metric. As expected, the scalar products $e_1 \ominus e_1$ and $e_1 \ominus e_2$ turn out to be 1 and 0 respectively.

$$e_1 \ominus e_1 == e_1 \vee \overline{e_1} == e_1 \vee (e_2 \wedge e_3) == (e_1 \wedge e_2 \wedge e_3) \vee 1 == \overline{1} \vee 1 == 1$$
$$e_1 \ominus e_2 == e_1 \vee \overline{e_2} == e_1 \vee (e_3 \wedge e_1) == (e_1 \wedge e_3 \wedge e_1) \vee 1 == 0 \vee 1 == 0$$

Using the Common Factor Axiom [1.13] with the common factor equal to 1, it is straightforward to see that inner products of identical basis 2-elements are unity.

$$(e_1 \wedge e_2) \ominus (e_1 \wedge e_2) == (e_1 \wedge e_2) \vee \overline{(e_1 \wedge e_2)} == (e_1 \wedge e_2) \vee e_3$$
$$== (e_1 \wedge e_2 \wedge e_3) \vee 1 == \overline{1} \vee 1 == 1$$

Inner products of non-identical basis 2-elements are zero.

$$(e_1 \wedge e_2) \ominus (e_2 \wedge e_3) == (e_1 \wedge e_2) \vee \overline{(e_2 \wedge e_3)} == (e_1 \wedge e_2) \vee e_1$$
$$== (e_1 \wedge e_2 \wedge e_1) \vee 1 == 0 \vee 1 == 0$$

If a basis 2-element contains a given basis 1-element, then their interior product is not zero.

$$(e_1 \wedge e_2) \ominus e_1 == (e_1 \wedge e_2) \vee \overline{e_1} == (e_1 \wedge e_2) \vee (e_2 \wedge e_3)$$
$$== (e_1 \wedge e_2 \wedge e_3) \vee e_2 == \overline{1} \vee e_2 == e_2$$

If a basis 2-element does *not* contain a given basis 1-element, then their interior product is zero:

$$(e_1 \wedge e_2) \ominus e_3 == (e_1 \wedge e_2) \vee \overline{e_3} == (e_1 \wedge e_2) \vee (e_1 \wedge e_2) == 0$$

Expanding interior products

To expand interior products, we will use the Interior Common Factor Theorem [6.49] developed in section 6.4. This theorem shows how an interior product of a simple element $\underset{m}{\alpha}$ with another, not necessarily simple element of equal or lower grade $\underset{k}{\beta}$, may be expressed as a linear combination of the ν $(= \binom{m}{k})$ essentially different factors $\underset{m-k}{\alpha_i}$ (of grade $m-k$) of the simple element of higher grade.

$$\underset{m}{\alpha} \ominus \underset{k}{\beta} == \sum_{i==1}^{\nu} \left(\underset{k}{\alpha_i} \ominus \underset{k}{\beta} \right) \underset{m-k}{\alpha_i}$$

$$\underset{m}{\alpha} == \underset{k}{\alpha_1} \wedge \underset{m-k}{\alpha_1} == \underset{k}{\alpha_2} \wedge \underset{m-k}{\alpha_2} == \cdots == \underset{k}{\alpha_\nu} \wedge \underset{m-k}{\alpha_\nu} \qquad 1.31$$

For example, the Interior Common Factor Theorem may be used to prove a relationship involving the interior product of a 1-element x with the exterior product of two factors, each of which may not be simple. This relationship and the special cases that derive from it find application throughout the algebra. See section 6.8 and formula [6.100].

$$\left(\underset{m}{\alpha} \wedge \underset{k}{\beta} \right) \ominus x == \left(\underset{m}{\alpha} \ominus x \right) \wedge \underset{k}{\beta} + (-1)^m \underset{m}{\alpha} \wedge \left(\underset{k}{\beta} \ominus x \right) \qquad 1.32$$

The Interior Common Factor Theorem may also be expressed in a more algorithmically powerful form in terms of the span and cospan of $\underset{m}{\alpha}$. See sections 2.12 and 6.4.

The interior product of a bivector and a vector

Suppose x is a vector and $B = x_1 \wedge x_2$ is a simple bivector. The interior product of the bivector B with the vector x is the vector $B \ominus x$. This can be expanded by the Interior Common Factor Theorem, or formula [1.32] to give:

$$B \ominus x == (x_1 \wedge x_2) \ominus x == (x \ominus x_1)\, x_2 - (x \ominus x_2)\, x_1$$

Since $B \ominus x$ is expressed as a linear combination of x_1 and x_2 it is clearly *contained in* the bivector B so that the exterior product of B with $B \ominus x$ is zero.

$$B \wedge (B \ominus x) == 0$$

The resulting vector $B \ominus x$ is also *orthogonal* to x. We can show this by taking its scalar product with x, and then using formula [1.24].

$$(B \ominus x) \ominus x == B \ominus (x \wedge x) == 0$$

If $\hat{B}$ is the unit bivector of B, the projection $x^\#$ of x onto B is given by

$$x^\# == -\hat{B} \ominus \left(\hat{B} \ominus x \right)$$

The component $x^{\perp}$ of x orthogonal to B is given by

$$x^{\perp} == \left(\hat{B} \wedge x\right) \ominus \hat{B}$$

It is easily shown that the sum of these two components is equal to x.

$$x == x^{\#} + x^{\perp}$$

In the diagram below we depict these relationships. But remember that the vectors and bivector are only docked in a *convenient* location.

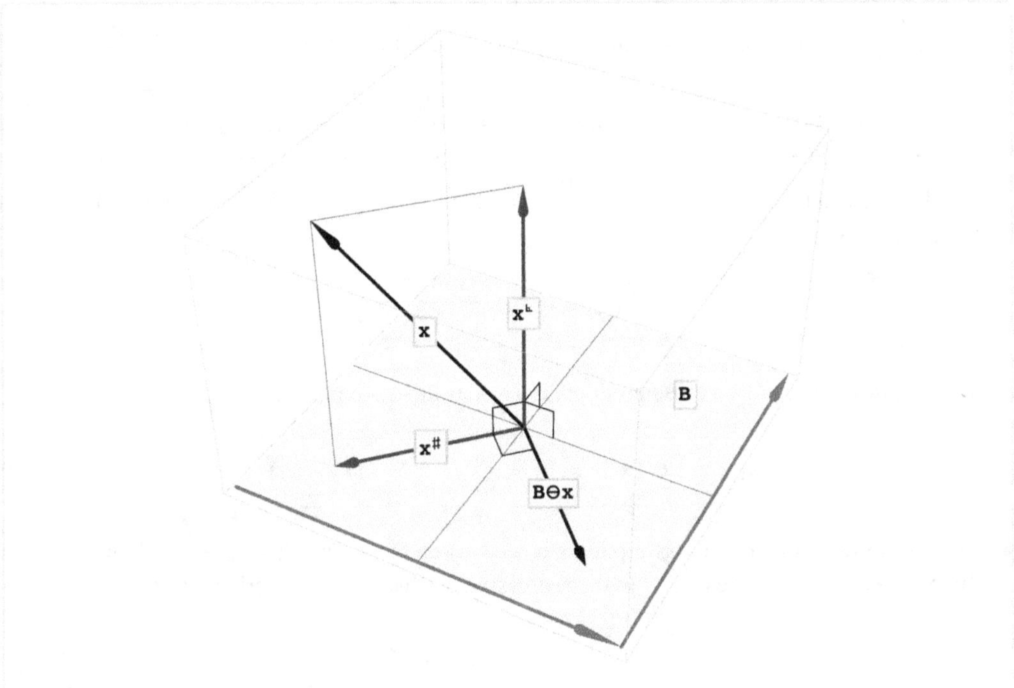

The interior product of a bivector with a vector

These concepts may easily be extended to geometric entities of higher grade. We will explore them further in Chapter 6, beginning in section 6.11.

The cross product

The cross or vector product of the three-dimensional vector calculus of Gibbs *et al.* [Gibbs 1928] corresponds to two operations in Grassmann's more general calculus. *Taking the cross-product of two vectors in three dimensions corresponds to taking the complement of their exterior product.* However, whilst the usual cross product formulation is valid only for vectors in three dimensions, the exterior product formulation is valid for elements of *any* grade in *any* number of dimensions. Therefore the opportunity exists to generalize the concept.

Because our generalization reduces to the usual definition under the usual circumstances, we take the liberty of continuing to refer to the generalized cross product as, simply, the cross product.

Let A and B be elements of any grade, then their cross product is denoted $A \times B$ and is defined as the complement of their exterior product. The cross product of an *m*-element and a *k*-element is thus an (*n*–(*m*+*k*))-element.

$$\mathbf{A} \times \mathbf{B} == \overline{\mathbf{A} \wedge \mathbf{B}} \qquad 1.33$$

This definition preserves the basic property of the cross product: that the cross product of two elements is an element orthogonal to both, and reduces to the usual notion for vectors in a three dimensional metric vector space. For 1-elements x_i the definition has the following consequences, independent of the dimension of the space.

- The triple cross product is a 1-element in any number of dimensions.

$$(x_1 \times x_2) \times x_3 == (x_1 \wedge x_2) \ominus x_3 == (x_3 \ominus x_1)\, x_2 - (x_3 \ominus x_2)\, x_1 \qquad 1.34$$

- The box product, or scalar triple product, is an (*n*–3)-element, and therefore a scalar only in three dimensions.

$$(x_1 \times x_2) \ominus x_3 == \overline{x_1 \wedge x_2 \wedge x_3} \qquad 1.35$$

- The scalar product of two cross products is a scalar in any number of dimensions.

$$(x_1 \times x_2) \ominus (x_3 \times x_4) == (x_1 \wedge x_2) \ominus (x_3 \wedge x_4) \qquad 1.36$$

- The cross product of two cross products is a (4–*n*)-element, and therefore a 1-element only in three dimensions. It corresponds to the regressive product of two exterior products.

$$(x_1 \times x_2) \times (x_3 \times x_4) == (x_1 \wedge x_2) \vee (x_3 \wedge x_4) \qquad 1.37$$

The cross product of the three-dimensional vector calculus requires the space to have a metric, since it is defined to be orthogonal to its factors. Equation [1.37] however, shows that in this particular case, the result does not explicitly require a metric.

1.7 Summary

Summary of operations

In this chapter we have briefly introduced the four fundamental operations which underpin the Grassmann algebra: the exterior product, the regressive product, the complement operation, and the interior product. They have been introduced in this sequence because each one depends on those preceding.

The *exterior product* is the first of the fundamental operations upon which all the others are based. It encodes the notion of linear independence in a way that enables higher order entities (for example, lines and planes) to be constructed from lower order ones (for example, points and vectors).

The *regressive product* was introduced as a true dual product operation to the exterior product. It is a true dual because its axioms can be derived from the axioms of the exterior product and *vice-versa*. The intersection of two lines in the plane was computed using the regressive product, previewing how it will be particularly powerful in Projective Geometry to compute intersections of any geometric entities in any space.

The exterior and regressive products, although they can build entities and intersect them, cannot measure or compare any of them unless they are congruent. For example, given two vectors which are not scalar multiples of each other, the two products do not lead to any invariant mechanism for deciding which is 'larger'. To do this we need to provide the algebra with more information with which to make its decisions. The complement operation introduces just such information.

The *complement* operation is a rule (or mapping) which, given an element, A say, of the algebra enables us to correspond an element $\overline{A}$, such that the exterior product of A with $\overline{A}$ is equal to a scalar multiple of the unit n-element of the space; and the regressive product of A with $\overline{A}$ is equal simply to this scalar multiple. If we take this scalar to be the square of a quantity which we call the *measure* or *magnitude* of A, we find a beautiful correspondence between magnitudes computed thus, and the commonly accepted magnitudes associated with geometric figures. For example the magnitude of a vector corresponds to its length, the magnitude of a bivector corresponds to the area of any of the parallelograms formed from its vectors, and the magnitude of a trivector corresponds to the volume of any of the parallelepipeds formed from its vectors.

The complement mapping can be defined on the elements of any basis chosen for the underlying linear space of the algebra and induced onto the rest of the algebra. That is, once we can measure basis elements, we can measure any entity of the algebra. This ability to measure in a way which conforms to our notions of physical measurement of geometric objects is so important that this scalar square of the magnitude of the element A is defined to be a new product: the *inner product* of A with itself - defined as the regressive product of A with its complement. It is straightforward then to extend this to general elements by defining the *interior product* of A with B as the regressive product of A with the complement of B.

Summary of objects

In the previous section we have discussed the fundamental *operations* of Grassmann algebra. However, an important contributor to the power of the algebra is that its *objects* may be endowed with different geometric interpretations.

0. A *multilinear algebra* like the Grassmann algebra is just the algebra you need when you have more than one independent object. Like all mathematics, it is devoid of interpretation, existing only as a set of objects, and a consistent set of operations and rules.

1. Interpreting all the basic objects of a multilinear algebra with just the exterior and regressive products as 'directed line segments with no location' gives us *vector geometry*.

2. If to vector geometry we add another object, and interpret it as the 'origin point', we get *projective geometry*.

3. If to vector geometry we add a rule for measuring and comparing the objects we get *metric vector geometry*.

4. If to projective geometry we add a rule for measuring and comparing the objects we get *metric projective geometry*.

5. Some physical phenomena can be modelled by one or more of these types of geometry by representing the physical 'objects' by geometric ones.

2 The Exterior Product

2.1 Introduction

The exterior product is the natural fundamental product operation for elements of a linear space. Although it is not a closed operation (that is, the product of two elements is not itself an element of the *same* linear space), the products it generates form a series of new linear spaces, the totality of which can be used to define a true algebra, which is closed.

The exterior product is naturally and fundamentally connected with the notion of linear dependence. Several 1-elements are linearly dependent if and only if their exterior product is zero. (Two common geometric interpretations of 1-elements are as points or vectors). All the properties of determinants flow naturally from the simple axioms of the exterior product. The notion of 'exterior' is equivalent to the notion of linear independence, since elements which are truly exterior to one another (that is, not lying in the same space) will have a non-zero product. An exterior product of 1-elements has a straightforward geometric interpretation as the multidimensional equivalent to the 1-elements from which it is constructed. And if the space possesses a metric, its measure or magnitude may be interpreted as a length, area, volume, or hypervolume according to the grade of the product.

However, the exterior product does not require the linear space to possess a metric. This is in direct contrast to the three-dimensional vector calculus in which the vector (or cross) product *does* require a metric, since it is defined as the vector *orthogonal* to its two factors, and orthogonality is a metric concept. Some of the results which use the cross product can equally well be cast in terms of the exterior product, thus avoiding unnecessary assumptions.

We start the chapter with an informal discussion of the exterior product, and then collect the axioms for exterior linear spaces together more formally into a set of 12 axioms which combine those of the underlying field and underlying linear space with the specific properties of the exterior product. In later chapters we will derive equivalent sets for the regressive and interior products, to which we will often refer.

Next, we pin down the linear space notions that we have introduced axiomatically by introducing a basis onto the underlying (primary) linear space and showing how this can induce a basis onto each of the other linear spaces generated by the exterior product. A constantly useful partner to a basis element of any of these exterior linear spaces is its cobasis element. The exterior product of a basis element with its cobasis element always gives the basis n-element of the algebra. We develop the notion of cobasis following that of basis.

The next three sections look at some standard topics in linear algebra from a Grassmannian viewpoint: determinants, cofactors, and the solution of systems of linear equations. We show that all the well-known properties of determinants, cofactors, and linear equations proceed directly from the properties of the exterior product.

The following two sections discuss two concepts dependent on the exterior product that have no direct counterpart in standard linear algebra: simplicity and exterior division. If an element is simple it can be factorized into 1-elements. If a simple element is divided by another simple element contained in it, the result is not unique. Both these properties will find application later in our explorations of geometry.

At the end of the chapter we introduce the concepts of span and cospan of a simple element, and the result that we can define multilinear forms involving the spans and cospans of an element which are invariant to any refactorization of the element. Following this we explore their application to calculating the union and intersection of two simple elements, and the factorization of a simple element. These applications involve only the exterior product. But we will continue to use the concepts of span, cospan and multilinear forms throughout the book, where they will play an important role in computable formulae for the expansion of expressions involving regressive and interior products.

◆ The *GrassmannAlgebra* application

As mentioned in the Preface, this book has a companion *Mathematica*-based application called *GrassmannAlgebra* to help in its explorations and take the tedium and potential for errors out of some of the longer computations. The application is free to download from the book's companion website (see the Preface). To assist those who may be reading the book while learning to use the application, some sections of the book will be dedicated to discussing conventions which are about to be used in the book. For those readers who are not interested in these computational issues, please ignore them. Doing so should not detract from understanding the rest of the material!

2.2 The Exterior Product

Basic properties of the exterior product

Let $\underset{0}{\Lambda}$ denote the field of real numbers. Its elements, called 0-*elements* (or scalars) will be denoted by a, b, c, … . Let $\underset{1}{\Lambda}$ denote a linear space of dimension n whose field is $\underset{0}{\Lambda}$. Its elements, called 1-*elements*, will be denoted by …, x, y, z.

We call $\underset{1}{\Lambda}$ a linear space rather than a vector space and its elements 1-elements rather than vectors because we will be interpreting its elements later as points as well as vectors.

A second linear space denoted $\underset{2}{\Lambda}$ may be constructed as the space of sums of exterior products of 1-elements taken two at a time. The *exterior product* operation is denoted by $\wedge$, and has the following properties:

$$a\,(x \wedge y) == (a\,x) \wedge y == x \wedge (a\,y) \tag{2.1}$$

$$x \wedge (y + z) == x \wedge y + x \wedge z \tag{2.2}$$

$$(y + z) \wedge x == y \wedge x + z \wedge x \tag{2.3}$$

$$x \wedge x == 0 \tag{2.4}$$

An important property of the exterior product (which is sometimes taken as an axiom) is the *anti-symmetry* of the product of two 1-elements.

$$x \wedge y == -y \wedge x \qquad 2.5$$

This may be proved from the *distributivity* and *nilpotency* axioms [2.2] to [2.4] since

$$(x + y) \wedge (x + y) == 0$$
$$\Longrightarrow \; x \wedge x + x \wedge y + y \wedge x + y \wedge y == 0$$
$$\Longrightarrow \; x \wedge y + y \wedge x == 0$$
$$\Longrightarrow \; x \wedge y == -y \wedge x$$

An element of $\underset{2}{\Lambda}$ will be called a *2-element* (of *grade* 2) and may be denoted by a kernel letter with a '2' written below. For example:

$$\underset{2}{\alpha} == x \wedge y + z \wedge w + \ldots$$

A *simple* 2-element is the exterior product of two 1-elements.

It is important to note the distinction between a 2-element and a simple 2-element (a distinction of no consequence in $\underset{1}{\Lambda}$, where all elements are simple). A 2-element is in general a sum of simple 2-elements, and is not generally expressible as the product of two 1-elements (except where $\underset{1}{\Lambda}$ is of dimension $n \leqslant 3$). The structure of $\underset{2}{\Lambda}$, whilst still that of a linear space, is thus richer than that of $\underset{1}{\Lambda}$, that is, it has more properties.

◆ Example: 2-elements in a three-dimensional space are simple

By way of example, we show that when $\underset{1}{\Lambda}$ is of dimension 3, every element of $\underset{2}{\Lambda}$ is simple.

Suppose that $\underset{1}{\Lambda}$ has basis e_1, e_2, e_3. Then a basis of $\underset{2}{\Lambda}$ is the set of all *essentially* different (linearly independent) products of basis elements of $\underset{1}{\Lambda}$ taken two at a time: $e_1 \wedge e_2$, $e_1 \wedge e_3$, $e_2 \wedge e_3$. (The product $e_1 \wedge e_2$ is not considered essentially different from $e_2 \wedge e_1$ in view of the anti-symmetry property).

Let a general 2-element A be expressed in terms of this basis as

$$A == a\, e_1 \wedge e_2 + b\, e_1 \wedge e_3 + c\, e_2 \wedge e_3$$

Without loss of generality, suppose $a \neq 0$. Then A can be recast in the form below, thus proving the proposition.

$$A == a \left(e_1 - \frac{c\, e_3}{a}\right) \wedge \left(e_2 + \frac{b\, e_3}{a}\right)$$

◆ *Historical Note*

In the *Ausdehnungslehre* of 1844 Grassmann denoted the exterior product of two symbols x and y by a simple concatenation *viz.* x y; whilst in the *Ausdehnungslehre* of 1862 he enclosed them in square brackets *viz.* [x y]. This notation has survived in the three-dimensional vector calculus as the 'box' product [x y z] used for the scalar triple product. Amongst other writers, Whitehead [1898] used the 1844 version whilst Forder [1941] and Cartan [1922] followed the 1862 version. Modern usage denotes the exterior product operation by the wedge ∧ (thus $x \wedge y$), possibly coined by Cartan.

Declaring scalar and vector symbols in *GrassmannAlgebra*

The *GrassmannAlgebra* software needs to know which symbols it will treat as scalar (0-element) symbols, and which symbols it will treat as vector (1-element) symbols. These lists must be distinct (disjoint), as they have different interpretations in the algebra. It loads with default values for lists of these symbols.

The default settings are:

```
ScalarSymbols    {a,b,c,d,e,f,g,h}
VectorSymbols    {p,q,r,s,t,u,v,w,x,y,z}
```

The current values are displayed dynamically on the *GrassmannAlgebra* Palette.

- *GrassmannAlgebra* uses the five-pointed star symbol ★ as the first character in each of its aliases to commonly used commands, objects or functions. This is to create short symbols which are unlikely to conflict with your own definitions. There is a button for it on the Palette.

- To declare extra scalar or vector symbols use `★★S` or `★★V` on the Palette. You can declare most expressions including patterns. For example to declare any subscripted versions of 𝔞, 𝔟, or 𝔠 as extra scalars, enter

  ```
  ★★S[𝔞_, 𝔟_, 𝔠_]
  ```

  ```
  {a, b, c, d, e, f, g, h, 𝔞_, 𝔟_, 𝔠_}
  ```

Now any subscripted versions will be treated as scalars

```
ScalarQ[𝔞1 a + 𝔞2 b + 𝔞3 c]
```

```
True
```

- You can reset all the default declarations by entering `★A`.

- For further information on Preferences, check the ? buttons in the Preferences section on the palette.

Entering exterior products

The exterior product is symbolized by the wedge '∧'. This may be entered either by typing `\[Wedge]`, by typing ESC^ESC, or by clicking on the ∧ button on the *GrassmannAlgebra* palette.

As an infix operator, `x∧y∧z∧⋯` is interpreted as `Wedge[x,y,z,⋯]`.

A quick way to enter `x∧y` is to type `x`, click on ∧ in the palette, then type `y`.

Other ways are to type `x`ESC^ESC`y`, or use the ■∧□ button on the palette.

- For further information on composing expressions, check the ? buttons in the Expression Composition section of the palette.

2.3 Exterior Linear Spaces

Composing *m*-elements

In the previous section the exterior product was introduced as an operation on the elements of the linear space $\underset{1}{\Lambda}$ to produce elements belonging to a new space $\underset{2}{\Lambda}$. Just as $\underset{2}{\Lambda}$ was composed from sums of exterior products of elements of $\underset{1}{\Lambda}$ two at a time, so $\underset{m}{\Lambda}$ may be composed from sums of exterior products of elements of $\underset{1}{\Lambda}$ m at a time.

A *simple m-element* is a product of the form:

$$x_1 \wedge x_2 \wedge \ldots \wedge x_m \qquad x_i \in \underset{1}{\Lambda}$$

An *m-element* is a linear combination of simple m-elements. It may be denoted by a kernel letter with an 'm' written below:

$$\underset{m}{\alpha} == a_1 \, (x_1 \wedge x_2 \wedge \ldots \wedge x_m) + a_2 \, (y_1 \wedge y_2 \wedge \ldots \wedge y_m) + \ldots \qquad a_i \in \underset{0}{\Lambda}$$

The number m is called the *grade* of the m-element.

The exterior linear space $\underset{0}{\Lambda}$ has essentially only one element in its basis, which we take as the unit 1. All other elements of $\underset{0}{\Lambda}$ are scalar multiples of this basis element.

The exterior linear space $\underset{n}{\Lambda}$ (where n is the dimension of $\underset{1}{\Lambda}$) also has essentially only one element in its basis: $e_1 \wedge e_2 \wedge \ldots \wedge e_n$. All other elements of $\underset{n}{\Lambda}$ are scalar multiples of this basis element.

Any m-element, where $m > n$, is zero, since there are no more than n independent elements available from which to compose it, and the nilpotency property causes it to vanish.

Composing elements automatically

Suppose you are working in a 3-space with the basis $\{e_1, e_2, e_3\}$. You can compose elements in this basis by using buttons on the Palette:

- A 1-element:

  $\star\mathbb{V}_a$

  $a_1 \, e_1 + a_2 \, e_2 + a_3 \, e_3$

- A 2-element:

  $\star\mathbb{B}_b$

  $b_1 \, e_1 \wedge e_2 + b_2 \, e_1 \wedge e_3 + b_3 \, e_2 \wedge e_3$

- A simple 2-element:

  $\star\mathbb{W}_c$

  $(c_1 \, e_1 + c_2 \, e_2 + c_3 \, e_3) \wedge (c_4 \, e_1 + c_5 \, e_2 + c_6 \, e_3)$

- *GrassmannAlgebra* will automatically add the new scalar symbols to the current ScalarSymbols.

  ScalarSymbols

  $\{a, b, c, d, e, f, g, h, a_1, a_2, a_3, b_1, b_2, b_3, c_1, c_2, c_3, c_4, c_5, c_6\}$

- For further information, check the **?** buttons in the Expression Composition section in the palette.

Spaces and congruence

In this book, the term *space* is another term for a Grassmann algebra. The term *n-space* is another term for a Grassmann algebra whose underlying linear space is of dimension *n*. By abuse of terminology we will refer to the dimension and basis of an *n*-space as that of its underlying linear space. In *GrassmannAlgebra* you can declare that you want to work in an *n*-space by entering $\star\mathcal{B}_n$, where *n* is a positive integer.

The *space of a simple m-element* is the *m*-space whose underlying linear space consists of all the 1-elements in the *m*-element. A basis of the space of a simple *m*-element may be taken as any set of *m* independent factors of the *m*-element. An *n*-space and the space of its *n*-element are identical.

A 1-element x is said *to belong to, to be contained in,* or simply, *to be in* a simple *m*-element A if and only if their exterior product is zero, that is $A \wedge x = 0$.

We may also say that x is *in the space of* A, or that A *defines* its own space.

A simple element B is said to be contained in another simple element A if and only if all 1-elements contained in B are contained in A.

We say that two elements are *congruent* if one is a scalar multiple of the other, and denote the relationship with the symbol ≡. For example we may write

$$A \equiv a\,A$$

We can therefore say that if two elements are congruent, then their spaces are the same.

When we have one element equal to a scalar multiple of another, $A = a\,B$ say, we may sometimes take the liberty of writing the scalar multiple as a quotient of the two elements.

$$a == \frac{A}{B}$$

These notions will be encountered many times in the rest of the book.

- An overview of the terminology used in this book may found as an appendix.

The associativity of the exterior product

The exterior product is associative in all groups of (adjacent) factors. For example:

$$(w \wedge x) \wedge y \wedge z == w \wedge (x \wedge y) \wedge z == (w \wedge x \wedge y) \wedge z == \ldots$$

Hence the brackets may be omitted altogether, and for elements A, B, and C of any grade we can write

$$(A \wedge B) \wedge C == A \wedge (B \wedge C) == A \wedge B \wedge C \qquad 2.6$$

From this associativity together with the anti-symmetric property of 1-elements it may be shown that *the exterior product is anti-symmetric for any two 1-element factors*. That is, a transposition of any two 1-element factors changes the sign of the product. For example:

$$w \wedge x \wedge y \wedge z == -\,y \wedge x \wedge w \wedge z == y \wedge z \wedge w \wedge x == \ldots$$

Furthermore, from the nilpotency axiom, *a product with two identical* 1-*element factors is zero*. For example:

```
w ∧ x ∧ y ∧ x == 0
```

◆ Example: Non-simple elements are not generally nilpotent

It should be noted that for *simple* elements `A`, `A ∧ A` is zero, but that non-simple elements do not necessarily possess this property, as the following example shows.

Suppose `w`, `x`, `y`, `z` are independent 1-elements. Then the following exterior product of identical 2-elements is not zero:

```
(w ∧ x + y ∧ z) ∧ (w ∧ x + y ∧ z) == 2 w ∧ x ∧ y ∧ z
```

Transforming exterior products

◆ Expanding exterior products

To expand an exterior product, you can use the function `GrassmannExpand` or its alias ⋆$\mathcal{E}$, which you can find on the *GrassmannAlgebra* palette. For example, to expand the product in the previous example you can enter:

```
X = ⋆ℰ[(w ∧ x + y ∧ z) ∧ (w ∧ x + y ∧ z)]
w ∧ x ∧ w ∧ x + w ∧ x ∧ y ∧ z + y ∧ z ∧ w ∧ x + y ∧ z ∧ y ∧ z
```

Note that `GrassmannExpand` does not simplify the expression.

◆ Simplifying exterior products

To simplify the expression you would use the `GrassmannSimplify` function. An alias for `GrassmannSimplify` is ⋆$\mathcal{S}$, which you can also find on the palette.

However in this case, you must first tell *GrassmannAlgebra* that you have chosen to work in a space of dimension 4 or greater, because you have an expression involving four independent 1-elements. You can declare a 4-space by entering the alias ⋆$\mathcal{B}_4$. (We will discuss the detail of declaring bases in a later section).

Now `GrassmannSimplify` will simplify the expanded product.

```
⋆ℬ4; ⋆𝒮[X]
2 w ∧ x ∧ y ∧ z
```

◆ Simplifying and expanding exterior products

To expand and then simplify the expression in one operation, you can use the single function `GrassmannExpandAndSimplify` (which has the alias ⋆$\mathcal{G}$).

```
⋆𝒢[(w ∧ x + y ∧ z) ∧ (w ∧ x + y ∧ z)]
2 w ∧ x ∧ y ∧ z
```

- Note that you can also use these functions to expand and/or simplify any Grassmann expression.

- For further information, check the ? buttons in the Expression Transformation section of the palette.

2.4 Axioms for Exterior Linear Spaces

Summary of axioms

The axioms for exterior linear spaces $\underset{m}{\Lambda}$ are summarized here for future reference. They are a composite of the requisite field, linear space, and exterior product axioms, and may be used to derive the properties discussed informally in the preceding sections. Composite statements are given in list form.

◆ **∧1: The sum of *m*-elements is itself an *m*-element**

$$\left\{\underset{m}{\alpha} \in \underset{m}{\Lambda},\ \underset{m}{\beta} \in \underset{m}{\Lambda}\right\} \Rightarrow \left\{\underset{m}{\alpha} + \underset{m}{\beta} \in \underset{m}{\Lambda}\right\} \tag{2.7}$$

◆ **∧2: Addition of *m*-elements is associative**

$$\left(\underset{m}{\alpha} + \underset{m}{\beta}\right) + \underset{m}{\gamma} == \underset{m}{\alpha} + \left(\underset{m}{\beta} + \underset{m}{\gamma}\right) \tag{2.8}$$

◆ **∧3: *m*-elements have an additive identity (zero element)**

$$\mathcal{E}xists\left[\underset{m}{0},\ \underset{m}{0} \in \underset{m}{\Lambda},\ \underset{m}{\alpha} == \underset{m}{0} + \underset{m}{\alpha}\right] \tag{2.9}$$

◆ **∧4: *m*-elements have an additive inverse**

$$\mathcal{E}xists\left[-\underset{m}{\alpha},\ -\underset{m}{\alpha} \in \underset{m}{\Lambda},\ \underset{m}{0} == \underset{m}{\alpha} + -\underset{m}{\alpha}\right] \tag{2.10}$$

◆ **∧5: Addition of *m*-elements is commutative**

$$\underset{m}{\alpha} + \underset{m}{\beta} == \underset{m}{\beta} + \underset{m}{\alpha} \tag{2.11}$$

◆ **∧6: The exterior product of an *m*-element and a *k*-element is an (*m*+*k*)-element**

$$\left\{\underset{m}{\alpha} \in \underset{m}{\Lambda},\ \underset{k}{\beta} \in \underset{k}{\Lambda}\right\} \Longrightarrow \left\{\underset{m}{\alpha} \wedge \underset{k}{\beta} \in \underset{m+k}{\Lambda}\right\} \qquad 2.12$$

◆ **∧7: The exterior product is associative**

$$\left(\underset{m}{\alpha} \wedge \underset{k}{\beta}\right) \wedge \underset{j}{\gamma} == \underset{m}{\alpha} \wedge \left(\underset{k}{\beta} \wedge \underset{j}{\gamma}\right) \qquad 2.13$$

◆ **∧8: There is a unit scalar which acts as an identity under the exterior product**

$$\mathcal{E}xists\left[1,\ 1 \in \underset{0}{\Lambda},\ \underset{m}{\alpha} == 1 \wedge \underset{m}{\alpha}\right] \qquad 2.14$$

◆ **∧9: Non-zero scalars have inverses with respect to the exterior product**

$$\mathcal{F}or\mathcal{A}ll\left[\alpha,\ \left(\alpha \in \underset{0}{\Lambda}\right) \bigwedge \left(\alpha \neq \underset{0}{0}\right),\ \mathcal{E}xists\left[\alpha^{-1},\ \alpha^{-1} \in \underset{0}{\Lambda},\ 1 == \alpha \wedge \alpha^{-1}\right]\right] \qquad 2.15$$

◆ **∧10: The exterior product of elements of odd grade is anti-commutative**

$$\underset{m}{\alpha} \wedge \underset{k}{\beta} == (-1)^{m\,k}\ \underset{k}{\beta} \wedge \underset{m}{\alpha} \qquad 2.16$$

◆ **∧11: Additive identities act as multiplicative zero elements under the exterior product**

$$\mathcal{E}xists\left[\underset{k}{0},\ \underset{k}{0} \in \underset{k}{\Lambda},\ \underset{k}{0} \wedge \underset{m}{\alpha} == \underset{k+m}{0}\right] \qquad 2.17$$

◆ **∧12: The exterior product is both left and right distributive under addition**

$$\begin{gathered}\left(\underset{m}{\alpha} + \underset{m}{\beta}\right) \wedge \underset{k}{\gamma} == \underset{m}{\alpha} \wedge \underset{k}{\gamma} + \underset{m}{\beta} \wedge \underset{k}{\gamma} \\ \underset{m}{\alpha} \wedge \left(\underset{k}{\beta} + \underset{k}{\gamma}\right) == \underset{m}{\alpha} \wedge \underset{k}{\beta} + \underset{m}{\alpha} \wedge \underset{k}{\gamma}\end{gathered} \qquad 2.18$$

◆ ∧13: Scalar multiplication commutes with the exterior product

$$\underset{m}{\alpha} \wedge \left(a \underset{k}{\beta}\right) == \left(a \underset{m}{\alpha}\right) \wedge \underset{k}{\beta} == a \left(\underset{m}{\alpha} \wedge \underset{k}{\beta}\right) \qquad 2.19$$

◆ A convention for the zeros

It can be seen from the above axioms that each of the linear spaces has its *own* zero. Apart from grade, these zeros all have the same properties, so that for simplicity in computations we will denote them all by the one symbol `0`.

However, this also means that we cannot computationally determine the grade of this symbol `0` from the symbol alone. In practice, we overcome this problem by defining the grade of `0` to be the (undefined) symbol `★0`.

Grassmann algebras

Under the foregoing axioms it may be directly shown that:

1. $\underset{0}{\Lambda}$ is a field.
2. The $\underset{m}{\Lambda}$ are linear spaces over $\underset{0}{\Lambda}$.
3. The direct sum of the $\underset{m}{\Lambda}$ is an algebra.

This algebra is called a *Grassmann algebra*. Its elements are sums of elements from the $\underset{m}{\Lambda}$, thus allowing closure over both addition and exterior multiplication.

A Grassmann algebra is also a linear space of dimension 2^n, where n is the dimension of the underlying linear space $\underset{1}{\Lambda}$.

◆ Example: The product of elements is also an element

The product of two elements of the algebra is also an element of the algebra. Here we expand and simplify the product using the alias `★G` for `GrassmannExpandAndSimplify`.

```
★G[(1 + 2 x + 3 x ∧ y) ∧ (2 + 3 x + 4 y ∧ z + 5 x ∧ y ∧ z)]
```

$2 + 7\,x + 6\,x \wedge y + 4\,y \wedge z + 13\,x \wedge y \wedge z$

On the nature of scalar multiplication

The anti-commutativity axiom ∧10 [2.16] for general elements states that

$$\underset{m}{\alpha} \wedge \underset{k}{\beta} == (-1)^{m\,k}\, \underset{k}{\beta} \wedge \underset{m}{\alpha}$$

If either of these factors is a scalar, a say, its grade is equal to zero and the axiom reduces to:

$$\underset{m}{\alpha} \wedge a == a \wedge \underset{m}{\alpha}$$

Since by axiom ∧6 each of these expressions is an m-element, we may permit the exterior product to subsume the normal field multiplication. Thus, if a is a scalar and A is an element of any grade, $a \wedge A$ is equivalent to $a\,A$. The latter (conventional) notation will usually be adopted.

$$a \wedge A \;==\; A \wedge a \;==\; a\,A \qquad a \in \underset{0}{\Lambda} \tag{2.20}$$

In the usual definitions of linear spaces no discussion is given on the nature of the product of a scalar and an element of the space. A notation is usually adopted (that is, the omission of the product sign) that leads one to suppose this product to be of the same nature as that between two scalars. From the axioms above it may be seen that both the product of a scalar and an element of the linear space *and the product of two scalars*, may be viewed as exterior products.

Factoring scalars

In *GrassmannAlgebra* scalars can be factored out of a product by `GrassmannSimplify` (alias ★*S*) so that they are collected at the beginning. For example:

```
★S[5 (2 x) ∧ (3 y) ∧ (b z)]

30 b x ∧ y ∧ z
```

If any of the factors in the product is scalar, it will also be collected at the beginning. Here a and b are scalars since they have been declared so by default.

```
★S[5 (2 x) ∧ (3 a) ∧ (b z)]

30 a b x ∧ z
```

`GrassmannSimplify` works with any Grassmann expression, or lists (to any level) of Grassmann expressions:

$$\star S\left[\begin{pmatrix} 1 + a \wedge b & a \wedge x \\ z \wedge b & 1 - z \wedge x \end{pmatrix}\right] \text{ // MatrixForm}$$

$$\begin{pmatrix} 1 + a\,b & a\,x \\ b\,z & 1 + x \wedge z \end{pmatrix}$$

Notice that `GrassmannSimplify` may rearrange the order of factors in exterior products. This ordering is necessary in order to compare expressions for equality.

Grassmann expressions

A *Grassmann expression* is any well-formed expression recognized by *GrassmannAlgebra* as a valid element of a Grassmann algebra. To check whether an expression is considered to be a Grassmann expression, you can use the function `GrassmannExpressionQ`.

```
GrassmannExpressionQ[1 + x ∧ y]

True
```

`GrassmannExpressionQ` also works on lists of expressions. Below we determine that whereas the product of a scalar a and a 1-element x is a valid element of the Grassmann algebra, the ordinary multiplication of two 1-elements x y is not.

```
GrassmannExpressionQ[{x ∧ y, a x, x y}]

{True, True, False}
```

- For further information, check the ? buttons in the Expression Analysis section of the palette.

Calculating the grade of a Grassmann expression

An m-element is a Grassmann expression that can be reduced to a sum of products, each of which is the exterior product of m 1-elements. Typical 1-elements are points and vectors.

The *grade* of an m-element is m.

At this stage we have not yet begun to use expressions for which the grade is other than obvious by inspection. However, as will be seen later, the grade will not always be obvious, especially when general Grassmann expressions, for example those involving Clifford or hypercomplex numbers, are being considered. A Clifford number may, for example, have components of different grade. (See Volume 2 for a discussion of Clifford and hypercomplex algebras).

In *GrassmannAlgebra* you can use the function `Grade` (alias `★G`) to calculate the grade of a Grassmann expression. `Grade` works with single Grassmann expressions or with lists (to any level) of expressions.

```
Grade[x ∧ y ∧ z]
```

```
3
```

```
Grade[{1, x, x ∧ y, x ∧ y ∧ z}]
```

```
{0, 1, 2, 3}
```

It will also calculate the grades of the elements in a more general expression, returning them in a list.

```
Grade[1 + x + x ∧ y + x ∧ y ∧ z]
```

```
{0, 1, 2, 3}
```

◆ The grade of zero

As mentioned in the summary of axioms above the zero symbol `0` will be used indiscriminately for the zero element of *any* of the exterior linear spaces. The grade of the symbol `0` is therefore ambiguous. In *GrassmannAlgebra*, whenever `Grade` encounters a `0` it will return the symbol `★0`. This symbol may be read as "the grade of zero". `Grade` also returns `★0` when it encounters an element whose notional grade is greater than the dimension of the currently declared space (and hence is zero).

As an example, we first declare a 3-space, then compute the grades of three elements. The first element `0` could be the zero of any exterior linear space, hence its grade is returned as `★0`. The second element, the unit scalar 1, is of grade 0 (this 0 is a true scalar!). The third element has a notional grade of 4, and hence reduces to `0` in the currently declared 3-space.

```
★𝓑3; Grade[{0, 1, x ∧ y ∧ z ∧ w}]
```

```
{★0, 0, ★0}
```

For the greater part of applications of the algebra, this technicality will not be encountered.

2.5 Bases

Bases for exterior linear spaces

Suppose e_1, e_2, ..., e_n is a basis for $\underset{1}{\Lambda}$. Then, as is well known, any element of $\underset{1}{\Lambda}$ may be expressed as a linear combination of these basis elements.

A basis for $\underset{2}{\Lambda}$ may be constructed from the e_i by assembling all the essentially different non-zero products $e_{i_1} \wedge e_{i_2}$ $(i_1, i_2 : 1, \ldots, n)$.

Two products are *essentially different* if they do not involve the same 1-elements. There are obviously $\binom{n}{2}$ such products, making $\binom{n}{2}$ also the dimension of $\underset{2}{\Lambda}$.

In general, a basis for $\underset{m}{\Lambda}$ may be constructed from the e_i by taking all the essentially different products $e_{i_1} \wedge e_{i_2} \wedge \ldots \wedge e_{i_m}$ $(i_1, i_2, \ldots, i_m : 1, \ldots, n)$. $\underset{m}{\Lambda}$ is thus $\binom{n}{m}$-dimensional.

Essentially different products are of course linearly independent.

Declaring a basis in *GrassmannAlgebra*

GrassmannAlgebra allows the setting up of an environment in which given symbols are declared basis elements of $\underset{1}{\Lambda}$.

When first loaded it sets up a default basis $\{e_1, e_2, e_3\}$. You can see this from the Current Basis pane on the palette.

This is the basis of a 3-dimensional linear space which may be interpreted as a 3-dimensional vector space if you wish. (We will be exploring interpretations other than the vectorial in Chapter 4).

To declare your own basis, enter `DeclareBasis[list]`. For example:

```
DeclareBasis[{i, j, k}]
```

{i, j, k}

Notice that the Current Basis pane of the palette now shows {i, j, k}.

`DeclareBasis` either takes a list (of your own basis elements) or a positive integer as its argument. By using the positive integer form you can simply declare a subscripted basis of any dimension you please.

```
DeclareBasis[8]
```

$\{e_1, e_2, e_3, e_4, e_5, e_6, e_7, e_8\}$

An optional argument gives you control over the kernel symbol used for the basis elements.

```
DeclareBasis[8, ϵ]
```

$\{\epsilon_1, \epsilon_2, \epsilon_3, \epsilon_4, \epsilon_5, \epsilon_6, \epsilon_7, \epsilon_8\}$

A convenient way to change your space to one of dimension n is by entering the symbol $\star\mathcal{B}_n$ with n a positive integer. The simplest way to enter this is from the palette using the $\star\mathcal{B}_\square$ symbol, and entering the integer into the placeholder □.

```
★𝓑4
{e1, e2, e3, e4}
```

You can return to the default basis by entering ★𝓑$_3$, or set all preferences back to their default values by entering `★A`.

- For further information, check the **?** buttons in the Preferences section of the palette.

Composing bases of exterior linear spaces

The function `GradeBasis[m]` composes a list of basis elements of $\underset{m}{\Lambda}$ (of grade m) arranged in standard order from the declared basis of $\underset{1}{\Lambda}$. For example, let us declare the basis to be that of a 3-dimensional space, and then compose the basis of $\underset{2}{\Lambda}$.

```
★𝓑3; GradeBasis[2]
{e1 ∧ e2, e1 ∧ e3, e2 ∧ e3}
```

`GradeBasis` is listable, so we could compose several bases at once.

```
GradeBasis[{0, 1, 3}]
{{1}, {e1, e2, e3}, {e1 ∧ e2 ∧ e3}}
```

To compose all the bases at once, enter `GradeBasis[]`.

```
GradeBasis[]
{{1}, {e1, e2, e3}, {e1 ∧ e2, e1 ∧ e3, e2 ∧ e3}, {e1 ∧ e2 ∧ e3}}
```

You can combine these into a single basis for the algebra by using *Mathematica's* inbuilt `Flatten` function.

```
Flatten[GradeBasis[]]
{1, e1, e2, e3, e1 ∧ e2, e1 ∧ e3, e2 ∧ e3, e1 ∧ e2 ∧ e3}
```

This is a basis of the Grassmann algebra whose underlying linear space is 3-dimensional.

Composing palettes of basis elements

If you would like to compose a palette of the basis elements of all the exterior linear spaces induced by the declared basis, you can use the *GrassmannAlgebra* command `BasisPalette`. For example suppose you are working in a four-dimensional space, entering `BasisPalette` would then give you the 16 basis elements of the corresponding Grassmann algebra.

```
★𝓑4; BasisPalette
```

Basis Palette

$\underset{0}{\Lambda}$	$\underset{1}{\Lambda}$	$\underset{2}{\Lambda}$	$\underset{3}{\Lambda}$	$\underset{4}{\Lambda}$
1	e_1	$e_1 \wedge e_2$	$e_1 \wedge e_2 \wedge e_3$	$e_1 \wedge e_2 \wedge e_3 \wedge e_4$
	e_2	$e_1 \wedge e_3$	$e_1 \wedge e_2 \wedge e_4$	
	e_3	$e_1 \wedge e_4$	$e_1 \wedge e_3 \wedge e_4$	
	e_4	$e_2 \wedge e_3$	$e_2 \wedge e_3 \wedge e_4$	
		$e_2 \wedge e_4$		
		$e_3 \wedge e_4$		

- You can 'float' any generated palette by using the `Float` command. A copy will be floated as a separate palette.

```
Float[BasisPalette];
```

- You can click on any of the buttons to get its contents pasted into your notebook. For example you can compose expressions of your choice by clicking on the relevant basis elements.

$$X = a + b\, e_1 + c\, e_1 \wedge e_3 + d\, e_1 \wedge e_3 \wedge e_4$$

Standard ordering

The *standard ordering* of the basis of $\underset{1}{\Lambda}$ is defined as the ordering of the elements in the declared list of basis elements. This ordering induces a natural standard ordering on the basis elements of $\underset{m}{\Lambda}$: if the basis elements of $\underset{1}{\Lambda}$ were letters of the alphabet arranged alphabetically, then the basis elements of $\underset{m}{\Lambda}$ would be words arranged alphabetically. Equivalently, if the basis elements were digits, the ordering would be numeric.

For example, if we take `{A,B,C,D}` as basis, we can see from its `BasisPalette` that the basis elements for each of the bases are arranged alphabetically.

```
DeclareBasis[{A, B, C, D}]; BasisPalette
```

Basis Palette

$\underset{0}{\Lambda}$	$\underset{1}{\Lambda}$	$\underset{2}{\Lambda}$	$\underset{3}{\Lambda}$	$\underset{4}{\Lambda}$
1	A	$A \wedge B$	$A \wedge B \wedge C$	$A \wedge B \wedge C \wedge D$
	B	$A \wedge C$	$A \wedge B \wedge D$	
	C	$A \wedge D$	$A \wedge C \wedge D$	
	D	$B \wedge C$	$B \wedge C \wedge D$	
		$B \wedge D$		
		$C \wedge D$		

Normally, *GrassmannAlgebra* does not permit numbers to be used as basis elements (since they are scalars), but in *Mathematica*, the double-struck number symbols are not considered numeric, so we can in this case use them as basis elements.

```
DeclareBasis[{1, 2, 3, 4}]; BasisPalette
```

Basis Palette

$\underset{0}{\Lambda}$	$\underset{1}{\Lambda}$	$\underset{2}{\Lambda}$	$\underset{3}{\Lambda}$	$\underset{4}{\Lambda}$
1	1	1∧2	1∧2∧3	1∧2∧3∧4
	2	1∧3	1∧2∧4	
	3	1∧4	1∧3∧4	
	4	2∧3	2∧3∧4	
		2∧4		
		3∧4		

Indexing basis elements of exterior linear spaces

Just as we can denote the basis elements of $\underset{1}{\Lambda}$ by indexed symbols e_i, $(i : 1, \cdots, n)$, so too can we denote the basis elements of $\underset{m}{\Lambda}$ by indexed symbols $\underset{m}{e_i}$, $(i : 1, \cdots, \binom{n}{m})$. This notation enables us to denote the basis elements of an unspecified exterior linear space.

For example, suppose we have a basis $\{e_1, e_2, e_3, e_4\}$ for $\underset{1}{\Lambda}$, then the standard basis for $\underset{2}{\Lambda}$ is:

```
★B4; GradeBasis[2]
```

$\{e_1 \wedge e_2, e_1 \wedge e_3, e_1 \wedge e_4, e_2 \wedge e_3, e_2 \wedge e_4, e_3 \wedge e_4\}$

We could set up a palette of correspondences by using *GrassmannAlgebra's* `Palettize`.

$$\text{Palettize}\left[\left\{\text{GradeBasis}[2], \left\{\underset{2}{e_1}, \underset{2}{e_2}, \underset{2}{e_3}, \underset{2}{e_4}, \underset{2}{e_5}, \underset{2}{e_6}\right\}\right\}\right]$$

$e_1 \wedge e_2$	$e_1 \wedge e_3$	$e_1 \wedge e_4$	$e_2 \wedge e_3$	$e_2 \wedge e_4$	$e_3 \wedge e_4$
$\underset{2}{e_1}$	$\underset{2}{e_2}$	$\underset{2}{e_3}$	$\underset{2}{e_4}$	$\underset{2}{e_5}$	$\underset{2}{e_6}$

We will find this more compact notation of some use in later theoretical derivations. Note however that *GrassmannAlgebra* is set up to accept only declarations for the basis of $\underset{1}{\Lambda}$ from which it generates the bases of the other exterior linear spaces of the algebra.

2.6 Cobases

Definition of a cobasis

The notion of *cobasis* of a basis will be conceptually and notationally useful in our development of later concepts. The cobasis of a basis of $\underset{m}{\Lambda}$ is simply the basis of $\underset{n-m}{\Lambda}$ which has been ordered in a special way relative to the basis of $\underset{m}{\Lambda}$.

Let $\{e_1, e_2, \ldots, e_n\}$ be a basis for $\underset{1}{\Lambda}$. The *cobasis element* associated with a basis element e_i is denoted $\underline{e_i}$ and is defined as the product of the remaining basis elements such that the exterior product of a basis element with its cobasis element is equal to the basis element of $\underset{n}{\Lambda}$.

$$e_i \wedge \underline{e_i} == e_1 \wedge e_2 \wedge \ldots \wedge e_n$$

That is:

$$\underline{e_i} == (-1)^{i-1} e_1 \wedge \ldots \wedge \square_i \wedge \ldots \wedge e_n \qquad 2.21$$

where $\square_i$ means that e_i is missing from the product.

The choice of the underbar notation to denote the cobasis may be viewed as a mnemonic to indicate that the element $\underline{e_i}$ is the basis element of $\underset{n}{\Lambda}$ with e_i 'struck out' from it.

Cobasis elements have similar properties to Euclidean complements, which are denoted with *overbars* (see Chapter 5: The Complement). However, it should be noted that the underbar denotes an element, and is not an invariant operation. For example: $\underline{a\, e_i}$ is *not* defined.

Suppose now we have a basis m-element of $\underset{m}{\Lambda}$:

$$\underset{m}{e_i} == e_{i_1} \wedge e_{i_2} \wedge \ldots \wedge e_{i_m}$$

We can denote its corresponding cobasis element as $\underline{\underset{m}{e_i}}$ and define it as the product of the remaining basis elements such that:

$$\underset{m}{e_i} \wedge \underline{\underset{m}{e_i}} == e_1 \wedge e_2 \wedge \ldots \wedge e_n \qquad 2.22$$

That is:

$$\underline{\underset{m}{e_i}} == (-1)^{K_m} e_1 \wedge \ldots \wedge \square_{i_1} \wedge \ldots \wedge \square_{i_m} \wedge \ldots \wedge e_n$$

$$K_m == \sum_{\gamma==1}^{m} i_\gamma + \frac{1}{2} m\,(m+1) \qquad 2.23$$

From the above definition it can be seen that the exterior product of a basis element with the cobasis element of another basis element is zero. We can write this succinctly in terms of the Kronecker delta as:

$$\underset{m}{e_i} \wedge \underline{\underset{m}{e_j}} == \delta_{ij}\, e_1 \wedge e_2 \wedge \ldots \wedge e_n \qquad 2.24$$

The cobasis of unity

The natural basis of $\underset{0}{\Lambda}$ is 1. The cobasis $\underline{1}$ of this basis element is defined by formula [2.22] as the product of the remaining basis elements such that:

$$1 \wedge \underline{1} == e_1 \wedge e_2 \wedge \ldots \wedge e_n$$

Thus:

$$\underline{1} == e_1 \wedge e_2 \wedge \dots \wedge e_n == \underset{n}{e} \qquad 2.25$$

Any of these three representations will be called the *basis n-element* of the space. $\underline{1}$ may also be described as the *cobasis of unity*. Note carefully that $\underline{1}$ may be different in different bases.

Composing palettes of cobasis elements

You can compose a palette of basis elements of the currently declared basis together with their corresponding cobasis elements by the command `CobasisPalette`. For the default basis $\{e_1, e_2, e_3\}$ this would give

```
★𝓑3; CobasisPalette
```

Cobasis Palette

Basis	Cobasis
1	$e_1 \wedge e_2 \wedge e_3$
e_1	$e_2 \wedge e_3$
e_2	$-(e_1 \wedge e_3)$
e_3	$e_1 \wedge e_2$
$e_1 \wedge e_2$	e_3
$e_1 \wedge e_3$	$-e_2$
$e_2 \wedge e_3$	e_1
$e_1 \wedge e_2 \wedge e_3$	1

If you want to compose a palette for another basis, first declare the basis.

```
DeclareBasis[{𝓘, 𝓙}]; CobasisPalette
```

Cobasis Palette

Basis	Cobasis
1	$\mathcal{I} \wedge \mathcal{J}$
$\mathcal{I}$	$\mathcal{J}$
$\mathcal{J}$	$-\mathcal{I}$
$\mathcal{I} \wedge \mathcal{J}$	1

The cobasis of a cobasis

Let $\underset{m}{e_i}$ be a basis element and $\underline{\underset{m}{e_i}}$ be its cobasis element. Then the cobasis element of this cobasis element is denoted $\underline{\underline{\underset{m}{e_i}}}$ and is defined as expected by the product of the remaining basis elements such that:

$$\underline{\underset{m}{e_i}} \wedge \underline{\underline{\underset{m}{e_j}}} == \delta_{ij}\, e_1 \wedge e_2 \wedge \dots \wedge e_n \qquad 2.26$$

The left-hand side may be rearranged to give:

$$\underline{e_i}_m \wedge \underline{\underline{e_j}}_m == (-1)^{m(n-m)} \underline{\underline{e_j}}_m \wedge \underline{e_i}_m$$

which, by comparison with the definition for the cobasis shows that:

$$\underline{\underline{e_i}}_m == (-1)^{m(n-m)} e_{i\,m} \qquad 2.27$$

That is, the cobasis element of the cobasis element of an element is (apart from a possible sign) equal to the element itself. We shall find this formula useful as we develop general formulae for the complement and interior products in later chapters.

2.7 Determinants

Determinants from exterior products

All the properties of determinants follow naturally from the properties of the exterior product. Indeed, it may be reasonably posited that the theory of determinants was a system for answering questions of linear dependence, developed only because Grassmann's work was ignored. Had Grassmann's work been more widely known, his simpler and more understandable approach would have rendered determinant theory of less interest.

The determinant

$$D == \begin{vmatrix} a_{11} & a_{12} & \cdots & a_{1\,n} \\ a_{21} & a_{22} & \cdots & a_{2\,n} \\ \vdots & \vdots & \cdots & \vdots \\ a_{n1} & a_{n2} & \cdots & a_{n\,n} \end{vmatrix}$$

may be calculated by considering each of its rows (or columns) as a 1-element. Here, we consider rows. Development by columns may be obtained *mutatis mutandis*. Introduce an arbitrary basis e_i in order to encode the position of the entry in the row, and let the rows be encoded as 1-elements α_i.

$$\alpha_i == a_{i1}\, e_1 + a_{i2}\, e_2 + \ldots + a_{in}\, e_n$$

Then form the exterior product of all the α_i. We can arrange them in rows to portray the effect of an array.

$$\begin{aligned} \alpha_1 \wedge \alpha_2 \wedge \ldots \wedge \alpha_n == \\ & (a_{11}\, e_1 + a_{12}\, e_2 + \ldots + a_{1\,n}\, e_n) \wedge \\ & (a_{21}\, e_1 + a_{22}\, e_2 + \ldots + a_{2\,n}\, e_n) \wedge \ldots \wedge \\ & (a_{i1}\, e_1 + a_{i2}\, e_2 + \ldots + a_{i\,n}\, e_n) \wedge \ldots \wedge \\ & (a_{n1}\, e_1 + a_{n2}\, e_2 + \ldots + a_{n\,n}\, e_n) \end{aligned}$$

The determinant D is then the coefficient of the resulting *n*-element.

$$\alpha_1 \wedge \alpha_2 \wedge \ldots \wedge \alpha_n == D\, e_1 \wedge e_2 \wedge \ldots \wedge e_n$$

Because this is a congruence relation, we can express the determinant D uniquely as

$$D == \frac{\alpha_1 \wedge \alpha_2 \wedge \ldots \wedge \alpha_n}{e_1 \wedge e_2 \wedge \ldots \wedge e_n} \qquad 2.28$$

Properties of determinants

All the well-known properties of determinants proceed directly from the properties of the exterior product.

◆ **1 The determinant changes sign on the interchange of any two rows.**

The exterior product is anti-symmetric in any two factors.

$$\ldots \wedge \alpha_i \wedge \ldots \wedge \alpha_j \wedge \ldots == -\ldots \wedge \alpha_j \wedge \ldots \wedge \alpha_i \wedge \ldots$$

◆ **2 The determinant is zero if two of its rows are equal.**

The exterior product is nilpotent.

$$\ldots \wedge \alpha_i \wedge \ldots \wedge \alpha_i \wedge \ldots == 0$$

◆ **3 The determinant is multiplied by a factor *c* if any row is multiplied by *c*.**

The exterior product is commutative with respect to scalars.

$$\alpha_1 \wedge \ldots \wedge (c\,\alpha_i) \wedge \ldots == c\,(\alpha_1 \wedge \ldots \wedge \alpha_i \wedge \ldots)$$

◆ **4 The determinant is equal to the sum of *p* determinants if each element of a given row is the sum of *p* terms.**

The exterior product is distributive with respect to addition.

$$\alpha_1 \wedge \ldots \wedge \left(\sum_i \alpha_i\right) \wedge \ldots == \sum_i (\alpha_1 \wedge \ldots \wedge \alpha_i \wedge \ldots)$$

◆ **5 The determinant is unchanged if to any row is added scalar multiples of other rows.**

The exterior product is unchanged if to any factor is added multiples of other factors.

$$\alpha_1 \wedge \ldots \wedge \left(\alpha_i + \sum_{j \neq i} c^j\,\alpha_j\right) \wedge \ldots == \alpha_1 \wedge \ldots \wedge \alpha_i \wedge \ldots$$

The Laplace expansion technique

The Laplace expansion technique is equivalent to the calculation of the exterior product in four stages:

1. Take the exterior product of any *m* of the α_i.
2. Take the exterior product of the remaining *n*–*m* of the α_i.
3. Take the exterior product of the results of the first two operations.
4. Adjust the sign to ensure the parity of the original ordering of the α_i is preserved.

Each of the first two operations produces an element with $\binom{n}{m} = \binom{n}{n-m}$ terms.

A generalization of the Laplace expansion technique is evident from the fact that the exterior product of the α_i may be effected in *any* grouping and sequence which facilitate the computation.

Calculating determinants

As an example take the following matrix. We can calculate its determinant with *Mathematica's* inbuilt `Det` function:

$$\text{Det}\left[\begin{pmatrix} a & 0 & 0 & 0 & 0 & 0 & 0 & 0 \\ 0 & 0 & 0 & c & 0 & 0 & 0 & a \\ 0 & b & 0 & 0 & 0 & f & 0 & 0 \\ 0 & 0 & 0 & 0 & e & 0 & 0 & 0 \\ 0 & 0 & e & 0 & 0 & 0 & g & 0 \\ 0 & f & 0 & d & 0 & c & 0 & 0 \\ g & 0 & 0 & 0 & 0 & 0 & 0 & h \\ 0 & 0 & 0 & 0 & d & 0 & b & 0 \end{pmatrix}\right]$$

$$a\, b^2\, c^2\, e^2\, h - a\, b\, c\, e^2\, f^2\, h$$

Alternatively, we can declare an 8-dimensional space and then expand and simplify the exterior product of the 8 rows or columns expressed as 1-elements. Here we use $\star\mathcal{B}_8$ to declare the 8-space and $\star\mathcal{G}$ to expand and simplify by *columns*.

$\star\mathcal{B}_8$;
$\star\mathcal{G}[((a\,e_1 + g\,e_7) \wedge (b\,e_3 + f\,e_6) \wedge (e\,e_5) \wedge (c\,e_2 + d\,e_6) \wedge (e\,e_4 + d\,e_8) \wedge (f\,e_3 + c\,e_6) \wedge (g\,e_5 + b\,e_8) \wedge (a\,e_2 + h\,e_7))]$

$$\left(a\, b^2\, c^2\, e^2\, h - a\, b\, c\, e^2\, f^2\, h\right) e_1 \wedge e_2 \wedge e_3 \wedge e_4 \wedge e_5 \wedge e_6 \wedge e_7 \wedge e_8$$

The determinant is then the coefficient of this basis 8-element.

- For maximum speed it is clearly better to use *Mathematica's* inbuilt `Det` wherever possible. If however you are using *GrassmannAlgebra*, it is usually faster to apply the generalized Laplace expansion technique. That is, the order in which the products are calculated is arranged with the objective of minimizing the number of resulting terms produced at each stage. You can do this by dividing the product up into groupings of factors where each grouping contains as few basis elements as possible in common with the others.

◆ *Historical Note*

Grassmann applied his *Ausdehnungslehre* to the theory of determinants and linear equations quite early in his work. Later, Cauchy published his technique of 'algebraic keys' which essentially duplicated Grassmann's results. To claim priority, Grassmann was led to publish his only paper in French, obviously directed at Cauchy: '*Sur les différents genres de multiplication*' ('On different types of multiplication') [1855]. For a complete treatise on the theory of determinants from a Grassmannian viewpoint see R. F. Scott [1880].

2.8 Cofactors

Cofactors from exterior products

The cofactor is an important concept in the usual approach to determinants. One often calculates a determinant by summing the products of the elements of a row by their corresponding cofactors. Cofactors divided by the determinant form the elements of an inverse matrix. In the subsequent development of the Grassmann algebra, particularly the development of the complement operation, we will find it useful to see how cofactors arise from exterior products.

Consider the product of n 1-elements introduced in the previous section:

$$\begin{aligned}\alpha_1 \wedge \alpha_2 \wedge \ldots \wedge \alpha_n =\!= \\ &(a_{11}\, e_1 + a_{12}\, e_2 + \ldots + a_{1\,n}\, e_n) \wedge \\ &(a_{21}\, e_1 + a_{22}\, e_2 + \ldots + a_{2\,n}\, e_n) \wedge \ldots \wedge \\ &(a_{i1}\, e_1 + a_{i2}\, e_2 + \ldots + a_{i\,n}\, e_n) \wedge \ldots \wedge \\ &(a_{n1}\, e_1 + a_{n2}\, e_2 + \ldots + a_{n\,n}\, e_n)\end{aligned}$$

Omitting the first factor α_1 gives

$$\begin{aligned}\alpha_2 \wedge \ldots \wedge \alpha_n =\!= \\ &(a_{21}\, e_1 + a_{22}\, e_2 + \ldots + a_{2\,n}\, e_n) \wedge \ldots \wedge \\ &(a_{i1}\, e_1 + a_{i2}\, e_2 + \ldots + a_{i\,n}\, e_n) \wedge \ldots \wedge \\ &(a_{n1}\, e_1 + a_{n2}\, e_2 + \ldots + a_{n\,n}\, e_n)\end{aligned}$$

Multiplying out the remaining $n-1$ factors results in an expression of the form

$$\begin{aligned}\alpha_2 \wedge \ldots \wedge \alpha_n =\!= \\ &\underline{a_{11}}\; e_2 \wedge e_3 \wedge \ldots \wedge e_n + \\ &\underline{a_{12}}\; (-\, e_1 \wedge e_3 \wedge \ldots \wedge e_n) + \ldots + \\ &\underline{a_{1\,n}}\; (-1)^{n-1}\, e_1 \wedge e_2 \wedge \ldots \wedge e_{n-1}\end{aligned}$$

Here, the signs attached to the basis $(n-1)$-elements have been specifically chosen so that together they correspond to the cobasis elements of e_1 to e_n. The underscored scalar coefficients have yet to be interpreted. Thus we can write:

$$\alpha_2 \wedge \ldots \wedge \alpha_n =\!= \underline{a_{11}}\,\underline{e_1} + \underline{a_{12}}\,\underline{e_2} + \ldots + \underline{a_{1\,n}}\,\underline{e_n}$$

If we now premultiply by the first factor α_1 we get a particularly symmetric form.

$$\begin{aligned}\alpha_1 \wedge \alpha_2 \wedge \ldots \wedge \alpha_n =\!= \\ &(a_{11}\, e_1 + a_{12}\, e_2 + \ldots + a_{1\,n}\, e_n) \wedge \\ &\left(\underline{a_{11}}\,\underline{e_1} + \underline{a_{12}}\,\underline{e_2} + \ldots + \underline{a_{1\,n}}\,\underline{e_n}\right)\end{aligned}$$

Multiplying this out and remembering that the exterior product of a basis element with the cobasis element of another basis element is zero, we get the sum:

$$\alpha_1 \wedge \alpha_2 \wedge \ldots \wedge \alpha_n =\!= \left(a_{11}\,\underline{a_{11}} + a_{12}\,\underline{a_{12}} + \ldots + a_{1\,n}\,\underline{a_{1\,n}}\right) e_1 \wedge e_2 \wedge \ldots \wedge e_n$$

But we have already seen that

$$\alpha_1 \wedge \alpha_2 \wedge \ldots \wedge \alpha_n == D\, e_1 \wedge e_2 \wedge \ldots \wedge e_n$$

Hence the determinant D can be expressed as the sum of products:

$$D == a_{11}\,\underline{a_{11}} + a_{12}\,\underline{a_{12}} + \ldots + a_{1n}\,\underline{a_{1n}}$$

showing that the $\underline{a_{ij}}$ are the *cofactors* of the a_{ij}.

Mnemonically we can visualize the $\underline{a_{ij}}$ as $(-1)^{i+j}$ times the determinant of the array with the row and column containing a_{ij} as 'struck out' by the underscore.

Of course, there is nothing special in the choice of the element α_1 about which to expand the determinant. We could have written the expansion in terms of any factor (row or column).

$$D == a_{i1}\,\underline{a_{i1}} + a_{i2}\,\underline{a_{i2}} + \ldots + a_{in}\,\underline{a_{in}} \qquad 2.29$$

It can be seen immediately that the product of the elements of a row with the cofactors of any *other* row is zero, since in the exterior product formulation the row must have been included in the calculation of the cofactors.

Summarizing these results for both row and column expansions we have:

$$\sum_{j=1}^{n} a_{ij}\,\underline{a_{kj}} == \sum_{j=1}^{n} a_{ji}\,\underline{a_{jk}} == D\,\delta_{ik} \qquad 2.30$$

In matrix terms of course this is equivalent to the standard results

$$A\,A_c^{T} == A^{T}\,A_c == D\,I$$

where A_c is the matrix of cofactors of the elements of A.

The Laplace expansion in cofactor form

We have already introduced the Laplace expansion technique in our discussion of determinants.

1. Take the exterior product of any *m* of the α_i.
2. Take the exterior product of the remaining *n–m* of the α_i.
3. Take the exterior product of the results of the first two operations.
4. Adjust the sign to ensure the parity of the original ordering of the α_i is preserved.

In this section we revisit the technique more specifically in the context of the cofactor. The results will be important in deriving later results.

Let $\underset{m}{e_i}$ be basis *m*-elements and $\underline{\underset{m}{e_i}}$ be their corresponding cobasis (*n–m*)-elements (i : 1, … ,ν and $\nu = \binom{n}{m}$). To fix ideas, suppose we expand a determinant about the first *m* rows:

$$(\alpha_1 \wedge \ldots \wedge \alpha_m) \wedge (\alpha_{m+1} \wedge \ldots \wedge \alpha_n)$$

$$== \left(a_1\,\underset{m}{e_1} + a_2\,\underset{m}{e_2} + \ldots + a_\nu\,\underset{m}{e_\nu}\right) \wedge \left(\underline{a_1}\,\underline{\underset{m}{e_1}} + \underline{a_2}\,\underline{\underset{m}{e_2}} + \ldots + \underline{a_\nu}\,\underline{\underset{m}{e_\nu}}\right)$$

Here, the a_i and $\underline{a_i}$ are simply the coefficients resulting from the partial expansions. As with basis 1-elements, the exterior product of a basis m-element with the cobasis element of another basis m-element is zero. That is:

$$\underset{m}{e_i} \wedge \underline{\underset{m}{e_j}} == \delta_{ij}\, e_1 \wedge e_2 \wedge \ldots \wedge e_n$$

Expanding the product and applying this simplification then yields

$$D == a_1\, \underline{a_1} + a_2\, \underline{a_2} + \ldots + a_\nu\, \underline{a_\nu}$$

This is the Laplace expansion. The a_i are minors and the $\underline{a_i}$ are their cofactors.

Transformations of cobases

In this section we show that if a basis of the underlying linear space is transformed by a transformation whose components are a_{ij}, then its cobasis is transformed by the induced transformation whose components are the cofactors of the a_{ij}. For simplicity in what follows, we use Einstein's summation convention in which a summation over repeated indices is understood.

Let a_{ij} be a transformation on the basis e_j to give the new basis ε_i. That is, $\varepsilon_i = a_{ij}\, e_j$.

Let $\underline{a_{ij}}$ be the corresponding transformation on the cobasis $\underline{e_j}$ to give the new cobasis $\underline{\varepsilon_i}$. That is, $\underline{\varepsilon_i} = \underline{a_{ij}}\, \underline{e_j}$.

Now take the exterior product of these two equations.

$$\varepsilon_i \wedge \underline{\varepsilon_k} == (a_{ip}\, e_p) \wedge \left(\underline{a_{kj}}\, \underline{e_j}\right) == a_{ip}\, \underline{a_{kj}}\, e_p \wedge \underline{e_j}$$

But the product of a basis element and its cobasis element is equal to the n-element of that basis.

$$\varepsilon_i \wedge \underline{\varepsilon_k} == \delta_{ik}\, \underset{n}{\varepsilon} \qquad\qquad e_p \wedge \underline{e_j} == \delta_{pj}\, \underset{n}{e}$$

Substituting in the previous equation gives

$$\delta_{ik}\, \underset{n}{\varepsilon} == a_{ip}\, \underline{a_{kj}}\, \delta_{pj}\, \underset{n}{e}$$

Using the properties of the Kronecker delta we can simplify the right side to give

$$\delta_{ik}\, \underset{n}{\varepsilon} == a_{ij}\, \underline{a_{kj}}\, \underset{n}{e}$$

We can now substitute $\underset{n}{\varepsilon} = D\, \underset{n}{e}$ where $D = \mathrm{Det}[a_{ij}]$ is the determinant of the transformation.

$$\delta_{ik}\, D == a_{ij}\, \underline{a_{kj}}$$

This is precisely the relationship [2.30] derived above for the expansion of a determinant in terms of cofactors. Hence we have shown that $\underline{a_{kj}} = \underline{a_{kj}}$.

$$\varepsilon_i == a_{ij}\, e_j \quad \Longleftrightarrow \quad \underline{\varepsilon_i} == \underline{a_{ij}}\, \underline{e_j} \qquad 2.31$$

In sum: If a basis of the underlying linear space is transformed by a transformation whose components are a_{ij}, then its cobasis is transformed by the induced transformation whose components are the cofactors $\underline{a_{ij}}$ of the a_{ij}.

2.9 Solution of Linear Equations

Grassmann's approach to solving linear equations

Because of its encapsulation of the properties of linear independence, Grassmann was able to use the exterior product to present a theory and formulae for the solution of linear equations of remarkable elegance.

Suppose m independent equations in n $(m \leq n)$ unknowns x_i.

$$\begin{array}{l} a_{11}\, x_1 + a_{12}\, x_2 + \cdots + a_{1\,n}\, x_n == a_1 \\ a_{21}\, x_1 + a_{22}\, x_2 + \cdots + a_{2\,n}\, x_n == a_2 \\ \vdots \qquad \vdots \quad \cdots \quad \vdots \qquad \vdots \\ a_{m1}\, x_1 + a_{m2}\, x_2 + \cdots + a_{m\,n}\, x_n == a_m \end{array}$$

Multiply these equations by $e_1, e_2, \cdots, e_m$ respectively and define

$$C_0 == a_1\, e_1 + a_2\, e_2 + \ldots + a_m\, e_m$$

$$C_i == a_{1\,i}\, e_1 + a_{2\,i}\, e_2 + \ldots + a_{m\,i}\, e_m$$

The C_0 and C_i are therefore 1-elements in a linear space of dimension m.

Adding the resulting equations then gives the system in the form

$$x_1\, C_1 + x_2\, C_2 + \ldots + x_n\, C_n == C_0 \qquad 2.32$$

To obtain an equation from which the unknowns x_i have been eliminated, it is only necessary to multiply the linear system through by the corresponding C_i.

If $m = n$ and a solution for x_i exists, it is obtained by eliminating $x_1, \ldots, x_{i-1}, x_{i+1}, \ldots, x_n$; that is, by multiplying the linear system through by $C_1 \wedge \ldots \wedge C_{i-1} \wedge C_{i+1} \wedge \ldots \wedge C_n$.

$$x_i\, C_i \wedge C_1 \wedge \ldots \wedge C_{i-1} \wedge C_{i+1} \wedge \ldots \wedge C_n == C_0 \wedge C_1 \wedge \ldots \wedge C_{i-1} \wedge C_{i+1} \wedge \ldots \wedge C_n$$

Because the exterior products are congruent we can express x_i as

$$x_i == \frac{C_0 \wedge (C_1 \wedge \ldots \wedge C_{i-1} \wedge C_{i+1} \wedge \ldots \wedge C_n)}{C_i \wedge (C_1 \wedge \ldots \wedge C_{i-1} \wedge C_{i+1} \wedge \ldots \wedge C_n)} \qquad 2.33$$

In this form we only have to calculate the $(n-1)$-element $C_1 \wedge \ldots \wedge C_{i-1} \wedge C_{i+1} \wedge \ldots \wedge C_n$ once.

An alternative form more reminiscent of Cramer's Rule is

$$x_i == \frac{C_1 \wedge \ldots \wedge C_{i-1} \wedge C_0 \wedge C_{i+1} \wedge \ldots \wedge C_n}{C_1 \wedge C_2 \wedge \ldots \wedge C_n} \qquad 2.34$$

All the well-known properties of solutions to systems of linear equations proceed directly from the properties of the exterior products of the C_i.

Example solution: 3 equations in 4 unknowns

Consider the following system of 3 equations in 4 unknowns.

$$a - 2b + 3c + 4d == 2$$

$$2a + 7c - 5d == 9$$

$$a + b + c + d == 8$$

To solve this system, first declare a basis of dimension at least equal to the number of equations. In this example, we declare a 4-space so that we can add another equation later.

```
DeclareBasis[4]
```

$\{e_1, e_2, e_3, e_4\}$

Next define a 1-element for each unknown which encodes its coefficients, and a 1-element which encodes the constants

$$C_a = e_1 + 2e_2 + e_3;$$
$$C_b = -2e_1 + e_3;$$
$$C_c = 3e_1 + 7e_2 + e_3;$$
$$C_d = 4e_1 - 5e_2 + e_3;$$
$$C_0 = 2e_1 + 9e_2 + 8e_3;$$

The system equation then becomes:

$$a\,C_a + b\,C_b + c\,C_c + d\,C_d == C_0$$

Suppose we wish to eliminate a and d thus giving a relationship between b and c. To accomplish this we multiply the system equation through by $C_a \wedge C_d$.

$$(a\,C_a + b\,C_b + c\,C_c + d\,C_d) \wedge C_a \wedge C_d == C_0 \wedge C_a \wedge C_d$$

Or, since the terms involving a and d will obviously be eliminated by their product with $C_a \wedge C_d$, we have more simply:

$$(b\,C_b + c\,C_c) \wedge C_a \wedge C_d == C_0 \wedge C_a \wedge C_d$$

This can be put in a form similar to that of equations [2.33] and [2.34] by dividing through by the right side.

$$\frac{(b\,C_b + c\,C_c) \wedge C_a \wedge C_d}{C_0 \wedge C_a \wedge C_d} == 1$$

By expanding and simplifying the expression we can obtain the relationship between b and c that we require.

$$\star\mathcal{G}\left[\frac{(b\,C_b + c\,C_c) \wedge C_a \wedge C_d}{C_0 \wedge C_a \wedge C_d}\right] == 1$$

$$\frac{1}{63}(27\,b - 29\,c) == 1$$

Example solution: 4 equations in 4 unknowns

Had we had a fourth equation, say $b - 3c + d = 7$, we could have simply added the new information for each coefficient.

$$C_a = e_1 + 2\,e_2 + e_3;$$
$$C_b = -2\,e_1 + e_3 + e_4;$$
$$C_c = 3\,e_1 + 7\,e_2 + e_3 - 3\,e_4;$$
$$C_d = 4\,e_1 - 5\,e_2 + e_3 + e_4;$$
$$C_0 = 2\,e_1 + 9\,e_2 + 8\,e_3 + 7\,e_4;$$

Now we can use equation [2.34] directly

$$b == \frac{C_a \wedge C_0 \wedge C_c \wedge C_d}{C_a \wedge C_b \wedge C_c \wedge C_d}$$

and calculate the result directly by simplifying:

$$b == \frac{\star\mathcal{G}[C_a \wedge C_0 \wedge C_c \wedge C_d]}{\star\mathcal{G}[C_a \wedge C_b \wedge C_c \wedge C_d]}$$

$$b == -\frac{30}{41}$$

2.10 Simplicity

The concept of simplicity

An important concept in the Grassmann algebra is that of *simplicity*. Earlier we introduced the concept informally. Now we will discuss it in more detail.

An element is *simple* if it is the exterior product of 1-elements. We extend this definition to scalars by defining all scalars to be simple. Clearly also, since any n-element always reduces to a product of 1-elements, all n-elements are simple. Thus we see immediately that all 0-elements, 1-elements, and n-elements are simple. In the next section we show that all $(n-1)$-elements are also simple.

In 2-space therefore, all 0-, 1-, and 2-elements are simple. In 3-space all 0-, 1-, 2- and 3-elements are simple. In higher dimensional spaces elements of grade $2 \leq m \leq n-2$ are therefore the only ones that may not be simple. In 4-space, the only elements that may not be simple are those of grade 2.

There is a straightforward way of testing the simplicity of 2-elements not shared by elements of higher grade. *A 2-element is simple if and only if its exterior square is zero*. (The exterior square of an element A is $A \wedge A$.) Since odd elements (elements of odd grade) anti-commute, the exterior square of odd elements will be zero, even if they are not simple. An even element of grade 4 or higher may be of the form of the exterior product of a 1-element with a non-simple 3-element: whence its exterior square is zero without its being simple.

We will return to a further discussion of simplicity from the point of view of factorization in Chapter 3: The Regressive Product, section 3.9.

All $(n-1)$-elements are simple

In general, $(n-1)$-elements are simple. We can show this as follows.

Consider first two simple $(n-1)$-elements. Since they can differ by at most *one* 1-element factor (otherwise they would together contain more than n independent factors), we can express them as $A \wedge x$ and $A \wedge y$, where A is another simple element of grade n–2. Summing these, and factoring out A, gives a simple (n–1)-element.

$$A \wedge x + A \wedge y == A \wedge (x + y)$$

Any (n–1)-element can be expressed as the sum of simple (n–1)-elements. We can therefore prove the general case from the above by supposing pairs of simple elements to be combined to form another simple element, until just one simple element remains.

Conditions for simplicity of a 2-element in a 4-space

Consider a simple 2-element in a 4-dimensional space. First we declare a 4-dimensional basis for the space, and then compose a general bivector B using the alias $\star\mathbb{B}_a$ for `ComposeBivector`.

$$\star\mathcal{B}_4;\ B = \star\mathbb{B}_a$$

$$a_1\, e_1 \wedge e_2 + a_2\, e_1 \wedge e_3 + a_3\, e_1 \wedge e_4 + a_4\, e_2 \wedge e_3 + a_5\, e_2 \wedge e_4 + a_6\, e_3 \wedge e_4$$

Since we are supposing B to be simple, $B \wedge B$ is zero. To see how this constrains the coefficients a_i of the terms of B we expand and simplify the product.

$$\star\mathcal{G}[B \wedge B]$$

$$(2\, a_3\, a_4 - 2\, a_2\, a_5 + 2\, a_1\, a_6)\, e_1 \wedge e_2 \wedge e_3 \wedge e_4$$

We thus see that the condition for a 2-element in a 4-dimensional space to be simple may be written:

$$a_3\, a_4 - a_2\, a_5 + a_1\, a_6 == 0 \qquad 2.35$$

Conditions for simplicity of a 2-element in a 5-space

The situation in 5-space is a little more complex. First we declare a 5-space, and then compose a general bivector.

$$\star\mathcal{B}_5;\ B = \star\mathbb{B}_a$$

$$a_1\, e_1 \wedge e_2 + a_2\, e_1 \wedge e_3 + a_3\, e_1 \wedge e_4 + a_4\, e_1 \wedge e_5 + a_5\, e_2 \wedge e_3 + a_6\, e_2 \wedge e_4 + a_7\, e_2 \wedge e_5 + a_8\, e_3 \wedge e_4 + a_9\, e_3 \wedge e_5 + a_{10}\, e_4 \wedge e_5$$

Next, we expand and simplify its exterior square.

$$B_2 = \star\mathcal{G}[B \wedge B]$$

$$\begin{aligned}&(2\, a_3\, a_5 - 2\, a_2\, a_6 + 2\, a_1\, a_8)\, e_1 \wedge e_2 \wedge e_3 \wedge e_4 +\\ &(2\, a_4\, a_5 - 2\, a_2\, a_7 + 2\, a_1\, a_9)\, e_1 \wedge e_2 \wedge e_3 \wedge e_5 +\\ &(2\, a_4\, a_6 - 2\, a_3\, a_7 + 2\, a_1\, a_{10})\, e_1 \wedge e_2 \wedge e_4 \wedge e_5 +\\ &(2\, a_4\, a_8 - 2\, a_3\, a_9 + 2\, a_2\, a_{10})\, e_1 \wedge e_3 \wedge e_4 \wedge e_5 +\\ &(2\, a_7\, a_8 - 2\, a_6\, a_9 + 2\, a_5\, a_{10})\, e_2 \wedge e_3 \wedge e_4 \wedge e_5\end{aligned}$$

For the bivector B to be simple we require this product B_2 to be zero, hence each of its five coefficients must be zero simultaneously. We can extract the coefficients by using *GrassmannAlgebra's* `ExtractCoefficients` function, and then thread them into a list of five equations.

```
Bc = Thread[ExtractCoefficients[GradeBasis[4]][B₂] == 0]
```

$\{2\,a_3\,a_5 - 2\,a_2\,a_6 + 2\,a_1\,a_8 == 0,\; 2\,a_4\,a_5 - 2\,a_2\,a_7 + 2\,a_1\,a_9 == 0,\; 2\,a_4\,a_6 - 2\,a_3\,a_7 + 2\,a_1\,a_{10} == 0,\; 2\,a_4\,a_8 - 2\,a_3\,a_9 + 2\,a_2\,a_{10} == 0,\; 2\,a_7\,a_8 - 2\,a_6\,a_9 + 2\,a_5\,a_{10} == 0\}$

Finally, we can solve these five equations with *Mathematica*'s inbuilt `Solve` function.

```
Solve[Bc, {a₈, a₉, a₁₀}]
```

$$\left\{\left\{a_8 \to -\frac{a_3\,a_5 - a_2\,a_6}{a_1},\; a_9 \to -\frac{a_4\,a_5 - a_2\,a_7}{a_1},\; a_{10} \to -\frac{a_4\,a_6 - a_3\,a_7}{a_1}\right\}\right\}$$

By inspection we see that these three solutions correspond to the first three equations. Consequently, from the generality of the expression for the bivector we expect *any* three equations to suffice. (To see this note that any three equations contain just one coefficient common to all of them (a_1 in the example above). The solution equations can be written down immediately for the coefficients multiplying this common one (a_8, a_9, and a_{10} in this example).

As a second example solution, take a_7 as the common coefficient. We then get solutions for a_2, a_3, and a_8 which satisfy all the equations.

In summary we can say that a 2-element in a 5-space is simple if any three of the following equations are satisfied:

$$\begin{aligned} a_3\,a_5 - a_2\,a_6 + a_1\,a_8 &== 0 \\ a_4\,a_5 - a_2\,a_7 + a_1\,a_9 &== 0 \\ a_4\,a_6 - a_3\,a_7 + a_1\,a_{10} &== 0 \\ a_4\,a_8 - a_3\,a_9 + a_2\,a_{10} &== 0 \\ a_7\,a_8 - a_6\,a_9 + a_5\,a_{10} &== 0 \end{aligned} \qquad 2.36$$

- The concept of simplicity will be explored in more detail in Chapter 3: The Regressive Product.

Factorizing simple elements from first principles

In Chapter 3: The Regressive Product, we will develop an algorithm for obtaining a factorization of a simple element. In this section we discuss factorization from first principles using only the exterior product. To make the discussion concrete, we take an example.

Suppose we have a 4-element A which we would like to factorize. A is expressed in terms of six independent 1-elements, and is known to be simple. Our working space must therefore be at least six-dimensional, but to exemplify the generality of the computation we choose a 7-space.

For reference purposes we begin with the element in factored form A_0 to assure ourselves that it is simple, but the example will take its expanded form A as starting point. The factorization we will obtain is of course unlikely to be of the same form as A_0.

```
A₀ = (2 x + 3 y) ∧ (u - v - x) ∧ (2 w - 5 z + y + u) ∧ (z - x);
```

```
★𝓑₇; A = ★𝒢[A₀]
```

$3\,u\wedge v\wedge x\wedge y + 2\,u\wedge v\wedge x\wedge z + 3\,u\wedge v\wedge y\wedge z + 6\,u\wedge w\wedge x\wedge y + 4\,u\wedge w\wedge x\wedge z + 6\,u\wedge w\wedge y\wedge z - 14\,u\wedge x\wedge y\wedge z - 6\,v\wedge w\wedge x\wedge y - 4\,v\wedge w\wedge x\wedge z - 6\,v\wedge w\wedge y\wedge z + 17\,v\wedge x\wedge y\wedge z + 6\,w\wedge x\wedge y\wedge z$

Clearly, a factor of A must be a linear combination of u, v, w, x, y and z. Let us call it f_1. The s_i are scalar coefficients which we declare as scalar symbols by entering ★★S[s_].

```
★★S[s_]; f1 = s1 u + s2 v + s3 w + s4 x + s5 y + s6 z;
```

Now if f_1 is to be a factor of A, its exterior product with A must be zero. Hence

```
A1 = ★𝒢[f1 ∧ A] == 0
```

```
(-6 s1 - 6 s2 + 3 s3) u∧v∧w∧x∧y +
  (-4 s1 - 4 s2 + 2 s3) u∧v∧w∧x∧z + (-6 s1 - 6 s2 + 3 s3) u∧v∧w∧y∧z +
  (17 s1 + 14 s2 + 3 s4 - 2 s5 + 3 s6) u∧v∧x∧y∧z +
  (6 s1 + 14 s3 + 6 s4 - 4 s5 + 6 s6) u∧w∧x∧y∧z +
  (6 s2 - 17 s3 - 6 s4 + 4 s5 - 6 s6) v∧w∧x∧y∧z == 0
```

Since the coefficients of each of these terms must be zero, we now have some equations of constraint between the coefficients of f_1. In this case it is easy to see that the first 3 are identical, so we can satisfy them by substituting $2\,s_1 + 2\,s_2$ for s_3.

```
A2 = (A1 /. s3 → 2 s1 + 2 s2) // Simplify
```

```
(17 s1 + 14 s2 + 3 s4 - 2 s5 + 3 s6)
  (u∧v∧x∧y∧z + 2 u∧w∧x∧y∧z - 2 v∧w∧x∧y∧z) == 0
```

Similarly we can solve for s_5 to satisfy the remaining constraint, substitute back into the expression for f_1 to get an expression for a generic factor of A. which we call f_2. (The extra factor 2 is just to simplify the resulting terms).

```
f2 =
 FullSimplify[2 (f1 /. s3 → 2 s1 + 2 s2 /. s5 → 1/2 (17 s1 + 14 s2 + 3 s4 + 3 s6))]
```

```
(2 u + 4 w + 17 y) s1 + 2 (v + 2 w + 7 y) s2 + (2 x + 3 y) s4 + (3 y + 2 z) s6
```

A can now be constructed from the exterior product of any four linearly independent versions of f_2. As an example, we create 1-elements x_i by letting s_i equal 1, and the remaining s_j equal 0.

```
x1 = 2 u + 4 w + 17 y;
x2 = 2 (v + 2 w + 7 y);
x3 = 2 x + 3 y;
x4 = 3 y + 2 z;
```

The exterior product of these x_i is

```
A3 = ★𝒢[x1 ∧ x2 ∧ x3 ∧ x4]
```

```
24 u∧v∧x∧y + 16 u∧v∧x∧z + 24 u∧v∧y∧z + 48 u∧w∧x∧y +
 32 u∧w∧x∧z + 48 u∧w∧y∧z - 112 u∧x∧y∧z - 48 v∧w∧x∧y -
 32 v∧w∧x∧z - 48 v∧w∧y∧z + 136 v∧x∧y∧z + 48 w∧x∧y∧z
```

We can verify that this is 8 times A, hence giving us a factorization.

```
★𝒢[A == ★𝒢[1/8 x1 ∧ x2 ∧ x3 ∧ x4]]
```

```
True
```

◆ Applying the process to a non-simple element

Consider a second case of a non-simple element. For example

```
Y = x ∧ y + u ∧ v;
```

If Y had a 1-element factor, it would have to be some linear combination of x, y, u, and v.

```
f = s1 x + s2 y + s3 u + s4 v;
```

If f is a factor of Y, its exterior product with Y must be zero.

```
★G[f ∧ Y]
```

$s_1\, u \wedge v \wedge x + s_2\, u \wedge v \wedge y + s_3\, u \wedge x \wedge y + s_4\, v \wedge x \wedge y$

Clearly this expression can only be zero if all the coefficients s_i are zero, which in turn implies f must be zero. Hence *in this case*, Y has no 1-element factors.

More generally, an element being non-simple does not imply that it has no 1-element factors. Here is a non-simple element with two 1-element factors.

```
Z = (x ∧ y + u ∧ v) ∧ w ∧ z;

f = s1 x + s2 y + s3 u + s4 v + s5 w + s6 z;

★G[f ∧ Z]
```

$-s_1\, u \wedge v \wedge w \wedge x \wedge z - s_2\, u \wedge v \wedge w \wedge y \wedge z + s_3\, u \wedge w \wedge x \wedge y \wedge z + s_4\, v \wedge w \wedge x \wedge y \wedge z$

Hence s_1, s_2, s_3, s_4 must be zero, leaving $f = s_5\, w + s_6\, z$ as a generic factor of Z, which in turn implies that $w \wedge z$ is a factor.

2.11 Exterior Division

The definition of an exterior quotient

It will be seen in Chapter 4: Geometric Interpretations, that one way of defining geometric entities like lines and planes is by the exterior quotient of two interpreted elements. Such a quotient does not yield a unique element. Indeed this is why it is useful for defining geometric objects, for they are composed of *sets* of elements.

We define the exterior quotient of a simple (*m*+*k*)-element $\underset{m+k}{\alpha}$ by another simple element $\underset{k}{\beta}$ which is contained in $\underset{m+k}{\alpha}$ to be the most general *m*-element contained in $\underset{m+k}{\alpha}$ ($\underset{m}{\mu}$, say) such that the exterior product of the quotient $\underset{m}{\mu}$ with the denominator $\underset{k}{\beta}$ yields the numerator $\underset{m+k}{\alpha}$.

$$\underset{m}{\mu} == \frac{\underset{m+k}{\alpha}}{\underset{k}{\beta}} \quad \Longrightarrow \quad \underset{m}{\mu} \wedge \underset{k}{\beta} == \underset{m+k}{\alpha} \qquad 2.37$$

Note the convention adopted for the order of the factors.

Division by a 1-element

Consider the quotient of a simple (*m*+1)-element $\underset{m+1}{\alpha}$ by a 1-element β. Since $\underset{m+1}{\alpha}$ contains β we can write it as

$$\underset{m+1}{\alpha} == \alpha_1 \wedge \alpha_2 \wedge \ldots \wedge \alpha_m \wedge \beta$$

However, we could also have written $\underset{m+1}{\alpha}$ in the more general form

$$\underset{m+1}{\alpha} == (\alpha_1 + t_1 \beta) \wedge (\alpha_2 + t_2 \beta) \wedge \ldots \wedge (\alpha_m + t_m \beta) \wedge \beta$$

where the t_i are arbitrary scalars. It is in this more general form that the numerator must be written before β can be 'divided out'. Thus the quotient may be written

$$\frac{\alpha_1 \wedge \alpha_2 \wedge \ldots \wedge \alpha_m \wedge \beta}{\beta} == (\alpha_1 + t_1 \beta) \wedge (\alpha_2 + t_2 \beta) \wedge \ldots \wedge (\alpha_m + t_m \beta) \qquad 2.38$$

Division by a *k*-element

Suppose now we have a quotient with a simple *k*-element denominator. Since the denominator is contained in the numerator, we can write the quotient as

$$\frac{(\alpha_1 \wedge \alpha_2 \wedge \ldots \wedge \alpha_m) \wedge (\beta_1 \wedge \beta_2 \wedge \ldots \wedge \beta_k)}{\beta_1 \wedge \beta_2 \wedge \ldots \wedge \beta_k}$$

To prepare for the 'dividing out' of the β factors we rewrite the numerator in the more general form

$$(\alpha_1 + t_{11} \beta_1 + \ldots + t_{1k} \beta_k) \wedge (\alpha_2 + t_{21} \beta_1 + \ldots + t_{2k} \beta_k) \wedge \ldots \wedge$$
$$(\alpha_m + t_{m1} \beta_1 + \ldots + t_{mk} \beta_k) \wedge (\beta_1 \wedge \beta_2 \wedge \ldots \wedge \beta_k)$$

where the t_{ij} are arbitrary scalars. Hence:

$$\begin{aligned}\frac{(\alpha_1 \wedge \alpha_2 \wedge \ldots \wedge \alpha_m) \wedge (\beta_1 \wedge \beta_2 \wedge \ldots \wedge \beta_k)}{\beta_1 \wedge \beta_2 \wedge \ldots \wedge \beta_k} &== (\alpha_1 + t_{11} \beta_1 + \ldots + t_{1k} \beta_k) \wedge \\ &(\alpha_2 + t_{21} \beta_1 + \ldots + t_{2k} \beta_k) \wedge \ldots \wedge \\ &(\alpha_m + t_{m1} \beta_1 + \ldots + t_{mk} \beta_k)\end{aligned} \qquad 2.39$$

In the special case of m equal to 1, this reduces to:

$$\frac{\alpha \wedge \beta_1 \wedge \beta_2 \wedge \ldots \wedge \beta_k}{\beta_1 \wedge \beta_2 \wedge \ldots \wedge \beta_k} == \alpha + t_1 \beta_1 + \ldots + t_k \beta_k \qquad 2.40$$

We will later see that this formula neatly defines a multiplane.

◆ Special cases

In the special cases where $\underset{m}{\alpha}$ or $\underset{k}{\beta}$ is a scalar, the results are unique.

$$\frac{a\, \beta_1 \wedge \beta_2 \wedge \ldots \wedge \beta_k}{\beta_1 \wedge \beta_2 \wedge \cdots \wedge \beta_k} == a \qquad 2.41$$

$$\frac{b\,\alpha_1 \wedge \alpha_2 \wedge \ldots \wedge \alpha_m}{b} == \alpha_1 \wedge \alpha_2 \wedge \ldots \wedge \alpha_m \qquad 2.42$$

Automating the division process

By using the *GrassmannAlgebra* function `ExteriorQuotient` you can automatically generate the expression resulting from the exterior quotient of two exterior products. For example, in a 3-space, the exterior quotient of the basis trivector $e_1 \wedge e_2 \wedge e_3$ by the basis bivector $e_1 \wedge e_2$ results in the variable vector $e_3 + e_1 \star t_1 + e_2 \star t_2$, where the coefficients $\star t_1$ and $\star t_2$ are automatically generated arbitrary scalar symbols. (*Mathematica* automatically puts these scalar symbols *after* the basis elements because of its internal ordering routine).

$$\star\mathcal{B}_3;\ \text{ExteriorQuotient}\left[\frac{e_1 \wedge e_2 \wedge e_3}{e_1 \wedge e_2}\right]$$

$$e_3 + e_1 \star t_1 + e_2 \star t_2$$

The exterior quotient of the basis trivector $e_1 \wedge e_2 \wedge e_3$ by the basis vector e_1 results in the variable bivector $(e_2 + e_1 \star t_1) \wedge (e_3 + e_1 \star t_2)$.

$$\text{ExteriorQuotient}\left[\frac{e_1 \wedge e_2 \wedge e_3}{e_1}\right]$$

$$(e_2 + e_1 \star t_1) \wedge (e_3 + e_1 \star t_2)$$

`ExteriorQuotient` requires that the factors in the denominator be *explicitly* displayed in the numerator.

$$\text{ExteriorQuotient}\left[\frac{(x - y) \wedge (y - z) \wedge (z - x)}{(z - x) \wedge (x - y)}\right]$$

$$y - z + (x - y) \star t_1 + (-x + z) \star t_2$$

More generally, you can use `ExteriorQuotient` on any Grassmann expression involving exterior quotients.

$$\text{ExteriorQuotient}\left[a\,\frac{e_1 \wedge e_2 \wedge e_3}{e_1} + b\,\frac{e_1 \wedge e_2 \wedge e_3}{e_1 \wedge e_2} + c\,\frac{e_1 \wedge e_2 \wedge e_3}{e_1 \wedge e_2 \wedge e_3}\right]$$

$$c + b\,(e_3 + e_1 \star t_1 + e_2 \star t_2) + a\,(e_2 + e_1 \star t_3) \wedge (e_3 + e_1 \star t_4)$$

2.12 Multilinear Forms

The span of a simple element

We have already introduced the notion of the *space of a simple m-element* in section 2.3. The space of a simple m-element is the m-space whose underlying linear space consists of all the 1-elements in the m-element. A 1-element is said to be *in* a simple m-element if and only if their exterior product is zero. The space of a simple m-element is a subspace of the n-space in which it resides. An n-space and the space of its n-element are thus identical.

In order to explore the space of a simple m-element we can of course choose as basis for its underlying linear space, *any* set of m independent 1-elements whose exterior product is congruent to the given *m*-element. Such sets *span* the space of the simple *m*-element, and by extension, we call each of them *a span of the m-element*.

For example, if `A = x ∧ y ∧ z`, then we can say that *a* span of `A` is `{x, y, z}`. But of course, `{x + y, y, z}` and `{a x, y, z}` are also spans of `A`.

Practically however, the simple *m*-elements we are working with will often be available to us in some known factored form, that is, as a given exterior product of *m* 1-elements. In these cases it is convenient for computational purposes to choose these factors to span the element, and, by abuse of terminology, call the list of these factors *the* span of the element. (We can read "*the* span" to mean "*the current working* span" of the element).

For example, if `A = x ∧ y ∧ z`, then we say that *the* span of `A` is `{x, y, z}`.

More generally, we define *the k-span of a simple m-element* as the list of all the essentially different exterior products of grade *k* formed from the span (or 1-span) of the element.

Given this preamble, we can now say that the most straightforward way to conceive of the span of a simple *m*-element is as the basis of an *m*-space, where the basis is the list of factors of the *m*-element in order. Similarly the *k*-span of a simple *m*-element may be conceived of as the basis of the concomitant exterior product space of grade *k*. And with this conception we can now introduce the *k-cospan* of the element as the cobasis of this space of grade *k*.

The *k*-span of a simple *m*-element has no natural notion of invariance attached to it, because the *same m*-element, factorized differently, will generate different *k*-spans. On the other hand, the *pair* comprising the *k*-span of an *m*-element and its concomitant *k*-cospan will, if present together appropriately in a *multilinear form*, generate results invariant to different factorizations. Many expressions in the Grassmann algebra are multilinear forms of this nature, and an understanding of their properties is instructive and useful. We will explore these forms in the sections to follow. But first, we solidify what these definitions mean by seeing how to compose a span.

Composing spans

In this section we discuss how to compose the span (1-span), *k*-span or *k*-cospan of a simple *m*-element. As a concrete case, let us take a 4-element `A`. The compositions are independent of the dimension of the currently declared space, but any computations with them will require the dimension of the space to be at least that of the *m*-element. First we compose `A`. (By using `ComposeSimpleForm`, the α_i are automatically added to the list of declared vector symbols.)

$$\star\mathcal{B}_6;\ \mathbf{A} = \mathbf{ComposeSimpleForm}\left[\underset{4}{\alpha},\ 1\right]$$

$$\alpha_1 \wedge \alpha_2 \wedge \alpha_3 \wedge \alpha_4$$

To compose the span of `A`, you can use the *GrassmannAlgebra* function `ComposeSpan[A][1]` or its alternative representation $\star\mathbb{S}_1$`[A]`. To compose the 2-span of `A` you can use `ComposeSpan[A][2]` or its alternative $\star\mathbb{S}_2$`[A]`.

$$\{\mathbf{ComposeSpan[A][1]},\ \star\mathbb{S}_2\mathbf{[A]}\}$$

$$\{\{\alpha_1, \alpha_2, \alpha_3, \alpha_4\}, \{\alpha_1 \wedge \alpha_2, \alpha_1 \wedge \alpha_3, \alpha_1 \wedge \alpha_4, \alpha_2 \wedge \alpha_3, \alpha_2 \wedge \alpha_4, \alpha_3 \wedge \alpha_4\}\}$$

To compose the cospan of `A`, you can use `ComposeCospan[A][1]` or $\star\mathbb{S}^1$`[A]`. To compose the 2-cospan of `A` you can use `ComposeCospan[A][2]` or $\star\mathbb{S}^2$`[A]`. (You can enter the 2 in $\star\mathbb{S}^2$ as either a superscript or a power.) Note that while a *k*-span is a list of *k*-elements, a *k*-cospan is a list of (*m*-*k*)-elements.

$$\{\mathbf{ComposeCospan[A][1]},\ \star\mathbb{S}^2[\mathbf{A}]\}$$

$$\{\{\alpha_2\wedge\alpha_3\wedge\alpha_4,\ -(\alpha_1\wedge\alpha_3\wedge\alpha_4),\ \alpha_1\wedge\alpha_2\wedge\alpha_4,\ -(\alpha_1\wedge\alpha_2\wedge\alpha_3)\},$$
$$\{\alpha_3\wedge\alpha_4,\ -(\alpha_2\wedge\alpha_4),\ \alpha_2\wedge\alpha_3,\ \alpha_1\wedge\alpha_4,\ -(\alpha_1\wedge\alpha_3),\ \alpha_1\wedge\alpha_2\}\}$$

Since the exterior product is `Listable` (like all Grassmann products in *GrassmannAlgebra*), the exterior product of a k-span and its k-cospan gives a list of 4-elements which simplify to 6 copies of `A`.

$$\star\mathbb{S}_2[\mathbf{A}]\wedge\star\mathbb{S}^2[\mathbf{A}]$$

$$\{\alpha_1\wedge\alpha_2\wedge\alpha_3\wedge\alpha_4,\ \alpha_1\wedge\alpha_3\wedge-(\alpha_2\wedge\alpha_4),\ \alpha_1\wedge\alpha_4\wedge\alpha_2\wedge\alpha_3,$$
$$\alpha_2\wedge\alpha_3\wedge\alpha_1\wedge\alpha_4,\ \alpha_2\wedge\alpha_4\wedge-(\alpha_1\wedge\alpha_3),\ \alpha_3\wedge\alpha_4\wedge\alpha_1\wedge\alpha_2\}$$

In the discussion of multilinear forms to follow we will usually use the alternative representation based on $\star\mathbb{S}$ since its compact nature makes the composition of the forms clearer. Further syntax for using $\star\mathbb{S}$ follows.

- You can compose a span or a cospan for an m-element entered only as an underscripted symbol `x`, say. `ComposeSpan` and `ComposeCospan` automatically define the m-element as the exterior product of m 1-element factors based on the symbol `x`.

$$\star\mathbb{S}_2\left[\underset{4}{\mathbf{x}}\right]$$

$$\{x_1\wedge x_2,\ x_1\wedge x_3,\ x_1\wedge x_4,\ x_2\wedge x_3,\ x_2\wedge x_4,\ x_3\wedge x_4\}$$

$$\star\mathbb{S}^2\left[\underset{4}{\mathbf{x}}\right]$$

$$\{x_3\wedge x_4,\ -(x_2\wedge x_4),\ x_2\wedge x_3,\ x_1\wedge x_4,\ -(x_1\wedge x_3),\ x_1\wedge x_2\}$$

- To compose the k-spans or k-cospans for several values of k you can enter the values of k enclosed in `★[ ]`. For example

$$\star\mathbb{S}_{\star[1,3]}[\mathbf{A}]$$

$$\{\{\alpha_1,\ \alpha_2,\ \alpha_3,\ \alpha_4\},\ \{\alpha_1\wedge\alpha_2\wedge\alpha_3,\ \alpha_1\wedge\alpha_2\wedge\alpha_4,\ \alpha_1\wedge\alpha_3\wedge\alpha_4,\ \alpha_2\wedge\alpha_3\wedge\alpha_4\}\}$$

$$\star\mathbb{S}^{\star[1,3]}[\mathbf{A}]$$

$$\{\{\alpha_2\wedge\alpha_3\wedge\alpha_4,\ -(\alpha_1\wedge\alpha_3\wedge\alpha_4),\ \alpha_1\wedge\alpha_2\wedge\alpha_4,\ -(\alpha_1\wedge\alpha_2\wedge\alpha_3)\},\ \{\alpha_4,\ -\alpha_3,\ \alpha_2,\ -\alpha_1\}\}$$

- If you want to multiply the elements of a span by scalar coefficients you can use *Mathematica's* inbuilt listability attribute of `Times`, and simply multiply the span by the list of coefficients (making sure that the two lists are the same length). The alias `★★S` stands for `DeclareExtraScalarSymbols`, and is here used to declare any subscripted `s` as a scalar.

$$\star\star\mathbf{S}[\mathbf{s\_}];\ \{s_1,\ s_2,\ s_3,\ s_4,\ s_5,\ s_6\}\ \star\mathbb{S}_2[\mathbf{A}]$$

$$\{s_1\,\alpha_1\wedge\alpha_2,\ s_2\,\alpha_1\wedge\alpha_3,\ s_3\,\alpha_1\wedge\alpha_4,\ s_4\,\alpha_2\wedge\alpha_3,\ s_5\,\alpha_2\wedge\alpha_4,\ s_6\,\alpha_3\wedge\alpha_4\}$$

- If you want to multiply the elements of a span by scalar coefficients and then sum them, you can use *Mathematica's* `Dot` (`.`) function.

$$\{s_1,\ s_2,\ s_3,\ s_4,\ s_5,\ s_6\}.\star\mathbb{S}_2[\mathbf{A}]$$

$$s_1\,\alpha_1\wedge\alpha_2+s_2\,\alpha_1\wedge\alpha_3+s_3\,\alpha_1\wedge\alpha_4+s_4\,\alpha_2\wedge\alpha_3+s_5\,\alpha_2\wedge\alpha_4+s_6\,\alpha_3\wedge\alpha_4$$

- If you want to use some other type of listable product between your elements (other than `Times`) and then sum them, you can use the *GrassmannAlgebra* alias ⋆Σ for *Mathematica's* `Total` function which simply sums the elements of any list.

```
⋆Σ[{s1, s2, s3, s4, s5, s6} ∧ ⋆S2[A]]
```

$s_1 \wedge \alpha_1 \wedge \alpha_2 + s_2 \wedge \alpha_1 \wedge \alpha_3 + s_3 \wedge \alpha_1 \wedge \alpha_4 + s_4 \wedge \alpha_2 \wedge \alpha_3 + s_5 \wedge \alpha_2 \wedge \alpha_4 + s_6 \wedge \alpha_3 \wedge \alpha_4$

- To pick out the *i*th element in any of these lists you can use *Mathematica's* inbuilt `Part` function by giving a subscript to the list in the form ⟦*i*⟧. For example

```
⋆S2[A]⟦3⟧
```

$\alpha_1 \wedge \alpha_4$

Thus, for example, we can write A as the exterior product of any two corresponding terms from a *k*-span and a *k*-cospan.

```
⋆G[A == ⋆S2[A]⟦3⟧ ∧ ⋆S^2[A]⟦3⟧]
```

True

Example: Refactorizations

Let us suppose we have a simple element A for which we have a factorization, and that we wish to obtain a factorization that has a given 1-element x (which we know belongs to A) as one of the factors. Suppose A is expressed in terms of the basis of an 8-space.

```
⋆B8; A = (e4 - 2 e7) ∧ (e2 + 3 e5) ∧ (e1 + 2 e2); x = e1 - 6 e5;
```

By taking the 2-span of A and multiplying each of the elements by x we get 3 candidates.

```
⋆S2[A] ∧ x
```

$\{(e_4 - 2 e_7) \wedge (e_2 + 3 e_5) \wedge (e_1 - 6 e_5),$
$(e_4 - 2 e_7) \wedge (e_1 + 2 e_2) \wedge (e_1 - 6 e_5), (e_2 + 3 e_5) \wedge (e_1 + 2 e_2) \wedge (e_1 - 6 e_5)\}$

Expanding and simplifying these shows that the first two candidates are congruent and the third is zero.

```
⋆G[⋆S2[A] ∧ x]
```

$\{-(e_1 \wedge e_2 \wedge e_4) + 2 e_1 \wedge e_2 \wedge e_7 + 3 e_1 \wedge e_4 \wedge e_5 + 6 e_1 \wedge e_5 \wedge e_7 +$
$6 e_2 \wedge e_4 \wedge e_5 + 12 e_2 \wedge e_5 \wedge e_7, -2 e_1 \wedge e_2 \wedge e_4 + 4 e_1 \wedge e_2 \wedge e_7 +$
$6 e_1 \wedge e_4 \wedge e_5 + 12 e_1 \wedge e_5 \wedge e_7 + 12 e_2 \wedge e_4 \wedge e_5 + 24 e_2 \wedge e_5 \wedge e_7, 0\}$

In expanded form, A was equal to the first candidate

```
⋆G[A]
```

$-(e_1 \wedge e_2 \wedge e_4) + 2 e_1 \wedge e_2 \wedge e_7 + 3 e_1 \wedge e_4 \wedge e_5 +$
$6 e_1 \wedge e_5 \wedge e_7 + 6 e_2 \wedge e_4 \wedge e_5 + 12 e_2 \wedge e_5 \wedge e_7$

Hence valid refactorizations which include the required factor x are:

$$A == (e_4 - 2 e_7) \wedge (e_2 + 3 e_5) \wedge (e_1 - 6 e_5)$$

$$A == \frac{1}{2} (e_4 - 2 e_7) \wedge (e_1 + 2 e_2) \wedge (e_1 - 6 e_5)$$

Multilinear forms

In the chapters to follow in this volume we introduce three new *Grassmann products* in addition to the exterior product: the regressive, interior and inner products. And in Volume 2 we will introduce the generalized, hypercomplex and Clifford products. In the computation of expressions involving these products we will often come across formulae which express the computation as a sum of terms where each term itself is a product. These sums must have the same natural *invariance* to different factorizations possessed by the original expression, and it turns out they are all examples of *multilinear forms*.

For our purposes, a typical multilinear form is a function $M[\alpha_1, \alpha_2, \ldots, \alpha_m]$ of 1-elements α_i which is linear in each of the α_i. That is,

$$M[\alpha_1, \alpha_2, \ldots, a_1 \beta_1 + a_2 \beta_2, \ldots, \alpha_m] = a_1 M[\alpha_1, \alpha_2, \ldots, \beta_1, \ldots, \alpha_m] + a_2 M[\alpha_1, \alpha_2, \ldots, \beta_2, \ldots, \alpha_m]$$

where the a_i are scalars and the β_i are also 1-elements. In this book however, because the M are usually viewed as products, we write them in their *infix* form, just as we have seen that *Mathematica* allows us to do with the exterior product: entering an expression in the argument form above displays it in its infix form.

$$\text{Wedge}[\alpha_1, \alpha_2, \ldots, a_1 \beta_1 + a_2 \beta_2, \ldots, \alpha_m] = a_1 \text{Wedge}[\alpha_1, \alpha_2, \ldots, \beta_1, \ldots, \alpha_m] + a_2 \text{Wedge}[\alpha_1, \alpha_2, \ldots, \beta_2, \ldots, \alpha_m]$$

$$\alpha_1 \wedge \alpha_2 \wedge \ldots \wedge (a_1 \beta_1 + a_2 \beta_2) \wedge \ldots \wedge \alpha_m = a_1 \alpha_1 \wedge \alpha_2 \wedge \ldots \wedge \beta_1 \wedge \ldots \wedge \alpha_m + a_2 \alpha_1 \wedge \alpha_2 \wedge \ldots \wedge \beta_2 \wedge \ldots \wedge \alpha_m$$

More specifically, we say that a product operation $\otimes$ is a *bilinear product* if it is both left and right distributive under addition, and permits scalars to be factored out.

$$\begin{gathered}(A + B) \otimes C = A \otimes C + B \otimes C \\ A \otimes (B + C) = A \otimes B + A \otimes C \\ A \otimes (a B) = (a A) \otimes B = a (A \otimes B)\end{gathered} \qquad 2.43$$

In the chapters to follow, the multilinear forms of interest may involve two, or sometimes three, different bilinear or multilinear product operations. For simplicity we will refer to any of these as *linear products*. To represent these we choose three initially undefined infix operators from *Mathematica's* library of symbols, $\otimes$, $\oplus$, and $\odot$ (`CircleTimes`, `CirclePlus`, `CircleDot`). We make them `Listable` (as all the Grassmann products are), and make them behave linearly within any functions we define intended to operate on expressions involving them.

As a beginning example, consider a simple 3-element $A = \alpha_1 \wedge \alpha_2 \wedge \alpha_3$ in a space of at least 3 dimensions, the three operations $\otimes$, $\oplus$, and $\odot$, and two arbitrary Grassmann expressions G_1, and G_2. From these we can form the expression

$$F = (G_1 \otimes \alpha_1) \odot (G_2 \oplus (\alpha_2 \wedge \alpha_3))$$

Now if we replace α_2, say, by $a_1 \beta_1 + a_2 \beta_2$, we can expand the expression to get a sum of two expressions.

$$\begin{aligned}F &= (G_1 \otimes \alpha_1) \odot (G_2 \oplus ((a_1 \beta_1 + a_2 \beta_2) \wedge \alpha_3)) \\ &= (G_1 \otimes \alpha_1) \odot (G_2 \oplus (a_1 \beta_1 \wedge \alpha_3 + a_2 \beta_2 \wedge \alpha_3)) \\ &= (G_1 \otimes \alpha_1) \odot (a_1 G_2 \oplus (\beta_1 \wedge \alpha_3) + a_2 G_2 \oplus (\beta_2 \wedge \alpha_3)) \\ &= a_1 (G_1 \otimes \alpha_1) \odot (G_2 \oplus (\beta_1 \wedge \alpha_3)) + a_2 (G_1 \otimes \alpha_1) \odot (G_2 \oplus (\beta_2 \wedge \alpha_3))\end{aligned}$$

Since the individual product operations are linear, the whole expression is multilinear: substituting for any of the G_i or α_i with a sum of terms will lead to a sum of expressions.

In later applications these linear products will of course be Grassmann products (exterior, regressive, interior, inner, generalized, hypercomplex or Clifford), so the multilinear form F will simply be a Grassmann expression (an element of the algebra under consideration).

For example, using the exterior, regressive and interior products we are able to write

$$(G_1 \ominus \alpha_1) \vee (G_2 \ominus ((a_1 \beta_1 + a_2 \beta_2) \wedge \alpha_3))$$
$$== a_1 (G_1 \ominus \alpha_1) \vee (G_2 \ominus (\beta_1 \wedge \alpha_3)) + a_2 (G_1 \ominus \alpha_1) \vee (G_2 \ominus (\beta_2 \wedge \alpha_3))$$

Defining m:k-forms

In this section we define in more detail the particular types of multilinear forms we will find useful later in the book. We will find it convenient to call them *m:k-forms* because, for any given simple *m*-element, there are forms of grade *k*, where *k* runs from 0 to *m*. To begin, consider a simple *m*-element A.

$$A == \alpha_1 \wedge \alpha_2 \wedge \ldots \wedge \alpha_k \wedge \alpha_{k+1} \wedge \ldots \wedge \alpha_m$$

The *k*-span and *k*-cospan of A are denoted $\star\mathbb{S}_k[A]$ and $\star\mathbb{S}^k[A]$. Each of these lists contain $\nu = \binom{m}{k}$ elements of grade *k* and grade *m–k* respectively. The *i*th element of any list is denoted with a subscripted $[\![i]\!]$.

Hence A can always be written as

$$A == \star\mathbb{S}_k[A]_{[\![i]\!]} \wedge \star\mathbb{S}^k[A]_{[\![i]\!]} \qquad 2.44$$

where A is a simple *m*-element, *k* can range from 0 to *m*, and *i* can range from 1 to $\nu = \binom{m}{k}$.

A generic *m:k-form* for A is denoted $F_{m,k}$ and can now be defined as

$$F_{m,k} == \sum_{i=1}^{\nu} \left(G_1 \oplus \star\mathbb{S}_k[A]_{[\![i]\!]}\right) \odot \left(G_2 \otimes \star\mathbb{S}^k[A]_{[\![i]\!]}\right)$$
$$A == \star\mathbb{S}_k[A]_{[\![i]\!]} \wedge \star\mathbb{S}^k[A]_{[\![i]\!]} \qquad 2.45$$

Here, A is a simple *m*-element, $0 \le k \le m$, G_1 and G_2 are *as yet undefined* Grassmann expressions, and $\oplus$, $\odot$, and $\otimes$ are *as yet undefined* linear products. The form is intended to be generic so that it includes all its special cases, for example

$$
\begin{array}{c}
\sum_{i=1}^{\nu}\left(G_1 \oplus \star\mathbb{S}_k[A]_{[\![i]\!]}\right) \odot \left(G_2 \otimes \star\mathbb{S}^k[A]_{[\![i]\!]}\right) \\
\sum_{i=1}^{\nu}\left(G_1 \oplus \star\mathbb{S}_k[A]_{[\![i]\!]}\right) \odot \left(\star\mathbb{S}^k[A]_{[\![i]\!]}\right) \\
\sum_{i=1}^{\nu}\left(\star\mathbb{S}_k[A]_{[\![i]\!]}\right) \odot \left(G_2 \otimes \star\mathbb{S}^k[A]_{[\![i]\!]}\right) \\
\sum_{i=1}^{\nu}\left(\star\mathbb{S}_k[A]_{[\![i]\!]}\right) \odot \left(\star\mathbb{S}^k[A]_{[\![i]\!]}\right)
\end{array}
\qquad 2.46
$$

Composing m:k-forms

All the product operations we will be dealing with are defined with the *Mathematica* attribute `Listable`. This means that to build up expressions, you only need to enter a product of lists (with the same number of elements in each), or the product of a single element and a list, to get a list of the products. To see how this works in building up forms, we can take the simple case of $A = \alpha_1 \wedge \alpha_2 \wedge \alpha_3$. Note that the exterior product is of highest precedence.

$$\star\star V[\alpha\_];\ A = \alpha_1 \wedge \alpha_2 \wedge \alpha_3;\ \{\star\mathbb{S}_1[A],\ \star\mathbb{S}^1[A]\}$$

$$\{\{\alpha_1, \alpha_2, \alpha_3\}, \{\alpha_2 \wedge \alpha_3, -(\alpha_1 \wedge \alpha_3), \alpha_1 \wedge \alpha_2\}\}$$

$$\{G_1 \oplus \star\mathbb{S}_1[A],\ G_2 \otimes \star\mathbb{S}^1[A]\}$$

$$\{\{G_1 \oplus \alpha_1, G_1 \oplus \alpha_2, G_1 \oplus \alpha_3\}, \{G_2 \otimes \alpha_2 \wedge \alpha_3, G_2 \otimes -(\alpha_1 \wedge \alpha_3), G_2 \otimes \alpha_1 \wedge \alpha_2\}\}$$

$$(G_1 \oplus \star\mathbb{S}_1[A]) \odot (G_2 \otimes \star\mathbb{S}^1[A])$$

$$\{(G_1 \oplus \alpha_1) \odot (G_2 \otimes \alpha_2 \wedge \alpha_3),\ (G_1 \oplus \alpha_2) \odot (G_2 \otimes -(\alpha_1 \wedge \alpha_3)),\ (G_1 \oplus \alpha_3) \odot (G_2 \otimes \alpha_1 \wedge \alpha_2)\}$$

If at any stage you want to sum the terms of a list, you can apply *Mathematica's* `Total` function, or its *GrassmannAlgebra* alias $\star\Sigma$.

$$\star\Sigma[(G_1 \oplus \star\mathbb{S}_1[A]) \odot (G_2 \otimes \star\mathbb{S}^1[A])]$$

$$(G_1 \oplus \alpha_1) \odot (G_2 \otimes \alpha_2 \wedge \alpha_3) + (G_1 \oplus \alpha_2) \odot (G_2 \otimes -(\alpha_1 \wedge \alpha_3)) + (G_1 \oplus \alpha_3) \odot (G_2 \otimes \alpha_1 \wedge \alpha_2)$$

Because all *GrassmannAlgebra's* product operations are defined to be listable you can compose the forms in [2.46] above without explicitly indexing the terms (i) and specifying their number (ν). All you need to do is eliminate the index and replace the summation sign with $\star\Sigma$.

$$
\begin{array}{ll}
\star\Sigma[(G_1 \oplus \star\mathbb{S}_k[A]) \odot (G_2 \otimes \star\mathbb{S}^k[A])] & \star\Sigma[(\star\mathbb{S}_k[A]) \odot (\star\mathbb{S}^k[A])] \\
\star\Sigma[(G_1 \oplus \star\mathbb{S}_k[A]) \odot (\star\mathbb{S}^k[A])] & \star\Sigma[(\star\mathbb{S}_k[A]) \odot (G_2 \otimes \star\mathbb{S}^k[A])]
\end{array}
\qquad 2.47
$$

For example if $A = \alpha_1 \wedge \alpha_2 \wedge \alpha_3$, the following two expressions give the same result. (Note that we must use *Mathematica*'s inbuilt summation sign.

$$\sum_{i=1}^{3}\left(G_1 \oplus \star\mathbb{S}_1[A]_{[\![i]\!]}\right) \odot \left(G_2 \otimes \star\mathbb{S}^1[A]_{[\![i]\!]}\right)$$

$$(G_1 \oplus \alpha_1) \odot (G_2 \otimes \alpha_2 \wedge \alpha_3) + (G_1 \oplus \alpha_2) \odot (G_2 \otimes -(\alpha_1 \wedge \alpha_3)) + (G_1 \oplus \alpha_3) \odot (G_2 \otimes \alpha_1 \wedge \alpha_2)$$

```
★Σ[(G₁ ⊕ ★S₁[A]) ⊙ (G₂ ⊗ ★S¹[A])]
```

$(G_1 \oplus \alpha_1) \odot (G_2 \otimes \alpha_2 \wedge \alpha_3) + (G_1 \oplus \alpha_2) \odot (G_2 \otimes -(\alpha_1 \wedge \alpha_3)) + (G_1 \oplus \alpha_3) \odot (G_2 \otimes \alpha_1 \wedge \alpha_2)$

Expanding and simplifying m:k-forms

If you want to expand and simplify an m:k-form, you can use the *GrassmannAlgebra* function `ExpandAndSimplifyForm` or its alias ★$\mathcal{F}$. This function distributes products over sums, and factors scalars out of terms. For example, consider the following form:

```
F = (G₁ ⊕ (a x + b y)) ⊙ (G₂ ⊗ (c z));
```

Note that, because there are a number of product operators with their own precedence, it is advisable to assure any expected behaviour by using parentheses in the input expression.

```
{ExpandAndSimplifyForm[F], ★ℱ[F]}
```

$\{a\,c\,(G_1 \oplus x) \odot (G_2 \otimes z) + b\,c\,(G_1 \oplus y) \odot (G_2 \otimes z),$
$a\,c\,(G_1 \oplus x) \odot (G_2 \otimes z) + b\,c\,(G_1 \oplus y) \odot (G_2 \otimes z)\}$

As another example let `A` be a 3-element and $F_{3,1}$ be a form based on it.

```
A = α₁ ∧ α₂ ∧ α₃; F₃,₁ = ★Σ[(G₁ ⊕ ★S₁[A]) ⊙ (G₂ ⊗ ★S¹[A])]
```

$(G_1 \oplus \alpha_1) \odot (G_2 \otimes \alpha_2 \wedge \alpha_3) + (G_1 \oplus \alpha_2) \odot (G_2 \otimes -(\alpha_1 \wedge \alpha_3)) + (G_1 \oplus \alpha_3) \odot (G_2 \otimes \alpha_1 \wedge \alpha_2)$

Now if α_1 is expressed as a linear combination of other 1-elements say, so that `A` is given by:

```
A = (a x + b y) ∧ α₂ ∧ α₃;
```

the form is now composed as

```
F₃,₁ = ★Σ[(G₁ ⊕ ★S₁[A]) ⊙ (G₂ ⊗ ★S¹[A])]
```

$(G_1 \oplus (a\,x + b\,y)) \odot (G_2 \otimes \alpha_2 \wedge \alpha_3) +$
$(G_1 \oplus \alpha_2) \odot (G_2 \otimes -((a\,x + b\,y) \wedge \alpha_3)) + (G_1 \oplus \alpha_3) \odot (G_2 \otimes (a\,x + b\,y) \wedge \alpha_2)$

Expanding and simplifying this new form with ★$\mathcal{F}$ now gives

```
★ℱ[F₃,₁]
```

$a\,(G_1 \oplus x) \odot (G_2 \otimes \alpha_2 \wedge \alpha_3) + b\,(G_1 \oplus y) \odot (G_2 \otimes \alpha_2 \wedge \alpha_3) - a\,(G_1 \oplus \alpha_2) \odot (G_2 \otimes x \wedge \alpha_3) -$
$b\,(G_1 \oplus \alpha_2) \odot (G_2 \otimes y \wedge \alpha_3) + a\,(G_1 \oplus \alpha_3) \odot (G_2 \otimes x \wedge \alpha_2) + b\,(G_1 \oplus \alpha_3) \odot (G_2 \otimes y \wedge \alpha_2)$

Developing invariant forms

Our objective in this section is to understand the construction of forms *which are invariant to precisely the same types of substitutions that leave an m-element invariant*. For example, a 3-element $A = \alpha_1 \wedge \alpha_2 \wedge \alpha_3$ is left invariant if to any of its 1-element factors is added a linear combination of its other 1-element factors. For example, `A` is unchanged if to α_1 we add a α_2.

```
A = α₁ ∧ α₂ ∧ α₃; Aₐ = A /. α₁ → α₁ + a α₂
```

$(\alpha_1 + a\,\alpha_2) \wedge \alpha_2 \wedge \alpha_3$

```
★𝒢[A == Aₐ]
```

```
True
```

Now consider the expression `F` involving the factors of `A` and make the same substitution.

$$F = (G_1 \otimes \alpha_1) \odot (G_2 \oplus (\alpha_2 \wedge \alpha_3));\ F_a = F\ /.\ \alpha_1 \to \alpha_1 + a\,\alpha_2$$

$$(G_1 \otimes (\alpha_1 + a\,\alpha_2)) \odot (G_2 \oplus \alpha_2 \wedge \alpha_3)$$

Applying $\star\mathcal{F}$ to F_a effects the expansion and extracts the scalar coefficient.

$$\star\mathcal{F}[F_a]$$

$$(G_1 \otimes \alpha_1) \odot (G_2 \oplus \alpha_2 \wedge \alpha_3) + a\,(G_1 \otimes \alpha_2) \odot (G_2 \oplus \alpha_2 \wedge \alpha_3)$$

If the form F had been invariant to this transformation as the 3-element A was, F_a would have simplified back again to F. But F_a now has an extra term which we notice displays α_2 twice, but in different parts of the product. This isolates the two occurrences of α_2 so that the antisymmetry of the exterior product cannot put their product to zero.

This suggests that we should *antisymmetrize* the original form F to create a new form F_b by adding a term featuring the same factors α_i of A but reordered to bring α_2 to first place. If we do this we find F_b is invariant to the substitution $\alpha_1 \to \alpha_1 + a\,\alpha_2$.

$$F_b = (G_1 \otimes \alpha_1) \odot (G_2 \oplus (\alpha_2 \wedge \alpha_3)) + (G_1 \otimes \alpha_2) \odot (G_2 \oplus (-(\alpha_1 \wedge \alpha_3)));$$

$$F_c = F_b\ /.\ \alpha_1 \to \alpha_1 + a\,\alpha_2$$

$$(G_1 \otimes \alpha_2) \odot (G_2 \oplus -((\alpha_1 + a\,\alpha_2) \wedge \alpha_3)) + (G_1 \otimes (\alpha_1 + a\,\alpha_2)) \odot (G_2 \oplus \alpha_2 \wedge \alpha_3)$$

$$\star\mathcal{F}[F_b == F_c]$$

True

Of course, if we want a form which is antisymmetric with respect to substitution of *all* of its factors α_i, we will need to include *all* the essentially different decompositions of the 3-element into a 1-element and a 2-element.

$$F_d = (G_1 \otimes \alpha_1) \odot (G_2 \oplus (\alpha_2 \wedge \alpha_3)) + (G_1 \otimes \alpha_2) \odot (G_2 \oplus (-(\alpha_1 \wedge \alpha_3))) + (G_1 \otimes \alpha_3) \odot (G_2 \oplus (\alpha_1 \wedge \alpha_2));$$

Now, replacing α_1 by $\alpha_1 + a\,\alpha_2 + b\,\alpha_3$ in this form leaves the form invariant to the substitution.

$$F_e = F_d\ /.\ \alpha_1 \to \alpha_1 + a\,\alpha_2 + b\,\alpha_3$$

$$(G_1 \otimes \alpha_2) \odot (G_2 \oplus -((\alpha_1 + a\,\alpha_2 + b\,\alpha_3) \wedge \alpha_3)) + (G_1 \otimes \alpha_3) \odot (G_2 \oplus (\alpha_1 + a\,\alpha_2 + b\,\alpha_3) \wedge \alpha_2) + (G_1 \otimes (\alpha_1 + a\,\alpha_2 + b\,\alpha_3)) \odot (G_2 \oplus \alpha_2 \wedge \alpha_3)$$

$$\star\mathcal{F}[F_d == F_e]$$

True

The form F_d may be recognized as the m:k-form $F_{3,1}$ introduced above in our discussion on expanding and simplifying m:k-forms. That is, when $A = \alpha_1 \wedge \alpha_2 \wedge \alpha_3$, $F_{3,1} = F_d$.

$$F_{3,1} = \star\Sigma\left[(G_1 \oplus \star\mathbb{S}_1[A]) \odot \left(G_2 \otimes \star\mathbb{S}^1[A]\right)\right]$$

$$(G_1 \oplus \alpha_1) \odot (G_2 \otimes \alpha_2 \wedge \alpha_3) + (G_1 \oplus \alpha_2) \odot (G_2 \otimes -(\alpha_1 \wedge \alpha_3)) + (G_1 \oplus \alpha_3) \odot (G_2 \otimes \alpha_1 \wedge \alpha_2)$$

The invariance of m:k-forms

To this point we have discussed simple *m*-elements A and the m:k-forms $F_{m,k}$ based on them.

$$A == \alpha_1 \wedge \alpha_2 \wedge \ldots \wedge \alpha_k \wedge \alpha_{k+1} \wedge \ldots \wedge \alpha_m$$

$$F_{m,k} == \star\Sigma\left[(G_1 \oplus \star\mathbb{S}_k[A]) \odot \left(G_2 \otimes \star\mathbb{S}^k[A]\right)\right]$$

Our interest is to show that *any transformation which leaves* A *invariant, also leaves any of its associated m:k-forms* $F_{m,k}$ *invariant*. These transformations are equivalent to refactorizations of the m-element: we want to show that *an m:k-form is unchanged no matter what factorization of* A *we use*.

For concreteness, let us take A to be a 3-element as we have in previous sections.

$$\mathbf{A} = \alpha_1 \wedge \alpha_2 \wedge \alpha_3;$$

A is of course invariant to any of its factorizations. Any factorization A′ of A is congruent to the exterior product of three independent linear combinations of the α_i.

$$\mathbf{A}' = (a_{1,1}\,\alpha_1 + a_{1,2}\,\alpha_2 + a_{1,3}\,\alpha_3) \wedge (a_{2,1}\,\alpha_1 + a_{2,2}\,\alpha_2 + a_{2,3}\,\alpha_3) \wedge (a_{3,1}\,\alpha_1 + a_{3,2}\,\alpha_2 + a_{3,3}\,\alpha_3);$$

Expanding and simplifying A′ gives, as expected, the original factorization A multiplied by the determinant of the coefficients. (But first we need to declare the scalar and vector symbols.)

$$\star\star\mathbf{S}[\mathbf{a\_}];\ \star\star\mathbf{V}[\alpha\_];\ \star\mathcal{G}[\mathbf{A}']$$

$$(-a_{1,3}\,a_{2,2}\,a_{3,1} + a_{1,2}\,a_{2,3}\,a_{3,1} + a_{1,3}\,a_{2,1}\,a_{3,2} - a_{1,1}\,a_{2,3}\,a_{3,2} - a_{1,2}\,a_{2,1}\,a_{3,3} + a_{1,1}\,a_{2,2}\,a_{3,3})\ \alpha_1 \wedge \alpha_2 \wedge \alpha_3$$

$$\mathbf{det} = \mathbf{Det}[\{\{a_{1,1}, a_{1,2}, a_{1,3}\}, \{a_{2,1}, a_{2,2}, a_{2,3}\}, \{a_{3,1}, a_{3,2}, a_{3,3}\}\}]$$

$$-a_{1,3}\,a_{2,2}\,a_{3,1} + a_{1,2}\,a_{2,3}\,a_{3,1} + a_{1,3}\,a_{2,1}\,a_{3,2} - a_{1,1}\,a_{2,3}\,a_{3,2} - a_{1,2}\,a_{2,1}\,a_{3,3} + a_{1,1}\,a_{2,2}\,a_{3,3}$$

Thus, any set of coefficients whose determinant is unity will provide a transformation to which A is invariant. Now let us compose the four 3:k-forms for A, and for A′.

$$\mathbf{F}_{3,k} = \star\mathcal{F}\left[\mathbf{Table}\left[\star\Sigma\left[(\mathbf{G}_1 \oplus \star\mathbb{S}_k[\mathbf{A}]) \odot (\mathbf{G}_2 \otimes \star\mathbb{S}^k[\mathbf{A}])\right], \{\mathbf{k}, 0, 3\}\right]\right]$$

$$\{(G_1 \oplus 1) \odot (G_2 \otimes \alpha_1 \wedge \alpha_2 \wedge \alpha_3),\ (G_1 \oplus \alpha_1) \odot (G_2 \otimes \alpha_2 \wedge \alpha_3) - (G_1 \oplus \alpha_2) \odot (G_2 \otimes \alpha_1 \wedge \alpha_3) + (G_1 \oplus \alpha_3) \odot (G_2 \otimes \alpha_1 \wedge \alpha_2),\ (G_1 \oplus \alpha_1 \wedge \alpha_2) \odot (G_2 \otimes \alpha_3) - (G_1 \oplus \alpha_1 \wedge \alpha_3) \odot (G_2 \otimes \alpha_2) + (G_1 \oplus \alpha_2 \wedge \alpha_3) \odot (G_2 \otimes \alpha_1),\ (G_1 \oplus \alpha_1 \wedge \alpha_2 \wedge \alpha_3) \odot (G_2 \otimes 1)\}$$

$$\mathbf{F'}_{3,k} = \left(\star\mathcal{F}\left[\mathbf{Table}\left[\star\Sigma\left[(\mathbf{G}_1 \oplus \star\mathbb{S}_k[\mathbf{A}']) \odot (\mathbf{G}_2 \otimes \star\mathbb{S}^k[\mathbf{A}'])\right], \{\mathbf{k}, 0, 3\}\right]\right] // \mathbf{Simplify}\right) /.\ \{\mathbf{Simplify}[\mathbf{det}] \to \mathbb{D}, \mathbf{Simplify}[-\mathbf{det}] \to -\mathbb{D}\}$$

$$\{\mathbb{D}\ (G_1 \oplus 1) \odot (G_2 \otimes \alpha_1 \wedge \alpha_2 \wedge \alpha_3),\ \mathbb{D}\ ((G_1 \oplus \alpha_1) \odot (G_2 \otimes \alpha_2 \wedge \alpha_3) - (G_1 \oplus \alpha_2) \odot (G_2 \otimes \alpha_1 \wedge \alpha_3) + (G_1 \oplus \alpha_3) \odot (G_2 \otimes \alpha_1 \wedge \alpha_2)),\ \mathbb{D}\ ((G_1 \oplus \alpha_1 \wedge \alpha_2) \odot (G_2 \otimes \alpha_3) - (G_1 \oplus \alpha_1 \wedge \alpha_3) \odot (G_2 \otimes \alpha_2) + (G_1 \oplus \alpha_2 \wedge \alpha_3) \odot (G_2 \otimes \alpha_1)),\ \mathbb{D}\ (G_1 \oplus \alpha_1 \wedge \alpha_2 \wedge \alpha_3) \odot (G_2 \otimes 1)\}$$

Here we have used *Mathematica's* inbuilt `Simplify` to extract the determinant of the transformation and denote it symbolically by $\mathbb{D}$. If we divide by $\mathbb{D}$, we can easily see that this list of forms is equal to the original list of forms.

$$\mathbf{F}_{3,k} == \frac{\mathbf{F'}_{3,k}}{\mathbb{D}}$$

```
True
```

Thus transformations (factorizations) which leave the 3-element A invariant, also leave all its 3:k-forms invariant.

The complete span of a simple element

In this section we discuss the notion of complete span and cospan and a form based on it. We will use the result derived for the form to develop alternative formulations of the General Product Formula in the next chapter.

The *complete span* of a simple m-element A is denoted $\star S_\star[A]$ and defined to be the ordered list of the k-spans of A, with k ranging from 0 to m.

Similarly the *complete cospan* of a simple m-element A is denoted $\star S^\star[A]$ and defined to be the ordered list of the k-cospans of A, with k ranging from 0 to m.

The complete span of the n-element of an algebra is simply the GrassmannBases of the algebra. Correspondingly, the complete cospan of the n-element of the algebra is the GrassmannCobases of the algebra.

For example, if A = u ∧ x ∧ y ∧ z and we are working in space of at least 4 dimensions (say, 6), the complete span and complete cospan of A are given by

$\star\mathcal{B}_6$; A = u ∧ x ∧ y ∧ z;

$\star S_\star[A]$

{{1}, {u, x, y, z}, {u ∧ x, u ∧ y, u ∧ z, x ∧ y, x ∧ z, y ∧ z},
{u ∧ x ∧ y, u ∧ x ∧ z, u ∧ y ∧ z, x ∧ y ∧ z}, {u ∧ x ∧ y ∧ z}}

$\star S^\star[A]$

{{u ∧ x ∧ y ∧ z}, {x ∧ y ∧ z, -(u ∧ y ∧ z), u ∧ x ∧ z, -(u ∧ x ∧ y)},
{y ∧ z, -(x ∧ z), x ∧ y, u ∧ z, -(u ∧ y), u ∧ x}, {z, -y, x, -u}, {1}}

Since the exterior product is Listable, simplifying the exterior product of a complete span with its complete cospan gives lists of the original m-element.

$S_1 = \star S\big[\star S_\star[A] \wedge \star S^\star[A]\big]$

{{u ∧ x ∧ y ∧ z}, {u ∧ x ∧ y ∧ z, u ∧ x ∧ y ∧ z, u ∧ x ∧ y ∧ z, u ∧ x ∧ y ∧ z},
{u ∧ x ∧ y ∧ z, u ∧ x ∧ y ∧ z, u ∧ x ∧ y ∧ z, u ∧ x ∧ y ∧ z, u ∧ x ∧ y ∧ z, u ∧ x ∧ y ∧ z},
{u ∧ x ∧ y ∧ z, u ∧ x ∧ y ∧ z, u ∧ x ∧ y ∧ z, u ∧ x ∧ y ∧ z}, {u ∧ x ∧ y ∧ z}}

Reversing the order by taking the exterior product of a complete cospan with a complete span give a similar result, but this time the exterior products of the k-cospan elements with the k-span elements include a sign $(-1)^{k(m-k)}$.

$S_2 = \star S\big[\star S^\star[A] \wedge \star S_\star[A]\big]$

{{u ∧ x ∧ y ∧ z},
{-(u ∧ x ∧ y ∧ z), -(u ∧ x ∧ y ∧ z), -(u ∧ x ∧ y ∧ z), -(u ∧ x ∧ y ∧ z)},
{u ∧ x ∧ y ∧ z, u ∧ x ∧ y ∧ z, u ∧ x ∧ y ∧ z, u ∧ x ∧ y ∧ z, u ∧ x ∧ y ∧ z, u ∧ x ∧ y ∧ z},
{-(u ∧ x ∧ y ∧ z), -(u ∧ x ∧ y ∧ z), -(u ∧ x ∧ y ∧ z), -(u ∧ x ∧ y ∧ z)},
{u ∧ x ∧ y ∧ z}}

The *list* of signs which differentiate this list from the previous one may be computed from the grades of the elements in the span (k) and cospan (m–k) lists. This is the list equivalent of the change of sign effected by interchanging the factors in any exterior product.

$(-1)^{\star G[\star S^\star[A]] \, \star G[\star S_\star[A]]}$

{{1}, {-1, -1, -1, -1}, {1, 1, 1, 1, 1, 1}, {-1, -1, -1, -1}, {1}}

If m (the grade of A) is even, as in the above example, then $k(m-k)$ will alternate between even and odd as k ranges from 0 to m. If m is odd then $k(m-k)$ will be even for all k, and hence there will be no change in sign. Thus we can construct a list of signs dependent only on the parity of m. Further, since the signs are repeated in each sublist, we only need to generate one candidate sign from each list, and then rely again on the definition of *Mathematica's* `Listable` property to propagate each individual sign across the elements of any list it is multiplied into. For example, the collection of signs in the previous result may be encoded by `{1, -1, 1, -1, 1}`. Multiplying this by S_2 above gives S_1 as intended.

```
{1, -1, 1, -1, 1} S2
```

```
{{u∧x∧y∧z}, {u∧x∧y∧z, u∧x∧y∧z, u∧x∧y∧z, u∧x∧y∧z},
 {u∧x∧y∧z, u∧x∧y∧z, u∧x∧y∧z, u∧x∧y∧z, u∧x∧y∧z, u∧x∧y∧z},
 {u∧x∧y∧z, u∧x∧y∧z, u∧x∧y∧z, u∧x∧y∧z}, {u∧x∧y∧z}}
```

One way of defining a generator `★rr` of lists of this form as a function of m is:

```
★r[m_] := Mod[Range[0, m], 2];
★rr[m_] := If[EvenQ[m], ★r[m], 0]
```

Here is a list of `★rr` for m from 1 to 6. Note that the previous example is encoded by `★rr[4]`.

```
{(-1)^★rr[1], (-1)^★rr[2], (-1)^★rr[3], (-1)^★rr[4], (-1)^★rr[5], (-1)^★rr[6]}
```

```
{1, {1, -1, 1}, 1, {1, -1, 1, -1, 1}, 1, {1, -1, 1, -1, 1, -1, 1}}
```

With this sign construction, we can now determine a relationship between products of complete spans and cospans. Below we verify the identity for m ranging from 1 to 5. Remember that `★S` will automatically generate a simple product from an underscripted symbol, and it is good practice if the symbol which is underscripted is undefined. So we clear our previous definition of `A`.

```
Clear[A];
Table[★S[★S★[A_m] ∧ ★S★[A_m] == (-1)^★rr[m] (★S★[A_m] ∧ ★S★[A_m])], {m, 1, 5}]
```

```
{True, True, True, True, True}
```

In the more general case of a *form* based on the complete span and cospan of a simple m-element $\underset{m}{A}$, we can obtain a similar identity providing we reverse *all* the lists, that is, reverse each list of elements *and* the list of these lists. To simplify this two-level reversing process we define a function `ReverseAll` with alias `★R`.

```
ReverseAll[s_List] := Reverse[Reverse /@ s]; ★R = ReverseAll;
```

$$\left(G_1 \oplus \star S_\star\left[\underset{m}{A}\right]\right) \odot \left(G_2 \otimes \star S^\star\left[\underset{m}{A}\right]\right) = \star R\left[(-1)^{\star rr[m]} \left(G_1 \oplus \star S^\star\left[\underset{m}{A}\right]\right) \odot \left(G_2 \otimes \star S_\star\left[\underset{m}{A}\right]\right)\right] \quad 2.48$$

Again we verify the identity for m ranging from 1 to 5.

```
Table[★ℱ[(G₁ ⊕ ★S⋆[A m]) ⊙ (G₂ ⊗ ★S*[A m])] ==
  ★ℱ[★R[(-1)^(★rr[m]) (G₁ ⊕ ★S*[A m]) ⊙ (G₂ ⊗ ★S⋆[A m])]], {m, 1, 5}]

{True, True, True, True, True}
```

As might be expected, the identity remains valid no matter to which form the sign change $(-1)^{\star rr[m]}$ or the ReverseAll operation is applied. Let

```
F₁[m_] := (G₁ ⊕ ★S⋆[A m]) ⊙ (G₂ ⊗ ★S*[A m])

F₂[m_] := (G₁ ⊕ ★S*[A m]) ⊙ (G₂ ⊗ ★S⋆[A m])
```

then the following identities hold

$$F_1[m] == \star R[(-1)^{\star rr[m]} F_2[m]] == (-1)^{\star rr[m]} \star R[F_2[m]] \qquad 2.49$$

$$F_2[m] == \star R[(-1)^{\star rr[m]} F_1[m]] == (-1)^{\star rr[m]} \star R[F_1[m]] \qquad 2.50$$

$$\star R[F_1[m]] == (-1)^{\star rr[m]} F_2[m] \qquad 2.51$$

$$\star R[F_2[m]] == (-1)^{\star rr[m]} F_1[m] \qquad 2.52$$

The Zero Form Theorem

An m:k-form is zero whenever the m-element on which it is based is zero.

The m:k-forms we are including here are those of the form shown in [2.47]. Since the complete span of an *m*-element includes the forms for all *k* from 0 to *m*, we will only need to verify for complete spans. As has been the case with all the properties of m:k-forms to this point, the complete span of a zero *m*-element will be zero independent of the dimension of the space in which it resides.

To fix ideas, suppose we take a non-zero 3-element A in a space of dimension at least 3.

```
★ℬ₄; ★★V[α_]; A = α₁ ∧ α₂ ∧ α₃;
```

The product ★S⋆[A] ⊙ ★S*[A] gives 4 lists of terms: the terms of the 3:k-forms for *k* from 0 to 3.

```
★S⋆[A] ⊙ ★S*[A]

{{1 ⊙ (α₁ ∧ α₂ ∧ α₃)}, {α₁ ⊙ (α₂ ∧ α₃), α₂ ⊙ (-(α₁ ∧ α₃)), α₃ ⊙ (α₁ ∧ α₂)},
 {(α₁ ∧ α₂) ⊙ α₃, (α₁ ∧ α₃) ⊙ (-α₂), (α₂ ∧ α₃) ⊙ α₁}, {(α₁ ∧ α₂ ∧ α₃) ⊙ 1}}
```

Now suppose that the α_i are no longer independent with α_3, say, equal to a linear combination of the other factors, thus making A zero.

```
A = α₁ ∧ α₂ ∧ (a α₁ + b α₂);
```

The product ★S⋆[A] ⊙ ★S*[A] now gives

⋆S⋆[A] ⊙ ⋆S*[A]

{{1 ⊙ ($\alpha_1 \wedge \alpha_2 \wedge$ (a α_1 + b α_2))},
{α_1 ⊙ ($\alpha_2 \wedge$ (a α_1 + b α_2)), α_2 ⊙ (- ($\alpha_1 \wedge$ (a α_1 + b α_2))), (a α_1 + b α_2) ⊙ ($\alpha_1 \wedge \alpha_2$)},
{($\alpha_1 \wedge \alpha_2$) ⊙ (a α_1 + b α_2), ($\alpha_1 \wedge$ (a α_1 + b α_2)) ⊙ (-α_2), ($\alpha_2 \wedge$ (a α_1 + b α_2)) ⊙ α_1},
{($\alpha_1 \wedge \alpha_2 \wedge$ (a α_1 + b α_2)) ⊙ 1}}

Expanding and simplifying these terms shows the pattern by which they will eventually sum to zero.

⋆$\mathcal{F}$[⋆S⋆[A] ⊙ ⋆S*[A]]

{{0}, {-a α_1 ⊙ ($\alpha_1 \wedge \alpha_2$), -b α_2 ⊙ ($\alpha_1 \wedge \alpha_2$), a α_1 ⊙ ($\alpha_1 \wedge \alpha_2$) + b α_2 ⊙ ($\alpha_1 \wedge \alpha_2$)},
{a ($\alpha_1 \wedge \alpha_2$) ⊙ α_1 + b ($\alpha_1 \wedge \alpha_2$) ⊙ α_2, -b ($\alpha_1 \wedge \alpha_2$) ⊙ α_2, -a ($\alpha_1 \wedge \alpha_2$) ⊙ α_1}, {0}}

The 3:0- and 3:3-forms are immediately zero as expected. The 3:1 and 3:2-forms are still displayed as lists of their component terms. However it is clear that these terms sum to zero.

◆ **Verification**

Let us take the forms shown in [2.47], define functions F, F_1, F_2, F_{12}, of an *m*-element A and verify that they are all zero for a few lower-grade values of a zero A.

F[A_] := ⋆Σ /@ ⋆$\mathcal{F}$[(⋆S⋆[A]) ⊙ (⋆S*[A])]
F_1[A_] := ⋆Σ /@ ⋆$\mathcal{F}$[($G_1 \oplus$ ⋆S⋆[A]) ⊙ (⋆S*[A])]
F_2[A_] := ⋆Σ /@ ⋆$\mathcal{F}$[(⋆S⋆[A]) ⊙ ($G_2 \otimes$ ⋆S*[A])]
F_{12}[A_] := ⋆Σ /@ ⋆$\mathcal{F}$[($G_1 \oplus$ ⋆S⋆[A]) ⊙ ($G_2 \otimes$ ⋆S*[A])]

⋆A; ⋆$\mathcal{B}_6$; ⋆⋆V[α_];

A = $\alpha_1 \wedge$ (a α_1); {F[A], F_1[A], F_2[A], F_{12}[A]}

{{0, 0, 0}, {0, 0, 0}, {0, 0, 0}, {0, 0, 0}}

A = $\alpha_1 \wedge \alpha_2 \wedge$ (a α_1 + b α_2); {F[A], F_1[A], F_2[A], F_{12}[A]}

{{0, 0, 0, 0}, {0, 0, 0, 0}, {0, 0, 0, 0}, {0, 0, 0, 0}}

A = $\alpha_1 \wedge \alpha_2 \wedge \alpha_3 \wedge$ (a α_1 + b α_2 + c α_3); {F[A], F_1[A], F_2[A], F_{12}[A]}

{{0, 0, 0, 0, 0}, {0, 0, 0, 0, 0}, {0, 0, 0, 0, 0}, {0, 0, 0, 0, 0}}

A = $\alpha_1 \wedge \alpha_2 \wedge \alpha_3 \wedge \alpha_4 \wedge$ (a α_1 + b α_2 + c α_3 + d α_4);
{F[A], F_1[A], F_2[A], F_{12}[A]}

{{0, 0, 0, 0, 0, 0}, {0, 0, 0, 0, 0, 0},
 {0, 0, 0, 0, 0, 0}, {0, 0, 0, 0, 0, 0}}

Zero Form formulae

To see the types of formulae that this theorem can yield, return to the case where A is zero but no factor is expressed as a linear combination of the others.

◆ 3 elements

$$A = \alpha_1 \wedge \alpha_2 \wedge \alpha_3;\ \{F[A], F_1[A], F_2[A], F_{12}[A]\}$$

$$\{\{1 \odot (\alpha_1 \wedge \alpha_2 \wedge \alpha_3),\ \alpha_1 \odot (\alpha_2 \wedge \alpha_3) - \alpha_2 \odot (\alpha_1 \wedge \alpha_3) + \alpha_3 \odot (\alpha_1 \wedge \alpha_2),$$
$$(\alpha_1 \wedge \alpha_2) \odot \alpha_3 - (\alpha_1 \wedge \alpha_3) \odot \alpha_2 + (\alpha_2 \wedge \alpha_3) \odot \alpha_1,$$
$$(\alpha_1 \wedge \alpha_2 \wedge \alpha_3) \odot 1\},\ \{(G_1 \oplus 1) \odot (\alpha_1 \wedge \alpha_2 \wedge \alpha_3),$$
$$(G_1 \oplus \alpha_1) \odot (\alpha_2 \wedge \alpha_3) - (G_1 \oplus \alpha_2) \odot (\alpha_1 \wedge \alpha_3) + (G_1 \oplus \alpha_3) \odot (\alpha_1 \wedge \alpha_2),$$
$$(G_1 \oplus \alpha_1 \wedge \alpha_2) \odot \alpha_3 - (G_1 \oplus \alpha_1 \wedge \alpha_3) \odot \alpha_2 + (G_1 \oplus \alpha_2 \wedge \alpha_3) \odot \alpha_1,$$
$$(G_1 \oplus \alpha_1 \wedge \alpha_2 \wedge \alpha_3) \odot 1\},$$
$$\{1 \odot (G_2 \otimes \alpha_1 \wedge \alpha_2 \wedge \alpha_3),\ \alpha_1 \odot (G_2 \otimes \alpha_2 \wedge \alpha_3) - \alpha_2 \odot (G_2 \otimes \alpha_1 \wedge \alpha_3) + \alpha_3 \odot (G_2 \otimes \alpha_1 \wedge \alpha_2),$$
$$(\alpha_1 \wedge \alpha_2) \odot (G_2 \otimes \alpha_3) - (\alpha_1 \wedge \alpha_3) \odot (G_2 \otimes \alpha_2) + (\alpha_2 \wedge \alpha_3) \odot (G_2 \otimes \alpha_1),$$
$$(\alpha_1 \wedge \alpha_2 \wedge \alpha_3) \odot (G_2 \otimes 1)\},\ \{(G_1 \oplus 1) \odot (G_2 \otimes \alpha_1 \wedge \alpha_2 \wedge \alpha_3),$$
$$(G_1 \oplus \alpha_1) \odot (G_2 \otimes \alpha_2 \wedge \alpha_3) - (G_1 \oplus \alpha_2) \odot (G_2 \otimes \alpha_1 \wedge \alpha_3) + (G_1 \oplus \alpha_3) \odot (G_2 \otimes \alpha_1 \wedge \alpha_2),$$
$$(G_1 \oplus \alpha_1 \wedge \alpha_2) \odot (G_2 \otimes \alpha_3) - (G_1 \oplus \alpha_1 \wedge \alpha_3) \odot (G_2 \otimes \alpha_2) + (G_1 \oplus \alpha_2 \wedge \alpha_3) \odot (G_2 \otimes \alpha_1),$$
$$(G_1 \oplus \alpha_1 \wedge \alpha_2 \wedge \alpha_3) \odot (G_2 \otimes 1)\}\}$$

We can thus write that for any three linearly dependent 1-elements, $\odot$, $\otimes$ and $\oplus$ arbitrary multilinear product operations, and G_1 and G_2 arbitrary Grassmann expressions:

$$\alpha_1 \wedge \alpha_2 \wedge \alpha_3 == 0 \implies \alpha_1 \odot (\alpha_2 \wedge \alpha_3) - \alpha_2 \odot (\alpha_1 \wedge \alpha_3) + \alpha_3 \odot (\alpha_1 \wedge \alpha_2) == 0 \tag{2.53}$$

$$\alpha_1 \wedge \alpha_2 \wedge \alpha_3 == 0 \implies (\alpha_1 \wedge \alpha_2) \odot \alpha_3 - (\alpha_1 \wedge \alpha_3) \odot \alpha_2 + (\alpha_2 \wedge \alpha_3) \odot \alpha_1 == 0 \tag{2.54}$$

$$\alpha_1 \wedge \alpha_2 \wedge \alpha_3 == 0 \implies (G_1 \oplus \alpha_1) \odot (\alpha_2 \wedge \alpha_3) - (G_1 \oplus \alpha_2) \odot (\alpha_1 \wedge \alpha_3) + (G_1 \oplus \alpha_3) \odot (\alpha_1 \wedge \alpha_2) == 0 \tag{2.55}$$

$$\alpha_1 \wedge \alpha_2 \wedge \alpha_3 == 0 \implies (G_1 \oplus \alpha_1 \wedge \alpha_2) \odot \alpha_3 - (G_1 \oplus \alpha_1 \wedge \alpha_3) \odot \alpha_2 + (G_1 \oplus \alpha_2 \wedge \alpha_3) \odot \alpha_1 == 0 \tag{2.56}$$

$$\alpha_1 \wedge \alpha_2 \wedge \alpha_3 == 0 \implies \alpha_1 \odot (G_2 \otimes \alpha_2 \wedge \alpha_3) - \alpha_2 \odot (G_2 \otimes \alpha_1 \wedge \alpha_3) + \alpha_3 \odot (G_2 \otimes \alpha_1 \wedge \alpha_2) == 0 \tag{2.57}$$

$$\alpha_1 \wedge \alpha_2 \wedge \alpha_3 == 0 \implies (\alpha_1 \wedge \alpha_2) \odot (G_2 \otimes \alpha_3) - (\alpha_1 \wedge \alpha_3) \odot (G_2 \otimes \alpha_2) + (\alpha_2 \wedge \alpha_3) \odot (G_2 \otimes \alpha_1) == 0 \tag{2.58}$$

$$\alpha_1 \wedge \alpha_2 \wedge \alpha_3 == 0 \implies (G_1 \oplus \alpha_1) \odot (G_2 \otimes \alpha_2 \wedge \alpha_3) - (G_1 \oplus \alpha_2) \odot (G_2 \otimes \alpha_1 \wedge \alpha_3) + (G_1 \oplus \alpha_3) \odot (G_2 \otimes \alpha_1 \wedge \alpha_2) == 0 \tag{2.59}$$

$$\alpha_1 \wedge \alpha_2 \wedge \alpha_3 == 0 \Longrightarrow (G_1 \oplus \alpha_1 \wedge \alpha_2) \odot (G_2 \otimes \alpha_3) - (G_1 \oplus \alpha_1 \wedge \alpha_3) \odot (G_2 \otimes \alpha_2) + (G_1 \oplus \alpha_2 \wedge \alpha_3) \odot (G_2 \otimes \alpha_1) == 0 \qquad 2.60$$

◇ Examples for ∧, ∨, and ⊖

For example, take the first two formulae and replace ⊙ successively by ∧, ∨, and ⊖ to show that the formulae are valid for the exterior, regressive and interior products - each being an example of a multilinear product.

```
A1 = α1 ⊙ (α2 ∧ α3) - α2 ⊙ (α1 ∧ α3) + α3 ⊙ (α1 ∧ α2) /. α2 → a α1 + b α3;
A2 = (α1 ∧ α2) ⊙ α3 - (α1 ∧ α3) ⊙ α2 + (α2 ∧ α3) ⊙ α1 /. α2 → a α1 + b α3;
AA = {A1, A2};
Table[★𝓑n; {★𝒢[AA /. a_⊙b_ :→ a ∧ b],
  ToScalarProducts[AA /. a_⊙b_ :→ a⊖b],
  ToCommonFactor[AA /. a_⊙b_ :→ a ∨ b]}, {n, 2, 4}]
```

```
{{{0, 0}, {0, 0}, {0, 0}},
 {{0, 0}, {0, 0}, {0, 0}}, {{0, 0}, {0, 0}, {0, 0}}}
```

◆ 4 elements

Here are a few examples of zero form formulae generated by a zero 4-element.

```
A = α1 ∧ α2 ∧ α3 ∧ α4; F[A]
```

```
{1⊙ (α1 ∧ α2 ∧ α3 ∧ α4),
 α1 ⊙ (α2 ∧ α3 ∧ α4) - α2 ⊙ (α1 ∧ α3 ∧ α4) + α3 ⊙ (α1 ∧ α2 ∧ α4) - α4 ⊙ (α1 ∧ α2 ∧ α3),
 (α1 ∧ α2) ⊙ (α3 ∧ α4) - (α1 ∧ α3) ⊙ (α2 ∧ α4) + (α1 ∧ α4) ⊙ (α2 ∧ α3) +
  (α2 ∧ α3) ⊙ (α1 ∧ α4) - (α2 ∧ α4) ⊙ (α1 ∧ α3) + (α3 ∧ α4) ⊙ (α1 ∧ α2),
 (α1 ∧ α2 ∧ α3) ⊙α4 - (α1 ∧ α2 ∧ α4) ⊙α3 + (α1 ∧ α3 ∧ α4) ⊙α2 - (α2 ∧ α3 ∧ α4) ⊙α1,
 (α1 ∧ α2 ∧ α3 ∧ α4) ⊙1}
```

$$\alpha_1 \wedge \alpha_2 \wedge \alpha_3 \wedge \alpha_4 == 0 \Longrightarrow \alpha_1 \odot (\alpha_2 \wedge \alpha_3 \wedge \alpha_4) - \alpha_2 \odot (\alpha_1 \wedge \alpha_3 \wedge \alpha_4) + \alpha_3 \odot (\alpha_1 \wedge \alpha_2 \wedge \alpha_4) - \alpha_4 \odot (\alpha_1 \wedge \alpha_2 \wedge \alpha_3) == 0 \qquad 2.61$$

$$\alpha_1 \wedge \alpha_2 \wedge \alpha_3 \wedge \alpha_4 == 0 \Longrightarrow (\alpha_1 \wedge \alpha_2) \odot (\alpha_3 \wedge \alpha_4) - (\alpha_1 \wedge \alpha_3) \odot (\alpha_2 \wedge \alpha_4) + (\alpha_1 \wedge \alpha_4) \odot (\alpha_2 \wedge \alpha_3) + (\alpha_2 \wedge \alpha_3) \odot (\alpha_1 \wedge \alpha_4) - (\alpha_2 \wedge \alpha_4) \odot (\alpha_1 \wedge \alpha_3) + (\alpha_3 \wedge \alpha_4) \odot (\alpha_1 \wedge \alpha_2) == 0 \qquad 2.62$$

$$\alpha_1 \wedge \alpha_2 \wedge \alpha_3 \wedge \alpha_4 == 0 \Longrightarrow (\alpha_1 \wedge \alpha_2 \wedge \alpha_3) \odot \alpha_4 - (\alpha_1 \wedge \alpha_2 \wedge \alpha_4) \odot \alpha_3 + (\alpha_1 \wedge \alpha_3 \wedge \alpha_4) \odot \alpha_2 - (\alpha_2 \wedge \alpha_3 \wedge \alpha_4) \odot \alpha_1 == 0 \qquad 2.63$$

◇ **Examples for ∧, ∨, and ⊖**

```
B1 = α₁ ⊙ (α₂ ∧ α₃ ∧ α₄) - α₂ ⊙ (α₁ ∧ α₃ ∧ α₄) + α₃ ⊙ (α₁ ∧ α₂ ∧ α₄) -
      α₄ ⊙ (α₁ ∧ α₂ ∧ α₃) /. α₂ → a α₁ + b α₃ + c α₄;
B2 = (α₁ ∧ α₂) ⊙ (α₃ ∧ α₄) - (α₁ ∧ α₃) ⊙ (α₂ ∧ α₄) + (α₁ ∧ α₄) ⊙ (α₂ ∧ α₃) +
      (α₂ ∧ α₃) ⊙ (α₁ ∧ α₄) - (α₂ ∧ α₄) ⊙ (α₁ ∧ α₃) + (α₃ ∧ α₄) ⊙ (α₁ ∧ α₂) /.
    α₂ → a α₁ + b α₃ + c α₄;
B3 = (α₁ ∧ α₂ ∧ α₃) ⊙ α₄ - (α₁ ∧ α₂ ∧ α₄) ⊙ α₃ + (α₁ ∧ α₃ ∧ α₄) ⊙ α₂ -
      (α₂ ∧ α₃ ∧ α₄) ⊙ α₁ /. α₂ → a α₁ + b α₃ + c α₄;
B = {B1, B2, B3};
Table[★𝓑n; {★𝒢[B /. a_⊙b_ :→ a ∧ b],
  ToScalarProducts[B /. a_⊙b_ :→ a⊖b],
  ToCommonFactor[B /. a_⊙b_ :→ a ∨ b]}, {n, 3, 5}]

{{{0, 0, 0}, {0, 0, 0}, {0, 0, 0}},
 {{0, 0, 0}, {0, 0, 0}, {0, 0, 0}}, {{0, 0, 0}, {0, 0, 0}, {0, 0, 0}}}
```

2.13 Unions and Intersections

Union and intersection as a multilinear form

One of the more immediate applications of multilinear forms that we can make at this stage is to the computing of unions and intersections of simple elements. To the author's knowledge, this approach was not discussed by Grassmann.

The *intersection* of two simple elements A and B is the element C of *highest grade* which is common to both A and B. Let

$$A == A^* \wedge C \qquad B == C \wedge B^* \qquad A^* \wedge C \wedge B^* \neq 0$$

The element $A^* \wedge C \wedge B^*$ will be called the *union* of A and B.

Two elements A and B are said to be *disjoint* if C is of grade 0, that is, a scalar: thus $A \wedge B$ is not zero. The formulae that we are going to develop still remain valid in this case, so that we will not need to consider it further.

The first thing to note is that defining the union and intersection only defines them up to congruence because they still remain valid if we multiply and divide the factors in the formulae by an arbitrary scalar.

$$A == \frac{A^*}{c} \wedge (c\, C) \qquad B == (c\, C) \wedge \frac{B^*}{c} \qquad \frac{A^*}{c} \wedge (c\, C) \wedge \frac{B^*}{c} \neq 0$$

Thus now, if $c\,C$ is the intersection, $\left(\frac{1}{c}\right) A^* \wedge C \wedge B^*$ is the union.

The observation that the union and intersection are only defined up to congruence leads us to consider how we might develop a formula for the union and intersection which is not dependent on an arbitrary congruence factor. Suppose ⊙ is a linear product as discussed in the previous section, not equal to the exterior product, but otherwise arbitrary. Then *if* the ⊙ product of A and B is defined to be equal to the ⊙ product of their union and intersection, the congruence factors cancel.

$$A \odot B == (A^* \wedge C) \odot (C \wedge B^*) == (A^* \wedge C \wedge B^*) \odot C \qquad 2.64$$

This formula can also be written in the alternative forms:

$$A \odot B == A \odot (C \wedge B^*) == (A \wedge B^*) \odot C$$
$$A \odot B == (A^* \wedge C) \odot B == (A^* \wedge B) \odot C$$

Suppose B is of grade m, and B^* is of grade k, then C is of grade $m-k$. To compute the union $A \wedge B^*$ we need to 'partition' B into such k and $m-k$ elements, and then find the maximal value of k (equal to k^*, say) for which $A \wedge B^*$ is not zero. To do this partitioning, B has to be simple, and we need to know some factorization of it. If we partition B using its span and cospan for various k and for the given factorization, then our formula becomes a special case of an m:k-form, and invariance of the result to the actual factorization of B is guaranteed.

Hence we write the formula for the $\odot$ product of the possibly intersecting simple elements A and B in terms of their union and intersection as

$$A \odot B == \star\Sigma\left[\left(A \wedge \star\mathbb{S}^{k^*}[B]\right) \odot \star\mathbb{S}_{k^*}[B]\right] \qquad 2.65$$

where k^* is the maximum grade for which the right hand side is not zero.

In formula [2.64] above we have chosen certain arrangements for the decomposition of the elements. Had the factors in any of these arrangements been interchanged, the results for the union and intersection would have changed by at most a sign. Our motivation for the precise arrangements above is that they correspond closely to the important Common Factor Axiom and Common Factor Theorem which we will introduce in the next chapter.

In the sections to follow, we will make these notions more concrete with some examples.

- It may be of interest to note that if $\odot$ is taken as the exterior product, the resulting formulation is still valid - though trivially so: if C is a scalar all terms reduce to $A \wedge B$. If C is of higher grade, all terms reduce to zero.

  $$A \wedge B == (A^* \wedge C) \wedge (C \wedge B^*) == (A^* \wedge C \wedge B^*) \wedge C$$

Where the intersection is evident

To see how this formula works, let us first take some simple examples where the two elements are expressed in such a way that their union and intersection are evident by inspection.

To enable the computations we first declare any subscripted α, β, or γ as vector symbols.

```
★★V[α_, β_, γ_];
```

◆ Example: 2:k-forms

Suppose A and B have a common element γ_1.

```
A = α1 ∧ α2 ∧ γ1; B = γ1 ∧ β1;
```

Initially we suppose that $A \wedge \beta_1$ is not zero. Formula [2.64] should then give us

$$(\alpha_1 \wedge \alpha_2 \wedge \gamma_1) \odot (\gamma_1 \wedge \beta_1) == (\alpha_1 \wedge \alpha_2 \wedge \gamma_1 \wedge \beta_1) \odot \gamma_1$$

If we simply compose a table of all the 2:k-forms for B (k equal to 0, 1, and 2 in this case) without simplifying the forms, we get the three forms

```
T = Table[★Σ[(A ∧ ★𝕊^k[B]) ⊙ ★𝕊_k[B]], {k, 0, 2}]; TableForm[T]
```

$(\alpha_1 \wedge \alpha_2 \wedge \gamma_1 \wedge \gamma_1 \wedge \beta_1) \odot 1$
$(\alpha_1 \wedge \alpha_2 \wedge \gamma_1 \wedge \beta_1) \odot \gamma_1 + (\alpha_1 \wedge \alpha_2 \wedge \gamma_1 \wedge -\gamma_1) \odot \beta_1$
$(\alpha_1 \wedge \alpha_2 \wedge \gamma_1 \wedge 1) \odot (\gamma_1 \wedge \beta_1)$

Clearly, when simplified, the first form ($k = 0$) is equal to zero, and the third form ($k = 2$) is equal to $A \odot B$. The second form ($k = 1$) reduces to a single term, giving us the expected expression for the union and intersection.

```
★ℱ[T] // TableForm
```

0
$-(\alpha_1 \wedge \alpha_2 \wedge \beta_1 \wedge \gamma_1) \odot \gamma_1$
$-(\alpha_1 \wedge \alpha_2 \wedge \gamma_1) \odot (\beta_1 \wedge \gamma_1)$

Note that ★ℱ has automatically put the factors in the forms into canonical order (based in this case on the alphabetical order of the symbols used). Rearrangement shows that we have indeed recovered the expression expected.

$-(\alpha_1 \wedge \alpha_2 \wedge \beta_1 \wedge \gamma_1) \odot \gamma_1 == (\alpha_1 \wedge \alpha_2 \wedge \gamma_1 \wedge \beta_1) \odot \gamma_1$

Suppose now that $A \wedge \beta_1$ is in fact zero. The formula still remains valid but now gives $(\alpha_1 \wedge \alpha_2 \wedge \gamma_1) \odot (\gamma_1 \wedge \beta_1)$ as the only non-zero form, whence we conclude that $\alpha_1 \wedge \alpha_2 \wedge \gamma_1$ is the union and $\gamma_1 \wedge \beta_1$ is the intersection. That is, $\gamma_1 \wedge \beta_1$ is contained in $\alpha_1 \wedge \alpha_2 \wedge \gamma_1$, or B is contained in A.

◆ Example: 3:k-forms

Had B also been a 3-element, we would have generated four forms.

$A = \alpha_1 \wedge \alpha_2 \wedge \gamma_1;\ B = \gamma_1 \wedge \beta_1 \wedge \beta_2;$

```
T = Table[★Σ[(A ∧ ★𝕊^k[B]) ⊙ ★𝕊_k[B]], {k, 0, 3}]
```

$\{(\alpha_1 \wedge \alpha_2 \wedge \gamma_1 \wedge \gamma_1 \wedge \beta_1 \wedge \beta_2) \odot 1,\ (\alpha_1 \wedge \alpha_2 \wedge \gamma_1 \wedge -(\gamma_1 \wedge \beta_2)) \odot \beta_1 +$
$(\alpha_1 \wedge \alpha_2 \wedge \gamma_1 \wedge \beta_1 \wedge \beta_2) \odot \gamma_1 + (\alpha_1 \wedge \alpha_2 \wedge \gamma_1 \wedge \gamma_1 \wedge \beta_1) \odot \beta_2,$
$(\alpha_1 \wedge \alpha_2 \wedge \gamma_1 \wedge -\beta_1) \odot (\gamma_1 \wedge \beta_2) + (\alpha_1 \wedge \alpha_2 \wedge \gamma_1 \wedge \beta_2) \odot (\gamma_1 \wedge \beta_1) +$
$(\alpha_1 \wedge \alpha_2 \wedge \gamma_1 \wedge \gamma_1) \odot (\beta_1 \wedge \beta_2),\ (\alpha_1 \wedge \alpha_2 \wedge \gamma_1 \wedge 1) \odot (\gamma_1 \wedge \beta_1 \wedge \beta_2)\}$

```
★ℱ[T]
```

$\{0,\ (\alpha_1 \wedge \alpha_2 \wedge \beta_1 \wedge \beta_2 \wedge \gamma_1) \odot \gamma_1,$
$-(\alpha_1 \wedge \alpha_2 \wedge \beta_1 \wedge \gamma_1) \odot (\beta_2 \wedge \gamma_1) + (\alpha_1 \wedge \alpha_2 \wedge \beta_2 \wedge \gamma_1) \odot (\beta_1 \wedge \gamma_1),$
$(\alpha_1 \wedge \alpha_2 \wedge \gamma_1) \odot (\beta_1 \wedge \beta_2 \wedge \gamma_1)\}$

The union and intersection is clearly the first non-zero form. It may be reordered to show the expression expected.

$(\alpha_1 \wedge \alpha_2 \wedge \beta_1 \wedge \beta_2 \wedge \gamma_1) \odot \gamma_1 == (\alpha_1 \wedge \alpha_2 \wedge \gamma_1 \wedge \beta_1 \wedge \beta_2) \odot \gamma_1$

◆ Example: Interchanging the order of A and B

We would not expect that interchanging the order of A and B in the form would make any difference to the results for the union and intersection, except perhaps for a change in sign.

```
A = α1 ∧ α2 ∧ γ1; B = γ1 ∧ β1;
★𝓕[Table[★Σ[(B ∧ ★𝕊^k[A]) ⊙ ★𝕊_k[A]], {k, 0, 2}]]
{0, -(α1 ∧ α2 ∧ β1 ∧ γ1) ⊙ γ1, -(α1 ∧ β1 ∧ γ1) ⊙ (α2 ∧ γ1) + (α2 ∧ β1 ∧ γ1) ⊙ (α1 ∧ γ1)}
```

◆ **Example: Interchanging the order of the span and cospan**

Neither would we expect that interchanging the order of the span $\star\mathbb{S}_k$ and cospan $\star\mathbb{S}^k$ functions in the form would make any difference to the results, except again perhaps for a change in sign. However this interchange does reverse the order of the forms, making the form displaying the union and intersection come before the zero form, rather than after it.

```
A = α1 ∧ α2 ∧ γ1; B = γ1 ∧ β1;
★𝓕[Table[★Σ[(A ∧ ★𝕊_k[B]) ⊙ ★𝕊^k[B]], {k, 0, 2}]] // TableForm
-(α1 ∧ α2 ∧ γ1) ⊙ (β1 ∧ γ1)
(α1 ∧ α2 ∧ β1 ∧ γ1) ⊙ γ1
0
```

Where the intersections is not evident

In the previous section we explored the application of the union and intersection formula in trivial cases where we could have more easily obtained the results by inspection. The usefulness of the formula becomes evident however when the intersection of A and B is not displayed explicitly.

◆ **Example: 2:k-forms**

Consider elements A and B with the common factor $e_1 + 2\,e_2$.

```
A == (e4 - 2 e7) ∧ (3 e5 + e2) ∧ (e1 + 2 e2)
B == (2 e3 - e2) ∧ (e1 + 2 e2)
```

Formula [2.65] requires only one of the elements to be in factored form. The factorization does not need to display the common factor. To test the formula we expand and simplify A, then expand and simplify B but then refactorize it into a factorization that does not display the intersection. (To do this we use the *GrassmannAlgebra* function ExteriorFactorize which we develop in the next chapter.)

```
★𝓑8; A = ★𝒢[(e4 - 2 e7) ∧ (3 e5 + e2) ∧ (e1 + 2 e2)]
-(e1 ∧ e2 ∧ e4) + 2 e1 ∧ e2 ∧ e7 + 3 e1 ∧ e4 ∧ e5 +
  6 e1 ∧ e5 ∧ e7 + 6 e2 ∧ e4 ∧ e5 + 12 e2 ∧ e5 ∧ e7
B = ExteriorFactorize[★𝒢[(2 e3 - e2) ∧ (e1 + 2 e2)]]
(e1 + 4 e3) ∧ (e2 - 2 e3)
```

In this case we know that a span of grade 1 will give us the result we are looking for.

$F_1 = \star\mathcal{F}\left[\star\Sigma\left[(A \wedge \star\mathbb{S}_1[B]) \odot \star\mathbb{S}^1[B]\right]\right]$

$2\ (e_1 \wedge e_2 \wedge e_3 \wedge e_4) \odot e_1 + 4\ (e_1 \wedge e_2 \wedge e_3 \wedge e_4) \odot e_2 -$
$4\ (e_1 \wedge e_2 \wedge e_3 \wedge e_7) \odot e_1 - 8\ (e_1 \wedge e_2 \wedge e_3 \wedge e_7) \odot e_2 - 3\ (e_1 \wedge e_2 \wedge e_4 \wedge e_5) \odot e_1 -$
$6\ (e_1 \wedge e_2 \wedge e_4 \wedge e_5) \odot e_2 - 6\ (e_1 \wedge e_2 \wedge e_5 \wedge e_7) \odot e_1 - 12\ (e_1 \wedge e_2 \wedge e_5 \wedge e_7) \odot e_2 +$
$6\ (e_1 \wedge e_3 \wedge e_4 \wedge e_5) \odot e_1 + 12\ (e_1 \wedge e_3 \wedge e_4 \wedge e_5) \odot e_2 + 12\ (e_1 \wedge e_3 \wedge e_5 \wedge e_7) \odot e_1 +$
$24\ (e_1 \wedge e_3 \wedge e_5 \wedge e_7) \odot e_2 + 12\ (e_2 \wedge e_3 \wedge e_4 \wedge e_5) \odot e_1 +$
$24\ (e_2 \wedge e_3 \wedge e_4 \wedge e_5) \odot e_2 + 24\ (e_2 \wedge e_3 \wedge e_5 \wedge e_7) \odot e_1 + 48\ (e_2 \wedge e_3 \wedge e_5 \wedge e_7) \odot e_2$

Of course, this is in expanded form. We would like to factorize it to display the union and intersection explicitly. We can do this with *GrassmannAlgebra's* `FactorizeForm`.

$F_2 = \text{FactorizeForm}[F_1]$

$$\left((e_1 - 6\, e_5) \wedge (e_2 + 3\, e_5) \wedge \left(e_3 + \frac{3\, e_5}{2}\right) \wedge (e_4 - 2\, e_7)\right) \odot (2\, e_1 + 4\, e_2)$$

The intersection displayed here is a scalar multiple of the factor that we introduced as our 'hidden' intersection in the A and B supplied to the formula, and is therefore a correct result as any congruent factor would be.

Although the union does not explicitly display the intersection, we can verify that the intersection is indeed a factor of it, and that it is also a factor of A and B as supplied to the formula. Let U be the union and J be the intersection.

$$U = (e_1 - 6\, e_5) \wedge (e_2 + 3\, e_5) \wedge \left(e_3 + \frac{3\, e_5}{2}\right) \wedge (e_4 - 2\, e_7);$$

$J = 2\, e_1 + 4\, e_2;$

Then the following are zero as expected.

$\star\mathcal{G}[\{U \wedge J,\ A \wedge J,\ B \wedge J\}]$

$\{0,\ 0,\ 0\}$

All the other factors of the original A and B are also easily verified to belong to U.

$\star\mathcal{G}[U \wedge \{(e_4 - 2\, e_7),\ (3\, e_5 + e_2),\ (2\, e_3 - e_2),\ (e_1 + 4\, e_3),\ (e_2 - 2\, e_3)\}]$

$\{0,\ 0,\ 0,\ 0,\ 0\}$

Intersection with a non-simple element

To this point we have assumed A is simple. But, provided the intersection with B is simple, Formula [2.65] will still yield results. The following examples demonstrate however, that one may need to pay special attention to their interpretation.

◆ Example 1

We take the simplest example with A non-simple, and a 1-element intersection γ_1, which does not occur in the non-simple 2-element factor of A.

$A = (\alpha_1 \wedge \alpha_2 + \alpha_3 \wedge \alpha_4) \wedge \alpha_5 \wedge \gamma_1;\quad B = \beta_1 \wedge \gamma_1;$

We assume that $A \wedge \beta_1$ is not zero. Since the intersection is a 1-element, and B is a 2-element, we can get the union and intersection from the 1-span and 1-cospan of B.

$\star\Sigma\left[(\mathbf{A}\wedge\star\mathbb{S}_1[\mathbf{B}])\odot\star\mathbb{S}^1[\mathbf{B}]\right]$

$((\alpha_1\wedge\alpha_2+\alpha_3\wedge\alpha_4)\wedge\alpha_5\wedge\gamma_1\wedge\beta_1)\odot\gamma_1+((\alpha_1\wedge\alpha_2+\alpha_3\wedge\alpha_4)\wedge\alpha_5\wedge\gamma_1\wedge\gamma_1)\odot(-\beta_1)$

Upon simplification, the second term is clearly zero, leading to the first term displaying the expected result.

◆ **Example 2**

Now let B contain one of the 1-element factors which occurs in the non-simple 2-element factor of A, say α_1. We assume that $A\wedge\alpha_1$ is not zero.

$\mathbf{A} = (\alpha_1\wedge\alpha_2+\alpha_3\wedge\alpha_4)\wedge\alpha_5\wedge\gamma_1;\ \mathbf{B} = \alpha_1\wedge\gamma_1;$

$\star\Sigma\left[(\mathbf{A}\wedge\star\mathbb{S}_1[\mathbf{B}])\odot\star\mathbb{S}^1[\mathbf{B}]\right]$

$((\alpha_1\wedge\alpha_2+\alpha_3\wedge\alpha_4)\wedge\alpha_5\wedge\gamma_1\wedge\alpha_1)\odot\gamma_1+((\alpha_1\wedge\alpha_2+\alpha_3\wedge\alpha_4)\wedge\alpha_5\wedge\gamma_1\wedge\gamma_1)\odot(-\alpha_1)$

Upon simplification, the second term is again zero, but the first term is now a form composed of *simple* elements.

- Thus we conclude for these cases that when A is not simple, but the intersection is known to be simple (Example 1) the union and intersection formula will give straightforwardly interpretable results. However if A is not simple, and the intersection is not known to be simple (Example 2) the formula may give misleading results.

Factorizing simple elements

In the previous section we have seen how one may obtain the union and intersection of two simple elements. We can now use these results to obtain a factorization of a simple element. For concreteness we take the example from the previous section where A is a simple 3-element in an 8-space. We will work with the expanded form of A, and pretend that we do not know any factorization of it.

$\star\mathcal{B}_8;\ \mathbf{A} = \star\mathcal{G}[(\mathbf{e}_4-2\,\mathbf{e}_7)\wedge(3\,\mathbf{e}_5+\mathbf{e}_2)\wedge(\mathbf{e}_1+2\,\mathbf{e}_2)]$

$-(e_1\wedge e_2\wedge e_4)+2\,e_1\wedge e_2\wedge e_7+3\,e_1\wedge e_4\wedge e_5+$
$6\,e_1\wedge e_5\wedge e_7+6\,e_2\wedge e_4\wedge e_5+12\,e_2\wedge e_5\wedge e_7$

Since A is a simple 3-element in an 8-space, we can always obtain a 1-element belonging to it by obtaining the union and intersection of A with a simple 6-element B using the formula involving the 5-span and 5-cospan of B.

But as we have seen in the sections above, it is really only the number of *distinct symbols* (hence independent 1-elements) that are important in computing unions and intersections. Thus we can simplify our computations by considering A as a 3-element in a 5-space with a basis of $\{e_1, e_2, e_4, e_5, e_7\}$ and B as a simple 3-element using the formula involving the 2-span and 2-cospan of B.

To obtain our first factor of A, suppose we take B as $e_1\wedge e_2\wedge e_4$.

$\mathbf{B}_1 = \mathbf{e}_1\wedge\mathbf{e}_2\wedge\mathbf{e}_4;$

$\mathbf{FactorizeForm}\left[\star\mathcal{F}\left[\star\Sigma\left[(\mathbf{A}\wedge\star\mathbb{S}_2[\mathbf{B}_1])\odot\star\mathbb{S}^2[\mathbf{B}_1]\right]\right]\right]$

$(e_1\wedge e_2\wedge e_4\wedge e_5\wedge e_7)\odot(6\,e_1+12\,e_2)$

The factor of interest, our first factor of A, is the intersection $6\, e_1 + 12\, e_2$. We can ignore the union as it is not of any significance in this procedure.

To obtain the rest of the factors we repeating the procedure for all the possible simple 3-elements constructed from $\{e_1, e_2, e_4, e_5, e_7\}$, that is, the 3-span of $e_1 \wedge e_2 \wedge e_4 \wedge e_5 \wedge e_7$.

```
BB = ★S3[e1 ∧ e2 ∧ e4 ∧ e5 ∧ e7]
```

$$\{e_1 \wedge e_2 \wedge e_4,\ e_1 \wedge e_2 \wedge e_5,\ e_1 \wedge e_2 \wedge e_7,\ e_1 \wedge e_4 \wedge e_5,\ e_1 \wedge e_4 \wedge e_7,\ e_1 \wedge e_5 \wedge e_7,\ e_2 \wedge e_4 \wedge e_5,\ e_2 \wedge e_4 \wedge e_7,\ e_2 \wedge e_5 \wedge e_7,\ e_4 \wedge e_5 \wedge e_7\}$$

```
(FactorizeForm[★ℱ[★Σ[(A ∧ ★S2[#]) ⊙ ★S^2[#]]]] &) /@ BB // TableForm
```

$$(e_1 \wedge e_2 \wedge e_4 \wedge e_5 \wedge e_7) \odot (6\, e_1 + 12\, e_2)$$
$$0$$
$$(e_1 \wedge e_2 \wedge e_4 \wedge e_5 \wedge e_7) \odot (3\, e_1 + 6\, e_2)$$
$$(e_1 \wedge e_2 \wedge e_4 \wedge e_5 \wedge e_7) \odot (2\, e_1 - 12\, e_5)$$
$$(e_1 \wedge e_2 \wedge e_4 \wedge e_5 \wedge e_7) \odot (6\, e_4 - 12\, e_7)$$
$$(e_1 \wedge e_2 \wedge e_4 \wedge e_5 \wedge e_7) \odot (-e_1 + 6\, e_5)$$
$$(e_1 \wedge e_2 \wedge e_4 \wedge e_5 \wedge e_7) \odot (2\, e_2 + 6\, e_5)$$
$$(e_1 \wedge e_2 \wedge e_4 \wedge e_5 \wedge e_7) \odot (-3\, e_4 + 6\, e_7)$$
$$(e_1 \wedge e_2 \wedge e_4 \wedge e_5 \wedge e_7) \odot (-e_2 - 3\, e_5)$$
$$(e_1 \wedge e_2 \wedge e_4 \wedge e_5 \wedge e_7) \odot (-e_4 + 2\, e_7)$$

By inspection we can see that there are 4 different factors. We can take the simplest form congruent to each. Clearly they are not independent.

```
AA = (e1 + 2 e2) ∧ (e1 - 6 e5) ∧ (e4 - 2 e7) ∧ (e2 + 3 e5); ★𝒢[AA]
```

$$0$$

But it is an easy matter to compute all the products of 3 different factors by taking the 3-span of AA.

```
★S3[AA] // TableForm
```

$$(e_1 + 2\, e_2) \wedge (e_1 - 6\, e_5) \wedge (e_4 - 2\, e_7)$$
$$(e_1 + 2\, e_2) \wedge (e_1 - 6\, e_5) \wedge (e_2 + 3\, e_5)$$
$$(e_1 + 2\, e_2) \wedge (e_4 - 2\, e_7) \wedge (e_2 + 3\, e_5)$$
$$(e_1 - 6\, e_5) \wedge (e_4 - 2\, e_7) \wedge (e_2 + 3\, e_5)$$

By inspection, the second factorization in the list, because it does not display all the symbols of A, cannot be congruent to A. The others are all candidates for a factorization of A, and all can be seen to be congruent to A when expanded and simplified.

```
★𝒢[★S3[AA]]
```

$$\{-2\, e_1 \wedge e_2 \wedge e_4 + 4\, e_1 \wedge e_2 \wedge e_7 + 6\, e_1 \wedge e_4 \wedge e_5 + 12\, e_1 \wedge e_5 \wedge e_7 + 12\, e_2 \wedge e_4 \wedge e_5 + 24\, e_2 \wedge e_5 \wedge e_7,\ 0,\ -(e_1 \wedge e_2 \wedge e_4) + 2\, e_1 \wedge e_2 \wedge e_7 + 3\, e_1 \wedge e_4 \wedge e_5 + 6\, e_1 \wedge e_5 \wedge e_7 + 6\, e_2 \wedge e_4 \wedge e_5 + 12\, e_2 \wedge e_5 \wedge e_7,\ -(e_1 \wedge e_2 \wedge e_4) + 2\, e_1 \wedge e_2 \wedge e_7 + 3\, e_1 \wedge e_4 \wedge e_5 + 6\, e_1 \wedge e_5 \wedge e_7 + 6\, e_2 \wedge e_4 \wedge e_5 + 12\, e_2 \wedge e_5 \wedge e_7\}$$

- In Chapter 3, we will explore an equivalent algorithm for factorizing simple elements, based on the regressive product as the multilinear product.

2.14 Summary

This chapter has introduced the exterior product. Its codification of linear dependence is the foundation of Grassmann's contribution to the mathematical and physical sciences. From it he was able to build an algebraic system widely applicable to much of the physical sciences.

We began the chapter by formally extending the axioms of a linear space to include the exterior product. We then showed how the exterior product leads naturally to an understanding of determinants and linear systems of equations. Although Grassmann was the first to develop an algebraic structure which naturally handled these now accepted denizens of linear algebra, we showed, in discussing the concepts of simplicity and exterior division, that his exterior product spaces (although themselves linear spaces) had the potential to be much richer than the simple linear space of linear algebra.

Towards the end of the chapter we introduced the notion of multilinear forms. Grassmann algebra is a linear algebra populated by a number of linear product operations, and it turns out that provided a product of elements takes a certain multilinear form, it can be 'expanded' in a manner invariant to the way in which the elements are represented. This invariance is critical to the tensor-geometric nature of the algebra where the entities being manipulated must be independent of their representation. As a first application of multilinear forms we explored an algorithm for computing the union and intersection of two simple elements or spaces. As a second application we looked at how we could obtain a factorization of a simple element.

The chapter has not yet introduced any geometric or physical interpretations for the elements of the algebra, as, being a mathematical system, its power lies in its potentially multiple interpretations, as I hope Chapter 1 has previewed. Nevertheless, I am acutely aware that comprehension is enhanced by context, particularly the geometric. I beg your indulgence that the next chapter is still devoid of any depiction of the entities discussed, but hope that Chapter 4: Geometric Interpretations and subsequent examples will redress this balance.

3 The Regressive Product

3.1 Introduction

Since the linear spaces $\underset{m}{\Lambda}$ and $\underset{n-m}{\Lambda}$ are of the same dimension $\binom{n}{m} = \binom{n}{n-m}$ and hence are isomorphic, the opportunity exists to define a product operation (called the regressive product) dual to the exterior product such that theorems involving exterior products of *m*-elements have duals involving regressive products of (*n*–*m*)-elements.

Very roughly speaking, if the exterior product is associated with the notion of *union*, then the regressive product is associated with the notion of *intersection*.

The regressive product appears to be little used in the recent literature. Grassmann's original development did not distinguish *notationally* between the exterior and regressive product operations. Instead he capitalized on the inherent duality and used a notation which, depending on the grade of the elements in the product, could only sensibly be *interpreted* by one or the other operation. This was a very elegant idea, but its subtlety may have been one of the reasons the notion has become lost. (See 'Grassmann's notation for the regressive product' in section 3.3.)

However, since the regressive product is a simple dual operation to the exterior product, an enticing and powerful symmetry is lost by ignoring it. We will find that its 'intersection' properties are a useful conceptual and algorithmic addition to non-metric geometry, and that its algebraic properties enable a firm foundation for the development of metric geometry. Some of the results which are usually proven in metric spaces via the inner and cross products can also be proven using the exterior and regressive products, thus showing that the result is independent of whether or not the space has a metric.

The approach adopted in this book is to distinguish between the two product operations by using different notations, just as Boolean algebra has its dual operations of union and intersection. We will find that this approach does not detract from the elegance of the results. We will also find that differentiating the two operations explicitly enhances the simplicity and power of the derivation of results.

Since the commonly accepted modern notation for the exterior product operation is the 'wedge' symbol $\wedge$, we will denote the regressive product operation by a 'vee' symbol $\vee$. Note however that this (unfortunately) does not correspond to the Boolean algebra usage of the 'vee' for union and the 'wedge' for intersection.

3.2 Duality

The notion of duality

In order to ensure that the regressive product is defined as an operation correctly dual to the exterior product, we give the defining axiom set for the regressive product the same formal symbolic structure as the axiom set for the exterior product. This may be accomplished by replacing $\wedge$ by $\vee$, and replacing the grades of elements and spaces by their complementary grades. The *complementary grade* of a grade m is defined in a linear space of n dimensions to be $n-m$.

$$\wedge \longrightarrow \vee \qquad \underset{m}{\alpha} \longrightarrow \underset{n-m}{\alpha} \qquad \underset{m}{\Lambda} \longrightarrow \underset{n-m}{\Lambda} \tag{3.1}$$

Note here that we are undertaking the construction of a mathematical structure and thus there is no specific mapping implied between individual elements at this stage. In Chapter 5: The Complement, we will introduce a mapping between the elements of $\underset{m}{\Lambda}$ and $\underset{n-m}{\Lambda}$ which will lead to the definitions of *complement* and *interior product*.

For concreteness, we take some examples.

Examples: Obtaining the dual of an axiom

◆ The dual of axiom ∧6

We begin with the exterior product axiom $\wedge 6$ [2.12].

The exterior product of an m-element and a k-element is an (m+k)-element.

$$\left\{\underset{m}{\alpha} \in \underset{m}{\Lambda},\ \underset{k}{\beta} \in \underset{k}{\Lambda}\right\} \Rightarrow \left\{\underset{m}{\alpha} \wedge \underset{k}{\beta} \in \underset{m+k}{\Lambda}\right\}$$

To form the dual of this axiom, replace $\wedge$ with $\vee$, and the grades of elements and spaces by their complementary grades.

$$\left\{\underset{n-m}{\alpha} \in \underset{n-m}{\Lambda},\ \underset{n-k}{\beta} \in \underset{n-k}{\Lambda}\right\} \Rightarrow \left\{\underset{n-m}{\alpha} \vee \underset{n-k}{\beta} \in \underset{n-(m+k)}{\Lambda}\right\}$$

Although this is indeed the dual of axiom $\wedge 6$, it is not necessary to display the grades of what are arbitrary elements in the more complex form specifically involving the dimension n of the space. It will be more convenient to display them as grades denoted by simple symbols like m and k as were the grades of the elements of the original axiom. To effect this transformation most expeditiously we first let $m' = n - m$, $k' = n - k$ to get

$$\left\{\underset{m'}{\alpha} \in \underset{m'}{\Lambda},\ \underset{k'}{\beta} \in \underset{k'}{\Lambda}\right\} \Rightarrow \left\{\underset{m'}{\alpha} \vee \underset{k'}{\beta} \in \underset{(m'+k')-n}{\Lambda}\right\}$$

The grade of the space to which the regressive product belongs is

$$n-(m+k) \;=\; n-((n-m')+(n-k')) \;=\; (m'+k')-n.$$

Finally, since the primes are no longer necessary we drop them. Then the final form of the axiom dual to $\wedge 6$, which we label $\vee 6$, becomes:

$$\left\{\underset{m}{\alpha} \in \underset{m}{\Lambda},\ \underset{k}{\beta} \in \underset{k}{\Lambda}\right\} \Rightarrow \left\{\underset{m}{\alpha} \vee \underset{k}{\beta} \in \underset{m+k-n}{\Lambda}\right\}$$

In words, this says that the regressive product of an m-element and a k-element is an $(m+k-n)$-element.

◆ The dual of axiom ∧8

Axiom $\wedge 8$ [2.14] says:

There is a unit scalar which acts as an identity under the exterior product.

$$\mathcal{E}xists\left[1,\ 1 \in \underset{0}{\Lambda},\ \underset{m}{\alpha} == 1 \wedge \underset{m}{\alpha}\right]$$

For simplicity we do not normally display designated scalars with an underscripted zero. However, the duality transformation will be clearer if we rewrite the axiom with $\underset{0}{1}$ in place of 1.

$$\mathcal{E}xists\left[\underset{0}{1},\ \underset{0}{1} \in \underset{0}{\Lambda},\ \underset{m}{\alpha} == \underset{0}{1} \wedge \underset{m}{\alpha}\right]$$

Replace $\wedge$ with $\vee$, and the grades of elements and spaces by their complementary grades.

$$\mathcal{E}xists\left[\underset{n}{1},\ \underset{n}{1} \in \underset{n}{\Lambda},\ \underset{n-m}{\alpha} == \underset{n}{1} \vee \underset{n-m}{\alpha}\right]$$

Replace *arbitrary* grades m with $n - m'$. An arbitrary grade is one without a specific value, like $0, 1$, or n.

$$\mathcal{E}xists\left[\underset{n}{1},\ \underset{n}{1} \in \underset{n}{\Lambda},\ \underset{m'}{\alpha} == \underset{n}{1} \vee \underset{m'}{\alpha}\right]$$

Drop the primes.

$$\mathcal{E}xists\left[\underset{n}{1},\ \underset{n}{1} \in \underset{n}{\Lambda},\ \underset{m}{\alpha} == \underset{n}{1} \vee \underset{m}{\alpha}\right]$$

In words, this says that there is a unit n-element which acts as an identity under the regressive product.

◆ The dual of axiom ∧10

Axiom ∧10 [2.16] says:

The exterior product of elements of odd grade is anti-commutative.

$$\underset{m}{\alpha} \wedge \underset{k}{\beta} == (-1)^{m k}\, \underset{k}{\beta} \wedge \underset{m}{\alpha}$$

Replace $\wedge$ with $\vee$, and the grades of elements by their complementary grades. Note that it is only the grades as shown in the underscripts which should be replaced, not those figuring elsewhere in the formula.

$$\underset{n-m}{\alpha} \vee \underset{n-k}{\beta} == (-1)^{m k}\, \underset{n-k}{\beta} \vee \underset{n-m}{\alpha}$$

Replace arbitrary grades m (wherever they occur in the formula) with $n - m'$, k with $n - k'$. (n is not an arbitrary grade.)

$$\underset{m'}{\alpha} \vee \underset{k'}{\beta} == (-1)^{(n-m')\,(n-k')}\, \underset{k'}{\beta} \vee \underset{m'}{\alpha}$$

Drop the primes.

$$\underset{m}{\alpha} \vee \underset{k}{\beta} == (-1)^{(n-m)\,(n-k)}\, \underset{k}{\beta} \vee \underset{m}{\alpha}$$

In words this says that the regressive product of elements of odd complementary grade is anti-commutative.

Summary: The duality transformation algorithm

The algorithm for the duality transformation may be summarized as follows:

1. Replace $\wedge$ with $\vee$, and the grades of elements and spaces by their complementary grades.
2. Replace arbitrary grades m with $n - m'$, k with $n - k'$. Drop the primes.

An arbitrary grade is one without a specific value, like 0, 1, or n.

3.3 Properties of the Regressive Product

Axioms for the regressive product

In this section we collect the results of applying the duality algorithm above to the exterior product axioms $\wedge 6$ to $\wedge 12$. Axioms $\wedge 1$ to $\wedge 5$ transform unchanged since there are no products involved.

◆ **$\vee$6: The regressive product of an *m*-element and a *k*-element is an (*m*+*k*–*n*)-element**

$$\left\{\underset{m}{\alpha} \in \underset{m}{\Lambda},\ \underset{k}{\beta} \in \underset{k}{\Lambda}\right\} \Rightarrow \left\{\underset{m}{\alpha} \vee \underset{k}{\beta} \in \underset{m+k-n}{\Lambda}\right\} \qquad 3.2$$

◆ **$\vee$7: The regressive product is associative**

$$\left(\underset{m}{\alpha} \vee \underset{k}{\beta}\right) \vee \underset{j}{\gamma} == \underset{m}{\alpha} \vee \left(\underset{k}{\beta} \vee \underset{j}{\gamma}\right) \qquad 3.3$$

◆ **$\vee$8: There is a unit *n*-element which acts as an identity under the regressive product**

$$\mathcal{E}xists\left[\underset{n}{1},\ \underset{n}{1} \in \underset{n}{\Lambda},\ \underset{m}{\alpha} == \underset{n}{1} \vee \underset{m}{\alpha}\right] \qquad 3.4$$

◆ **$\vee$9: Non-zero *n*-elements have inverses with respect to the regressive product**

$$\mathcal{F}or\mathcal{A}ll\left[\alpha,\ \left(\alpha \in \underset{n}{\Lambda}\right) \bigwedge \left(\alpha \neq \underset{n}{0}\right),\ \mathcal{E}xists\left[\frac{1}{\alpha},\ \frac{1}{\alpha} \in \underset{n}{\Lambda},\ \underset{n}{1} == \alpha \vee \frac{1}{\alpha}\right]\right] \qquad 3.5$$

- **∨10: The regressive product of elements of odd complementary grade is anti-commutative**

$$\underset{m}{\alpha} \vee \underset{k}{\beta} == (-1)^{(n-k)\,(n-m)}\, \underset{k}{\beta} \vee \underset{m}{\alpha} \tag{3.6}$$

- **∨11: Additive identities act as multiplicative zero elements under the regressive product**

$$\mathcal{E}xists\left[\underset{k}{0},\ \underset{k}{0} \in \underset{k}{\Lambda},\ \underset{k}{0} \vee \underset{m}{\alpha} == \underset{k+m-n}{0}\right] \tag{3.7}$$

- **∨12: The regressive product is both left and right distributive under addition**

$$\left(\underset{m}{\alpha} + \underset{m}{\beta}\right) \vee \underset{k}{\gamma} == \underset{m}{\alpha} \vee \underset{k}{\gamma} + \underset{m}{\beta} \vee \underset{k}{\gamma}$$

$$\underset{m}{\alpha} \vee \left(\underset{k}{\beta} + \underset{k}{\gamma}\right) == \underset{m}{\alpha} \vee \underset{k}{\beta} + \underset{m}{\alpha} \vee \underset{k}{\gamma} \tag{3.8}$$

- **∨13: Scalar multiplication commutes with the regressive product**

$$\underset{m}{\alpha} \vee \left(a\, \underset{k}{\beta}\right) == \left(a\, \underset{m}{\alpha}\right) \vee \underset{k}{\beta} == a \left(\underset{m}{\alpha} \vee \underset{k}{\beta}\right) \tag{3.9}$$

The unit *n*-element

- **The unit *n*-element is congruent to any basis *n*-element**

The duality algorithm has generated a unit *n*-element $\underset{n}{1}$ which acts as the multiplicative identity for the regressive product, just as the unit scalar 1 (or $\underset{0}{1}$) acts as the multiplicative identity for the exterior product (axioms ∧8 and ∨8).

We have already seen that any basis of $\underset{n}{\Lambda}$ contains only one element. If a basis of $\underset{1}{\Lambda}$ is $\{e_1, e_2, \ldots, e_n\}$, then the single basis element of $\underset{n}{\Lambda}$ is equal to $e_1 \wedge e_2 \wedge \ldots \wedge e_n$. If the basis of $\underset{1}{\Lambda}$ is changed by an arbitrary (non-singular) linear transformation, then the basis of $\underset{n}{\Lambda}$ changes by a scalar factor which is the determinant of the transformation. Any basis of $\underset{n}{\Lambda}$ may therefore be expressed as a scalar multiple of some given basis, say $e_1 \wedge e_2 \wedge \ldots \wedge e_n$. Hence we can therefore also express $e_1 \wedge e_2 \wedge \ldots \wedge e_n$ as some scalar multiple $\star c$ of $\underset{n}{1}$. (We denote the scalar in this form so that *GrassmannAlgebra* can understand its special role in later computations.)

$$e_1 \wedge e_2 \wedge \ldots \wedge e_n \;==\; \star c \underset{n}{1} \qquad 3.10$$

Defining $\underset{n}{1}$ any more specifically than this is normally done by imposing a metric onto the space. This we do in Chapter 5: The Complement, and Chapter 6: The Interior Product. It turns out then that $\underset{n}{1}$ is the n-element whose *measure* (magnitude, volume) *is unity*.

On the other hand, for geometric application in spaces without a metric, for example the calculation of intersections of lines, planes, and hyperplanes, it is inconsequential that we only know $\underset{n}{1}$ up to congruence, because we will see that *if an element defines a geometric entity then any element congruent to it will define the same geometric entity.*

All n-elements are congruent. Hence we can also immediately say that any n-element is some scalar factor of the unit n-element.

$$\underset{n}{\alpha} \;==\; a \underset{n}{1} \qquad 3.11$$

◆ The unit n-element is idempotent under the regressive product

By putting $\underset{m}{\alpha}$ equal to $\underset{n}{1}$ in the regressive product axiom $\vee 8$ above we have immediately that:

$$\underset{n}{1} \vee \underset{n}{1} \;==\; \underset{n}{1} \qquad 3.12$$

◆ The regressive product of an n-element with the unit 1 is a scalar

By putting $\underset{m}{\alpha}$ equal to 1 in the regressive product axiom $\vee 8$ we also have that:

$$\underset{n}{1} \vee 1 \;==\; 1 \vee \underset{n}{1} \;==\; 1 \qquad 3.13$$

By multiplying through by the scalar a

$$a \left(\underset{n}{1} \vee 1\right) \;==\; \left(a \underset{n}{1}\right) \vee 1 \;==\; \underset{n}{\alpha} \vee 1$$

we obtain that

$$\underset{n}{\alpha} \vee 1 \;==\; 1 \vee \underset{n}{\alpha} \;==\; a \qquad\qquad \underset{n}{\alpha} \;==\; a \underset{n}{1} \qquad 3.14$$

◆ The regressive product with an n-element

The regressive product of an m-element with an n-element is congruent to the m-element.

$$\underset{n}{\alpha} \vee \underset{m}{\beta} \;==\; \left(a \underset{n}{1}\right) \vee \underset{m}{\beta} \;==\; a \left(\underset{n}{1} \vee \underset{m}{\beta}\right) \;==\; a \underset{m}{\beta}$$

$$\underset{n}{\alpha} \vee \underset{m}{\beta} == \underset{m}{\beta} \vee \underset{n}{\alpha} == a \underset{m}{\beta} \qquad \underset{n}{\alpha} == a \underset{n}{1} \tag{3.15}$$

◆ *n*-elements allow a sort of associativity

By taking the exterior product of the previous equation with a new element, we can show that regressive products with an *n*-element allow a sort of associativity with the exterior product.

$$\left(\underset{n}{\alpha} \vee \underset{m}{\beta}\right) \wedge \underset{j}{\gamma} == \left(a \underset{m}{\beta}\right) \wedge \underset{j}{\gamma} == a \left(\underset{m}{\beta} \wedge \underset{j}{\gamma}\right) == \underset{n}{\alpha} \vee \left(\underset{m}{\beta} \wedge \underset{j}{\gamma}\right)$$

$$\left(\underset{n}{\alpha} \vee \underset{m}{\beta}\right) \wedge \underset{j}{\gamma} == \underset{n}{\alpha} \vee \left(\underset{m}{\beta} \wedge \underset{j}{\gamma}\right) \tag{3.16}$$

◆ Division by *n*-elements

An equation may be 'divided' through by an *n*-element under the regressive product.

$$\underset{n}{\beta} \vee \underset{m}{\alpha} == \underset{n}{\beta} \vee \underset{m}{\gamma} \implies \underset{m}{\alpha} == \underset{m}{\gamma} \tag{3.17}$$

We can easily see this by putting $\underset{n}{\beta} = b \underset{n}{1}$, and using axiom $\vee 8$.

The inverse of an *n*-element

Axiom $\vee 9$ says that every (non-zero) *n*-element has an inverse with respect to the regressive product, and the unit *n*-element.

$$\text{ForAll}\left[\alpha, \left(\alpha \in \underset{n}{\Lambda}\right) \bigwedge \left(\alpha \neq \underset{n}{0}\right), \text{Exists}\left[\frac{1}{\alpha}, \frac{1}{\alpha} \in \underset{n}{\Lambda}, \underset{n}{1} == \alpha \vee \frac{1}{\alpha}\right]\right]$$

Suppose we have an *n*-element $\underset{n}{\alpha}$ expressed in terms of some basis. Then, according to formula [3.11] we can express it as a scalar multiple (a, say) of the unit *n*-element $\underset{n}{1}$.

We may then write the inverse $\underset{n}{\alpha}^{-1}$ of $\underset{n}{\alpha}$ with respect to the regressive product as the scalar multiple $\frac{1}{a}$ of $\underset{n}{1}$.

$$\underset{n}{\alpha} == a \underset{n}{1} \iff \underset{n}{\alpha}^{-1} == \frac{1}{a} \underset{n}{1}$$

We can see this by taking the regressive product of $a \underset{n}{1}$ with $\frac{1}{a} \underset{n}{1}$:

$$\underset{n}{\alpha} \vee \underset{n}{\alpha}^{-1} == \left(a \underset{n}{1}\right) \vee \left(\frac{1}{a} \underset{n}{1}\right) == \underset{n}{1} \vee \underset{n}{1} == \underset{n}{1}$$

$$\underset{n}{\alpha} == a \underset{n}{1} \iff \underset{n}{\alpha}^{-1} == \frac{1}{a} \underset{n}{1} \tag{3.18}$$

If $\underset{n}{\alpha}$ is now expressed in terms of a basis we have

$$\underset{n}{\alpha} == b\, e_1 \wedge e_2 \wedge \cdots \wedge e_n == b \star c \underset{n}{1}$$

Hence $\underset{n}{\alpha}^{-1}$ can be written as

$$\underset{n}{\alpha}^{-1} == \frac{1}{b} \frac{1}{\star c} \underset{n}{1} == \frac{1}{b} \frac{1}{\star c^2} e_1 \wedge e_2 \wedge \cdots \wedge e_n$$

$$\underset{n}{\alpha} == b\, e_1 \wedge e_2 \wedge \cdots \wedge e_n \quad \Leftrightarrow \quad \underset{n}{\alpha}^{-1} == \frac{1}{b} \frac{1}{\star c^2} e_1 \wedge e_2 \wedge \cdots \wedge e_n \qquad 3.19$$

Grassmann's notation for the regressive product

In Grassmann's *Ausdehnungslehre* of 1862 Section 95 (translated by Lloyd C. Kannenberg) he states:

> If *q* and *r* are the orders of two magnitudes A and B, and *n* that of the principal domain, then the order of the product [A B] is *first* equal to *q+r* if *q+r* is smaller than *n*, and *second* equal to *q+r–n* if *q+r* is greater than or equal to *n*.

Translating this into the terminology used in this book we have:

> If *q* and *r* are the *grades* of two *elements* A and B, and *n* that of the *underlying linear space*, then the *grade* of the product [A B] is *first* equal to *q+r* if *q+r* is smaller than *n*, and *second* equal to *q+r–n* if *q+r* is greater than or equal to *n*.

Translating this further into the notation used in this book we have:

$$\begin{array}{lll} [A\,B] \Leftrightarrow \underset{p}{A} \wedge \underset{q}{B} & & p + q < n \\ [A\,B] \Leftrightarrow \underset{p}{A} \vee \underset{q}{B} & & p + q \geq n \end{array} \qquad 3.20$$

Grassmann called the product [AB] in the first case a *progressive exterior* product, and in the second case a *regressive exterior* product. In current terminology, which we adopt in this book, the *progressive exterior product* is called simply the *exterior product*, and the *regressive exterior product* is called simply the *regressive product*.

In the equivalence above, Grassmann has opted to define the product of two elements whose grades sum to the dimension *n* of the space as *regressive*, and thus a *scalar*. However, the more explicit notation that we have adopted identifies that some definition is still required for the *progressive* (exterior) product of two such elements.

The advantage of denoting the two products differently enables us to correctly define the exterior product of two elements whose grades sum to *n*, *as an n-element*. Grassmann by his choice of notation has decided to define it as a scalar. In modern terminology this is equivalent to confusing scalars with pseudo-scalars. A separate notation for the two products thus avoids this tensorially invalid confusion.

We can see how then, in not being explicitly denoted, the regressive product may have become lost.

3.4 The Grassmann Duality Principle

The dual of a dual

The duality of the axiom sets for the exterior and regressive products is completed by requiring that the dual of a dual of an axiom be the axiom itself. The dual of a regressive product axiom may be obtained by applying the following algorithm:

1. Replace $\vee$ with $\wedge$, and the grades of elements and spaces by their complementary grades.
2. Replace arbitrary grades *m* with $n-m'$, *k* with $n-k'$. Drop the primes.

This differs from the algorithm for obtaining the dual of an exterior product axiom only in the replacement of $\vee$ with $\wedge$ instead of *vice versa*.

It is easy to see that applying this algorithm to the regressive product axioms generates the original exterior product axiom set.

We can combine both transformation algorithms by restating them as:

1. Replace each product operation by its dual operation, and the grades of elements and spaces by their complementary grades.
2. Replace arbitrary grades *m* with $n-m'$, *k* with $n-k'$. Drop the primes.

Since this algorithm applies to both sets of axioms, it also applies to any theorem. Thus to each theorem involving exterior or regressive products corresponds a dual theorem obtained by applying the algorithm. We call this the Grassmann Duality Principle (or, where the context is clear, simply the Duality Principle).

The Grassmann Duality Principle

To every theorem involving exterior and regressive products, a dual theorem may be obtained by:
1. Replacing each product operation by its dual operation, and the grades of elements and spaces by their complementary grades.
2. Replacing arbitrary grades *m* with $n-m'$, *k* with $n-k'$, then dropping the primes.

In each of the following examples we state a theorem and then its dual. We can recover the original theorem as the dual of this dual. Checking theorems using *GrassmannAlgebra's* `Dual` function requires the dimension of the space to be entered as `★n`.

◆ Example 1

The exterior product of two elements is zero if the sum of their grades is greater than the dimension of the linear space.

$$\left\{\underset{m}{\alpha}\wedge\underset{k}{\beta} == 0,\ m+k-n>0\right\}$$

The dual theorem states that the regressive product of two elements is zero if the sum of their grades is less than the dimension of the linear space.

$$\left\{\underset{m}{\alpha} \vee \underset{k}{\beta} == 0,\ n - (k + m) > 0\right\}$$

◆ Example 2

The exterior square of a *simple* element is zero unless it is a scalar.

$$\left\{\underset{m}{\alpha} \wedge \underset{m}{\alpha} == 0,\ m \neq 0\right\}$$

The dual theorem states the regressive square of a simple element is zero unless it is an *n*-element.

$$\left\{\underset{m}{\alpha} \vee \underset{m}{\alpha} == 0,\ n - m \neq 0\right\}$$

◆ Example 3

The exterior square of an element (simple or non-simple) is zero if it is of odd grade.

$$\left\{\underset{m}{\alpha} \wedge \underset{m}{\alpha} == 0,\ m \in \{\text{OddPositiveIntegers}\}\right\}$$

The dual theorem states that the regressive square of an element is zero if its complementary grade is odd.

$$\left\{\underset{m}{\alpha} \vee \underset{m}{\alpha} == 0,\ (n - m) \in \{\text{OddPositiveIntegers}\}\right\}$$

◆ Example 4

The exterior product of unity with itself any number of times remains unity.

$$1 \wedge 1 \wedge \ldots \wedge 1 \wedge \underset{m}{\alpha} == \underset{m}{\alpha}$$

The dual theorem states the corresponding fact for the regressive product of unit *n*-elements $\underset{n}{1}$.

$$\underset{n}{1} \vee \underset{n}{1} \vee \ldots \vee \underset{n}{1} \vee \underset{m}{\alpha} == \underset{m}{\alpha}$$

Using the *GrassmannAlgebra* function Dual

The algorithm of the Duality Principle has been encapsulated in the function `Dual` in *GrassmannAlgebra*. `Dual` takes an expression or list of expressions which comprise an axiom or theorem and generates the list of dual expressions by transforming them according to the replacement rules of the Duality Principle.

In order that *GrassmannAlgebra* can understand the special nature of the symbol used to represent the dimension *n* of the underlying linear space we use the special symbol `★n` instead whenever any computations are being done.

We have adopted *Mathematica's* syntax for the logical quantifiers of `Exists` and `ForAll`, but to enhance readability we have purposely frustrated its output form for them. This is signaled by their being written in script.

◆ Example 1: The Dual of axiom ∧10

Our first examples are to show how `Dual` may be used to develop dual axioms. For example, to take the dual of axiom $\wedge$10 simply enter:

$$\textbf{Dual}\left[\underset{m}{\alpha}\wedge\underset{k}{\beta} == (-1)^{m\,k}\,\underset{k}{\beta}\wedge\underset{m}{\alpha}\right]$$

$$\underset{m}{\alpha}\vee\underset{k}{\beta} == (-1)^{(\star n-k)\,(\star n-m)}\,\underset{k}{\beta}\vee\underset{m}{\alpha}$$

◆ Example 2: The Dual of axiom $\wedge$8

Again, to take the `Dual` of axiom $\wedge$8 enter:

$$\textbf{Dual}\left[\mathcal{E}xists\left[1,\ 1\in\underset{0}{\Lambda},\ \underset{m}{\alpha} == 1\wedge\underset{m}{\alpha}\right]\right]$$

$$\mathcal{E}xists\left[\underset{\star n}{1},\ \underset{\star n}{1}\in\underset{\star n}{\Lambda},\ \underset{m}{\alpha} == \underset{\star n}{1}\vee\underset{m}{\alpha}\right]$$

◆ Example 3: The Dual of axiom $\wedge$9

To apply `Dual` to an axiom involving more than one statement, collect the statements in a list. For example, the dual of axiom $\wedge$9 is obtained as:

$$\textbf{Dual}\left[\mathcal{F}or\mathcal{A}ll\left[\alpha,\ \left(\alpha\in\underset{0}{\Lambda}\right)\bigwedge\left(\alpha\neq\underset{0}{0}\right),\ \mathcal{E}xists\left[\alpha^{-1},\ \alpha^{-1}\in\underset{0}{\Lambda},\ 1 == \alpha\wedge\alpha^{-1}\right]\right]\right]$$

$$\mathcal{F}or\mathcal{A}ll\left[\alpha,\ \alpha\in\underset{\star n}{\Lambda}\ \&\&\ \alpha\neq\underset{\star n}{0},\ \mathcal{E}xists\left[\frac{1}{\alpha},\ \frac{1}{\alpha}\in\underset{\star n}{\Lambda},\ \underset{\star n}{1} == \alpha\vee\frac{1}{\alpha}\right]\right]$$

◆ Example 4: The Dual of the Common Factor Axiom

Our fourth example is to generate the dual of the *Common Factor Axiom*. This axiom will be introduced in the next section. Remember that the dimension of the space must be entered as ★n in order for `Dual` to recognize it as such.

$$\textbf{CommonFactorAxiom} = \left\{\left(\underset{m}{\alpha}\wedge\underset{j}{\mu}\right)\vee\left(\underset{k}{\beta}\wedge\underset{j}{\mu}\right) == \left(\underset{m}{\alpha}\wedge\underset{k}{\beta}\wedge\underset{j}{\mu}\right)\vee\underset{j}{\mu},\ m+k+j-\star n == 0\right\};$$

```
Dual[CommonFactorAxiom]
```

$$\left\{\left(\underset{m}{\alpha}\vee\underset{j}{\mu}\right)\wedge\left(\underset{k}{\beta}\vee\underset{j}{\mu}\right) == \left(\underset{m}{\alpha}\vee\underset{k}{\beta}\vee\underset{j}{\mu}\right)\wedge\underset{j}{\mu},\ 2\star n-j-k-m == 0\right\}$$

We can verify that the dual of the dual of the axiom is the axiom itself.

```
CommonFactorAxiom == Dual[Dual[CommonFactorAxiom]]
```

```
True
```

◆ Example 5: The Dual of a formula

Our fifth example is to obtain the dual of one of the product formulae derived in a later section of this chapter. Again, the dimension of the space must be entered as ★n.

$$\textbf{F} = \left(\underset{m}{\alpha}\wedge\underset{k}{\beta}\right)\vee\underset{\star n-1}{x} == \left(\underset{m}{\alpha}\vee\underset{\star n-1}{x}\right)\wedge\underset{k}{\beta} + (-1)^{m}\,\underset{m}{\alpha}\wedge\left(\underset{k}{\beta}\vee\underset{\star n-1}{x}\right);$$

```
Dual[F]
```

$$(\underset{m}{\alpha} \vee \underset{k}{\beta}) \wedge x == (-1)^{\star n-m} \underset{m}{\alpha} \vee (\underset{k}{\beta} \wedge x) + (\underset{m}{\alpha} \wedge x) \vee \underset{k}{\beta}$$

Again, we can verify that the dual of the dual of the formula is the formula itself.

```
F == Dual[Dual[F]]
```

```
True
```

◆ Setting up your own input to the Dual function

- You can use the *GrassmannAlgebra* function `Dual` with any expression involving exterior and regressive products of graded, vector or scalar symbols.
- You can include multiple statements by combining them in a list.
- The statements should not include any expressions which will evaluate when entered.
- The dimension of the underlying linear space should be denoted as `★n`.
- Avoid using scalar or vector symbols as (underscripted) grades if they appear other than as underscripts elsewhere in the expression.

3.5 The Common Factor Axiom

Motivation

Although the axiom sets for the exterior and regressive products have been posited, it still remains to propose an axiom explicitly relating the two types of products.

The axiom set for the regressive product, which we have created above simply as a dual axiom set to that of the exterior product, asserts that in an *n*-space the regressive product of an *m*-element and a *k*-element is an *(m+k–n)*-element. Our first criterion for the new axiom then, is that it can be used to compute this new element.

Earlier we have also remarked on some enticing correspondences between the dual exterior and regressive products and the dual union and intersection operations of a Boolean algebra. We can already see that the (non-zero) exterior product of simple elements is an element whose space is the 'union' of the spaces of its factors. Our second criterion then, is that the new axiom enables the 'dual' of this property: the (non-zero) regressive product of simple elements is an element whose space is the 'intersection' of the spaces of its factors.

This criterion leads us first to return to Chapter 2: Section 2.13 Unions and Intersections, where, with an arbitrary multilinear product operation (which we denoted ⊙), we were able to compute the union and intersection of simple elements. In the cases explored the simple elements were of arbitrary grade. We saw there that given simple elements $A^* \wedge C$ and $C \wedge B^*$, we could obtain their union and intersection as (congruent to) $A^* \wedge C \wedge B^*$ and C respectively.

$$A \odot B \;==\; (A^* \wedge C) \odot (C \wedge B^*) \;==\; (A^* \wedge C \wedge B^*) \odot C$$

Since ⊙ is an arbitrary multilinear product, we can satisfy the second criterion by positing the same form for the regressive product.

$$A \vee B == (A^* \wedge C) \vee (C \wedge B^*) == (A^* \wedge C \wedge B^*) \vee C \qquad 3.21$$

What additional requirements on this would one need in order to satisfy the first criterion, that is, that the axiom enables a result which does not involve the regressive product? Axiom $\vee 8$ [3.4] gives us the clue.

$$\mathcal{E}xists\left[\underset{n}{1}, \underset{n}{1} \in \underset{n}{\Lambda}, \underset{m}{\alpha} == \underset{n}{1} \vee \underset{m}{\alpha}\right]$$

If the grade of $A^* \wedge C \wedge B^*$ is equal to the dimension of the space n, then since all n-elements are congruent we can write $A^* \wedge C \wedge B^* = c\,\underset{n}{1}$, where c is some scalar, showing that the right hand side is congruent to the common factor (intersection) C.

$$(A^* \wedge C) \vee (C \wedge B^*) == (A^* \wedge C \wedge B^*) \vee C == \left(c\,\underset{n}{1}\right) \vee C == c\,C \equiv C$$

An axiom which satisfies both of our original criteria is then

$$\{(A^* \wedge C) \vee (C \wedge B^*) == (A^* \wedge C \wedge B^*) \vee C,\ \ \mathrm{Grade}[A^* \wedge C \wedge B^*] == n\}$$

Using graded symbols, we could also write this in either of the following forms

$$\left\{\left(\underset{m}{\alpha} \wedge \underset{j}{\mu}\right) \vee \left(\underset{j}{\mu} \wedge \underset{k}{\beta}\right) == \left(\underset{m}{\alpha} \wedge \underset{j}{\mu} \wedge \underset{k}{\beta}\right) \vee \underset{j}{\mu},\ \ m + k + j - n == 0\right\}$$

$$\left\{\left(\underset{m}{\alpha} \wedge \underset{j}{\mu}\right) \vee \left(\underset{k}{\beta} \wedge \underset{j}{\mu}\right) == \left(\underset{m}{\alpha} \wedge \underset{k}{\beta} \wedge \underset{j}{\mu}\right) \vee \underset{j}{\mu},\ \ m + k + j - n == 0\right\}$$

As we will see, an axiom of this form works very well. Indeed, it turns out to be one of the fundamental underpinnings of the algebra. We call it the Common Factor Axiom and explore it further below.

◆ ***Historical Note***

The approach we have adopted in this chapter of treating the common factor relation as an axiom is effectively the same as Grassmann used in his first *Ausdehnungslehre* (1844) but differs from the approach that he used in his second *Ausdehnungslehre* (1862). (See Chapter 3, Section 5 in Kannenberg.) In the 1862 version Grassmann *proves* this relation from another which is (almost) the same as the Complement Axiom that we introduce in Chapter 5: The Complement. Whitehead [1898], and other writers in the Grassmannian tradition follow his 1862 approach.

The relation which Grassmann used in the 1862 *Ausdehnungslehre* is in effect equivalent to assuming the space has a Euclidean metric (his *Ergänzung* or *supplement*). However the Common Factor Axiom does not depend on the space having a metric; that is, it is completely independent of any correspondence we set up between $\underset{m}{\Lambda}$ and $\underset{n-m}{\Lambda}$. Hence we would rather not adopt an approach which introduces an unnecessary constraint, especially since we want to show later that the *Ausdehnungslehre* is easily extended to metrics more general than the Euclidean.

The Common Factor Axiom

We begin completely afresh in positing the axiom. Let $\underset{m}{\alpha}$, $\underset{k}{\beta}$, and $\underset{j}{\mu}$ be *simple* elements with $m+k+j = n$, where n is the dimension of the space. Then the Common Factor Axiom states that:

$$\left\{\left(\underset{m}{\alpha}\wedge\underset{j}{\mu}\right)\vee\left(\underset{k}{\beta}\wedge\underset{j}{\mu}\right) == \left(\underset{m}{\alpha}\wedge\underset{k}{\beta}\wedge\underset{j}{\mu}\right)\vee\underset{j}{\mu},\ m+k+j-n == 0\right\} \qquad 3.22$$

Thus, the regressive product of two elements $\underset{m}{\alpha}\wedge\underset{j}{\mu}$ and $\underset{k}{\beta}\wedge\underset{j}{\mu}$ with a common factor $\underset{j}{\mu}$ is equal to the regressive product of the 'union' of the elements $\underset{m}{\alpha}\wedge\underset{k}{\beta}\wedge\underset{j}{\mu}$ with the common factor $\underset{j}{\mu}$ (their 'intersection').

Since the union $\underset{m}{\alpha}\wedge\underset{k}{\beta}\wedge\underset{j}{\mu}$ is an *n*-element, we can write it as some scalar factor c, say, of the unit *n*-element: $\underset{m}{\alpha}\wedge\underset{k}{\beta}\wedge\underset{j}{\mu} = c\,\underset{n}{1}$. Hence by axiom $\vee 8$ we derive immediately that the regressive product of two elements $\underset{m}{\alpha}\wedge\underset{j}{\mu}$ and $\underset{k}{\beta}\wedge\underset{j}{\mu}$ with a common factor $\underset{j}{\mu}$ is congruent to that factor.

$$\left\{\left(\underset{m}{\alpha}\wedge\underset{j}{\mu}\right)\vee\left(\underset{k}{\beta}\wedge\underset{j}{\mu}\right) \equiv \underset{j}{\mu},\ m+k+j-n == 0\right\} \qquad 3.23$$

It is easy to see that by using the anti-commutativity axiom $\wedge 10$ [2.16], that the axiom may be arranged in any of a number of alternative forms, the most useful of which are:

$$\left\{\left(\underset{m}{\alpha}\wedge\underset{j}{\mu}\right)\vee\left(\underset{j}{\mu}\wedge\underset{k}{\beta}\right) == \left(\underset{m}{\alpha}\wedge\underset{j}{\mu}\wedge\underset{k}{\beta}\right)\vee\underset{j}{\mu},\ m+k+j-n == 0\right\} \qquad 3.24$$

$$\left\{\left(\underset{j}{\mu}\wedge\underset{m}{\alpha}\right)\vee\left(\underset{j}{\mu}\wedge\underset{k}{\beta}\right) == \left(\underset{j}{\mu}\wedge\underset{m}{\alpha}\wedge\underset{k}{\beta}\right)\vee\underset{j}{\mu},\ m+k+j-n == 0\right\} \qquad 3.25$$

And since for the regressive product, any *n*-element commutes with any other element, we can always rewrite the right hand side with the common factor first.

$$\left\{\left(\underset{m}{\alpha}\wedge\underset{j}{\mu}\right)\vee\left(\underset{j}{\mu}\wedge\underset{k}{\beta}\right) == \underset{j}{\mu}\vee\left(\underset{m}{\alpha}\wedge\underset{j}{\mu}\wedge\underset{k}{\beta}\right),\ m+k+j-n == 0\right\} \qquad 3.26$$

Extension of the Common Factor Axiom to general elements

The axiom has been stated for simple elements. In this section we show that it remains valid for general (possibly non-simple) elements, *provided that the common factor remains simple*.

Consider two simple elements $\underset{m}{\alpha_1}$ and $\underset{m}{\alpha_2}$. Then the Common Factor Axiom can be written for each as:

$$\left(\underset{m}{\alpha_1}\wedge\underset{j}{\mu}\right)\vee\left(\underset{k}{\beta}\wedge\underset{j}{\mu}\right) == \left(\underset{m}{\alpha_1}\wedge\underset{k}{\beta}\wedge\underset{j}{\mu}\right)\vee\underset{j}{\mu}$$

$$\left(\underset{m}{\alpha_2}\wedge\underset{j}{\mu}\right)\vee\left(\underset{k}{\beta}\wedge\underset{j}{\mu}\right) == \left(\underset{m}{\alpha_2}\wedge\underset{k}{\beta}\wedge\underset{j}{\mu}\right)\vee\underset{j}{\mu}$$

Adding these two equations and using the distributivity of $\wedge$ and $\vee$ gives:

$$\left(\left(\underset{m}{\alpha_1}+\underset{m}{\alpha_2}\right)\wedge\underset{j}{\mu}\right)\vee\left(\underset{k}{\beta}\wedge\underset{j}{\mu}\right) == \left(\left(\underset{m}{\alpha_1}+\underset{m}{\alpha_2}\right)\wedge\underset{k}{\beta}\wedge\underset{j}{\mu}\right)\vee\underset{j}{\mu}$$

Extending this process, we see that the formula remains true for arbitrary $\underset{m}{\alpha}$ and $\underset{k}{\beta}$, *providing* $\underset{j}{\mu}$ *is simple*.

$$\left\{\left(\underset{m}{\alpha}\wedge\underset{j}{\mu}\right)\vee\left(\underset{k}{\beta}\wedge\underset{j}{\mu}\right) == \left(\underset{m}{\alpha}\wedge\underset{k}{\beta}\wedge\underset{j}{\mu}\right)\vee\underset{j}{\mu} \equiv \underset{j}{\mu},\quad m+k+j-n==0\right\} \qquad 3.27$$

This is an extended version of the Common Factor Axiom. It states that: *the regressive product of two arbitrary elements containing a simple common factor is congruent to that factor.*

For applications involving computations in a non-metric space, particularly those with a geometric interpretation, we will see that the congruence form is not restrictive. Indeed, it will be quite elucidating. For more general applications in metric spaces we will see that the associated scalar factor is no longer arbitrary but is determined by the metric imposed.

Special cases of the Common Factor Axiom

In this section we list some special cases of the Common Factor Axiom that we will find useful later. We assume, without explicitly stating, that the common factor is simple.

If there is no common factor (other than the scalar `1`), then the axiom reduces to:

$$\left\{\underset{m}{\alpha}\vee\underset{k}{\beta} == \left(\underset{m}{\alpha}\wedge\underset{k}{\beta}\right)\vee 1 \in \underset{0}{\Lambda},\quad m+k-n==0\right\} \qquad 3.28$$

This can be rewritten in terms of three disjoint elements by putting $\underset{k}{\beta}$ equal to $\underset{k}{\beta}\wedge\underset{j}{\mu}$.

$$\left\{\underset{m}{\alpha}\vee\left(\underset{k}{\beta}\wedge\underset{j}{\mu}\right) == \left(\underset{m}{\alpha}\wedge\underset{k}{\beta}\wedge\underset{j}{\mu}\right)\vee 1 \in \underset{0}{\Lambda},\quad m+k+j-n==0\right\} \qquad 3.29$$

By rewriting $\underset{k}{\beta}$ as $\underset{j}{\mu}$, and $\underset{m}{\alpha}$ as $\underset{m}{\alpha}\wedge\underset{k}{\beta}$ we get a similar result. We can then combine this with the above to get a sort of associativity for products which are scalar.

$$\left\{\left(\underset{m}{\alpha}\wedge\underset{k}{\beta}\right)\vee\underset{j}{\mu} == \underset{m}{\alpha}\vee\left(\underset{k}{\beta}\wedge\underset{j}{\mu}\right) \in \underset{0}{\Lambda},\quad m+k+j-n==0\right\} \qquad 3.30$$

Dual versions of the Common Factor Axiom

As discussed earlier in the chapter, dual versions of formulae can be obtained automatically by using the *GrassmannAlgebra* function `Dual`.

The dual Common Factor Axiom is:

$$\left\{\left(\underset{m}{\alpha} \vee \underset{j}{\mu}\right) \wedge \left(\underset{k}{\beta} \vee \underset{j}{\mu}\right) == \left(\underset{m}{\alpha} \vee \underset{k}{\beta} \vee \underset{j}{\mu}\right) \wedge \underset{j}{\mu} \in \underset{j}{\Lambda}, \quad m + k + j - 2\,n == 0\right\} \qquad 3.31$$

The duals of the special cases in the previous section are:

$$\left\{\underset{m}{\alpha} \wedge \underset{k}{\beta} == \left(\underset{m}{\alpha} \vee \underset{k}{\beta}\right) \wedge \underset{n}{1} == \left(\underset{m}{\alpha} \vee \underset{k}{\beta}\right) \underset{n}{1} \in \underset{n}{\Lambda}, \quad m + k - n == 0\right\} \qquad 3.32$$

$$\left\{\underset{m}{\alpha} \wedge \left(\underset{k}{\beta} \vee \underset{j}{\mu}\right) == \left(\underset{m}{\alpha} \vee \underset{k}{\beta} \vee \underset{j}{\mu}\right) \wedge \underset{n}{1} \in \underset{n}{\Lambda}, \quad m + k + j - 2\,n == 0\right\} \qquad 3.33$$

$$\left\{\left(\underset{m}{\alpha} \vee \underset{k}{\beta}\right) \wedge \underset{j}{\mu} == \underset{m}{\alpha} \wedge \left(\underset{k}{\beta} \vee \underset{j}{\mu}\right) \in \underset{n}{\Lambda}, \quad m + k + j - 2\,n == 0\right\} \qquad 3.34$$

Taking the regressive product of formula [3.32] with an element $\underset{j}{\mu}$ of arbitrary grade gives

$$\left(\underset{m}{\alpha} \wedge \underset{k}{\beta}\right) \vee \underset{j}{\mu} == \left(\underset{m}{\alpha} \vee \underset{k}{\beta}\right) \underset{n}{1} \vee \underset{j}{\mu} == \left(\underset{m}{\alpha} \vee \underset{k}{\beta}\right) \underset{j}{\mu} == \left(\underset{m}{\alpha} \vee \underset{k}{\beta}\right) \wedge \underset{j}{\mu}$$

$$\left\{\left(\underset{m}{\alpha} \wedge \underset{k}{\beta}\right) \vee \underset{j}{\mu} == \left(\underset{m}{\alpha} \vee \underset{k}{\beta}\right) \wedge \underset{j}{\mu} == \left(\underset{m}{\alpha} \vee \underset{k}{\beta}\right) \underset{j}{\mu} \in \underset{j}{\Lambda}, \quad m + k - n == 0\right\} \qquad 3.35$$

Application of the Common Factor Axiom

We now work through an example to illustrate how the Common Factor Axiom might be applied. In most cases however, results will be obtained more effectively using the Common Factor Theorem. This will be discussed in the section to follow.

Suppose we have two general 2-elements X and Y in a 3-space and we wish to find a formula for their 1-element intersection Z. Because X and Y are in a 3-space, we are assured that they are simple.

$$X == x_1\, e_1 \wedge e_2 + x_2\, e_1 \wedge e_3 + x_3\, e_2 \wedge e_3$$

$$Y == y_1\, e_1 \wedge e_2 + y_2\, e_1 \wedge e_3 + y_3\, e_2 \wedge e_3$$

We calculate Z as the regressive product of X and Y:

$$\begin{aligned} Z &== X \vee Y \\ &== (x_1\, e_1 \wedge e_2 + x_2\, e_1 \wedge e_3 + x_3\, e_2 \wedge e_3) \vee (y_1\, e_1 \wedge e_2 + y_2\, e_1 \wedge e_3 + y_3\, e_2 \wedge e_3) \end{aligned}$$

Expanding this product, and remembering (see section 3.4) that the regressive product of identical basis 2-elements is zero, we obtain:

$$\begin{aligned} Z == & (x_1\, e_1 \wedge e_2) \vee (y_2\, e_1 \wedge e_3) + (x_1\, e_1 \wedge e_2) \vee (y_3\, e_2 \wedge e_3) \\ & + (x_2\, e_1 \wedge e_3) \vee (y_1\, e_1 \wedge e_2) + (x_2\, e_1 \wedge e_3) \vee (y_3\, e_2 \wedge e_3) \\ & + (x_3\, e_2 \wedge e_3) \vee (y_1\, e_1 \wedge e_2) + (x_3\, e_2 \wedge e_3) \vee (y_2\, e_1 \wedge e_3) \end{aligned}$$

In a 3-space, regressive products of 2-elements are anti-commutative since

$$(-1)^{(n-m)(n-k)} = (-1)^{(3-2)(3-2)} = -1$$

Hence we can collect pairs of terms with the same factors by making the corresponding sign change:

$$\begin{aligned} Z \;==\; & (x_1\, y_2 - x_2\, y_1)\; (e_1 \wedge e_2) \vee (e_1 \wedge e_3) \\ & + (x_1\, y_3 - x_3\, y_1)\; (e_1 \wedge e_2) \vee (e_2 \wedge e_3) \\ & + (x_2\, y_3 - x_3\, y_2)\; (e_1 \wedge e_3) \vee (e_2 \wedge e_3) \end{aligned}$$

We can now apply the Common Factor Axiom to each of these regressive products:

$$\begin{aligned} Z \;==\; & (x_1\, y_2 - x_2\, y_1)\; (e_1 \wedge e_2 \wedge e_3) \vee e_1 \\ & + (x_1\, y_3 - x_3\, y_1)\; (e_1 \wedge e_2 \wedge e_3) \vee e_2 \\ & + (x_2\, y_3 - x_3\, y_2)\; (e_1 \wedge e_2 \wedge e_3) \vee e_3 \end{aligned}$$

Finally, by putting $e_1 \wedge e_2 \wedge e_3$ equal to $\star c\, \underset{n}{1}$, we have Z expressed as a 1-element:

$$Z \;==\; \star c\; ((x_1\, y_2 - x_2\, y_1)\, e_1 + (x_1\, y_3 - x_3\, y_1)\, e_2 + (x_2\, y_3 - x_3\, y_2)\, e_3)$$

Thus, in sum, we have the general congruence relation for the intersection of two 2-elements in a 3-space.

$$\begin{aligned} & (x_1\, e_1 \wedge e_2 + x_2\, e_1 \wedge e_3 + x_3\, e_2 \wedge e_3) \vee (y_1\, e_1 \wedge e_2 + y_2\, e_1 \wedge e_3 + y_3\, e_2 \wedge e_3) \\ & \equiv (x_1\, y_2 - x_2\, y_1)\, e_1 + (x_1\, y_3 - x_3\, y_1)\, e_2 + (x_2\, y_3 - x_3\, y_2)\, e_3 \end{aligned} \qquad 3.36$$

The result on the right hand side *almost* looks as if it could have been obtained from a cross-product operation. However the cross-product requires the space to have a metric, while this does not. We will see in Chapter 6 how, once we have introduced a metric, we can transform this into the formula for the cross product. This is an example of how the regressive product can generate results, independent of whether the space has a metric; whereas the usual vector algebra, in using the cross-product, must assume that it does.

◆ Check by calculating the exterior products

We can check that Z is indeed a common element to X and Y by determining if the exterior product of Z with each of X and Y is zero. We compose the expressions for X, Y, and Z, and then use the alias ⋆𝒢 of `GrassmannExpandAndSimplify` to do the check.

GrassmannAlgebra already knows that the special congruence symbol ⋆c is scalar, and `ComposeBivector` or its alias ⋆𝔹□ will automatically declare the scalar coefficients as `ScalarSymbols`.

```
⋆𝓑3; X = ⋆𝔹x
```

$x_1\, e_1 \wedge e_2 + x_2\, e_1 \wedge e_3 + x_3\, e_2 \wedge e_3$

```
Y = ⋆𝔹y
```

$y_1\, e_1 \wedge e_2 + y_2\, e_1 \wedge e_3 + y_3\, e_2 \wedge e_3$

```
Z = ⋆c ((x1 y2 - x2 y1) e1 + (x1 y3 - x3 y1) e2 + (x2 y3 - x3 y2) e3);
Expand[⋆𝒢[{Z ∧ X, Z ∧ Y}]]
```

{0, 0}

When the common factor is not simple

Let us take a precautionary example to show what happens when we have a factor that looks like a common factor but it is not simple. When an m-element is simple, it only needs m independent 1-elements to express it. When an m-element is not simple, it requires at least $m+2$ independent 1-elements to express it. The simplest example of a non-simple element is the 2-element C, say, equal to $u \wedge v + x \wedge y$, where u, v, x and y are independent 1-elements.

In a 4-space we only have 4 independent 1-elements available. Let X and Y be 3-elements with C *apparently* a common factor. First, suppose we simplify them before applying the Common Factor Axiom. Note that a 3-element in a 4-space is necessarily simple (see section 2.10).

$$X == u \wedge C == u \wedge (u \wedge v + x \wedge y) == u \wedge x \wedge y$$
$$Y == C \wedge x == (u \wedge v + x \wedge y) \wedge x == u \wedge v \wedge x$$

$$X \vee Y == (u \wedge x \wedge y) \vee (u \wedge v \wedge x) == (y \wedge u \wedge x) \vee (-u \wedge x \wedge v)$$

The Common Factor Axiom gives us the correct result for the simplified factors because the simplified factors had a simple common factor. But of course it is not C.

$$X \vee Y == (y \wedge u \wedge x \wedge v) \vee (x \wedge u)$$

Now if we *formally* (that is, only looking at the form), (and thus incorrectly) apply the Common Factor Axiom *without* simplifying we get

$$X \vee Y == (u \wedge C) \vee (C \wedge x) == (u \wedge C \wedge x) \vee C$$

Substituting back for C on the right hand side gives zero, which is of course an incorrect result.

- *In sum* The Common Factor Axiom gives the correct result (as expected) when applied to terms resulting from the complete expansion of a regressive product, since these terms will either have simple common factors, or be zero. On the other hand it will not necessarily give the correct result (as expected) when the symbol for the common factor takes on a non-simple value.

The Common Factor Axiom is only valid when the common factor is simple.

3.6 The Common Factor Theorem

Development of the Common Factor Theorem

The Common Factor Theorem which we are about to develop is one of the most important in the Grassmann algebra as it is the source of many formulae. We will also see in Chapter 6 that it has a counterpart which forms the principal expansion theorem for interior products.

The example above in the section Application of the Common Factor Axiom applied the Common Factor Axiom to two elements by expanding all the terms in their regressive product, applying the Common Factor Axiom to each of the terms, and then factoring the result. This is always possible. But in situations where there is a large number of terms, it may not be very efficient. The Common Factor Theorem will enable more efficient solutions.

Let us begin by returning to the motivation of the Common Factor Axiom where we recast the union and intersection formula from Chapter 2 to become the axiom. The union and intersection formula was:

$$A \vee B == (A^* \wedge C) \vee (C \wedge B^*) == (A^* \wedge C \wedge B^*) \vee C$$

Here, the grade of $A^* \wedge C \wedge B^*$ is equal to the dimension of the space. Since an n-element will commute with an element of any grade, the term on the right hand side can always be rewritten as $C \vee (A^* \wedge C \wedge B^*)$. This latter form is often convenient from a mnemonic point of view, and we may make such exchanges without further comment. Thus

$$A \vee B == (A^* \wedge C) \vee (C \wedge B^*) == (A^* \wedge C \wedge B^*) \vee C == C \vee (A^* \wedge C \wedge B^*) \qquad 3.37$$

There are two different forms of the Common Factor Theorem which we will be developing: the A form and the B form. It is useful to have both forms at hand, since depending on the case, one of them is likely to be more computationally efficient than the other; and formulae resulting from each, while ultimately yielding the same result, give distinctly different forms to their expressions. The A form requires A to be simple and obtains the common factor C from the factors of A by spanning A (that is, composing an appropriate span and cospan of A (see section 2.12)). The B form requires B to be simple and obtains the common factor C from the factors of B by spanning B.

◆ The A form

The basic formula for the A form begins with the Common Factor Axiom written as

$$A \vee B == (A^* \wedge C) \vee B == (A^* \wedge B) \vee C \qquad 3.38$$

Let us now display the grades explicitly. Let A be of grade m, B be of grade k, and C be of grade j. A^* is then of grade $m-j = n-k$.

$$\underset{m}{A} \vee \underset{k}{B} == \left(\underset{m-j}{A^*} \wedge \underset{j}{C}\right) \vee \underset{k}{B} == \left(\underset{m-j}{A^*} \wedge \underset{k}{B}\right) \vee \underset{j}{C}$$

The crux of the development comes from the intrinsic relationship [2.44] between A, its span, and its cospan: that is, for any i and s, A is equal to the exterior product of the ith s-span element and the corresponding ith s-cospan element.

$$A == \left(\star S_s[A]_{[[i]]}\right) \wedge \left(\star S^s[A]_{[[i]]}\right)$$

In our case, we want s to be $m-j$. So we can write

$$A == \underset{m-j}{A^*} \wedge \underset{j}{C} == \left(\star S_{m-j}[A]_{[[i]]}\right) \wedge \left(\star S^{m-j}[A]_{[[i]]}\right)$$

We now focus on the right hand side of the Common Factor Axiom. If, for a given value of i, we were to replace $\underset{m-j}{A^*}$ by $\star S_{m-j}[A]_{[[i]]}$ and $\underset{j}{C}$ by $\star S^{m-j}[A]_{[[i]]}$, we would get the term

$$\left(\star S_{m-j}[A]_{[[i]]} \wedge \underset{k}{B}\right) \vee \star S^{m-j}[A]_{[[i]]}$$

Clearly, this term may be different for different values of i, since, for example, wherever the span element of A ($\star S_{m-j}[A]_{[[i]]}$) contains a factor of the common factor it will be zero, since B will contain the same factor. In other cases it may be non-zero.

As it turns out, the common factor needs the *sum* of these terms. We will see why in the next section when we discuss the proof of the theorem.

$$\underset{m}{\mathbf{A}} \vee \underset{k}{\mathbf{B}} == \sum_{i=1}^{\nu} \left(\star\mathbb{S}_{m-j}\left[\underset{m}{\mathbf{A}}\right]_{[[i]]} \wedge \underset{k}{\mathbf{B}} \right) \vee \star\mathbb{S}^{m-j}\left[\underset{m}{\mathbf{A}}\right]_{[[i]]} \qquad 3.39$$

The index i ranges from 1 to ν, where ν is equal to $\binom{m}{j}$ (or equivalently $\binom{m}{m-j}$).

Because the exterior and regressive products have *Mathematica's* `Listable` attribute, we do not actually need to index the list elements and then sum them. If we write ★Σ as a shorthand for *Mathematica's* inbuilt `Total` function, which sums the elements of a list, we can write the A form of the Common Factor Theorem in the computable form as:

$$\underset{m}{\mathbf{A}} \vee \underset{k}{\mathbf{B}} == \star\Sigma\left[\left(\star\mathbb{S}_{m-j}\left[\underset{m}{\mathbf{A}}\right] \wedge \underset{k}{\mathbf{B}} \right) \vee \star\mathbb{S}^{m-j}\left[\underset{m}{\mathbf{A}}\right] \right] \qquad 3.40$$

To make these ideas more concrete we now apply the formula to some simple examples before proceeding to a proof in the following section.

◆ Example 1

Suppose we are working in a 5-space and `A` and `B` are expressed in such a way as to display their common factor `C`. We declare the extra vector symbols with the alias ★★V. We also turn on the `ShowPrecedence` capability (alias ★P) to make the output groupings of exterior and regressive products easier to read.

$\star\mathcal{B}_5;\ \star\star\mathbf{V}[\alpha\_, \beta\_, \gamma\_];\ \star\mathbf{P};\ \mathbf{A} = \alpha_1 \wedge \gamma_1 \wedge \gamma_2;\ \mathbf{B} = \gamma_1 \wedge \gamma_2 \wedge \beta_1 \wedge \beta_2;$

Thus, n is 5, m (the grade of `A`) is 3, k (the grade of `B`) is 4, and j (the grade of `C`) is 2.

For reference, the 1-span and 1-cospan of `A` are

$\{\star\mathbb{S}_1[\mathbf{A}], \star\mathbb{S}^1[\mathbf{A}]\}$

$\{\{\alpha_1, \gamma_1, \gamma_2\}, \{\gamma_1 \wedge \gamma_2, -(\alpha_1 \wedge \gamma_2), \alpha_1 \wedge \gamma_1\}\}$

Now if we simply evaluate the formula, we notice that two of the three terms are zero due to repeated factors in the exterior product.

$\mathbf{F} = \star\Sigma[(\star\mathbb{S}_1[\mathbf{A}] \wedge \mathbf{B}) \vee \star\mathbb{S}^1[\mathbf{A}]]$

$(\alpha_1 \wedge \gamma_1 \wedge \gamma_2 \wedge \beta_1 \wedge \beta_2) \vee (\gamma_1 \wedge \gamma_2) +$
$(\gamma_1 \wedge \gamma_1 \wedge \gamma_2 \wedge \beta_1 \wedge \beta_2) \vee (-(\alpha_1 \wedge \gamma_2)) + (\gamma_2 \wedge \gamma_1 \wedge \gamma_2 \wedge \beta_1 \wedge \beta_2) \vee (\alpha_1 \wedge \gamma_1)$

Simplifying then gives us the result.

$\star\mathcal{G}[\mathbf{F}]$

$(\gamma_1 \wedge \gamma_2) \vee (\alpha_1 \wedge \beta_1 \wedge \beta_2 \wedge \gamma_1 \wedge \gamma_2)$

Because the second factor in the regressive product is an n-element, we can immediately write this as congruent to the common factor.

$(\gamma_1 \wedge \gamma_2) \vee (\alpha_1 \wedge \beta_1 \wedge \beta_2 \wedge \gamma_1 \wedge \gamma_2) \equiv \gamma_1 \wedge \gamma_2$

◆ Example 2

We now take the same example as we did above when we multiplied out the regressive product term by term (Application of the Common Factor Axiom, section 3.5). `X` and `Y` are both 2-

elements in a 3-space, and we wish to compute a 1-element Z common to them (remember Z can only be computed up to congruence).

$$\star\mathcal{B}_3;\ \{X, Y\} = \{\star\mathbb{B}_x, \star\mathbb{B}_y\}$$

$$\{x_1\, e_1 \wedge e_2 + x_2\, e_1 \wedge e_3 + x_3\, e_2 \wedge e_3,\ y_1\, e_1 \wedge e_2 + y_2\, e_1 \wedge e_3 + y_3\, e_2 \wedge e_3\}$$

To apply the Common Factor Theorem in the form above, we need one of the bivectors, say X, in factored form. In a 3-space, every bivector is simple and we can write X in the form:

$$X = (x_1\, e_1 - e_3\, x_3) \wedge \left(e_2 + \frac{e_3\, x_2}{x_1}\right);$$

(Note that to compute the span of an element, the element needs to be a simple product of 1-elements - not a scalar multiple of such a product. The *GrassmannAlgebra* function ⋆S ignores any such scalar factors. In applications where only a result up to congruence is required, this is entirely satisfactory. Otherwise if we wish to include the scalar factor, we will need to multiply it into one of the other factors.)

Applying formula [3.40] gives:

$$Z = \star\Sigma\left[(\star S_1[X] \wedge Y) \vee \star S^1[X]\right]$$

$$\left(\left(e_2 + \frac{e_3\, x_2}{x_1}\right) \wedge (y_1\, e_1 \wedge e_2 + y_2\, e_1 \wedge e_3 + y_3\, e_2 \wedge e_3)\right) \vee (-e_1\, x_1 + e_3\, x_3) +$$
$$\left((e_1\, x_1 - e_3\, x_3) \wedge (y_1\, e_1 \wedge e_2 + y_2\, e_1 \wedge e_3 + y_3\, e_2 \wedge e_3)\right) \vee \left(e_2 + \frac{e_3\, x_2}{x_1}\right)$$

Simplifying:

$$\star\mathcal{G}[Z]$$

$$(-x_2\, y_1 + x_1\, y_2)\, e_1 \vee (e_1 \wedge e_2 \wedge e_3) +$$
$$(-x_3\, y_1 + x_1\, y_3)\, e_2 \vee (e_1 \wedge e_2 \wedge e_3) + (-x_3\, y_2 + x_2\, y_3)\, e_3 \vee (e_1 \wedge e_2 \wedge e_3)$$

The *n*-element in this case is $e_1 \wedge e_2 \wedge e_3$. Factoring it out gives:

$$Z == ((x_1\, y_2 - x_2\, y_1)\, e_1 + (x_1\, y_3 - x_3\, y_1)\, e_2 + (x_2\, y_3 - x_3\, y_2)\, e_3) \vee e_1 \wedge e_2 \wedge e_3$$

Writing this in congruence form gives us the common factor required, and corroborates the result [3.36] originally obtained by applying the Common Factor Axiom to each term of the full expansion.

$$Z \equiv (x_1\, y_2 - x_2\, y_1)\, e_1 + (x_1\, y_3 - x_3\, y_1)\, e_2 + (x_2\, y_3 - x_3\, y_2)\, e_3$$

Proof of the Common Factor Theorem

Consider a regressive product $\underset{m}{A} \vee \underset{k}{B}$ where $\underset{m}{A}$ is given as a simple product of 1-element factors and $m + k = n + j,\ j > 0$. Then $\underset{m}{A} \vee \underset{k}{B}$ is either zero, or, by the Common Factor Axiom, has a common factor $\underset{j}{C}$. We assume it is not zero. We then express $\underset{m}{A}$ and $\underset{k}{B}$ in terms of $\underset{j}{C}$ as we have done previously.

$$\underset{m}{A} == \underset{m-j}{A^*} \wedge \underset{j}{C} \qquad \underset{k}{B} == \underset{j}{C} \wedge \underset{k-j}{B^*}$$

Let the $(m-j)$-span and $(m-j)$-cospan of $\underset{m}{A}$ be written as

$$\star\mathbb{S}_{m-j}\left[\underset{m}{A}\right] == \left\{\underset{m-j}{A_1}, \underset{m-j}{A_2}, \ldots, \underset{m-j}{A_\nu}\right\} \qquad \star\mathbb{S}^{m-j}\left[\underset{m}{A}\right] == \left\{\underset{j}{A_1}, \underset{j}{A_2}, \ldots, \underset{j}{A_\nu}\right\}$$

then $\underset{m}{A}$ can be written in any of the forms

$$\underset{m}{A} == \underset{m-j}{A_1} \wedge \underset{j}{A_1} == \underset{m-j}{A_2} \wedge \underset{j}{A_2} == \ldots == \underset{m-j}{A_\nu} \wedge \underset{j}{A_\nu}$$

Here ν is equal to $\binom{m}{j}$, or equivalently $\binom{m}{m-j}$.

Since the common factor $\underset{j}{C}$ is a *j*-element belonging to $\underset{m}{A}$ it can be expressed as a linear combination of the cospan elements of $\underset{m}{A}$.

$$\underset{j}{C} == a_1 \underset{j}{A_1} + a_2 \underset{j}{A_2} + \ldots + a_\nu \underset{j}{A_\nu}$$

The exterior product of the *i*th (*m-j*)-span element $\underset{m-j}{A_i}$ with $\underset{j}{C}$ can be expanded to give

$$\underset{m-j}{A_i} \wedge \underset{j}{C} == \underset{m-j}{A_i} \wedge \left(a_1 \underset{j}{A_1} + a_2 \underset{j}{A_2} + \ldots + a_\nu \underset{j}{A_\nu}\right)$$

$$== a_1 \left(\underset{m-j}{A_i} \wedge \underset{j}{A_1}\right) + a_2 \left(\underset{m-j}{A_i} \wedge \underset{j}{A_2}\right) + \ldots + a_i \left(\underset{m-j}{A_i} \wedge \underset{j}{A_i}\right) + \ldots + a_\nu \left(\underset{m-j}{A_i} \wedge \underset{j}{A_\nu}\right)$$

And indeed, because by definition, the exterior product of an (*m-j*)-span element (say, $\underset{m-j}{A_2}$) and any of the *non-corresponding* cospan elements (say, $\underset{j}{A_3}$) is zero, we can write for any *i* that

$$\underset{m-j}{A_i} \wedge \underset{j}{C} == a_i \left(\underset{m-j}{A_i} \wedge \underset{j}{A_i}\right) == a_i \underset{m}{A}$$

Now return to the Common Factor Axiom in the form

$$Z == \underset{m}{A} \vee \underset{k}{B} == \left(\underset{m-j}{A^*} \wedge \underset{j}{C}\right) \vee \left(\underset{j}{C} \wedge \underset{k-j}{B^*}\right) == \left(\underset{m-j}{A^*} \wedge \underset{j}{C} \wedge \underset{k-j}{B^*}\right) \vee \underset{j}{C} == \left(\underset{m}{A} \wedge \underset{k-j}{B^*}\right) \vee \underset{j}{C}$$

Substituting for the common factor in the right hand side gives

$$Z == \left(\underset{m}{A} \wedge \underset{k-j}{B^*}\right) \vee \underset{j}{C} == \left(\underset{m}{A} \wedge \underset{k-j}{B^*}\right) \vee \left(a_1 \underset{j}{A_1} + a_2 \underset{j}{A_2} + \ldots + a_\nu \underset{j}{A_\nu}\right)$$

By the distributivity of the exterior and regressive products, and their behaviour with scalars, the scalar factors can be transferred and attached to $\underset{m}{A}$.

$$Z == \left(\left(a_1 \underset{m}{A}\right) \wedge \underset{k-j}{B^*}\right) \vee \underset{j}{A_1} + \left(\left(a_2 \underset{m}{A}\right) \wedge \underset{k-j}{B^*}\right) \vee \underset{j}{A_2} + \ldots + \left(\left(a_\nu \underset{m}{A}\right) \wedge \underset{k-j}{B^*}\right) \vee \underset{j}{A_\nu}$$

But we have shown earlier that

$$a_i \underset{m}{A} == \underset{m-j}{A_i} \wedge \underset{j}{C}$$

Hence substituting gives

$$Z == \left(\underset{m-j}{A_1} \wedge \underset{j}{C} \wedge \underset{k-j}{B^*}\right) \vee \underset{j}{A_1} + \left(\underset{m-j}{A_2} \wedge \underset{j}{C} \wedge \underset{k-j}{B^*}\right) \vee \underset{j}{A_2} + \ldots + \left(\underset{m-j}{A_\nu} \wedge \underset{j}{C} \wedge \underset{k-j}{B^*}\right) \vee \underset{j}{A_\nu}$$

Finally, a further substitution of $\underset{k}{B}$ for $\underset{j}{C} \wedge \underset{k-j}{B^*}$ gives the final result.

$$\underset{m}{A} \vee \underset{k}{B} == \left(\underset{m-j}{A_1} \wedge \underset{k}{B}\right) \vee \underset{j}{A_1} + \left(\underset{m-j}{A_2} \wedge \underset{k}{B}\right) \vee \underset{j}{A_2} + \ldots + \left(\underset{m-j}{A_\nu} \wedge \underset{k}{B}\right) \vee \underset{j}{A_\nu}$$
$$== \sum_{i==1}^{\nu} \left(\underset{m-j}{A_i} \wedge \underset{k}{B}\right) \vee \underset{j}{A_i} \tag{3.41}$$

The A and B forms of the Common Factor Theorem

We have just proved what we will now call the A form of the Common Factor Theorem (because it works by spanning the *first* factor in the regressive product). An analogous formula may be obtained *mutatis mutandis* by factoring $\underset{k}{B}$ rather than $\underset{m}{A}$. Hence $\underset{k}{B}$ must now be simple. This formula we will call the B form of the Common Factor Theorem.

In the case where both $\underset{m}{A}$ and $\underset{k}{B}$ are simple but not of the same grade, the form which decomposes the element of lower grade will generate the least number of terms and be more computationally efficient. If both $\underset{m}{A}$ and $\underset{k}{B}$ are simple, but not in factored form, the Common Factor Theorems can still be applied to the simple component terms of the product. Multiple regressive products may be treated by successive applications of the theorem in either of its forms.

◆ The A form of the Common Factor Theorem

$$\left\{\underset{m}{A} \vee \underset{k}{B} == \sum_{i==1}^{\nu} \left(\underset{m-j}{A_i} \wedge \underset{k}{B}\right) \vee \underset{j}{A_i}, \quad m - j + k - n == 0,\right.$$
$$\left.\underset{m}{A} == \underset{m-j}{A_1} \wedge \underset{j}{A_1} == \underset{m-j}{A_2} \wedge \underset{j}{A_2} == \ldots == \underset{m-j}{A_\nu} \wedge \underset{j}{A_\nu}, \quad \nu == \binom{m}{j}\right\} \tag{3.42}$$

As we have already seen, this is equivalent to the computationally oriented form.

$$\underset{m}{A} \vee \underset{k}{B} == \sum_{i==1}^{\nu} \left(\star\mathbb{S}_{m-j}\left[\underset{m}{A}\right]_{[[i]]} \wedge \underset{k}{B}\right) \vee \star\mathbb{S}^{m-j}\left[\underset{m}{A}\right]_{[[i]]}$$

$$\star\mathbb{S}_{m-j}\left[\underset{m}{A}\right] == \left\{\underset{m-j}{A_1}, \underset{m-j}{A_2}, \ldots, \underset{m-j}{A_\nu}\right\} \qquad \star\mathbb{S}^{m-j}\left[\underset{m}{A}\right] == \left\{\underset{j}{A_1}, \underset{j}{A_2}, \ldots, \underset{j}{A_\nu}\right\}$$

Or, using *Mathematica's* inbuilt Listability:

$$\underset{m}{A} \vee \underset{k}{B} == \star\Sigma\left[\left(\star\mathbb{S}_{m-j}\left[\underset{m}{A}\right] \wedge \underset{k}{B}\right) \vee \star\mathbb{S}^{m-j}\left[\underset{m}{A}\right]\right] \tag{3.43}$$

◆ The B form of the Common Factor Theorem

$$\left\{\underset{m}{A} \vee \underset{k}{B} == \sum_{i=1}^{\nu} \left(\underset{m}{A} \wedge \underset{k-j}{B_i}\right) \vee \underset{j}{B_i}, \quad m - j + k - n == 0, \quad \underset{k}{B} == \underset{j}{B_1} \wedge \underset{k-j}{B_1} == \underset{j}{B_2} \wedge \underset{k-j}{B_2} == \ldots == \underset{j}{B_\nu} \wedge \underset{k-j}{B_\nu}, \quad \nu == \binom{k}{j}\right\} \qquad 3.44$$

The computationally oriented form is

$$\underset{m}{A} \vee \underset{k}{B} == \sum_{i=1}^{\nu} \left(\underset{m}{A} \wedge \star S^j\left[\underset{k}{B}\right]_{[[i]]}\right) \vee \star S_j\left[\underset{k}{B}\right]_{[[i]]}$$

$$\star S_j\left[\underset{k}{B}\right] == \left\{\underset{j}{B_1}, \underset{j}{B_2}, \ldots, \underset{j}{B_\nu}\right\} \qquad \star S^j\left[\underset{k}{B}\right] == \left\{\underset{k-j}{B_1}, \underset{k-j}{B_2}, \ldots, \underset{k-j}{B_\nu}\right\}$$

Or, using *Mathematica's* inbuilt Listability:

$$\underset{m}{A} \vee \underset{k}{B} == \star\Sigma\left[\left(\underset{m}{A} \wedge \star S^j\left[\underset{k}{B}\right]\right) \vee \star S_j\left[\underset{k}{B}\right]\right] \qquad 3.45$$

Example: The decomposition of a 1-element

The special case of the regressive product of a 1-element β with an n-element $\underset{n}{\alpha}$ enables us to decompose β directly in terms of the factors of $\underset{n}{\alpha}$. The A form of the Common Factor Theorem [3.42] gives:

$$\underset{n}{\alpha} \vee \beta == \sum_{i=1}^{n} \left(\underset{n-1}{\alpha_i} \wedge \beta\right) \vee \alpha_i$$

where:

$$\underset{n}{\alpha} == \alpha_1 \wedge \alpha_2 \wedge \ldots \wedge \alpha_n$$

$$== (-1)^{n-i} (\alpha_1 \wedge \alpha_2 \wedge \ldots \wedge \square_i \wedge \ldots \wedge \alpha_n) \wedge \alpha_i == \underset{n-1}{\alpha_i} \wedge \alpha_i$$

The symbol $\square_i$ means that the ith factor is missing from the product. Substituting in the Common Factor Theorem gives:

$$\underset{n}{\alpha} \vee \beta == \sum_{i=1}^{n} (-1)^{n-i} ((\alpha_1 \wedge \alpha_2 \wedge \ldots \wedge \square_i \wedge \ldots \wedge \alpha_n) \wedge \beta) \vee \alpha_i$$

$$== \sum_{i=1}^{n} (\alpha_1 \wedge \alpha_2 \wedge \ldots \wedge \alpha_{i-1} \wedge \beta \wedge \alpha_{i+1} \wedge \ldots \wedge \alpha_n) \vee \alpha_i$$

Hence the decomposition formula becomes

$$(\alpha_1 \wedge \alpha_2 \wedge \ldots \wedge \alpha_n) \vee \beta == \sum_{i=1}^{n} (\alpha_1 \wedge \alpha_2 \wedge \ldots \wedge \alpha_{i-1} \wedge \beta \wedge \alpha_{i+1} \wedge \ldots \wedge \alpha_n) \vee \alpha_i \qquad 3.46$$

Writing this out in full shows that we can expand the expression simply by interchanging β successively with each of the factors of $\alpha_1 \wedge \alpha_2 \wedge \ldots \wedge \alpha_n$, and summing the results.

$$(\alpha_1 \wedge \alpha_2 \wedge \ldots \wedge \alpha_n) \vee \beta == (\beta \wedge \alpha_2 \wedge \ldots \wedge \alpha_n) \vee \alpha_1 + (\alpha_1 \wedge \beta \wedge \ldots \wedge \alpha_n) \vee \alpha_2 + \ldots + (\alpha_1 \wedge \alpha_2 \wedge \ldots \wedge \beta) \vee \alpha_n \qquad 3.47$$

We can make the result express more explicitly the decomposition of β in terms of the α_i by writing each of the n-elements as a scalar multiple of the unit n-element, and using [3.4], [3.9] and the results of section 2.11.

$$\beta == \sum_{i=1}^{n} \frac{\alpha_1 \wedge \alpha_2 \wedge \ldots \wedge \alpha_{i-1} \wedge \beta \wedge \alpha_{i+1} \wedge \ldots \wedge \alpha_n}{\alpha_1 \wedge \alpha_2 \wedge \ldots \wedge \alpha_n} \alpha_i \qquad 3.48$$

The coefficients of this expression are equivalent to that which would have been obtained from Grassmann's approach to solving linear equations [2.32] with $C_0 = \beta$ and $C_i = \alpha_i$.

Example: Applying the Common Factor Theorem

In this section we show how the Common Factor Theorem generally leads to a more efficient computation of results than repeated application of the Common Factor Axiom, particularly when done manually, since there are fewer terms in the Common Factor Theorem expansion, and many are evidently zero by inspection.

Again, we take the problem of finding the 1-element common to two 2-elements in a 3-space. This time, however, we take a numerical example and suppose that we know the factors of at least one of the 2-elements. Let:

$$\xi_1 = 3\, e_1 \wedge e_2 + 2\, e_1 \wedge e_3 + 3\, e_2 \wedge e_3;$$
$$\xi_2 = (5\, e_2 + 7\, e_3) \wedge e_1;$$

We keep ξ_1 fixed initially and apply the Common Factor Theorem to rewrite the regressive product as a sum of the two products, each due to one of the essentially different rearrangements of ξ_2:

$$\xi_1 \vee \xi_2 == (\xi_1 \wedge e_1) \vee (5\, e_2 + 7\, e_3) - (\xi_1 \wedge (5\, e_2 + 7\, e_3)) \vee e_1$$

The next step is to expand out the exterior products with ξ_1. Often this step can be done by inspection, since all products with a repeated basis element factor will be zero.

$$\xi_1 \vee \xi_2 == (3\, e_2 \wedge e_3 \wedge e_1) \vee (5\, e_2 + 7\, e_3) - (10\, e_1 \wedge e_3 \wedge e_2 + 21\, e_1 \wedge e_2 \wedge e_3) \vee e_1$$

Factorizing out the 3-element $e_1 \wedge e_2 \wedge e_3$ then gives

$$\xi_1 \vee \xi_2 == (e_1 \wedge e_2 \wedge e_3) \vee (-11\, e_1 + 15\, e_2 + 21\, e_3)$$
$$== \star c\, (-11\, e_1 + 15\, e_2 + 21\, e_3) \equiv -11\, e_1 + 15\, e_2 + 21\, e_3$$

The scalar $\star c$ is called the *congruence factor*. The congruence factor has already been defined in formula [3.10] above as the connection between the current basis n-element and the unit n-element $\underset{n}{1}$.

$$e_1 \wedge e_2 \wedge \cdots \wedge e_n == \star c\, \underset{n}{1}$$

Since this connection cannot be defined unless a metric is introduced, the congruence factor remains arbitrary at this stage. Although such an arbitrariness may appear at first sight to be disadvantageous, it is on the contrary, highly elucidating. In application to the computing of unions and intersections of spaces, perhaps under a geometric interpretation where they represent lines, planes and hyperplanes, the notion of congruence becomes central, since spaces can only be determined up to congruence.

Thus all factors common to both ξ_1 and ξ_2 are congruent to $-11\,e_1 + 15\,e_2 + 21\,e_3$.

◆ Check by calculating the exterior products

We can check that this result is correct by taking its exterior product with each of ξ_1 and ξ_2 and simplifying. We should get zero in both cases, indicating that the factor determined is indeed common to both original 2-elements.

```
★B3; ★G[{ξ1, ξ2} ∧ (-11 e1 + 15 e2 + 21 e3)]
{0, 0}
```

Automating the application of the Common Factor Theorem

The Common Factor Theorem can be computed in *GrassmannAlgebra* by using either the A or B form in its computable form using the appropriate span and cospan. However *GrassmannAlgebra* also has a more direct function `ToCommonFactor` which will attempt to handle multiple regressive products in an optimum way.

◆ Using ToCommonFactor

`ToCommonFactor` reduces any regressive products in a Grassmann expression to their common factor form.

For example in the case explored above we have:

```
ξ1 = 3 e1 ∧ e2 + 2 e1 ∧ e3 + 3 e2 ∧ e3;
ξ2 = (5 e2 + 7 e3) ∧ e1;
★A; ToCommonFactor[ξ1 ∨ ξ2]
★c (-11 e1 + 15 e2 + 21 e3)
```

In this example, ξ_1 and ξ_2 are expressed in terms of basis elements. `ToCommonFactor` can also apply the Common Factor Theorem to more general elements. For example, if ξ_1 and ξ_2 are expressed as symbolic bivectors, the common factor expansion could still be performed.

```
ξ1 = x ∧ y; ξ2 = u ∧ v;
ToCommonFactor[ξ1 ∨ ξ2]
★c (y ⟨u ∧ v ∧ x⟩ - x ⟨u ∧ v ∧ y⟩)
```

The expression $\langle u \wedge v \wedge x \rangle$ represents the *coefficient* of $u \wedge v \wedge x$ when $u \wedge v \wedge x$ is expressed in terms of the *current basis n*-element (in this case 3-element).

As an example, we use the *GrassmannAlgebra* function `ComposeBasisForm` to compose an expression for $u \wedge v \wedge x$ in terms of basis elements, and then expand and simplify the result.

```
X = ComposeBasisForm[u ∧ v ∧ x]
(e1 u1 + e2 u2 + e3 u3) ∧ (e1 v1 + e2 v2 + e3 v3) ∧ (e1 x1 + e2 x2 + e3 x3)
```

```
⋆𝒢[X]
```

$$(-u_3 v_2 x_1 + u_2 v_3 x_1 + u_3 v_1 x_2 - u_1 v_3 x_2 - u_2 v_1 x_3 + u_1 v_2 x_3)\, e_1 \wedge e_2 \wedge e_3$$

Hence in this case $\langle u \wedge v \wedge x \rangle$ is given by

$$\langle \mathbf{u} \wedge \mathbf{v} \wedge \mathbf{x} \rangle == (-u_3 v_2 x_1 + u_2 v_3 x_1 + u_3 v_1 x_2 - u_1 v_3 x_2 - u_2 v_1 x_3 + u_1 v_2 x_3)$$

◆ Using **ToCommonFactorA** and **ToCommonFactorB**

If you wish explicitly to apply the A form of the Common Factor Theorem, you can use `ToCommonFactorA`. Or if the B form, you can use `ToCommonFactorB`. The results will be equal, but they may look different.

We illustrate by taking the example in the previous section and expanding it both ways. Again define

```
ξ1 = x ∧ y; ξ2 = u ∧ v;
ToCommonFactorA[ξ1 ∨ ξ2]
```

$$\star c\ (y \langle u \wedge v \wedge x \rangle - x \langle u \wedge v \wedge y \rangle)$$

```
ToCommonFactorB[ξ1 ∨ ξ2]
```

$$\star c\ (-v \langle u \wedge x \wedge y \rangle + u \langle v \wedge x \wedge y \rangle)$$

To see most directly that these are equal, we express the vectors in terms of basis elements.

```
Z = ComposeBasisForm[ξ1 ∨ ξ2]
```

$$((e_1 x_1 + e_2 x_2 + e_3 x_3) \wedge (e_1 y_1 + e_2 y_2 + e_3 y_3)) \vee ((e_1 u_1 + e_2 u_2 + e_3 u_3) \wedge (e_1 v_1 + e_2 v_2 + e_3 v_3))$$

Applying the A form of the Common Factor Theorem gives

```
ToCommonFactorA[Z]
```

$$\begin{aligned}\star c\ (&e_1\ (u_3 v_1 x_2 y_1 - u_1 v_3 x_2 y_1 - u_2 v_1 x_3 y_1 + \\ &\quad u_1 v_2 x_3 y_1 - u_3 v_1 x_1 y_2 + u_1 v_3 x_1 y_2 + u_2 v_1 x_1 y_3 - u_1 v_2 x_1 y_3) + \\ &e_2\ (u_3 v_2 x_2 y_1 - u_2 v_3 x_2 y_1 - u_3 v_2 x_1 y_2 + u_2 v_3 x_1 y_2 - u_2 v_1 x_3 y_2 + \\ &\quad u_1 v_2 x_3 y_2 + u_2 v_1 x_2 y_3 - u_1 v_2 x_2 y_3) + \\ &e_3\ (u_3 v_2 x_3 y_1 - u_2 v_3 x_3 y_1 - u_3 v_1 x_3 y_2 + u_1 v_3 x_3 y_2 - u_3 v_2 x_1 y_3 + \\ &\quad u_2 v_3 x_1 y_3 + u_3 v_1 x_2 y_3 - u_1 v_3 x_2 y_3))\end{aligned}$$

The resulting expression is identical to that calculated by the B form.

```
ToCommonFactorB[Z] == ToCommonFactorA[Z]
True
```

◆ Expressions involving non-decomposable elements

The *GrassmannAlgebra* function `ToCommonFactor` uses the A or B form heuristically. When a common factor can be calculated by using either the A or B form, the A form is used by default. If an expansion is not possible by using the A form, `ToCommonFactor` tries the B form.

For example, consider the regressive product of two symbolic 2-elements in a 3-space. Attempting to compute the common factor fails, since neither factor is expressed as a product of 1-elements.

```
★A; ToCommonFactor[α ∨ β]
                   2   2
```

$\underset{2}{\alpha} \vee \underset{2}{\beta}$

However, if either one of the factors is so expressed, then a formula for the common factor can be generated. In the case below, the results are the same; remember that ★c is an *arbitrary* scalar factor.

```
{ToCommonFactor[α ∨ (x ∧ y)], ToCommonFactor[(x ∧ y) ∨ α]}
                2                                      2
```

$\left\{\star c \left(-y \left\langle x \wedge \underset{2}{\alpha}\right\rangle + x \left\langle y \wedge \underset{2}{\alpha}\right\rangle\right), \star c \left(y \left\langle x \wedge \underset{2}{\alpha}\right\rangle - x \left\langle y \wedge \underset{2}{\alpha}\right\rangle\right)\right\}$

◆ Example: The regressive product of three 3-elements in a 4-space

`ToCommonFactor` can find the common factor of any number of elements. As an example, we take the regressive product of three 3-elements in a 4-space. This is a 1-element, since $3+3+3-2\times 4=1$. Remember also that a 3-element in a 4-space is necessarily simple (see section 2.10).

To demonstrate this we first declare a 4-space, and then use the *GrassmannAlgebra* function `ComposeBasisForm` to compose the required product in basis form.

```
★B4; X = ComposeBasisForm[α ∨ β ∨ γ]
                          3   3   3
```

$(\alpha_{3,1}\, e_1 \wedge e_2 \wedge e_3 + \alpha_{3,2}\, e_1 \wedge e_2 \wedge e_4 + \alpha_{3,3}\, e_1 \wedge e_3 \wedge e_4 + \alpha_{3,4}\, e_2 \wedge e_3 \wedge e_4) \vee$
$(\beta_{3,1}\, e_1 \wedge e_2 \wedge e_3 + \beta_{3,2}\, e_1 \wedge e_2 \wedge e_4 + \beta_{3,3}\, e_1 \wedge e_3 \wedge e_4 + \beta_{3,4}\, e_2 \wedge e_3 \wedge e_4) \vee$
$(\gamma_{3,1}\, e_1 \wedge e_2 \wedge e_3 + \gamma_{3,2}\, e_1 \wedge e_2 \wedge e_4 + \gamma_{3,3}\, e_1 \wedge e_3 \wedge e_4 + \gamma_{3,4}\, e_2 \wedge e_3 \wedge e_4)$

The common factor is then determined up to congruence by `ToCommonFactor`.

```
Xc = ToCommonFactor[X]
```

$\star c^2$
$(e_1 (-\alpha_{3,3}\,\beta_{3,2}\,\gamma_{3,1} + \alpha_{3,2}\,\beta_{3,3}\,\gamma_{3,1} + \alpha_{3,3}\,\beta_{3,1}\,\gamma_{3,2} - \alpha_{3,1}\,\beta_{3,3}\,\gamma_{3,2} - \alpha_{3,2}\,\beta_{3,1}\,\gamma_{3,3} +$
$\alpha_{3,1}\,\beta_{3,2}\,\gamma_{3,3}) + e_2 (-\alpha_{3,4}\,\beta_{3,2}\,\gamma_{3,1} + \alpha_{3,2}\,\beta_{3,4}\,\gamma_{3,1} +$
$\alpha_{3,4}\,\beta_{3,1}\,\gamma_{3,2} - \alpha_{3,1}\,\beta_{3,4}\,\gamma_{3,2} - \alpha_{3,2}\,\beta_{3,1}\,\gamma_{3,4} + \alpha_{3,1}\,\beta_{3,2}\,\gamma_{3,4}) +$
$e_3 (-\alpha_{3,4}\,\beta_{3,3}\,\gamma_{3,1} + \alpha_{3,3}\,\beta_{3,4}\,\gamma_{3,1} + \alpha_{3,4}\,\beta_{3,1}\,\gamma_{3,3} - \alpha_{3,1}\,\beta_{3,4}\,\gamma_{3,3} -$
$\alpha_{3,3}\,\beta_{3,1}\,\gamma_{3,4} + \alpha_{3,1}\,\beta_{3,3}\,\gamma_{3,4}) + e_4 (-\alpha_{3,4}\,\beta_{3,3}\,\gamma_{3,2} + \alpha_{3,3}\,\beta_{3,4}\,\gamma_{3,2} +$
$\alpha_{3,4}\,\beta_{3,2}\,\gamma_{3,3} - \alpha_{3,2}\,\beta_{3,4}\,\gamma_{3,3} - \alpha_{3,3}\,\beta_{3,2}\,\gamma_{3,4} + \alpha_{3,2}\,\beta_{3,3}\,\gamma_{3,4}))$

Note that the congruence factor ★c is to the second power. This is because the original expression contained the regressive product operator twice: one ★c effectively stands in for each basis 4-element that is produced during the calculation (3+3+3 = 4+4+1).

It is easy to check that this result is indeed a common factor by taking its exterior product with each of the 3-elements.

```
Expand[★G[Xc ∧ List @@ X]]
```

```
{0, 0, 0}
```

A special form of the Common Factor Theorem

We now derive a special form of the Common Factor Theorem which will be especially useful in Chapter 5 in defining the complement, and in Chapter 6 in deriving the Interior Common Factor Theorem.

We start with the Common Factor Theorem in the A form [3.42].

$$\underset{m}{A} \vee \underset{k}{B} == \sum_{i==1}^{\nu} \left(\underset{m-j}{A_i} \wedge \underset{k}{B} \right) \vee \underset{j}{A_i}$$

$$\underset{m}{A} == \underset{m-j}{A_1} \wedge \underset{j}{A_1} == \underset{m-j}{A_2} \wedge \underset{j}{A_2} == \ldots == \underset{m-j}{A_\nu} \wedge \underset{j}{A_\nu}$$

where $\underset{m}{\alpha}$ is simple, $j = m+k-n$, and $\nu = \binom{m}{j}$.

One of the dual forms [3.32] of the Common Factor Axiom is:

$$\left\{ \underset{m}{\alpha} \wedge \underset{k}{\beta} == \left(\underset{m}{\alpha} \vee \underset{k}{\beta} \right) \wedge \underset{n}{1} == \left(\underset{m}{\alpha} \vee \underset{k}{\beta} \right) \underset{n}{1} \in \underset{n}{\Lambda}, \quad m+k-n == 0 \right\}$$

which allows us to write the bracketed n-elements in the sum as

$$\underset{m-j}{A_i} \wedge \underset{k}{B} == \left(\underset{m-j}{A_i} \vee \underset{k}{B} \right) \underset{n}{1}$$

Hence a special form of the Common Factor Theorem can be written

$$\underset{m}{A} \vee \underset{k}{B} == \sum_{i==1}^{\nu} \left(\underset{m-j}{A_i} \vee \underset{k}{B} \right) \underset{j}{A_i}$$

$$\underset{m}{A} == \underset{m-j}{A_1} \wedge \underset{j}{A_1} == \underset{m-j}{A_2} \wedge \underset{j}{A_2} == \ldots == \underset{m-j}{A_\nu} \wedge \underset{j}{A_\nu}$$

$$\left\{ \begin{array}{ll} \underset{m}{A} \vee \underset{k}{B} == \sum_{i==1}^{\nu} \left(\underset{m-j}{A_i} \vee \underset{k}{B} \right) \underset{j}{A_i}, & m-j+k-n == 0, \\ \underset{m}{A} == \underset{m-j}{A_1} \wedge \underset{j}{A_1} == \underset{m-j}{A_2} \wedge \underset{j}{A_2} == \ldots == \underset{m-j}{A_\nu} \wedge \underset{j}{A_\nu}, & \nu == \binom{m}{j} \end{array} \right\} \quad 3.49$$

A more explicit form may be obtained by writing a simple element $\underset{m}{A}$ in factored form.

$$(\alpha_1 \wedge \alpha_2 \wedge \cdots \wedge \alpha_m) \vee \underset{k}{B} ==$$
$$\sum_{i_1 \ldots i_\mu} \left(\left(\alpha_{i_1} \wedge \ldots \wedge \alpha_{i_\mu} \right) \vee \underset{k}{B} \right) (-1)^{K_\mu} \alpha_1 \wedge \ldots \wedge \square_{i_1} \wedge \ldots \wedge \square_{i_\mu} \wedge \ldots \wedge \alpha_m$$
$$K_\mu == \sum_{i==1}^{\mu} i_\gamma + \frac{1}{2} \mu (\mu+1) \qquad \mu == m-j == n-k \quad 3.50$$

where $\square_j$ means α_j is missing from the product.

3.7 The Regressive Product of Simple Elements

The regressive product of simple elements

The regressive product of simple elements is simple.

To show this, consider the (non-zero) regressive product $\underset{m}{\alpha} \vee \underset{k}{\beta}$, where $\underset{m}{\alpha}$ and $\underset{k}{\beta}$ are simple, $m+k \geq n$. The 1-element factors of $\underset{m}{\alpha}$ and $\underset{k}{\beta}$ must then have a common subspace of dimension $m+k-n = j$. Let $\underset{j}{\mu}$ be a simple j-element which spans this common subspace. We can then write:

$$\underset{m}{\alpha} \vee \underset{k}{\beta} == \left(\underset{m-j}{\alpha} \wedge \underset{j}{\mu}\right) \vee \left(\underset{k-j}{\beta} \wedge \underset{j}{\mu}\right) == \left(\underset{m-j}{\alpha} \wedge \underset{k-j}{\beta} \wedge \underset{j}{\mu}\right) \vee \underset{j}{\mu} \equiv \underset{j}{\mu}$$

The Common Factor Axiom then shows us that since $\underset{j}{\mu}$ is simple, then so is the original product of simple elements $\underset{m}{\alpha} \vee \underset{k}{\beta}$.

The regressive product of (n–1)-elements

Since we have shown in section 2.10 that all (n–1)-elements are simple, and in the previous section that the regressive product of simple elements is simple, it follows immediately that the regressive product of any number of (n–1)-elements is simple.

◆ Example: The regressive product of two 3-elements in a 4-space is simple

As an example of the foregoing result we calculate the regressive product of two 3-elements in a 4-space. We begin by declaring a 4-space and composing the regressive product of two general 3-elements.

★𝓑₄; Z = ComposeBasisForm[x ∨ y] (with $\underset{3}{x} \vee \underset{3}{y}$)

$$(x_{3,1}\, e_1 \wedge e_2 \wedge e_3 + x_{3,2}\, e_1 \wedge e_2 \wedge e_4 + x_{3,3}\, e_1 \wedge e_3 \wedge e_4 + x_{3,4}\, e_2 \wedge e_3 \wedge e_4) \vee (y_{3,1}\, e_1 \wedge e_2 \wedge e_3 + y_{3,2}\, e_1 \wedge e_2 \wedge e_4 + y_{3,3}\, e_1 \wedge e_3 \wedge e_4 + y_{3,4}\, e_2 \wedge e_3 \wedge e_4)$$

The common 2-element is:

Zc = ToCommonFactor[Z]

$$\star c\, ((-x_{3,2}\, y_{3,1} + x_{3,1}\, y_{3,2})\, e_1 \wedge e_2 + (-x_{3,3}\, y_{3,1} + x_{3,1}\, y_{3,3})\, e_1 \wedge e_3 + (-x_{3,3}\, y_{3,2} + x_{3,2}\, y_{3,3})\, e_1 \wedge e_4 + (-x_{3,4}\, y_{3,1} + x_{3,1}\, y_{3,4})\, e_2 \wedge e_3 + (-x_{3,4}\, y_{3,2} + x_{3,2}\, y_{3,4})\, e_2 \wedge e_4 + (-x_{3,4}\, y_{3,3} + x_{3,3}\, y_{3,4})\, e_3 \wedge e_4)$$

We can show that this 2-element is simple by confirming that its exterior product with itself is zero. (This technique was discussed in section 2.10).

★𝒢[Zc ∧ Zc] // Expand

0

Regressive products leading to scalar results

A formula particularly useful in its interior product form to be derived later in Chapter 6 is obtained by application of the Common Factor Theorem to the regressive product of two elements: the first a simple m-element ($\underset{m}{\mathrm{A}}$, say), and the second an element which has the *form* we would get if we take the dual of $\underset{m}{\mathrm{A}}$ - a regressive product of m (n-1)-elements ($\underset{n-m}{\mathrm{B}}$, say). Since the regressive product of m (n-1)-elements is of grade n-m, $\underset{m}{\mathrm{A}} \vee \underset{n-m}{\mathrm{B}}$ is a scalar. But the regressive product of any single factor of $\underset{m}{\mathrm{A}}$ (α_i, say) with any single factor of $\underset{n-m}{\mathrm{B}}$ ($\underset{n-1}{\beta_i}$, say) is also a scalar. We will show that $\underset{m}{\mathrm{A}} \vee \underset{n-m}{\mathrm{B}}$ can be determined as the determinant of the $\alpha_i \vee \underset{n-1}{\beta_j}$.

$$\underset{m}{\mathbf{A}} \vee \underset{n-m}{\mathbf{B}} == (\alpha_1 \wedge \ldots \wedge \alpha_m) \vee \left(\underset{n-1}{\beta_1} \vee \ldots \vee \underset{n-1}{\beta_m}\right)$$

$$== \left((\alpha_1 \wedge \ldots \wedge \alpha_m) \vee \underset{n-1}{\beta_j}\right) \vee \left((-1)^{j-1} \underset{n-1}{\beta_1} \vee \ldots \vee \square_j \vee \ldots \vee \underset{n-1}{\beta_m}\right)$$

Now apply the Common Factor Theorem A [3.42].

$$\underset{m}{\mathbf{A}} \vee \underset{n-m}{\mathbf{B}} == \left(\sum_i (-1)^{i-1} \left(\alpha_i \vee \underset{n-1}{\beta_j}\right) (\alpha_1 \wedge \ldots \wedge \square_i \wedge \ldots \wedge \alpha_m)\right) \vee \left((-1)^{j-1} \underset{n-1}{\beta_1} \vee \ldots \vee \square_j \vee \ldots \vee \underset{n-1}{\beta_m}\right)$$

$$== \sum_i (-1)^{i+j} \left(\alpha_i \vee \underset{n-1}{\beta_j}\right) \left((\alpha_1 \wedge \ldots \wedge \square_i \wedge \ldots \wedge \alpha_m) \vee \left(\underset{n-1}{\beta_1} \vee \ldots \vee \square_j \vee \ldots \vee \underset{n-1}{\beta_m}\right)\right)$$

By repeating this process one obtains finally that:

$$(\alpha_1 \wedge \ldots \wedge \alpha_m) \vee \left(\underset{n-1}{\beta_1} \vee \ldots \vee \underset{n-1}{\beta_m}\right) == \mathrm{Det}\left[\alpha_i \vee \underset{n-1}{\beta_j}\right] \qquad 3.51$$

Here $\mathrm{Det}\left[\alpha_i \vee \underset{n-1}{\beta_j}\right]$ is the determinant of the matrix whose elements are the scalars $\alpha_i \vee \underset{n-1}{\beta_j}$.

This determinant formula is of central importance in the computation of inner products to be discussed in Chapter 6.

◆ **Example**

For the case $m = 2$ we have:

$$(\alpha_1 \wedge \alpha_2) \vee \left(\underset{n-1}{\beta_1} \vee \underset{n-1}{\beta_2}\right) == \left(\alpha_1 \vee \underset{n-1}{\beta_1}\right) \left(\alpha_2 \vee \underset{n-1}{\beta_2}\right) - \left(\alpha_1 \vee \underset{n-1}{\beta_2}\right) \left(\alpha_2 \vee \underset{n-1}{\beta_1}\right)$$

The cobasis form of the Common Factor Axiom

The Common Factor Axiom has a significantly suggestive form when written in terms of cobasis elements (see section 2.6). This form will later help us extend the definition of the interior product to arbitrary elements.

We start with three basis elements of Λ whose exterior product is equal to the basis n-element. The basis n-element may also be expressed as the cobasis $\underline{1}$ of the unit element 1 of $\underset{0}{\Lambda}$.

$$\underset{m}{\mathbf{e_i}} \wedge \underset{k}{\mathbf{e_j}} \wedge \underset{p}{\mathbf{e_s}} == \mathbf{e_1} \wedge \mathbf{e_2} \wedge \cdots \wedge \mathbf{e_n} == \underline{1} == \star\mathrm{c}\,\underset{n}{1}$$

The Common Factor Axiom can be written for these basis elements as:

$$\left(\underset{m}{e_i} \wedge \underset{p}{e_s}\right) \vee \left(\underset{k}{e_j} \wedge \underset{p}{e_s}\right) = \left(\underset{m}{e_i} \wedge \underset{k}{e_j} \wedge \underset{p}{e_s}\right) \vee \underset{p}{e_s}, \qquad m + k + p = n \tag{3.52}$$

From the definition of cobasis elements we have that:

$$\underset{m}{e_i} \wedge \underset{p}{e_s} = (-1)^{m k}\, \underline{\underset{k}{e_j}} \qquad \underset{k}{e_j} \wedge \underset{p}{e_s} = \underline{\underset{m}{e_i}}$$

$$\underset{p}{e_s} = \underline{\underset{m}{e_i} \wedge \underset{k}{e_j}} \qquad \underset{m}{e_i} \wedge \underset{k}{e_j} \wedge \underset{p}{e_s} = \underline{1}$$

Substituting these four elements into the Common Factor Axiom above gives:

$$(-1)^{m k}\, \underline{\underset{k}{e_j}} \vee \underline{\underset{m}{e_i}} = \underline{1} \vee \left(\underline{\underset{m}{e_i} \wedge \underset{k}{e_j}}\right) = \star c\; \underline{\underset{m}{e_i} \wedge \underset{k}{e_j}}$$

Or, more symmetrically, by interchanging the first two factors:

$$\underline{\underset{m}{e_i}} \vee \underline{\underset{k}{e_j}} \equiv \underline{\underset{m}{e_i} \wedge \underset{k}{e_j}} \tag{3.53}$$

Thus, given any two basis elements of the Grassmann algebra Λ, the regressive product of their cobasis elements is congruent to the cobasis element of their exterior product.

It can be seen that in this form the Common Factor Axiom does not specifically display the common factor, and is valid for all basis elements, independent of their grades.

The regressive product of cobasis elements

In Chapter 5 we will have cause to calculate the regressive product of cobasis elements. From formula [3.54] below we will have an instance of the fact that the regressive product of $(n-1)$-elements is simple, and we will determine that simple element.

First, consider basis elements e_1 and e_2 of an n-space and their cobasis elements $\underline{e_1}$ and $\underline{e_2}$. The regressive product of $\underline{e_1}$ and $\underline{e_2}$ is given by:

$$\underline{e_1} \vee \underline{e_2} = (e_2 \wedge e_3 \wedge \ldots \wedge e_n) \vee (-\, e_1 \wedge e_3 \wedge \ldots \wedge e_n)$$

Applying the Common Factor Axiom [3.22] enables us to write

$$\underline{e_1} \vee \underline{e_2} = (e_1 \wedge e_2 \wedge e_3 \wedge \ldots \wedge e_n) \vee (\, e_3 \wedge \ldots \wedge e_n)$$

We can write this either in the form already derived in the section above:

$$\underline{e_1} \vee \underline{e_2} = \underline{1} \vee \left(\underline{e_1 \wedge e_2}\right)$$

or, by writing $\underline{1}$ equal to $\star c\ \underset{n}{1}$ as

$$\underline{e_1} \vee \underline{e_2} = \star c\ \left(\underline{e_1 \wedge e_2}\right)$$

Taking the regressive product of this equation with $\underline{e_3}$ and using [3.53] gives

$$\underline{e_1} \vee \underline{e_2} \vee \underline{e_3} = \star c\ \left(\underline{e_1 \wedge e_2}\right) \vee \underline{e_3} = \star c\ \underline{1} \vee \left(\underline{e_1 \wedge e_2 \wedge e_3}\right) = \star c^2\ \left(\underline{e_1 \wedge e_2 \wedge e_3}\right)$$

Continuing this process, we arrive finally at the result that the regressive product of cobasis elements of basis 1-elements is congruent to the cobasis element of their exterior product.

$$\underline{e_1} \vee \underline{e_2} \vee \dots \vee \underline{e_m} == \star c^{m-1}\, \underline{e_1 \wedge e_2 \wedge \dots \wedge e_m} \qquad 3.54$$

A special case which we will have occasion to use in Chapter 5 is where the result reduces to a 1-element. That is, the regressive product of *n*–1 cobasis elements of basis 1-elements is congruent to the remaining 1-element.

$$(-1)^{j-1}\, \underline{e_1} \vee \underline{e_2} \vee \dots \vee \square_j \vee \dots \vee \underline{e_n} == (-1)^{n-1} \star c^{n-2}\, e_j \qquad 3.55$$

3.8 Expressing an Element in another Basis

Expressing an element in terms of another basis

The Common Factor Theorem can be used to express a *k*-element $\underset{k}{\beta}$ in terms of a new basis with basis *n*-element $\underset{n}{\varepsilon}$ by expanding the product $\underset{n}{\varepsilon} \vee \underset{k}{\beta}$. The advantage of this method is that it does not require the calculation of the inverse of the basis transformation.

Let the new basis be $\{\varepsilon_1, \varepsilon_2, \dots, \varepsilon_n\}$ and let $\underset{n}{\varepsilon}$ be equal to $\varepsilon_1 \wedge \varepsilon_2 \wedge \dots \wedge \varepsilon_n$. Then the A form of the Common Factor Theorem [3.42] permits us to write:

$$\underset{n}{\varepsilon} \vee \underset{k}{\beta} == \sum_{i=1}^{\nu} \left(\underset{n-k}{\varepsilon_i} \wedge \underset{k}{\beta}\right) \vee \underset{k}{\varepsilon_i}$$

where $\nu = \binom{n}{k}$ and

$$\underset{n}{\varepsilon} == \underset{n-k}{\varepsilon_1} \wedge \underset{k}{\varepsilon_1} == \underset{n-k}{\varepsilon_2} \wedge \underset{k}{\varepsilon_2} == \dots == \underset{n-k}{\varepsilon_\nu} \wedge \underset{k}{\varepsilon_\nu}$$

We can visualize how the formula operates by writing $\underset{n}{\varepsilon}$ and $\underset{k}{\beta}$ as simple products and then exchanging the $\underset{k}{\varepsilon_i}$ with $\underset{k}{\beta}$ in all the essentially different ways possible whilst always retaining the original ordering.

To make this more concrete, suppose *n* is 5 and *k* is 2:

$$\begin{aligned}
&(\varepsilon_1 \wedge \varepsilon_2 \wedge \varepsilon_3 \wedge \varepsilon_4 \wedge \varepsilon_5) \vee (\beta_1 \wedge \beta_2) \\
&== (\varepsilon_1 \wedge \varepsilon_2 \wedge \varepsilon_3 \wedge \beta_1 \wedge \beta_2) \vee (\varepsilon_4 \wedge \varepsilon_5) \\
&\quad + (\varepsilon_1 \wedge \varepsilon_2 \wedge \beta_1 \wedge \varepsilon_4 \wedge \beta_2) \vee (\varepsilon_3 \wedge \varepsilon_5) \\
&\quad + (\varepsilon_1 \wedge \varepsilon_2 \wedge \beta_1 \wedge \beta_2 \wedge \varepsilon_5) \vee (\varepsilon_3 \wedge \varepsilon_4) \\
&\quad + (\varepsilon_1 \wedge \beta_1 \wedge \varepsilon_3 \wedge \varepsilon_4 \wedge \beta_2) \vee (\varepsilon_2 \wedge \varepsilon_5) \\
&\quad + (\varepsilon_1 \wedge \beta_1 \wedge \varepsilon_3 \wedge \beta_2 \wedge \varepsilon_5) \vee (\varepsilon_2 \wedge \varepsilon_4) \\
&\quad + (\varepsilon_1 \wedge \beta_1 \wedge \beta_2 \wedge \varepsilon_4 \wedge \varepsilon_5) \vee (\varepsilon_2 \wedge \varepsilon_3) \\
&\quad + (\beta_1 \wedge \varepsilon_2 \wedge \varepsilon_3 \wedge \varepsilon_4 \wedge \beta_2) \vee (\varepsilon_1 \wedge \varepsilon_5) \\
&\quad + (\beta_1 \wedge \varepsilon_2 \wedge \varepsilon_3 \wedge \beta_2 \wedge \varepsilon_5) \vee (\varepsilon_1 \wedge \varepsilon_4) \\
&\quad + (\beta_1 \wedge \varepsilon_2 \wedge \beta_2 \wedge \varepsilon_4 \wedge \varepsilon_5) \vee (\varepsilon_1 \wedge \varepsilon_3) \\
&\quad + (\beta_1 \wedge \beta_2 \wedge \varepsilon_3 \wedge \varepsilon_4 \wedge \varepsilon_5) \vee (\varepsilon_1 \wedge \varepsilon_2)
\end{aligned}$$

Now let the new n-elements on the right hand side be written as scalar factors a_i times the new basis n-element $\underset{n}{\varepsilon}$.

$$\underset{n-k}{\varepsilon_i} \wedge \underset{k}{\beta} == a_i \underset{n}{\varepsilon}$$

Substituting gives

$$\underset{n}{\varepsilon} \vee \underset{k}{\beta} == \sum_{i=1}^{\nu} \left(a_i \underset{n}{\varepsilon}\right) \vee \underset{k}{\varepsilon_i} == \underset{n}{\varepsilon} \vee \left(\sum_{i=1}^{\nu} a_i \underset{k}{\varepsilon_i}\right)$$

Since $\underset{n}{\varepsilon}$ is an n-element, we can 'divide' through by it (see section 2.11) and also replace the scalar factors a_i from their definition

$$a_i == \frac{\underset{n-k}{\varepsilon_i} \wedge \underset{k}{\beta}}{\underset{n}{\varepsilon}}$$

to give finally

$$\underset{k}{\beta} == \sum_{i=1}^{\nu} \left(\frac{\underset{n-k}{\varepsilon_i} \wedge \underset{k}{\beta}}{\underset{n}{\varepsilon}}\right) \underset{k}{\varepsilon_i} \qquad \nu == \binom{n}{k} \tag{3.56}$$

◆ Example: Expressing a 1-element in a new basis

In section 3.6, formula [3.48] we have already seen an example for k equal to 1.

$$\beta == \sum_{i=1}^{n} \frac{\varepsilon_1 \wedge \varepsilon_2 \wedge \ldots \wedge \varepsilon_{i-1} \wedge \beta \wedge \varepsilon_{i+1} \wedge \ldots \wedge \varepsilon_n}{\varepsilon_1 \wedge \varepsilon_2 \wedge \ldots \wedge \varepsilon_n} \varepsilon_i$$

Using the computable form of the Common Factor Theorem

We can use the computable form of the Common Factor Theorem A [3.43] to express an element in terms of a new basis.

$$\underset{m}{A} \vee \underset{k}{B} == \star\Sigma\left[\left(\star\mathbb{S}_{m-j}\left[\underset{m}{A}\right] \wedge \underset{k}{B}\right) \vee \star\mathbb{S}^{m-j}\left[\underset{m}{A}\right]\right] \qquad m - j + k - n == 0$$

First, put $\underset{m}{A}$ equal to $\underset{n}{\varepsilon}$ and $\underset{k}{B}$ equal to $\underset{k}{\beta}$.

$$\underset{n}{\varepsilon} \vee \underset{k}{\beta} == \star\Sigma\left[\left(\star\mathbb{S}_{n-k}\left[\underset{n}{\varepsilon}\right] \wedge \underset{k}{\beta}\right) \vee \star\mathbb{S}^{n-k}\left[\underset{n}{\varepsilon}\right]\right]$$

Since the first term of the regressive product on both sides of the equation is an n-element, we can write

$$\underset{k}{\beta} == \star\Sigma\left[\left(\frac{\star\mathbb{S}_{n-k}\left[\underset{n}{\varepsilon}\right] \wedge \underset{k}{\beta}}{\underset{n}{\varepsilon}}\right) \star\mathbb{S}^{n-k}\left[\underset{n}{\varepsilon}\right]\right] \tag{3.57}$$

This formula is not directly computable however, because the first factor must be computed with $\underset{n}{\varepsilon}$ and $\underset{k}{\beta}$ expressed in the old basis, while the second factor is computed with $\underset{n}{\varepsilon}$ in the new basis.

This is because, since $\underset{k}{\beta}$ is expressed in terms of the old basis, we need also to have the other factors in the quotient in the same basis. An example may make this clearer.

◆ Example: A 2-element in a new 5-basis

Suppose we have a 2-element $\underset{2}{\beta}$ expressed in terms of basis elements e_i and a new basis $\{\varepsilon_1, \varepsilon_2, \varepsilon_3, \varepsilon_4, \varepsilon_5\}$ with n-element $\underset{5}{\varepsilon}$ equal to $\varepsilon_1 \wedge \varepsilon_2 \wedge \varepsilon_3 \wedge \varepsilon_4 \wedge \varepsilon_5$.

- Step 1: Compute the second factor with $\underset{n}{\varepsilon}$ in the new basis:

$$\mathcal{E} = \star\mathbb{S}^{5-2}\left[\underset{5}{\varepsilon}\right]$$

$$\{\varepsilon_4 \wedge \varepsilon_5, -(\varepsilon_3 \wedge \varepsilon_5), \varepsilon_3 \wedge \varepsilon_4, \varepsilon_2 \wedge \varepsilon_5, -(\varepsilon_2 \wedge \varepsilon_4), \varepsilon_2 \wedge \varepsilon_3, -(\varepsilon_1 \wedge \varepsilon_5), \varepsilon_1 \wedge \varepsilon_4, -(\varepsilon_1 \wedge \varepsilon_3), \varepsilon_1 \wedge \varepsilon_2\}$$

- Step 2: Compute the first factor with $\underset{n}{\varepsilon}$ and $\underset{k}{\beta}$ expressed in the old basis:

$$\star\mathcal{B}_5;\ \underset{5}{\varepsilon} = (2\,e_1 + 3\,e_3) \wedge (5\,e_3 - e_4) \wedge (e_1 - e_3) \wedge (e_2 + e_3 + e_5) \wedge -e_2;$$

$$\underset{2}{\beta} = e_1 \wedge e_2 - 2\,e_1 \wedge e_4 + 3\,e_2 \wedge e_3 - 4\,e_2 \wedge e_5 + 5\,e_3 \wedge e_5;$$

$$C = \star\mathcal{G}\left[\frac{\star\mathbb{S}_{5-2}\left[\underset{5}{\varepsilon}\right] \wedge \underset{2}{\beta}}{\underset{5}{\varepsilon}}\right]$$

$$\left\{-4, \frac{27}{5}, -2, 0, 0, -\frac{6}{5}, -\frac{11}{5}, 1, -2, \frac{2}{5}\right\}$$

- Step 3: Multiply the first factor by the second factor and add the terms to get the final expression for $\underset{2}{\beta}$.

$$\underset{2}{\beta'} = \star\Sigma[C\,\mathcal{E}]$$

$$\frac{2\,\varepsilon_1 \wedge \varepsilon_2}{5} + 2\,\varepsilon_1 \wedge \varepsilon_3 + \varepsilon_1 \wedge \varepsilon_4 + \frac{11\,\varepsilon_1 \wedge \varepsilon_5}{5} - \frac{6\,\varepsilon_2 \wedge \varepsilon_3}{5} - 2\,\varepsilon_3 \wedge \varepsilon_4 - \frac{27\,\varepsilon_3 \wedge \varepsilon_5}{5} - 4\,\varepsilon_4 \wedge \varepsilon_5$$

- You can check that this is indeed a correct expression of $\underset{2}{\beta}$ by substituting back for the ε_i.

$$\star\mathcal{G}\left[\underset{2}{\beta} == \underset{2}{\beta'} /. \{\varepsilon_1 \to 2\,e_1 + 3\,e_3, \varepsilon_2 \to 5\,e_3 - e_4, \varepsilon_3 \to e_1 - e_3, \varepsilon_4 \to e_2 + e_3 + e_5, \varepsilon_5 \to -e_2\}\right]$$

```
True
```

Automating the process

It is straightforward to collect the steps we have made into a computational procedure for general elements. In the interests of simplicity this function is not robustified to take in to account all special cases, so we need to give a few provisos for its application.

Let us call the function ToNewBasis. Its arguments A and B should be given in terms of the currently declared (old) basis. A is the *new* n-element, and must be given in factored form with as many factors as the dimension of the currently declared basis (and should not be zero). Any symbolic scalars must be declared as ScalarSymbols. Also enter a symbol with which the new basis elements (that is, the factors of A in the new basis) are to be named (for example ε).

```
ToNewBasis[A_, B_, ε_] := Module[{ℰ, C, n, k}, n = Grade[A]; k = Grade[B];
  ℰ = ★𝕊^(n-k)[ε_n]; C = ★𝒢[★𝕊_(n-k)[A] ∧ B / A]; ★Σ[C ℰ]]
```

- Here is a numeric example using ToNewBasis.

```
★ℬ5; A = (2 e1 + 3 e3) ∧ (5 e3 - e4) ∧ (e1 - e3) ∧ (e1 + e2 + e3 + e4);
B = 2 e1 ∧ e2 ∧ e3 - 3 e1 ∧ e3 ∧ e4 - 2 e2 ∧ e3 ∧ e4;
B′ = ToNewBasis[A, B, ε]
```

$$-\frac{1}{5}\,\varepsilon_1\wedge\varepsilon_2\wedge\varepsilon_3+\frac{2}{5}\,\varepsilon_1\wedge\varepsilon_2\wedge\varepsilon_4+\frac{2}{5}\,\varepsilon_1\wedge\varepsilon_3\wedge\varepsilon_4+\frac{4}{5}\,\varepsilon_2\wedge\varepsilon_3\wedge\varepsilon_4$$

You can check by replacing the new basis elements with the old ones and simplifying.

```
Simplify[B == ★𝒢[B′ /. ε_i_ :→ A[[i]]]]
```

True

- Here is a symbolic example using ToNewBasis.

```
★ℬ4; A = (a e1 + (1 - a) e3) ∧ (c e3 - e4) ∧ (e1 - e3) ∧ (- e2);
B = b e1 ∧ e2 + d e3 ∧ e4;
B′ = ToNewBasis[A, B, ε]
```

$$-d\,\varepsilon_1\wedge\varepsilon_2-b\,\varepsilon_1\wedge\varepsilon_4-a\,d\,\varepsilon_2\wedge\varepsilon_3+(-b+a\,b)\,\varepsilon_3\wedge\varepsilon_4$$

We might also note that the original expression of B showed that it was not simple. It is straightforward to confirm that in the new basis it is also, as expected, not simple - since its exterior square is not zero (see section 2.10).

```
Simplify[★𝒢[B′ ∧ B′]]
```

$$2\,b\,d\,\varepsilon_1\wedge\varepsilon_2\wedge\varepsilon_3\wedge\varepsilon_4$$

The symmetric expansion of a 1-element in terms of another basis

For the case $k = 1$ we have seen in [3.46] and [3.48] that $\underset{n}{\alpha}\vee\beta$ may be expanded by the Common Factor Theorem to give the equivalent forms:

$$(\alpha_1\wedge\alpha_2\wedge\ldots\wedge\alpha_n)\vee\beta == \sum_{i=1}^{n}(\alpha_1\wedge\alpha_2\wedge\ldots\wedge\alpha_{i-1}\wedge\beta\wedge\alpha_{i+1}\wedge\ldots\wedge\alpha_n)\vee\alpha_i$$

$$\beta == \sum_{i=1}^{n}\frac{\alpha_1\wedge\alpha_2\wedge\ldots\wedge\alpha_{i-1}\wedge\beta\wedge\alpha_{i+1}\wedge\ldots\wedge\alpha_n}{\alpha_1\wedge\alpha_2\wedge\ldots\wedge\alpha_n}\,\alpha_i$$

or, in terms of the mnemonic expansion in the first section:

$$(\alpha_1 \wedge \alpha_2 \wedge \alpha_3 \wedge \ldots \wedge \alpha_n) \vee \beta$$
$$= (\beta \wedge \alpha_2 \wedge \alpha_3 \wedge \ldots \wedge \alpha_n) \vee \alpha_1$$
$$+ (\alpha_1 \wedge \beta \wedge \alpha_3 \wedge \ldots \wedge \alpha_n) \vee \alpha_2$$
$$+ (\alpha_1 \wedge \alpha_2 \wedge \beta \wedge \ldots \wedge \alpha_n) \vee \alpha_3$$
$$+ \ldots$$
$$+ (\alpha_1 \wedge \alpha_2 \wedge \alpha_3 \wedge \ldots \wedge \beta) \vee \alpha_n$$

Putting β equal to α_0, this may be written more symmetrically as:

$$\sum_{i=0}^{n} (-1)^i (\alpha_0 \wedge \alpha_1 \wedge \ldots \wedge \square_i \wedge \ldots \wedge \alpha_n) \vee \alpha_i = 0 \qquad 3.58$$

For example, suppose $\alpha_0, \alpha_1, \alpha_2, \alpha_3$ are four dependent 1-elements which span a 3-space, then the formula reduces to the identity:

$$(\alpha_1 \wedge \alpha_2 \wedge \alpha_3) \vee \alpha_0 - (\alpha_0 \wedge \alpha_2 \wedge \alpha_3) \vee \alpha_1 + (\alpha_0 \wedge \alpha_1 \wedge \alpha_3) \vee \alpha_2 - (\alpha_0 \wedge \alpha_1 \wedge \alpha_2) \vee \alpha_3 = 0 \qquad 3.59$$

It might be noted that formula [3.58] is a simple case of the more general result developed in section 2.12 for general multilinear product operations of which the regressive product is one. The particular case of [3.59] corresponds to formula [2.63] in which there is a linear dependence between the α_i.

$$\alpha_1 \wedge \alpha_2 \wedge \alpha_3 \wedge \alpha_4 = 0 \quad \Longrightarrow$$

$$(\alpha_1 \wedge \alpha_2 \wedge \alpha_3) \odot \alpha_4 - (\alpha_1 \wedge \alpha_2 \wedge \alpha_4) \odot \alpha_3 + (\alpha_1 \wedge \alpha_3 \wedge \alpha_4) \odot \alpha_2 - (\alpha_2 \wedge \alpha_3 \wedge \alpha_4) \odot \alpha_1$$
$$= 0$$

3.9 Factorization of Simple Elements

Factorization using the regressive product

The Common Factor Axiom asserts that in an *n*-space, the regressive product of a simple *m* element with a simple (*n*–*m*+1)-element will give either zero, or a 1-element belonging to them both, and hence a factor of the *m*-element. If *m* such factors can be obtained which are independent, their product will therefore constitute (apart from a scalar factor easily determined) a factorization of the *m*-element.

Let $\underset{m}{\alpha}$ be the simple element to be factorized. Choose first an (*n*–*m*)-element $\underset{n-m}{\beta}$ whose product with $\underset{m}{\alpha}$ is non-zero. Next, choose a set of *m* independent 1-elements β_j whose products with $\underset{n-m}{\beta}$ are also non-zero.

A factorization α_j of $\underset{m}{\alpha}$ is then obtained from:

$$\underset{m}{\alpha} = a\, \alpha_1 \wedge \alpha_2 \wedge \cdots \wedge \alpha_m \qquad \alpha_j = \underset{m}{\alpha} \vee \left(\beta_j \wedge \underset{n-m}{\beta}\right) \qquad 3.60$$

The scalar factor a may be determined simply by equating any two corresponding terms of the original element and the factorized version.

Note that no factorization is unique. Had different β_j been chosen, a different factorization would have been obtained. Nevertheless, any one factorization may be obtained from any other by adding multiples of the factors to each factor.

If an element is simple, then the exterior product of the element with itself is zero. The converse, however, is not true in general for elements of grade higher than 2, for it only requires the element to have just one simple factor to make the product with itself zero (see section 2.10).

If the method is applied to the factorization of a non-simple element, the result will still be a simple element. Thus an element may be tested to see if it is simple by applying the method of this section: if the factorization is not equivalent to the original element, the hypothesis of simplicity has been violated.

◆ Example: Factorization of a simple 2-element

Suppose we have a 2-element $\underset{2}{\alpha}$ which we wish to show is simple and which we wish to factorize.

$$\underset{2}{\alpha} == v \wedge w + v \wedge x + v \wedge y + v \wedge z + w \wedge z + x \wedge z + y \wedge z$$

There are 5 independent 1-elements in the expression for $\underset{2}{\alpha}$: v, w, x, y, z, hence we can choose *n* to be 5. Next, we choose $\underset{n-m}{\beta}$ ($= \underset{3}{\beta}$) arbitrarily as $x \wedge y \wedge z$, β_1 as v, and β_2 as w. Our two factors then become:

$$\alpha_1 \equiv \underset{2}{\alpha} \vee \left(\beta_1 \wedge \underset{3}{\beta}\right) == \underset{2}{\alpha} \vee (v \wedge x \wedge y \wedge z)$$

$$\alpha_2 \equiv \underset{2}{\alpha} \vee \left(\beta_2 \wedge \underset{3}{\beta}\right) == \underset{2}{\alpha} \vee (w \wedge x \wedge y \wedge z)$$

The Common Factor Theorem permits us to write for arbitrary 1-elements ξ and ψ:

$$(\xi \wedge \psi) \vee (v \wedge x \wedge y \wedge z) == (\xi \wedge v \wedge x \wedge y \wedge z) \vee \psi - (\psi \wedge v \wedge x \wedge y \wedge z) \vee \xi$$

Applying this expansion to each of the terms of $\underset{2}{\alpha}$ gives for α_1:

$$\alpha_1 \equiv -(w \wedge v \wedge x \wedge y \wedge z) \vee v + (w \wedge v \wedge x \wedge y \wedge z) \vee z \equiv v - z$$

Similarly for α_2 we have the same formula, except that w replaces v.

$$(\xi \wedge \psi) \vee (w \wedge x \wedge y \wedge z) == (\xi \wedge w \wedge x \wedge y \wedge z) \vee \psi - (\psi \wedge w \wedge x \wedge y \wedge z) \vee \xi$$

Applying this to each of the terms of $\underset{2}{\alpha}$ gives for α_2:

$$\alpha_2 \equiv (v \wedge w \wedge x \wedge y \wedge z) \vee w + (v \wedge w \wedge x \wedge y \wedge z) \vee x + (v \wedge w \wedge x \wedge y \wedge z) \vee y + (v \wedge w \wedge x \wedge y \wedge z) \vee z \equiv w + x + y + z$$

Hence the factorization is congruent to:

$$\alpha_1 \wedge \alpha_2 \equiv (v - z) \wedge (w + x + y + z)$$

Verification by expansion of this product shows that this is indeed a factorization of the original element.

$$\star\mathcal{G}[v\wedge w + v\wedge x + v\wedge y + v\wedge z + w\wedge z + x\wedge z + y\wedge z == (v - z)\wedge(w + x + y + z)]$$

True

Factorizing elements expressed in terms of basis elements

We now take the special case where the element to be factorized is expressed in terms of basis elements. This will enable us to develop formulae from which we can write down the factorization of an element (almost) by inspection.

The development is most clearly apparent from a specific example, but one that is general enough to cover the general concept. Consider a 3-element A in a 5-space, where we suppose the coefficients to be such as to ensure the simplicity of the element.

$$\begin{aligned} A == {} & a_1\, e_1\wedge e_2\wedge e_3 + a_2\, e_1\wedge e_2\wedge e_4 + a_3\, e_1\wedge e_2\wedge e_5 + a_4\, e_1\wedge e_3\wedge e_4 + \\ & a_5\, e_1\wedge e_3\wedge e_5 + a_6\, e_1\wedge e_4\wedge e_5 + a_7\, e_2\wedge e_3\wedge e_4 + a_8\, e_2\wedge e_3\wedge e_5 + \\ & a_9\, e_2\wedge e_4\wedge e_5 + a_{10}\, e_3\wedge e_4\wedge e_5 \end{aligned}$$

Choose $\beta_1 = e_1$, $\beta_2 = e_2$, $\beta_3 = e_3$, $\underset{2}{\beta} = e_4\wedge e_5$, then we can write each of the factors $\beta_i\wedge\underset{2}{\beta}$ as a cobasis element.

$$\beta_1\wedge\underset{2}{\beta} == e_1\wedge e_4\wedge e_5 == \underline{e_2\wedge e_3}$$

$$\beta_2\wedge\underset{2}{\beta} == e_2\wedge e_4\wedge e_5 == -\underline{e_1\wedge e_3}$$

$$\beta_3\wedge\underset{2}{\beta} == e_3\wedge e_4\wedge e_5 == \underline{e_1\wedge e_2}$$

Consider the cobasis element $\underline{e_2\wedge e_3}$, and a typical term of $A\vee\underline{e_2\wedge e_3}$, which we write as $(a\, e_i\wedge e_j\wedge e_k)\vee\underline{e_2\wedge e_3}$. The Common Factor Theorem (section 3.6) tells us that this product is zero if $e_i\wedge e_j\wedge e_k$ does not contain $e_2\wedge e_3$. We can thus simplify the product $A\vee\underline{e_2\wedge e_3}$ by dropping out the terms of A which do not contain $e_2\wedge e_3$. Thus:

$$A\vee\underline{e_2\wedge e_3} == (a_1\, e_1\wedge e_2\wedge e_3 + a_7\, e_2\wedge e_3\wedge e_4 + a_8\, e_2\wedge e_3\wedge e_5)\vee\underline{e_2\wedge e_3}$$

Furthermore, the Common Factor Theorem applied to a typical term $(e_2\wedge e_3\wedge e_i)\vee\underline{e_2\wedge e_3}$ of the expansion in which *both* $e_2\wedge e_3$ and $\underline{e_2\wedge e_3}$ occur yields a 1-element congruent to the remaining basis 1-element e_i in the product. This effectively cancels out (up to congruence) the product $e_2\wedge e_3$ from the original term.

$$(e_2\wedge e_3\wedge e_i)\vee\underline{e_2\wedge e_3} == \left((e_2\wedge e_3)\wedge(\underline{e_2\wedge e_3})\right)\vee e_i \equiv e_i$$

Thus we can further reduce $A\vee\underline{e_2\wedge e_3}$ to give

$$\alpha_1 \equiv A\vee\underline{e_2\wedge e_3} \equiv a_1\, e_1 + a_7\, e_4 + a_8\, e_5$$

Similarly we can determine the other factors:

$$\alpha_2 \equiv A\vee -\underline{e_1\wedge e_3} \equiv a_1\, e_2 - a_4\, e_4 - a_5\, e_5$$

$$\alpha_3 \equiv A\vee\underline{e_1\wedge e_2} \equiv a_1\, e_3 + a_2\, e_4 + a_3\, e_5$$

It is clear from inspecting the product of the first terms in each 1-element that the product requires a scalar divisor of a_1^2. The final result A' for the factored form of A is then

$$A' == \frac{1}{a_1^2}\alpha_1 \wedge \alpha_2 \wedge \alpha_3$$

$$== \frac{1}{a_1^2}(a_1 e_1 + a_7 e_4 + a_8 e_5) \wedge (a_1 e_2 - a_4 e_4 - a_5 e_5) \wedge (a_1 e_3 + a_2 e_4 + a_3 e_5)$$

◆ Verification and derivation of conditions for simplicity

We verify the factorization by multiplying out the factors and comparing the result with the original expression. When we do this we obtain a result which still requires some conditions to be met: those ensuring the original element is simple.

First we declare a 5-dimensional basis and compose the form of A as a general 3-element (this automatically declares the coefficients a_i to be scalar).

```
★𝓑5; A = ComposeMElement[3, a]
```

$$a_1 e_1 \wedge e_2 \wedge e_3 + a_2 e_1 \wedge e_2 \wedge e_4 + a_3 e_1 \wedge e_2 \wedge e_5 + a_4 e_1 \wedge e_3 \wedge e_4 + a_5 e_1 \wedge e_3 \wedge e_5 + a_6 e_1 \wedge e_4 \wedge e_5 + a_7 e_2 \wedge e_3 \wedge e_4 + a_8 e_2 \wedge e_3 \wedge e_5 + a_9 e_2 \wedge e_4 \wedge e_5 + a_{10} e_3 \wedge e_4 \wedge e_5$$

Expanding and simplifying the factored form A′ of A we found in the example above gives

$$A' = \star\mathcal{G}\left[\frac{1}{a_1^2}(a_1 e_1 + a_7 e_4 + a_8 e_5) \wedge (a_1 e_2 - a_4 e_4 - a_5 e_5) \wedge (a_1 e_3 + a_2 e_4 + a_3 e_5)\right]$$

$$a_1 e_1 \wedge e_2 \wedge e_3 + a_2 e_1 \wedge e_2 \wedge e_4 + a_3 e_1 \wedge e_2 \wedge e_5 + a_4 e_1 \wedge e_3 \wedge e_4 + a_5 e_1 \wedge e_3 \wedge e_5 + \left(-\frac{a_3 a_4}{a_1} + \frac{a_2 a_5}{a_1}\right) e_1 \wedge e_4 \wedge e_5 + a_7 e_2 \wedge e_3 \wedge e_4 + a_8 e_2 \wedge e_3 \wedge e_5 + \left(-\frac{a_3 a_7}{a_1} + \frac{a_2 a_8}{a_1}\right) e_2 \wedge e_4 \wedge e_5 + \left(-\frac{a_5 a_7}{a_1} + \frac{a_4 a_8}{a_1}\right) e_3 \wedge e_4 \wedge e_5$$

For this factored expression to be congruent to the original general expression we clearly need to apply the simplicity conditions for a general 3-element in a 5-space. Once this is done we retrieve the original 3-element A with which we began. The simplicity conditions can be expressed by the following rules, which we apply to the expression A′. We will discuss them in the next section.

$$A == A' \;/.\; \left\{\left(-\frac{a_3 a_4}{a_1} + \frac{a_2 a_5}{a_1}\right) \to a_6,\ \left(-\frac{a_3 a_7}{a_1} + \frac{a_2 a_8}{a_1}\right) \to a_9,\ \left(-\frac{a_5 a_7}{a_1} + \frac{a_4 a_8}{a_1}\right) \to a_{10}\right\}$$

```
True
```

- In sum: A 3-element in a 5-space of the form:

$$a_1 e_1 \wedge e_2 \wedge e_3 + a_2 e_1 \wedge e_2 \wedge e_4 + a_3 e_1 \wedge e_2 \wedge e_5 + a_4 e_1 \wedge e_3 \wedge e_4 + a_5 e_1 \wedge e_3 \wedge e_5 + a_6 e_1 \wedge e_4 \wedge e_5 + a_7 e_2 \wedge e_3 \wedge e_4 + a_8 e_2 \wedge e_3 \wedge e_5 + a_9 e_2 \wedge e_4 \wedge e_5 + a_{10} e_3 \wedge e_4 \wedge e_5$$

whose coefficients are constrained by the relations:

$$a_2 a_5 - a_3 a_4 - a_1 a_6 == 0$$
$$a_2 a_8 - a_3 a_7 - a_1 a_9 == 0$$
$$a_4 a_8 - a_5 a_7 - a_1 a_{10} == 0$$

is simple, and has a factorization of

$$\frac{1}{a_1^2}\,(a_1 e_1 + a_7 e_4 + a_8 e_5) \wedge (a_1 e_2 - a_4 e_4 - a_5 e_5) \wedge (a_1 e_3 + a_2 e_4 + a_3 e_5)$$

This factorization is of course not unique.

The factorization algorithm

In this section we take the development of the previous sections and extract an algorithm for writing down a factorization by inspection.

Suppose we have a *simple* m-element expressed as a sum of terms, each of which is of the form $a\,x \wedge y \wedge \ldots \wedge z$, where a denotes a scalar (which may be unity) and the x, y, ..., z are symbols denoting 1-element factors.

◇ To obtain a 1-element factor of the *m*-element

- Select an $(m-1)$-element belonging to at least one of the terms, for example $y \wedge \ldots \wedge z$.
- Drop any terms not containing the selected $(m-1)$-element.
- Factor this $(m-1)$-element from the resulting expression and eliminate it.
- The 1-element remaining is a 1-element factor of the m-element.

◇ To factorize the *m*-element

- Select an m-element belonging to at least one of the terms, for example $x \wedge y \wedge \ldots \wedge z$.
- Create m different $(m-1)$-elements by dropping a different 1-element factor each time. The sign of the result is not important, since a scalar factor will be determined in the last step.
- Obtain m independent 1-element factors corresponding to each of these $(m-1)$-elements.
- The original m-element is *congruent* to the exterior product of these 1-element factors.
- Compare this product to the original m-element to obtain the correct scalar factor and hence the final factorization.

◆ Example 1: Factorizing a 2-element in a 4-space

Suppose we have a 2-element A in a 4-space, and we wish to apply this algorithm to obtain a factorization $A' = a\,\alpha_1 \wedge \alpha_2$. We have already seen that such an element is in general *not* simple. We may however use the preceding algorithm to obtain the simplicity conditions on the coefficients.

$$A == a_1 e_1 \wedge e_2 + a_2 e_1 \wedge e_3 + a_3 e_1 \wedge e_4 + a_4 e_2 \wedge e_3 + a_5 e_2 \wedge e_4 + a_6 e_3 \wedge e_4$$

- Select a 2-element belonging to at least one of the terms, say $e_1 \wedge e_2$.
- Drop e_2 to create e_1. Then drop e_1 to create e_2.
- Select e_1.
- Drop the terms $a_4 e_2 \wedge e_3 + a_5 e_2 \wedge e_4 + a_6 e_3 \wedge e_4$ since they do not contain e_1.
- Factor e_1 from $a_1 e_1 \wedge e_2 + a_2 e_1 \wedge e_3 + a_3 e_1 \wedge e_4$ and eliminate it to give factor

  $\alpha_1 = a_1 e_2 + a_2 e_3 + a_3 e_4$.
- Select e_2.
- Drop the terms $a_2 e_1 \wedge e_3 + a_3 e_1 \wedge e_4 + a_6 e_3 \wedge e_4$ since they do not contain e_2.
- Factor e_2 from $a_1 e_1 \wedge e_2 + a_4 e_2 \wedge e_3 + a_5 e_2 \wedge e_4$ and eliminate it to give factor

$\alpha_2 = -a_1 e_1 + a_4 e_3 + a_5 e_4.$

- The exterior product of these 1-element factors is

$$\alpha_1 \wedge \alpha_2 == (a_1 e_2 + a_2 e_3 + a_3 e_4) \wedge (-a_1 e_1 + a_4 e_3 + a_5 e_4) ==$$
$$a_1 \left(a_1 e_1 \wedge e_2 + a_2 e_1 \wedge e_3 + a_3 e_1 \wedge e_4 + a_4 e_2 \wedge e_3 + a_5 e_2 \wedge e_4 + \left(\frac{-a_3 a_4 + a_2 a_5}{a_1}\right) e_3 \wedge e_4\right)$$

- Comparing this product to the original 2-element A gives the final factorization as

$$A' == \frac{1}{a_1} (a_1 e_2 + a_2 e_3 + a_3 e_4) \wedge (-a_1 e_1 + a_4 e_3 + a_5 e_4)$$

provided that the simplicity conditions on the coefficients are satisfied:

$$a_1 a_6 + a_3 a_4 - a_2 a_5 == 0$$

- In sum: A 2-element in a 4-space may be factorized if and only if a condition on its coefficients is satisfied.

$$\begin{aligned} a_1 e_1 \wedge e_2 + a_2 e_1 \wedge e_3 + a_3 e_1 \wedge e_4 + a_4 e_2 \wedge e_3 + a_5 e_2 \wedge e_4 + a_6 e_3 \wedge e_4 \\ == \frac{1}{a_1} (a_1 e_1 - a_4 e_3 - a_5 e_4) \wedge (a_1 e_2 + a_2 e_3 + a_3 e_4) \\ \Leftrightarrow \quad a_3 a_4 - a_2 a_5 + a_1 a_6 == 0 \end{aligned} \qquad 3.61$$

Again, this factorization is not unique.

◆ Example 2: Factorizing a 3-element in a 5-space

This time we have a numerical example of a 3-element in 5-space. We wish to determine if the element is simple, and if so, to obtain a factorization of it. To achieve this, we first *assume* that it is simple, and apply the factorization algorithm. We will verify its simplicity (or its non-simplicity) by comparing the results of this process to the original element.

$$A == -3 e_1 \wedge e_2 \wedge e_3 - 4 e_1 \wedge e_2 \wedge e_4 + 12 e_1 \wedge e_2 \wedge e_5 + 3 e_1 \wedge e_3 \wedge e_4 - 3 e_1 \wedge e_3 \wedge e_5 + 8 e_1 \wedge e_4 \wedge e_5 + 6 e_2 \wedge e_3 \wedge e_4 - 18 e_2 \wedge e_3 \wedge e_5 + 12 e_3 \wedge e_4 \wedge e_5$$

- Select a 3-element belonging to at least one of the terms, say $e_1 \wedge e_2 \wedge e_3$.
- Drop e_1 to create $e_2 \wedge e_3$. Drop e_2 to create $e_1 \wedge e_3$. Drop e_3 to create $e_1 \wedge e_2$.
- Select $e_2 \wedge e_3$.
- Drop the terms not containing it, factor it from the remainder, and eliminate it to give α_1.

$$\alpha_1 \equiv -3 e_1 + 6 e_4 - 18 e_5 \equiv -e_1 + 2 e_4 - 6 e_5$$

- Select $e_1 \wedge e_3$.
- Drop the terms not containing it, factor it from the remainder, and eliminate it to give α_2.

$$\alpha_2 \equiv 3 e_2 + 3 e_4 - 3 e_5 \equiv e_2 + e_4 - e_5$$

- Select $e_1 \wedge e_2$.
- Drop the terms not containing it, factor it from the remainder, and eliminate it to give α_3.

$$\alpha_3 \equiv -3 e_3 - 4 e_4 + 12 e_5$$

- The exterior product of these 1-element factors is:

$$\alpha_1 \wedge \alpha_2 \wedge \alpha_3 == (-e_1 + 2 e_4 - 6 e_5) \wedge (e_2 + e_4 - e_5) \wedge (-3 e_3 - 4 e_4 + 12 e_5)$$

- Expanding and simplifying this factorization gives:

$\star\mathcal{B}_5;\ \star\mathcal{G}[(-e_1 + 2 e_4 - 6 e_5) \wedge (e_2 + e_4 - e_5) \wedge (-3 e_3 - 4 e_4 + 12 e_5)]$

$3 e_1 \wedge e_2 \wedge e_3 + 4 e_1 \wedge e_2 \wedge e_4 - 12 e_1 \wedge e_2 \wedge e_5 - 3 e_1 \wedge e_3 \wedge e_4 + 3 e_1 \wedge e_3 \wedge e_5 - 8 e_1 \wedge e_4 \wedge e_5 - 6 e_2 \wedge e_3 \wedge e_4 + 18 e_2 \wedge e_3 \wedge e_5 - 12 e_3 \wedge e_4 \wedge e_5$

- Comparing this product to the original 3-element A verifies a final factorization as:

$$A' == (e_1 - 2 e_4 + 6 e_5) \wedge (e_2 + e_4 - e_5) \wedge (-3 e_3 - 4 e_4 + 12 e_5)$$

This factorization is, of course, not unique. For example, a slightly simpler factorization could be obtained by subtracting twice the first factor from the third factor to obtain:

$$A' == (e_1 - 2 e_4 + 6 e_5) \wedge (e_2 + e_4 - e_5) \wedge (-3 e_3 - 2 e_1)$$

Factorization of (*n*–1)-elements

The foregoing method may be used to prove constructively that any (*n*–1)-element is simple by obtaining a factorization and subsequently verifying its validity. Let the general (*n*–1)-element be

$$\underset{n-1}{\alpha} == \sum_{i=1}^{n} a_i \underline{e_i}, \qquad a_1 \neq 0$$

where the $\underline{e_i}$ are the cobasis elements (see section 2.6) of the e_i.

Choose $\underset{1}{\beta} = e_1$ and $\beta_j = e_j$ and apply the Common Factor Theorem to obtain (for j ≠1):

$$\underset{n-1}{\alpha} \vee (e_1 \wedge e_j) == \left(\underset{n-1}{\alpha} \wedge e_j\right) \vee e_1 - \left(\underset{n-1}{\alpha} \wedge e_1\right) \vee e_j$$

$$== \left(\sum_{i=1}^{n} a_i \underline{e_i} \wedge e_j\right) \vee e_1 - \left(\sum_{i=1}^{n} a_i \underline{e_i} \wedge e_1\right) \vee e_j$$

$$== (-1)^{n-1} \left(a_j \underset{n}{e} \vee e_1 - a_1 \underset{n}{e} \vee e_j\right)$$

$$== (-1)^{n-1} \underset{n}{e} \vee (a_j e_1 - a_1 e_j)$$

Factors of $\underset{n-1}{\alpha}$ are therefore of the form $a_j e_1 - a_1 e_j$, $j \neq 1$, so that $\underset{n-1}{\alpha}$ is congruent to:

$$\underset{n-1}{\alpha} \equiv (a_2 e_1 - a_1 e_2) \wedge (a_3 e_1 - a_1 e_3) \wedge \ldots \wedge (a_n e_1 - a_1 e_n)$$

The result may be summarized as follows:

$$a_1 \underline{e_1} + a_2 \underline{e_2} + \ldots + a_n \underline{e_n} \equiv (a_2 e_1 - a_1 e_2) \wedge (a_3 e_1 - a_1 e_3) \wedge \ldots \wedge (a_n e_1 - a_1 e_n), \quad a_1 \neq 0 \qquad 3.62$$

$$\sum_{i=1}^{n} a_i \underline{e_i} \equiv \bigwedge_{i=2}^{n} (a_i e_1 - a_1 e_i), \qquad a_1 \neq 0 \qquad 3.63$$

By putting equation [3.63] in its equality form rather than its congruence form, we have

$$\sum_{i=1}^{n} a_i \underline{e_i} == (-1)^{n-1} \frac{1}{a_1{}^{n-2}} \bigwedge_{i=2}^{n} (a_i\, e_1 - a_1\, e_i), \qquad a_1 \neq 0 \tag{3.64}$$

◆ Example: Verifying the formula

The left and right sides of the formula [3.64] are easily coded in *GrassmannAlgebra* as functions of the dimension *n* of the space.

```
L[n_] := (★𝓑n; a1^(n-2) Σ_{i=1}^{n} ai CobasisElement[ei])

R[n_] := (-1)^(n-1) Wedge @@ Table[ai e1 - a1 ei, {i, 2, n}]
```

This enables us to tabulate the formula for different dimensions. For example

```
Table[L[n] == R[n], {n, 3, 5}]
```

$\{a_1 (a_3\, e_1 \wedge e_2 - a_2\, e_1 \wedge e_3 + a_1\, e_2 \wedge e_3) == (a_2\, e_1 - a_1\, e_2) \wedge (a_3\, e_1 - a_1\, e_3),$
$a_1^2 (-a_4\, e_1 \wedge e_2 \wedge e_3 + a_3\, e_1 \wedge e_2 \wedge e_4 - a_2\, e_1 \wedge e_3 \wedge e_4 + a_1\, e_2 \wedge e_3 \wedge e_4) ==$
$-((a_2\, e_1 - a_1\, e_2) \wedge (a_3\, e_1 - a_1\, e_3) \wedge (a_4\, e_1 - a_1\, e_4)),$
$a_1^3 (a_5\, e_1 \wedge e_2 \wedge e_3 \wedge e_4 - a_4\, e_1 \wedge e_2 \wedge e_3 \wedge e_5 + a_3\, e_1 \wedge e_2 \wedge e_4 \wedge e_5 -$
$a_2\, e_1 \wedge e_3 \wedge e_4 \wedge e_5 + a_1\, e_2 \wedge e_3 \wedge e_4 \wedge e_5) ==$
$(a_2\, e_1 - a_1\, e_2) \wedge (a_3\, e_1 - a_1\, e_3) \wedge (a_4\, e_1 - a_1\, e_4) \wedge (a_5\, e_1 - a_1\, e_5)\}$

$$a_3\, e_1 \wedge e_2 - a_2\, e_1 \wedge e_3 + a_1\, e_2 \wedge e_3 == \frac{1}{a_1} (a_2\, e_1 - a_1\, e_2) \wedge (a_3\, e_1 - a_1\, e_3) \tag{3.65}$$

$$\begin{gathered} a_4\, e_1 \wedge e_2 \wedge e_3 - a_3\, e_1 \wedge e_2 \wedge e_4 + a_2\, e_1 \wedge e_3 \wedge e_4 - a_1\, e_2 \wedge e_3 \wedge e_4 \\ == \frac{1}{a_1^2} (a_2\, e_1 - a_1\, e_2) \wedge (a_3\, e_1 - a_1\, e_3) \wedge (a_4\, e_1 - a_1\, e_4) \end{gathered} \tag{3.66}$$

$$\begin{gathered} a_5\, e_1 \wedge e_2 \wedge e_3 \wedge e_4 - a_4\, e_1 \wedge e_2 \wedge e_3 \wedge e_5 + \\ a_3\, e_1 \wedge e_2 \wedge e_4 \wedge e_5 - a_2\, e_1 \wedge e_3 \wedge e_4 \wedge e_5 + a_1\, e_2 \wedge e_3 \wedge e_4 \wedge e_5 \\ == \frac{1}{a_1^3} (a_2\, e_1 - a_1\, e_2) \wedge (a_3\, e_1 - a_1\, e_3) \wedge (a_4\, e_1 - a_1\, e_4) \wedge (a_5\, e_1 - a_1\, e_5) \end{gathered} \tag{3.67}$$

We can also verify the formula for different dimensions by expanding and simplifying the equations (after first declaring the a_i as scalars).

```
★★S[a_]; ★𝒢[Table[L[n] == R[n], {n, 3, 10}]]
```

```
{True, True, True, True, True, True, True, True}
```

Factorizing simple *m*-elements

For *m*-elements known to be simple we can use the *GrassmannAlgebra* function `ExteriorFactorize` to obtain a factorization. Note that the result obtained from `ExteriorFactorize` will be only one factorization out of the generally infinitely many possibilities. However, every factorization may be obtained from any other by adding scalar multiples of each factor to the other factors.

We have shown above that every (*n*–1)-element in an *n*-space is simple. Applying `ExteriorFactorize` to a series of general (*n*–1)-elements in spaces of 4, 5, and 6 dimensions corroborates this result (although the form it generates is slightly different).

In each case we first declare a basis of the requisite dimension by entering $\star\mathcal{B}_n$;, and then create a general (*n*–1)-element A with scalars a_i using the *GrassmannAlgebra* function `ComposeMElement`.

◆ Factorizing a 3-element in a 4-space

$\star\mathcal{B}_4$; A = ComposeMElement[3, a]

$$a_1\, e_1 \wedge e_2 \wedge e_3 + a_2\, e_1 \wedge e_2 \wedge e_4 + a_3\, e_1 \wedge e_3 \wedge e_4 + a_4\, e_2 \wedge e_3 \wedge e_4$$

ExteriorFactorize[A]

$$a_1 \left(e_1 + \frac{a_4\, e_4}{a_1}\right) \wedge \left(e_2 - \frac{a_3\, e_4}{a_1}\right) \wedge \left(e_3 + \frac{a_2\, e_4}{a_1}\right)$$

◆ Factorizing a 4-element in a 5-space

$\star\mathcal{B}_5$; A = ComposeMElement[4, a]

$$a_1\, e_1 \wedge e_2 \wedge e_3 \wedge e_4 + a_2\, e_1 \wedge e_2 \wedge e_3 \wedge e_5 + a_3\, e_1 \wedge e_2 \wedge e_4 \wedge e_5 + a_4\, e_1 \wedge e_3 \wedge e_4 \wedge e_5 + a_5\, e_2 \wedge e_3 \wedge e_4 \wedge e_5$$

ExteriorFactorize[A]

$$a_1 \left(e_1 - \frac{a_5\, e_5}{a_1}\right) \wedge \left(e_2 + \frac{a_4\, e_5}{a_1}\right) \wedge \left(e_3 - \frac{a_3\, e_5}{a_1}\right) \wedge \left(e_4 + \frac{a_2\, e_5}{a_1}\right)$$

◆ Factorizing a 5-element in a 6-space

$\star\mathcal{B}_6$; A = ComposeMElement[5, a]

$$a_1\, e_1 \wedge e_2 \wedge e_3 \wedge e_4 \wedge e_5 + a_2\, e_1 \wedge e_2 \wedge e_3 \wedge e_4 \wedge e_6 + a_3\, e_1 \wedge e_2 \wedge e_3 \wedge e_5 \wedge e_6 + a_4\, e_1 \wedge e_2 \wedge e_4 \wedge e_5 \wedge e_6 + a_5\, e_1 \wedge e_3 \wedge e_4 \wedge e_5 \wedge e_6 + a_6\, e_2 \wedge e_3 \wedge e_4 \wedge e_5 \wedge e_6$$

ExteriorFactorize[A]

$$a_1 \left(e_1 + \frac{a_6\, e_6}{a_1}\right) \wedge \left(e_2 - \frac{a_5\, e_6}{a_1}\right) \wedge \left(e_3 + \frac{a_4\, e_6}{a_1}\right) \wedge \left(e_4 - \frac{a_3\, e_6}{a_1}\right) \wedge \left(e_5 + \frac{a_2\, e_6}{a_1}\right)$$

◆ The regressive product of two 3-elements in a 4-space

Let X and Y be 3-elements in a 4-space.

$\star$A; $\star\mathcal{B}_4$;
X = ComposeMElement[3, x]

$$x_1\, e_1 \wedge e_2 \wedge e_3 + x_2\, e_1 \wedge e_2 \wedge e_4 + x_3\, e_1 \wedge e_3 \wedge e_4 + x_4\, e_2 \wedge e_3 \wedge e_4$$

```
Y = ComposeMElement[3, y]
```

$y_1\, e_1 \wedge e_2 \wedge e_3 + y_2\, e_1 \wedge e_2 \wedge e_4 + y_3\, e_1 \wedge e_3 \wedge e_4 + y_4\, e_2 \wedge e_3 \wedge e_4$

We can obtain the 2-element common factor of these two 3-elements by applying `ToCommonFactor` to their regressive product.

```
Z = ToCommonFactor[X ∨ Y]
```

$$\star c\, ((-x_2\, y_1 + x_1\, y_2)\, e_1 \wedge e_2 + (-x_3\, y_1 + x_1\, y_3)\, e_1 \wedge e_3 + (-x_3\, y_2 + x_2\, y_3)\, e_1 \wedge e_4 + (-x_4\, y_1 + x_1\, y_4)\, e_2 \wedge e_3 + (-x_4\, y_2 + x_2\, y_4)\, e_2 \wedge e_4 + (-x_4\, y_3 + x_3\, y_4)\, e_3 \wedge e_4)$$

Applying `ExteriorFactorize` to this common factor and collecting terms shows that it is simple and can be factorized.

```
ExteriorFactorize[Z] //. a_ e_i_ + b_ e_i_ :→ Together[a + b] e_i
```

$$\star c\, (-x_2\, y_1 + x_1\, y_2) \left(e_1 + \frac{e_3\, (-x_4\, y_1 + x_1\, y_4)}{x_2\, y_1 - x_1\, y_2} + \frac{e_4\, (x_4\, y_2 - x_2\, y_4)}{-x_2\, y_1 + x_1\, y_2}\right) \wedge \left(e_2 + \frac{e_3\, (x_3\, y_1 - x_1\, y_3)}{x_2\, y_1 - x_1\, y_2} + \frac{e_4\, (-x_3\, y_2 + x_2\, y_3)}{-x_2\, y_1 + x_1\, y_2}\right)$$

Because `Z` is a 2-element we can of course also prove its simplicity by expanding and simplifying its exterior square to show that it is zero.

```
Expand[★𝒢[Z ∧ Z]]
```

0

Factorizing contingently simple *m*-elements

If `ExteriorFactorize` is applied to an element which is not simple, the element will simply be returned. For example, in a 4-space, the 2-element $x \wedge y + u \wedge v$ is not simple.

```
★𝔅4; ExteriorFactorize[x ∧ y + u ∧ v]
```

$u \wedge v + x \wedge y$

If `ExteriorFactorize` is applied to an element which may be simple *conditional* on the values taken by some of its scalar symbol coefficients, the conditional result will be returned.

◆ Contingently factorizing a 2-element in a 4-space

```
★𝔅4; A = ComposeMElement[2, a]
```

$a_1\, e_1 \wedge e_2 + a_2\, e_1 \wedge e_3 + a_3\, e_1 \wedge e_4 + a_4\, e_2 \wedge e_3 + a_5\, e_2 \wedge e_4 + a_6\, e_3 \wedge e_4$

```
A′ = ExteriorFactorize[A]
```

$$\text{If}\left[\{a_3\, a_4 - a_2\, a_5 + a_1\, a_6 == 0\},\ a_1 \left(e_1 - \frac{a_4\, e_3}{a_1} - \frac{a_5\, e_4}{a_1}\right) \wedge \left(e_2 + \frac{a_2\, e_3}{a_1} + \frac{a_3\, e_4}{a_1}\right)\right]$$

This *Mathematica* `If` function has syntax `If[C,F]`, where `C` is a list of constraints on the scalar symbols of the expression, and `F` is the factorization if the constraints are satisfied. Hence the above may be read: *if* the simplicity condition is satisfied, *then* the factorization is as given, *else* the element is not factorisable.

Instead of a list of constraints, we can also recast the list of constraints in predicate form by applying `And` to the list. (In this example, where there is only one constraint, the effect is simply to remove the braces from the constraint.)

```
A′ = A′ /. c_List :→ And @@ c
```

$$\text{If}\left[a_3\,a_4 - a_2\,a_5 + a_1\,a_6 == 0,\ a_1\left(e_1 - \frac{a_4\,e_3}{a_1} - \frac{a_5\,e_4}{a_1}\right) \wedge \left(e_2 + \frac{a_2\,e_3}{a_1} + \frac{a_3\,e_4}{a_1}\right)\right]$$

If we are able to assert the condition required, then the 2-element is indeed simple and the factorization is valid. Substituting this condition into the predicate form of the `If` statement, yields true for the predicate, hence the factorization is returned.

$$\mathbf{A' \;/.\; a_6 \to \frac{a_2\,a_5 - a_3\,a_4}{a_1}}$$

$$a_1\left(e_1 - \frac{a_4\,e_3}{a_1} - \frac{a_5\,e_4}{a_1}\right) \wedge \left(e_2 + \frac{a_2\,e_3}{a_1} + \frac{a_3\,e_4}{a_1}\right)$$

◆ Contingently factorizing a 2-element in a 5-space

```
★𝓑5; A = ComposeMElement[2, a]
```

$$a_1\,e_1 \wedge e_2 + a_2\,e_1 \wedge e_3 + a_3\,e_1 \wedge e_4 + a_4\,e_1 \wedge e_5 + a_5\,e_2 \wedge e_3 + a_6\,e_2 \wedge e_4 + a_7\,e_2 \wedge e_5 + a_8\,e_3 \wedge e_4 + a_9\,e_3 \wedge e_5 + a_{10}\,e_4 \wedge e_5$$

```
A′ = ExteriorFactorize[A]
```

$$\text{If}\left[\{a_3\,a_5 - a_2\,a_6 + a_1\,a_8 == 0,\ a_4\,a_5 - a_2\,a_7 + a_1\,a_9 == 0,\ a_4\,a_6 - a_3\,a_7 + a_1\,a_{10} == 0\},\ a_1\left(e_1 - \frac{a_5\,e_3}{a_1} - \frac{a_6\,e_4}{a_1} - \frac{a_7\,e_5}{a_1}\right) \wedge \left(e_2 + \frac{a_2\,e_3}{a_1} + \frac{a_3\,e_4}{a_1} + \frac{a_4\,e_5}{a_1}\right)\right]$$

In this case of a 2-element in a 5-space, we have a list of three constraints, which we can turn into a predicate by applying `And` to the list.

```
A′ = A′ /. c_List :→ And @@ c
```

$$\text{If}\left[a_3\,a_5 - a_2\,a_6 + a_1\,a_8 == 0\ \&\&\ a_4\,a_5 - a_2\,a_7 + a_1\,a_9 == 0\ \&\&\ a_4\,a_6 - a_3\,a_7 + a_1\,a_{10} == 0,\ a_1\left(e_1 - \frac{a_5\,e_3}{a_1} - \frac{a_6\,e_4}{a_1} - \frac{a_7\,e_5}{a_1}\right) \wedge \left(e_2 + \frac{a_2\,e_3}{a_1} + \frac{a_3\,e_4}{a_1} + \frac{a_4\,e_5}{a_1}\right)\right]$$

Now, if we assert the constraints, we get the simple factored expression.

$$\mathbf{A' \;/.\; \left\{a_8 \to -\frac{a_3\,a_5 - a_2\,a_6}{a_1},\ a_9 \to -\frac{a_4\,a_5 - a_2\,a_7}{a_1},\ a_{10} \to -\frac{a_4\,a_6 - a_3\,a_7}{a_1}\right\}}$$

$$a_1\left(e_1 - \frac{a_5\,e_3}{a_1} - \frac{a_6\,e_4}{a_1} - \frac{a_7\,e_5}{a_1}\right) \wedge \left(e_2 + \frac{a_2\,e_3}{a_1} + \frac{a_3\,e_4}{a_1} + \frac{a_4\,e_5}{a_1}\right)$$

◆ Elements expressed in terms of independent vector symbols

`ExteriorFactorize` will also contingently factorize *m*-elements expressed in terms of (independent) vector symbols. For example, here is a 2-element in a 4-space expressed in terms of independent vector symbols `x`, `y`, `z`, `w`, and scalars `a` and `b`.

```
★𝓑4; A = 2 (x∧y) + b (y∧z) - 5 (x∧w) + 7 a (y∧w) - 4 (z∧w);
ExteriorFactorize[A]
```

$$\text{If}\left[\{8 + 5\,b == 0\},\ 5\left(w - \frac{2\,y}{5}\right) \wedge \left(x - \frac{7\,a\,y}{5} + \frac{4\,z}{5}\right)\right]$$

Determining if an element is simple

To determine if an element is simple, we can use the *GrassmannAlgebra* function `SimpleQ`. If an element is simple, `SimpleQ` returns `True`. If it is not simple, it returns `False`. If it may be simple conditional on the values taken by some of its scalar coefficients, `SimpleQ` returns the conditions required.

We take some of the examples discussed in the previous section on factorization.

◆ For simple *m*-elements

◇ *A 4-element in a 5-space*

```
★𝓑5; A = ComposeMElement[4, a]
```

$a_1\, e_1 \wedge e_2 \wedge e_3 \wedge e_4 + a_2\, e_1 \wedge e_2 \wedge e_3 \wedge e_5 +$
$a_3\, e_1 \wedge e_2 \wedge e_4 \wedge e_5 + a_4\, e_1 \wedge e_3 \wedge e_4 \wedge e_5 + a_5\, e_2 \wedge e_3 \wedge e_4 \wedge e_5$

```
SimpleQ[A]
```

```
True
```

◆ For non-simple *m*-elements

If $x \wedge y \wedge u \wedge v \neq 0$, then `SimpleQ[x ∧ y + u ∧ v]` returns `False` as expected.

```
SimpleQ[x ∧ y + u ∧ v]
```

```
False
```

◆ For contingently simple *m*-elements

If `SimpleQ` is applied to an element which may be simple *conditional* on the values taken by some of its scalar coefficients, the conditional result will be returned.

◇ *A 2-element in a 4-space*

```
★𝓑4; A = ComposeMElement[2, a]
```

$a_1\, e_1 \wedge e_2 + a_2\, e_1 \wedge e_3 + a_3\, e_1 \wedge e_4 + a_4\, e_2 \wedge e_3 + a_5\, e_2 \wedge e_4 + a_6\, e_3 \wedge e_4$

```
SimpleQ[A]
```

`If[{`$a_3\, a_4 - a_2\, a_5 + a_1\, a_6 == 0$`}, True]`

If we are able to assert the condition required, that $a_3\, a_4 - a_2\, a_5 + a_1\, a_6 == 0$, then the 2-element is indeed simple.

$$\text{SimpleQ}\left[A\ /.\ a_6 \to -\frac{a_3\, a_4 - a_2\, a_5}{a_1}\right]$$

```
True
```

◇ *A 2-element in a 5-space*

```
★𝓑5; A = ComposeMElement[2, a]
```

$a_1\, e_1 \wedge e_2 + a_2\, e_1 \wedge e_3 + a_3\, e_1 \wedge e_4 + a_4\, e_1 \wedge e_5 + a_5\, e_2 \wedge e_3 +$
$a_6\, e_2 \wedge e_4 + a_7\, e_2 \wedge e_5 + a_8\, e_3 \wedge e_4 + a_9\, e_3 \wedge e_5 + a_{10}\, e_4 \wedge e_5$

```
SimpleQ[A]
```

```
If[{a3 a5 - a2 a6 + a1 a8 == 0,
   a4 a5 - a2 a7 + a1 a9 == 0, a4 a6 - a3 a7 + a1 a10 == 0}, True]
```

In this case of a 2-element in a 5-space, we have three conditions on the coefficients to satisfy before being able to assert that the element is simple.

◆ **For general *m*-elements**

`SimpleQ` will also check *m*-elements expressed in terms of (independent) vector symbols. The grade *m* must be equal to, or less than, the dimension of the declared space.

```
★𝓑10; A = 2 (x ∧ y) + b (y ∧ z) - 5 (x ∧ w) + 7 a (y ∧ w) - 4 (z ∧ w);
```

```
SimpleQ[A]
```

```
If[{8 + 5 b == 0}, True]
```

3.10 Product Formulae for Regressive Products

The Product Formula

The Common Factor Theorem forms the basis for many of the important formulae relating the various products of the Grassmann algebra. In this section, some of the more basic formulae which involve just the exterior and regressive products will be developed. These in turn will be shown in Chapter 6: The Interior Product to have their counterparts in terms of exterior and interior products. The formulae are usually directed at obtaining alternative expressions (or expansions) for an element, or for products of elements.

The first formula to be developed (which forms a basis for much of the rest) is an expansion for the regressive product of an (*n*–1)-element with the exterior product of two arbitrary elements. We call this (and its dual) *the* (basic) Product Formula.

Let $\underset{n-1}{x}$ be an (*n*–1)-element, then we wish to show that

$$\left(\underset{m}{\alpha} \wedge \underset{k}{\beta}\right) \vee \underset{n-1}{x} == \left(\underset{m}{\alpha} \vee \underset{n-1}{x}\right) \wedge \underset{k}{\beta} + (-1)^{m} \underset{m}{\alpha} \wedge \left(\underset{k}{\beta} \vee \underset{n-1}{x}\right) \qquad 3.68$$

To prove [3.68], suppose initially that $\underset{m}{\alpha}$ and $\underset{k}{\beta}$ are simple and can be expressed as:

$$\underset{m}{\alpha} == \alpha_1 \wedge \ldots \wedge \alpha_m \qquad \underset{k}{\beta} == \beta_1 \wedge \ldots \wedge \beta_k$$

Applying the Common Factor Theorem gives

$$\left(\underset{m}{\alpha}\wedge\underset{k}{\beta}\right)\vee\underset{n-1}{\mathbf{x}} == ((\alpha_1\wedge\ldots\wedge\alpha_m)\wedge(\beta_1\wedge\ldots\wedge\beta_k))\vee\underset{n-1}{\mathbf{x}} ==$$

$$\sum_{i=1}^{m}(-1)^{i-1}\left(\alpha_i\wedge\underset{n-1}{\mathbf{x}}\right)\vee((\alpha_1\wedge\ldots\wedge\square_i\wedge\ldots\wedge\alpha_m)\wedge(\beta_1\wedge\ldots\wedge\beta_k))$$

$$+\sum_{j=1}^{k}(-1)^{m+j-1}\left(\beta_j\wedge\underset{n-1}{\mathbf{x}}\right)\vee\left((\alpha_1\wedge\ldots\wedge\alpha_m)\wedge\left(\beta_1\wedge\ldots\wedge\square_j\wedge\ldots\wedge\beta_k\right)\right)$$

Since the $\alpha_i\wedge\underset{n-1}{x}$ and $\beta_j\wedge\underset{n-1}{x}$ are *n*-elements, we can rearrange the parentheses in the terms of the sums by using the property [3.16] of *n*-elements that they behave with regressive products just like scalars behave with exterior products: $\underset{n}{\omega}\vee\left(\underset{r}{\mu}\wedge\underset{s}{\nu}\right)\rightarrow\left(\underset{n}{\omega}\vee\underset{r}{\mu}\right)\wedge\underset{s}{\nu}$. So the equation becomes

$$\left(\underset{m}{\alpha}\wedge\underset{k}{\beta}\right)\vee\underset{n-1}{\mathbf{x}} == ((\alpha_1\wedge\ldots\wedge\alpha_m)\wedge(\beta_1\wedge\ldots\wedge\beta_k))\vee\underset{n-1}{\mathbf{x}} ==$$

$$\sum_{i=1}^{m}(-1)^{i-1}\left(\left(\alpha_i\wedge\underset{n-1}{\mathbf{x}}\right)\vee(\alpha_1\wedge\ldots\wedge\square_i\wedge\ldots\wedge\alpha_m)\right)\wedge(\beta_1\wedge\ldots\wedge\beta_k)$$

$$+\sum_{j=1}^{k}(-1)^{m+j-1}(-1)^{m(k-1)}\left(\left(\beta_j\wedge\underset{n-1}{\mathbf{x}}\right)\vee(\beta_1\wedge\ldots\wedge\square_j\wedge\ldots\wedge\beta_k)\right)\wedge(\alpha_1\wedge\ldots\wedge\alpha_m)$$

Reapplying the Common Factor Theorem in reverse enables us to condense the sums back to:

$$\left(\underset{m}{\alpha}\vee\underset{n-1}{\mathbf{x}}\right)\wedge(\beta_1\wedge\ldots\wedge\beta_k)+(-1)^{mk}\left(\underset{k}{\beta}\vee\underset{n-1}{\mathbf{x}}\right)\wedge(\alpha_1\wedge\ldots\wedge\alpha_m)$$

A reordering of the second term gives the final form.

$$\left(\underset{m}{\alpha}\vee\underset{n-1}{\mathbf{x}}\right)\wedge(\beta_1\wedge\ldots\wedge\beta_k)+(-1)^{m}(\alpha_1\wedge\ldots\wedge\alpha_m)\wedge\left(\underset{k}{\beta}\vee\underset{n-1}{\mathbf{x}}\right)$$

This result may be extended in a straightforward manner to the case where $\underset{m}{\alpha}$ and $\underset{k}{\beta}$ are not simple: since a non-simple element may be expressed as the sum of simple terms, and the formula is valid for each term, then by addition it can be shown to be valid for the sum.

We can calculate the dual of this formula by applying the *GrassmannAlgebra* function `Dual`. (Note that we use `★n` instead of `n` in order to let the `Dual` function know of its special meaning as the dimension of the space.)

$$\mathbf{Dual}\left[\left(\underset{m}{\alpha}\wedge\underset{k}{\beta}\right)\vee\underset{\star n-1}{\mathbf{x}} == \left(\underset{m}{\alpha}\vee\underset{\star n-1}{\mathbf{x}}\right)\wedge\underset{k}{\beta}+(-1)^{m}\underset{m}{\alpha}\wedge\left(\underset{k}{\beta}\vee\underset{\star n-1}{\mathbf{x}}\right)\right]$$

$$\left(\underset{m}{\alpha}\vee\underset{k}{\beta}\right)\wedge\mathbf{x} == (-1)^{\star n-m}\underset{m}{\alpha}\vee\left(\underset{k}{\beta}\wedge\mathbf{x}\right)+\left(\underset{m}{\alpha}\wedge\mathbf{x}\right)\vee\underset{k}{\beta}$$

$$\left(\underset{m}{\alpha}\vee\underset{k}{\beta}\right)\wedge\mathbf{x} == \left(\underset{m}{\alpha}\wedge\mathbf{x}\right)\vee\underset{k}{\beta}+(-1)^{n-m}\underset{m}{\alpha}\vee\left(\underset{k}{\beta}\wedge\mathbf{x}\right) \qquad 3.69$$

Note that here x is a 1-element.

Deriving Product Formulae

If x is of a grade higher than 1, then similar relations hold, but with extra terms on the right-hand side. For example, if we replace x by $x_1 \wedge x_2$ and note that:

$$\left(\underset{m}{\alpha} \vee \underset{k}{\beta}\right) \wedge (x_1 \wedge x_2) == \left(\left(\underset{m}{\alpha} \vee \underset{k}{\beta}\right) \wedge x_1\right) \wedge x_2$$

then the right-hand side may be expanded by applying the Product Formula [3.69] twice to obtain:

$$\left(\underset{m}{\alpha} \vee \underset{k}{\beta}\right) \wedge (x_1 \wedge x_2) == \left(\underset{m}{\alpha} \wedge (x_1 \wedge x_2)\right) \vee \underset{k}{\beta} + \underset{m}{\alpha} \vee \left(\underset{k}{\beta} \wedge (x_1 \wedge x_2)\right) + (-1)^{n-m} \left(\left(\underset{m}{\alpha} \wedge x_2\right) \vee \left(\underset{k}{\beta} \wedge x_1\right) - \left(\underset{m}{\alpha} \wedge x_1\right) \vee \left(\underset{k}{\beta} \wedge x_2\right)\right) \qquad 3.70$$

Each successive application doubles the number of terms. We started with two terms on the right hand side of the basic Product Formula [3.69]. By applying it again we obtain a Product Formula with four terms on the right hand side. The next application would give us eight terms as shown in the Product Formula below.

$$\left(\underset{m}{\alpha} \vee \underset{k}{\beta}\right) \wedge (x_1 \wedge x_2 \wedge x_3) == \left(\underset{m}{\alpha} \wedge x_1\right) \vee \left(\underset{k}{\beta} \wedge x_2 \wedge x_3\right) - \left(\underset{m}{\alpha} \wedge x_2\right) \vee \left(\underset{k}{\beta} \wedge x_1 \wedge x_3\right) + \left(\underset{m}{\alpha} \wedge x_3\right) \vee \left(\underset{k}{\beta} \wedge x_1 \wedge x_2\right) + \left(\underset{m}{\alpha} \wedge x_1 \wedge x_2 \wedge x_3\right) \vee \underset{k}{\beta} + (-1)^{n-m} \left(\underset{m}{\alpha} \vee \left(\underset{k}{\beta} \wedge x_1 \wedge x_2 \wedge x_3\right) + \left(\underset{m}{\alpha} \wedge x_1 \wedge x_2\right) \vee \left(\underset{k}{\beta} \wedge x_3\right) - \left(\underset{m}{\alpha} \wedge x_1 \wedge x_3\right) \vee \left(\underset{k}{\beta} \wedge x_2\right) + \left(\underset{m}{\alpha} \wedge x_2 \wedge x_3\right) \vee \left(\underset{k}{\beta} \wedge x_1\right)\right) \qquad 3.71$$

Thus the Product Formula for $\left(\underset{m}{\alpha} \vee \underset{k}{\beta}\right) \wedge \left(x_1 \wedge x_2 \wedge \ldots \wedge x_j\right)$ would lead to 2^j terms.

Because this derivation process is simply the repeated application of a formula, we can get *Mathematica* to do it for us automatically.

Deriving Product Formulae automatically

By doing the previous derivations by hand, we identify three basic steps.

1. Devise the rule to be applied at each step
2. Apply the rule as many times as required
3. Simplify the signs of the terms.

We discuss the coding of each of these steps in turn. The code is explanatory only. The final function `DeriveProductFormula` is built in to *GrassmannAlgebra* and is protected, so entering it will not in fact work.

◇ 1. Devise the rule to be applied at each step

```
DeriveProductFormulaOnce[F_, x_] : GrassmannExpand[F ∧ x] /.
  ((a_. * (u_ ∨ v_)) ∧ z_ :→
    a * ((u ∧ z) ∨ v) + (-1) ^ (★n - RawGrade[u]) * a * (u ∨ (v ∧ z)));
```

`DeriveProductFormulaOnce` takes an existing Product Formula `F`, together with a new 1-element `x`, and expands out their exterior product using `GrassmannExpand`. Each of the resulting terms will be of the form shown by the pattern `(a_.*(u_∨v_))∧z_`, where `u_`, `v_`, and `z_` *may* be exterior products, and `a_` *may* be a scalar coefficient.

Each term is now transformed into two terms with the forms `a*((u∧z)∨v)` and `(-1)^(★n-RawGrade[u])*a*(u∨(v∧z)))`.

Here `★n` is the (symbolic) dimension of the underlying linear space, and the function `RawGrade` computes the grade of an element without consideration given to the dimension of the currently declared space. Both these features enable the formula derivation to apply to an element of any grade.

◇ 2. Apply the rule as many times as required

```
DeriveProductFormula[(A_ ∨ B_) ∧ X_] := (A ∨ B) ∧ ComposeSimpleForm[X, 1] ==
   DPFSimplify[Fold[DeriveProductFormulaOnce, A ∨ B,
     {ComposeSimpleForm[X, 1]} /. Wedge → Sequence]];
```

`DeriveProductFormula` takes the left hand side of the original Product Formula `(A∨B)∧X`, and uses *Mathematica*'s `Fold` function to repeatedly apply `DeriveProductFormulaOnce` to the previous result, each time including a new 1-element factor of `X` (or the form resulting from applying `ComposeSimpleForm` to `X`).

◇ 3. Simplify the signs of the terms.

`DPFSimplify` is a set of rules which simplify any products of the form $(-1)^{a+b\,\mathrm{m}+c\,\star\mathrm{n}}$ which arise in the derivation. Here, `a`, `b`, and `c` may be odd or even integers. Symbol `m` may be symbolic. Symbol `★n` is symbolic.

◆ Examples: Testing the code

We can test out the code by seeing if we get the same results as the formulae derived by hand above. It is convenient to use *GrassmannAlgebra's* facility to show the precedences of the various product combinations with extra parentheses. This can be turned on by entering `ShowPrecedence` and turned off by entering `HidePrecedence`. It can also be facilitated by the check-box on the Palette.

$$\star\mathbf{A};\ \mathbf{DeriveProductFormula}\left[\left(\underset{m}{\alpha}\vee\underset{k}{\beta}\right)\wedge\mathbf{x}\right]$$

$$\left(\underset{m}{\alpha}\vee\underset{k}{\beta}\right)\wedge\mathbf{x} == (-1)^{\star n-m}\,\underset{m}{\alpha}\vee\left(\underset{k}{\beta}\wedge\mathbf{x}\right)+\left(\underset{m}{\alpha}\wedge\mathbf{x}\right)\vee\underset{k}{\beta}$$

DeriveProductFormula$\left[\left(\underset{m}{\alpha} \vee \underset{k}{\beta}\right) \wedge \underset{2}{x}\right]$

$$\left(\underset{m}{\alpha} \vee \underset{k}{\beta}\right) \wedge x_1 \wedge x_2 == \underset{m}{\alpha} \vee \left(\underset{k}{\beta} \wedge x_1 \wedge x_2\right) + (-1)^{\star n-m}\left(-\left(\left(\underset{m}{\alpha} \wedge x_1\right) \vee \left(\underset{k}{\beta} \wedge x_2\right)\right) + \left(\underset{m}{\alpha} \wedge x_2\right) \vee \left(\underset{k}{\beta} \wedge x_1\right)\right) + \left(\underset{m}{\alpha} \wedge x_1 \wedge x_2\right) \vee \underset{k}{\beta}$$

DeriveProductFormula$\left[\left(\underset{m}{\alpha} \vee \underset{k}{\beta}\right) \wedge \underset{3}{x}\right]$

$$\left(\underset{m}{\alpha} \vee \underset{k}{\beta}\right) \wedge x_1 \wedge x_2 \wedge x_3 == \left(\underset{m}{\alpha} \wedge x_1\right) \vee \left(\underset{k}{\beta} \wedge x_2 \wedge x_3\right) - \left(\underset{m}{\alpha} \wedge x_2\right) \vee \left(\underset{k}{\beta} \wedge x_1 \wedge x_3\right) + \left(\underset{m}{\alpha} \wedge x_3\right) \vee \left(\underset{k}{\beta} \wedge x_1 \wedge x_2\right) + (-1)^{\star n-m}\left(\underset{m}{\alpha} \vee \left(\underset{k}{\beta} \wedge x_1 \wedge x_2 \wedge x_3\right) + \left(\underset{m}{\alpha} \wedge x_1 \wedge x_2\right) \vee \left(\underset{k}{\beta} \wedge x_3\right) - \left(\underset{m}{\alpha} \wedge x_1 \wedge x_3\right) \vee \left(\underset{k}{\beta} \wedge x_2\right) + \left(\underset{m}{\alpha} \wedge x_2 \wedge x_3\right) \vee \left(\underset{k}{\beta} \wedge x_1\right)\right) + \left(\underset{m}{\alpha} \wedge x_1 \wedge x_2 \wedge x_3\right) \vee \underset{k}{\beta}$$

DeriveProductFormula$\left[\left(\underset{m}{\alpha} \vee \underset{k}{\beta}\right) \wedge \underset{4}{x}\right]$

$$\left(\underset{m}{\alpha} \vee \underset{k}{\beta}\right) \wedge x_1 \wedge x_2 \wedge x_3 \wedge x_4 == \underset{m}{\alpha} \vee \left(\underset{k}{\beta} \wedge x_1 \wedge x_2 \wedge x_3 \wedge x_4\right) + \left(\underset{m}{\alpha} \wedge x_1 \wedge x_2\right) \vee \left(\underset{k}{\beta} \wedge x_3 \wedge x_4\right) - \left(\underset{m}{\alpha} \wedge x_1 \wedge x_3\right) \vee \left(\underset{k}{\beta} \wedge x_2 \wedge x_4\right) + \left(\underset{m}{\alpha} \wedge x_1 \wedge x_4\right) \vee \left(\underset{k}{\beta} \wedge x_2 \wedge x_3\right) + \left(\underset{m}{\alpha} \wedge x_2 \wedge x_3\right) \vee \left(\underset{k}{\beta} \wedge x_1 \wedge x_4\right) - \left(\underset{m}{\alpha} \wedge x_2 \wedge x_4\right) \vee \left(\underset{k}{\beta} \wedge x_1 \wedge x_3\right) + \left(\underset{m}{\alpha} \wedge x_3 \wedge x_4\right) \vee \left(\underset{k}{\beta} \wedge x_1 \wedge x_2\right) + (-1)^{\star n-m}\left(-\left(\left(\underset{m}{\alpha} \wedge x_1\right) \vee \left(\underset{k}{\beta} \wedge x_2 \wedge x_3 \wedge x_4\right)\right) + \left(\underset{m}{\alpha} \wedge x_2\right) \vee \left(\underset{k}{\beta} \wedge x_1 \wedge x_3 \wedge x_4\right) - \left(\underset{m}{\alpha} \wedge x_3\right) \vee \left(\underset{k}{\beta} \wedge x_1 \wedge x_2 \wedge x_4\right) + \left(\underset{m}{\alpha} \wedge x_4\right) \vee \left(\underset{k}{\beta} \wedge x_1 \wedge x_2 \wedge x_3\right) - \left(\underset{m}{\alpha} \wedge x_1 \wedge x_2 \wedge x_3\right) \vee \left(\underset{k}{\beta} \wedge x_4\right) + \left(\underset{m}{\alpha} \wedge x_1 \wedge x_2 \wedge x_4\right) \vee \left(\underset{k}{\beta} \wedge x_3\right) - \left(\underset{m}{\alpha} \wedge x_1 \wedge x_3 \wedge x_4\right) \vee \left(\underset{k}{\beta} \wedge x_2\right) + \left(\underset{m}{\alpha} \wedge x_2 \wedge x_3 \wedge x_4\right) \vee \left(\underset{k}{\beta} \wedge x_1\right)\right) + \left(\underset{m}{\alpha} \wedge x_1 \wedge x_2 \wedge x_3 \wedge x_4\right) \vee \underset{k}{\beta}$$

Computing the General Product Formula

If we look carefully at the form of the product formulae for different grades j of $\underset{j}{x}$ we can see that each of them is a specific case of a more general explicit formula. We will call this more general explicit formula the General Product Formula. We derive it below by deconstructing the result for $\underset{4}{x}$, and then confirm its identity to the original iteratively derived form in the first few cases.

$$\mathbf{F_1} = \mathbf{DeriveProductFormula}\left[\left(\underset{m}{\alpha} \vee \underset{k}{\beta}\right) \wedge \underset{4}{\mathbf{x}}\right]$$

$$\left(\underset{m}{\alpha} \vee \underset{k}{\beta}\right) \wedge \mathbf{x_1} \wedge \mathbf{x_2} \wedge \mathbf{x_3} \wedge \mathbf{x_4} == \underset{m}{\alpha} \vee \left(\underset{k}{\beta} \wedge \mathbf{x_1} \wedge \mathbf{x_2} \wedge \mathbf{x_3} \wedge \mathbf{x_4}\right) +$$
$$\left(\underset{m}{\alpha} \wedge \mathbf{x_1} \wedge \mathbf{x_2}\right) \vee \left(\underset{k}{\beta} \wedge \mathbf{x_3} \wedge \mathbf{x_4}\right) - \left(\underset{m}{\alpha} \wedge \mathbf{x_1} \wedge \mathbf{x_3}\right) \vee \left(\underset{k}{\beta} \wedge \mathbf{x_2} \wedge \mathbf{x_4}\right) +$$
$$\left(\underset{m}{\alpha} \wedge \mathbf{x_1} \wedge \mathbf{x_4}\right) \vee \left(\underset{k}{\beta} \wedge \mathbf{x_2} \wedge \mathbf{x_3}\right) + \left(\underset{m}{\alpha} \wedge \mathbf{x_2} \wedge \mathbf{x_3}\right) \vee \left(\underset{k}{\beta} \wedge \mathbf{x_1} \wedge \mathbf{x_4}\right) -$$
$$\left(\underset{m}{\alpha} \wedge \mathbf{x_2} \wedge \mathbf{x_4}\right) \vee \left(\underset{k}{\beta} \wedge \mathbf{x_1} \wedge \mathbf{x_3}\right) + \left(\underset{m}{\alpha} \wedge \mathbf{x_3} \wedge \mathbf{x_4}\right) \vee \left(\underset{k}{\beta} \wedge \mathbf{x_1} \wedge \mathbf{x_2}\right) +$$
$$(-1)^{\star n-m} \left(-\left(\left(\underset{m}{\alpha} \wedge \mathbf{x_1}\right) \vee \left(\underset{k}{\beta} \wedge \mathbf{x_2} \wedge \mathbf{x_3} \wedge \mathbf{x_4}\right)\right) + \left(\underset{m}{\alpha} \wedge \mathbf{x_2}\right) \vee \left(\underset{k}{\beta} \wedge \mathbf{x_1} \wedge \mathbf{x_3} \wedge \mathbf{x_4}\right) -\right.$$
$$\left(\underset{m}{\alpha} \wedge \mathbf{x_3}\right) \vee \left(\underset{k}{\beta} \wedge \mathbf{x_1} \wedge \mathbf{x_2} \wedge \mathbf{x_4}\right) + \left(\underset{m}{\alpha} \wedge \mathbf{x_4}\right) \vee \left(\underset{k}{\beta} \wedge \mathbf{x_1} \wedge \mathbf{x_2} \wedge \mathbf{x_3}\right) -$$
$$\left(\underset{m}{\alpha} \wedge \mathbf{x_1} \wedge \mathbf{x_2} \wedge \mathbf{x_3}\right) \vee \left(\underset{k}{\beta} \wedge \mathbf{x_4}\right) + \left(\underset{m}{\alpha} \wedge \mathbf{x_1} \wedge \mathbf{x_2} \wedge \mathbf{x_4}\right) \vee \left(\underset{k}{\beta} \wedge \mathbf{x_3}\right) - \left(\underset{m}{\alpha} \wedge \mathbf{x_1} \wedge \mathbf{x_3} \wedge \mathbf{x_4}\right) \vee$$
$$\left.\left(\underset{k}{\beta} \wedge \mathbf{x_2}\right) + \left(\underset{m}{\alpha} \wedge \mathbf{x_2} \wedge \mathbf{x_3} \wedge \mathbf{x_4}\right) \vee \left(\underset{k}{\beta} \wedge \mathbf{x_1}\right)\right) + \left(\underset{m}{\alpha} \wedge \mathbf{x_1} \wedge \mathbf{x_2} \wedge \mathbf{x_3} \wedge \mathbf{x_4}\right) \vee \underset{k}{\beta}$$

The first thing we note is that apart possibly from the sign $(-1)^{\star n-m}$, the terms on the right hand side are all products of the form

$$\left(\underset{m}{\alpha} \wedge \underset{4-r}{\mathbf{x_i}}\right) \vee \left(\underset{k}{\beta} \wedge \underset{r}{\mathbf{x_i}}\right) \qquad \underset{j}{\mathbf{x}} == \underset{r}{\mathbf{x_i}} \wedge \underset{4-r}{\mathbf{x_i}}$$

This means we can compute them using the *r*-span and *r*-cospan of $\underset{j}{x}$. For *j* equal to 4, *r* will range from 0 to 4. Thus we have

$$\left(\underset{m}{\alpha} \wedge \star\mathbb{S}^0\left[\underset{4}{\mathbf{x}}\right]\right) \vee \left(\underset{k}{\beta} \wedge \star\mathbb{S}_0\left[\underset{4}{\mathbf{x}}\right]\right)$$

$$\left\{\left(\underset{m}{\alpha} \wedge \mathbf{x_1} \wedge \mathbf{x_2} \wedge \mathbf{x_3} \wedge \mathbf{x_4}\right) \vee \left(\underset{k}{\beta} \wedge 1\right)\right\}$$

$$\left(\underset{m}{\alpha} \wedge \star\mathbb{S}^1\left[\underset{4}{\mathbf{x}}\right]\right) \vee \left(\underset{k}{\beta} \wedge \star\mathbb{S}_1\left[\underset{4}{\mathbf{x}}\right]\right)$$

$$\left\{\left(\underset{m}{\alpha} \wedge \mathbf{x_2} \wedge \mathbf{x_3} \wedge \mathbf{x_4}\right) \vee \left(\underset{k}{\beta} \wedge \mathbf{x_1}\right),\ \left(\underset{m}{\alpha} \wedge \left(-(\mathbf{x_1} \wedge \mathbf{x_3} \wedge \mathbf{x_4})\right)\right) \vee \left(\underset{k}{\beta} \wedge \mathbf{x_2}\right),\right.$$
$$\left.\left(\underset{m}{\alpha} \wedge \mathbf{x_1} \wedge \mathbf{x_2} \wedge \mathbf{x_4}\right) \vee \left(\underset{k}{\beta} \wedge \mathbf{x_3}\right),\ \left(\underset{m}{\alpha} \wedge \left(-(\mathbf{x_1} \wedge \mathbf{x_2} \wedge \mathbf{x_3})\right)\right) \vee \left(\underset{k}{\beta} \wedge \mathbf{x_4}\right)\right\}$$

$$\left(\underset{m}{\alpha} \wedge \star\mathbb{S}^2\left[\underset{4}{\mathbf{x}}\right]\right) \vee \left(\underset{k}{\beta} \wedge \star\mathbb{S}_2\left[\underset{4}{\mathbf{x}}\right]\right)$$

$$\left\{\left(\underset{m}{\alpha} \wedge \mathbf{x_3} \wedge \mathbf{x_4}\right) \vee \left(\underset{k}{\beta} \wedge \mathbf{x_1} \wedge \mathbf{x_2}\right),\ \left(\underset{m}{\alpha} \wedge \left(-(\mathbf{x_2} \wedge \mathbf{x_4})\right)\right) \vee \left(\underset{k}{\beta} \wedge \mathbf{x_1} \wedge \mathbf{x_3}\right),\right.$$
$$\left(\underset{m}{\alpha} \wedge \mathbf{x_2} \wedge \mathbf{x_3}\right) \vee \left(\underset{k}{\beta} \wedge \mathbf{x_1} \wedge \mathbf{x_4}\right),\ \left(\underset{m}{\alpha} \wedge \mathbf{x_1} \wedge \mathbf{x_4}\right) \vee \left(\underset{k}{\beta} \wedge \mathbf{x_2} \wedge \mathbf{x_3}\right),$$
$$\left.\left(\underset{m}{\alpha} \wedge \left(-(\mathbf{x_1} \wedge \mathbf{x_3})\right)\right) \vee \left(\underset{k}{\beta} \wedge \mathbf{x_2} \wedge \mathbf{x_4}\right),\ \left(\underset{m}{\alpha} \wedge \mathbf{x_1} \wedge \mathbf{x_2}\right) \vee \left(\underset{k}{\beta} \wedge \mathbf{x_3} \wedge \mathbf{x_4}\right)\right\}$$

$$\left(\underset{m}{\alpha} \wedge \star\mathbb{S}^3\left[\underset{4}{\mathbf{x}}\right]\right) \vee \left(\underset{k}{\beta} \wedge \star\mathbb{S}_3\left[\underset{4}{\mathbf{x}}\right]\right)$$

$$\{(\underset{m}{\alpha} \wedge x_4) \vee (\underset{k}{\beta} \wedge x_1 \wedge x_2 \wedge x_3),\ (\underset{m}{\alpha} \wedge (-x_3)) \vee (\underset{k}{\beta} \wedge x_1 \wedge x_2 \wedge x_4),$$
$$(\underset{m}{\alpha} \wedge x_2) \vee (\underset{k}{\beta} \wedge x_1 \wedge x_3 \wedge x_4),\ (\underset{m}{\alpha} \wedge (-x_1)) \vee (\underset{k}{\beta} \wedge x_2 \wedge x_3 \wedge x_4)\}$$

$$\left(\underset{m}{\alpha} \wedge \star\mathbb{S}^4\left[\underset{4}{\mathbf{x}}\right]\right) \vee \left(\underset{k}{\beta} \wedge \star\mathbb{S}_4\left[\underset{4}{\mathbf{x}}\right]\right)$$

$$\{(\underset{m}{\alpha} \wedge 1) \vee (\underset{k}{\beta} \wedge x_1 \wedge x_2 \wedge x_3 \wedge x_4)\}$$

But all these terms can be computed at once using the complete span $\star\mathbb{S}_\star$ and cospan $\star\mathbb{S}^\star$.

$$\left(\underset{m}{\alpha} \wedge \star\mathbb{S}^\star\left[\underset{4}{\mathbf{x}}\right]\right) \vee \left(\underset{k}{\beta} \wedge \star\mathbb{S}_\star\left[\underset{4}{\mathbf{x}}\right]\right)$$

$$\{\{(\underset{m}{\alpha} \wedge x_1 \wedge x_2 \wedge x_3 \wedge x_4) \vee (\underset{k}{\beta} \wedge 1)\},$$
$$\{(\underset{m}{\alpha} \wedge x_2 \wedge x_3 \wedge x_4) \vee (\underset{k}{\beta} \wedge x_1),\ (\underset{m}{\alpha} \wedge (-(x_1 \wedge x_3 \wedge x_4))) \vee (\underset{k}{\beta} \wedge x_2),$$
$$(\underset{m}{\alpha} \wedge x_1 \wedge x_2 \wedge x_4) \vee (\underset{k}{\beta} \wedge x_3),\ (\underset{m}{\alpha} \wedge (-(x_1 \wedge x_2 \wedge x_3))) \vee (\underset{k}{\beta} \wedge x_4)\},$$
$$\{(\underset{m}{\alpha} \wedge x_3 \wedge x_4) \vee (\underset{k}{\beta} \wedge x_1 \wedge x_2),\ (\underset{m}{\alpha} \wedge (-(x_2 \wedge x_4))) \vee (\underset{k}{\beta} \wedge x_1 \wedge x_3),$$
$$(\underset{m}{\alpha} \wedge x_2 \wedge x_3) \vee (\underset{k}{\beta} \wedge x_1 \wedge x_4),\ (\underset{m}{\alpha} \wedge x_1 \wedge x_4) \vee (\underset{k}{\beta} \wedge x_2 \wedge x_3),$$
$$(\underset{m}{\alpha} \wedge (-(x_1 \wedge x_3))) \vee (\underset{k}{\beta} \wedge x_2 \wedge x_4),\ (\underset{m}{\alpha} \wedge x_1 \wedge x_2) \vee (\underset{k}{\beta} \wedge x_3 \wedge x_4)\},$$
$$\{(\underset{m}{\alpha} \wedge x_4) \vee (\underset{k}{\beta} \wedge x_1 \wedge x_2 \wedge x_3),\ (\underset{m}{\alpha} \wedge (-x_3)) \vee (\underset{k}{\beta} \wedge x_1 \wedge x_2 \wedge x_4),$$
$$(\underset{m}{\alpha} \wedge x_2) \vee (\underset{k}{\beta} \wedge x_1 \wedge x_3 \wedge x_4),\ (\underset{m}{\alpha} \wedge (-x_1)) \vee (\underset{k}{\beta} \wedge x_2 \wedge x_3 \wedge x_4)\},$$
$$\{(\underset{m}{\alpha} \wedge 1) \vee (\underset{k}{\beta} \wedge x_1 \wedge x_2 \wedge x_3 \wedge x_4)\}\}$$

Now we need to multiply the terms by a scalar factor which is 1 when the grade of the span is even, and $(-1)^{\star n-m}$ when it is odd. Since the lists of terms in the collection above are arranged according to the grades of their spans, we then need a list whose elements alternate between 1 and $(-1)^{\star n-m}$. To construct this list, we first define a function `★r` which returns an alternating list of `1`s and `0`s of the correct length $j+1$, multiply this list by `★n - m`, and then use it as a power. *Mathematica's* inbuilt `Listable` attributes of `Times` and `Power` will again return the result we want.

```
★r[j_] := Mod[Range[0, j], 2]
```

$$(-1)^{\star r[4]\,(\star n-m)}$$

$$\{1,\ (-1)^{-m+\star n},\ 1,\ (-1)^{-m+\star n},\ 1\}$$

Multiplying by this list of scalar factors now gives us the terms we are looking for.

$$(-1)^{\star r[4]\,(\star n-m)} \left(\underset{m}{\alpha} \wedge \star S^{\star}\left[\underset{4}{x}\right]\right) \vee \left(\underset{k}{\beta} \wedge \star S_{\star}\left[\underset{4}{x}\right]\right)$$

$$\{\{(\underset{m}{\alpha} \wedge x_1 \wedge x_2 \wedge x_3 \wedge x_4) \vee (\underset{k}{\beta} \wedge 1)\},$$
$$\{(-1)^{\star n-m} (\underset{m}{\alpha} \wedge x_2 \wedge x_3 \wedge x_4) \vee (\underset{k}{\beta} \wedge x_1),\ (-1)^{\star n-m} (\underset{m}{\alpha} \wedge (-(x_1 \wedge x_3 \wedge x_4))) \vee (\underset{k}{\beta} \wedge x_2),\ (-1)^{\star n-m} (\underset{m}{\alpha} \wedge x_1 \wedge x_2 \wedge x_4) \vee (\underset{k}{\beta} \wedge x_3),$$
$$(-1)^{\star n-m} (\underset{m}{\alpha} \wedge (-(x_1 \wedge x_2 \wedge x_3))) \vee (\underset{k}{\beta} \wedge x_4)\},$$
$$\{(\underset{m}{\alpha} \wedge x_3 \wedge x_4) \vee (\underset{k}{\beta} \wedge x_1 \wedge x_2),\ (\underset{m}{\alpha} \wedge (-(x_2 \wedge x_4))) \vee (\underset{k}{\beta} \wedge x_1 \wedge x_3),$$
$$(\underset{m}{\alpha} \wedge x_2 \wedge x_3) \vee (\underset{k}{\beta} \wedge x_1 \wedge x_4),\ (\underset{m}{\alpha} \wedge x_1 \wedge x_4) \vee (\underset{k}{\beta} \wedge x_2 \wedge x_3),$$
$$(\underset{m}{\alpha} \wedge (-(x_1 \wedge x_3))) \vee (\underset{k}{\beta} \wedge x_2 \wedge x_4),\ (\underset{m}{\alpha} \wedge x_1 \wedge x_2) \vee (\underset{k}{\beta} \wedge x_3 \wedge x_4)\},$$
$$\{(-1)^{\star n-m} (\underset{m}{\alpha} \wedge x_4) \vee (\underset{k}{\beta} \wedge x_1 \wedge x_2 \wedge x_3),\ (-1)^{\star n-m} (\underset{m}{\alpha} \wedge (-x_3)) \vee (\underset{k}{\beta} \wedge x_1 \wedge x_2 \wedge x_4),$$
$$(-1)^{\star n-m} (\underset{m}{\alpha} \wedge x_2) \vee (\underset{k}{\beta} \wedge x_1 \wedge x_3 \wedge x_4),\ (-1)^{\star n-m} (\underset{m}{\alpha} \wedge (-x_1)) \vee (\underset{k}{\beta} \wedge x_2 \wedge x_3 \wedge x_4)\},$$
$$\{(\underset{m}{\alpha} \wedge 1) \vee (\underset{k}{\beta} \wedge x_1 \wedge x_2 \wedge x_3 \wedge x_4)\}\}$$

Using $\star\Sigma$ to flatten the list and then sum the terms gives the final result.

$$F_2 = \left(\underset{m}{\alpha} \vee \underset{k}{\beta}\right) \wedge (x_1 \wedge x_2 \wedge x_3 \wedge x_4) ==$$
$$\star\Sigma\left[(-1)^{\star r[4]\,(\star n-m)} \left(\underset{m}{\alpha} \wedge \star S^{\star}\left[\underset{4}{x}\right]\right) \vee \left(\underset{k}{\beta} \wedge \star S_{\star}\left[\underset{4}{x}\right]\right)\right]$$

$$(\underset{m}{\alpha} \vee \underset{k}{\beta}) \wedge x_1 \wedge x_2 \wedge x_3 \wedge x_4 ==$$
$$(\underset{m}{\alpha} \wedge 1) \vee (\underset{k}{\beta} \wedge x_1 \wedge x_2 \wedge x_3 \wedge x_4) + (-1)^{\star n-m} (\underset{m}{\alpha} \wedge (-x_1)) \vee (\underset{k}{\beta} \wedge x_2 \wedge x_3 \wedge x_4) +$$
$$(-1)^{\star n-m} (\underset{m}{\alpha} \wedge x_2) \vee (\underset{k}{\beta} \wedge x_1 \wedge x_3 \wedge x_4) + (-1)^{\star n-m} (\underset{m}{\alpha} \wedge (-x_3)) \vee (\underset{k}{\beta} \wedge x_1 \wedge x_2 \wedge x_4) +$$
$$(-1)^{\star n-m} (\underset{m}{\alpha} \wedge x_4) \vee (\underset{k}{\beta} \wedge x_1 \wedge x_2 \wedge x_3) + (\underset{m}{\alpha} \wedge (-(x_1 \wedge x_3))) \vee (\underset{k}{\beta} \wedge x_2 \wedge x_4) +$$
$$(\underset{m}{\alpha} \wedge (-(x_2 \wedge x_4))) \vee (\underset{k}{\beta} \wedge x_1 \wedge x_3) + (-1)^{\star n-m} (\underset{m}{\alpha} \wedge (-(x_1 \wedge x_2 \wedge x_3))) \vee (\underset{k}{\beta} \wedge x_4) +$$
$$(-1)^{\star n-m} (\underset{m}{\alpha} \wedge (-(x_1 \wedge x_3 \wedge x_4))) \vee (\underset{k}{\beta} \wedge x_2) + (\underset{m}{\alpha} \wedge x_1 \wedge x_2) \vee (\underset{k}{\beta} \wedge x_3 \wedge x_4) +$$
$$(\underset{m}{\alpha} \wedge x_1 \wedge x_4) \vee (\underset{k}{\beta} \wedge x_2 \wedge x_3) + (\underset{m}{\alpha} \wedge x_2 \wedge x_3) \vee (\underset{k}{\beta} \wedge x_1 \wedge x_4) +$$
$$(\underset{m}{\alpha} \wedge x_3 \wedge x_4) \vee (\underset{k}{\beta} \wedge x_1 \wedge x_2) + (-1)^{\star n-m} (\underset{m}{\alpha} \wedge x_1 \wedge x_2 \wedge x_4) \vee (\underset{k}{\beta} \wedge x_3) +$$
$$(-1)^{\star n-m} (\underset{m}{\alpha} \wedge x_2 \wedge x_3 \wedge x_4) \vee (\underset{k}{\beta} \wedge x_1) + (\underset{m}{\alpha} \wedge x_1 \wedge x_2 \wedge x_3 \wedge x_4) \vee (\underset{k}{\beta} \wedge 1)$$

Applying `GrassmannExpandAndSimplify` will simplify the signs and allow us to confirm the result is identical to that produced by `DeriveProductFormula`.

```
★𝒢[F₁] === ★𝒢[F₂]
```

```
True
```

◆ The General Product Formula

To summarize the results above: we can take the Product Formula derived in the previous section and rewrite it in its computational form using the notions of complete span and complete cospan. The computational form relies on the fact that in *GrassmannAlgebra* the exterior and regressive products have an inbuilt `Listable` attribute. (This attribute means that a product of lists is automatically converted to a list of products.)

$$\left(\underset{m}{\alpha} \vee \underset{k}{\beta}\right) \wedge \underset{j}{x} == \star\Sigma\left[(-1)^{\star r[j]\,(\star n-m)} \left(\underset{m}{\alpha} \wedge \star\mathbb{S}^{\star}\left[\underset{j}{x}\right]\right) \vee \left(\underset{k}{\beta} \wedge \star\mathbb{S}_{\star}\left[\underset{j}{x}\right]\right)\right] \qquad 3.72$$

In this formula, $\star\mathbb{S}_{\star}\left[\underset{j}{x}\right]$ and $\star\mathbb{S}^{\star}\left[\underset{j}{x}\right]$ are the complete span and complete cospan of $\underset{j}{x}$.

The sign function $\star r$ just creates a list of elements alternating between 1 and $(-1)^{\star n-m}$. For example, for *j* equal to 8:

$$(-1)^{\star r[8]\,(\star n-m)}$$

$$\left\{1,\ (-1)^{-m+\star n},\ 1,\ (-1)^{-m+\star n},\ 1,\ (-1)^{-m+\star n},\ 1,\ (-1)^{-m+\star n},\ 1\right\}$$

Comparing the two forms of the Product Formula

◆ Encapsulating the computable formula

We can easily encapsulate this computable formula in a function `GeneralProductFormula` for easy comparison with the results from our initial iterative derivation using `DeriveProductFormula`. This is built in to *GrassmannAlgebra* and is protected, so entering it will not work.

```
GeneralProductFormula[(A_ ∨ B_) ∧ X_] := (A ∨ B) ∧ ComposeSimpleForm[X, 1] ==
    ★Σ[(-1)^(★r[★G[X]] (★n-★G[A])) (A ∧ ★S^★[X]) ∨ (B ∧ ★S_★[X])];
★r[j_] := Mod[Range[0, j], 2];
```

◆ Comparing results from `DeriveProductFormula` with `GeneralProductFormula`.

We can compare the results obtained by our two formulations. `DeriveProductFormula` invoked the successive application of a basic formula. `GeneralProductFormula` invoked an explicit formula for the general case. Below we show the two formulations give the same result for all simple elements from grades 1 to 10.

$\star\mathcal{B}_{10}$;

$$\text{Table}\left[\star\mathcal{G}\left[\text{DeriveProductFormula}\left[\left(\underset{m}{\alpha} \vee \underset{k}{\beta}\right) \wedge \underset{j}{x}\right]\right] ===\ \star\mathcal{G}\left[\text{GeneralProductFormula}\left[\left(\underset{m}{\alpha} \vee \underset{k}{\beta}\right) \wedge \underset{j}{x}\right]\right],\ \{j, 1, 10\}\right]$$

{True, True, True, True, True, True, True, True, True, True}

The invariance of the General Product Formula

In Chapter 2 section 2.12, we saw that the m:k-forms based on a simple factorized *m*-element A were, like the *m*-element itself, independent of the factorization. A typical such form is

$$\star\Sigma\left[\left(G_1 \oplus \star\mathbb{S}_k[A]\right) \odot \left(G_2 \otimes \star\mathbb{S}^k[A]\right)\right]$$

As we have seen in its development, the General Product Formula is composed of a number of m:k-forms, and, as expected, shows the same invariance. For example, suppose we generate the formula F_1 for a 3-element A equal to $x \wedge y \wedge z$:

$$F_1 = \text{GeneralProductFormula}\left[\left(\underset{m}{\alpha} \vee \underset{k}{\beta}\right) \wedge (x \wedge y \wedge z)\right]$$

$$\left(\underset{m}{\alpha} \vee \underset{k}{\beta}\right) \wedge x \wedge y \wedge z == \left(\underset{m}{\alpha} \wedge x\right) \vee \left(\underset{k}{\beta} \wedge y \wedge z\right) - \left(\underset{m}{\alpha} \wedge y\right) \vee \left(\underset{k}{\beta} \wedge x \wedge z\right) + \left(\underset{m}{\alpha} \wedge z\right) \vee \left(\underset{k}{\beta} \wedge x \wedge y\right) +$$
$$(-1)^{\star n-m}\left(\underset{m}{\alpha} \vee \left(\underset{k}{\beta} \wedge x \wedge y \wedge z\right) + \left(\underset{m}{\alpha} \wedge x \wedge y\right) \vee \left(\underset{k}{\beta} \wedge z\right) - \left(\underset{m}{\alpha} \wedge x \wedge z\right) \vee \left(\underset{k}{\beta} \wedge y\right) + \left(\underset{m}{\alpha} \wedge y \wedge z\right) \vee \left(\underset{k}{\beta} \wedge x\right)\right) + \left(\underset{m}{\alpha} \wedge x \wedge y \wedge z\right) \vee \underset{k}{\beta}$$

We can express the 3-element as a product of different 1-element factors by adding to any given factor, scalar multiples of the other factors. For example

$$F_2 = \text{GeneralProductFormula}\left[\left(\underset{m}{\alpha} \vee \underset{k}{\beta}\right) \wedge (x \wedge y \wedge (z + a\,x + b\,y))\right]$$

$$\left(\underset{m}{\alpha} \vee \underset{k}{\beta}\right) \wedge x \wedge y \wedge (a\,x + b\,y + z) == \left(\underset{m}{\alpha} \wedge x\right) \vee \left(\underset{k}{\beta} \wedge y \wedge (a\,x + b\,y + z)\right) - \left(\underset{m}{\alpha} \wedge y\right) \vee \left(\underset{k}{\beta} \wedge x \wedge (a\,x + b\,y + z)\right) +$$
$$\left(\underset{m}{\alpha} \wedge (a\,x + b\,y + z)\right) \vee \left(\underset{k}{\beta} \wedge x \wedge y\right) + (-1)^{\star n-m}\left(\underset{m}{\alpha} \vee \left(\underset{k}{\beta} \wedge x \wedge y \wedge (a\,x + b\,y + z)\right) + \right.$$
$$\left(\underset{m}{\alpha} \wedge x \wedge y\right) \vee \left(\underset{k}{\beta} \wedge (a\,x + b\,y + z)\right) - \left(\underset{m}{\alpha} \wedge x \wedge (a\,x + b\,y + z)\right) \vee \left(\underset{k}{\beta} \wedge y\right) +$$
$$\left.\left(\underset{m}{\alpha} \wedge y \wedge (a\,x + b\,y + z)\right) \vee \left(\underset{k}{\beta} \wedge x\right)\right) + \left(\underset{m}{\alpha} \wedge x \wedge y \wedge (a\,x + b\,y + z)\right) \vee \underset{k}{\beta}$$

Applying `GrassmannExpandAndSimplify` to these expressions shows that they are equal.

$$\star\mathcal{G}[F_1 == F_2]$$

True

Alternative forms for the General Product Formula

If we start from the General Product Formula we can rearrange it to interchange the span and cospan, so that the span elements become associated with $\underset{m}{\alpha}$ and the cospan elements become associated with $\underset{k}{\beta}$. Of course, there will need to be some associated changes in the signs of the terms as well. We start with the General Product Formula in the form developed above.

$$\left(\underset{m}{\alpha} \vee \underset{k}{\beta}\right) \wedge \underset{j}{x} == \star\Sigma\left[(-1)^{\star r[j]\,(\star n-m)}\left(\underset{m}{\alpha} \wedge \star\mathbb{S}^{\star}\left[\underset{j}{x}\right]\right) \vee \left(\underset{k}{\beta} \wedge \star\mathbb{S}_{\star}\left[\underset{j}{x}\right]\right)\right]$$

In section 2.12, we have shown that a form such as

$$\left(\underset{m}{\alpha}\wedge\star S^{\star}\left[\underset{j}{x}\right]\right)\vee\left(\underset{k}{\beta}\wedge\star S_{\star}\left[\underset{j}{x}\right]\right)$$

can be rewritten in the form

$$(-1)^{\star rr[j]}\ \star R\left[\left(\underset{m}{\alpha}\wedge\star S_{\star}\left[\underset{j}{x}\right]\right)\vee\left(\underset{k}{\beta}\wedge\star S^{\star}\left[\underset{j}{x}\right]\right)\right]$$

Thus we can write the General Product Formula in the alternative 'B' form

$$\left(\underset{m}{\alpha}\vee\underset{k}{\beta}\right)\wedge\underset{j}{x} == \star\Sigma\left[(-1)^{\star r[j]\ (\star n-m)}\ (-1)^{\star rr[j]}\ \star R\left[\left(\underset{m}{\alpha}\wedge\star S_{\star}\left[\underset{j}{x}\right]\right)\vee\left(\underset{k}{\beta}\wedge\star S^{\star}\left[\underset{j}{x}\right]\right)\right]\right] \qquad 3.73$$

We have discussed the definitions of ★r, ★rr and ★R when we introduced the notion of the complete span of a simple element at the end of section 2.12. We repeat their definition below for reference.

```
★r[j_] := Mod[Range[0, j], 2];
★rr[j_] := If[EvenQ[j], ★r[j], 0];
ReverseAll[s_List] := Reverse[Reverse /@ s]; ★R = ReverseAll;
```

◆ Encapsulating the second computable formula

Just as we encapsulated the first computable formula, we can similarly encapsulate this alternative one. We call the function for generating it `GeneralProductFormulaB`.

```
GeneralProductFormulaB[(A_ ∨ B_) ∧ X_] := (A ∨ B) ∧ ComposeSimpleForm[X, 1] ==
    ★Σ[(-1)^(★r[★G[X]] (★n-★G[A])) (-1)^★rr[★G[X]] ★R[(A ∧ ★S★[X]) ∨ (B ∧ ★S★[X])]];
```

◆ Comparing results from `DerivoProductFormula` with `GeneralProductFormulaB`.

Again comparing the results of `DeriveProductFormula` with this alternative formula verifies their equivalence for values of j from 1 to 10.

```
★𝓑10;
Table[★𝒢[DeriveProductFormula[(α_m ∨ β_k) ∧ x_j]] ===
    ★𝒢[GeneralProductFormulaB[(α_m ∨ β_k) ∧ x_j]], {j, 1, 10}]
```

```
{True, True, True, True, True, True, True, True, True, True}
```

The Decomposition Formula

Recall the General Product Formula [3.72] in its computable form.

$$\left(\underset{m}{\alpha} \vee \underset{k}{\beta}\right) \wedge \underset{j}{x} == \star\Sigma\left[(-1)^{\star r[j]\,(\star n-m)} \left(\underset{m}{\alpha} \wedge \star S^{\star}\left[\underset{j}{x}\right]\right) \vee \left(\underset{k}{\beta} \wedge \star S_{\star}\left[\underset{j}{x}\right]\right)\right]$$

Putting k equal to $\star n-m$ permits the left-hand side to be expressed as a scalar multiple of $\underset{j}{x}$ and hence expresses a type of 'decomposition' of $\underset{j}{x}$ into components.

$$\underset{j}{x} == \star\Sigma\left[(-1)^{\star r[j]\,(\star n-m)} \frac{\left(\underset{m}{\alpha} \wedge \star S^{\star}\left[\underset{j}{x}\right]\right) \vee \left(\underset{\star n-m}{\beta} \wedge \star S_{\star}\left[\underset{j}{x}\right]\right)}{\underset{m}{\alpha} \vee \underset{\star n-m}{\beta}}\right] \qquad 3.74$$

If the element to be decomposed is a 1-element, that is $j = 1$, there are just two terms, reducing the decomposition formula to:

$$x == \star\mathcal{G}\left[\star\Sigma\left[(-1)^{\star r[1]\,(\star n-m)} \frac{\left(\underset{m}{\alpha} \wedge \star S^{\star}[x]\right) \vee \left(\underset{\star n-m}{\beta} \wedge \star S_{\star}[x]\right)}{\underset{m}{\alpha} \vee \underset{\star n-m}{\beta}}\right]\right]$$

$$x == \frac{(-1)^{\star n+m}\, \underset{m}{\alpha} \vee \left(\underset{\star n-m}{\beta} \wedge x\right) + \left(\underset{m}{\alpha} \wedge x\right) \vee \underset{\star n-m}{\beta}}{\underset{m}{\alpha} \vee \underset{\star n-m}{\beta}}$$

From this it is clear that we can absorb the sign by interchanging the order of the factors x and $\underset{\star n-m}{\beta}$.

$$x == \frac{\underset{m}{\alpha} \vee \left(x \wedge \underset{\star n-m}{\beta}\right)}{\underset{m}{\alpha} \vee \underset{\star n-m}{\beta}} + \frac{\left(\underset{m}{\alpha} \wedge x\right) \vee \underset{\star n-m}{\beta}}{\underset{m}{\alpha} \vee \underset{\star n-m}{\beta}} \qquad 3.75$$

In fact, it turns out that we can absorb the sign in the general case also by interchanging the factors $\star S_{\star}[x]$ and $\underset{\star n-m}{\beta}$. This is important because the interior product forms of the General Product Formula which we will develop later in Chapter 6 would not be expected (intuitively) to involve the dimension of the space in an explicit sign. The crux of this considerable simplification is that *the grades of* $\underset{m}{\alpha}$ *and* $\underset{\star n-m}{\beta}$ *must be complementary*. Thus we can write

$$\underset{j}{x} == \star\Sigma\left[\frac{\left(\underset{m}{\alpha} \wedge \star S^{\star}\left[\underset{j}{x}\right]\right) \vee \left(\star S_{\star}\left[\underset{j}{x}\right] \wedge \underset{\star n-m}{\beta}\right)}{\underset{m}{\alpha} \vee \underset{\star n-m}{\beta}}\right] \qquad 3.76$$

To verify this we compare the two formulations by computing the original form and then reversing the order of the exterior product involving $\underset{\star n-m}{\beta}$. For simplicity we ignore the denominator for the moment.

```
J₁[j_] := ★𝒢[★Σ[(-1)^(★r[j] (★n-m)) (α_m ∧ ★S*[x_j]) ∨ (β_{★n-m} ∧ ★S_★[x_j])] /.
    β_{★n-m} ∧ z_ :→ (-1)^(★G[z] (★n-m)) z ∧ β_{★n-m}]
```

```
J₂[j_] := ★𝒢[★Σ[(α_m ∧ ★S*[x_j]) ∨ (★S_★[x_j] ∧ β_{★n-m})]]
```

```
★ℬ₁₀; Table[J₁[j] == J₂[j], {j, 1, 6}]
{True, True, True, True, True, True}
```

In section 4.10 we will explore some of the geometric significance of these formulae. They will also find application in Chapter 6: The Interior Product.

Exploration: Dual forms of the General Product Formulae

You can obtain the dual forms of the various specific product formulae by applying the `Dual` function to them. Below we simply display the duals of the first three specific cases.

◆ First Product Formula

```
F₁ = GeneralProductFormula[(α_m ∨ β_k) ∧ x]
```

$$(\underset{m}{\alpha} \vee \underset{k}{\beta}) \wedge x == (-1)^{\star n-m}\, \underset{m}{\alpha} \vee (\underset{k}{\beta} \wedge x) + (\underset{m}{\alpha} \wedge x) \vee \underset{k}{\beta}$$

```
Dual[F₁]
```

$$(\underset{m}{\alpha} \wedge \underset{k}{\beta}) \vee \underset{-1+\star n}{x} == (-1)^{m}\, \underset{m}{\alpha} \wedge (\underset{k}{\beta} \vee \underset{-1+\star n}{x}) + (\underset{m}{\alpha} \vee \underset{-1+\star n}{x}) \wedge \underset{k}{\beta}$$

◆ Second Product Formula

```
F₂ = GeneralProductFormula[(α_m ∨ β_k) ∧ x_2]
```

$$(\underset{m}{\alpha} \vee \underset{k}{\beta}) \wedge x_1 \wedge x_2 == \underset{m}{\alpha} \vee (\underset{k}{\beta} \wedge x_1 \wedge x_2) + (-1)^{\star n-m} \left(-\left((\underset{m}{\alpha} \wedge x_1) \vee (\underset{k}{\beta} \wedge x_2) \right) + (\underset{m}{\alpha} \wedge x_2) \vee (\underset{k}{\beta} \wedge x_1) \right) + (\underset{m}{\alpha} \wedge x_1 \wedge x_2) \vee \underset{k}{\beta}$$

```
Dual[F₂]
```

$$(\underset{m}{\alpha} \wedge \underset{k}{\beta}) \vee \underset{-1+\star n}{x_1} \vee \underset{-1+\star n}{x_2} == \underset{m}{\alpha} \wedge (\underset{k}{\beta} \vee \underset{-1+\star n}{x_1} \vee \underset{-1+\star n}{x_2}) + (-1)^{m} \left(-\left((\underset{m}{\alpha} \vee \underset{-1+\star n}{x_1}) \wedge (\underset{k}{\beta} \vee \underset{-1+\star n}{x_2}) \right) + (\underset{m}{\alpha} \vee \underset{-1+\star n}{x_2}) \wedge (\underset{k}{\beta} \vee \underset{-1+\star n}{x_1}) \right) + (\underset{m}{\alpha} \vee \underset{-1+\star n}{x_1} \vee \underset{-1+\star n}{x_2}) \wedge \underset{k}{\beta}$$

◆ Third Product Formula

$$\mathbf{F_3 = GeneralProductFormula}\left[\left(\underset{m}{\alpha}\vee\underset{k}{\beta}\right)\wedge\underset{3}{\mathbf{x}}\right]$$

$$\left(\underset{m}{\alpha}\vee\underset{k}{\beta}\right)\wedge x_1\wedge x_2\wedge x_3 ==$$

$$\left(\underset{m}{\alpha}\wedge x_1\right)\vee\left(\underset{k}{\beta}\wedge x_2\wedge x_3\right)-\left(\underset{m}{\alpha}\wedge x_2\right)\vee\left(\underset{k}{\beta}\wedge x_1\wedge x_3\right)+\left(\underset{m}{\alpha}\wedge x_3\right)\vee\left(\underset{k}{\beta}\wedge x_1\wedge x_2\right)+$$

$$(-1)^{\star n-m}\left(\underset{m}{\alpha}\vee\left(\underset{k}{\beta}\wedge x_1\wedge x_2\wedge x_3\right)+\left(\underset{m}{\alpha}\wedge x_1\wedge x_2\right)\vee\left(\underset{k}{\beta}\wedge x_3\right)-\left(\underset{m}{\alpha}\wedge x_1\wedge x_3\right)\vee\left(\underset{k}{\beta}\wedge x_2\right)+\right.$$

$$\left.\left(\underset{m}{\alpha}\wedge x_2\wedge x_3\right)\vee\left(\underset{k}{\beta}\wedge x_1\right)\right)+\left(\underset{m}{\alpha}\wedge x_1\wedge x_2\wedge x_3\right)\vee\underset{k}{\beta}$$

$$\mathbf{Dual[F_3]}$$

$$\left(\underset{m}{\alpha}\wedge\underset{k}{\beta}\right)\vee\underset{-1+\star n}{x_1}\vee\underset{-1+\star n}{x_2}\vee\underset{-1+\star n}{x_3} ==$$

$$\left(\underset{m}{\alpha}\vee\underset{-1+\star n}{x_1}\right)\wedge\left(\underset{k}{\beta}\vee\underset{-1+\star n}{x_2}\vee\underset{-1+\star n}{x_3}\right)-\left(\underset{m}{\alpha}\vee\underset{-1+\star n}{x_2}\right)\wedge\left(\underset{k}{\beta}\vee\underset{-1+\star n}{x_1}\vee\underset{-1+\star n}{x_3}\right)+$$

$$\left(\underset{m}{\alpha}\vee\underset{-1+\star n}{x_3}\right)\wedge\left(\underset{k}{\beta}\vee\underset{-1+\star n}{x_1}\vee\underset{-1+\star n}{x_2}\right)+(-1)^{m}\left(\underset{m}{\alpha}\wedge\left(\underset{k}{\beta}\vee\underset{-1+\star n}{x_1}\vee\underset{-1+\star n}{x_2}\vee\underset{-1+\star n}{x_3}\right)+\right.$$

$$\left(\underset{m}{\alpha}\vee\underset{-1+\star n}{x_1}\vee\underset{-1+\star n}{x_2}\right)\wedge\left(\underset{k}{\beta}\vee\underset{-1+\star n}{x_3}\right)-\left(\underset{m}{\alpha}\vee\underset{-1+\star n}{x_1}\vee\underset{-1+\star n}{x_3}\right)\wedge\left(\underset{k}{\beta}\vee\underset{-1+\star n}{x_2}\right)+$$

$$\left.\left(\underset{m}{\alpha}\vee\underset{-1+\star n}{x_2}\vee\underset{-1+\star n}{x_3}\right)\wedge\left(\underset{k}{\beta}\vee\underset{-1+\star n}{x_1}\right)\right)+\left(\underset{m}{\alpha}\vee\underset{-1+\star n}{x_1}\vee\underset{-1+\star n}{x_2}\vee\underset{-1+\star n}{x_3}\right)\wedge\underset{k}{\beta}$$

The double sum form of the General Product Formula

Although not directly computable, the following double sum form of the General Product Formula may be shown to be equivalent to the computable form.

$$\left(\underset{m}{\alpha}\vee\underset{k}{\beta}\right)\wedge\underset{j}{\mathbf{x}} == \sum_{r==0}^{j}(-1)^{r\,(n-m)}\sum_{i==1}^{\nu}\left(\underset{m}{\alpha}\wedge\underset{j-r}{\mathbf{x}_i}\right)\vee\left(\underset{k}{\beta}\wedge\underset{r}{\mathbf{x}_i}\right) \qquad 3.77$$

$$\underset{j}{\mathbf{x}} == \underset{r}{\mathbf{x}_1}\wedge\underset{j-r}{\mathbf{x}_1} == \underset{r}{\mathbf{x}_2}\wedge\underset{j-r}{\mathbf{x}_2} == \ldots == \underset{r}{\mathbf{x}_\nu}\wedge\underset{j-r}{\mathbf{x}_\nu}$$

The index ν is $\binom{j}{r}$.

Note that the formula is, in effect, the sum of $j+1$ sums A_r, indexed by the parameter r.

$$\left(\underset{m}{\alpha}\vee\underset{k}{\beta}\right)\wedge\underset{j}{\mathbf{x}} == \sum_{r==0}^{j}A_r$$

The most straightforward way of using the formula for a given value of j is to compose each of these sums separately, and then substitute more meaningful symbols for the factors of $\underset{j}{\mathbf{x}}$. However, the notation for the decomposition of $\underset{j}{\mathbf{x}}$ into the various exterior products is not unambiguous, and care should be taken using the formula in this form.

◆ Example: j = 1

$$\left(\underset{m}{\alpha} \vee \underset{k}{\beta}\right) \wedge \underset{1}{x} == A_0 + A_1$$

Compose the A_r:

$$A_0 == (-1)^{0\,(n-m)} \left(\underset{m}{\alpha} \wedge \underset{1-0}{x_1}\right) \vee \left(\underset{k}{\beta} \wedge \underset{0}{x_1}\right) \qquad \underset{1}{x} == \underset{0}{x_1} \wedge \underset{1-0}{x_1} \qquad \nu == 1$$

$$A_1 == (-1)^{1\,(n-m)} \left(\underset{m}{\alpha} \wedge \underset{1-1}{x_1}\right) \vee \left(\underset{k}{\beta} \wedge \underset{1}{x_1}\right) \qquad \underset{1}{x} == \underset{1}{x_1} \wedge \underset{1-1}{x_1} \qquad \nu == 1$$

$$A_0 + A_1 == \left(\underset{m}{\alpha} \wedge \underset{1-0}{x_1}\right) \vee \left(\underset{k}{\beta} \wedge \underset{0}{x_1}\right) + (-1)^{n-m} \left(\underset{m}{\alpha} \wedge \underset{1-1}{x_1}\right) \vee \left(\underset{k}{\beta} \wedge \underset{1}{x_1}\right)$$

Substitute more meaningful symbols:

$$\underset{0}{x_1} \longrightarrow 1 \qquad \underset{1}{x_1} \longrightarrow x$$

Finally yielding:

$$\left(\underset{m}{\alpha} \vee \underset{k}{\beta}\right) \wedge x == \left(\underset{m}{\alpha} \wedge x\right) \vee \left(\underset{k}{\beta} \wedge 1\right) + (-1)^{n-m} \left(\underset{m}{\alpha} \wedge 1\right) \vee \left(\underset{k}{\beta} \wedge x\right)$$

◆ Example: j = 2

$$\left(\underset{m}{\alpha} \vee \underset{k}{\beta}\right) \wedge \underset{2}{x} == A_0 + A_1 + A_2$$

Compose the A_r:

$$A_0 == (-1)^{0\,(n-m)} \left(\underset{m}{\alpha} \wedge \underset{2-0}{x_1}\right) \vee \left(\underset{k}{\beta} \wedge \underset{0}{x_1}\right) \qquad \underset{2}{x} == \underset{0}{x_1} \wedge \underset{2-0}{x_1} \qquad \nu == 1$$

$$A_1 == (-1)^{1\,(n-m)} \left(\left(\underset{m}{\alpha} \wedge \underset{2-1}{x_1}\right) \vee \left(\underset{k}{\beta} \wedge \underset{1}{x_1}\right) + \left(\underset{m}{\alpha} \wedge \underset{2-1}{x_2}\right) \vee \left(\underset{k}{\beta} \wedge \underset{1}{x_2}\right)\right)$$

$$\underset{2}{x} == \underset{1}{x_1} \wedge \underset{2-1}{x_1} == \underset{1}{x_2} \wedge \underset{2-1}{x_2} \qquad \nu == 2$$

$$A_2 == (-1)^{2\,(n-m)} \left(\underset{m}{\alpha} \wedge \underset{2-2}{x_1}\right) \vee \left(\underset{k}{\beta} \wedge \underset{2}{x_1}\right) \qquad \underset{2}{x} == \underset{2}{x_1} \wedge \underset{2-2}{x_1} \qquad \nu == 1$$

$$A_0 + A_1 + A_2 == \left(\underset{m}{\alpha} \wedge \underset{2-0}{x_1}\right) \vee \left(\underset{k}{\beta} \wedge \underset{0}{x_1}\right)$$

$$+ (-1)^{n-m} \left(\left(\underset{m}{\alpha} \wedge \underset{2-1}{x_1}\right) \vee \left(\underset{k}{\beta} \wedge \underset{1}{x_1}\right) + \left(\underset{m}{\alpha} \wedge \underset{2-1}{x_2}\right) \vee \left(\underset{k}{\beta} \wedge \underset{1}{x_2}\right)\right) + \left(\underset{m}{\alpha} \wedge \underset{2-2}{x_1}\right) \vee \left(\underset{k}{\beta} \wedge \underset{2}{x_1}\right)$$

Substitute more meaningful symbols:

$$\underset{0}{x_1} \longrightarrow 1 \qquad \underset{1}{x_1} \longrightarrow x \qquad \underset{2-1}{x_1} \longrightarrow y \qquad \underset{1}{x_2} \longrightarrow y \qquad \underset{2-1}{x_2} \longrightarrow -x \qquad \underset{2}{x_1} \longrightarrow x \wedge y$$

Finally yielding:

$$\left(\underset{m}{\alpha}\vee\underset{k}{\beta}\right)\wedge(\mathbf{x}\wedge\mathbf{y}) == \left(\underset{m}{\alpha}\wedge(\mathbf{x}\wedge\mathbf{y})\right)\vee\left(\underset{k}{\beta}\wedge 1\right)$$
$$+(-1)^{n-m}\left(\left(\underset{m}{\alpha}\wedge\mathbf{y}\right)\vee\left(\underset{k}{\beta}\wedge\mathbf{x}\right)+\left(\underset{m}{\alpha}\wedge(-\mathbf{x})\right)\vee\left(\underset{k}{\beta}\wedge\mathbf{y}\right)\right)+\left(\underset{m}{\alpha}\wedge 1\right)\vee\left(\underset{k}{\beta}\wedge(\mathbf{x}\wedge\mathbf{y})\right)$$

3.11 Summary

This chapter has introduced the regressive product as a true *dual* to the exterior product. This means that to every theorem `T` involving exterior and regressive products there corresponds a dual theorem `Dual[T]` such that `T = Dual[Dual[T]]`.

But although the regressive product axioms assert that the regressive product of an *m*-element with a *k*-element is an (*m*+*k*–*n*)-element (where *n* is the dimension of the underlying linear space), the axiom sets for the exterior and regressive products alone do not provide a mechanism for deriving such an element explicitly. A further explicit axiom involving *both* exterior and regressive products is necessary. We called this axiom the Common Factor Axiom and motivated it by a combination of algebraic and geometric argument. If the exterior product is viewed as a sort of 'union' of independent elements, the regressive product may be viewed as a sort of 'intersection'. Because Grassmann considered only Euclidean spaces, used the same notation for both exterior and regressive products, and equated scalars and pseudo-scalars, the Common Factor Axiom was effectively hidden in his notation.

The Common Factor Axiom was then extended to prove one of the most important formulae in the Grassmann algebra, the Common Factor Theorem. This theorem enables the regressive product of any two arbitrary elements of the algebra to be computed in an effective manner. It will be shown in Chapter 5: The Complement that if the underlying linear space is endowed with a metric, then the result is specific, and depends on the metric for its precise value. Otherwise, if there is no metric, the element is specific only up to congruence (that is, up to an arbitrary scalar factor). It was then shown how the Common Factor Theorem could be used in the factorization of simple elements.

Finally, it was shown how the Common Factor Theorem leads to a suite of formulae, called product formulae. These formulae expand expressions involving an exterior product of an element with a regressive product of elements, or, involving a regressive product of an element with an exterior product of elements. In Chapter 6 we will show how these lead naturally to formulae where the regressive products are replaced by interior products. These interior product forms of the product formulae find application throughout the Grassmann algebra.

The next chapter is an interlude in the development of the algebraic fundamentals. In it we begin to explore one of Grassmann algebra's most enticing interpretations: geometry. But at this stage we will only be discussing non-metric geometry. Chapters 5 and 6 to follow will develop the algebra's metric concepts, and thus complete the fundamentals required for subsequent applications and geometric interpretations in metric space.

4 Geometric Interpretations

4.1 Introduction

In Chapter 2, the exterior product operation was introduced onto the elements of a linear space $\underset{1}{\Lambda}$, enabling the construction of a series of new linear spaces $\underset{m}{\Lambda}$ possessing new types of elements. In this chapter, the elements of $\underset{1}{\Lambda}$ will be *interpreted*, some as *vectors*, some as *points*. This will be done by singling out one particular element of $\underset{1}{\Lambda}$ and conferring upon it the *interpretation* of *origin point*. All the other elements of the linear space then divide into two categories. Those that involve the origin point will be called (weighted) *points*, and those that do not will be called *vectors*. As this distinction is developed in succeeding sections it will be seen to be both illuminating and consistent with accepted notions. Vectors and points will be called *geometric interpretations* of the elements of $\underset{1}{\Lambda}$.

Some of the more important consequences however, of the distinction between vectors and points arise from the distinctions thereby generated in the higher grade spaces $\underset{m}{\Lambda}$. It will be shown that a simple element of $\underset{m}{\Lambda}$ takes on two interpretations. The first, that of a *multivector* (or *m-vector*) is when the *m*-element can be expressed in terms of vectors alone. The second, that of a *bound multivector* is when the *m*-element requires both points and vectors to express it. These simple interpreted elements will be found useful for defining geometric entities such as lines and planes and their higher dimensional analogues known as *multiplanes*. Unions and intersections of multiplanes may then be calculated straightforwardly by using the bound multivectors which define them. A multivector may be visualized as a 'free' entity with no location. A bound multivector may be visualized as 'bound' *through* a location in space.

We will show that Projective Geometry is nothing but Grassmann algebra interpreted in this way.

It is not only simple interpreted elements which will be found useful in applications however. In Volume 2 of this work, a basis for a theory of mechanics is developed in which a principal quantity (for example, a system of forces, the momentum of a system of particles, the velocity of a rigid body) may be represented by a general interpreted 2-element, that is, by the *sum* of a bound vector and a bivector.

In the literature of the nineteenth century, wherever vectors and points were considered together, vectors were introduced as point differences. When it is required to designate physical quantities it is not satisfactory that all vectors should arise as the differences of points. In later literature, this problem appeared to be overcome by designating points by their position vectors alone, making vectors the fundamental entities [Gibbs 1886]. This approach is not satisfactory either, since by excluding points much of the power of the calculus for dealing with free and located entities *together* is excluded. In this book we do not require that vectors be defined in terms of points, but rather, propose a difference of interpretation between the origin element and those elements not involving the origin. This approach permits the existence of points and vectors together without the vectors necessarily arising as point differences.

In this chapter, as in the preceding chapters, $\underset{1}{\Lambda}$ and the spaces $\underset{m}{\Lambda}$ do not yet have a metric. That is, there is no way of calculating a measure or magnitude associated with an element. The interpretation discussed therefore may also be supposed non-metric. In the next chapter a metric will be introduced onto the uninterpreted spaces and the consequences of this for the interpreted elements developed.

In summary then, it is the aim of this chapter to set out the distinct non-metric geometric interpretations of m-elements brought about by the interpretation of one specific element of $\underset{1}{\Lambda}$ as an origin point.

4.2 Geometrically Interpreted 1-Elements

Vectors

◆ Depicting vectors

The most common current geometric interpretation for an element of a linear space is that of a *vector*. We suppose in this chapter that the linear space does not have a metric (that is, we cannot calculate magnitudes). Such a vector may be endowed with the geometric properties of *direction* and *sense,* but *not location* and *not magnitude*. They will be graphically *depicted* (in the usual way) by a directed and sensed line segment thus:

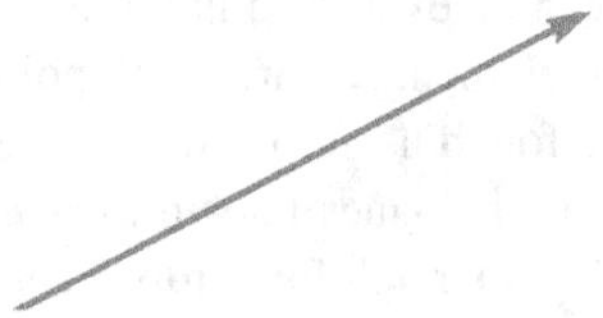

The usual depiction of a vector. But this vector is located nowhere!

This depiction is unsatisfactory in that the line segment has a definite location and length whilst the vector it is depicting is supposed to possess neither of these properties. One way perhaps to depict an *unlocated* entity is to show it in *many* locations.

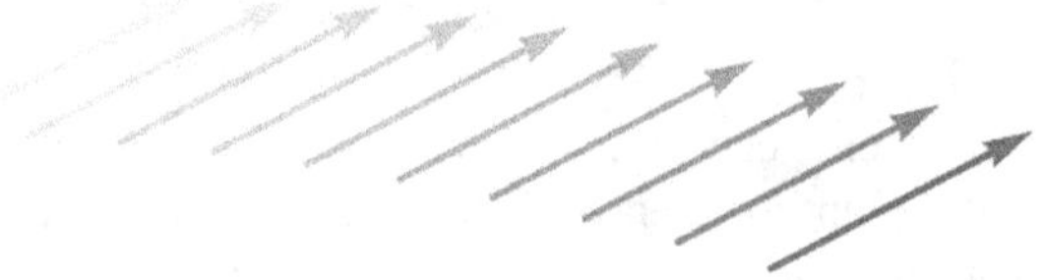

One way to depict an unlocated entity is to show it in *many* locations

Another way to emphasize that a vector has no location is show it on a dynamic graphic. If you are viewing this notebook live with the *GrassmannAlgebra* package loaded, you can use your mouse to manipulate the vector below to reinforce that *any* of the vector's locations that retain its direction and sense are equally satisfactory depictions for it. This *dynamic* depiction is a somewhat more faithful way of depicting something without locating it. But it is still not fully satisfactory, since we are depicting the property of *no-location* by using *any location*.

Another way is to allow it to be moveable to *any* location

A linear space, all of whose elements are interpreted as vectors, will be called a *vector space*.

◆ **Depicting the addition of vectors**

The geometric interpretation of the addition operation of a vector space is the triangle (or parallelogram) rule. To effect a sum of two vectors using the triangle rule in this 'unlocated' view of vectors, you would slide them (always in a direction parallel to themselves) from where you had each of them conveniently docked, until their tails touched, make the sum in the standard textbook manner, then park the result anywhere you please.

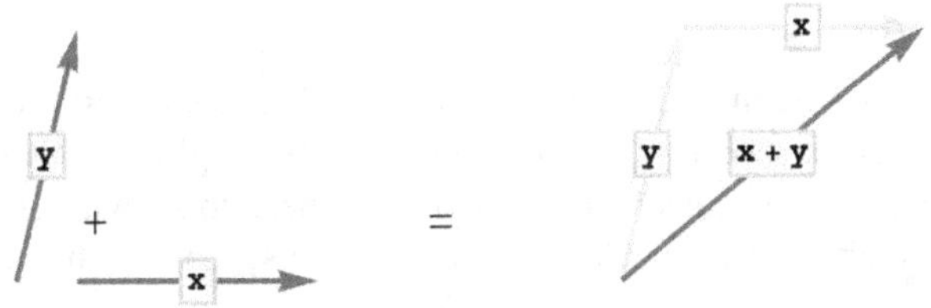

The triangle rule for the addition of two vectors

- From this point onwards *we will always show vectors docked in a visually* ***convenient*** *location*, often one that is suggestive of the operation we wish to perform. Of course as discussed above, this does *not* mean it is actually located there.

◆ **Comparing vectors**

In a metric space we can define the magnitude of a vector and hence compare vectors by comparing their magnitudes even if they are not in the same direction. In spaces where no metric has been imposed, as are the spaces we are discussing in this chapter, we cannot define the magnitude of a vector, and hence we cannot compare it with another vector in a different direction.

However, we can compare vectors in the same direction. Vectors in the same direction are *congruent*. That is, one is a scalar multiple (not zero) of the other. This scalar multiple is called the *congruence factor*.

Thus vectors $a\,x$ and $b\,x$ are congruent with congruence factor $\frac{a}{b}$ or $\frac{b}{a}$.

Most often it is not the magnitude of the congruence factor that is important, but its sign. If it is positive, the vectors may be said to have the same *orientation*. If it is negative the vectors may be said to have an opposite orientation. Orientation is thus a *relative* concept. It applies in the same way to elements of any grade.

For the special case of vectors however, the term *sense* is generally used instead of orientation. Let us depict the sense of the vector x as in the figure below with the arrow in the direction shown. If we multiply x by –1 we will obtain a vector –x, the additive inverse of x. We say vectors x and –x have *opposite* sense. Arrowheads then enable us to distinguish a vector from its congruent partner of opposite sense.

Below we depict three vectors with congruence factors (relative to x) of 1, -1 and 2.

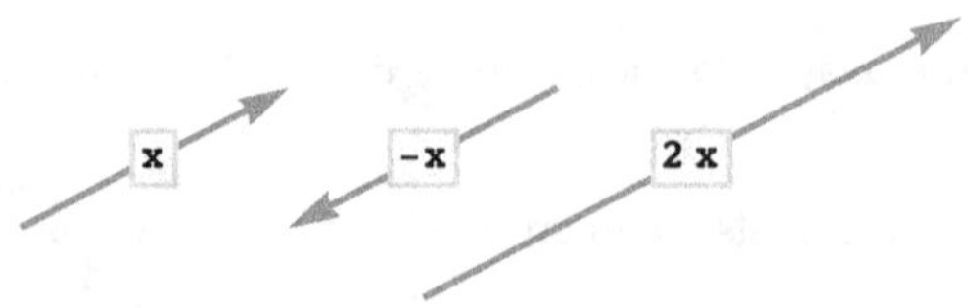

The effect of multiplying the vector x by factors 1, –1, and 2

Points

◆ The origin

In order to describe *position* in space it is necessary to have a *reference point*. This point is usually called the *Origin*.

Rather than the standard technique of implicitly assuming the origin and working only with vectors to describe position, we find it important for later applications to *augment the vector space with the origin as a new element* to create a new linear space with one more element in its basis. For reasons which will appear later, such a linear space will be called a *bound vector space*.

The only difference between the origin element and the vector elements of the linear space is their *interpretation*. The origin element is interpreted as a point. We will denote it in *GrassmannAlgebra* by ⋆𝕆, which you can access from the palette or type as ESC*5ESC ESCdsOESC (a five-star followed by a double-struck capital O).

In order to describe position it is necessary to have a reference point

◆ The sum of the origin and a vector is a point

The bound vector space in addition to its vectors and its origin now possesses a new set of elements requiring interpretation: those formed from the sum of the origin and a vector.

```
P == ⋆𝕆 + x
```

It is these elements that will be used to describe position and that we will call *points*. The vector x is called the *position vector* of the point P. A position vector is just like any other vector and is therefore, of course, located nowhere (even though we may show it docked in a conveniently suggestive position!).

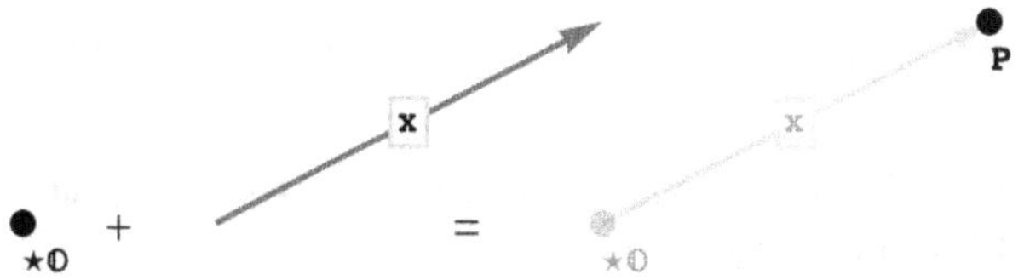

The sum of the origin and a vector is a point

◆ The difference of two points is a vector

It follows immediately from this definition of a point that *the difference of two points is a vector*.

$$P - Q == (\star O + p) - (\star O + q) == p - q == x$$

Thus *the difference of two points is not a point*.

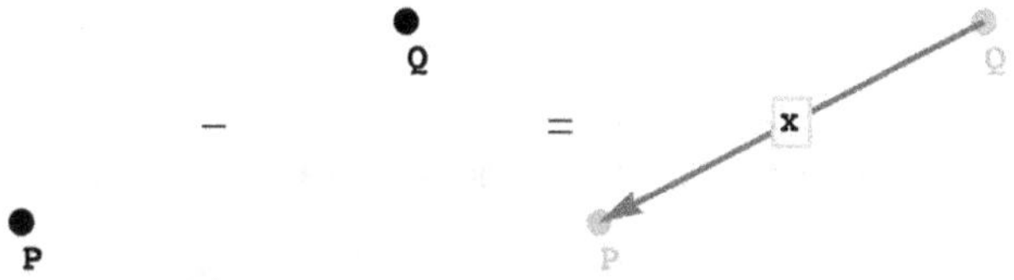

The difference of two points is a vector

Remember that the bound vector space we are discussing does not yet have a metric. That is, the distance between two points (the magnitude of the vector equal to the point difference) is not meaningful. However, the *relative* distances between points *on the same line* can be measured since the point differences are congruent vectors.

◆ The sum of a point and a vector is a point

The *sum of a point and a vector* is another point.

$$Q == P + y == (\star O + x) + y == \star O + (x + y)$$

We depict this by conveniently docking the tails of the vectors at the points.

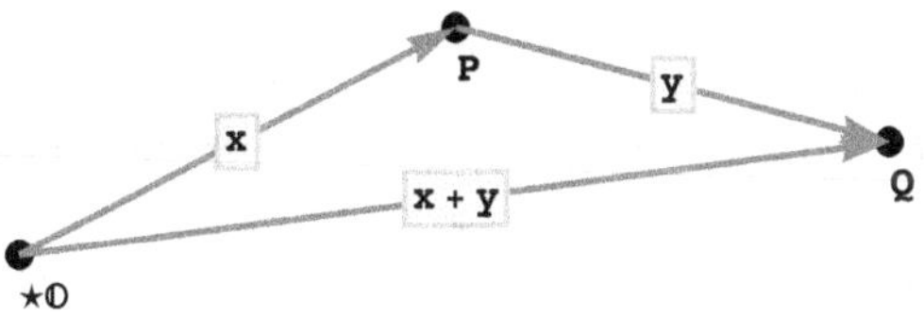

The sum of a point and a vector is another point

and so a vector may be viewed as a *carrier of points*. That is, the addition of a vector to a point carries or transforms it to another point. (See the Historical Note below.)

◆ The sum of two points is a weighted point

A scalar multiple m of a point P (of the form m P or m ∧ P) will be called a *weighted point* with *weight* m.

The *sum of two points* is not quite another point. Rather, it is a weighted point with weight 2, situated mid-way between the two points.

$$P_1 + P_2 == (\star O + x_1) + (\star O + x_2)$$

$$== 2 \star O + (x_1 + x_2) == 2\left(\star O + \frac{1}{2}(x_1 + x_2)\right) == 2 P_c$$

Wherever pictorially feasible we will show weighted points with their weights attached to their names and a size change to distinguish them from the 'pure' points on the same graphic.

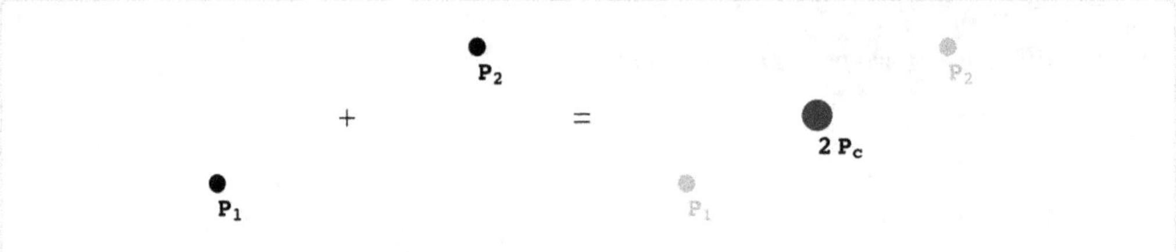

The sum of two points is a point of weight 2 located mid-way between them

Similarly, the *sum of n points* is a point with weight *n*, situated at the centre of mass (centre of gravity) of the points.

◆ The sum of weighted points

Weighted points are summed just like a set of point masses is deemed equivalent to their total mass located at their centre of mass. For example the sum of two weighted points with weights (masses) m_1 and m_2 is equal to the weighted point with weight $(m_1 + m_2)$ located on the line between them. The location is the point P_c about which the 'mass moments' $l_1\, m_1$ and $l_2\, m_2$ are equal.

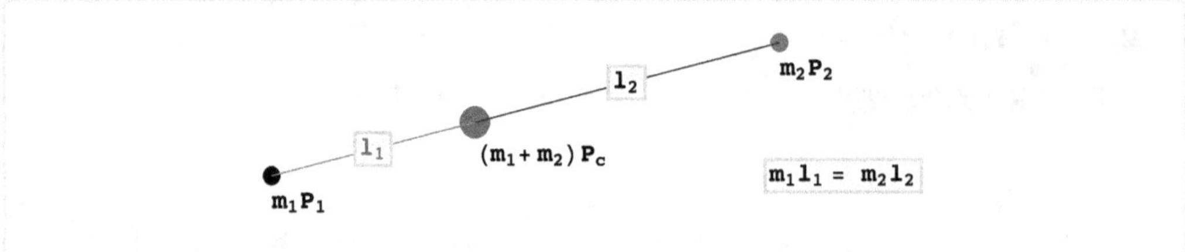

The sum of two weighted points is a weighted point between them

In the more general case we have

$$\left(\sum m_i\right) P_c == \sum (m_i\, P_i)$$

We can also easily express this equation in terms of the position vectors of the weighted points.

$$\sum (m_i\, P_i) == \sum (m_i\, (\star\mathbb{O} + x_i)) == \left(\sum m_i\right)\left(\star\mathbb{O} + \frac{\sum (m_i\, x_i)}{\sum m_i}\right)$$

Thus the sum of a number of mass-weighted points $\Sigma\,(m_i\, P_i)$ is equivalent to the centre of gravity $\star\mathbb{O} + \frac{\Sigma\,(m_i\, x_i)}{\Sigma\, m_i}$ weighted by the total mass $\Sigma\, m_i$. A numerical example is included later in this section.

As will be seen in Section 4.4 below, a weighted point may also be viewed as a bound scalar.

◆ ***Historical Note***

Sir William Rowan Hamilton in his *Lectures on Quaternions* [Hamilton 1853] was the first to coin the term vector as a 'carrier' of points.

... I regard the symbol B–A as denoting "the *step* from B to A": namely, that step by making which, from the given point A, we should reach or arrive at the sought point B; and so determine, generate, mark or *construct* that point. This step, (which we always suppose to be a straight line) may also in my opinion be properly called a *vector*; or more fully, it may be called "the vector of the point B from the point A": because it may be considered as having for its office, function, work, task or business, to transport or *carry* (in Latin *vehere*) a movable point, from the given or initial position A, to the sought or final position B.

Declaring a basis for a bound vector space

Any geometry that we do with points in *GrassmannAlgebra* will require us to declare the origin $\star\mathbb{O}$ as one of the elements of the basis of the space. We have already seen that a shorthand way of declaring a basis $\{e_1, e_2, \ldots, e_n\}$ is by entering $\star\mathcal{B}_n$. Declaring the augmented basis $\{\star\mathbb{O}, e_1, e_2, \ldots, e_n\}$ can be accomplished by entering the alias $\star\mathcal{P}_n$ (most easily from the palette). These are 5-star-script-capital letters subscripted with the integer *n* denoting the desired 'vectorial' dimension of the space. For example, entering $\star\mathcal{P}_2$ gives a basis for the plane.

$\star\mathcal{P}_2$

$\{\star\mathbb{O}, e_1, e_2\}$

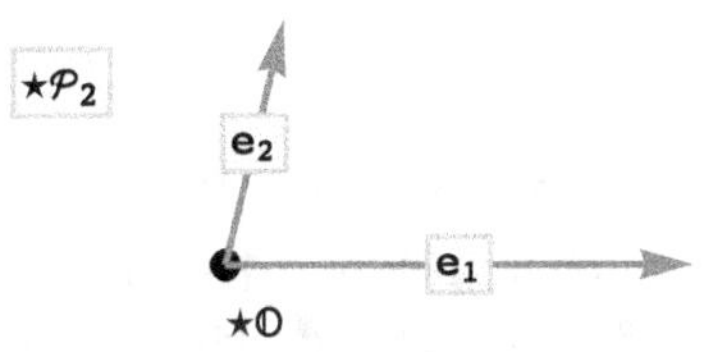

Declaring a basis for the plane

We do not depict the basis vectors at right angles because the space does not yet have a metric, and orthogonality is a metric concept.

As always, you can confirm your currently declared basis by checking the status pane at the top of the palette.

We may often precede a calculation with one of these `DeclareBasis` aliases followed by a semi-colon. This accomplishes the declaration of the basis but for brevity suppresses the confirming output. For example, below we declare a bound 3-space, and then compose a palette of the basis of the algebra constructed on it.

```
★𝒫3; BasisPalette
```

Basis Palette

Λ_0	Λ_1	Λ_2	Λ_3	Λ_4
1	★O	★O ∧ e_1	★O ∧ e_1 ∧ e_2	★O ∧ e_1 ∧ e_2 ∧ e_3
	e_1	★O ∧ e_2	★O ∧ e_1 ∧ e_3	
	e_2	★O ∧ e_3	★O ∧ e_2 ∧ e_3	
	e_3	e_1 ∧ e_2	e_1 ∧ e_2 ∧ e_3	
		e_1 ∧ e_3		
		e_2 ∧ e_3		

Of course, you can declare your own bound vector space basis if you wish. For example if you want the vector basis elements to be i, j, and k, you could enter

```
DeclareBasis[{i, j, k, ★O}]
```

```
{★O, i, j, k}
```

(DeclareBasis will always rearrange the ordering to make the origin ★O come first.)

Composing vectors and points

◆ Composing vectors

Once you have declared a basis for your bound vector space, say ★$\mathcal{P}_3$,

```
★𝒫3
```

{★O, e_1, e_2, e_3}

you can quickly compose any vectors in this basis with ComposeVector or its alias ★$\mathbb{V}_\square$. The placeholder □ is used for entering the symbol upon which you want the coefficients to be based. Here we choose the symbol a.

V_a = ★$\mathbb{V}_a$

$a_1 e_1 + a_2 e_2 + a_3 e_3$

ComposeVector automatically declares the a_i to be scalar symbols.

If you enter ★$\mathbb{V}_\square$, you get an expression with placeholders in which you can enter your own coefficients just by clicking on any placeholder, and tabbing through them.

★$\mathbb{V}_\square$

$\square\, e_1 + \square\, e_2 + \square\, e_3$

Note however that in this case if you want any symbolic coefficients you enter to be recognized as scalar symbols, you would need to declare them as extra scalar symbols, say by using ★★S.

◆ Composing points

Similarly, to compose a point you can use ★$\mathbb{P}_\square$.

$P_b = \star\mathbb{P}_b$

$\star\mathbb{O} + b_1\, e_1 + b_2\, e_2 + b_3\, e_3$

$\star\mathbb{P}_\square$

$\star\mathbb{O} + \square\, e_1 + \square\, e_2 + \square\, e_3$

Now you can use the points and vectors you have composed in expressions. For example, here we add P_b and $2\,V_a$, then use `GrassmannSimplify` (alias $\star S$) to collect their coefficients.

$P_b + 2\, V_a\ //\ \star S$

$\star\mathbb{O} + (2\, a_1 + b_1)\, e_1 + (2\, a_2 + b_2)\, e_2 + (2\, a_3 + b_3)\, e_3$

Example: Calculation of the centre of mass

Suppose a space with basis $\{\star\mathbb{O}, e_1, e_2, e_3\}$ and a set of masses situated at points P_i. It is required to find their centre of mass. First declare the basis, then enter the mass points.

$\star\mathcal{P}_3$

$\{\star\mathbb{O}, e_1, e_2, e_3\}$

$M_1 = 2\, P_1;\quad P_1 = \star\mathbb{O} + e_1 + 3\, e_2 - 4\, e_3;$
$M_2 = 4\, P_2;\quad P_2 = \star\mathbb{O} + 2\, e_1 - e_2 - 2\, e_3;$
$M_3 = 7\, P_3;\quad P_3 = \star\mathbb{O} - 5\, e_1 + 3\, e_2 - 6\, e_3;$
$M_4 = 5\, P_4;\quad P_4 = \star\mathbb{O} + 4\, e_1 + 2\, e_2 - 9\, e_3;$

We simply add the mass-weighted points.

$$M = \sum_{i=1}^{4} M_i$$

$$5\,(\star\mathbb{O} + 4\, e_1 + 2\, e_2 - 9\, e_3) + 7\,(\star\mathbb{O} - 5\, e_1 + 3\, e_2 - 6\, e_3) + 2\,(\star\mathbb{O} + e_1 + 3\, e_2 - 4\, e_3) + 4\,(\star\mathbb{O} + 2\, e_1 - e_2 - 2\, e_3)$$

Simplifying this gives a weighted point with weight 18, the scalar attached to the origin.

$M = \star\mathcal{G}[M]$

$18\,\star\mathbb{O} - 5\, e_1 + 33\, e_2 - 103\, e_3$

To take the weight out as a factor, that is, expressing the result in the form *mass* × *point*, we can use `ToWeightedPointForm`.

`M = ToWeightedPointForm[M]`

$$18\left(\star\mathbb{O} - \frac{5\, e_1}{18} + \frac{11\, e_2}{6} - \frac{103\, e_3}{18}\right)$$

(`ToPointForm` will remove the weight altogether).

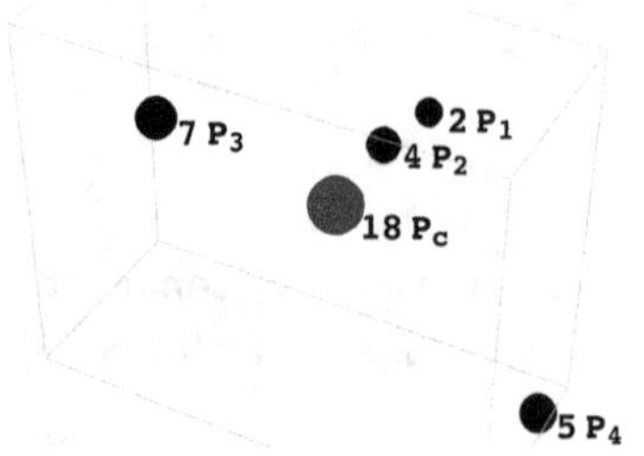

The centre of mass of four weighted points

Thus the total mass is 18 situated at the point

$$P_c == \star\mathcal{O} - \frac{5\,e_1}{18} + \frac{11\,e_2}{6} - \frac{103\,e_3}{18}$$

If you are reading this notebook in *Mathematica*, you can rotate this graphic, and check the positions of the points by hovering your mouse over their symbols.

4.3 Geometrically Interpreted 2-Elements

Simple geometrically interpreted 2-elements

It has been seen in Chapter 2 that the linear space $\underset{2}{\Lambda}$ may be generated from $\underset{1}{\Lambda}$ by the exterior product operation. In the preceding section the elements of $\underset{1}{\Lambda}$ have been given two geometric interpretations: that of a *vector* and that of a *point*. These interpretations in turn generate various other interpretations for the elements of $\underset{2}{\Lambda}$.

In $\underset{2}{\Lambda}$ there are at first sight three possibilities for simple elements:

1. $x \wedge y$ (vector by vector)
2. $P \wedge x$ (point by vector)
3. $P \wedge Q$ (point by point)

However, $P \wedge Q$ may be expressed as $P \wedge (Q - P)$ which is the product of a point and a vector, and thus reduces to the second case.

There are thus two *simple* interpreted elements in $\underset{2}{\Lambda}$:

1. $x \wedge y$ (*the simple bivector*)
2. $P \wedge x$ (*the bound vector*)

◆ *A note on terminology*

The term *bound* as in the phrase "the *bound vector* $P \wedge x$" indicates that the vector x is conceived of as bound *through* the point P, rather than *to* the point P, since the latter conception would give the incorrect impression that the vector was located *at* the point P. By adhering to the terminology *bound through*, we get a slightly more correct impression of the 'freedom' that the vector enjoys.

The term *bound vector space* will be used to denote a vector space to whose basis has been added an origin point. This should be read as "bound *vector-space*", rather than "*bound-vector* space".

The term *bound m-space* will be used to denote an *m*-dimensional vector space to whose basis has been added an origin point.

Bivectors

◆ Depicting bivectors

Earlier in this chapter, a vector was depicted graphically by a directed and sensed line segment supposed to be located nowhere in particular.

In like manner we depict a simple bivector by depicting its two component vectors supposed located nowhere in particular. But to indicate that they are part of the same exterior product, the vectors are depicted *conveniently* docked head to tail *in the order of the product*, and 'joined' by a somewhat transparent parallelogram.

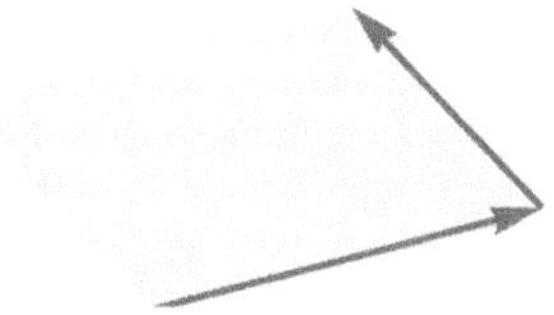

Depicting a simple bivector. This bivector is located nowhere.

This bivector defines a planar 2-*direction*, the precise 2-dimensional analog of the unlocated vector. Its vectors are located nowhere, and *it* is located nowhere. The parallelogram should be viewed as an artifact which is used to visually tie the two vectors together in the exterior product. We motivate this choice by noting that if this bivector were in a metric space, the area of the parallelogram would correspond to the magnitude of the bivector. We will discuss this further in Chapter 6.

If we change the sign of both vectors in the product, we do not change the bivector, but we get a different depiction.

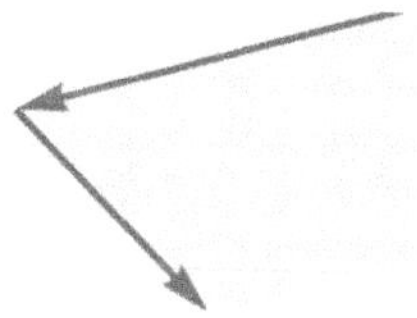

Changing the signs of the both vectors gives the same bivector

Any bivector congruent to this bivector (that is, a scalar multiple of it) will represent the same 2-direction. Reversing the order of the factors in a bivector is equivalent to multiplying it by `-1`, producing an 'opposite orientation'.

Congruent bivectors with different orientations

This parallelogram depiction of the simple bivector is still misleading in a further major respect that did not arise for vectors. It incorrectly suggests a specific *shape* of the parallelogram. Indeed, since the bivector $x \wedge y$ may also be expressed as $x \wedge (x + y)$, another valid depiction of this simple bivector would be a parallelogram with sides constructed from vectors x and $x + y$.

$$x \wedge y == x \wedge (x + y)$$

Or, more generally, for any scalar multiplier a:

$$x \wedge y == x \wedge (a\, x + y)$$

This is true for a space of any dimension (greater than 1).

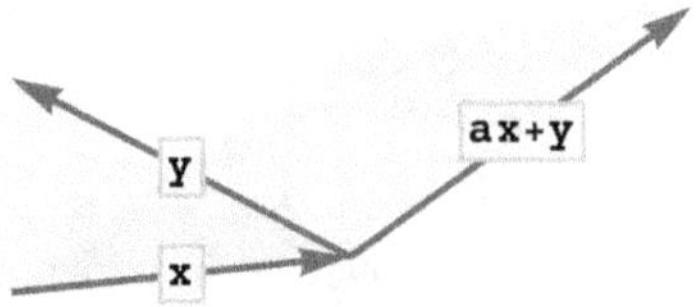

A bivector displaying its ability to change shape.
The bivectors $x \wedge y$ and $x \wedge (ax+y)$ are equal with equal areas.

However in what follows, just as we will usually depict a bivector docked in a *location* most convenient for the discussion at hand, so too we will usually depict it in the *shape* most convenient for the discussion at hand, usually one with the simplest factors.

A *sum of simple bivectors* is called a *bivector*. In two and three dimensions all bivectors are simple. This will have important consequences for our exploration of mechanics in Volume 2.

Earlier in the chapter it was remarked that a vector may be viewed as a 'carrier' of points. Analogously, a simple bivector may be viewed as a *carrier of bound vectors*. This view will be more fully explored in the next section.

◆ Bivectors in a metric space

In the following chapter a metric will be introduced onto $\underset{1}{\Lambda}$ from which a metric is induced onto $\underset{2}{\Lambda}$. This will permit the definition of the measure of a vector (its length) and the measure of the simple bivector (its area).

The measure of a simple bivector is geometrically interpreted as the *area of the parallelogram* formed by its two vectors. However, as demonstrated above, the bivector can be expressed in terms of an infinite number of pairs of vectors. Despite this, the simple geometric fact that they have the same base and height shows that parallelograms formed from all of them have the same area. For example, the areas of the parallelograms in the previous figure are the same. Thus the area definition of the measure of the bivector is truly an invariant measure. From this point of view the parallelogram depiction in a metric space is correctly *suggestive*, although the parallelogram is not of fixed shape.

In Chapter 6 it will be shown that this parallelogram-area notion of measure is simply the 2-dimensional case of a very much more general notion of measure applicable to entities of any grade.

Bound vectors

In mechanics the concept of force is paramount. In Volume 2 we will show that a force may be represented by a bound vector, and that a system of forces may be represented by a sum of bound vectors.

It has already been shown that a bound vector may be expressed either as the product of a point with a vector or as the product of two points.

$$P \wedge x == P \wedge (Q - P) == P \wedge Q \qquad Q == P + x$$

The bound vector in the form $P \wedge x$ *defines* a line through the point P in the direction of x. Similarly, in the form $P \wedge Q$ it defines the line through P and Q. These are of course the same line.

Here we use the word 'define' in the context discussed in the topic *Spaces and congruence* in section 2.3. To say that a bound vector L 'defines a line' means that the line may be defined as the set of all points P which belong to the space of L, that is, such that $L \wedge P = 0$. A consequence of this definition is that any other bound vector congruent to L defines the same line.

We depict a bound vector as located in its line in either of two ways, by:

1. A single point and a vector
2. Two points and their vector difference

In both cases the vector indicates the order of the factors in the exterior product.

Point∧vector depiction

Point∧point depiction

Two different depictions of a bound vector in its line

These graphical depictions of the bound vector are each misleading in their own way.

The first (point-vector) depiction suggests that the vector lies in the line. Since the vector is located nowhere, it is certainly not bound *to* the point. However as a component of the bound vector, we will often find it convenient to imagine it docked in the line, and to speak of it as bound *through* the point. And the point, again as a component of the bound vector, can be anywhere in the line since if P and P^* are any two points in the line, $P - P^*$ is a vector of the same direction as x and the bound vector can be expressed as either $P \wedge x$ or $P^* \wedge x$.

$$(P - P^*) \wedge x == 0 \qquad \Longrightarrow \qquad P \wedge x == P^* \wedge x$$

Hence, although a lone vector is located nowhere, and a lone point is immovably fixed, *the bound vector formed by the exterior product of the two is bound to a line through the point in the direction of the vector*. And to add to this representational conundrum, the bound vector has no specific location in the line. The following figure depicts a bound vector docked in two different locations in the line.

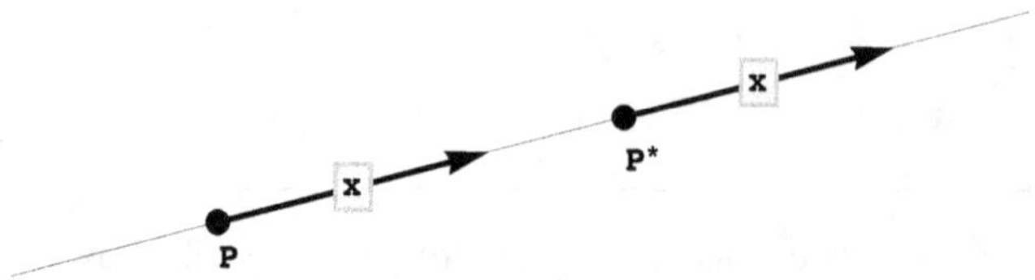

A bound vector displaying its ability to slide along its line

The second (point-point) depiction suggests that the depicted points are of specific importance over other points in the line. However, *any* pair of points in the line with the same vector difference may be used to express the bound vector. Let

$$x == Q - P == Q^* - P^*$$

then

$$P \wedge x == P^* \wedge x \quad \Longrightarrow \quad P \wedge (Q - P) == P^* \wedge (Q^* - P^*) \quad \Longrightarrow \quad P \wedge Q == P^* \wedge Q^*$$

A typical physical entity which has many of the properties of the bound vector is *force*. It is well known that a force has a 'line of action' and has the same effect on a body no matter how it is applied as long as it remains in this line.

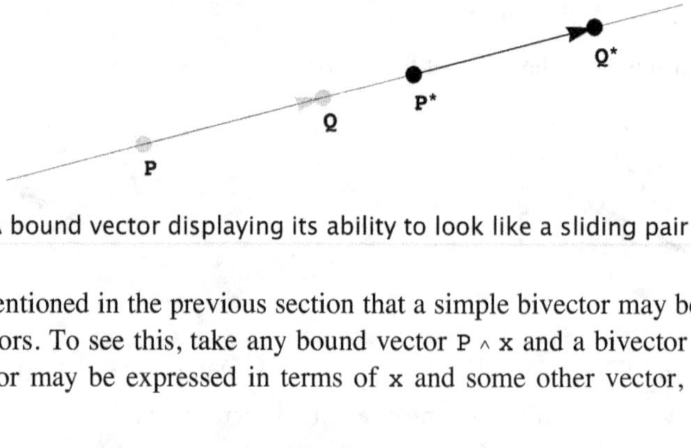

A bound vector displaying its ability to look like a sliding pair of points

It has been mentioned in the previous section that a simple bivector may be viewed as a 'carrier' of bound vectors. To see this, take any bound vector $P \wedge x$ and a bivector whose space contains x. The bivector may be expressed in terms of x and some other vector, y say, yielding $y \wedge x$. Thus:

$$P \wedge x + y \wedge x == (P + y) \wedge x == P^* \wedge x$$

In the plane, the addition of any bivector to any bound vector will shift the bound vector since they will always have a vector in common.

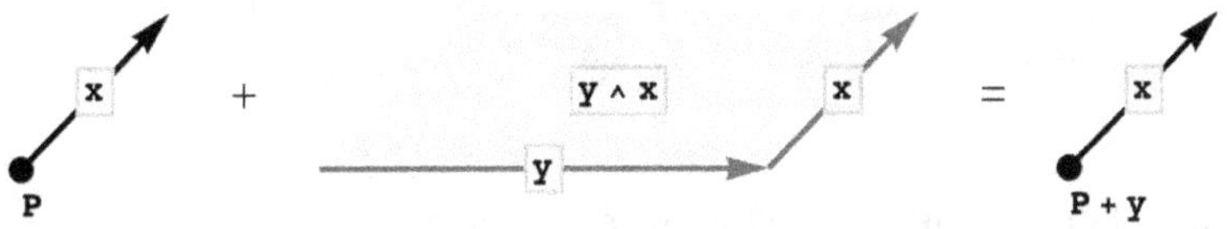

The bivector as a carrier of bound vectors

The geometric interpretation of the addition of such a simple bivector to a bound vector is then similar to that for the addition of a vector to a point, that is, a shift in position of the bound element.

Composing bivectors and bound vectors

◆ Composing bivectors

Once you have declared a basis for your bound vector space, say $\star\mathcal{P}_4$,

$\star\mathcal{P}_4$

$\{\star\mathbb{O}, e_1, e_2, e_3, e_4\}$

you can use quickly compose any bivectors in this basis with `ComposeBivector` or its alias $\star\mathbb{B}_\square$. The placeholder □ is used for entering the symbol upon which you want the coefficients to be based. Here we choose α.

$\mathbf{B}_\alpha = \star\mathbb{B}_\alpha$

$\alpha_1\, e_1 \wedge e_2 + \alpha_2\, e_1 \wedge e_3 + \alpha_3\, e_1 \wedge e_4 + \alpha_4\, e_2 \wedge e_3 + \alpha_5\, e_2 \wedge e_4 + \alpha_6\, e_3 \wedge e_4$

Just as with vectors and points, if you enter $\star\mathbb{B}_\square$, you get an expression with placeholders in which you can enter your own coefficients. (But if you want to do further manipulation, make sure they are declared as scalar symbols, say by using ★★S).

$\star\mathbb{B}_\square$

$\square\, e_1 \wedge e_2 + \square\, e_1 \wedge e_3 + \square\, e_1 \wedge e_4 + \square\, e_2 \wedge e_3 + \square\, e_2 \wedge e_4 + \square\, e_3 \wedge e_4$

◆ Composing simple bivectors

In a 4-space, bivectors are not necessarily simple. If you want to compose a simple bivector, you can use `ComposeSimpleBivector` or its alias $\star\mathbb{W}_\square$.

$\mathbf{B}_\beta = \star\mathbb{W}_\beta$

$(e_1\, \beta_1 + e_2\, \beta_2 + e_3\, \beta_3 + e_4\, \beta_4) \wedge (e_1\, \beta_5 + e_2\, \beta_6 + e_3\, \beta_7 + e_4\, \beta_8)$

◆ Composing bound vectors

Similarly, to compose a bound vector you can use `ComposeBoundVector`, or its alias $\star\mathbb{L}_\square$. The symbol $\mathbb{L}$ has been chosen because as we have seen, bound vectors are often used to define lines.

$\mathbf{L}_1 = \star\mathbb{L}_q$

$(\star\mathbb{O} + e_1\, q_1 + e_2\, q_2 + e_3\, q_3 + e_4\, q_4) \wedge (e_1\, q_5 + e_2\, q_6 + e_3\, q_7 + e_4\, q_8)$

The sum of two parallel bound vectors

Suppose we have two bound vectors $\mathbb{B}_1$ and $\mathbb{B}_2$, and we wish to explore how their sum might be expressed. In this section we explore the special case where they are parallel. The results are valid in a space of any dimension.

◆ The sum of two parallel bound vectors whose vectors are equal but of opposite sense

Consider a space of any dimension. Let $\mathbb{B}_1 = P \wedge x$ and $\mathbb{B}_2 = Q \wedge y$ be two bound vectors. Then their sum $\mathbb{B}$ is

$$\mathbb{B} == \mathbb{B}_1 + \mathbb{B}_2 == P \wedge x + Q \wedge y$$

In the case the bound vectors are parallel we can write y equal to a x (where a is a scalar) so that

$$\mathbb{B} == P \wedge x + Q \wedge y == P \wedge x + Q \wedge (a\, x) == (P + a\, Q) \wedge x$$

In the special case that y is equal to $-x$ (that is, a is -1), we have that $\mathbb{B}$ reduces to a *bivector*.

$$\mathbb{B} == (P - Q) \wedge x$$

In our discussion of mechanics in Volume 2, we will see that this is the algebraic equivalent of two oppositely directed forces of equal magnitude reducing to a couple.

The sum of two bound vectors, whose vectors are equal but of opposite sense, is a bivector

◆ The sum of two parallel bound vectors is in general a bound vector parallel to them

Supposing now that a is not -1, then $P + a\, Q$ is a weighted point of weight $1 + a$ situated on the line joining P and Q. We can shift this weight to the vectorial term (so that the bound vector again becomes the product of a (pure) point and a vector) by writing the bound vector $\mathbb{B}$ as

$$\mathbb{B} == R \wedge ((1 + a)\, x) \qquad R == \frac{P + a\, Q}{1 + a}$$

In mechanics, this is equivalent to two parallel forces reducing to a resultant force parallel to them. If the two parallel forces are of equal magnitude (a is equal to 1), the resultant force passes through a point mid-way between them.

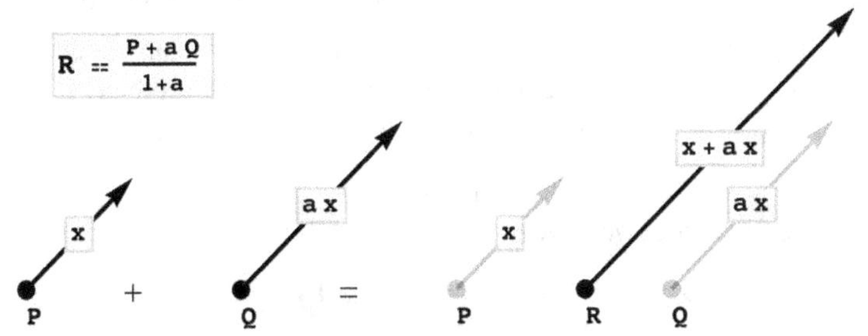

The sum of two parallel bound vectors is a bound vector parallel to them

The sum of two non-parallel bound vectors

We now turn to the more general case *where we assume the bound vectors are not parallel*. In a space of any dimension we can always express a sum of bound vectors in terms of an arbitrary point, R say, as the sum of a bound vector through R and a bivector B.

$$\mathbb{B} == P \wedge x + Q \wedge y == R \wedge (x + y) + (P - R) \wedge x + (Q - R) \wedge y == R \wedge (x + y) + B$$

Because our two bivectors are not parallel, the sum of their vectors cannot be zero. Hence the first term is a new bound vector which is not zero. The remaining terms are simple bivectors. In the general dimensional case, this sum of bivectors B may not be reducible to a simple bivector. In 3-dimensional space, the sum of bivectors B can always be reduced to a single simple bivector. In the plane, a point R can always be found which makes the sum B zero, leaving us with the result that the sum of two bound vectors in the plane (whose sum of vectors is not zero) can always be reduced to a single bound vector.

◆ The sum of two non-parallel bound vectors in the plane

In the plane, the point R is in fact the point of intersection of the lines of the bound vectors. We have seen earlier in the chapter that a bound vector is independent of the point used to express it, provided that the point lies on its line. We can thus imagine 'sliding' each bound vector along its line until its point reaches the point of intersection with the line of the other bound vector. We can then take the common point out of the sum as a factor, and we are left with the vector of the resultant being the sum of the vectors of the bound vectors.

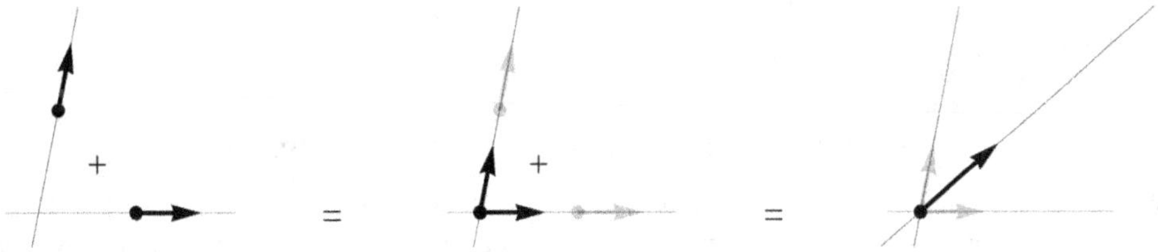

The sum of two non-parallel bound vectors in the plane
is a bound vector through their point of intersection

This result applies of course, not only to bound vectors in the plane, but to any pair of intersecting bound vectors in a space of any dimensions, since together they form a planar subspace.

As we might expect, given any two bound vectors *in the plane*, we can check if they intersect or are parallel by extracting a common factor from their regressive product. If they intersect, the common factor will be their point of intersection. If they are parallel, it will be their common vector.

- For example, if they intersect we have

```
★P2; B1 = (★O + a x + b y) ∧ x; B2 = (★O + c x + d y) ∧ y;
Simplify[ToCommonFactor[B1 ∨ B2]]
```

```
★c (★O + c x + b y) ⟨★O ∧ x ∧ y⟩
```

Whence their point of intersection can be extracted as

```
★O + c x + b y
```

(The expression ⟨★O∧x∧y⟩ represents the *coefficient* of ★O∧x∧y when ★O∧x∧y is expressed in terms of the *current basis* n-element (in this case 3-element). See the examples on using `ToCommonFactor` in section 3.6.)

◆ The sum of two non-parallel bound vectors in 3-space

If the lines of the two bound vectors in 3-space intersect, then the bound vectors together belong to a plane and the results of the previous section apply. We are thus left to explore the remaining case where the bivectors are neither parallel nor intersecting. In this case, the exterior product of the bound vectors will not be zero.

As we have seen above, the sum of two bound vectors in a space of any number of dimensions, may always be reduced to the sum of a bound vector and a bivector. The resultant bound vector can be chosen through an arbitrary point, but its vector must be the sum of the vectors of the original two bound vectors. The bivector will depend on the choice of the point defining the resultant bound vector. And in 3-space this bivector is guaranteed to be simple.

$$\mathbb{B} == P \wedge x + Q \wedge y == R \wedge (x + y) + B$$

To visualize how this works, we take a simple example. Define the two bound vectors as

$$\mathbb{L} == \star\mathbb{O} \wedge e_1 \qquad \mathbb{M} == (\star\mathbb{O} + 2\,e_2) \wedge e_3$$

Choose a convenient point through which to define the resultant bound vector, say half way between the points used to define the component bound vectors.

$$\mathbb{N} == (\star\mathbb{O} + e_2) \wedge (e_1 + e_3)$$

Then the bivector of the sum will be

$$B == \mathbb{L} + \mathbb{M} - \mathbb{N}$$
$$== \star\mathbb{O} \wedge e_1 + (\star\mathbb{O} + 2\,e_2) \wedge e_3 - (\star\mathbb{O} + e_2) \wedge (e_1 + e_3) == e_2 \wedge (e_3 - e_1)$$

The following depictions give two different views of the same summation. The original component bound vectors $\mathbb{L}$ and $\mathbb{M}$ are shown in light grey, the resultant bivector B in dark grey, and the resultant bound vector $\mathbb{N}$ in black.

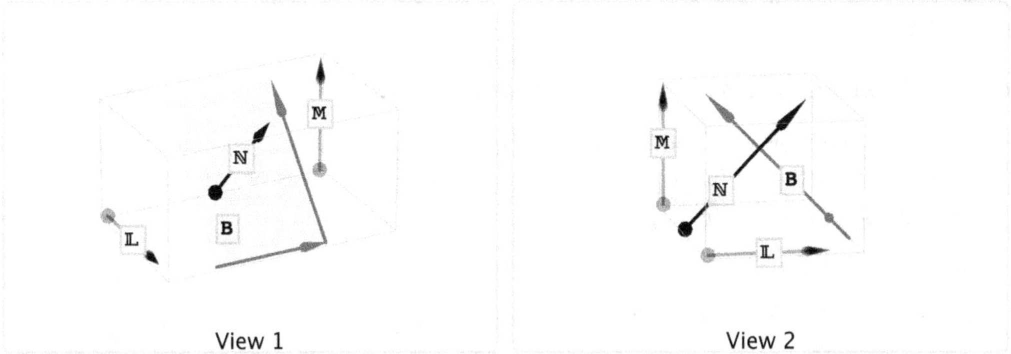

The sum of non–parallel bound vectors 𝕃 and 𝕄 gives a bound vector ℕ and a bivector B

We shall see in the next section that the sum of *any* number of bound vectors in 3-space may be reduced to the sum of a single bound vector and a single (simple) bivector.

- In Volume 2 we will see that bound vectors may be corresponded to forces and bivectors to moments. If the resultant bound vector force is chosen through a point which makes it and its bivector moment orthogonal, such a system of forces is called a *wrench*.

Sums of bound vectors

◆ Sums of bound vectors in the plane

We have seen in the sections above that the sum of two bound vectors in the plane (bound vector 2-space) is either a bound vector or a bivector or zero. We have also seen that the sum of a bound vector and a bivector is another bound vector.

A sum of any number of bound vectors in the plane, therefore, is clearly also reducible to a bound vector, a bivector, or zero; for we can simply add them two at a time.

◆ Sums of bound vectors in a bound vector 3-space

In a bound vector 3-space we can present the same sort of argument. The sum of two bound vectors is either a bound vector, a simple bivector, the sum of a bound vector and a simple bivector, or zero. In addition, the sum of two simple bivectors in 3-space is itself a simple bivector.

Thus again, we can add our bound vectors two at a time, and will end up with a bound vector, a simple bivector, the sum of a bound vector and a simple bivector, or zero.

In a bound vector 3-space, the sum of a bound vector and a bivector can also always be recast back into the sum of two bound vectors by adding and subtracting a bound vector.

$$\mathbf{P} \wedge \mathbf{x} + \mathbf{y} \wedge \mathbf{z} == (\mathbf{P} \wedge \mathbf{x} - \mathbf{P} \wedge \mathbf{z}) + (\mathbf{P} \wedge \mathbf{z} + \mathbf{y} \wedge \mathbf{z}) == \mathbf{P} \wedge (\mathbf{x} - \mathbf{z}) + (\mathbf{P} + \mathbf{y}) \wedge \mathbf{z}$$

◆ Sums of bound vectors in a bound vector *n*-space

Sums of bound vectors in a bound vector *n*-space ($n > 3$) differ from sums in a bound vector 3-space only in one aspect. The bivector may not be simple.

If the bivector is simple, a sum of bound vectors in *n*-space has essentially the same properties as one in 3-space.

◆ A criterion for a sum of bound vectors to be reducible to a single bound vector

A *non-zero* sum of bound vectors S can be reduced to a single bound vector only in the case that the vector of the bound vector is a factor of the bivector. That is, S can be expressed in the form

$$S == P \wedge x + y \wedge x == (P + y) \wedge x$$

In this case it is clear that the exterior square of S is zero.

$$S \wedge S == ((P + y) \wedge x) \wedge ((P + y) \wedge x) == 0$$

Suppose now that the vector of the bound vector is *not* a factor of the bivector. Then we have

$$S == P \wedge x + y \wedge z$$

where x, y, and z are independent. The exterior square of S is then

$$S \wedge S == (P \wedge x + y \wedge z) \wedge (P \wedge x + y \wedge z) == 2\, P \wedge x \wedge y \wedge z$$

This forms a straightforward test to see if the reduction of the sum to a single bound vector can be effected. If the exterior square is zero, the sum can be reduced to a single bound vector. If the exterior square is not zero, then the sum cannot be reduced to a single bound vector. We have already seen this property of exterior squares of 2-elements in our discussion of simplicity in Chapter 2, section 2.10.

◆ Sums of bound vectors: Summary

A *non-zero sum of bound vectors* $\Sigma P_i \wedge x_i$ (except in the case $\Sigma x_i = 0$) may always be reduced to the *sum of a bound vector and a bivector (or zero)*, since, by choosing an *arbitrary* point P, $\Sigma P_i \wedge x_i$ may always be written in the form:

$$\sum P_i \wedge x_i == P \wedge \left(\sum x_i\right) + \sum (P_i - P) \wedge x_i \qquad 4.1$$

The first term on the right hand side is a bound vector. The second term is a bivector. This decomposition is clearly valid in a space of any dimension.

If $\Sigma x_i = 0$ then the sum is a bivector. In spaces of vector dimension greater than 3, the bivector may not be simple.

If $\Sigma (P_i - P) \wedge x_i = 0$ for some P, then the sum is a bound vector.

Alternatively, if $(\Sigma P_i \wedge x_i) \wedge (\Sigma P_i \wedge x_i) = 0$, then the sum is a bound vector.

This transformation is of fundamental importance in our exploration of mechanics in Volume 2.

Example: Reducing a sum of bound vectors

Here is an example of a sum of bound vectors and its reduction to the sum of a bound vector and a bivector. We begin by declaring a bound vector space of three vector dimensions, so in this case the bivector is guaranteed to be simple.

```
★A; ★𝒫3;
```

Next, we define and enter the four bound vectors.

$$L_1 = P_1 \wedge x_1; \quad P_1 = \star O + e_1 + 3\,e_2 - 4\,e_3; \quad x_1 = e_1 - e_3;$$

$$L_2 = P_2 \wedge x_2; \quad P_2 = \star O + 2\,e_1 - e_2 - 2\,e_3; \quad x_2 = e_1 - e_2 + e_3;$$

$$L_3 = P_3 \wedge x_3; \quad P_3 = \star O - 5\,e_1 + 3\,e_2 - 6\,e_3; \quad x_3 = 2\,e_1 + 3\,e_2;$$

$$L_4 = P_4 \wedge x_4; \quad P_4 = \star O + 4\,e_1 + 2\,e_2 - 9\,e_3; \quad x_4 = 5\,e_3;$$

The sum of the four bound vectors is:

$$\mathbb{B} = \sum_{i=1}^{4} L_i$$

$$(\star O + 4\,e_1 + 2\,e_2 - 9\,e_3) \wedge (5\,e_3) + (\star O - 5\,e_1 + 3\,e_2 - 6\,e_3) \wedge (2\,e_1 + 3\,e_2) + (\star O + e_1 + 3\,e_2 - 4\,e_3) \wedge (e_1 - e_3) + (\star O + 2\,e_1 - e_2 - 2\,e_3) \wedge (e_1 - e_2 + e_3)$$

By expanding these products, simplifying and collecting terms, we obtain the sum of a bound vector (through the origin) and a bivector. We can use `GrassmannExpandAndSimplify` (or its alias $\star\mathcal{G}$) to do the computations for us.

$$\mathbb{B} = \star\mathcal{G}[\mathbb{B}]$$

$$\star O \wedge (4\,e_1 + 2\,e_2 + 5\,e_3) - 25\,e_1 \wedge e_2 + 39\,e_1 \wedge e_3 + 22\,e_2 \wedge e_3$$

We could just as well have expressed this 2-element as bound through (for example) the point $\star O + e_1$. To do this, we simply add $e_1 \wedge (4\,e_1 + 2\,e_2 + 5\,e_3)$ to the bound vector and subtract it from the bivector to get:

$$(\star O + e_1) \wedge (4\,e_1 + 2\,e_2 + 5\,e_3) - 27\,e_1 \wedge e_2 + 34\,e_1 \wedge e_3 + 22\,e_2 \wedge e_3$$

We can of course also express the bivector in factored form in many ways. Using `ExteriorFactorize` (discussed in section 3.9) gives a default result.

$$\text{ExteriorFactorize}[-27\,e_1 \wedge e_2 + 34\,e_1 \wedge e_3 + 22\,e_2 \wedge e_3]$$

$$-27\left(e_1 + \frac{22\,e_3}{27}\right) \wedge \left(e_2 - \frac{34\,e_3}{27}\right)$$

Finally, we can check to see if the sum can be reduced to a bound vector by calculating its exterior square.

$$\star\mathcal{G}[\mathbb{B} \wedge \mathbb{B}]$$

$$-230\,\star O \wedge e_1 \wedge e_2 \wedge e_3$$

Since this is not zero, the reduction of the sum to a single bound vector is not possible in this case.

4.4 Geometrically Interpreted *m*-Elements

Types of geometrically interpreted *m*-elements

In $\underset{m}{\Lambda}$ the situation is analogous to that in $\underset{2}{\Lambda}$. A simple product of *m* points and vectors may always be reduced to the product of a point and a simple (*m*–1)-vector by subtracting one of the points from all of the others. For example, take the exterior product of three points and two vectors. By subtracting the first point, say, from the other two, we can cast the product into the form of a bound simple 4-vector.

$$P \wedge P_1 \wedge P_2 \wedge x \wedge y \;==\; P \wedge (P_1 - P) \wedge (P_2 - P) \wedge x \wedge y$$

There are thus only two *simple* interpreted elements in $\underset{m}{\Lambda}$:

1. $x_1 \wedge x_2 \wedge \ldots \wedge x_m$ (the simple m-vector).

2. $P \wedge x_2 \wedge \ldots \wedge x_m$ (the bound simple (m–1)-vector).

A sum of simple m-vectors is called an m-vector.

If $\underset{m}{\alpha}$ is a (not necessarily simple) m-vector, then $P \wedge \underset{m}{\alpha}$ is called a *bound m-vector*.

A sum of bound m-vectors may always be reduced to the sum of a bound m-vector and an (m+1)-vector (with the proviso that either or both may be zero).

These interpreted elements and their relationships will be discussed further in the following sections.

The *m*-vector

The simple m-vector, or *multivector*, is the multidimensional equivalent of the vector. As with a vector, it does not have the property of location. The m-dimensional vector space of a simple m-vector may be used to define the multidimensional *direction* of the m-vector.

If we had access to m-dimensional paper, we would depict the simple m-vector, just as we did for the bivector, by depicting its vectors *conveniently* docked head to tail *in the order of the product*, and 'joined' by a somewhat transparent m-dimensional parallelepiped.

In practice of course, we will only be depicting vectors, bivectors and trivectors. A trivector may be depicted by its three vectors in order joined by a parallelepiped.

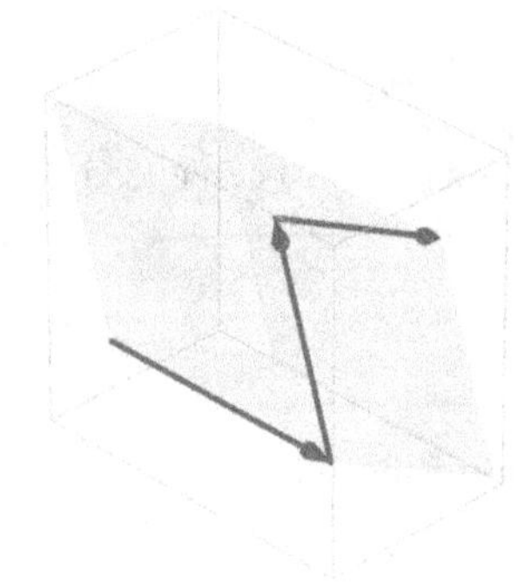

Depicting a simple trivector. A trivector is located nowhere.

The orientation is given by the order of the factors in the simple m-vector. An interchange of any two factors produces an m-vector of opposite orientation. By the anti-symmetry of the exterior product, there are just two distinct orientations.

In section 6.6, it will be shown that the measure of a trivector may be geometrically interpreted as the volume of this parallelepiped. However, the depiction of the simple trivector in the manner above suffers from similar defects to those already described for the bivector: namely, it incorrectly suggests a specific location and shape of the parallelepiped.

A simple m-vector may also be viewed as a *carrier* of bound simple $(m-1)$-vectors in a manner analogous to that already described for the simple bivector as a carrier of bound vectors. Thus, a simple trivector may be viewed as a carrier for bound simple bivectors. (See below for a depiction of this.)

A sum of simple m-vectors (that is, an m-vector) is not necessarily reducible to a simple m-vector, except in $\underset{0}{\Lambda}$, $\underset{1}{\Lambda}$, $\underset{n-1}{\Lambda}$, and $\underset{n}{\Lambda}$.

The bound m-vector

The exterior product of a point and an m-vector is called a *bound m-vector*. Note that it belongs to $\underset{m+1}{\Lambda}$. A sum of bound m-vectors is not necessarily a bound m-vector. However, it may in general be reduced to the sum of a bound m-vector and an $(m+1)$-vector as follows:

$$\sum P_i \wedge \underset{m}{\alpha_i} == \sum (P + \beta_i) \wedge \underset{m}{\alpha_i} == P \wedge \sum \underset{m}{\alpha_i} + \sum \beta_i \wedge \underset{m}{\alpha_i} \qquad 4.2$$

The first term on the right hand side is a bound m-vector (providing that the sum of the $\underset{m}{\alpha_i}$ is not zero); and the second term is an $(m+1)$-vector.

When $m = 0$, a bound 0-vector or *bound scalar* $P \wedge a$ ($= a\,P$) is seen to be equivalent to a weighted point.

When $m = n$ (n being the dimension of the underlying *vector space*), any bound n-vector is but a scalar multiple of the basis $(n+1)$-element. (The default basis $(n+1)$-element is denoted $\star\mathbb{O} \wedge e_1 \wedge e_2 \wedge \ldots \wedge e_n$.)

The graphical depiction of bound simple m-vectors presents even greater difficulties than those already discussed for bound vectors. As in the case of the bound vector, the point used to express the bound simple m-vector is not unique.

In section 4.6 we will see how bound simple m-vectors may be used to define multiplanes.

Bound simple m-vectors expressed by points

A bound simple m-vector may always be expressed as a product of $m+1$ points.
Let $P_i = P_0 + x_i$, then:

$$P_0 \wedge x_1 \wedge x_2 \wedge \ldots \wedge x_m$$
$$== P_0 \wedge (P_1 - P_0) \wedge (P_2 - P_0) \wedge \ldots \wedge (P_m - P_0)$$
$$== P_0 \wedge P_1 \wedge P_2 \wedge \ldots \wedge P_m$$

Conversely, as we have already seen, a product of $m+1$ points may always be expressed as the product of a point and a simple m-vector by subtracting one of the points from all of the others.

The *m-vector of a bound simple m-vector* $P_0 \wedge P_1 \wedge P_2 \wedge \ldots \wedge P_m$ may thus be expressed in terms of these points as:

$$(P_1 - P_0) \wedge (P_2 - P_0) \wedge \ldots \wedge (P_m - P_0)$$

A particularly symmetrical formula results from the expansion of this product reducing it to a form no longer showing preference for P_0.

$$(P_1 - P_0) \wedge (P_2 - P_0) \wedge \ldots \wedge (P_m - P_0) = \sum_{i=0}^{m} (-1)^i P_0 \wedge P_1 \wedge \ldots \wedge \square_i \wedge \ldots \wedge P_m \qquad 4.3$$

Here $\square_i$ denotes deletion of the factor P_i from the product.

Bound simple bivectors

In this section, by way of example, we discuss the special case of the bound simple 2-vector, or bound simple bivector.

A bound bivector is a 3-element. Any exterior product of three 1-elements, in which at least one of them is a point, is a bound simple bivector. (For brevity in this section, we will suppose all the bound bivectors discussed are bound simple bivectors.)

Consider three points in a space of any number of dimensions:

$$P$$
$$Q = P + x$$
$$R = P + y$$

We can construct the *same* bound bivector $\mathbb{B}$ from these points in a number of ways

$$\mathbb{B} = P \wedge Q \wedge R = Q \wedge R \wedge P = R \wedge P \wedge Q$$

By substituting for R and expanding, we see that $\mathbb{B}$ can also be expressed as

$$\mathbb{B} = P \wedge Q \wedge y = Q \wedge y \wedge P = y \wedge P \wedge Q$$

Substituting again for Q and expanding gives

$$\mathbb{B} = P \wedge x \wedge y = x \wedge y \wedge P = y \wedge P \wedge x$$

Thus we have nine (inessentially different) ways of constructing or denoting the same bound bivector. From here on we will usually only discuss bound bivectors in any of the three *conceptually* distinct forms in which points are denoted before vectors:

$$\mathbb{B} = P \wedge Q \wedge R = P \wedge Q \wedge y = P \wedge x \wedge y$$

We depict each of these forms in a slightly different way: in each case emphasizing the entities involved in the form, but in all cases showing the order of factors in the product by the chain of arrows. (These graphical conventions will also apply to our discussion of bound trivectors later.)

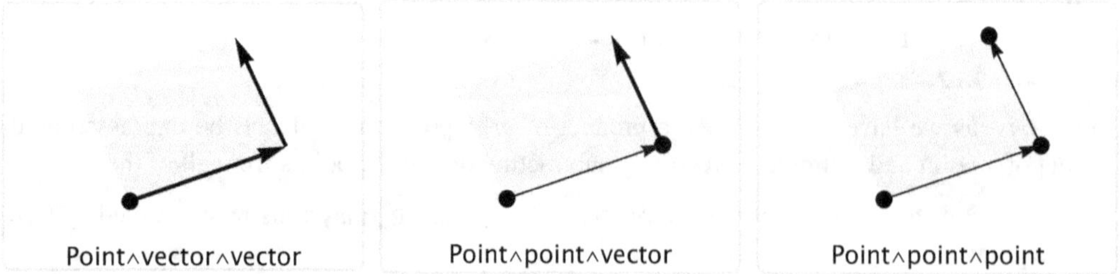

Three different depictions of a bound bivector

The bound bivector $\mathbb{B}$ defines a *plane* through the points P, Q and R in the directions of x and y.

Here we use the word 'define' in the context discussed in the appendix Terminology and Glossary. To say that a bound bivector $\mathbb{B}$ 'defines a plane' means that the plane may be defined as the set of all points P which belong to the space of $\mathbb{B}$, that is, such that $\mathbb{B} \wedge P = 0$. A consequence of this definition is that any other bound bivector congruent to $\mathbb{B}$ defines the same plane.

These graphical depictions of the bound bivector are each misleading in their own way. Since the bound bivector *defines* a plane, we can conceive of it as lying *in* the plane. However, just as a bound vector can lie anywhere in its line, *a bound bivector can lie anywhere in its plane*.

The following diagram depicts a bound bivector docked in three different locations in its plane. The depictions each have the same vectors, but differ in the point of the plane they use.

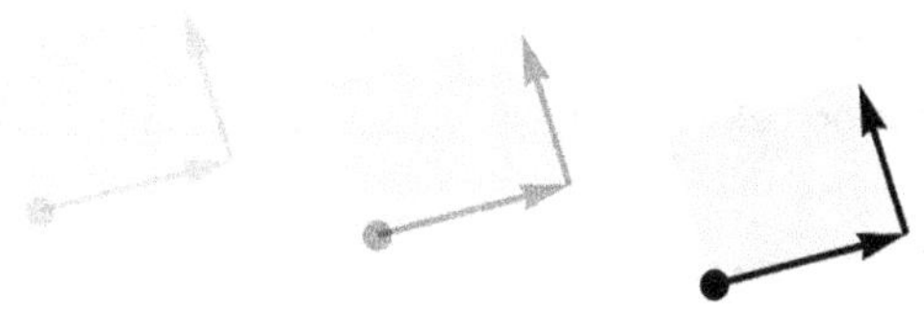

A bound bivector displaying its ability to slide around in its plane

Hence, although a lone bivector is located nowhere, and a lone point is immovably fixed, *the bound bivector formed by the exterior product of the two is bound to a plane through the point in the direction of the bivector*.

It has been shown earlier that a simple bivector may be viewed as a 'carrier' of bound vectors. In a similar manner, a simple trivector may be viewed as a 'carrier' of bound bivectors. To see this, take any bound bivector $P \wedge y \wedge z$ and a trivector $x \wedge y \wedge z$. Thus:

$$P \wedge y \wedge z + x \wedge y \wedge z == (P + x) \wedge y \wedge z$$

The geometric interpretation of the addition of such a simple trivector to a bound bivector is then similar to that for the addition of a vector to a point, that is, a shift in position of the bound element.

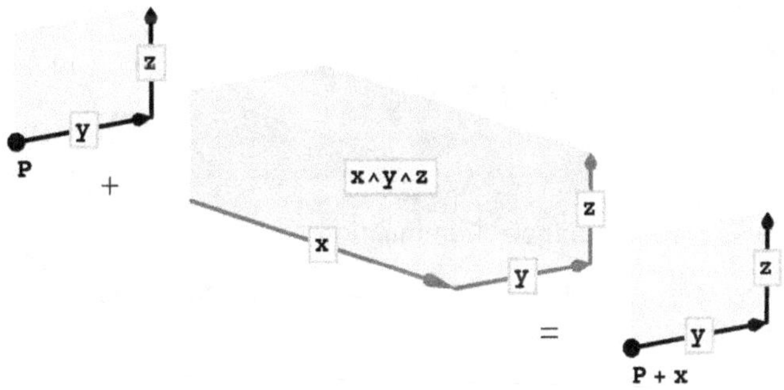

The trivector as a carrier of bound bivectors

Composing *m*-vectors and bound *m*-vectors

◆ Composing *m*-vectors

Suppose that we are interested in a bound vector space of 3 (vector) dimensions.

```
★𝒫3
{★O, e1, e2, e3}
```

Even though we are in a bound vector space we can still can compose *m*-vectors with `ComposeMVector`. In a bound 3-space, we have three possible *m*-vectors. Here we base the coefficients on the symbol a.

```
Grid[Table[{m, ComposeMVector[m, a]}, {m, 1, 3}], Alignment → Left]
```

1 $a_1\, e_1 + a_2\, e_2 + a_3\, e_3$
2 $a_1\, e_1 \wedge e_2 + a_2\, e_1 \wedge e_3 + a_3\, e_2 \wedge e_3$
3 $a\, e_1 \wedge e_2 \wedge e_3$

◆ Composing simple *m*-vectors

Similarly we can compose simple *m*-vectors with `ComposeSimpleMVector`. If we wish, we can use a placeholder to enable later input of specific coefficients.

```
Grid[Table[{m, ComposeSimpleMVector[m, □]}, {m, 1, 3}],
  Alignment → Left]
```

1 $\Box\, e_1 + \Box\, e_2 + \Box\, e_3$
2 $(\Box\, e_1 + \Box\, e_2 + \Box\, e_3) \wedge (\Box\, e_1 + \Box\, e_2 + \Box\, e_3)$
3 $(\Box\, e_1 + \Box\, e_2 + \Box\, e_3) \wedge (\Box\, e_1 + \Box\, e_2 + \Box\, e_3) \wedge (\Box\, e_1 + \Box\, e_2 + \Box\, e_3)$

◆ Composing bound simple *m*-vectors

```
Grid[Table[{m, ComposeMPlaneElement[m, □]}, {m, 1, 3}],
  Alignment → Left]
```

1 $(\star\mathbb{O} + \Box\, e_1 + \Box\, e_2 + \Box\, e_3) \wedge (\Box\, e_1 + \Box\, e_2 + \Box\, e_3)$
2 $(\star\mathbb{O} + \Box\, e_1 + \Box\, e_2 + \Box\, e_3) \wedge (\Box\, e_1 + \Box\, e_2 + \Box\, e_3) \wedge (\Box\, e_1 + \Box\, e_2 + \Box\, e_3)$
3 $(\star\mathbb{O} + \Box\, e_1 + \Box\, e_2 + \Box\, e_3) \wedge (\Box\, e_1 + \Box\, e_2 + \Box\, e_3) \wedge$
$(\Box\, e_1 + \Box\, e_2 + \Box\, e_3) \wedge (\Box\, e_1 + \Box\, e_2 + \Box\, e_3)$

◆ Composing bound *m*-vectors

Of course you can also combine composition functions.

```
Grid[Table[{m, ★ℙ□ ∧ ComposeMVector[m, □]}, {m, 1, 3}],
  Alignment → Left]
```

1 $(\star\mathbb{O} + \Box\, e_1 + \Box\, e_2 + \Box\, e_3) \wedge (\Box\, e_1 + \Box\, e_2 + \Box\, e_3)$
2 $(\star\mathbb{O} + \Box\, e_1 + \Box\, e_2 + \Box\, e_3) \wedge (\Box\, e_1 \wedge e_2 + \Box\, e_1 \wedge e_3 + \Box\, e_2 \wedge e_3)$
3 $(\star\mathbb{O} + \Box\, e_1 + \Box\, e_2 + \Box\, e_3) \wedge (\Box\, e_1 \wedge e_2 \wedge e_3)$

4.5 Geometrically Interpreted Spaces

Vector and point spaces

The space of a simple non-zero m-element A has been defined in section 2.3 as the set of 1-elements x whose exterior product with A is zero: $\{x : A \wedge x = 0\}$.

If x is interpreted as a vector v, then the *vector space* of A is defined as $\{v : A \wedge v = 0\}$.

If x is interpreted as a point P, then the *point space* of A is defined as $\{P : A \wedge P = 0\}$.

The point space of a simple m-vector is empty.

The vector space of a simple m-vector is an m-dimensional vector space. Conversely, the m-dimensional vector space may be said to be *defined* by the m-vector.

The point space of a bound simple m-vector is called an *m-plane* (sometimes *multiplane*). Thus the point space of a bound vector is a 1-plane (or line) and the point space of a bound simple bivector is a 2-plane (or, simply, a plane). The m-plane will be said to be *defined* by the bound simple m-vector.

The vector space of a bound simple m-vector is an m-dimensional vector space.

The geometric interpretation for the notion of set inclusion is taken as 'to lie in'. Thus for example, a point may be said *to lie in* an m-plane.

The point and vector spaces of bound simple m-vectors are tabulated below.

Point and vector spaces of bound simple m–vectors

m	Bound Simple m–Vector	Point Space	Vector Space
0	bound scalar	point	
1	bound vector	line	1–dimensional vector space
2	bound simple bivector	plane	2–dimensional vector space
m	bound simple m–vector	m–plane	m–dimensional vector space
$n-1$	bound $(n-1)$–vector	hyperplane	$(n-1)$–dimensional vector space
n	bound n–vector	n–plane	n–dimensional vector space

Two congruent bound simple m-vectors $P \wedge A$ and $c\, P \wedge A$ define the same m-plane. Thus, for example if A is a scalar, the point $\star\mathbb{O} + x$ and the weighted point $2\,(\star\mathbb{O} + x)$ define the same point.

◆ *A note on bound simple m-vectors as m-planes*

We have defined an m-plane as a set of points *defined by* a bound simple m-vector. It will often turn out however to be more convenient and conceptually fruitful to work with m-planes *as if* they were the bound simple m-vectors which define them. This is in fact the approach taken by Grassmann and the early workers in the Grassmannian tradition (for example, [Whitehead 1898], [Hyde 1884-1906] and [Forder 1941]). This will be satisfactory provided that the equality relationship we define for m-planes is that of *congruence* rather than the more specific *equals*.

Thus we may, when speaking in a geometric context, refer to a bound simple *m*-vector as an *m*-plane and *vice versa*. Hence in saying Π is an *m*-plane, we are also saying that all a Π (where a is a scalar factor, not zero) is the same *m*-plane.

Coordinate spaces

The *coordinate spaces* of a Grassmann algebra are the spaces defined by the basis elements.

The *coordinate m-spaces* are the spaces defined by the basis elements of $\underset{m}{\Lambda}$. For example, if $\underset{1}{\Lambda}$ has basis $\{\star\mathbb{O}, e_1, e_2, e_3\}$, that is, we are working in a bound vector 3-space, then the Grassmann algebra it generates has basis:

```
★P3; Flatten[GrassmannBases]
```

$\{1, \star\mathbb{O}, e_1, e_2, e_3, \star\mathbb{O}\wedge e_1, \star\mathbb{O}\wedge e_2, \star\mathbb{O}\wedge e_3, e_1\wedge e_2, e_1\wedge e_3, e_2\wedge e_3, \star\mathbb{O}\wedge e_1\wedge e_2, \star\mathbb{O}\wedge e_1\wedge e_3, \star\mathbb{O}\wedge e_2\wedge e_3, e_1\wedge e_2\wedge e_3, \star\mathbb{O}\wedge e_1\wedge e_2\wedge e_3\}$

Each one of these basis elements defines a coordinate space. Most familiar are the coordinate *m*-planes. The basis 1-planes $\star\mathbb{O}\wedge e_1$, $\star\mathbb{O}\wedge e_2$, $\star\mathbb{O}\wedge e_3$ define the coordinate axes, while the basis 2-planes $\star\mathbb{O}\wedge e_1\wedge e_2$, $\star\mathbb{O}\wedge e_1\wedge e_3$, $\star\mathbb{O}\wedge e_2\wedge e_3$ define the coordinate planes. Additionally however there are the coordinate vectors e_1, e_2, e_3 and the coordinate bivectors $e_1\wedge e_2, e_1\wedge e_3, e_2\wedge e_3$.

Perhaps less familiar is the fact that there are no coordinate *m*-planes in a vector space, but rather simply coordinate *m*-vectors.

Geometric dependence

In Chapter 2 the notion of dependence was discussed for elements of a linear space. Non-zero 1-elements are said to be dependent if and only if their exterior product is zero.

If the elements concerned have been endowed with a geometric interpretation, the notion of dependence takes on an additional geometric interpretation, as the following table shows. The x_i are vectors and the P_i are points.

$x_1\wedge x_2 = 0$ $P_1\wedge P_2 = 0$	x_1, x_2 are parallel (co–directional) P_1, P_2 are coincident
$x_1\wedge x_2\wedge x_3 = 0$ $P_1\wedge P_2\wedge P_3 = 0$	x_1, x_2, x_3 are co–2–directional (or parallel) P_1, P_2, P_3 are collinear (or coincident)
$x_1\wedge \ldots \wedge x_m = 0$ $P_1\wedge \ldots \wedge P_m = 0$	$x_1, \ldots, x_m$ are co–k–directional, $k < m$ $P_1, \ldots, P_m$ are co–k–planar, $k < m-1$

Thus, co-directionality and coincidence, while being distinct interpretive geometric notions, are equivalent algebraically.

Geometric duality

The concept of duality introduced in Chapter 3 is most striking when interpreted geometrically. Suppose:

P defines a point
L defines a line

π defines a plane
V defines a 3-plane

In what follows we tabulate the dual relationships of these bound or 'positioned' entities to each other.

◆ **Duality in a plane**

In a plane there are just three types of bound geometric entity: points, lines and planes. In the table below we can see that in the plane, points and lines are 'dual' entities, and planes and scalars are 'dual' entities, because their definitions convert under the application of the Duality Principle (see section 3.2).

$L \equiv P_1 \wedge P_2$	$P \equiv L_1 \vee L_2$
$\pi \equiv P_1 \wedge P_2 \wedge P_3$	$1 \equiv L_1 \vee L_2 \vee L_3$
$\pi \equiv L \wedge P$	$1 \equiv P \vee L$

◆ **Duality in a 3-plane**

In the 3-plane there are just four types of bound geometric entity: points, lines, planes and 3-planes. In the table below we can see that in the 3-plane, lines are self-dual, points and planes are now dual, and scalars are now dual to 3-planes.

$L \equiv P_1 \wedge P_2$	$L \equiv \pi_1 \vee \pi_2$
$\pi \equiv P_1 \wedge P_2 \wedge P_3$	$P \equiv \pi_1 \vee \pi_2 \vee \pi_3$
$V \equiv P_1 \wedge P_2 \wedge P_3 \wedge P_4$	$1 \equiv \pi_1 \vee \pi_2 \vee \pi_3 \vee \pi_4$
$\pi \equiv L \wedge P$	$P \equiv L \vee \pi$
$V \equiv L_1 \wedge L_2$	$1 \equiv L_1 \vee L_2$
$V \equiv \pi \wedge P$	$1 \equiv P \vee \pi$

◆ **Duality in an *n*-plane**

From these cases the types of relationships in higher dimensions may be composed straightforwardly. For example, if P defines a point and H defines a hyperplane ((*n*–1)-plane), then we have the dual formulations:

$H \equiv P_1 \wedge P_2 \wedge \ldots \wedge P_{n-1}$	$P \equiv H_1 \vee H_2 \vee \ldots \vee H_{n-1}$

4.6 *m*-Planes

In this section we summarize three different ways we can represent an *m*-plane: the first in terms of a simple exterior product of points, or points and vectors; the second as an *m*-vector; and the third as an exterior quotient.

Note carefully that an *m*-plane is strictly *a set of points*, but that since the properties of the Grassmann algebra *representation* of an *m*-plane as a bound *m*-vector are so faithfully useful in their geometric manipulations, we will often abuse the terminology by saying that "an *m*-plane *is* a bound *m*-vector", rather than "an *m*-plane *is represented by* a bound *m*-vector", or "an *m*-plane *is defined by* a bound *m*-vector".

m-planes defined by points or bound m-vectors

Grassmann and those who wrote in the style of the *Ausdehnungslehre* considered the point more fundamental than the vector for exploring geometry. This approach indeed has its merits. An *m*-plane is quite straightforwardly defined and expressed as the (space of the) exterior product of *m*+1 points or its equivalent in terms of its (vectorial) point differences.

$$\Pi \;==\; P_0 \wedge P_1 \wedge P_2 \wedge \ldots \wedge P_m$$

$$\Pi \;==\; P_0 \wedge (P_1 - P_0) \wedge (P_2 - P_0) \wedge \ldots \wedge (P_m - P_0)$$

$$\Pi \;==\; P_0 \wedge x_1 \wedge x_2 \wedge \ldots \wedge x_m$$

m-planes defined by *m*-vectors

Consider a bound simple *m*-vector Π.

$$\Pi \;==\; P_0 \wedge x_1 \wedge x_2 \wedge \ldots \wedge x_m$$

Its *m*-plane is the set of points P such that $P \wedge \Pi = 0$.

$$P \wedge P_0 \wedge x_1 \wedge x_2 \wedge \ldots \wedge x_m \;==\; 0$$

This equation is equivalent to the statement: *there exist scalars* a, a_0, a_i, *not all zero, such that*:

$$a\,P + a_0\,P_0 + \sum a_i\,x_i \;==\; 0$$

And since this is only possible if $a = -a_0$ (since for the sum to be zero, it must be a sum of vectors) then we can rewrite the condition as:

$$a\,(P - P_0) + \sum a_i\,x_i \;==\; 0$$

or equivalently:

$$(P - P_0) \wedge x_1 \wedge x_2 \wedge \ldots \wedge x_m \;==\; 0$$

We are thus lead to the following alternative definition of an *m*-plane: an *m*-plane defined by the bound simple *m*-vector Π is the set of points P

$$\{P \;:\; (P - P_0) \wedge x_1 \wedge x_2 \wedge \ldots \wedge x_m \;==\; 0\}$$

This is of course equivalent to the usual definition of an *m*-plane. That is, since the vectors $P - P_0, x_1, x_2, \ldots, x_m$ are dependent, then for scalar parameters t_i we have

$$P - P_0 \;==\; t_1\,x_1 + t_2\,x_2 + \ldots + t_m\,x_m \qquad 4.4$$

m-planes as exterior quotients

The alternative definition of an *m*-plane developed above shows that an *m*-plane may be defined as the set of points P such that

$$P \wedge x_1 \wedge x_2 \wedge \ldots \wedge x_m \;==\; P_0 \wedge x_1 \wedge x_2 \wedge \ldots \wedge x_m$$

'Solving' for P and noting from section 2.11 that the quotient of an (*m*+1)-element by an *m*-element contained in it is a 1-element with *m* arbitrary scalar parameters, we can write

$$P = \frac{P_0 \wedge x_1 \wedge x_2 \wedge \ldots \wedge x_m}{x_1 \wedge x_2 \wedge \ldots \wedge x_m} = P_0 + t_1 x_1 + t_2 x_2 + \ldots + t_m x_m$$

$$P = \frac{P_0 \wedge x_1 \wedge x_2 \wedge \ldots \wedge x_m}{x_1 \wedge x_2 \wedge \ldots \wedge x_m} \qquad 4.5$$

Computing exterior quotients

In section 2.11 we introduced the *GrassmannAlgebra* function `ExteriorQuotient`. In what follows we use `ExteriorQuotient` to compute the expressions for various *m*-planes in a bound 3-space.

Note that all the 1-elements in the numerator must be declared vector symbols, so we need to declare any points we use.

```
★A; ★P3; DeclareExtraVectorSymbols[P];
```

◆ Points

The exterior quotient of a weighted point by its weight gives (as would be expected), the (pure) point.

$$\text{ExteriorQuotient}\left[\frac{P \wedge a}{a}\right]$$

P

◆ Points in a line

The exterior quotient of a bound vector with its vector gives all the points in the line defined by the bound vector, parametrized by the scalar $\star t_1$.

$$\text{ExteriorQuotient}\left[\frac{P \wedge x}{x}\right]$$

$P + x \star t_1$

We can see that any other point used to define the bound vector gives effectively the same parametrization.

$$\text{ExteriorQuotient}\left[\frac{(P + a\, x) \wedge x}{x}\right]$$

$P + a\, x + x \star t_1$

◆ Points in a plane

The exterior quotient of a bound simple bivector with its simple bivector gives all the points in the plane defined by the bound simple bivector, parametrized by the scalars $\star t_1$ and $\star t_2$.

$$\text{ExteriorQuotient}\left[\frac{P \wedge x \wedge y}{x \wedge y}\right]$$

$P + x \star t_1 + y \star t_2$

◆ Points in a 3-plane

The exterior quotient of a bound simple trivector with its simple trivector gives all the points in the 3-plane defined by the bound simple trivector, parametrized by the scalars $\star t_1$, $\star t_2$ and $\star t_3$.

$$\text{ExteriorQuotient}\left[\frac{P\wedge x\wedge y\wedge z}{x\wedge y\wedge z}\right]$$

$$P + x\,\star t_1 + y\,\star t_2 + z\,\star t_3$$

◆ Lines in a plane

The exterior quotient of a bound simple bivector with a vector in it gives a set of bound vectors. These bound vectors define lines in the plane of the bound simple bivector, parametrized by the scalars $\star t_1$ and $\star t_2$. They are characterized by the fact that their exterior product with the vector gives the original bound simple bivector.

$$\text{ExteriorQuotient}\left[\frac{P\wedge x\wedge y}{y}\right]$$

$$(P + y\,\star t_1)\wedge(x + y\,\star t_2)$$

◆ Lines in a 3-plane

The exterior quotient of a bound simple trivector with a bivector in it gives a set of bound vectors. These bound vectors define lines in the 3-plane of the bound simple trivector, parametrized by the scalars $\star t_1$, $\star t_2$, $\star t_3$ and $\star t_4$. They are characterized by the fact that their exterior product with the bivector gives the original bound simple trivector.

$$\text{ExteriorQuotient}\left[\frac{P\wedge x\wedge y\wedge z}{y\wedge z}\right]$$

$$(P + y\,\star t_1 + z\,\star t_2)\wedge(x + y\,\star t_3 + z\,\star t_4)$$

◆ Planes in a 3-plane

The exterior quotient of a bound simple trivector with a vector in it gives a set of bound bivectors. These bound bivectors define planes in the 3-plane of the bound simple trivector, parametrized by the scalars $\star t_1$, $\star t_2$ and $\star t_3$. They are characterized by the fact that their exterior product with the vector gives the original bound simple trivector.

$$\text{ExteriorQuotient}\left[\frac{P\wedge x\wedge y\wedge z}{z}\right]$$

$$(P + z\,\star t_1)\wedge(x + z\,\star t_2)\wedge(y + z\,\star t_3)$$

The *m*-vector of a bound *m*-vector

◆ Applying the ∂ operator

We can define an operator ∂ which takes a simple (*m*+1)-element and converts it to an *m*-element form

$$\partial\,(P_0\wedge P_1\wedge P_1\wedge\ldots\wedge P_m) == (P_1 - P_0)\wedge(P_2 - P_0)\wedge\ldots\wedge(P_m - P_0)$$

The interesting property of this operation is that when it is applied twice, the result is zero.

Operationally, $\bar{\partial}^2 = 0$. For example:

$$\partial(P) == 1$$

$$\partial(P_0 \wedge P_1) == P_1 - P_0$$
$$\partial(P_1 - P_0) == 1 - 1 == 0$$

$$\partial(P_0 \wedge P_1 \wedge P_2) == P_1 \wedge P_2 + P_2 \wedge P_0 + P_0 \wedge P_1$$
$$\partial(P_1 \wedge P_2 + P_2 \wedge P_0 + P_0 \wedge P_1) == P_2 - P_1 + P_0 - P_2 + P_1 - P_0 == 0$$

Remember that $P_1 \wedge P_2 + P_2 \wedge P_0 + P_0 \wedge P_1$ is simple since it may be expressed as

$$(P_1 - P_0) \wedge (P_2 - P_0) == (P_2 - P_1) \wedge (P_0 - P_1) == (P_0 - P_2) \wedge (P_1 - P_2)$$

This property of nilpotence is shared by the boundary operator of algebraic topology and the exterior derivative. Furthermore, if a product with a given 1-element is considered an operation, then the exterior, regressive and interior products are all likewise nilpotent.

◆ Taking the interior product with the origin

In Chapter 6 we will show that under the hybrid metric we will adopt for bound vector spaces (in which the origin is orthogonal to all vectors), taking the interior product of any *bound* element with the origin ⋆O has the same effect as the ∂ operator.

Applied to a bound *m*-vector, this operator generates the *m*-vector, thus creating a 'free' entity from a bound one. Applied a second time to the result gives zero, since the *m*-vector is already free.

◆ Composing the first co-span

Another way of generating the *m*-vector from a bound *m*-vector is to compose the first cospan (see section 2.12) of the bound *m*-vector, and then sum the elements.

Suppose we have a bound trivector Ω in a bound *m*-space ($m \geq 3$)

$$\Omega = P_0 \wedge P_1 \wedge P_2 \wedge P_3;$$

Declare the P_i as 'vector' symbols (that is, as 1-elements), and sum the elements of its first cospan.

$$\star\mathcal{P}_5;\ \star\star V[P\_];\ A = \star\Sigma\left[\star\mathbb{S}^1[\Omega]\right]$$

$$-(P_0 \wedge P_1 \wedge P_2) + P_0 \wedge P_1 \wedge P_3 - P_0 \wedge P_2 \wedge P_3 + P_1 \wedge P_2 \wedge P_3$$

This can be seen to be a 'free' *m*-vector by factorizing it and noting that it is the exterior product of point differences only.

```
ExteriorFactorize[A]
```

$$-((P_0 - P_3) \wedge (P_1 - P_3) \wedge (P_2 - P_3))$$

- Note carefully that we cannot simply apply the same cospan operation a second time to A, since as discussed in section 2.12, the operation is not defined for sums. However, since it is defined for simple products, we can apply it to each of the products of points in A, whence we obtain the same nilpotent behaviour as the ∂ operator.

$$-\star\Sigma\left[\star\mathbb{S}^1[P_0 \wedge P_1 \wedge P_2]\right] + \star\Sigma\left[\star\mathbb{S}^1[P_0 \wedge P_1 \wedge P_3]\right] - \star\Sigma\left[\star\mathbb{S}^1[P_0 \wedge P_2 \wedge P_3]\right] + \star\Sigma\left[\star\mathbb{S}^1[P_1 \wedge P_2 \wedge P_3]\right]$$

0

4.7 Line Coordinates

We have already seen that lines are defined by bound vectors independent of the dimension of the space. We now look at the types of coordinate descriptions we can use to define lines in bound spaces (multiplanes) of various dimensions.

For simplicity of exposition we refer to a bound vector as 'a line', rather than as 'defining a line', and to a bound (simple) bivector as 'a plane', rather than as 'defining a plane'.

Lines in a plane

To explore lines in a plane, we first declare the basis of the plane: $\star\mathcal{P}_2$.

$\star\mathcal{P}_2$

$\{\star\mathbb{O}, e_1, e_2\}$

A line can be written in several forms. The most intuitive form perhaps is as a product of two points $\star\mathbb{O} + x$ and $\star\mathbb{O} + y$ where x and y are position vectors.

$$L \equiv (\star\mathbb{O} + x) \wedge (\star\mathbb{O} + y)$$

Note that we have used the congruence symbol '≡' since any scalar multiple of this product will define the same line.

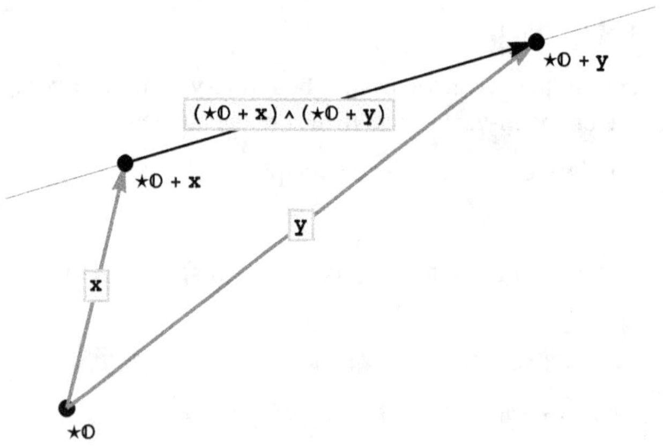

A line defined by the product of two points

We can automatically generate a basis form for each of the points in the plane by using the *GrassmannAlgebra* function ComposePoint, or its alias form $\star\mathbb{P}_{\square}$.

$$L \equiv \star\mathbb{P}_x \wedge \star\mathbb{P}_y$$

$$L \equiv (\star\mathbb{O} + e_1\, x_1 + e_2\, x_2) \wedge (\star\mathbb{O} + e_1\, y_1 + e_2\, y_2)$$

Or, we can express the line as the product of any point in it and a vector parallel to it. For example:

$$L \equiv (\star\mathbb{O} + x) \wedge (y - x) \equiv (\star\mathbb{O} + y) \wedge (y - x)$$

Often, the most convenient way to express a line is as the product of a point and a vector. Informally we adopt this as the 'canonical' form.

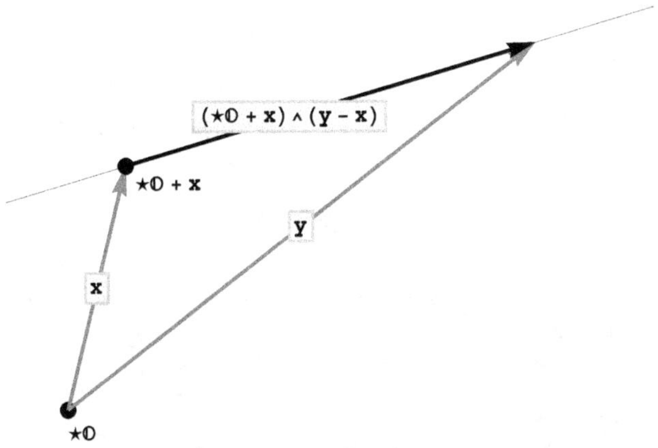

A line defined by the product of a point and a vector

If we want to compose a line element in this form, we can also use `ComposeLineElement` or its alias $\star\mathbb{L}_\square$. By leaving the placeholder unfilled we get a template for entering our own scalar coefficients.

$\star\mathbb{L}_\square$

$(\star\mathbb{O} + \square\, e_1 + \square\, e_2) \wedge (\square\, e_1 + \square\, e_2)$

Alternatively, we can express L without specific reference to points in it. For example:

$$\mathbf{L} \equiv \star\mathbb{O} \wedge (a\, e_1 + b\, e_2) + c\, e_1 \wedge e_2$$

The first term $\star\mathbb{O} \wedge (a\, e_1 + b\, e_2)$ is a vector bound through the origin, and hence defines a line through the origin. The second term $c\, e_1 \wedge e_2$ is a bivector whose addition represents a shift in the line parallel to itself, away from the origin.

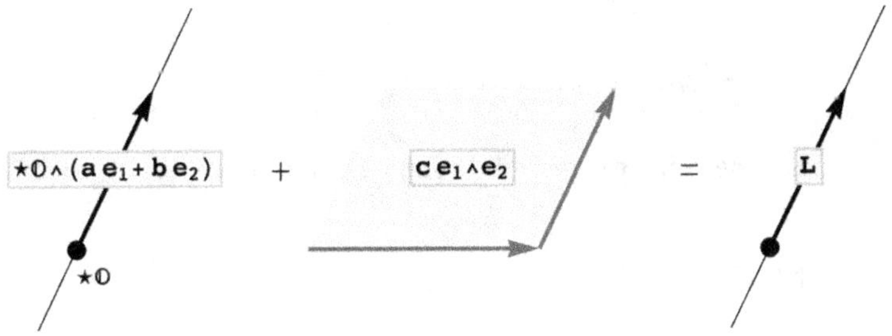

Components of a line in the plane expressed in basis form

We know that this can indeed represent a line since we can factorize it into any of the following three forms.

- A line of gradient $\frac{b}{a}$ through the point with coordinate $\frac{c}{b}$ on the $\star\mathbb{O} \wedge e_1$ axis.

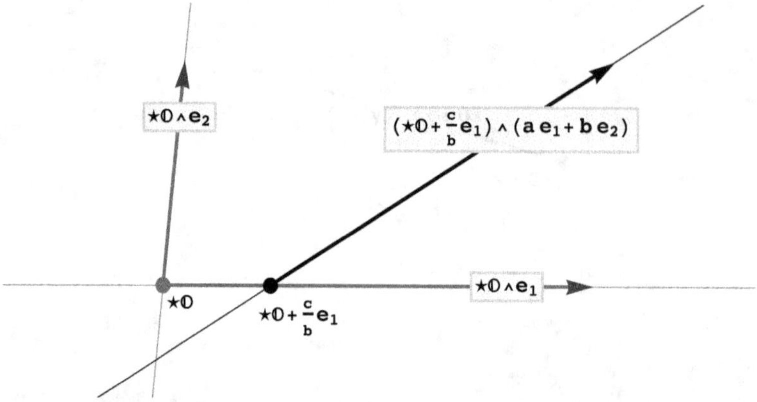

A line in the plane expressed through a point on the $\star\mathbb{O} \wedge e_1$ axis

$$L \equiv \left(\star\mathbb{O} + \frac{c}{b}\, e_1\right) \wedge (a\, e_1 + b\, e_2)$$

- A line of gradient $\frac{b}{a}$ through the point with coordinate $-\frac{c}{a}$ on the $\star\mathbb{O} \wedge e_2$ axis.

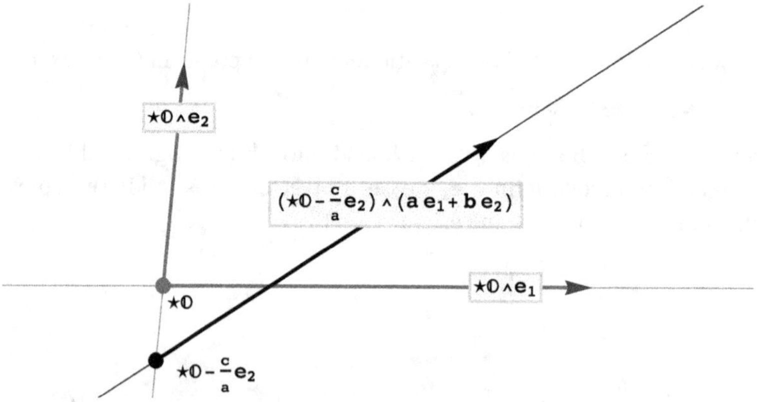

A line in the plane expressed through a point on the $\star\mathbb{O} \wedge e_2$ axis

$$L \equiv \left(\star\mathbb{O} - \frac{c}{a}\, e_2\right) \wedge (a\, e_1 + b\, e_2)$$

- Or, a line through both points.

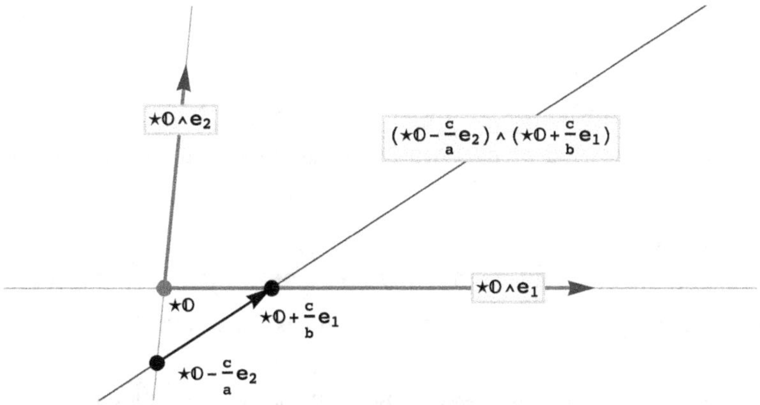

A line in the plane expressed through points on both axes

$$L \equiv \frac{a\,b}{c}\left(\star\mathbb{O}-\frac{c}{a}e_2\right)\wedge\left(\star\mathbb{O}+\frac{c}{b}e_1\right)$$

Of course the scalar factor $\frac{a\,b}{c}$ is inessential so we can just as well say:

$$L \equiv \left(\star\mathbb{O}-\frac{c}{a}e_2\right)\wedge\left(\star\mathbb{O}+\frac{c}{b}e_1\right)$$

◆ **Information required to express a line in a plane**

The expression of the line above in terms of a pair of points requires the four coordinates of the points. Expressed without specific reference to points, we seem to need three parameters. However, the last expression shows, as expected, that it is really only two parameters that are necessary (*viz* $y = m\,x + c$).

Lines in a 3-plane

Our normal notion of three-dimensional space corresponds to a 3-plane. Lines in a 3-plane have the same form when expressed in coordinate-free notation as they do in a plane (2-plane). Remember that a 3-plane is a bound vector 3-space whose basis may be chosen as 3 independent vectors and a point, or equivalently as 4 independent points. For example, we can still express a line in a 3-plane in any of the following equivalent forms.

$$L \equiv (\star\mathbb{O}+x)\wedge(\star\mathbb{O}+y)$$
$$L \equiv (\star\mathbb{O}+x)\wedge(y-x)$$
$$L \equiv (\star\mathbb{O}+y)\wedge(y-x)$$
$$L \equiv \star\mathbb{O}\wedge(y-x)+x\wedge y$$

Here, x and y are independent vectors in the 3-plane.

The coordinate form however will appear somewhat different to that in the 2-plane case. To explore this, we redeclare the basis with $\star\mathcal{P}_3$, and use `ComposePoint` in its alias form to define a line `L` as the product of 2 points.

$\star\mathcal{P}_3$

$\{\star\mathbb{O}, e_1, e_2, e_3\}$

$L = \star\mathbb{P}_x \wedge \star\mathbb{P}_y$

$(\star\mathbb{O} + e_1 x_1 + e_2 x_2 + e_3 x_3) \wedge (\star\mathbb{O} + e_1 y_1 + e_2 y_2 + e_3 y_3)$

We can expand and simplify this product with `GrassmannExpandAndSimplify` or its alias $\star\mathcal{G}$.

$L = \star\mathcal{G}[L]$

$$\star\mathbb{O} \wedge (e_1 (-x_1 + y_1) + e_2 (-x_2 + y_2) + e_3 (-x_3 + y_3)) + (-x_2 y_1 + x_1 y_2) e_1 \wedge e_2 + (-x_3 y_1 + x_1 y_3) e_1 \wedge e_3 + (-x_3 y_2 + x_2 y_3) e_2 \wedge e_3$$

The scalar coefficients in this expression are sometimes called the *Plücker coordinates* of the line.

Alternatively, we can express L in terms of basis elements, but without specific reference to points or vectors in it. For example:

$L = \star\mathbb{O} \wedge (a e_1 + b e_2 + c e_3) + d e_1 \wedge e_2 + e e_2 \wedge e_3 + f e_1 \wedge e_3;$

The first term $\star\mathbb{O} \wedge (a e_1 + b e_2 + c e_3)$ is a vector bound through the origin, and hence defines a line through the origin. The second term $d e_1 \wedge e_2 + e e_2 \wedge e_3 + f e_1 \wedge e_3$ is a bivector whose addition represents a shift in the line parallel to itself, away from the origin. In order to effect this shift, however, it is necessary that the bivector contain the vector $a e_1 + b e_2 + c e_3$. Hence there will be some constraint on the coefficients d, e, and f. To determine this we only need to determine the condition that the exterior product of the vector and the bivector is zero.

$\star\mathcal{G}[(a e_1 + b e_2 + c e_3) \wedge (d e_1 \wedge e_2 + e e_2 \wedge e_3 + f e_1 \wedge e_3)] == 0$

$(c d + a e - b f) e_1 \wedge e_2 \wedge e_3 == 0$

Alternatively, this constraint amongst the coefficients could have been obtained by noting that in order to be a line, L must be simple, hence the exterior product with itself must be zero.

$\star\mathcal{G}[L \wedge L] == 0$

$(2 c d + 2 a e - 2 b f) \star\mathbb{O} \wedge e_1 \wedge e_2 \wedge e_3 == 0$

Thus the constraint that the coefficients must obey in order for a general bound vector of the form L to be a line in a 3-plane is that:

$c d + a e - b f == 0$

This constraint is sometimes referred to as the *Plücker identity*.

Given this constraint, and supposing neither a, b nor c to be zero, we can factorize the line into any of the following forms:

$$L == \left(\star\mathbb{O} + \frac{f}{c} e_1 + \frac{e}{c} e_2\right) \wedge (a e_1 + b e_2 + c e_3)$$

$$L == \left(\star\mathbb{O} - \frac{d}{a} e_2 - \frac{f}{a} e_3\right) \wedge (a e_1 + b e_2 + c e_3)$$

$$L == \left(\star\mathbb{O} + \frac{d}{b} e_1 - \frac{e}{b} e_3\right) \wedge (a e_1 + b e_2 + c e_3)$$

Each of these forms represents a line in the direction of $a e_1 + b e_2 + c e_3$ and intersecting a coordinate plane. For example, the first form intersects the $\star\mathbb{O} \wedge e_1 \wedge e_2$ coordinate plane in the point $\star\mathbb{O} + \frac{f e_1}{c} + \frac{e e_2}{c}$ with coordinates $(\frac{f}{c}, \frac{e}{c}, 0)$.

The most compact form, in terms of the number of scalar parameters used, is when L is expressed as the product of two points, each of which lies in a coordinate plane.

$$L = \frac{a\,c}{f}\left(\star\mathbb{O} - \frac{d}{a}\,e_2 - \frac{f}{a}\,e_3\right) \wedge \left(\star\mathbb{O} + \frac{f}{c}\,e_1 + \frac{e}{c}\,e_2\right);$$

We can verify that this formulation gives us the original form of the line by expanding and simplifying the product and substituting the constraint relation previously obtained.

```
★G[Expand[★E[★G[L] /. (c d + a e → f b)]]]
```

$$\star\mathbb{O} \wedge (a\,e_1 + b\,e_2 + c\,e_3) + d\,e_1 \wedge e_2 + f\,e_1 \wedge e_3 + e\,e_2 \wedge e_3$$

Similar expressions may be obtained for L in terms of points lying in the other coordinate planes.

To summarize, there are three possibilities in a 3-plane, corresponding to there being three different pairs of coordinate 2-planes in the 3-plane.

$$\begin{aligned} L &\equiv (\star\mathbb{O} + x_1\,e_1 + x_2\,e_2) \wedge (\star\mathbb{O} + y_2\,e_2 + y_3\,e_3) \equiv P \wedge Q \\ L &\equiv (\star\mathbb{O} + x_1\,e_1 + x_2\,e_2) \wedge (\star\mathbb{O} + z_1\,e_1 + z_3\,e_3) \equiv P \wedge R \\ L &\equiv (\star\mathbb{O} + y_2\,e_2 + y_3\,e_3) \wedge (\star\mathbb{O} + z_1\,e_1 + z_3\,e_3) \equiv Q \wedge R \end{aligned} \qquad 4.6$$

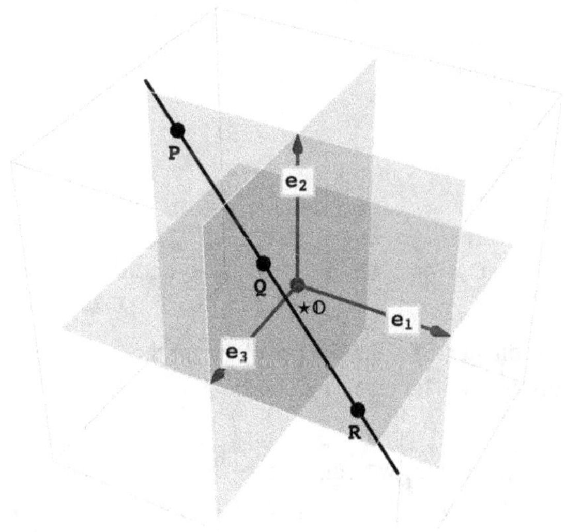

A line in a 3-plane through points in the coordinate planes

◆ Information required to express a line in a 3-plane

As with a line in a 2-plane, we find that a line in a 3-plane is expressed with the minimum number of parameters by expressing it as the product of two points, each in one of the coordinate planes. In this form, there are just 4 independent scalar parameters (coordinates) required to express the line.

◆ Determining the point of intersection with the third coordinate plane

Suppose we have a line expressed as the exterior product of points in two coordinate planes, say $\star\mathbb{O} \wedge e_1 \wedge e_2$ and $\star\mathbb{O} \wedge e_2 \wedge e_3$.

```
L = (★O + x₁ e₁ + x₂ e₂) ∧ (★O + y₂ e₂ + y₃ e₃);
```

We can always find the point on the line in the third coordinate plane $\star\mathbb{O} \wedge e_3 \wedge e_1$ by computing the intersection of the line and the third plane.

To use *GrassmannAlgebra* for this, we first need to declare the space, and the scalar symbols.

```
★P₃; DeclareExtraScalarSymbols[{x₁, x₂, y₂, y₃}];
```

The intersection of the line and the third coordinate plane is given by finding their common factor.

```
F = ToCommonFactor[L ∨ (★O ∧ e₁ ∧ e₃)]
```

$$\star c\,(\star\mathbb{O}\,(x_2 - y_2) - e_1\,x_1\,y_2 + e_3\,x_2\,y_3)$$

The point we need is then

```
R = ToPointForm[F]
```

$$\star\mathbb{O} - \frac{e_1\,x_1\,y_2}{x_2 - y_2} + \frac{e_3\,x_2\,y_3}{x_2 - y_2}$$

We can confirm that this point is both on the line and on the coordinate plane by simplifying their exterior products.

```
{★S[★G[R ∧ L]], ★G[R ∧ ★O ∧ e₁ ∧ e₃]}
```

{0, 0}

Lines in a 4-plane

Lines in a 4-plane have the same form when expressed in coordinate-free notation as they do in any multiplane.

To obtain the Plücker coordinates of a line in a 4-plane, express the line as the exterior product of two points and multiply it out. The resulting coefficients of the basis elements are the Plücker coordinates of the line.

Additionally, from the results above, we can expect that a line in a 4-plane may be expressed with the least number of scalar parameters as the exterior product of two points, each point lying in one of the coordinate 3-planes. For example, the expression for the line as the product of the points in the coordinate 3-planes $\star\mathbb{O} \wedge e_1 \wedge e_2 \wedge e_3$ and $\star\mathbb{O} \wedge e_2 \wedge e_3 \wedge e_4$ is

```
L = (★O + x₁ e₁ + x₂ e₂ + x₃ e₃) ∧ (★O + y₂ e₂ + y₃ e₃ + y₄ e₄)
```

Lines in an *m*-plane

The formulae below summarize some of the expressions for defining a line, valid in a multiplane of any dimension.

Coordinate-free expressions may take any of a number of forms. For example:

$$\begin{aligned}
L &\equiv (\star\mathbb{O} + x) \wedge (\star\mathbb{O} + y) \\
L &\equiv (\star\mathbb{O} + x) \wedge (y - x) \\
L &\equiv (\star\mathbb{O} + y) \wedge (y - x) \\
L &\equiv \star\mathbb{O} \wedge (y - x) + x \wedge y
\end{aligned} \tag{4.7}$$

A line can be expressed in terms of the 2*m* coordinates of any two points on it.

$$L \equiv (\star\mathbb{O} + x_1\, e_1 + x_2\, e_2 + \ldots + x_m\, e_m) \wedge (\star\mathbb{O} + y_1\, e_1 + y_2\, e_2 + \ldots + y_m\, e_m) \tag{4.8}$$

When multiplied out, the expression for the line takes a form explicitly displaying the Plücker coordinates of the line.

$$\begin{aligned}
L \equiv \star\mathbb{O} \wedge ((y_1 - x_1)\, e_1 + (y_2 - x_2)\, e_2 + \ldots + (y_m - x_m)\, e_m) \\
+ (x_1\, y_2 - x_2\, y_1)\, e_1 \wedge e_2 + (x_1\, y_3 - x_3\, y_1)\, e_1 \wedge e_3 + \\
(x_1\, y_4 - x_4\, y_1)\, e_1 \wedge e_4 + \ldots + (x_{m-1}\, y_m - x_m\, y_{m-1})\, e_{m-1} \wedge e_m
\end{aligned} \tag{4.9}$$

Alternatively, a line in a coordinate *m*-plane $\star\mathbb{O} \wedge e_1 \wedge e_2 \wedge \ldots \wedge e_m$ can be expressed in terms of its intersections with two of its coordinate (*m*–1)-planes, $\star\mathbb{O} \wedge e_1 \wedge \ldots \wedge \square_i \wedge \ldots \wedge e_m$ and $\star\mathbb{O} \wedge e_1 \wedge \ldots \wedge \square_j \wedge \ldots \wedge e_m$ say. The notation $\square_i$ means that the *i*th element or term is missing.

$$L \equiv (\star\mathbb{O} + x_1\, e_1 + \cdots + \square_i + \cdots + x_m\, e_m) \wedge \left(\star\mathbb{O} + y_1\, e_1 + \cdots + \square_j + \cdots + y_m\, e_m\right) \tag{4.10}$$

This formulation indicates that a line in *m*-space has at most 2(*m*–1) independent parameters required to describe it.

It also implies that in the special case when the line lies in one of the coordinate (*m*–1)-spaces, it can be even more economically expressed as the product of two points, each lying in one of the coordinate (*m*–2)-spaces contained in the (*m*–1)-space. And so on.

Exploration: Line simplicity

In what follows we will explore in more detail how the requirement for a line to be a simple bound vector (in a space of any dimension) can lead us to its Plücker coordinates.

◆ Lines in space

Consider a general 2-element in a bound 3-space.

```
★𝒫3; A = ComposeMElement[2, a]
```

$$a_1 \star\mathbb{O} \wedge e_1 + a_2 \star\mathbb{O} \wedge e_2 + a_3 \star\mathbb{O} \wedge e_3 + a_4\, e_1 \wedge e_2 + a_5\, e_1 \wedge e_3 + a_6\, e_2 \wedge e_3$$

We can attempt to factorize this using `ExteriorFactorize`.

```
ExteriorFactorize[A]
```

$$\text{If}\left[\{a_3\, a_4 - a_2\, a_5 + a_1\, a_6 == 0\},\ a_1 \left(\star\mathbb{O} - \frac{a_4\, e_2}{a_1} - \frac{a_5\, e_3}{a_1}\right) \wedge \left(e_1 + \frac{a_2\, e_2}{a_1} + \frac{a_3\, e_3}{a_1}\right)\right]$$

This says that the given factorization is possible only if and only if the condition $a_3 a_4 - a_2 a_5 + a_1 a_6 == 0$ on the coefficients of A is met.

There is a straightforward way of testing the simplicity of 2-elements not shared by elements of higher grade. *A 2-element is simple if and only if its exterior square is zero*. (The exterior square of A is $A \wedge A$. See section 2.10.) In our case, we can write A as $A = \star\mathbb{O} \wedge \beta + B$, where β is a vector, and B is a bivector.

In the case of a 2-element $A = \star\mathbb{O} \wedge \beta + B$ in a bound space then, we require

$$\mathbf{A \wedge A == (\star\mathbb{O} \wedge \beta + B) \wedge (\star\mathbb{O} \wedge \beta + B) == 2 \star\mathbb{O} \wedge \beta \wedge B == 0}$$

Here $B \wedge B$ is zero because it is a bivector in a vector 3-space.

Thus we have the reduced requirement only that $\beta \wedge B = 0$. In the case of a bound 3-space

$$\beta = a_1 e_1 + a_2 e_2 + a_3 e_3;$$
$$B = a_4 e_1 \wedge e_2 + a_5 e_1 \wedge e_3 + a_6 e_2 \wedge e_3;$$

Simplifying $\beta \wedge B$ gives

$$\star\mathcal{G}[\beta \wedge B]$$

$$(a_3 a_4 - a_2 a_5 + a_1 a_6)\, e_1 \wedge e_2 \wedge e_3$$

Hence by these means we arrive at the same simplicity condition we obtained initially by using `ExteriorFactorize`.

At this point we should note that these results have not involved any metric concepts. The simple requirement that $\beta \wedge B$ be zero, has led to the required constraint.

On the other hand, if we require from the start that $\beta \wedge B = 0$, that is B contains the factor β, then we can write $B = \alpha \wedge \beta$, and so write A immediately as the product of a point and a vector.

$$A == \star\mathbb{O} \wedge \beta + \alpha \wedge \beta == (\star\mathbb{O} + \alpha) \wedge \beta$$

The vector α may be visualized as the position vector of a point on the line, and the vector β as a vector in the direction of the line. The coefficients of $\alpha \wedge \beta$ are the same as the coefficients of the cross product $\alpha \times \beta$. The cross product is the complement of the exterior product, and is orthogonal to it. (The cross product and orthogonality are metric concepts which will be introduced in the next chapters. See particularly section 6.10.)

The Plücker coordinates of a line in space are often given by the coefficients of the two vectors α and $\alpha \times \beta$. Composing α and β in basis form, and then calculating their exterior product displays the coefficients of $\alpha \times \beta$.

$$\{\alpha = \star\mathbb{V}_a,\ \beta = \star\mathbb{V}_b,\ \star\mathcal{G}[\alpha \wedge \beta]\}$$

$$\{a_1 e_1 + a_2 e_2 + a_3 e_3,\ b_1 e_1 + b_2 e_2 + b_3 e_3,$$
$$(-a_2 b_1 + a_1 b_2)\, e_1 \wedge e_2 + (-a_3 b_1 + a_1 b_3)\, e_1 \wedge e_3 + (-a_3 b_2 + a_2 b_3)\, e_2 \wedge e_3\}$$

Then with the mapping

$$\{e_1 \wedge e_2,\ e_1 \wedge e_3,\ e_1 \wedge e_2\} \longrightarrow \{e_1,\ -e_2,\ e_3\}$$

the Plücker line coordinates are given by

$$\{a_1,\ a_2,\ a_3,\ (a_2 b_3 - a_3 b_2),\ (a_3 b_1 - a_1 b_3),\ (a_1 b_2 - a_2 b_1)\}$$

And as we would expect, the first three components as a vector are orthogonal to the second three.

```
{a1, a2, a3}.{(a2 b3 - a3 b2), (a3 b1 - a1 b3), (a1 b2 - a2 b1)} // Expand
```

0

Despite this common practice of developing the Plücker coordinates in terms of dot and cross products, we can see from the Grassmannian approach that we began with, that it is neither necessary, nor particularly elucidating. The coordinates of a line do not need to rely on metric notions. And if the geometry is projective, they perhaps should not.

◆ Lines in 4-space

Now consider a general 2-element in a bound 4-space.

```
★P4; A = ComposeMElement[2, a]
```

$$a_1 ★\mathbb{O} \wedge e_1 + a_2 ★\mathbb{O} \wedge e_2 + a_3 ★\mathbb{O} \wedge e_3 + a_4 ★\mathbb{O} \wedge e_4 + a_5 e_1 \wedge e_2 + a_6 e_1 \wedge e_3 + a_7 e_1 \wedge e_4 + a_8 e_2 \wedge e_3 + a_9 e_2 \wedge e_4 + a_{10} e_3 \wedge e_4$$

Factorizing gives

```
ExteriorFactorize[A]
```

$$\text{If}\left[\{a_3 a_5 - a_2 a_6 + a_1 a_8 == 0,\ a_4 a_5 - a_2 a_7 + a_1 a_9 == 0,\ a_4 a_6 - a_3 a_7 + a_1 a_{10} == 0\},\ a_1 \left(★\mathbb{O} - \frac{a_5 e_2}{a_1} - \frac{a_6 e_3}{a_1} - \frac{a_7 e_4}{a_1}\right) \wedge \left(e_1 + \frac{a_2 e_2}{a_1} + \frac{a_3 e_3}{a_1} + \frac{a_4 e_4}{a_1}\right)\right]$$

On the other hand, adopting the method of the previous section, we can write

```
β = a1 e1 + a2 e2 + a3 e3 + a4 e4;
B = a5 e1 ∧ e2 + a6 e1 ∧ e3 + a7 e1 ∧ e4 + a8 e2 ∧ e3 + a9 e2 ∧ e4 + a10 e3 ∧ e4;
```

Simplifying $\beta \wedge \text{B}$ gives

```
★G[β ∧ B]
```

$$(a_3 a_5 - a_2 a_6 + a_1 a_8)\ e_1 \wedge e_2 \wedge e_3 + (a_4 a_5 - a_2 a_7 + a_1 a_9)\ e_1 \wedge e_2 \wedge e_4 + (a_4 a_6 - a_3 a_7 + a_1 a_{10})\ e_1 \wedge e_3 \wedge e_4 + (a_4 a_8 - a_3 a_9 + a_2 a_{10})\ e_2 \wedge e_3 \wedge e_4$$

Again, we arrive at the same conditions, except that with this approach we have one extra one. However, it is easy to show that the number of independent conditions is still only three. Clearly, because this product will always be a trivector, in a space of n vector dimensions, it will always have $c = n(n\text{-}1)(n\text{-}2)/6$ components, and hence spawn c conditions.

On the other hand, if we require from the start that $\beta \wedge \text{B} = 0$, that is B contains the factor β, then we can write $\text{B} = \alpha \wedge \beta$, and so write A immediately as the product of a point and a vector, just as we did for the line in 3-space.

```
A == ★O ∧ β + α ∧ β == (★O + α) ∧ β
```

The vector α may be visualized as the position vector of a point on the line, and the vector β as a vector in the direction of the line. The coefficients of $\alpha \wedge \beta$ cannot be expressed as the coefficients of the cross product as before since the vector space is now 4-dimensional.

The Plücker coordinates of a line in 4-space must now be given by the coefficients of the vector α and the bivector $\alpha \wedge \beta$. If

```
α = ★Va; β = ★Vb;
```

Then the Plücker line coordinates are given by

```
First /@ExtractLinearCombinations[α + ★G[α ∧ β]]
```

$\{a_1, a_2, a_3, a_4, -a_2 b_1 + a_1 b_2, -a_3 b_1 + a_1 b_3, -a_4 b_1 + a_1 b_4, -a_3 b_2 + a_2 b_3, -a_4 b_2 + a_2 b_4, -a_4 b_3 + a_3 b_4\}$

In a bound 4-space then, the Grassmannian approach to determining Plücker line coordinates is the same as that in a bound 3-space, and indeed in a bound space of any dimensions. On the other hand, the standard approach developed through the dot and cross products of vector 3-space does not generalize to higher dimensions.

4.8 Plane Coordinates

We have already seen that planes are defined by simple bound bivectors independent of the dimension of the space. We now look at the types of coordinate descriptions we can use to define planes in bound spaces (multiplanes) of various dimensions.

Planes in a 3-plane

A plane Π in a 3-plane can be written in several forms. The most intuitive form perhaps is as a product of three non-collinear points P, Q, and R.

$$\mathbf{P} == \star\mathbb{O} + \mathbf{x}$$
$$\mathbf{Q} == \star\mathbb{O} + \mathbf{y}$$
$$\mathbf{R} == \star\mathbb{O} + \mathbf{z}$$

$$\Pi \equiv \mathbf{P} \wedge \mathbf{Q} \wedge \mathbf{R} == (\star\mathbb{O} + \mathbf{x}) \wedge (\star\mathbb{O} + \mathbf{y}) \wedge (\star\mathbb{O} + \mathbf{z})$$

Here, x, y and z are the position vectors of P, Q, and R.

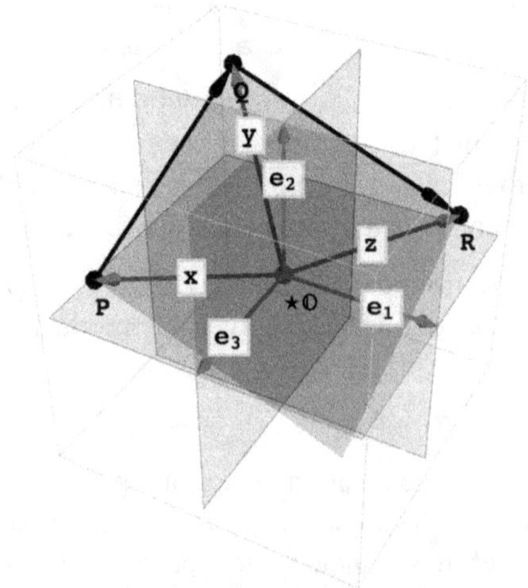

A 2-plane in a 3-plane through three points

Or, we can express it as the product of any two different points in it and a vector parallel to it (but not in the direction of the line joining the two points). For example:

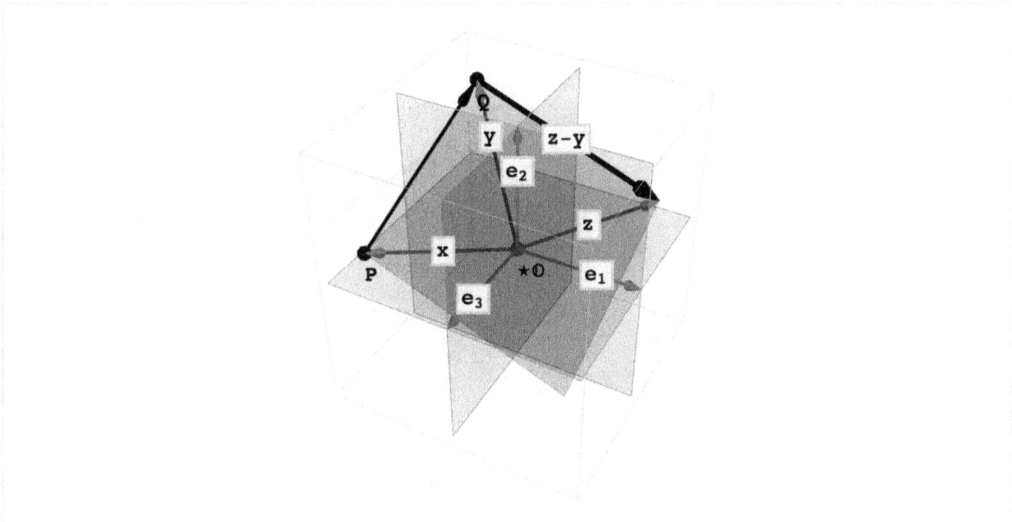

A 2-plane in a 3-plane through two points and a vector

$\Pi \equiv (\star\mathbb{O} + \mathbf{x}) \wedge (\star\mathbb{O} + \mathbf{y}) \wedge (\mathbf{z} - \mathbf{y})$

Or, we can express it as the product of any point in it and any two independent vectors parallel to it. For example:

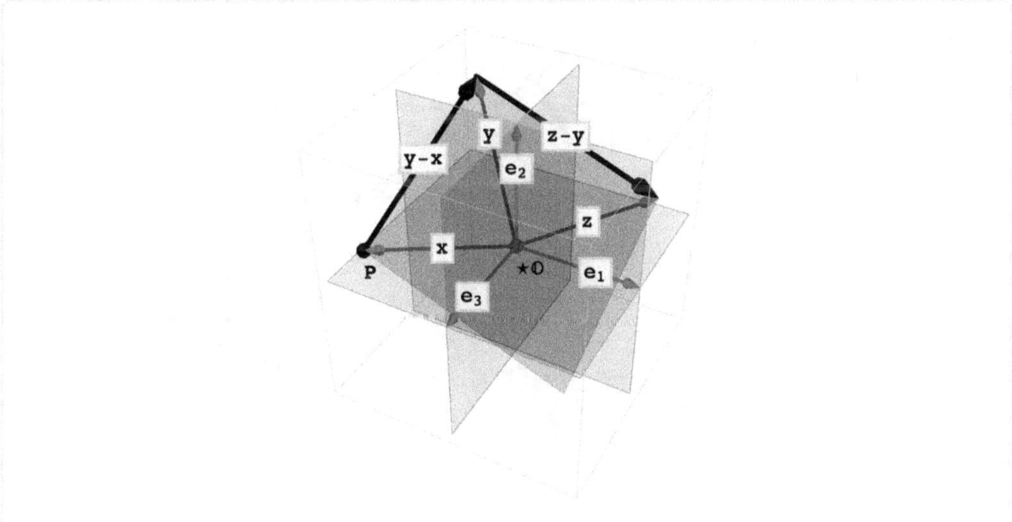

A 2-plane in a 3-plane through one point and two vectors

$\Pi \equiv (\star\mathbb{O} + \mathbf{x}) \wedge (\mathbf{y} - \mathbf{x}) \wedge (\mathbf{z} - \mathbf{y})$

Or, we can express it as the product of any point in it and any line in it not through the point. For example:

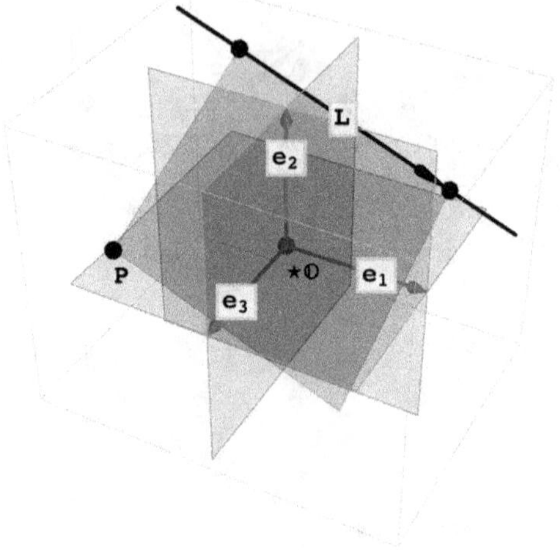

A 2-plane in a 3-plane through a point and a line

$\Pi \equiv P \wedge L$

Or, we can express it as the product of any line in it and any vector parallel to it (but not parallel to the line). For example:

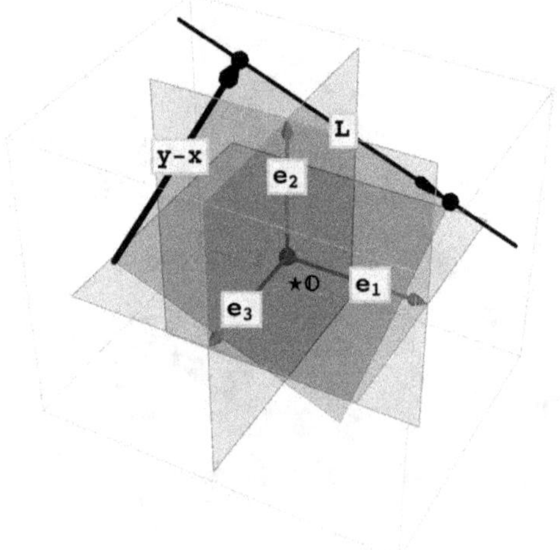

A 2-plane in a 3-plane through a vector and a line

$\Pi \equiv (y - x) \wedge L$

Given a basis, we can always express the plane in terms of the coordinates of the points or vectors in the expressions above. However the form which requires the least number of coordinates is that which expresses the plane as the exterior product of its three points of intersection with the coordinate axes.

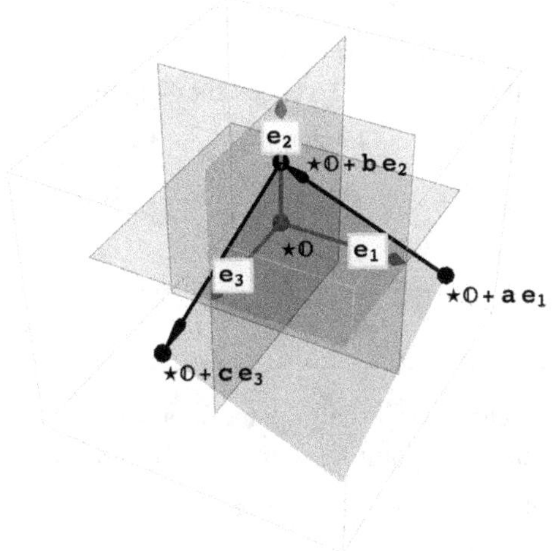

A 2-plane in a 3-plane through three points on the coordinate axes

$$\Pi \equiv (\star\mathbb{O} + a\, e_1) \wedge (\star\mathbb{O} + b\, e_2) \wedge (\star\mathbb{O} + c\, e_3)$$

If the plane is parallel to one of the coordinate axes, say $\star\mathbb{O} \wedge e_3$, it may be expressed as:

$$\Pi \equiv (\star\mathbb{O} + a\, e_1) \wedge (\star\mathbb{O} + b\, e_2) \wedge e_3$$

Whereas, if it is parallel to two of the coordinate axes, say $\star\mathbb{O} \wedge e_2$ and $\star\mathbb{O} \wedge e_3$, it may be expressed as:

$$\Pi \equiv (\star\mathbb{O} + a\, e_1) \wedge e_2 \wedge e_3$$

If we wish to express a plane as the exterior product of its intersection points with the coordinate axes, we first determine its points of intersection with the axes and then take the exterior product of the resulting points. This leads to the following identity:

$$\Pi \equiv (\Pi \vee (\star\mathbb{O} \wedge e_1)) \wedge (\Pi \vee (\star\mathbb{O} \wedge e_2)) \wedge (\Pi \vee (\star\mathbb{O} \wedge e_3))$$

◆ Example: To express a plane in terms of its intersections with the coordinate axes

Suppose we have a plane in a 3-plane defined by three points.

$\star\mathcal{P}_3$; $\Pi = (\star\mathbb{O} + e_1 + 2\, e_2 + 5\, e_3) \wedge (\star\mathbb{O} - e_1 + 9\, e_2) \wedge (\star\mathbb{O} - 7\, e_1 + 6\, e_2 + 4\, e_3)$;

To express this plane in terms of its intersections with the coordinate axes we calculate the intersection points with the axes.

```
ToCommonFactor[Π ∨ (★O ∧ e1)]
```

$\star c\, (13\, \star\mathbb{O} + 329\, e_1)$

```
ToCommonFactor[Π ∨ (★O ∧ e2)]
```

$\star c\, (38\, \star\mathbb{O} + 329\, e_2)$

```
ToCommonFactor[Π ∨ (★O ∧ e3)]
```

$\star c\, (48\, \star\mathbb{O} + 329\, e_3)$

We then take the product of these points (ignoring the weights) to form the plane.

$$\Pi \equiv \left(\star\mathbb{O} + \frac{329\, e_1}{13}\right) \wedge \left(\star\mathbb{O} + \frac{329\, e_2}{38}\right) \wedge \left(\star\mathbb{O} + \frac{329\, e_3}{48}\right)$$

To verify that this is indeed the same plane, we can check to see if these points are in the original plane. Using the listability of $\star\mathcal{G}$, we can check them all at once.

$$\star\mathcal{G}\left[\Pi \wedge \left\{\left(\star\mathbb{O} + \frac{329\, e_1}{13}\right), \left(\star\mathbb{O} + \frac{329\, e_2}{38}\right), \left(\star\mathbb{O} + \frac{329\, e_3}{48}\right)\right\}\right]$$

{0, 0, 0}

Planes in a 4-plane

From the results above, we can expect that a plane in a 4-plane is most economically expressed as the product of three points, each point lying in one of the coordinate 2-planes. For example:

$$\Pi \equiv (\star\mathbb{O} + x_1\, e_1 + x_2\, e_2) \wedge (\star\mathbb{O} + y_2\, e_2 + y_3\, e_3) \wedge (\star\mathbb{O} + z_3\, e_3 + z_4\, e_4)$$

(This is analogous to expressing a line in a 3-plane as the product of two points, each point lying in one of the coordinate 2-planes.)

If a plane is expressed in any other form, we can express it in the form above by first determining its points of intersection with the coordinate planes and then taking the exterior product of the resulting points. This leads to the following identity:

$$\Pi \equiv (\Pi \vee (\star\mathbb{O} \wedge e_1 \wedge e_2)) \wedge (\Pi \vee (\star\mathbb{O} \wedge e_2 \wedge e_3)) \wedge (\Pi \vee (\star\mathbb{O} \wedge e_3 \wedge e_4))$$

Planes in an *m*-plane

A plane in an *m*-plane is most economically expressed as the product of three points, each point lying in one of the coordinate (*m*–2)-planes.

$$\begin{aligned}\Pi \equiv\ & \left(\star\mathbb{O} + x_1\, e_1 + \ldots + \underline{x_{i_1}\, e_{i_1}} + \ldots + \underline{x_{i_2}\, e_{i_2}} + \ldots + x_m\, e_m\right) \wedge \\ & \left(\star\mathbb{O} + y_1\, e_1 + \ldots + \underline{y_{i_3}\, e_{i_3}} + \ldots + \underline{y_{i_4}\, e_{i_4}} + \ldots + y_m\, e_m\right) \wedge \\ & \left(\star\mathbb{O} + z_1\, e_1 + \ldots + \underline{z_{i_5}\, e_{i_5}} + \ldots + \underline{z_{i_6}\, e_{i_6}} + \ldots + z_m\, e_m\right)\end{aligned}$$

Here the notation $\underline{x_i\, e_i}$ means that the term is *missing* from the sum.

This formulation indicates that a plane in an *m*-plane needs at most 3(*m*–2) independent scalar parameters required to describe it.

The coordinates of geometric entities

In the sections above we have discussed some common geometric entities, and various ways in which they can be represented by coordinates in a given basis system. In this section we make some general observations about the coordinates of geometric entities from a Grassmannian point of view.

We can define a *geometric entity* as an element of an exterior product space which has a specific geometric interpretation. Examples we have already met include points, lines, planes, *m*-planes, vectors, bivectors, and *m*-vectors.

These entities are characterized by the fact that they are *simple*. That is, they may be factorized into a product of 1-elements. The 1-element factors are, of course, not unique. However, this is not a shortcoming. Indeed, the power of such a representation for computations owes much to this non-uniqueness. For example, if we know two lines intersect in a point P, we can represent them immediately and expressly as products involving P (say, $P \wedge Q$ and $P \wedge R$).

If an element is not simple, then it is not a candidate for geometric entities of this sort. (We will look at geometric entities represented by non-simple elements in Volume 2.) Apart from 0, 1, *n*-1 and *n*-elements, an *m*-element of an *n*-space is not necessarily simple. Thus in the plane (a three-dimensional linear space), *n* is equal to 3, and all entities (scalars, points, lines and planes) are simple. In space, *n* is equal to 4, and hence 2-elements are the only entities that potentially are not simple. This means that only a subset of the 2-elements of space (the simple 2-elements) can represent lines.

These comments extend to a space of any larger number of dimensions. In particular, (*n*-1)-elements in a space of *n* dimensions are simple. Hence any (*n*-1)-element represents a hyperplane. For other elements (apart from the trivial 0 and *n*-elements), only a subset of them can represent geometric entities like lines and planes. This subset may be defined by a set of constraints between the coefficients of the element which ensures that it be simple.

If we are constructing a geometric entity as the exterior product of known independent points, or as the regressive product of hyperplanes, then we are assured that the result will be simple. In the examples of the previous section, we have seen several cases of this approach.

4.9 Calculation of Intersections

The intersection of two lines in a plane

Suppose we wish to find the point P of intersection of two lines $L_1$and L_2 in a plane. We have seen in the previous section how we could express a line in a plane as the exterior product of two points, and that these points could be taken as the points of intersection of the line with the coordinate axes.

First declare the basis of the plane, and then define the lines.

```
★𝒫2;
L1 = (★O + a e1) ∧ (★O + b e2);
L2 = (★O + c e1) ∧ (★O + d e2);
```

To find the point of intersection of these two lines, we first find their common factor. The common factor of two elements can be found by applying ToCommonFactor to their regressive product. Remember that common factors are only determinable up to congruence (that is, to within an arbitrary scalar multiple). Hence the common factor we determine will, in general, be a weighted point.

```
wP = ToCommonFactor[L1 ∨ L2]
```

```
★c ((b c - a d) ★O + a c (b - d) e1 + b (-a + c) d e2)
```

We can convert this weighted point to a form which separates the weight from the point by applying ToWeightedPointForm.

ToWeightedPointForm[wP]

$$\star c\,(b\,c - a\,d)\left(\star\mathbb{O} - \frac{a\,c\,(b-d)\,e_1}{-b\,c + a\,d} - \frac{b\,(a-c)\,d\,e_2}{b\,c - a\,d}\right)$$

Or, if we are only interested in the point, we can apply `ToPointForm`.

P = ToPointForm[wP]

$$\star\mathbb{O} - \frac{a\,c\,(b-d)\,e_1}{-b\,c + a\,d} - \frac{b\,(a-c)\,d\,e_2}{b\,c - a\,d}$$

To verify that this point lies in both lines, we can take its exterior product with each of the lines and show the results to be zero.

$\star\mathcal{G}[\{L_1 \wedge P,\ L_2 \wedge P\}]$ // Simplify

{0, 0}

In the special case in which the lines are parallel, that is $b\,c - a\,d = 0$, their intersection is no longer a point, but a *vector defining their common direction*.

The intersection of a line and a plane in a 3-plane

Suppose we wish to find the point P of intersection of a line L and a plane Π in a 3-plane. We express the line as the exterior product of two points in two coordinate planes, and the plane as the exterior product of the points of intersection of the plane with the coordinate axes.

First declare the basis of the 3-plane, and then define the line and plane.

$\star\mathcal{P}_3$;
$L = (\star\mathbb{O} + a\,e_1 + b\,e_2) \wedge (\star\mathbb{O} + c\,e_2 + d\,e_3)$;
$\Pi = (\star\mathbb{O} + e\,e_1) \wedge (\star\mathbb{O} + f\,e_2) \wedge (\star\mathbb{O} + g\,e_3)$;

To find the point of intersection of the line with the plane, we first determine their common factor (a weighted point) by applying `ToCommonFactor` to their regressive product, and then drop the weight by applying `ToPointForm`.

P = ToPointForm[ToCommonFactor[$L \vee \Pi$]]

$$\star\mathbb{O} - \frac{a\,e\,(d\,f + (c-f)\,g)\,e_1}{-d\,e\,f + (b\,e - c\,e + a\,f)\,g} + \frac{f\,(b\,e\,(d-g) + c\,(-a+e)\,g)\,e_2}{d\,e\,f - (b\,e - c\,e + a\,f)\,g} - \frac{d\,(b\,e + (a-e)\,f)\,g\,e_3}{d\,e\,f - (b\,e - c\,e + a\,f)\,g}$$

To verify that this point lies in both the line and the plane, we can take its exterior product with each of the line and the plane and show the results to be zero.

$\star\mathcal{S}[\star\mathcal{G}[\{L \wedge P,\ \Pi \wedge P\}]]$

{0, 0}

In the special case in which the line is parallel to the plane, their intersection is no longer a point, but a vector defining their common direction. When the line lies in the plane, the result from the calculation will be zero.

The intersection of two planes in a 3-plane

Suppose we wish to find the line L of intersection of two planes Π_1 and Π_2 in a 3-plane.

First declare the basis of the 3-plane, and then define the planes.

```
★𝒫3;
Π1 = (★O + a e1) ∧ (★O + b e2) ∧ (★O + c e3);
Π2 = (★O + e e1) ∧ (★O + f e2) ∧ (★O + g e3);
```

To find the line of intersection of the two planes, we first determine their common factor (a line) by applying `ToCommonFactor` to their regressive product, and then (if we wish) drop the (arbitrary) congruence factor ★c by setting it equal to unity.

```
L = ToCommonFactor[Π1 ∨ Π2] /. ★c → 1
```

$$\star\mathbb{O}\wedge(a\,e\,(c\,f-b\,g)\,e_1+b\,f\,(-c\,e+a\,g)\,e_2+c\,(b\,e-a\,f)\,g\,e_3)+a\,b\,e\,f\,(-c+g)\,e_1\wedge e_2+a\,c\,e\,(b-f)\,g\,e_1\wedge e_3+b\,c\,(-a+e)\,f\,g\,e_2\wedge e_3$$

This is of course, a 2-element in a 4-space, and hence is not necessarily simple. However we can verify in a number of ways that this result is indeed simple, and therefore does define a line. The simplest way is to apply the *GrassmannAlgebra* function `SimpleQ`.

```
SimpleQ[L]
```

```
True
```

Perhaps more convincing is to factorize L so as to display it as the product of a point and a vector.

```
L′ = ExteriorFactorize[L]
```

$$a\,e\,(c\,f-b\,g)\left(\star\mathbb{O}+\frac{b\,f\,(-c+g)\,e_2}{-c\,f+b\,g}+\frac{c\,(b-f)\,g\,e_3}{-c\,f+b\,g}\right)\wedge\left(e_1-\frac{b\,f\,(c\,e-a\,g)\,e_2}{a\,c\,e\,f-a\,b\,e\,g}+\frac{c\,(b\,e-a\,f)\,g\,e_3}{a\,c\,e\,f-a\,b\,e\,g}\right)$$

We can extract the point and vector as specific expressions by applying the *GrassmannAlgebra* function `ExtractArguments`.

```
{P, V} = ExtractArguments[L′]
```

$$\left\{\star\mathbb{O}+\frac{b\,f\,(-c+g)\,e_2}{-c\,f+b\,g}+\frac{c\,(b-f)\,g\,e_3}{-c\,f+b\,g},\;e_1-\frac{b\,f\,(c\,e-a\,g)\,e_2}{a\,c\,e\,f-a\,b\,e\,g}+\frac{c\,(b\,e-a\,f)\,g\,e_3}{a\,c\,e\,f-a\,b\,e\,g}\right\}$$

To verify that both the point and the vector lie in both planes, we can take their exterior product with each of the planes and show the results to be zero.

```
Simplify[★𝒢[{Π1 ∧ P, Π1 ∧ V, Π2 ∧ P, Π2 ∧ V}]]
```

```
{0, 0, 0, 0}
```

In the special case in which the planes are parallel, their intersection is no longer a line, but a *bivector defining their common 2-direction.*

Example: The osculating plane to a curve

◇ The problem

Show that the osculating planes at any three points to the curve defined by:

$$P == \star\mathbb{O} + \nu\, e_1 + \nu^2\, e_2 + \nu^3\, e_3$$

intersect at a point coplanar with these three points. Here, ν is a scalar parametrizing the curve.

◇ The solution

The osculating plane Π to the curve at the point P is given by $\Pi = P \wedge P' \wedge P''$. P' and P'' are the first and second derivatives of P with respect to ν.

```
P = ★O + ν e1 + ν^2 e2 + ν^3 e3;

P' = e1 + 2 ν e2 + 3 ν^2 e3;

P'' = 2 e2 + 6 ν e3;

Π = P ∧ P' ∧ P'';
```

We can declare the space to be a 3-plane and subscripts of the parameter ν to be scalar, and then use `GrassmannSimplify` to derive the expression for the osculating plane as a function of ν. (We divide out a common multiple of 2 to make the resulting expressions a little simpler.)

```
★P3; DeclareExtraScalarSymbols[{ν, ν_}]; Π = ★G[Π / 2]
```

$$\star\mathbb{O} \wedge \left(e_1 \wedge e_2 + 3\,\nu\, e_1 \wedge e_3 + 3\,\nu^2\, e_2 \wedge e_3\right) + \nu^3\, e_1 \wedge e_2 \wedge e_3$$

Now select any three points on the curve P_1, P_2, and P_3.

```
P1 = ★O + ν1 e1 + ν1^2 e2 + ν1^3 e3;
P2 = ★O + ν2 e1 + ν2^2 e2 + ν2^3 e3;
P3 = ★O + ν3 e1 + ν3^2 e2 + ν3^3 e3;
```

The osculating planes at these three points are:

```
{Π1, Π2, Π3} = {Π /. ν → ν1, Π /. ν → ν2, Π /. ν → ν3}
```

$$\{\star\mathbb{O} \wedge \left(e_1 \wedge e_2 + 3\,\nu_1\, e_1 \wedge e_3 + 3\,\nu_1^2\, e_2 \wedge e_3\right) + \nu_1^3\, e_1 \wedge e_2 \wedge e_3,$$
$$\star\mathbb{O} \wedge \left(e_1 \wedge e_2 + 3\,\nu_2\, e_1 \wedge e_3 + 3\,\nu_2^2\, e_2 \wedge e_3\right) + \nu_2^3\, e_1 \wedge e_2 \wedge e_3,$$
$$\star\mathbb{O} \wedge \left(e_1 \wedge e_2 + 3\,\nu_3\, e_1 \wedge e_3 + 3\,\nu_3^2\, e_2 \wedge e_3\right) + \nu_3^3\, e_1 \wedge e_2 \wedge e_3\}$$

The (weighted) point of intersection of these three planes may be obtained by calculating their common factor. The congruence factor $\star c$ appears squared because there are two regressive product operations.

```
wP = Simplify[ToCommonFactor[Π1 ∨ Π2 ∨ Π3]]
```

$$-3\,\star c^2\,(\nu_1 - \nu_2)\,(\nu_1 - \nu_3)\,(\nu_2 - \nu_3)$$
$$(e_1\,(\nu_1 + \nu_2 + \nu_3) + 3\,(\star\mathbb{O} + e_3\,\nu_1\,\nu_2\,\nu_3) + e_2\,(\nu_2\,\nu_3 + \nu_1\,(\nu_2 + \nu_3)))$$

The scalar coefficient attached to the origin factors out of the expression so that we can extract the point of intersection Q from the resulting weighted point as

```
Q = ToPointForm[wP]
```

$$\star\mathbb{O} + e_3\, \nu_1\, \nu_2\, \nu_3 + \frac{1}{3}\, e_1\, (\nu_1 + \nu_2 + \nu_3) + \frac{1}{3}\, e_2\, (\nu_2\, \nu_3 + \nu_1\, (\nu_2 + \nu_3))$$

Finally, to confirm that this point of intersection Q is coplanar with the points P_1, P_2, and P_3, we compute their exterior product.

```
Simplify[★G[P₁ ∧ P₂ ∧ P₃ ∧ Q]]
```

0

This proves the original assertion.

4.10 Decomposition into Components

The shadow

In section 3.10 we developed the Decomposition Formula [3.76] which expressed a *j*-element as a sum of components, the element being decomposed with respect to a pair of elements $\underset{m}{\alpha}$ and $\underset{n-m}{\beta}$ which together span the whole space.

$$\underset{j}{x} = \star\Sigma\left[\frac{\left(\underset{m}{\alpha} \wedge \star\mathbb{S}^{\star}\left[\underset{j}{x}\right]\right) \vee \left(\star\mathbb{S}_{\star}\left[\underset{j}{x}\right] \wedge \underset{\star n-m}{\beta}\right)}{\underset{m}{\alpha} \vee \underset{\star n-m}{\beta}}\right]$$

The first and last terms of this decomposition are

$$\underset{j}{x} = \frac{\underset{m}{\alpha} \vee \left(\underset{j}{x} \wedge \underset{n-m}{\beta}\right)}{\underset{m}{\alpha} \vee \underset{n-m}{\beta}} + \ldots + \frac{\left(\underset{m}{\alpha} \wedge \underset{j}{x}\right) \vee \underset{n-m}{\beta}}{\underset{m}{\alpha} \vee \underset{n-m}{\beta}}$$

Grassmann called the first term the *shadow of* $\underset{j}{x}$ *on* $\underset{m}{\alpha}$ *excluding* $\underset{n-m}{\beta}$.

It can be seen that the last term can be rearranged as the *shadow of* $\underset{j}{x}$ *on* $\underset{n-m}{\beta}$ *excluding* $\underset{m}{\alpha}$.

$$\frac{\left(\underset{m}{\alpha} \wedge \underset{j}{x}\right) \vee \underset{n-m}{\beta}}{\underset{m}{\alpha} \vee \underset{n-m}{\beta}} = \frac{\underset{n-m}{\beta} \vee \left(\underset{j}{x} \wedge \underset{m}{\alpha}\right)}{\underset{n-m}{\beta} \vee \underset{m}{\alpha}}$$

If $j = 1$, the decomposition formula reduces to the sum of just two components, x_α and x_β, where x_α lies in $\underset{m}{\alpha}$ and x_β lies in $\underset{n-m}{\beta}$.

$$x = x_\alpha + x_\beta = \frac{\underset{m}{\alpha} \vee \left(x \wedge \underset{n-m}{\beta}\right)}{\underset{m}{\alpha} \vee \underset{n-m}{\beta}} + \frac{\underset{n-m}{\beta} \vee \left(x \wedge \underset{m}{\alpha}\right)}{\underset{n-m}{\beta} \vee \underset{m}{\alpha}} \qquad 4.11$$

Another variant on this decomposition formula is formula [3.48] derived in Chapter 3.

$$\mathbf{x} == \sum_{i==1}^{n} \frac{\alpha_1 \wedge \alpha_2 \wedge \ldots \wedge \alpha_{i-1} \wedge \mathbf{x} \wedge \alpha_{i+1} \wedge \ldots \wedge \alpha_n}{\alpha_1 \wedge \alpha_2 \wedge \ldots \wedge \alpha_n} \alpha_i$$

We now explore these formulae with a number of geometric examples, beginning with the simplest case of decomposition in a 2-space.

Decomposition in a 2-space

◆ Decomposition of a vector in a bivector

Suppose we have a vector 2-space defined by the bivector $\alpha \wedge \beta$. We wish to decompose a vector x in the space to give one component in α and the other in β. Applying the decomposition formula [4.11], and noting that $\mathbf{x} \wedge \beta$ and $\mathbf{x} \wedge \alpha$ are bivectors while $\mathbf{x} \vee \beta$ and $\mathbf{x} \vee \alpha$ are scalars, we can use formula [3.32] with n equal to 2 (making $\underset{2}{1}$ the unit n-element), together with axiom $\vee 8$ [3.4] to give:

$$\mathbf{x}_\alpha == \frac{\alpha \vee (\mathbf{x} \wedge \beta)}{\alpha \vee \beta} == \frac{\alpha \vee \left((\mathbf{x} \vee \beta) \underset{2}{1} \right)}{\alpha \vee \beta}$$

$$== \left(\frac{\mathbf{x} \vee \beta}{\alpha \vee \beta} \right) \left(\alpha \vee \underset{2}{1} \right) == \left(\frac{\mathbf{x} \vee \beta}{\alpha \vee \beta} \right) \alpha == \left(\frac{\mathbf{x} \wedge \beta}{\alpha \wedge \beta} \right) \alpha$$

$$\mathbf{x}_\beta == \frac{\beta \vee (\mathbf{x} \wedge \alpha)}{\beta \vee \alpha} == \frac{\beta \vee \left((\mathbf{x} \vee \alpha) \underset{2}{1} \right)}{\beta \vee \alpha}$$

$$== \left(\frac{\mathbf{x} \vee \alpha}{\beta \vee \alpha} \right) \left(\beta \vee \underset{2}{1} \right) == \left(\frac{\mathbf{x} \vee \alpha}{\beta \vee \alpha} \right) \beta == \left(\frac{\alpha \wedge \mathbf{x}}{\alpha \wedge \beta} \right) \beta$$

$$\star\mathcal{B}_2 : \quad \mathbf{x} == \mathbf{x}_\alpha + \mathbf{x}_\beta == \left(\frac{\mathbf{x} \wedge \beta}{\alpha \wedge \beta} \right) \alpha + \left(\frac{\alpha \wedge \mathbf{x}}{\alpha \wedge \beta} \right) \beta \qquad 4.12$$

(To see that the coefficient $\frac{\mathbf{x} \vee \beta}{\alpha \vee \beta}$ can be replaced by $\frac{\mathbf{x} \wedge \beta}{\alpha \wedge \beta}$, replace x by a linear combination of α and β.)

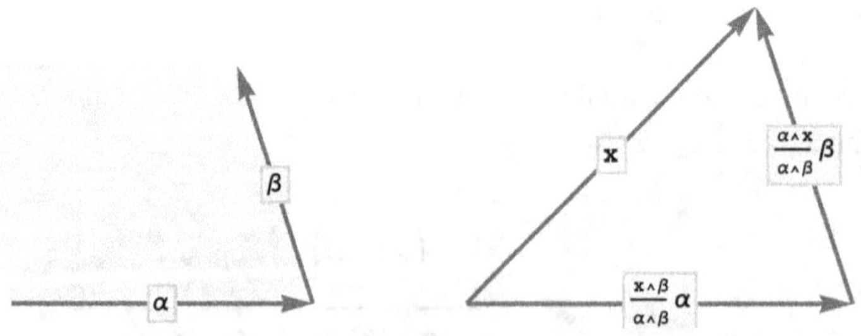

Decomposition of a vector in a bivector into two components

The coefficients of α and β are scalars showing that $\mathbf{x}_\alpha$ is congruent to α and $\mathbf{x}_\beta$ is congruent to

β. If each of the three vectors is expressed in basis form, we can determine these scalars more specifically. For example:

$$x == x_1 e_1 + x_2 e_2; \quad \alpha == a_1 e_1 + a_2 e_2; \quad \beta == b_1 e_1 + b_2 e_2;$$

$$\frac{x \wedge \beta}{\alpha \wedge \beta} == \frac{(x_1 e_1 + x_2 e_2) \wedge (b_1 e_1 + b_2 e_2)}{(a_1 e_1 + a_2 e_2) \wedge (b_1 e_1 + b_2 e_2)}$$

$$== \frac{(x_1 b_2 - x_2 b_1)\, e_1 \wedge e_2}{(a_1 b_2 - a_2 b_1)\, e_1 \wedge e_2} == \frac{(x_1 b_2 - x_2 b_1)}{(a_1 b_2 - a_2 b_1)}$$

Finally then we can express the original vector x as the required sum of two components, one in α and one in β.

$$x == \frac{(x_1 b_2 - x_2 b_1)}{(a_1 b_2 - a_2 b_1)} \alpha + \frac{(x_1 a_2 - x_2 a_1)}{(b_1 a_2 - b_2 a_1)} \beta$$

We can easily check that the right hand side does indeed reduce to x by composing x, α, and β, and then simplifying.

$$\star\mathcal{B}_2;\ \star\mathbb{V}_x;\ \alpha = \star\mathbb{V}_a;\ \beta = \star\mathbb{V}_b;$$

$$\frac{(x_1 b_2 - x_2 b_1)}{(a_1 b_2 - a_2 b_1)} \alpha + \frac{(x_1 a_2 - x_2 a_1)}{(b_1 a_2 - b_2 a_1)} \beta \ \text{// Simplify}$$

$$e_1 x_1 + e_2 x_2$$

◆ Decomposition of a point in a line

These same calculations apply *mutatis mutandis* to decomposing a point P into two component weighted points congruent to two given points Q and R in a line. For variety here, we take a more general and less coordinate-based approach.

In formula [4.12] let

$$x \to P \qquad \alpha \to Q \qquad \beta \to R$$

then the decomposition becomes

$$\star\mathcal{P}_1: \quad P == P_Q + P_R == \left(\frac{P \wedge R}{Q \wedge R}\right) Q + \left(\frac{Q \wedge P}{Q \wedge R}\right) R \qquad 4.13$$

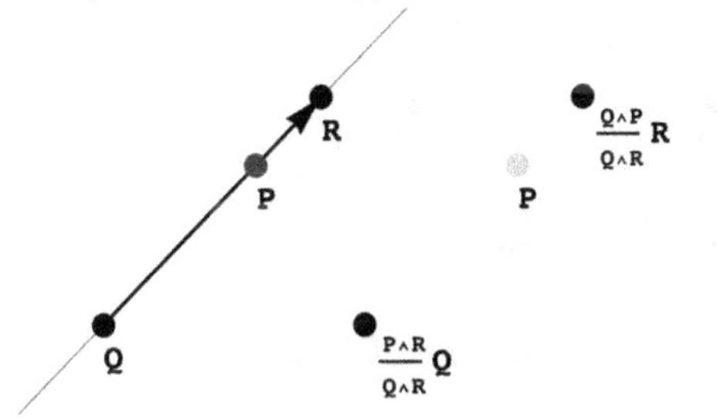

Decomposition of a point in a line into two weighted points

Because P is a point, we would expect the sum of its two component weights to be unity.

$$\frac{P \wedge R}{Q \wedge R} + \frac{Q \wedge P}{Q \wedge R} == \frac{P \wedge (R - Q)}{Q \wedge R}$$

We can see immediately that this is indeed so because adding them gives the same bound vector in the numerator and the denominator. We can either note this equivalence from our previous discussion of bound vectors, or do a substitution. For a substitution, let r and q be scalars, and v be a vector in (parallel to) the line.

$$Q == P - q\,v \qquad R == P + r\,v$$

$$\frac{P \wedge R}{Q \wedge R} == \frac{P \wedge (P + r\,v)}{(P - q\,v) \wedge (P + r\,v)} == \frac{r}{r + q}\,\frac{P \wedge v}{P \wedge v} == \frac{r}{r + q}$$

$$\frac{Q \wedge P}{Q \wedge R} == \frac{(P - q\,v) \wedge P}{(P - q\,v) \wedge (P + r\,v)} == \frac{q}{r + q}\,\frac{P \wedge v}{P \wedge v} == \frac{q}{r + q}$$

Hence, in these terms, the final decomposition reduces to

$$P == \left(\frac{r}{r + q}\right) Q + \left(\frac{q}{r + q}\right) R$$

◆ Decomposition of a vector in a line

Again, these same calculations apply *mutatis mutandis* to the situation above. Although the result turns out to be trivial in terms of our understanding that a vector may be expressed as the difference of two points, we include the example to emphasize that the decomposition formula applies easily to either bound or free entities. Let

$$P \to x$$

then the decomposition becomes

$$\star\mathcal{P}_1 : \quad x == x_Q + x_R == \left(\frac{x \wedge R}{Q \wedge R}\right) Q + \left(\frac{Q \wedge x}{Q \wedge R}\right) R \qquad 4.14$$

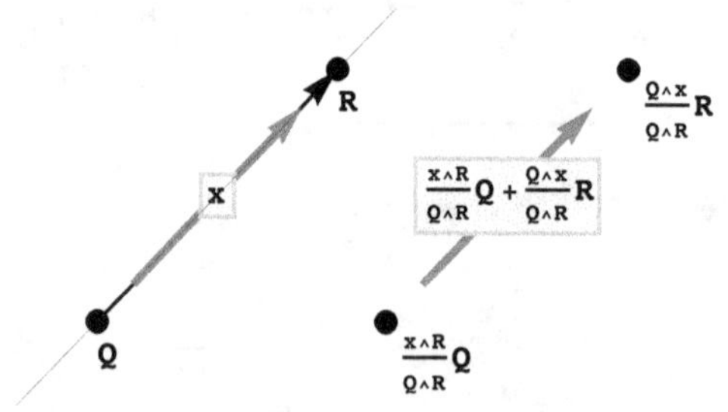

Decomposition of a vector in a line into two points of opposite weight

Note that while x is a vector, its two components are weighted points.

Because x is a vector, we would expect the sum of its two component weights to be *zero*.

$$\frac{x \wedge R}{Q \wedge R} + \frac{Q \wedge x}{Q \wedge R} == \frac{x \wedge (R - Q)}{Q \wedge R}$$

Again, we can see immediately that this is so because adding them gives a zero bivector in the numerator. For a substitution, let a be a scalar; then R - Q is a vector in (parallel to) the line.

$$x == a\,(R - Q)$$

$$\frac{x \wedge R}{Q \wedge R} == \frac{a\,(R - Q) \wedge R}{Q \wedge R} == -a\,\frac{Q \wedge R}{Q \wedge R} == -a$$

$$\frac{Q \wedge x}{Q \wedge R} == \frac{Q \wedge a\,(R - Q)}{Q \wedge R} == a\,\frac{Q \wedge R}{Q \wedge R} == a$$

Hence, in these terms, the decomposition reduces to our original definition for x.

$$x == -a\,Q + a\,R$$

Decomposition in a 3-space

◆ Decomposition of a vector in a trivector

Suppose we have a vector 3-space represented by the trivector $\alpha \wedge \beta \wedge \gamma$. We wish to decompose a vector x in this 3-space to give one component in $\alpha \wedge \beta$ and the other in γ. Applying the decomposition formula [4.11] gives:

$$x_{\alpha\wedge\beta} == \frac{(\alpha \wedge \beta) \vee (x \wedge \gamma)}{(\alpha \wedge \beta) \vee \gamma} \qquad x_\gamma == \frac{\gamma \vee (x \wedge \alpha \wedge \beta)}{\gamma \vee (\alpha \wedge \beta)}$$

Because $x \wedge \alpha \wedge \beta$ is a 3-element we can use formula [3.35] to see that the component x_γ can be written as a scalar multiple of γ where the scalar is expressed either as a ratio of regressive products (scalars) or exterior products (*n*-elements).

$$x_\gamma == \left(\frac{x \vee (\alpha \wedge \beta)}{\gamma \vee (\alpha \wedge \beta)}\right) \gamma == \left(\frac{\alpha \wedge \beta \wedge x}{\alpha \wedge \beta \wedge \gamma}\right) \gamma$$

$$\star\mathcal{B}_3 : \quad x == x_{\alpha\wedge\beta} + x_\gamma == \frac{(\alpha \wedge \beta) \vee (x \wedge \gamma)}{\alpha \wedge \beta \vee \gamma} + \left(\frac{\alpha \wedge \beta \wedge x}{\alpha \wedge \beta \wedge \gamma}\right) \gamma \qquad 4.15$$

The component $x_{\alpha\wedge\beta}$ will be a linear combination of α and β. To show this we can expand the expression above for $x_{\alpha\wedge\beta}$ using the Common Factor Axiom.

$$x_{\alpha\wedge\beta} == \frac{(\alpha \wedge \beta) \vee (x \wedge \gamma)}{(\alpha \wedge \beta) \vee \gamma} == \frac{(\alpha \wedge x \wedge \gamma) \vee \beta}{(\alpha \wedge \beta) \vee \gamma} - \frac{(\beta \wedge x \wedge \gamma) \vee \alpha}{(\alpha \wedge \beta) \vee \gamma}$$

Rearranging these two terms into a similar form to that derived for x_γ gives:

$$x_{\alpha\wedge\beta} == \left(\frac{x \wedge \beta \wedge \gamma}{\alpha \wedge \beta \wedge \gamma}\right) \alpha + \left(\frac{\alpha \wedge x \wedge \gamma}{\alpha \wedge \beta \wedge \gamma}\right) \beta$$

Of course we could have obtained this decomposition directly by using the results of the decomposition formula [3.48] for decomposing a 1-element into a linear combination of the factors of an *n*-element.

$$\star\mathcal{B}_3 : \quad x == x_\alpha + x_\beta + x_\gamma == \left(\frac{x \wedge \beta \wedge \gamma}{\alpha \wedge \beta \wedge \gamma}\right) \alpha + \left(\frac{\alpha \wedge x \wedge \gamma}{\alpha \wedge \beta \wedge \gamma}\right) \beta + \left(\frac{\alpha \wedge \beta \wedge x}{\alpha \wedge \beta \wedge \gamma}\right) \gamma \qquad 4.16$$

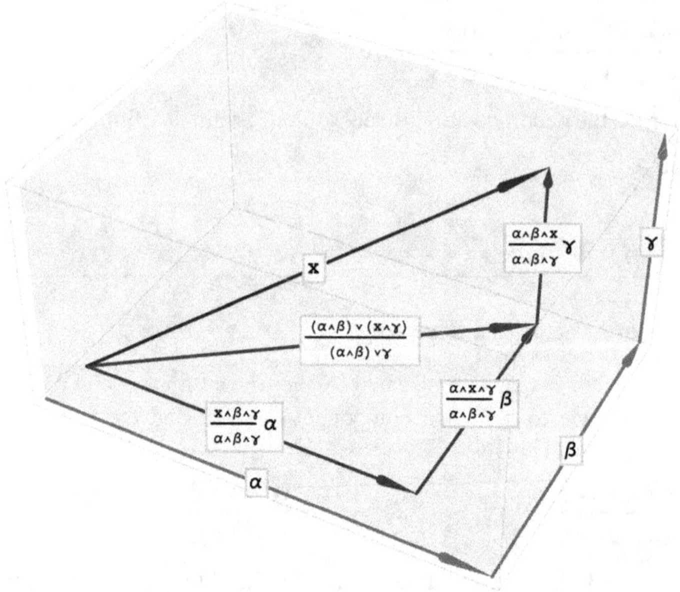

Decomposition of a vector in a trivector into three components

◆ Decomposition of a point in a plane

Suppose we have a plane represented by the bound bivector $Q \wedge R \wedge S$, where Q, R and S are points. We wish to decompose a point P in this plane in two ways. The first, as weighted points at Q, R, and S. The second as a weighted point at S and a weighted point in the line represented by $Q \wedge R$. Applying the decomposition formula [3.48] gives:

$$\star\mathcal{P}_2 : \quad P == P_Q + P_R + P_S == \left(\frac{P \wedge R \wedge S}{Q \wedge R \wedge S}\right) Q + \left(\frac{Q \wedge P \wedge S}{Q \wedge R \wedge S}\right) R + \left(\frac{Q \wedge R \wedge P}{Q \wedge R \wedge S}\right) S \qquad 4.17$$

This formula immediately displays P as a sum of weighted points situated at Q, R and S.

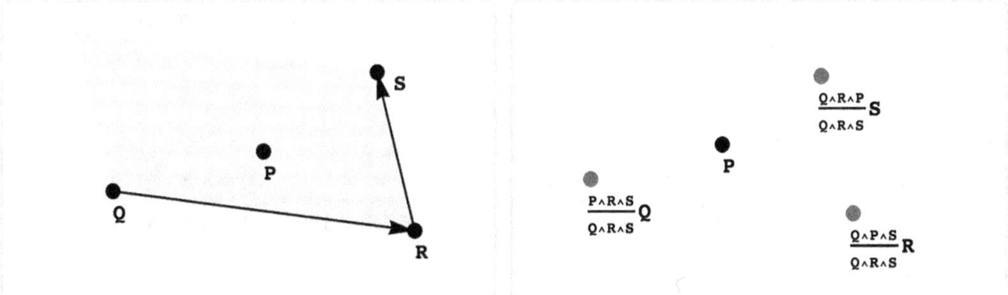

A point P in a bound bivector — P decomposed into 3 weighted points

Decomposition of a point in a bound bivector into three weighted points

Alternatively, we can group these terms two at a time to obtain a weighted point in the corresponding line. For example, adding the first two terms together gives us the point which we denote $P_{Q \wedge R}$.

$$\mathbf{P}_{\mathbf{Q}\wedge\mathbf{R}} == \left(\frac{\mathbf{P}\wedge\mathbf{R}\wedge\mathbf{S}}{\mathbf{Q}\wedge\mathbf{R}\wedge\mathbf{S}}\right)\mathbf{Q} + \left(\frac{\mathbf{Q}\wedge\mathbf{P}\wedge\mathbf{S}}{\mathbf{Q}\wedge\mathbf{R}\wedge\mathbf{S}}\right)\mathbf{R}$$

This grouping process is depicted below under the interpretation that the elements are points. Remember however, that these formulae are inherently independent of their interpretation; or, are also valid under other interpretations. For example, where some of the points are interpreted as vectors.

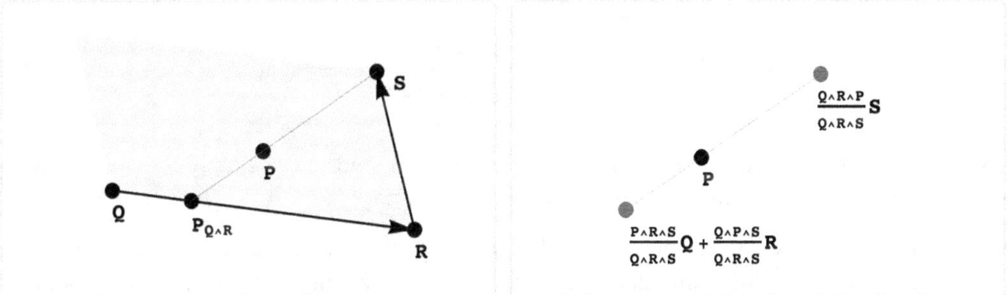

A point P in a bound bivector — P decomposed into 2 weighted points

Decomposition of a point in a bound bivector into two weighted points

Decomposition in a 4-space

◆ Decomposition of a vector in a 4-vector

Suppose we have a vector 4-space represented by the trivector $\alpha \wedge \beta \wedge \gamma \wedge \delta$ and we wish to decompose a vector x in this 4-space into two vectors, one in $\alpha \wedge \beta$ and one in $\gamma \wedge \delta$.

$$\mathbf{x} == \mathbf{x}_{\alpha\wedge\beta} + \mathbf{x}_{\gamma\wedge\delta}$$

We can apply formula [4.11] to get

$$\star\mathcal{B}_4 : \mathbf{x} == \mathbf{x}_{\alpha\wedge\beta} + \mathbf{x}_{\gamma\wedge\delta} == \frac{(\alpha\wedge\beta)\vee(\mathbf{x}\wedge\gamma\wedge\delta)}{(\alpha\wedge\beta)\vee(\gamma\wedge\delta)} + \frac{(\gamma\wedge\delta)\vee(\mathbf{x}\wedge\alpha\wedge\beta)}{(\alpha\wedge\beta)\vee(\gamma\wedge\delta)} \qquad 4.18$$

Or, we can apply formula [3.48] to get

$$\mathbf{x} == \left(\frac{\mathbf{x}\wedge\beta\wedge\gamma\wedge\delta}{\alpha\wedge\beta\wedge\gamma\wedge\delta}\right)\alpha + \left(\frac{\alpha\wedge\mathbf{x}\wedge\gamma\wedge\delta}{\alpha\wedge\beta\wedge\gamma\wedge\delta}\right)\beta + \left(\frac{\alpha\wedge\beta\wedge\mathbf{x}\wedge\delta}{\alpha\wedge\beta\wedge\gamma\wedge\delta}\right)\gamma + \left(\frac{\alpha\wedge\beta\wedge\gamma\wedge\mathbf{x}}{\alpha\wedge\beta\wedge\gamma\wedge\delta}\right)\delta$$

Hence we can write

$$\mathbf{x}_{\alpha\wedge\beta} == \frac{(\alpha\wedge\beta)\vee(\mathbf{x}\wedge\gamma\wedge\delta)}{(\alpha\wedge\beta)\vee(\gamma\wedge\delta)} == \left(\frac{\mathbf{x}\wedge\beta\wedge\gamma\wedge\delta}{\alpha\wedge\beta\wedge\gamma\wedge\delta}\right)\alpha + \left(\frac{\alpha\wedge\mathbf{x}\wedge\gamma\wedge\delta}{\alpha\wedge\beta\wedge\gamma\wedge\delta}\right)\beta$$

$$\mathbf{x}_{\gamma\wedge\delta} == \frac{(\gamma\wedge\delta)\vee(\mathbf{x}\wedge\alpha\wedge\beta)}{(\alpha\wedge\beta)\vee(\gamma\wedge\delta)} == \left(\frac{\alpha\wedge\beta\wedge\mathbf{x}\wedge\delta}{\alpha\wedge\beta\wedge\gamma\wedge\delta}\right)\gamma + \left(\frac{\alpha\wedge\beta\wedge\gamma\wedge\mathbf{x}}{\alpha\wedge\beta\wedge\gamma\wedge\delta}\right)\delta$$

As we have previously remarked, we can rearrange these components in whatever combinations or interpretations (as points or vectors) we require.

Decomposition of a point or vector in an *n*-space

Decomposition of a point or vector in a space of *n* dimensions is generally most directly accomplished by using the decomposition formula [3.48]. For example, providing at least one of the α_i is a point, the decomposition of a point P can be written:

$$\mathbf{P} == \sum_{i=1}^{n} \frac{\alpha_1\wedge\alpha_2\wedge\ldots\wedge\alpha_{i-1}\wedge\mathbf{P}\wedge\alpha_{i+1}\wedge\ldots\wedge\alpha_n}{\alpha_1\wedge\alpha_2\wedge\ldots\wedge\alpha_n}\,\alpha_i \qquad 4.19$$

As we have already seen in the examples above, the components are simply arranged in whatever combinations are required to achieve the decomposition.

- In summary, it is worth reemphasizing that, although the decomposition of elements depicts geometrically quite differently depending on whether the elements are interpreted as points or vectors, *the formulae are identical*. As with all the formulae of the Grassmann algebra, the only difference is the *interpretation*.

4.11 Projective Space

The relationship between Projective and Grassmann geometries

In perusing this chapter, you may well have remarked the similarities between the Grassmann geometry described, and the geometry of projective space. In this section, we will briefly explore these parallels. We will find that there are no *essential* differences, and that the inessential differences are ones simply of definition or interpretation.

The apparent differences stem in the main from the way in which the *intersection* of two geometric elements is treated in Projective Geometry on the one hand, and Grassmann geometry on the other. For example, in the plane, both geometries will say that two lines always intersect. In

Projective geometry the lines intersect in either a point or a *point-at-infinity*. In Grassmann geometry the lines intersect in either a point or a *vector*. We see here that there is no essential difference, but an inessential difference of definition or interpretation: a vector and a point-at-infinity are essentially the same object.

The following table compares the two geometries.

Comparing projective and Grassmann geometries

Grade	Projective Geometry	Grassmann Geometry
1	point	point
1	point at infinity	vector
2	line	bound vector
2	line at infinity	bivector
3	plane	bound bivector
3	plane at infinity	trivector
m	multiplane	bound $(m-1)$–vector
m	multiplane at infinity	m–vector

The advantage of the Grassmann geometry interpretation however, is that theorems may often be proved algebraically (and hence computationally). We shall see examples of this later in this section.

The intersection of two lines in a plane

In the Grassmann approach to the geometry of the plane, two (distinct) points define a line, and two (non-parallel) lines intersect in a common point. If the lines are parallel, they intersect in a common *vector* (that is, their common factor is a vector). Thus, as we intimated in the introduction to this section, we need only rename this vector as a *point-at-infinity* to recover the projective geometry of 2-space.

But this correspondence is not as arbitrary as it might first appear. It turns out that the result of computing the intersection of two lines approaching parallelism in the Grassmann algebra gives us a result which can be viewed on the one hand as a point approaching infinity, and on the other hand as a weighted point approaching a vector. That is, the point of intersection of two lines moves off towards infinity as the lines approach parallelism, and in the limit becomes a finite vector.

To see this we need only apply the methods previously developed for the calculation of intersections to two such lines. For simplicity, and without loss of generality, we can represent the two lines as follows:

- L_1 by a bound vector through a point P, in the direction of the vector x; and
- L_2 by a bound vector through the point $P + y$, with the vector of the line differing in direction from x by a scalar multiple ϵ of y.

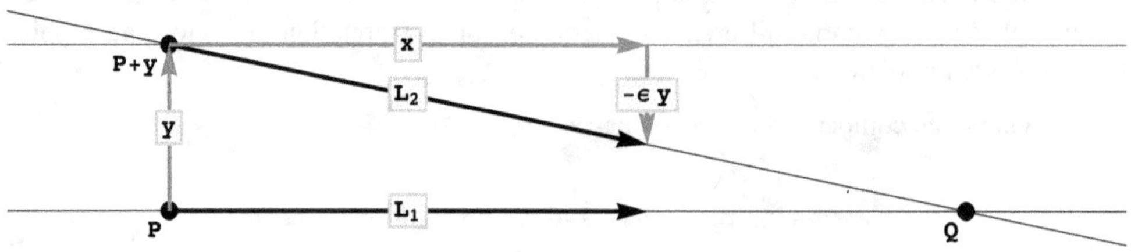

The intersection of two lines as they approach parallelism

Declare a bound 2-space, ϵ a scalar symbol, P a vector symbol, and define L_1 and L_2.

```
★𝒫2; ★★S[ϵ]; ★★V[P];
L1 = P ∧ x;
L2 = (P + y) ∧ (x - ϵ y);
```

We obtain the point of intersection *up to congruence* by applying ToCommonFactor to the regressive product of the lines.

```
ToCommonFactor[L1 ∨ L2]
```

```
★c (-x ⟨P ∧ x ∧ y⟩ - P ϵ ⟨P ∧ x ∧ y⟩)
```

Since the congruence factor ★c, and ⟨P ∧ x ∧ y⟩ (the coefficient of the exterior product in any basis) are all fixed scalars, we can say that *the intersection is congruent to the weighted point* wQ given by

$$wQ \equiv \epsilon P + x$$

By factoring out the scalar factor ϵ, we can also say that *the intersection is congruent to the point* Q given by

$$Q \equiv P + \frac{x}{\epsilon}$$

As ϵ approaches zero, the lines approach parallelism, the point of intersection Q has an ever-increasing position vector $\frac{x}{\epsilon}$, causing its position to approach infinity, and the weighted point wQ has an ever-decreasing influence from the point P, causing it to approach the (ultimately) common vector x.

It may be remarked that in this calculation, the point of intersection does not involve the vector y which specifies the position of L_2. In general then, we may conclude that the resulting vector and hence the resulting point-at-infinity is the same for all lines parallel to L_1.

The line at infinity in a plane

A basis for the plane involves the origin ★O and two basis vectors, which we denote, say, by e_1 and e_2. We can now repeat the previous computation for each of the basis vectors to obtain the weighted points

$$wQ_1 \equiv \epsilon \star O + e_1 \qquad wQ_2 \equiv \epsilon \star O + e_2$$

We can now see how the plane may be said to contain a line approaching a *line-at-infinity* if we define it by the product of these two weighted points.

$$L_\infty \equiv wQ_1 \wedge wQ_2 \equiv (\epsilon \star\mathbb{O} + e_1) \wedge (\epsilon \star\mathbb{O} + e_2)$$

This line has the property that in the limit, it contains all the points-at-infinity of the plane. We can show this by taking its exterior product with a third arbitrary point approaching a point-at-infinity, and obtain an element that becomes zero in the limit as ϵ approaches zero.

$$\star\mathcal{G}[(\epsilon \star\mathbb{O} + e_1) \wedge (\epsilon \star\mathbb{O} + e_2) \wedge (\epsilon \star\mathbb{O} + a\, e_1 + b\, e_2)]$$

$$(\epsilon - a\,\epsilon - b\,\epsilon)\, \star\mathbb{O} \wedge e_1 \wedge e_2$$

In fact, in the limit, *the line-at-infinity is the bivector of the plane*. We can see this by letting ϵ approach zero in the expression for L_∞ or its expansion:

$$\star\mathcal{G}[(\epsilon \star\mathbb{O} + e_1) \wedge (\epsilon \star\mathbb{O} + e_2)]$$

$$\star\mathbb{O} \wedge (-\epsilon\, e_1 + \epsilon\, e_2) + e_1 \wedge e_2$$

Projective 3-space

In projective 3-space, all the above concepts of the projective plane carry over naturally. We can see the parallels between Projective Geometry and Grassmann Geometry as follows:

Projective Geometry: Non-parallel planes intersect in lines. Parallel planes intersect in a line-at-infinity. All the lines-at-infinity belong to a single plane-at-infinity.

Grassmann Geometry: Non-parallel planes intersect in lines. Parallel planes intersect in a bivector. All the bivectors belong to a single trivector.

Homogeneous coordinates

◆ The plane

As we have seen, a point P in the plane may be represented *up to congruence* by

$$P = a_0 \star\mathbb{O} + a_1\, e_1 + a_2\, e_2;$$

The elements of the triplet $\{a_0, a_1, a_2\}$ are called the *homogeneous coordinates of the point* P. In projective geometry, the triplet $\{a_0, a_1, a_2\}$ and the triplet $\{k\, a_0, k\, a_1, k\, a_2\}$, where k is a scalar, are defined to represent the same point. In the Grassmann approach, this corresponds to weighted points with different weights also defining the same point.

The cobasis to the basis of the plane is

$$\star\mathcal{P}_2;\ \textbf{Cobasis}$$

$$\{e_1 \wedge e_2, -(\star\mathbb{O} \wedge e_2), \star\mathbb{O} \wedge e_1\}$$

A line in the plane can be represented up to congruence by a linear combination of these cobasis elements

$$L = b_0\, e_1 \wedge e_2 + b_1\, (-\star\mathbb{O} \wedge e_2) + b_2 \star\mathbb{O} \wedge e_1;$$

The elements of the triplet $\{b_0, b_1, b_2\}$ are called the *homogeneous coordinates of the line* L.

The equation to the line may then be obtained from the condition that a point lies in the line, that is, their exterior product is zero.

$$P \wedge L == 0$$

On simplifying the product, we recover the equation of the line in homogeneous coordinates ($a_0\, b_0 + a_1\, b_1 + a_2\, b_2 = 0$) by viewing the line as fixed and the point as variable.

$$\star\star S[a\_, b\_];\ \star\mathcal{G}[P \wedge L] == 0$$

$$(a_0\, b_0 + a_1\, b_1 + a_2\, b_2)\ \star\mathbb{O} \wedge e_1 \wedge e_2 == 0$$

However, we can also view the point as fixed, and the line as variable, wherein the same form of equation describes a variable line through a fixed point.

◆ Points and planes in a bound vector 3-space

A point P in a bound vector 3-space may be represented up to congruence by

$$\star\mathcal{P}_3;\ P = a_0 \star\mathbb{O} + a_1\, e_1 + a_2\, e_2 + a_3\, e_3;$$

The elements of the quartet $\{a_0, a_1, a_2, a_3\}$ are called the *homogeneous coordinates of the point* P.

The cobasis to the basis of the a bound vector 3-space is

```
Cobasis
```

$$\{e_1 \wedge e_2 \wedge e_3,\ -(\star\mathbb{O} \wedge e_2 \wedge e_3),\ \star\mathbb{O} \wedge e_1 \wedge e_3,\ -(\star\mathbb{O} \wedge e_1 \wedge e_2)\}$$

Hence now it is planes in the 3-space that can be represented up to congruence by a linear combination of these cobasis elements

$$\Pi = b_0\, e_1 \wedge e_2 \wedge e_3 + b_1\,(-\star\mathbb{O} \wedge e_2 \wedge e_3) + b_2 \star\mathbb{O} \wedge e_1 \wedge e_3 + b_3\,(-\star\mathbb{O} \wedge e_1 \wedge e_2);$$

The elements of the quartet $\{b_0, b_1, b_2, b_3\}$ are called the *homogeneous coordinates of the plane* Π.

The equation to the plane may then be obtained from the condition that a point lies in the plane, that is, their exterior product is zero.

$$P \wedge \Pi == 0$$

On simplifying the product, we recover the equation of the plane in homogeneous coordinates by viewing the plane as fixed and the point as variable.

$$\star\mathcal{G}[P \wedge \Pi] == 0$$

$$(a_0\, b_0 + a_1\, b_1 + a_2\, b_2 + a_3\, b_3)\ \star\mathbb{O} \wedge e_1 \wedge e_2 \wedge e_3 == 0$$

And as in the case of the plane, we can also view the point as fixed, and the plane as variable, wherein the same form of equation describes a variable plane through a fixed point.

◆ Points and hyperplanes in a bound vector *n*-space

The cases of points and lines in a bound vector 2-space (the plane), and points and planes in a bound vector 3-space (space) generalize in an obvious manner to points and hyperplanes in a bound vector *n*-space.

The reason that the relationships here are so straightforward is because points and hyperplanes are simple and have the same number of basis elements.

◆ *m*-planes in a bound vector *n*-space

This is not, in general, the case for elements of intermediate grade. For example, in a bound vector 3-space, 2-elements are not necessarily simple. Hence lines in a bound vector 3-space have no such simple form as a linear sum of the basis 2-elements. We can easily compose such a form and check for its simplicity.

```
★𝒫3; A = ComposeMElement[2, a]
```

$a_1 \star\mathbb{O} \wedge e_1 + a_2 \star\mathbb{O} \wedge e_2 + a_3 \star\mathbb{O} \wedge e_3 + a_4\, e_1 \wedge e_2 + a_5\, e_1 \wedge e_3 + a_6\, e_2 \wedge e_3$

```
SimpleQ[A]
```

$\text{If}[\{a_3\, a_4 - a_2\, a_5 + a_1\, a_6 == 0\}, \text{True}]$

Thus, such a form will only be simple, and thus able to represent a line if the coefficients are constrained by the relation

$$a_3\, a_4 - a_2\, a_5 + a_1\, a_6 == 0$$

- In this particular case the element A can be written as the sum of a bound vector and a bivector.

$$A == \star\mathbb{O} \wedge (a_1\, e_1 + a_2\, e_2 + a_3\, e_3) + (a_4\, e_1 \wedge e_2 + a_5\, e_1 \wedge e_3 + a_6\, e_2 \wedge e_3)$$
$$== \star\mathbb{O} \wedge x + B$$

Geometrically, this sum is simple if the vector of the bound vector lies in the bivector (so that the effect of the bivector is simply to shift the bound vector to another location). The condition for this is equivalent to the one which we required for simplicity of A.

$$\star\mathcal{G}[(a_1\, e_1 + a_2\, e_2 + a_3\, e_3) \wedge (a_4\, e_1 \wedge e_2 + a_5\, e_1 \wedge e_3 + a_6\, e_2 \wedge e_3)] == 0$$

$$(a_3\, a_4 - a_2\, a_5 + a_1\, a_6)\, e_1 \wedge e_2 \wedge e_3 == 0$$

Duality

The concept of duality in Grassmann algebra corresponds closely to the concept of duality in projective geometry. For example, consider the expression for a line as the exterior product of two points.

$$L == P \wedge Q$$

This may be read variously as:

The points P and Q *lie on* the line L.
The line L *passes through* the points P and Q.
The points P and Q *join* to form the line L.
The line L is the *join* of the points P and Q.

In a plane, the dual expression to "*a line is the exterior product of two points*", is "*a point is the regressive product of two lines*".

$$P == L \vee M$$

This may be read variously as:

The lines L and M *pass through* the point P.
The point P *lies on* the lines L and M.
The lines L and M *intersect* to form the point P.
The point P is the *intersection* of the lines L and M.

The *concurrency* of three points is defined by

$$P \wedge Q \wedge R == 0$$

The *collinearity* of three lines is defined by

```
L ∨ M ∨ N == 0
```

Duality extends naturally to a space of any dimension. Remember that, just as in projective geometry, these entities and expressions are only defined up to congruence.

Desargues' theorem

As an example of how Grassmann algebra can be used in Projective Geometry, we prove Desargues' theorem. It should be noted that the methods used are simply applications of the exterior and regressive products, and as are all the applications in this chapter, entirely non-metric. This is in contradistinction to 'proofs' of the theorem using the dot and cross products of 3-dimensional vector algebra, which are metric concepts.

> If two triangles have corresponding vertices joined by concurrent lines, then the intersections of corresponding sides are collinear. That is, if PU, QV, and RW all pass through one point O, then the intersections X = PQ.UV, Y = PR.UW and Z = QR.VW all lie on one line.

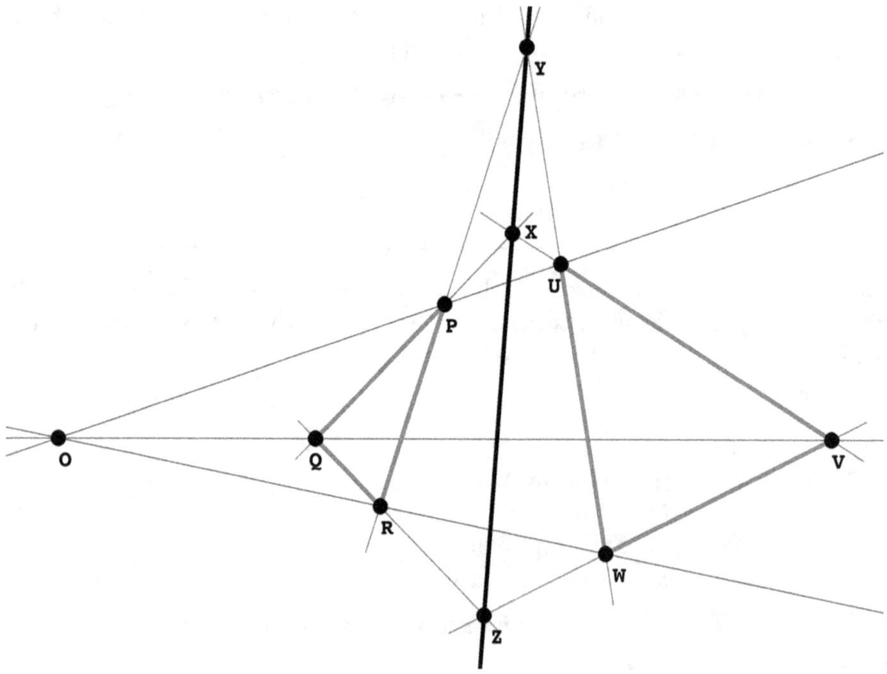

Desargues' theorem

We begin by declaring that we wish to work in the plane.

```
★P₂;
```

For simplicity, and without loss of generality, we may take the fixed point O as the origin ★O. We can then define the points P, Q, and R simply by

```
P = ★O + p; Q = ★O + q; R = ★O + r;
```

And since U, V, and W are on the same lines (respectively) through O as are P, Q, and R, we can define U, V, and W using scalar multiples of the position vectors of P, Q, and R.

```
U = ★O + a p; V = ★O + b q; W = ★O + c r;
```

To confirm that these triplets of points are collinear, we can compute their exterior product.

```
★G[{★O ∧ P ∧ U, ★O ∧ Q ∧ V, ★O ∧ R ∧ W}]
```

```
{0, 0, 0}
```

The next stage is to compute the intersections X, Y, and Z. We do this by using `ToCommonFactor`.

```
X = ToCommonFactor[(P ∧ Q) ∨ (U ∧ V)]
```

```
★c ((a - b) ★O ⟨★O ∧ p ∧ q⟩ - a (-1 + b) p ⟨★O ∧ p ∧ q⟩ + (-1 + a) b q ⟨★O ∧ p ∧ q⟩)
```

Here `X` is a weighted point We can extract the associated point by using `ToPointForm`. (Remember that ⟨★O ∧ p ∧ q⟩ represents the scalar coefficient of ★O ∧ p ∧ q in any basis that might be chosen, and ★c is the scalar congruence factor.)

```
X = ToPointForm[X]
```

$$\star O + \frac{(a - a\, b)\, p}{a - b} + \frac{(-1 + a)\, b\, q}{a - b}$$

Similarly, we can proceed with the other two intersections.

```
Y = ToCommonFactor[(P ∧ R) ∨ (U ∧ W)] // ToPointForm
```

$$\star O + \frac{(a - a\, c)\, p}{a - c} + \frac{(-1 + a)\, c\, r}{a - c}$$

```
Z = ToCommonFactor[(Q ∧ R) ∨ (V ∧ W)] // ToPointForm
```

$$\star O + \frac{(b - b\, c)\, q}{b - c} + \frac{(-1 + b)\, c\, r}{b - c}$$

We can now check whether these points are collinear by taking their exterior product. The first level of simplification gives

```
F = ★G[X ∧ Y ∧ Z]
```

$$\star O \wedge \left(\left(\frac{(-1 + a)\, a\, b\, (-1 + c)}{(a - b)\, (a - c)} - \frac{a\, (-1 + b)\, b\, (-1 + c)}{(a - b)\, (b - c)} - \frac{a\, b\, (-1 + c)^2}{(a - c)\, (-b + c)}\right) p \wedge q + \left(\frac{(-1 + a)\, (a - a\, b)\, c}{(a - b)\, (a - c)} + \frac{a\, (-1 + b)^2\, c}{(a - b)\, (b - c)} + \frac{a\, (-1 + b)\, (-1 + c)\, c}{(a - c)\, (-b + c)}\right) p \wedge r + \left(\frac{(-1 + a)^2\, b\, c}{(a - b)\, (a - c)} + \frac{(1 - a)\, (-1 + b)\, b\, c}{(a - b)\, (b - c)} + \frac{(1 - a)\, b\, (-1 + c)\, c}{(a - c)\, (-b + c)}\right) q \wedge r\right)$$

On simplifying the scalar coefficients we get zero as expected, thus proving the theorem.

```
★S[Simplify[F]]
```

```
0
```

Pappus' theorem

As a second example, we prove Pappus' theorem. The methods are essentially the same as used for Desargues' theorem in the previous section.

> Given two sets of collinear points P, Q, R, and U, V, W, then the intersection points X, Y, Z of line pairs PV and QU, PW and RU, QW and RV, are collinear.

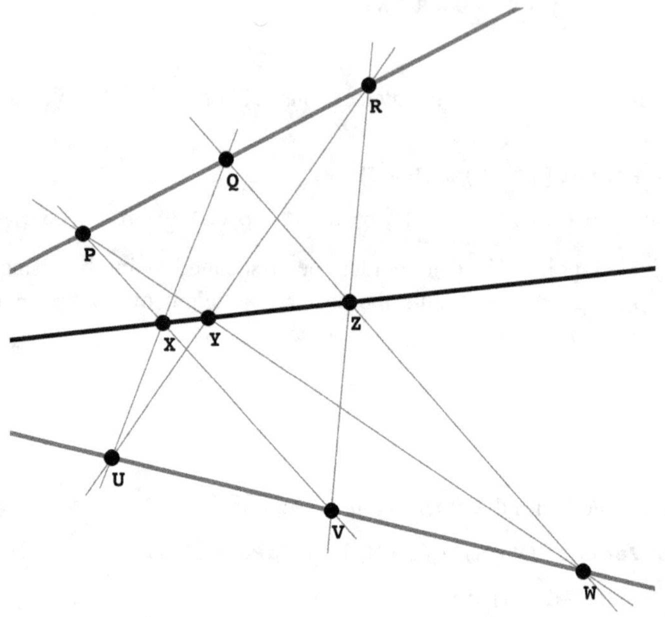

Pappus' theorem

We begin as before, by declaring that we wish to work in the plane.

```
★𝒫2;
```

As our example, we take the more general case in which the two lines through the two sets of collinear points are not parallel. For simplicity, and without loss of generality, we take their intersection point as the origin. We can then define the points P, Q, R, and U, V, W by means of two vectors p and u; and four scalars a, b, c, d.

```
P = ★O + p;
Q = ★O + a p;
R = ★O + b p;
U = ★O + u;
V = ★O + c u;
W = ★O + d u;
```

By definition these triplets of points are collinear.

The next stage is to compute the intersections X, Y, and Z. We do this by using `ToCommonFactor` and `ToPointForm` as in the previous section.

```
X = ToCommonFactor[(P ∧ V) ∨ (Q ∧ U)] // ToPointForm
```

$$★O + \frac{a(-1+c)\,p}{-1+a\,c} + \frac{(-1+a)\,c\,u}{-1+a\,c}$$

```
Y = ToCommonFactor[(P ∧ W) ∨ (R ∧ U)] // ToPointForm
```

$$\star\mathbb{O} + \frac{b\,(-1+d)\,p}{-1+b\,d} + \frac{(-1+b)\,d\,u}{-1+b\,d}$$

```
Z = ToCommonFactor[(Q ∧ W) ∨ (R ∧ V)] // ToPointForm
```

$$\star\mathbb{O} + \frac{a\,b\,(c-d)\,p}{a\,c-b\,d} + \frac{(a-b)\,c\,d\,u}{a\,c-b\,d}$$

We can now check whether these points are collinear by taking their exterior product. A value of zero thus proving the theorem.

```
★G[X ∧ Y ∧ Z] // Simplify
```

0

Projective *n*-space

A projective *n*-space is essentially, a bound vector *n*-space, that is, a Grassmann algebra whose underlying linear space is of *n*+1 dimensions; and in which one of the basis elements is interpreted as a point, say the origin, and the other *n* basis elements are interpreted as vectors. In projective *n*-space ($n \geq 2$), the concepts of the projective plane carry over naturally.

Non-parallel (*n*–1)-planes intersect in (*n*–2)-planes. Parallel (*n*–1)-planes may be said to intersect in an (*n*–2)-plane-at-infinity. All the (*n*–2)-planes-at-infinity belong to a single (*n*–1)-plane-at-infinity (which is in fact the *n*-vector of the space).

Here, a 0-plane is a point, a 1-plane is a line, a 2-plane is a plane, and so on. Two (*n*–1)-planes may be said to be parallel if their (*n*–1)-directions are the same (that is, their (*n*–1)-vectors are congruent).

We can see how this works more concretely by thinking in terms of the actual bound elements representing these geometric objects.

We can define two (*n*–1)-planes Π and Θ by

$$\Pi \equiv P \wedge \underset{n-1}{\alpha} \qquad \Theta \equiv Q \wedge \underset{n-1}{\beta}$$

where P and Q are points and $\underset{n-1}{\alpha}$ and $\underset{n-1}{\beta}$ are (*n*–1)-vectors.

The Common Factor Theorem shows that Π and Θ will have a common (*n*–1)-element, $\underset{n-1}{C}$ say. If $\underset{n-1}{\alpha}$ is congruent to $\underset{n-1}{\beta}$, then $\underset{n-1}{C}$ will be congruent to them both, and will be an (*n*–1)-vector. If $\underset{n-1}{\alpha}$ is not congruent to $\underset{n-1}{\beta}$, then $\underset{n-1}{C}$ will be of the form $X = R \wedge \underset{n-2}{V}$, thus representing an (*n*–2)-plane.

Projective space terminology is just simply equivalent to *defining* $\underset{n-1}{C}$ as an (*n*–2)-plane-at-infinity. All the $\underset{n-1}{C}$ from all the possible intersections, being (*n*–1)-vectors, belong to the *n*-vector of the space. In projective space terminology: an (*n*–1)-plane-at-infinity.

- Thus we can see that Projective Geometry differs in no essential way from the geometry which is a natural consequence of Grassmann's algebra. The superficial differences, as is the case with many mathematical systems covering the same applications, reside in definition or interpretation.

4.12 Projection

Central projection

In this section we consider the notion of projection of an element onto (or into) another element. To fix ideas however, we begin by discussing *central projection* in a bound vector space: lines are drawn *from* a centre of projection (a point), *through* an object entity, to *intersect* an image hyperplane in the object image. Sometimes we say that the object is projected *onto* the image hyperplane.

Simple cases are the projection of a point Q from a centre P onto a line $\mathbb{L}$ in the plane; and the projection of a point Q from a centre P onto a plane Π in a bound 3-space.

The projection R_i of a point Q_i from a centre of projection P onto a line $\mathbb{L}$ in the plane is achieved by passing a line through P and Q_i to intersect $\mathbb{L}$ in R_i. Given P, Q_i and $\mathbb{L}$, R_i may be calculated as the point congruent to the weighted point given by $(P \wedge Q_i) \vee \mathbb{L}$.

$$R_i \equiv (P \wedge Q_i) \vee \mathbb{L}$$

In the special case that the line $P \wedge Q_i$ is parallel to $\mathbb{L}$, the projection (intersection) R_i is their common direction (vector) rather than their common point.

Now suppose we are in bound 3-space and we wish to project three distinct points Q_i onto a plane Π as shown in the diagram below. We can easily show that if the Q_i lie on a line, then so do their projections R_i in Π.

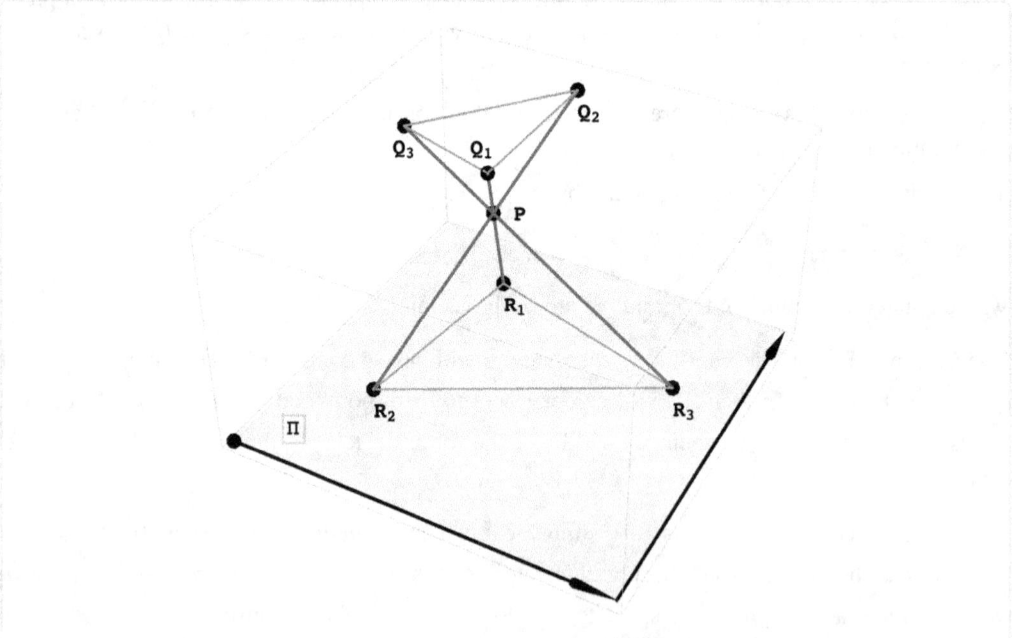

Central projection of points onto a plane

Supose that

$$Q_1 \wedge Q_2 \wedge Q_3 == 0$$
$$R_1 \equiv (P \wedge Q_1) \vee \Pi$$
$$R_2 \equiv (P \wedge Q_2) \vee \Pi$$
$$R_3 \equiv (P \wedge Q_3) \vee \Pi$$

then we wish to show that

$$A == R_1 \wedge R_2 \wedge R_3 == 0$$

To do this we first observe that the R_i are in a form immediately expandable by the Common Factor Theorem [3.42].

$$(P \wedge Q_i) \vee \Pi == (P \wedge \Pi) \vee Q_i - (Q_i \wedge \Pi) \vee P$$

Since $P \wedge \Pi$ and $Q_i \wedge \Pi$ are of grade 4 (equal to the dimension of the space), the right hand side can be written as a linear combination of P and Q_i. To do this we use formula [3.15] from our discussion of regressive products in Chapter 3.

$$\underset{n}{\alpha} \vee \underset{m}{\beta} == \underset{m}{\beta} \vee \underset{n}{\alpha} == a \underset{m}{\beta}$$

Applying this formula to the present case allows us to write

$$R_i \equiv (P \wedge \Pi) \vee Q_i - (Q_i \wedge \Pi) \vee P == p\, Q_i - q_i\, P$$

Here, p and q_i are scalars. This result makes sense since R_i is on the line joining P and Q_i, and hence must be a linear combination of them.

Substituting for the R_i in $R_1 \wedge R_2 \wedge R_3$ then gives

$$A == ((P \wedge Q_1) \vee \Pi) \wedge ((P \wedge Q_2) \vee \Pi) \wedge ((P \wedge Q_3) \vee \Pi)$$
$$== (p\, Q_1 - q_1\, P) \wedge (p\, Q_2 - q_2\, P) \wedge (p\, Q_3 - q_3\, P)$$

Multiplying this out simplifies it to

$$A == p^2 (p\, Q_1 \wedge Q_2 \wedge Q_3 - q_1\, P \wedge Q_2 \wedge Q_3 + q_2\, P \wedge Q_1 \wedge Q_3 - q_3\, P \wedge Q_1 \wedge Q_2)$$

Now if we apply the Common Factor Theorem [3.42] to $(P \wedge Q_1 \wedge Q_2 \wedge Q_3) \vee \Pi$, we get

$$(P \wedge Q_1 \wedge Q_2 \wedge Q_3) \vee \Pi$$
$$== (P \wedge \Pi) \vee (Q_1 \wedge Q_2 \wedge Q_3) - (Q_1 \wedge \Pi) \vee (P \wedge Q_2 \wedge Q_3) +$$
$$(Q_2 \wedge \Pi) \vee (P \wedge Q_1 \wedge Q_3) - (Q_3 \wedge \Pi) \vee (P \wedge Q_1 \wedge Q_2)$$

By comparing this to the previous expression for A we see that we can now rewrite A as

$$(P \wedge \Pi) \vee (P \wedge \Pi) \vee (P \wedge Q_1 \wedge Q_2 \wedge Q_3) \vee \Pi$$

Thus we have the final result that

$$((P \wedge Q_1) \vee \Pi) \wedge ((P \wedge Q_2) \vee \Pi) \wedge ((P \wedge Q_3) \vee \Pi) == (P \wedge \Pi) \vee (P \wedge \Pi) \vee (P \wedge Q_1 \wedge Q_2 \wedge Q_3) \vee \Pi \qquad 4.20$$

From this result it is clear that if $Q_1 \wedge Q_2 \wedge Q_3$ is zero then so is A, thus proving that if three points lie on a line then so do their projections onto a plane.

Thus we have assured ourselves of the intuitively obvious result: to project a *line* onto a plane, all we need to do is to project any two points onto the plane, and then form a line through them.

A general projection formula

Formula [4.20] above may be generalized to a bound vector n-space as follows:

$$\begin{gathered} R_1 \wedge R_2 \wedge \ldots \wedge R_m \equiv \\ ((P \wedge Q_1) \vee \mathbb{H}) \wedge ((P \wedge Q_2) \vee \mathbb{H}) \wedge \ldots \wedge ((P \wedge Q_m) \vee \mathbb{H}) \\ == p^{m-1} (P \wedge Q_1 \wedge Q_2 \wedge \ldots \wedge Q_m) \vee \mathbb{H} \\ P \wedge \mathbb{H} == p \underset{n+1}{1} \end{gathered} \qquad 4.21$$

Here, P is the centre of projection, the Q_i are the points to be projected, and the R_i are their images in the hyperplane $\mathbb{H}$. Remember, $\mathbb{H}$ is a bound (n–1)-vector in the bound vector n-space.

$$R_i \equiv (P \wedge Q_i) \vee \mathbb{H}$$

Note carefully that the $(P \wedge Q_i) \vee \mathbb{H}$ are *weighted* points, so that the image points R_i are only defined by this formula up to congruence.

◆ Projection of a point

In the case of a single point Q_1 projected from P onto a hyperplane $\mathbb{H}$, the formula reduces to an identity in which each term is congruent to the image point.

$$R_1 \equiv (P \wedge Q_1) \vee \mathbb{H} == (P \wedge Q_1) \vee \mathbb{H}$$

This formula is valid in a space of any number of dimensions.

For example, in a bound vector 1-space (a line), a hyperplane $\mathbb{H}$ is a point in the line. The points P and Q_1 are also on the line (and indeed define the line). The projection of Q_1 onto the hyperplane is then just the hyperplane itself. Hence in this special case, the image R_1 coincides with the hyperplane.

In a bound vector 2-space (a plane), a hyperplane is a line in the plane. The points P and Q_1 define a line $P \wedge Q_1$ which intersects the line in R_1.

◆ Projection of a line

In the case of two points Q_1 and Q_2 projected from P onto a hyperplane $\mathbb{H}$, the formula shows that the line $Q_1 \wedge Q_2$ is projected to give the image line $R_1 \wedge R_2$.

$$R_1 \wedge R_2 \equiv ((P \wedge Q_1) \vee \mathbb{H}) \wedge ((P \wedge Q_2) \vee \mathbb{H}) == p (P \wedge Q_1 \wedge Q_2) \vee \mathbb{H}$$

This line is also given by the intersection of the plane $P \wedge Q_1 \wedge Q_2$ with the hyperplane $\mathbb{H}$.

This formula is also valid in a space of any number of dimensions.

However, in a bound vector 1-space it tells us that if we attempt to project a line onto a hyperplane (a point in this case) we get a zero result, indicating that it is not geometrically meaningful.

On the other hand, in a bound vector 2-space (a plane), a hyperplane is a line in the plane. The image points R_1 and R_2 are on the line (and indeed define it). The projection of $Q_1 \wedge Q_2$ onto the line is then just the line itself. Hence in this special case, the image $R_1 \wedge R_2$ coincides with the hyperplane.

In a bound vector 3-space, a hyperplane is a plane. The points P, $Q_1$and Q_2 define a plane $P \wedge Q_1 \wedge Q_2$ which intersects it in the line $R_1 \wedge R_2$.

◆ Projection of a bound bivector

Just as lines or bound vectors can be projected, so can higher grade elements. Below we depict the projection of a bound bivector $Q_1 \wedge Q_2 \wedge Q_3$ through a point P onto a plane Π to form the bound bivector $R_1 \wedge R_2 \wedge R_3$. As always, we can view bound bivectors as defining planes, or view planes as being defined by bound bivectors.

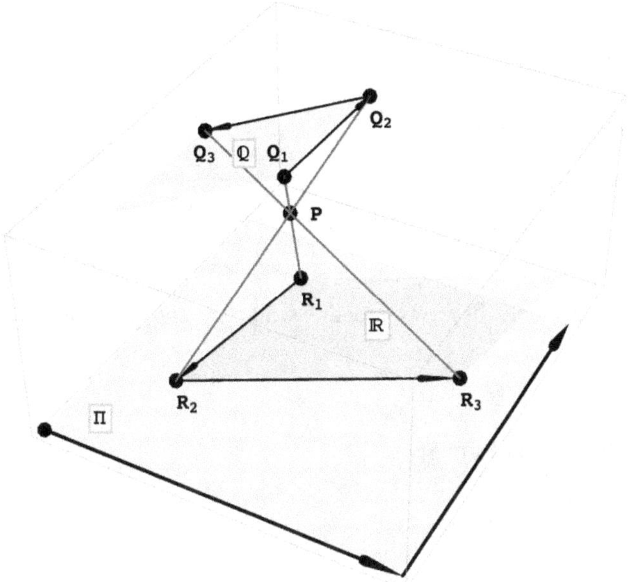

Central projection of a bound bivector onto a plane

Computing the image of a central projection

To compute the central projection C of a simple m-element $\underset{m}{Q}$ equal to $Q_1 \wedge Q_2 \wedge \ldots \wedge Q_m$ onto a hyperplane $\mathbb{H}$, we can take the general projection formula [4.21], and apply `ToCommonFactor`. Note that we can immediately ignore the scalars p, since we are only interested in determining the image up to congruence.

$$C \equiv (P \wedge Q_1 \wedge Q_2 \wedge \ldots \wedge Q_m) \vee \mathbb{H} == \left(P \wedge \underset{m}{Q}\right) \vee \mathbb{H} \qquad 4.22$$

First, we declare any vector symbols.

```
★A; ★★V[P, Q_];
```

◆ The projection of a point onto a line in the plane

After declaring that we want to work in the plane we can get a purely symbolic result immediately by applying `ToCommonFactor`, providing we tell *GrassmannAlgebra* that the hyperplane is of grade 2.

$$\star\mathcal{P}_2;\ \mathbf{C} \equiv \mathbf{ToCommonFactor}\left[(\mathbf{P}\wedge\mathbf{Q_1})\vee\underset{2}{\mathbb{L}}\right]$$

$$C \equiv \star c\left(-P\left\langle Q_1\wedge\underset{2}{\mathbb{L}}\right\rangle+\left\langle P\wedge\underset{2}{\mathbb{L}}\right\rangle Q_1\right)$$

Remember that the angle bracket expressions are scalar quantities. $\langle z\rangle$ is the scalar coefficient of z when z is expressed in the current basis. See section 3.6.

Since the the congruence factor $\star c$ is of no consequence in finding the image point we can ignore it by using `CongruenceSimplify` (or its alias $\star\mathcal{C}$) instead of `ToCommonFactor`.

$$\mathbf{C} \equiv \star\mathcal{C}\left[(\mathbf{P}\wedge\mathbf{Q_1})\vee\underset{2}{\mathbb{L}}\right]$$

$$C \equiv -P\left\langle Q_1\wedge\underset{2}{\mathbb{L}}\right\rangle+\left\langle P\wedge\underset{2}{\mathbb{L}}\right\rangle Q_1$$

$$\mathbf{C} \equiv -\mathbf{P}\left\langle \mathbf{Q_1}\wedge\underset{2}{\mathbb{L}}\right\rangle+\left\langle \mathbf{P}\wedge\underset{2}{\mathbb{L}}\right\rangle \mathbf{Q_1} \qquad 4.23$$

The expressions within angle brackets are the scalars p and q_1 which figured in our initial explorations.

$$\mathbf{P}\wedge\underset{2}{\mathbb{L}} == \mathbf{p}\,\underset{3}{\mathbf{1}} \qquad \mathbf{Q_1}\wedge\underset{2}{\mathbb{L}} == \mathbf{q_1}\,\underset{3}{\mathbf{1}}$$

Writing C again in this shorter form gives

$$\mathbf{C} \equiv -\mathbf{q_1}\,\mathbf{P}+\mathbf{p}\,\mathbf{Q_1}$$

Now, P and Q_1 are points, so we can write each of them as the sum of the origin and its position vector, and substitute back in the expression for C.

$$\mathbf{P} == \star\mathbb{O}+\mathbf{x} \qquad \mathbf{Q_1} == \star\mathbb{O}+\mathbf{z}$$

$$\mathbf{C} \equiv -\mathbf{q_1}\,(\star\mathbb{O}+\mathbf{x})+\mathbf{p}\,(\star\mathbb{O}+\mathbf{z}) == (\mathbf{p}-\mathbf{q_1})\,\star\mathbb{O}+(\mathbf{p}\,\mathbf{z}-\mathbf{q_1}\,\mathbf{x})$$

Because this is a congruence relation for C, we can write C finally as the point.

$$\mathbf{C} == \star\mathbb{O}+\frac{\mathbf{p}\,\mathbf{z}-\mathbf{q_1}\,\mathbf{x}}{\mathbf{p}-\mathbf{q_1}}$$

◆ The projection of a point onto a hyperplane in a space of arbitrary dimension

It is straightforward to deduce that the same formula holds for the projection of a point Q through a point P onto a hyperplane $\mathbb{H}$ in a space of arbitrary dimension.

$$\mathbf{C} \equiv -\mathbf{P}\,\langle \mathbf{Q}\wedge\mathbb{H}\rangle+\langle \mathbf{P}\wedge\mathbb{H}\rangle\,\mathbf{Q} \qquad 4.24$$

◆ The projection of a line onto a plane in bound 3-space

The computation is analogous to that for the projection of a point onto a line in the plane.

$$\star\mathcal{P}_3;\ \mathbf{C} \equiv \star\mathcal{C}\left[(\mathbf{P}\wedge\mathbf{Q_1}\wedge\mathbf{Q_2})\vee\underset{3}{\Pi}\right]$$

$$C \equiv \left\langle Q_2\wedge\underset{3}{\Pi}\right\rangle P\wedge Q_1-\left\langle Q_1\wedge\underset{3}{\Pi}\right\rangle P\wedge Q_2+\left\langle P\wedge\underset{3}{\Pi}\right\rangle Q_1\wedge Q_2$$

$$C \equiv \left\langle Q_2 \wedge \underset{3}{\Pi} \right\rangle P \wedge Q_1 - \left\langle Q_1 \wedge \underset{3}{\Pi} \right\rangle P \wedge Q_2 + \left\langle P \wedge \underset{3}{\Pi} \right\rangle Q_1 \wedge Q_2 \qquad 4.25$$

From the general projection formula [4.21], we would expect to obtain this result also by projecting each point.

$$C \equiv \star C\left[\left((P \wedge Q_1) \vee \underset{3}{\Pi}\right) \wedge \left((P \wedge Q_2) \vee \underset{3}{\Pi}\right)\right]$$

$$C \equiv \left\langle P \wedge \underset{3}{\Pi} \right\rangle \left\langle Q_2 \wedge \underset{3}{\Pi} \right\rangle P \wedge Q_1 - \left\langle P \wedge \underset{3}{\Pi} \right\rangle \left\langle Q_1 \wedge \underset{3}{\Pi} \right\rangle P \wedge Q_2 + \left\langle P \wedge \underset{3}{\Pi} \right\rangle^2 Q_1 \wedge Q_2$$

By factoring out the scalar factor $\left\langle P \wedge \underset{3}{\Pi} \right\rangle$, we see that, up to congruence, the results are indeed the same.

Examples in coordinate form

◆ Algebraic example: Projection of a point onto a line in the plane

Let us consider a plane and project point Q from point P onto a line $\mathbb{L}$ in the plane. We shall give the entities coordinates, and obtain an expression for R as a function of $\mathbb{L}$, P and Q from the formula we have developed.

$$\star\mathcal{P}_2;\ \{P = \star\mathbb{P}_p,\ Q = \star\mathbb{P}_q,\ \mathbb{L} = \star\mathbb{L}_1\}$$

$$\{\star\mathbb{O} + e_1\, p_1 + e_2\, p_2,\ \star\mathbb{O} + e_1\, q_1 + e_2\, q_2,\ (\star\mathbb{O} + e_1\, l_1 + e_2\, l_2) \wedge (e_1\, l_3 + e_2\, l_4)\}$$

$$R = \mathbf{ToPointForm}[\star C[(P \wedge Q) \vee \mathbb{L}]]$$

$$\star\mathbb{O} + \frac{e_1\, ((l_2\, l_3 - l_1\, l_4)\, (p_1 - q_1) + l_3\, (p_2\, q_1 - p_1\, q_2))}{l_4\, (-p_1 + q_1) + l_3\, (p_2 - q_2)} + \frac{e_2\, ((l_2\, l_3 - l_1\, l_4)\, (p_2 - q_2) + l_4\, (p_2\, q_1 - p_1\, q_2))}{l_4\, (-p_1 + q_1) + l_3\, (p_2 - q_2)}$$

◆ Algebraic example: Projection of a point onto a plane in space

Following the same procedure as the previous example, let us consider a bound 3-space and project point Q from point P onto a plane Π in the space.

$$\star\mathcal{P}_3;\ \{P = \star\mathbb{P}_p,\ Q = \star\mathbb{P}_q,\ \Pi = \star\Pi_\pi\}$$

$$\{\star\mathbb{O} + e_1\, p_1 + e_2\, p_2 + e_3\, p_3,\ \star\mathbb{O} + e_1\, q_1 + e_2\, q_2 + e_3\, q_3,\ (\star\mathbb{O} + e_1\, \pi_1 + e_2\, \pi_2 + e_3\, \pi_3) \wedge (e_1\, \pi_4 + e_2\, \pi_5 + e_3\, \pi_6) \wedge (e_1\, \pi_7 + e_2\, \pi_8 + e_3\, \pi_9)\}$$

$$R = \star C[(P \wedge Q) \vee \Pi]$$

$$\begin{aligned}
&\star\mathbb{O}\, ((\pi_6\, \pi_8 - \pi_5\, \pi_9)\, (p_1 - q_1) + \\
&\quad (\pi_6\, \pi_7 - \pi_4\, \pi_9)\, (-p_2 + q_2) + (\pi_5\, \pi_7 - \pi_4\, \pi_8)\, (p_3 - q_3)) + \\
&e_1\, ((\pi_3\, (-\pi_5\, \pi_7 + \pi_4\, \pi_8) + \pi_2\, (\pi_6\, \pi_7 - \pi_4\, \pi_9) + \pi_1\, (-\pi_6\, \pi_8 + \pi_5\, \pi_9))\, (-p_1 + q_1) + \\
&\quad (\pi_6\, \pi_7 - \pi_4\, \pi_9)\, (-p_2\, q_1 + p_1\, q_2) + (\pi_5\, \pi_7 - \pi_4\, \pi_8)\, (p_3\, q_1 - p_1\, q_3)) + \\
&e_2\, ((\pi_3\, (-\pi_5\, \pi_7 + \pi_4\, \pi_8) + \pi_2\, (\pi_6\, \pi_7 - \pi_4\, \pi_9) + \pi_1\, (-\pi_6\, \pi_8 + \pi_5\, \pi_9))\, (-p_2 + q_2) + \\
&\quad (\pi_6\, \pi_8 - \pi_5\, \pi_9)\, (-p_2\, q_1 + p_1\, q_2) + (\pi_5\, \pi_7 - \pi_4\, \pi_8)\, (p_3\, q_2 - p_2\, q_3)) + \\
&e_3\, ((\pi_3\, (-\pi_5\, \pi_7 + \pi_4\, \pi_8) + \pi_2\, (\pi_6\, \pi_7 - \pi_4\, \pi_9) + \pi_1\, (-\pi_6\, \pi_8 + \pi_5\, \pi_9))\, (-p_3 + q_3) + \\
&\quad (\pi_6\, \pi_8 - \pi_5\, \pi_9)\, (-p_3\, q_1 + p_1\, q_3) + (\pi_6\, \pi_7 - \pi_4\, \pi_9)\, (p_3\, q_2 - p_2\, q_3))
\end{aligned}$$

Explicitly expressing this weighted point in point form leads to a rather cumbersome expression. However, as always, the point form may be read directly by dividing the weighted point by the coefficient of the origin $\star\mathbb{O}$. We repeat this example numerically below.

◆ Numerical example: Projection of a point onto a plane in space

We repeat the example above, but this time using numerical coordinates for the entities. We can compose the numerically defined entities most effectively by using the composition functions with blank placeholders for the coordinates. Beginning to fill these in automatically transforms the initial output into an input cell in which we fill the placeholders with numbers by tabbing through them.

$\star\mathcal{P}_3$; $\{\star\mathbb{P}_\square, \star\mathbb{P}_\square, \star\Pi_\square\}$

$\{P = \star\mathbb{O} + 1\, e_1 + 2\, e_2 + 3\, e_3,\ Q = \star\mathbb{O} + 4\, e_1 - 5\, e_2 + 6\, e_3,$
$\Pi = (\star\mathbb{O} - 7\, e_1 + 8\, e_2 + 9\, e_3) \wedge (9\, e_1 - 8\, e_2 + 7\, e_3) \wedge (6\, e_1 + 5\, e_2 - 4\, e_3)\};$

R = ToPointForm[$\star\mathcal{C}$[(P ∧ Q) ∨ Π]]

$$\star\mathbb{O} - \frac{479\, e_1}{46} + \frac{1317\, e_2}{46} - \frac{387\, e_3}{46}$$

◆ Numerical example: Projection of a line onto a plane in space

To project a line onto a plane in space, we can either project two points from the line onto the plane and then construct the image line from their projections; or, we can project the line directly.

$\{P = \star\mathbb{O} + 1\, e_1 + 2\, e_2 + 3\, e_3,\ Q_1 = \star\mathbb{O} + 4\, e_1 - 5\, e_2 + 6\, e_3,\ Q_2 = \star\mathbb{O} + 9\, e_1 - 7\, e_2 + 3\, e_3,$
$\Pi = (\star\mathbb{O} - 7\, e_1 + 8\, e_2 + 9\, e_3) \wedge (9\, e_1 - 8\, e_2 + 7\, e_3) \wedge (6\, e_1 + 5\, e_2 - 4\, e_3)\};$

◇ Projecting the line by projecting two points

$R_a = \star\mathcal{C}[((P \wedge Q_1) \vee \Pi) \wedge ((P \wedge Q_2) \vee \Pi)]$

$\star\mathbb{O} \wedge (-31\,500\, e_1 - 2\,727\,900\, e_2 + 2\,286\,900\, e_3) +$
$29\,307\,600\, e_1 \wedge e_2 - 24\,078\,600\, e_1 \wedge e_3 + 42\,525\,000\, e_2 \wedge e_3$

◇ Projecting the line directly

$R_b = \star\mathcal{C}[(P \wedge Q_1 \wedge Q_2) \vee \Pi]$

$\star\mathbb{O} \wedge (-30\, e_1 - 2598\, e_2 + 2178\, e_3) + 27\,912\, e_1 \wedge e_2 - 22\,932\, e_1 \wedge e_3 + 40\,500\, e_2 \wedge e_3$

One way to see that R_a and R_b define the same line is to use `ExteriorFactorize` to express each as a product of a point and a vector.

ExteriorFactorize[R_a]

$$-31\,500 \left(\star\mathbb{O} + \frac{4652\, e_2}{5} - \frac{3822\, e_3}{5}\right) \wedge \left(e_1 + \frac{433\, e_2}{5} - \frac{363\, e_3}{5}\right)$$

ExteriorFactorize[R_b]

$$-30 \left(\star\mathbb{O} + \frac{4652\, e_2}{5} - \frac{3822\, e_3}{5}\right) \wedge \left(e_1 + \frac{433\, e_2}{5} - \frac{363\, e_3}{5}\right)$$

General projection

Although we have discussed specific examples in the preceding sections, the process by which we derived them transform, *mutatis mutandis* for the central projection of any bound simple element onto a hyperplane in a bound space of any dimension.

Furthermore, although the formulae have been derived within the specific geometric context of bound elements (points, lines, planes, ...), they are still valid formulae of the algebra *independent* of this interpretation. For example, provided we respect the grades of the elements involved, all the formulae possess a valid interpretation in which all the points are 1-elements, and the hyperplane is an (n-1)-element. To illustrate what we mean, we copy the preceding coordinate examples, and modify them *mutatis mutandis* to interpret them in terms of m-elements only.

◆ Algebraic example: Projection of a 1-element onto a 2-element in a 3-space

Let us consider a 3-space and project a 1-element β from a 1-element α onto a 2-element B to form an image ρ.

★A; {α = ★V_a, β = ★V_b, B = ★B_c}

$\{a_1\, e_1 + a_2\, e_2 + a_3\, e_3,\ b_1\, e_1 + b_2\, e_2 + b_3\, e_3,\ c_1\, e_1 \wedge e_2 + c_2\, e_1 \wedge e_3 + c_3\, e_2 \wedge e_3\}$

ρ = ToCommonFactor[(α ∧ β) ∨ B]

$$\star c\ ((a_3\, b_1\, c_1 - a_1\, b_3\, c_1 - a_2\, b_1\, c_2 + a_1\, b_2\, c_2)\, e_1 + (a_3\, b_2\, c_1 - a_2\, b_3\, c_1 - a_2\, b_1\, c_3 + a_1\, b_2\, c_3)\, e_2 + (a_3\, b_2\, c_2 - a_2\, b_3\, c_2 - a_3\, b_1\, c_3 + a_1\, b_3\, c_3)\, e_3)$$

If we interpret these elements vectorially, we can say that the projection of a vector β from another vector α onto a bivector B in 3-space is equal to the intersection (or common vector, or direction) of the bivector $\alpha \wedge \beta$ and the bivector B.

◆ Numeric example: Projection of a 1-element onto a 3-element in a 4-space

Here, we consider a 4-space and project a 1-element β from a 1-element α onto a 3-element T to form an image ρ.

★𝔅₄; {★V□, ★V□, ComposeMVector[3, □]}

$$\{\alpha = e_1 + 2\, e_2 + 3\, e_3 + 4\, e_4,\ \beta = 5\, e_1 + 6\, e_2 + 7\, e_3 + 8\, e_4,\ T = 3\, e_1 \wedge e_2 \wedge e_3 - 2\, e_1 \wedge e_2 \wedge e_4 + 7\, e_1 \wedge e_3 \wedge e_4 - 5\, e_2 \wedge e_3 \wedge e_4\};$$

ρ = ToCommonFactor[(α ∧ β) ∨ T]

$\star c\ (-80\, e_1 - 12\, e_2 + 56\, e_3 + 124\, e_4)$

Interpreting this result vectorially, we can say that the projection of a vector β from another vector α onto a trivector T in 4-space is equal to the intersection (or common vector, or direction) of the bivector $\alpha \wedge \beta$ and the bivector B.

◆ Numerical example: Projection of a simple 2-element onto a 3-element in 4-space

Let us now project a simple 2-element B from a 1-element α onto a 3-element T to form an image F.

★𝔅₄; {★V□, ★W□ ComposeMVector[3, □]}

```
{α = e1 + 2 e2 + 3 e3 + 4 e4, B = (e1 + e2 + 9 e3 + 8 e4) ∧ (7 e1 + 6 e2 + 5 e3 + 4 e4),
  T = 3 e1 ∧ e2 ∧ e3 - 2 e1 ∧ e2 ∧ e4 + 7 e1 ∧ e3 ∧ e4 - 5 e2 ∧ e3 ∧ e4};
F = ToCommonFactor[(α ∧ B) ∨ T]
```

$$\star c\,(-296\, e_1 \wedge e_2 + 688\, e_1 \wedge e_3 + 232\, e_1 \wedge e_4 - 104\, e_2 \wedge e_3 - 424\, e_2 \wedge e_4 + 904\, e_3 \wedge e_4)$$

We can see that this result is simple by applying `ExteriorFactorize`.

```
ExteriorFactorize[F]
```

$$-296\,\star c \left(e_1 - \frac{13\, e_3}{37} - \frac{53\, e_4}{37}\right) \wedge \left(e_2 - \frac{86\, e_3}{37} - \frac{29\, e_4}{37}\right)$$

Again, interpreting this result vectorially, we can say that the projection of a bivector B from a vector α onto a trivector T in 4-space is equal to the intersection (or common bivector, or 2-direction) of the trivector $\alpha \wedge B$ and the trivector T.

◆ Projection of curves and surfaces

From the forgoing examples, it is straightforward to see that if we allow a point to move, then its projected image on a hyperplane will move correspondingly. Hence any series of points will yield a corresponding series of image points on the hyperplane. If we can parametrize the moving point then the image may be traced correspondingly. The space in which this takes place may be of any number of dimensions, and the number of parameters adjustable to make the point move may also be any number. In particular, there is no reason why the coordinates of the moving point should not depend on more independent parameters than the space has dimensions.

As an example, take the point Q to be in motion around a circle. A simple parametrization would be a uniform motion as a function of angle θ.

```
Q = ★O + Cos[θ] e1 + Sin[θ] e2
```

In the plane, this motion may be projected through a point onto a hyperplane of the space (a line) giving an oscillating image in the line forming a segment of it. In space (bound vector 3-space) this image projected onto the hyperplane of the space (a plane) will be a conic section of some type (depending on the inclination of the plane to the circle). For a plane through coordinates a, b and c on the coordinate axes we can easily obtain the image of Q as the point

$$R == \star\mathbb{O} + \frac{a\,(c\,\mathrm{Cos}[\theta] + b\,(-c + \mathrm{Sin}[\theta]))\, e_1}{-b\,c + a\,c\,\mathrm{Cos}[\theta] + a\,b\,\mathrm{Sin}[\theta]} + \frac{(-1 + a)\, b\, c\,\mathrm{Cos}[\theta]\, e_2}{-b\,c + a\,c\,\mathrm{Cos}[\theta] + a\,b\,\mathrm{Sin}[\theta]} + \frac{(-1 + a)\, b\, c\,\mathrm{Sin}[\theta]\, e_3}{-b\,c + a\,c\,\mathrm{Cos}[\theta] + a\,b\,\mathrm{Sin}[\theta]}$$

Parallel projection

We have left the discussion of parallel projection until last because it is just another interpretation of the formulae already developed. Parallel projection is simply central projection with the centre of projection P replaced by a direction of projection v.

- As we have seen, we can project points (which may be the points of an entity):

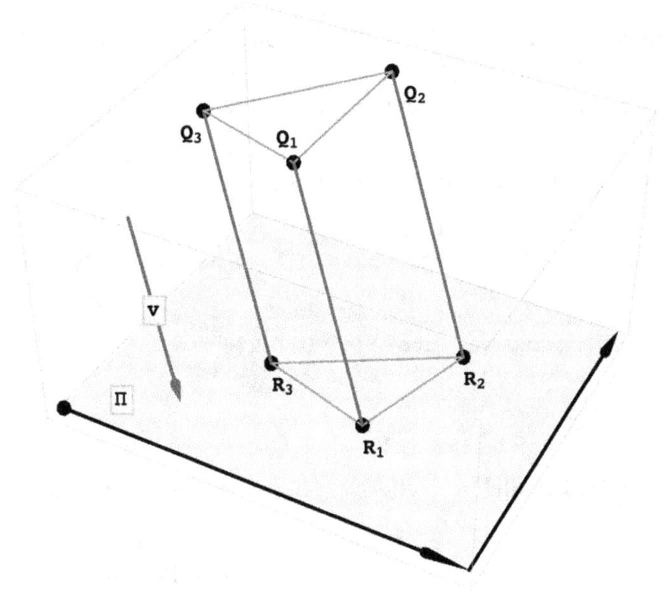

Parallel projection of points onto a plane

- Or, we can project entities (composed of points):

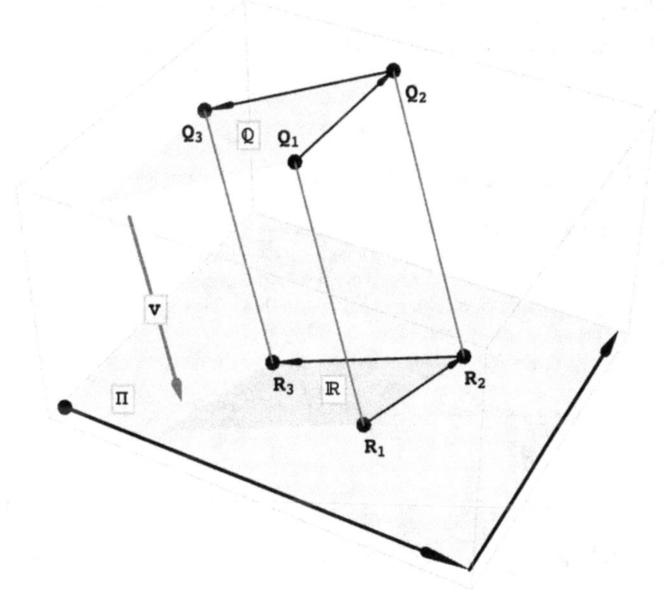

Parallel projection of a bound bivector onto a plane

We take two simple examples to confirm the notions involved.

◆ Numerical example: Parallel projection of a point onto a plane in space

Here we take the numerical example for the central projection of a point onto a plane in space, but replace the centre of projection P by a direction vector v. In this trivial case involving one point only, if we make the direction vector equal to P - Q, we should get the same result as for central projection since

$$(\mathbf{v} \wedge \mathbf{Q}) \vee \Pi \;==\; ((\mathbf{P} - \mathbf{Q}) \wedge \mathbf{Q}) \vee \Pi \;==\; (\mathbf{P} \wedge \mathbf{Q}) \vee \Pi$$

$\star\mathcal{P}_3$; $\{\mathbf{P} = \star\mathbb{O} + 1\,e_1 + 2\,e_2 + 3\,e_3,\ \mathbf{Q} = \star\mathbb{O} + 4\,e_1 - 5\,e_2 + 6\,e_3,$
$\quad\Pi = (\star\mathbb{O} - 7\,e_1 + 8\,e_2 + 9\,e_3) \wedge (9\,e_1 - 8\,e_2 + 7\,e_3) \wedge (6\,e_1 + 5\,e_2 - 4\,e_3)\};$

$\mathbf{v} = \mathbf{P} - \mathbf{Q};$

$\mathbf{R} = \mathbf{ToPointForm}[\mathbf{ToCommonFactor}[(\mathbf{v} \wedge \mathbf{Q}) \vee \Pi]]$

$$\star\mathbb{O} - \frac{479\,e_1}{46} + \frac{1317\,e_2}{46} - \frac{387\,e_3}{46}$$

◆ Numerical example: Parallel projection of a line onto a plane in space

Again we take the numerical example for the central projection of a line onto a plane in space, but replace the centre of projection P by a direction vector v. Because there are now two points to be projected we will generally get a different result.

$\{\mathbf{v} = e_1 + 5\,e_2 - 3\,e_3,\ \mathbf{Q}_1 = \star\mathbb{O} + 4\,e_1 - 5\,e_2 + 6\,e_3,\ \mathbf{Q}_2 = \star\mathbb{O} + 9\,e_1 - 7\,e_2 + 3\,e_3,$
$\quad\Pi = (\star\mathbb{O} - 7\,e_1 + 8\,e_2 + 9\,e_3) \wedge (9\,e_1 - 8\,e_2 + 7\,e_3) \wedge (6\,e_1 + 5\,e_2 - 4\,e_3)\};$

◇ Projecting the line by projecting two points

We can either express the projected line as the line through two projected points:

$\mathbf{R}_a = \mathbf{ToPointForm}[\mathbf{ToCommonFactor}[((\mathbf{v} \wedge \mathbf{Q}_1) \vee \Pi) \wedge ((\mathbf{v} \wedge \mathbf{Q}_2) \vee \Pi)]]$

$$-\star c^2 \left(\star\mathbb{O} + \frac{229\,e_1}{9} + \frac{677\,e_2}{9} - \frac{139\,e_3}{3}\right) \wedge \left(\star\mathbb{O} + \frac{293\,e_1}{18} + \frac{1015\,e_2}{18} - \frac{185\,e_3}{6}\right)$$

Or alternately, we can obtain it by using CongruenceSimplify.

$\mathbf{R}_a = \star\mathcal{C}[((\mathbf{v} \wedge \mathbf{Q}_1) \vee \Pi) \wedge ((\mathbf{v} \wedge \mathbf{Q}_2) \vee \Pi)]$

$$\star\mathbb{O} \wedge (106\,920\,e_1 + 219\,672\,e_2 - 180\,792\,e_3) - 2\,453\,328\,e_1 \wedge e_2 + 353\,808\,e_1 \wedge e_3 - 3\,421\,440\,e_2 \wedge e_3$$

◇ Projecting the line directly

Again using CongruenceSimplify we obtain

$\mathbf{R}_b = \star\mathcal{C}[(\mathbf{v} \wedge \mathbf{Q}_1 \wedge \mathbf{Q}_2) \vee \Pi]$

$$\star\mathbb{O} \wedge (-990\,e_1 - 2034\,e_2 + 1674\,e_3) + 22\,716\,e_1 \wedge e_2 - 3276\,e_1 \wedge e_3 + 31\,680\,e_2 \wedge e_3$$

Factorizing both R_a and R_b into point-vector form shows that they are congruent.

$\mathbf{ExteriorFactorize}[\mathbf{R}_a]$

$$106\,920 \left(\star\mathbb{O} + \frac{1262\,e_2}{55} - \frac{182\,e_3}{55}\right) \wedge \left(e_1 + \frac{113\,e_2}{55} - \frac{93\,e_3}{55}\right)$$

ExteriorFactorize[R_b]

$$-990\left(\star\mathbb{O}+\frac{1262\,e_2}{55}-\frac{182\,e_3}{55}\right)\wedge\left(e_1+\frac{113\,e_2}{55}-\frac{93\,e_3}{55}\right)$$

The generality of the projection formulae

If we look back at our projection formulae, we can see that there is a symmetry evident between the projecting element, and the projected element. We get the *same result* in the formula $(A \wedge B) \vee \mathbb{H}$ whether we consider projecting A from B, or B from A. (There may be a possible sign difference, but this is ignorable if we are only interested in projective results.)

For example, the following two cases give the same result.

- The projection of a bivector B from a vector p onto a trivector T in 4-space is equal to the intersection (or common bivector, or 2-direction) of the trivector $p \wedge B$ and the trivector T.
- The projection of a vector p from a bivector B onto a trivector T in 4-space is equal to the intersection (or common bivector, or 2-direction) of the trivector $B \wedge p$ and the trivector T.

More generally, we can see that because we have framed our discussion of projection in terms of the intersection of two elements, *any intersection of two simple elements* $(A \wedge B) \vee C$ *may be reinterpreted as the projection from one element* A, *through a second element* B *to form an image on (in) a third element* C.

Even more generally, since the elements A, B and C are simple, we may group their factors in any of possibly many ways. The projective *interpretations* of the regroupings will be different, but the image element for all of them will be the same.

We will meet the concept of projection again in Chapter 6, section [6.15]. In this case we will explore the implications of the direction of projection being determined by its orthogonality to the image element.

4.13 Regions of Space

Regions of space

Some geometric applications require an assessment of the *region* in which a point is located. Once we have a test which determines if a point is inside or outside a region, we can then *define* the region as the set of points which satisfy the test.

In this section we begin with the very simplest of regions, and build up to more complex ones of higher dimension. A closed region of dimension r is bounded by regions of dimension $r-1$. A closed region is often called a *polytope*, a generalization of the notion of polygon and polyhedron.

Regions of a plane

Consider a line L in a plane. The line will divide the plane into two regions (half-planes). A half-plane region is an *open region* if it does not contain any points of the line.

Any point P will be located in either one of the half-planes or in the line itself.

We can form the exterior product $P \wedge L$. This exterior product will vary in weight as P varies over the regions, *but will only change sign* when P crosses the line. The product (and hence the weight) is of course zero when P is on the line itself.

Suppose now we have a *reference point* Q, *not in the line*, and we want to find if P and Q are on the same side of the line. All we need to do is to take the ratio of $P \wedge L$ to $Q \wedge L$. If the ratio is positive, P and Q are on the same side. If the ratio is negative, they are on opposite sides. If the ratio is zero, P is in the line.

Note that we can take the ratio of $P \wedge L$ to $Q \wedge L$ because they are both bound n-vectors (in this case n is 2 because we are working in the plane - a bound vector 2-space). See section 2.11.

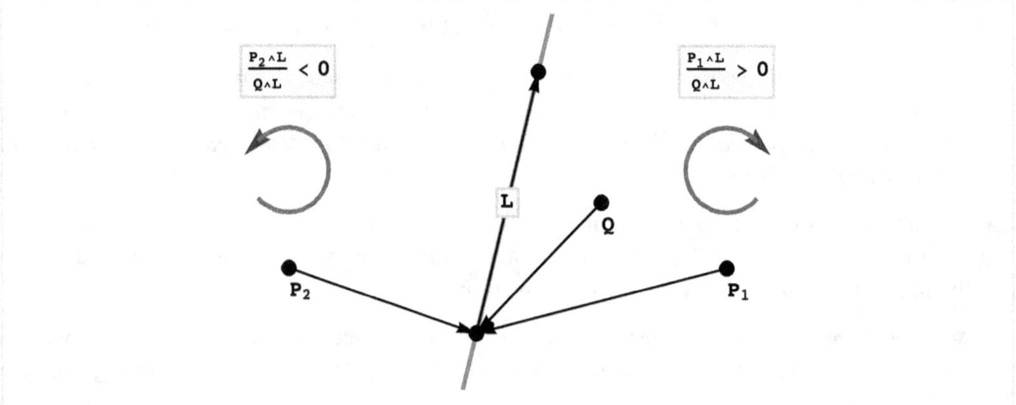

Regions of a plane divided by a line L

As can be seen from the figure above, the reason this occurs is because the ordering of the points in the product $P \wedge L$ is clockwise on one side of the line, and anti-clockwise on the other. (Remember that the line L itself can be expressed as the product of any two points on it.) Because the line L is in both the numerator and denominator of the ratio, the ratio is invariant with respect to how the line is expressed (for example, which two points are chosen to express it).

If we know in which region the point Q is located, we can therefore determine the region in which the point P is located. In the sections below we will see how this simple result can be extended in a boolean manner to regions of higher dimension, and regions with more complex boundaries.

◆ Example: Regions of a plane divided by a line

As shown in the graphic below, let L be a line through the origin $\star\mathbb{O}$ in the direction of the basis vector e_2, the reference point Q be at the point $\star\mathbb{O} + e_1$, and point P_1 and P_2 at the points $\star\mathbb{O} + a\, e_1 + e_2$ and $\star\mathbb{O} - b\, e_1 + e_2$ respectively. The scalar coefficients a and b are strictly positive.

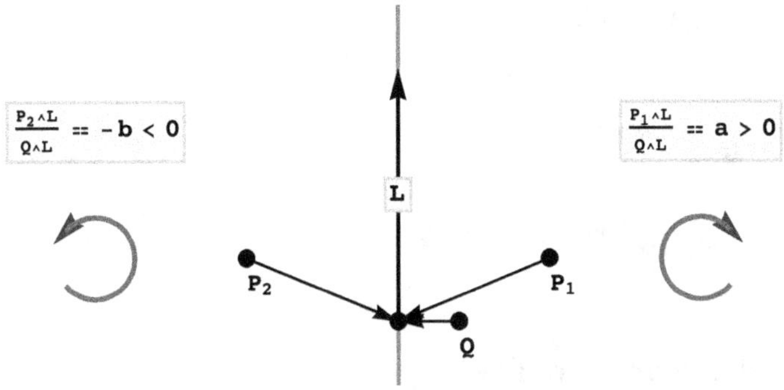

Example: Regions of a plane divided by a line

The products of each of the three points with the line are:

$$Q \wedge L == (\star O + e_1) \wedge \star O \wedge e_2 == -\star O \wedge e_1 \wedge e_2$$

$$P_1 \wedge L == (\star O + a\, e_1 + e_2) \wedge \star O \wedge e_2 == -a \star O \wedge e_1 \wedge e_2$$

$$P_2 \wedge L == (\star O - b\, e_1 + e_2) \wedge \star O \wedge e_2 == b \star O \wedge e_1 \wedge e_2$$

The ratios of $P_1 \wedge L$ and $P_2 \wedge L$ with the reference product $Q \wedge L$ are:

$$\frac{P_1 \wedge L}{Q \wedge L} == a > 0 \qquad \frac{P_2 \wedge L}{Q \wedge L} == -b < 0$$

Thus we conclude that P_1 is on the same side of the line L as Q, while P_2 is on the opposite side.

Regions of a line

Before we explore more complex regions and higher dimensions, let us take a step back to consider a line as the complete space (a bound 1-space), and a point R in the line as dividing it into two regions. The point R now takes the role of the line L in the previous example. Suppose the 1-space has basis $\{\star O, e_1\}$, and for simplicity take R to be the origin $\star O$.

Following the pattern of the previous example, we let the reference point Q be the point $\star O + e_1$, and points P_1 and P_2 be the points $\star O + a\, e_1$ and $\star O - b\, e_1$ respectively. Here, as above, the scalar coefficients a and b are strictly positive.

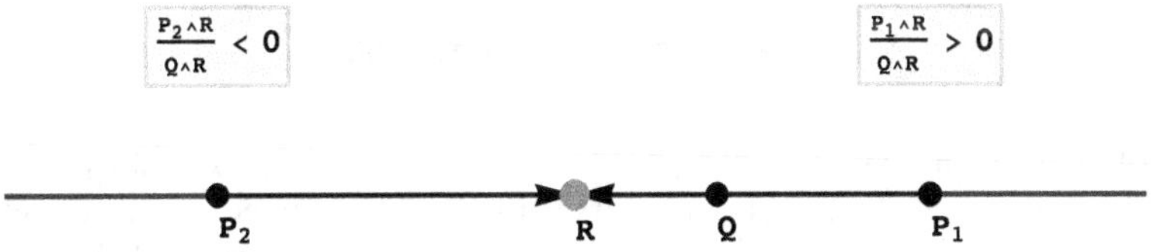

Example: Regions of a line divided by a point R

The products of each of these three points with the dividing point R are:

$$Q \wedge R == (\star O + e_1) \wedge \star O == -\star O \wedge e_1$$

$$P_1 \wedge R == (\star 0 + a\, e_1) \wedge \star 0 == -a \star 0 \wedge e_1$$

$$P_2 \wedge R == (\star 0 - b\, e_1) \wedge \star 0 == b \star 0 \wedge e_1$$

The quotients of $P_1 \wedge R$ and $P_2 \wedge R$ with the reference product $Q \wedge R$ are:

$$\frac{P_1 \wedge R}{Q \wedge R} == a > 0 \qquad \frac{P_2 \wedge R}{Q \wedge R} == -b < 0$$

Thus we conclude that P_1 is on the same side of the dividing point R as Q, while P_2 is on the opposite side.

Planar regions defined by two lines

Suppose now, we have two intersecting lines L_1 and L_2 in the plane and a known reference point Q, not on either line. We wish to find in which quadrant defined by the intersecting lines that a given point lies. In order to explore this we define the regions by four nominal test points P_1, P_2, P_3, P_4, one in each quadrant, and for both lines L_1 and L_2, making a pair of ratios for each point. For a given region, represented by the point P_i we can form the ratio as we did above for each of the lines L_j and portray them as follows:

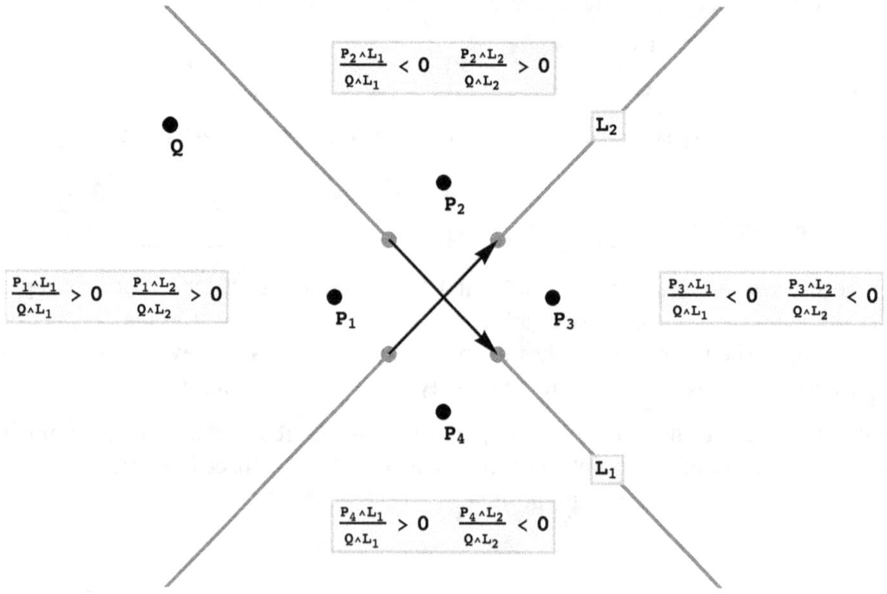

Planar regions defined by two lines

For easier comprehension, the graphic shows the two lines as if they are at right angles, and the points as if they are equidistant from the intersection, but remember that all the explorations in this chapter are non-metric, and so angles and distances at this stage are undefined.

We can summarize the region definitions more succinctly in a table.

Region	L_1	L_2
P_1	$\frac{P_1 \wedge L_1}{Q \wedge L_1} > 0$	$\frac{P_1 \wedge L_2}{Q \wedge L_2} > 0$
P_2	$\frac{P_2 \wedge L_1}{Q \wedge L_1} < 0$	$\frac{P_2 \wedge L_2}{Q \wedge L_2} > 0$
P_3	$\frac{P_3 \wedge L_1}{Q \wedge L_1} < 0$	$\frac{P_3 \wedge L_2}{Q \wedge L_2} < 0$
P_4	$\frac{P_4 \wedge L_1}{Q \wedge L_1} > 0$	$\frac{P_4 \wedge L_2}{Q \wedge L_2} < 0$

Planar regions defined by two lines

◆ Example: Regions of a plane divided by two lines

To see how this works, suppose we know that some point Q is in some quadrant, and we would like to know which quadrant another given point P is in. We simply calculate the two ratios and compare their results to the table. We do this for a numerical example below.

First define the lines and the point Q:

```
L₁ = (★O + e₂) ∧ (★O + 2 e₁ - e₂);
L₂ = (★O - e₂) ∧ (★O + 2 e₁ + e₂);
Q = ★O - 4 e₁ + 3 e₂;
```

To see if a point P is in the same quadrant as the reference point Q, we simply check that the signs of both products $\frac{P \wedge L_1}{Q \wedge L_1}$ and $\frac{P \wedge L_2}{Q \wedge L_2}$ are positive. Below we write a function CheckP to take a random point P and calculate the value of the combined predicate for it.

```
CheckP :=
 Module[{P}, P = ★O + RandomReal[{-5, 5}] e₁ + RandomReal[{-5, 5}] e₂;
  {P, ★G[P ∧ L₁] / ★G[Q ∧ L₁] > 0 && ★G[P ∧ L₂] / ★G[Q ∧ L₂] > 0}]
```

Below we check 10 random points and find that three of them return True for CheckP and are thus in the same quadrant as Q.

```
Table[CheckP, {10}] // TableForm
```

$\star\mathcal{O} + 2.50516\, e_1 - 0.477195\, e_2$	False
$\star\mathcal{O} - 4.78253\, e_1 + 4.11593\, e_2$	True
$\star\mathcal{O} - 0.511969\, e_1 + 2.70047\, e_2$	False
$\star\mathcal{O} - 2.45479\, e_1 + 1.84245\, e_2$	True
$\star\mathcal{O} + 2.7643\, e_1 - 2.33078\, e_2$	False
$\star\mathcal{O} + 3.40323\, e_1 - 4.45599\, e_2$	False
$\star\mathcal{O} + 3.68059\, e_1 - 4.49096\, e_2$	False
$\star\mathcal{O} - 1.29608\, e_1 + 1.51291\, e_2$	True
$\star\mathcal{O} + 3.90768\, e_1 - 1.83592\, e_2$	False
$\star\mathcal{O} + 2.52696\, e_1 - 0.82049\, e_2$	False

Planar regions defined by three lines

Suppose now, we have three intersecting lines L_1, L_2 and L_3 in the plane and a known reference point Q, not on any line. We wish to find in which region defined by the intersecting lines that a given point lies. We will assume the lines are not parallel and do not all intersect in a single point. We define the regions by seven points $P_1, P_2, P_3, P_4, P_5, P_6, P_7$, one in each region, and for each of the three lines L_1, L_2 and L_3, making a triplet of ratios for each point.

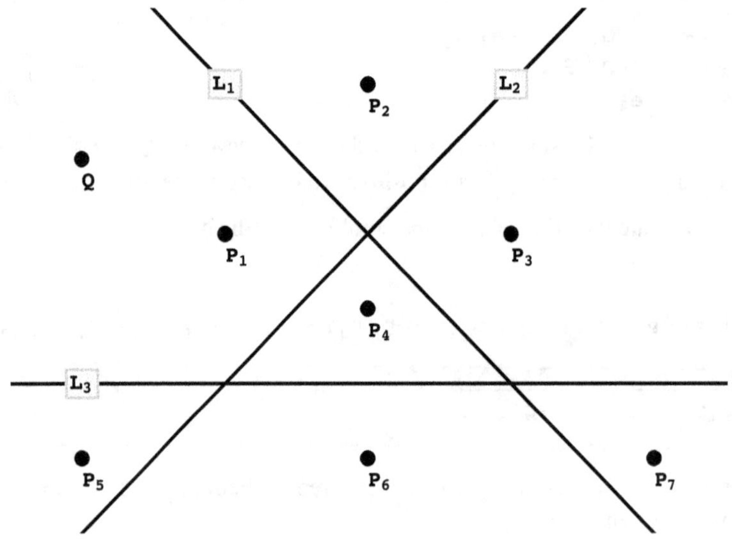

Planar regions defined by three lines

For a given region, represented by the point P_i we can form the ratio as we did above for each of the lines L_j and tabulate the regions as follows:

Region	L_1	L_2	L_3
P_1	$\frac{P_1 \wedge L_1}{Q \wedge L_1} > 0$	$\frac{P_1 \wedge L_2}{Q \wedge L_2} > 0$	$\frac{P_1 \wedge L_3}{Q \wedge L_3} > 0$
P_2	$\frac{P_2 \wedge L_1}{Q \wedge L_1} < 0$	$\frac{P_2 \wedge L_2}{Q \wedge L_2} > 0$	$\frac{P_2 \wedge L_3}{Q \wedge L_3} > 0$
P_3	$\frac{P_3 \wedge L_1}{Q \wedge L_1} < 0$	$\frac{P_3 \wedge L_2}{Q \wedge L_2} < 0$	$\frac{P_3 \wedge L_3}{Q \wedge L_3} > 0$
P_4	$\frac{P_4 \wedge L_1}{Q \wedge L_1} > 0$	$\frac{P_4 \wedge L_2}{Q \wedge L_2} < 0$	$\frac{P_4 \wedge L_3}{Q \wedge L_3} > 0$
P_5	$\frac{P_5 \wedge L_1}{Q \wedge L_1} > 0$	$\frac{P_5 \wedge L_2}{Q \wedge L_2} > 0$	$\frac{P_5 \wedge L_3}{Q \wedge L_3} < 0$
P_6	$\frac{P_6 \wedge L_1}{Q \wedge L_1} > 0$	$\frac{P_6 \wedge L_2}{Q \wedge L_2} < 0$	$\frac{P_6 \wedge L_3}{Q \wedge L_3} < 0$
P_7	$\frac{P_7 \wedge L_1}{Q \wedge L_1} < 0$	$\frac{P_7 \wedge L_2}{Q \wedge L_2} < 0$	$\frac{P_7 \wedge L_3}{Q \wedge L_3} < 0$

Planar regions defined by three lines

◆ Example: Regions of a plane divided by three lines

Here are some actual lines and a reference point `Q`:

```
L1 = (★O + e2) ∧ (★O + 2 e1 - e2);
L2 = (★O - e2) ∧ (★O + 2 e1 + e2);
L3 = (★O - e1 - 2 e2) ∧ (★O + 3 e1 - 2 e2);
Q = ★O - 3 e1 + e2;
```

To see if a point `P` is in the centre triangle (the region in which P_4 is in), we check that the signs of the products are +, —, +, according to the table above, and hence write the predicate `CheckP4` for this triangular region as

```
CheckP4 :=
 Module[{P}, P = ★O + RandomReal[{-3, 3}] e1 + RandomReal[{-3, 3}] e2;
  {P, ★G[P ∧ L1] / ★G[Q ∧ L1] > 0 && ★G[P ∧ L2] / ★G[Q ∧ L2] < 0 && ★G[P ∧ L3] / ★G[Q ∧ L3] > 0}]
```

Below we check 10 random points and find that only one of them returns `True` for `CheckP4` and is thus in the centre triangle.

```
Table[CheckP4, {10}] // TableForm
```

$\star\mathbb{O} - 0.830005\, e_1 - 0.848417\, e_2$	False
$\star\mathbb{O} - 2.94884\, e_1 - 2.22242\, e_2$	False
$\star\mathbb{O} + 2.77172\, e_1 + 0.3699\, e_2$	False
$\star\mathbb{O} + 0.13623\, e_1 - 0.109629\, e_2$	False
$\star\mathbb{O} - 2.48096\, e_1 - 1.96646\, e_2$	False
$\star\mathbb{O} - 2.17117\, e_1 + 0.0203384\, e_2$	False
$\star\mathbb{O} + 1.02258\, e_1 + 1.51379\, e_2$	False
$\star\mathbb{O} + 0.506026\, e_1 - 0.54325\, e_2$	True
$\star\mathbb{O} - 1.15566\, e_1 + 0.842319\, e_2$	False
$\star\mathbb{O} + 2.25353\, e_1 - 1.15175\, e_2$	False

Creating a pentagonal region

In the previous example the centre triangle was a *bounded region* (while the remaining regions were unbounded). This suggests that we ought to be able to define the interior and exterior of more complex polgons (and polytopes).

In this section we explore a simple example of a pentagon in the plane defined by 5 (non-parallel) lines. Each line is defined by the exterior product of two points. If we know the *vertices* of the pentagon, then the most obvious way to construct these lines is as products of pairs of adjacent vertices. However, it is just as easy if all we know are the lines forming the *sides* of the pentagon. And it is more general since a line can be constructed as the product of any two points on a side or its extension.

In the section which follows this one, we will extend this example to look at the construction of a 5-star. To make use of common geometry in the two examples, we will define the sides of the pentagon in this example by the vertices of the 5-star (as shown below). We locate the reference point Q in the middle of the pentagon.

Each half-plane in the diagram, on the side of its line which includes the reference point Q, is coloured an opaque grey. Such half-planes which overlap therefore show themselves in a darker grey. There are four shades of grey shown. The pentagonal region is the overlap of all five half-planes, and hence shows in the darkest grey. The lightest shade is the overlap of two half-planes.

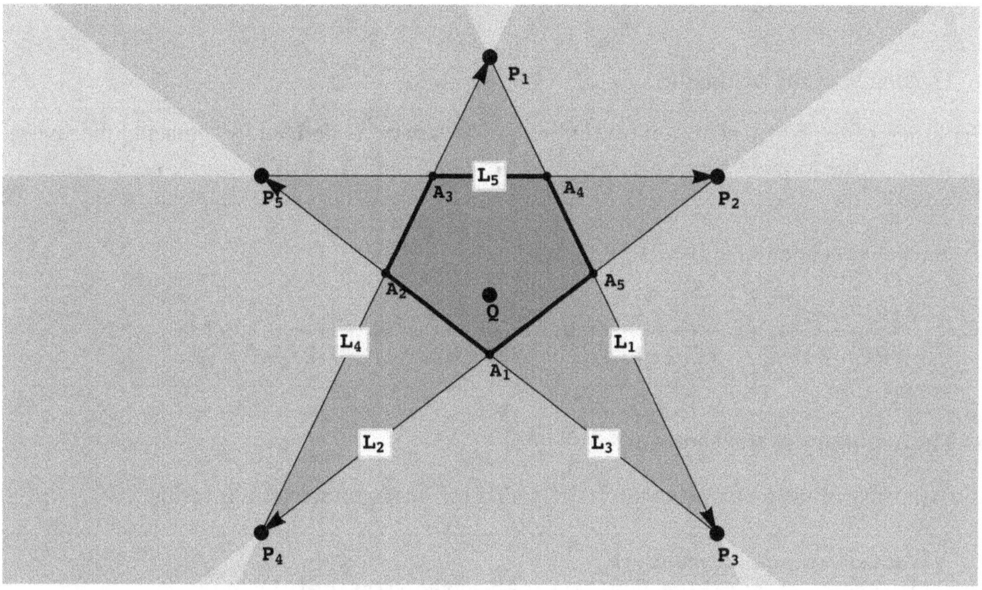

Creating a pentagonal region by intersecting the half-planes which include the point Q

To be specific, we define the points and lines involved.

```
Q = ★O;
P1 = ★O + 2 e2;
P2 = ★O + 2 e1 + e2;
P3 = ★O + 2 e1 - 2 e2;
P4 = ★O - 2 e1 - 2 e2;
P5 = ★O - 2 e1 + e2;
L1 = P1 ∧ P3;
L2 = P2 ∧ P4;
L3 = P3 ∧ P5;
L4 = P4 ∧ P1;
L5 = P5 ∧ P2;
```

◆ The pentagon in the star

The pentagon internal to the 5-star is the region (including the boundary) defined by the set of points `P` such that the predicate `FiveStarPentagon` below is `True`.

```
FiveStarPentagon[P_] :=
  And[★G[P ∧ L1]/★G[Q ∧ L1] ≥ 0, ★G[P ∧ L2]/★G[Q ∧ L2] ≥ 0,
   ★G[P ∧ L3]/★G[Q ∧ L3] ≥ 0, ★G[P ∧ L4]/★G[Q ∧ L4] ≥ 0, ★G[P ∧ L5]/★G[Q ∧ L5] ≥ 0];
```

For example, the reference point `Q` belongs to the pentagon.

```
FiveStarPentagon[Q]
True
```

◆ The inside of the pentagon

The inside of the 5-star pentagon (excluding the boundary) is defined by replacing the sign ≥ by the sign > in the formula above.

```
FiveStarPentagonInside[P_] := And[★G[P∧L1]/★G[Q∧L1] > 0,
   ★G[P∧L2]/★G[Q∧L2] > 0, ★G[P∧L3]/★G[Q∧L3] > 0, ★G[P∧L4]/★G[Q∧L4] > 0, ★G[P∧L5]/★G[Q∧L5] > 0];
```

◆ The boundary of the pentagon

The boundary of the 5-star pentagon is defined as the pentagon without its inside.

```
FiveStarPentagonBoundary[P_] :=
 FiveStarPentagon[P] && Not[FiveStarPentagonInside[P]]
```

To check, we choose three points one inside, one on, and one outside the pentagon boundary. For the point on the boundary we choose the point `M`, say, mid-way between P_2 and P_5. For the other two we perturb the position of `M` by a small vertical displacement ϵ. As expected, only the point `M` on the boundary returns `True`.

```
M = (P2 + P5) / 2;  ϵ = .000001 e2;
FiveStarPentagonBoundary /@ {M - ϵ, M, M + ϵ}
{False, True, False}
```

Note that the pentagon boundary cannot be defined by a Boolean expression of predicates in which the ratios are zero, because the lines from which the boundary is formed extend beyond the boundary.

◆ The outside of the pentagon

The plane outside the 5-star pentagon is defined simply by the negation of the `FiveStarPentagon` predicate. Since `FiveStarPentagon` includes the boundary, `FiveStarPentagonOutside` excludes it.

```
FiveStarPentagonOutside[P_] := Not[FiveStarPentagon[P]]
```

For example, the vertices of the 5-star are outside the pentagon.

```
FiveStarPentagonOutside[#] & /@ {P1, P2, P3, P4, P5}
{True, True, True, True, True}
```

◆ The vertices of the pentagon

Although in this example it has been unnecessary to know the vertices of the pentagon, we can of course obtain these points by applying `ToCommonFactor` to each of the pairs of intersecting lines.

```
{A5, A1, A2, A3, A4} = ToPointForm[ToCommonFactor[#]] & /@
  {L1 ∨ L2, L2 ∨ L3, L3 ∨ L4, L4 ∨ L5, L5 ∨ L1}
```

$$\left\{\star\mathbb{O} + \frac{10\,e_1}{11} + \frac{2\,e_2}{11},\ \star\mathbb{O} - \frac{e_2}{2},\ \star\mathbb{O} - \frac{10\,e_1}{11} + \frac{2\,e_2}{11},\ \star\mathbb{O} - \frac{e_1}{2} + e_2,\ \star\mathbb{O} + \frac{e_1}{2} + e_2\right\}$$

We confirm that these are on the boundary of the pentagon:

```
FiveStarPentagonBoundary /@ {A5, A1, A2, A3, A4}
```

```
{True, True, True, True, True}
```

Creating a 5-star region

We now define the 5-star region itself. That is, a Boolean function of the point `P` which will yield `True` if `P` is inside the 5-star or on its boundary, and `False` otherwise. The Boolean function will of course also contain, as parameters, the reference point `Q` and the lines defining the 5-star.

The vertices of the 5-star are the same as the points we chose in the pentagon example in the previous section, and the edges of the 5-star are segments of the lines we chose.

There are many ways of creating a Boolean expression to define the 5-star. The most conceptually straightforward way is to use either a disjunction or conjunction of sub-regions - each of which is itself defined by a Boolean expression. A disjunction (Boolean `Or`) would be used when the region being defined is composed of one *or* other of the sub-regions. A conjunction (Boolean `And`) would be used when the region being defined is the overlap of the sub-regions.

We avoid discussing solutions requiring knowledge of the vertices of the inner pentagon, as these may not be known (although of course, as shown in the previous section, they can be easily calculated using the regressive product). However, in the diagram below we name them as points A_i (opposite to P_i) in order to explain our construction.

◆ Building the formula

◇ The triangle

We view the 5-star as two polygons: the triangle $P_2\ P_5\ A_1$, and the quadrilateral $P_1\ P_3\ A_1\ P_4$. The triangular region can be defined as the intersection of the three half-planes defined by the lines L_2, L_3, L_5, and which include the point Q.

$$\text{Triangle[P\_] := And}\left[\frac{\star\mathcal{G}[P \wedge L_2]}{\star\mathcal{G}[Q \wedge L_2]} \geq 0,\ \frac{\star\mathcal{G}[P \wedge L_3]}{\star\mathcal{G}[Q \wedge L_3]} \geq 0,\ \frac{\star\mathcal{G}[P \wedge L_5]}{\star\mathcal{G}[Q \wedge L_5]} \geq 0\right]$$

This region is shown in the darkest grey in the following figure.

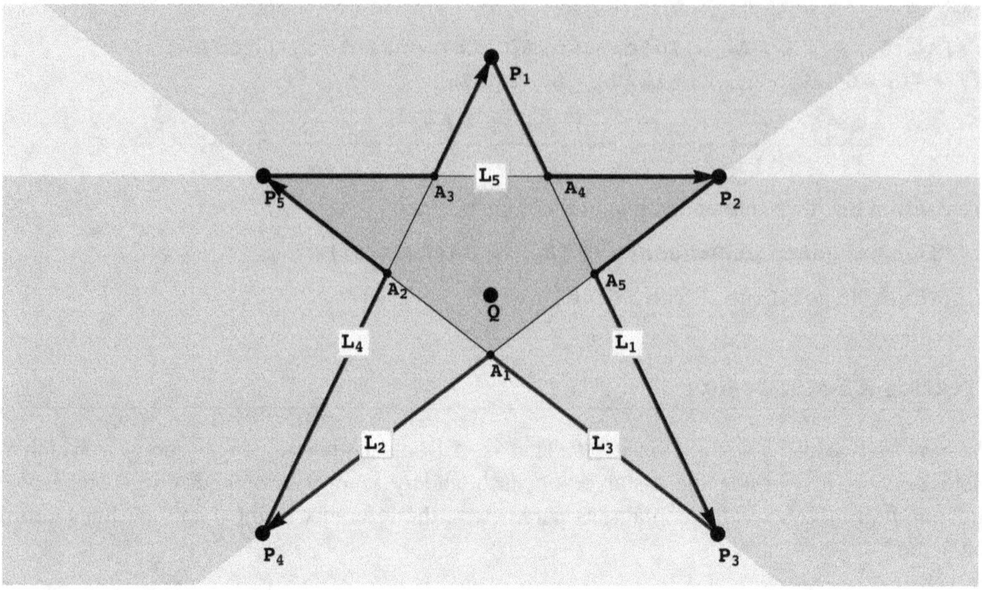

Creating a 5–star region Stage I: Creating the triangular region $P_2P_5A_1$

◇ The quadrilateral

The region defined by the quadrilateral can be viewed as the intersection of the (infinite) region below the lines L_1 and L_4, and the (infinite) region above the lines L_2 and L_3. The region below the lines L_1 and L_4 can be defined by the conjunction of the regions on the positive side (relative to Q) of the lines L_1 and L_4.

$$\texttt{BelowL1L4[P\_] := And}\left[\frac{\star\mathcal{G}[P \wedge L_1]}{\star\mathcal{G}[Q \wedge L_1]} \geq 0,\ \frac{\star\mathcal{G}[P \wedge L_4]}{\star\mathcal{G}[Q \wedge L_4]} \geq 0\right]$$

This shows the first region from which the quadrilateral is constructed.

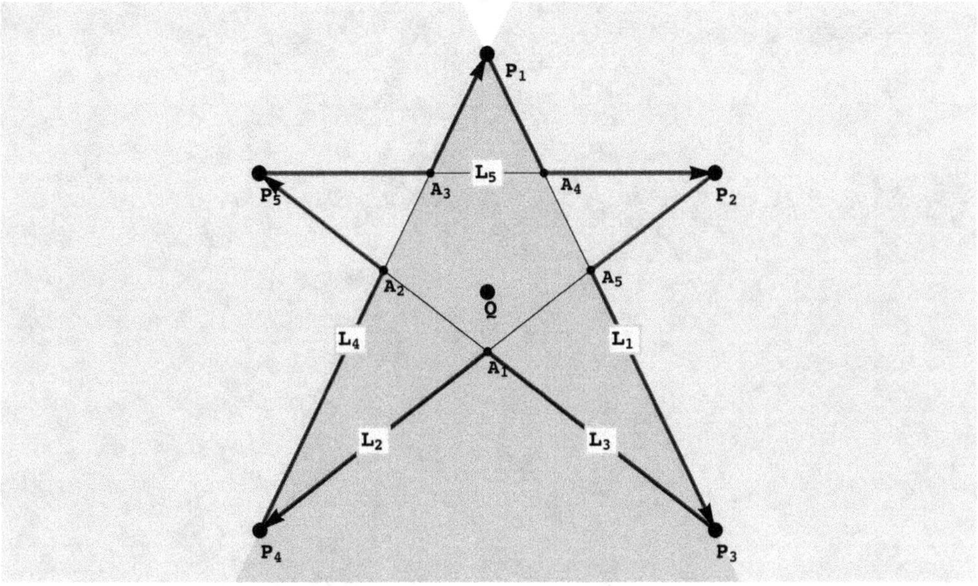

Creating a 5-star region Stage IIa: Creating the region under the lines L_1 and L_4

The region above the lines L_2 and L_3 can be defined by the *disjunction* of the regions on the positive side (relative to Q) of the lines L_2 and L_3. This region is shown in the figure below with various opacities of grey.

$$\text{AboveL2L3}[\text{P\_}] := \text{Or}\left[\frac{\star\mathcal{G}[\text{P} \wedge \text{L}_2]}{\star\mathcal{G}[\text{Q} \wedge \text{L}_2]} \geq 0,\ \frac{\star\mathcal{G}[\text{P} \wedge \text{L}_3]}{\star\mathcal{G}[\text{Q} \wedge \text{L}_3]} \geq 0\right]$$

This shows the second region from which the quadrilateral is constructed.

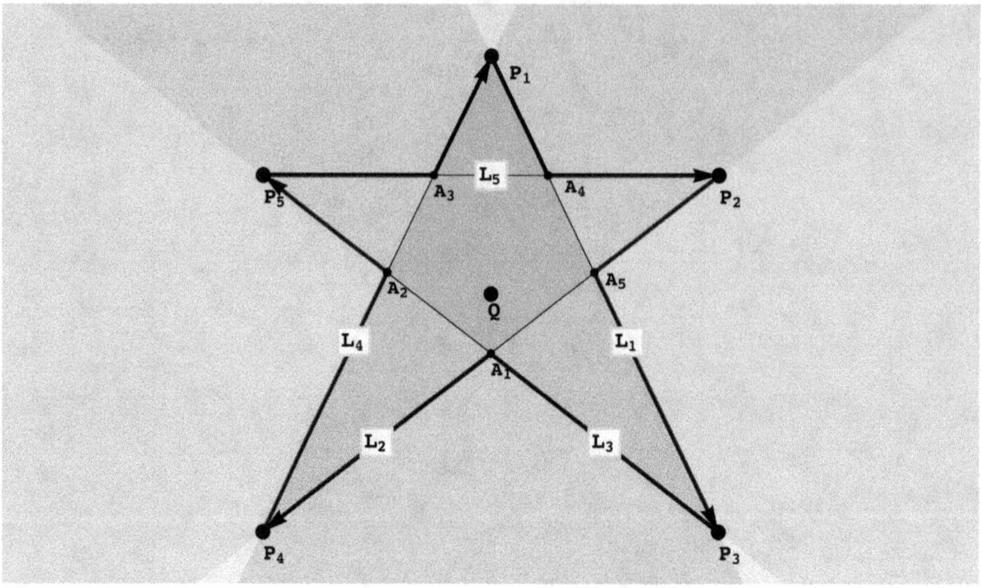

Creating a 5-star region Stage IIb: Creating the region above the lines L_2 and L_3

The quadrilateral is then the conjunction of these two regions.

```
Quadrilateral[P_] := And[BelowL1L4[P], AboveL2L3[P]]
```

◇ The 5-star

And the 5-star is the disjunction of the triangle and the quadrilateral.

```
FiveStar[P_] := Or[Triangle[P], Quadrilateral[P]]
```

◆ Testing the formula

To test this formula, we can choose test points on, inside and outside the 5-star. The vertices are on the 5-star, so should all yield `True`.

```
FiveStar /@ {P1, P2, P3, P4, P5}
```

```
{True, True, True, True, True}
```

And so also are the vertices of the interior pentagon explored in the previous section.

```
FiveStar /@ {A5, A1, A2, A3, A4}
```

```
{True, True, True, True, True}
```

We can add small perturbations to these vertices to create points just outside the 5-star. These should all yield `False`.

```
δ = .000001 e₁; ϵ = .000001 e₂;
{FiveStar /@ {P₁ + ϵ, P₂ + ϵ, P₃ + ϵ, P₄ + ϵ, P₅ + ϵ},
 FiveStar /@ {A₁ - ϵ, A₂ - δ, A₃ + ϵ, A₄ + ϵ, A₅ + δ}}
```

```
{{False, False, False, False, False},
 {False, False, False, False, False}}
```

Perturbations to create points just inside the 5-star should all yield `True`.

```
{FiveStar /@ {P₁ - ϵ, P₂ - 2 δ - ϵ, P₃ - δ + ϵ, P₄ + δ + ϵ, P₅ + 2 δ - ϵ},
 FiveStar /@ {A₁ + ϵ, A₂ + δ, A₃ - ϵ, A₄ - ϵ, A₅ - δ}}
```

```
{{True, True, True, True, True}, {True, True, True, True, True}}
```

◆ Defining a general 5-star region formula

To define a general *explicit* formula for a 5-star region, all we need to do is to collect the regions into one expression.

```
FiveStar[P_, Q_, L_List] :=
 Or[And[★𝒢[P ∧ L[[2]]]/★𝒢[Q ∧ L[[2]]] ≥ 0, ★𝒢[P ∧ L[[3]]]/★𝒢[Q ∧ L[[3]]] ≥ 0, ★𝒢[P ∧ L[[5]]]/★𝒢[Q ∧ L[[5]]] ≥ 0],
  And[And[★𝒢[P ∧ L[[1]]]/★𝒢[Q ∧ L[[1]]] ≥ 0, ★𝒢[P ∧ L[[4]]]/★𝒢[Q ∧ L[[4]]] ≥ 0],
   Or[★𝒢[P ∧ L[[2]]]/★𝒢[Q ∧ L[[2]]] ≥ 0, ★𝒢[P ∧ L[[3]]]/★𝒢[Q ∧ L[[3]]] ≥ 0]]]
```

This formula is valid for any set of five lines intersecting to form the 5-star, and any reference point `Q` inside the 5-star region. We test the formula with the points just inside the vertices of the pentagon as above.

```
FiveStar[#, Q, {L₁, L₂, L₃, L₄, L₅}] & /@ {A₁ + ϵ, A₂ + δ, A₃ - ϵ, A₄ - ϵ, A₅ - δ}
```

```
{True, True, True, True, True}
```

Creating a 5-star pyramid

Having defined a region in the plane, we can easily extend it into three dimensions by choosing points in the third dimension, and converting all the lines to planes by multiplying them by one or another of these points.

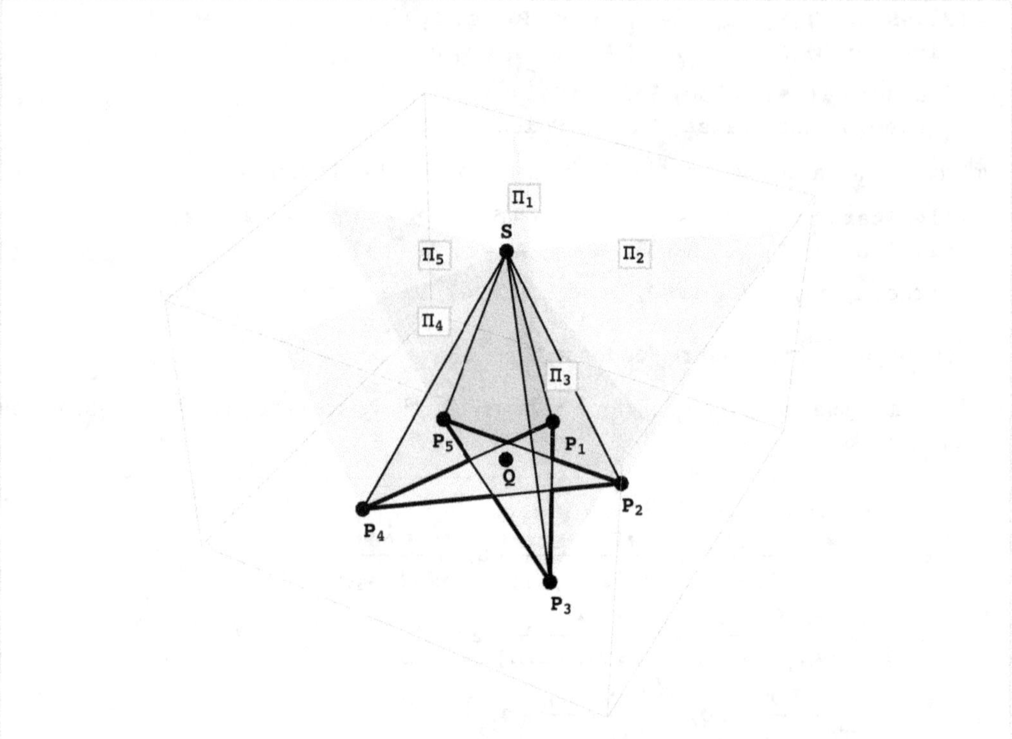

Creating a 5−star pyramid

To create a three-dimensionalized pyramidal-type 5-star we only need to take a single point S above the origin, and extend each of the lines through S to form five intersecting planes. The formula (predicate) for this is given by

FiveStar[P, Q, L ∧ S]

Here, L is the *list* of lines forming the original 5-star. Since the exterior product operation Wedge is Listable, taking the exterior product of this list of lines with the point S gives the corresponding list of planes through S.

We then truncate the (infinite) pyramidal cone thus formed by a sixth plane equivalent to the plane of the original planar 5-star by adding an extra predicate term. We define the plane Π as the exterior product of any three of the original 5-star vertices.

$\Pi = \mathbf{P}_1 \wedge \mathbf{P}_2 \wedge \mathbf{P}_3;$

We must also redefine the location of the reference point Q. Since the original point Q is now on the boundary of the new three dimensional region, we will need to relocate it to an interior point.

The required predicate for the truncating plane is then

$$\frac{\star\mathcal{G}[\mathbf{P} \wedge \Pi]}{\star\mathcal{G}[\mathbf{Q} \wedge \Pi]} \geq 0$$

We can now define the solid region FiveStar3D as the conjunction of these two regions.

```
FiveStar3D[P_] := And[FiveStar[P, Q, L∧S], ★𝒢[P∧Π]/★𝒢[Q∧Π] ≥ 0]
```

Here is our original data for the planar 5-star:

```
P₁ = ★O + 2 e₂;
P₂ = ★O + 2 e₁ + e₂;
P₃ = ★O + 2 e₁ - 2 e₂;
P₄ = ★O - 2 e₁ - 2 e₂;
P₅ = ★O - 2 e₁ + e₂;
L₁ = P₁ ∧ P₃;
L₂ = P₂ ∧ P₄;
L₃ = P₃ ∧ P₅;
L₄ = P₄ ∧ P₁;
L₅ = P₅ ∧ P₂;
L = {L₁, L₂, L₃, L₄, L₅};
```

We now change to a bound 3-space, and define the two new points, and the new plane.

```
★𝒫3;
Q = ★O + e₃;
S = ★O + 2 e₃;
Π = P₁ ∧ P₂ ∧ P₃;
```

To verify the formula, we can first check that the 5-star pyramid contains all its defining points.

```
FiveStar3D /@ {P₁, P₂, P₃, P₄, P₅, S}
```

```
{True, True, True, True, True, True}
```

We can also check that points clustered sufficiently close to the origin are inside the region.

```
Table[FiveStar3D[★O + RandomReal[0.5] e₁ + RandomReal[0.5] e₂ +
    RandomReal[0.5] e₃], {10}]
```

```
{True, True, True, True, True, True, True, True, True, True}
```

We can add small perturbations to the vertices to create points just outside the 5-star. These should all yield `False`.

```
ϵ = .000001 e₃; FiveStar3D /@ ({P₁, P₂, P₃, P₄, P₅, S} + ϵ)
```

```
{False, False, False, False, False, False}
```

Summary

- In a bound 1-space, a hyperplane is a point.
  In a bound 2-space, a hyperplane is a line.
  In a bound 3-space, a hyperplane is a plane.
  In a bound *n*-space, a hyperplane is an $(n-1)$-plane.

- Hyperplanes divide their space into *regions*.

- If two points `P` and `Q` lie on the *same* side of a hyperplane `H`, then the ratio `R` of the exterior products of the two points with the hyperplane is *positive*:

$$R == \frac{P \wedge \mathbb{H}}{Q \wedge \mathbb{H}} > 0$$

- If P and Q lie on *opposite* sides of H, then R is *negative*:

$$R == \frac{P \wedge \mathbb{H}}{Q \wedge \mathbb{H}} < 0$$

- If P lies *on* H, then R is *zero*:

$$R == \frac{P \wedge \mathbb{H}}{Q \wedge \mathbb{H}} == 0$$

- The point Q may be considered as a *reference point*, and the region in which it is located as the *reference region*.

- The point P may be considered as a *variable point*, and the region in which it is located as the *region of interest*.

- Regions of space may be defined by Boolean functions of the predicates $R > 0$, $R = 0$, $R < 0$, $R \geq 0$, and $R \leq 0$.

4.14 Geometric Constructions

Geometric expressions

In this chapter, we have been exploring non-metric geometry involving *geometric entities*: points, lines, planes, ..., hyperplanes. In this section we will explore expressions, called *geometric expressions*, involving exterior and regressive products of these entities - the exterior product to extend them, and the regressive product to intersect them. When geometric expressions are equated to zero they becomes geometric equations, leading to corresponding algebraic equations for curves and surfaces of various degrees.

One of the most enticing features of Grassmann algebra applied to geometry which we will also explore in this section, is the idea that some geometric expressions may not only be geometrically *depicted*, but may also double as *prescriptions for their geometric construction*.

The types of geometric expressions we will explore are exemplified by that leading to the equation to a conic section in the plane (discussed in the next section).

$$P \wedge P_1 \wedge (L_1 \vee (P_0 \wedge (L_2 \vee (P \wedge P_2)))) == 0$$

Here, the factors P_0, P_1, P_2, L_1 and L_2 are fixed (constant) entities defining the parameters of the conic section. P is considered the independent variable, and since it occurs twice, the expression represents an equation of the second degree. (If P had occurred *m* times in the expression, then the equation would have been of degree *m*.)

The reduction of a geometric expression to an algebraic equation for a curve or surface requires that the expression be either a scalar, or a bound *n*-vector. The expression above is in fact the exterior product of three points, which reduces to a 3-element in the plane (a bound vector 2-space whose underlying linear space is of 3 dimensions). Each time a regressive product is simplified to its common factor, for example when two lines intersect, the result is in general a *weighted* point multiplied by an arbitrary scalar congruence factor (symbolized ⋆c in *GrassmannAlgebra*). The zero on the right hand side of the equation means that we can effectively ignore the weights and congruence factors, and visualize each regressive product as resulting simply in a point. In intermediate computations however, it is most effective to retain these weights to avoid coordinates with denominators.

For a geometric expression to result in a bound n-vector it will be equivalent to the exterior product of $n+1$ points (3 points for the plane, 4 points for 3-space). If the product is zero, one possibility is that the points all belong to a hyperplane in the space, that is, collinear in the plane, coplanar in 3-space.

For a geometric expression to result in a scalar it will be equivalent to the regressive product of $n+1$ hyperplanes (3 lines for the plane, 4 planes for 3-space). If the product is zero one possibility is that the hyperplanes all intersect in a point in the space, that is, three lines intersecting in one point in the plane, four planes intersecting in one point in 3-space.

◆ *Historical Note*

Grassmann discusses geometric constructions in Chapter 5 (Applications to Geometry) in his *Ausdehnungslehre* of 1862, and also in *Crelle's Journal* volumes XXXI, XXXVI and LII. Probably the most comprehensive additional treatment is given by Alfred North Whitehead in Book IV of his *Treatise on Universal Algebra* (Chapter IV: Descriptive Geometry and Chapter V: Descriptive Geometry of Conics and Cubics) 1898.

Geometric equations for lines and planes

Before embarking on exploring the more general geometric expressions, we take this opportunity to first review the simplest cases of geometric equations, those defining lines in the plane, and planes in space.

◆ The equation of a line in the plane

Suppose we have a fixed line L through two points in the plane, and we want to derive its usual algebraic equation. Let the line L be defined by the points P_a and P_b.

$$\star\mathcal{P}_2;\ P_a = \star\mathbb{P}_a;\ P_b = \star\mathbb{P}_b;\ L = P_a \wedge P_b$$

$$(\star\mathbb{O} + a_1 e_1 + a_2 e_2) \wedge (\star\mathbb{O} + b_1 e_1 + b_2 e_2)$$

The *geometric equation* to this line is $P \wedge L = 0$.

$$P = \star\mathbb{P}_x;\ P \wedge L == 0$$

$$(\star\mathbb{O} + e_1 x_1 + e_2 x_2) \wedge (\star\mathbb{O} + a_1 e_1 + a_2 e_2) \wedge (\star\mathbb{O} + b_1 e_1 + b_2 e_2) == 0$$

The *algebraic equation* to this line is obtained by simplifying this 3-element and extracting its scalar coefficient by dividing through by the basis 3-element of the space

$$\frac{\star\mathcal{G}[P \wedge L]}{\star\mathbb{O} \wedge e_1 \wedge e_2} == 0$$

$$-a_2 b_1 + a_1 b_2 + a_2 x_1 - b_2 x_1 - a_1 x_2 + b_1 x_2 == 0$$

Clearly this equation can be rearranged in the more usual forms

$$(a_2 - b_2)\ x_1 + (b_1 - a_1)\ x_2 + (a_1 b_2 - a_2 b_1) == 0$$

$$x_2 == \left(\frac{b_2 - a_2}{b_1 - a_1}\right) x_1 + \frac{a_2 b_1 - a_1 b_2}{b_1 - a_1}$$

The first of these is often cast in the 'hyperplane' form.

$$A x_1 + B x_2 + C == 0$$

◆ The equation of a plane in space

To determine the equation of a plane in a bound vector 3-space we follow the same process.

$$\star\mathcal{P}_3;\ P_a = \star\mathbb{P}_a;\ P_b = \star\mathbb{P}_b;\ P_c = \star\mathbb{P}_c;\ P = \star\mathbb{P}_x;\ \Pi = P_a \wedge P_b \wedge P_c;\ P \wedge \Pi == 0$$

$$(\star\mathbb{O} + e_1 x_1 + e_2 x_2 + e_3 x_3) \wedge (\star\mathbb{O} + a_1 e_1 + a_2 e_2 + a_3 e_3) \wedge (\star\mathbb{O} + b_1 e_1 + b_2 e_2 + b_3 e_3) \wedge (\star\mathbb{O} + c_1 e_1 + c_2 e_2 + c_3 e_3) == 0$$

Simplifying this expression gives us the equation to a plane in a bound vector 3-space in the more recognizable form

$$A + B x_1 + C x_2 + D x_3 == 0$$

We can use the *Mathematica* function `Collect` for collecting the coefficients of the algebraic equation.

$$\text{Collect}\left[\frac{\star\mathcal{G}[P \wedge \Pi]}{\star\mathbb{O} \wedge e_1 \wedge e_2 \wedge e_3}, \{x_1, x_2, x_3\}\right] == 0$$

$$-a_3 b_2 c_1 + a_2 b_3 c_1 + a_3 b_1 c_2 - a_1 b_3 c_2 - a_2 b_1 c_3 + a_1 b_2 c_3 + (a_3 b_2 - a_2 b_3 - a_3 c_2 + b_3 c_2 + a_2 c_3 - b_2 c_3)\, x_1 + (-a_3 b_1 + a_1 b_3 + a_3 c_1 - b_3 c_1 - a_1 c_3 + b_1 c_3)\, x_2 + (a_2 b_1 - a_1 b_2 - a_2 c_1 + b_2 c_1 + a_1 c_2 - b_1 c_2)\, x_3 == 0$$

The same process applied to a hyperplane in a bound vector *n*-space yields, as expected:

$$C + C_1 x_1 + C_2 x_2 + C_2 x_3 + \ldots + C_n x_n == 0$$

The geometric equation of a conic section in the plane

The previous section introduced the simplest cases of geometric equations: those of linear elements. We now turn our attention to quadratic elements, exploring how their equations may be generated by constraining the position of a point in a more general manner than we used for the linear elements.

The idea is that an equation of the general form $P \wedge \mathbb{H} = 0$ still holds, where P is a variable point, and $\mathbb{H}$ is a hyperplane in the space. But for higher degree curves, $\mathbb{H}$ will itself need to be dependent on the point P. Thus for a conic section, P will occur twice in the product, leading ultimately to an equation of the second degree in its coordinates.

The hyperplane in a bound vector 2-space is a line, so in this section we are looking for ways in which this line ($\mathbb{L}_a$, say) can be *nontrivially* dependent on P.

In what follows, we will use double struck symbols $\mathbb{L}_i$ for line parameters which will not appear in the final expression. We will also use the congruence relation when defining intersections of lines, since these normally lead to weighted points (although, as we have previously mentioned, the weights are inconsequential in the final result). Thus we begin with

$$P \wedge \mathbb{L}_a == 0$$

If P is a simple exterior factor of $\mathbb{L}_a$, the equation is true for all P, leading us nowhere.

How can $\mathbb{L}_a$ depend linearly on P, without P being an (exterior) factor of $\mathbb{L}_a$? The answer lies in the regressive product. Geometrically, a line is most meaningfully constructed as an exterior product of points (P_1 and Q_1, say). Hence

$$\mathbb{L}_a == P_1 \wedge Q_1$$

So one of these points must depend on P. Suppose it is Q_1. Thus in turn, Q_1 must be a regressive product, since in the plane the regressive product of two lines yields a point. Thus

$$Q_1 \equiv L_1 \vee \mathbb{L}_b$$

Again, allowing P to be a simple exterior factor of one of these lines leads to a trivial dependence, so we leave one of them constant (L_1, say) and replace the other, $\mathbb{L}_b$ by an exterior product of points, neither of which is P.

$$\mathbb{L}_b == P_0 \wedge Q_2$$

As in the previous steps, we leave one factor constant, and express the other as a product (in this case regressive).

$$Q_2 \equiv L_2 \vee \mathbb{L}_c$$

Finally, we are linearly far enough away from P on our chain of lines through points to come back and intersect it non-trivially.

$$\mathbb{L}_c == P \wedge P_2$$

We can now collect all these steps together

$$0 == P \wedge \mathbb{L}_a$$
$$\mathbb{L}_a == P_1 \wedge Q_1$$
$$Q_1 \equiv L_1 \vee \mathbb{L}_b$$
$$\mathbb{L}_b == P_0 \wedge Q_2$$
$$Q_2 \equiv L_2 \vee \mathbb{L}_c$$
$$\mathbb{L}_c == P \wedge P_2$$

and eliminate $\mathbb{L}_a$, $\mathbb{L}_b$, $\mathbb{L}_c$, Q_1 and Q_2 to give the final equation for the conic section.

$$P \wedge P_1 \wedge (L_1 \vee (P_0 \wedge (L_2 \vee (P \wedge P_2)))) == 0 \qquad 4.26$$

Note that since interchanging the order of factors in an exterior or regressive product changes at most its sign, and that this change is inconsequential to the final result, we can if we wish, rewrite the geometric equation in any of a number of ways, in particular, in reverse order.

$$((((P_2 \wedge P) \vee L_2) \wedge P_0) \vee L_1) \wedge P_1 \wedge P == 0 \qquad 4.27$$

The geometric equation as a prescription to construct

This geometric equation, either as a series of steps, or in its final form can be used to guide the construction of points on a conic section. Thus, the equation can be viewed as a *prescription to construct* a conic section. We begin with the second form of the equation above, and work from left to right.

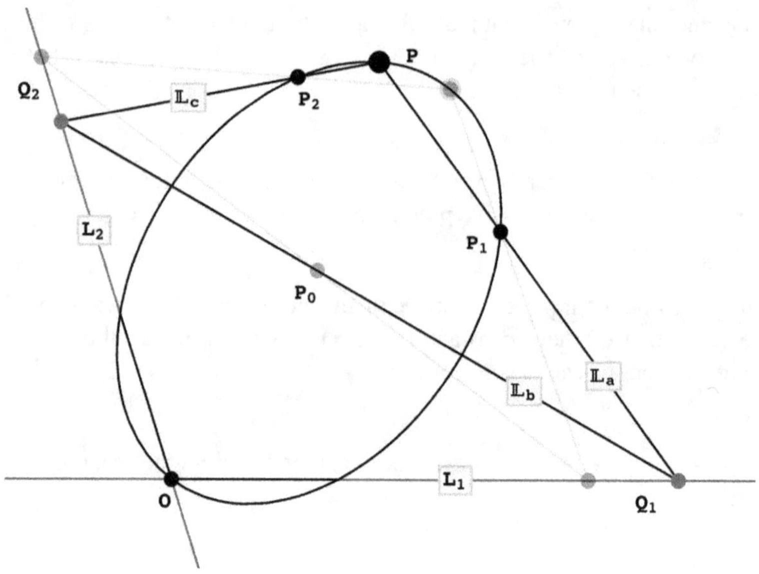

Constructing a conic section through 5 points

1. Construct the fixed points and lines: P_0, P_1, P_2, L_1, and L_2.
2. Construct a line $\mathbb{L}_c$ through P_2 to intersect line L_2 in point Q_2.
3. Construct a line $\mathbb{L}_b$ through Q_2 and P_0 to intersect line L_1 in point Q_1.
4. Construct a line $\mathbb{L}_a$ through Q_1 and P_1 to intersect line $\mathbb{L}_c$ in point P.
5. Repeat the construction by drawing a new line $\mathbb{L}_c$ through P_2.

We do of course still need to verify that this formula and construction leads to a scalar second degree equation in two variables. This we do in the next section.

The algebraic equation of a conic section in the plane

To construct the algebraic equation for the conic section we essentially follow the same process as we did with the geometric construction. In simplifying the regressive products, it will be convenient to use the *GrassmannAlgebra* function `CongruenceSimplify` (through its shorthand $\star C$). `CongruenceSimplify` takes a nested expression involving exterior and regressive products, and simplifies it, ignoring any scalar congruence factors $\star c$.

First, define the variable point P.

$$\star\mathcal{P}_2;\ \star\star S[\xi, \psi];\ P = \star O + \xi\, e_1 + \psi\, e_2;$$

1. Define the fixed points and lines: P_0, P_1, P_2, L_1, and L_2.

$$\star\star S[\{a, b, c, d, j, k, p, q, r, s\}];$$

$$P_0 = \star O + j\, e_1 + k\, e_2;$$
$$P_1 = \star O + a\, e_1 + b\, e_2;$$
$$P_2 = \star O + c\, e_1 + d\, e_2;$$
$$L_1 = (\star O + p\, e_1) \wedge (\star O + q\, e_2);$$
$$L_2 = (\star O + r\, e_1) \wedge (\star O + s\, e_2);$$

2. Construct a line $\mathbb{L}_c$ through P and P_2 to intersect line L_2 in (weighted) point Q_2.

$$Q_2 = \star\mathcal{C}[L_2 \vee (P \wedge P_2)]$$

$$\star\mathbb{O}\,(-d\,r - c\,s + s\,\xi + r\,\psi) + r\,((-d+s)\,\xi + c\,(-s+\psi))\,e_1 + s\,(d\,(-r+\xi) + (-c+r)\,\psi)\,e_2$$

3. Construct a line $\mathbb{L}_b$ through Q_2 and P_0 to intersect line L_1 in (weighted) point Q_1.

$$Q_1 = \star\mathcal{C}[L_1 \vee (P_0 \wedge Q_2)]$$

$$\begin{aligned}
&\star\mathbb{O}\,(-p\,(k\,(d\,r + c\,s - s\,\xi - r\,\psi) + s\,(d\,(-r+\xi) + (-c+r)\,\psi)) - \\
&\qquad q\,(j\,(d\,r + c\,s - s\,\xi - r\,\psi) + r\,((-d+s)\,\xi + c\,(-s+\psi)))) + \\
&\quad p\,(j\,q\,s\,\xi + k\,r\,s\,\xi - q\,r\,s\,\xi - d\,((k-q)\,r\,\xi + j\,(q\,r - r\,s + s\,\xi)) + \\
&\qquad j\,q\,r\,\psi - j\,r\,s\,\psi + c\,(-(k-q)\,r\,(s-\psi) + j\,s\,(-q+\psi)))\,e_1 + \\
&\quad q\,(k\,r\,((d-s)\,\xi + c\,(s-\psi)) + j\,s\,(d\,(-r+\xi) + (-c+r)\,\psi) - \\
&\qquad p\,(k\,(d\,r + c\,s - s\,\xi - r\,\psi) + s\,(d\,(-r+\xi) + (-c+r)\,\psi)))\,e_2
\end{aligned}$$

4. Construct a line $\mathbb{L}_a$ through Q_1 and P_1 to intersect line $\mathbb{L}_c$ in point P. Q_1, P_1 and P are thus on the same line, and the left hand side of the equation (which we call $\mathcal{E}_1$) becomes

$$\mathcal{E}_1 = \star\mathcal{C}[P \wedge P_1 \wedge Q_1]$$

$$\begin{aligned}
&(-b\,p\,(j\,q\,s\,\xi + k\,r\,s\,\xi - q\,r\,s\,\xi - d\,((k-q)\,r\,\xi + j\,(q\,r - r\,s + s\,\xi)) + \\
&\qquad j\,q\,r\,\psi - j\,r\,s\,\psi + c\,(-(k-q)\,r\,(s-\psi) + j\,s\,(-q+\psi))) + \\
&\quad a\,q\,(k\,r\,((d-s)\,\xi + c\,(s-\psi)) + j\,s\,(d\,(-r+\xi) + (-c+r)\,\psi) - \\
&\qquad p\,(k\,(d\,r + c\,s - s\,\xi - r\,\psi) + s\,(d\,(-r+\xi) + (-c+r)\,\psi))) + \\
&\quad \psi\,(p\,(j\,q\,s\,\xi + k\,r\,s\,\xi - q\,r\,s\,\xi - d\,((k-q)\,r\,\xi + j\,(q\,r - r\,s + s\,\xi)) + \\
&\qquad j\,q\,r\,\psi - j\,r\,s\,\psi + c\,(-(k-q)\,r\,(s-\psi) + j\,s\,(-q+\psi))) - \\
&\qquad a\,(-p\,(k\,(d\,r + c\,s - s\,\xi - r\,\psi) + s\,(d\,(-r+\xi) + (-c+r)\,\psi)) - \\
&\qquad\quad q\,(j\,(d\,r + c\,s - s\,\xi - r\,\psi) + r\,((-d+s)\,\xi + c\,(-s+\psi))))) - \\
&\quad \xi\,(q\,(k\,r\,((d-s)\,\xi + c\,(s-\psi)) + j\,s\,(d\,(-r+\xi) + (-c+r)\,\psi) - \\
&\qquad p\,(k\,(d\,r + c\,s - s\,\xi - r\,\psi) + s\,(d\,(-r+\xi) + (-c+r)\,\psi))) - \\
&\qquad b\,(-p\,(k\,(d\,r + c\,s - s\,\xi - r\,\psi) + s\,(d\,(-r+\xi) + (-c+r)\,\psi)) - \\
&\qquad\quad q\,(j\,(d\,r + c\,s - s\,\xi - r\,\psi) + r\,((-d+s)\,\xi + c\,(-s+\psi)))))) \star\mathbb{O} \wedge e_1 \wedge e_2
\end{aligned}$$

Dividing out the basis n-element and collecting the coefficients of ξ and ψ with *GrassmannAlgebra's* `ExpandAndCollect` gives the final equation for the conic.

$$\mathcal{E}_2 = \mathbf{ExpandAndCollect}\left[\frac{\mathcal{E}_1}{\star\mathbb{O} \wedge e_1 \wedge e_2}, \{\xi, \psi\}, 2\right] == 0$$

$$\begin{aligned}
&b\,d\,j\,p\,q\,r - a\,d\,k\,p\,q\,r + b\,c\,j\,p\,q\,s - a\,c\,k\,p\,q\,s - \\
&\quad b\,d\,j\,p\,r\,s + b\,c\,k\,p\,r\,s - a\,d\,j\,q\,r\,s + a\,c\,k\,q\,r\,s - b\,c\,p\,q\,r\,s + a\,d\,p\,q\,r\,s + \\
&\quad (-b\,d\,j\,q\,r + a\,d\,k\,q\,r - b\,d\,p\,q\,r + d\,k\,p\,q\,r + b\,d\,j\,p\,s - b\,c\,k\,p\,s - b\,c\,j\,q\,s + \\
&\qquad a\,d\,j\,q\,s - a\,d\,p\,q\,s - b\,j\,p\,q\,s + a\,k\,p\,q\,s + c\,k\,p\,q\,s + b\,d\,p\,r\,s - b\,k\,p\,r\,s + \\
&\qquad b\,c\,q\,r\,s + d\,j\,q\,r\,s - a\,k\,q\,r\,s - c\,k\,q\,r\,s + b\,p\,q\,r\,s - d\,p\,q\,r\,s)\,\xi + \\
&\quad (b\,d\,q\,r - d\,k\,q\,r - b\,d\,p\,s + b\,k\,p\,s + b\,j\,q\,s - d\,j\,q\,s + d\,p\,q\,s - k\,p\,q\,s - b\,q\,r\,s + k\,q\,r\,s)\,\xi^2 + \\
&\quad (-b\,c\,k\,p\,r + a\,d\,k\,p\,r + a\,d\,j\,q\,r - a\,c\,k\,q\,r + b\,c\,p\,q\,r - b\,j\,p\,q\,r - \\
&\qquad d\,j\,p\,q\,r + a\,k\,p\,q\,r - b\,c\,j\,p\,s + a\,c\,k\,p\,s + a\,c\,p\,q\,s - c\,j\,p\,q\,s - a\,d\,p\,r\,s + \\
&\qquad b\,j\,p\,r\,s + d\,j\,p\,r\,s - c\,k\,p\,r\,s - a\,c\,q\,r\,s + a\,j\,q\,r\,s - a\,p\,q\,r\,s + c\,p\,q\,r\,s)\,\psi + \\
&\quad (b\,k\,p\,r - d\,k\,p\,r - b\,c\,q\,r - a\,d\,q\,r + b\,j\,q\,r + c\,k\,q\,r + d\,p\,q\,r - k\,p\,q\,r + b\,c\,p\,s + a\,d\,p\,s - \\
&\qquad d\,j\,p\,s - a\,k\,p\,s - a\,j\,q\,s + c\,j\,q\,s - c\,p\,q\,s + j\,p\,q\,s - b\,p\,r\,s + k\,p\,r\,s + a\,q\,r\,s - j\,q\,r\,s)\,\xi\,\psi + \\
&\quad (-a\,k\,p\,r + c\,k\,p\,r + a\,c\,q\,r - a\,j\,q\,r - c\,p\,q\,r + j\,p\,q\,r - a\,c\,p\,s + c\,j\,p\,s + a\,p\,r\,s - j\,p\,r\,s)\,\psi^2 == 0
\end{aligned}$$

To check if this gives the correct form, we can substitute in some values for the coordinates of the points and lines.

```
ℰ₂ /. {a → 2, b → 1, c → 1, d → 2, j → 1, k → 1, p → 1, q → 0, r → 1, s → 3}
```

$-3 + 6\xi - 3\xi^2 + 2\xi\psi - \psi^2 == 0$

This is an ellipse.

```
ContourPlot[-3 + 6 ξ - 3 ξ^2 + 2 ξ ψ - ψ^2 == 0, {ξ, 0, 3}, {ψ, 0, 3},
  GridLines → Automatic, ImageSize → {3 × 72, 3 × 72}]
```

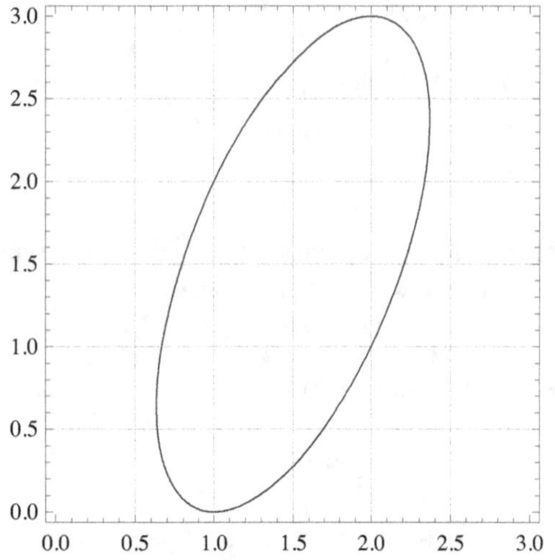

An alternative geometric equation of a conic section in the plane

We can derive an alternative formula which is a little more symmetric. As in the previous derivation, we still take the fundamental form of the equation to be the collinearity of three points.

$$P_0 \wedge Q_1 \wedge Q_2 == 0$$

But this time, P_0 is a fixed point while Q_1 and Q_2 are functions of the variable point P. Because we are looking for a more symmetric formula, let us suppose the form for Q_i is the same as that for Q_2 in the previous formulation.

$$Q_i \equiv L_i \vee (P \wedge P_i)$$

Here, the L_i and P_i are fixed. The Q_i lie on the lines L_i.

The equation becomes

$$P_0 \wedge (L_1 \vee (P \wedge P_1)) \wedge (L_2 \vee (P \wedge P_2)) == 0 \qquad 4.28$$

Thus we can imagine two fixed lines L_1 and L_2 and three fixed points P_0, P_1, and P_2. The prescription to construct the conic section is as follows:

1. Construct the fixed points and lines: P_0, P_1, P_2, L_1, and L_2.
2. Construct a line through the point P_0 to intersect L_1 and L_2 in Q_1 and Q_2 respectively.
3. Construct a line through Q_1 and P_1.
4. Construct a line through Q_2 and P_2.

The first line through P_0, Q_1 and Q_2 satisfies the condition that $P_0 \wedge Q_1 \wedge Q_2 = 0$.

The second and third lines through Q_i and P_i satisfy the conditions that $Q_i \wedge P \wedge P_i = 0$. Thus their intersection gives the point P.

We now have two formulae purporting to describe a conic section: the formula above, and the formula we first derived.

$$\mathcal{S}_1 == \mathbf{P} \wedge \mathbf{P_1} \wedge (\mathbf{L_1} \vee (\mathbf{P_0} \wedge (\mathbf{L_2} \vee (\mathbf{P} \wedge \mathbf{P_2}))))$$

$$\mathcal{S}_2 == \mathbf{P_0} \wedge (\mathbf{L_1} \vee (\mathbf{P} \wedge \mathbf{P_1})) \wedge (\mathbf{L_2} \vee (\mathbf{P} \wedge \mathbf{P_2}))$$

To see if they are the same, we could choose a basis, substitute symbolic coordinates, and check that the same second degree equation results. An alternative method is to prove the identity of the formulae from the Common Factor Theorem by applying `CongruenceSimplify` ($\star\mathcal{C}$) to the symbolic expressions in the formulae. To indicate that the symbols for the lines represent elements of grade 2 we underscript them with their grade. To indicate that the symbols for the points represent elements of grade 1 we declare them as vector symbols. All previous definitions should be cleared first.

```
★A; ★P2; DeclareExtraVectorSymbols[{P, P0, P1, P2}];
```

A typical computation might then be to find a weighted intersection point:

$$\star\mathcal{C}\left[\underset{2}{\mathbf{L_1}} \vee (\mathbf{P} \wedge \mathbf{P_1})\right]$$

$$P \left\langle P_1 \wedge \underset{2}{L_1} \right\rangle - \left\langle P \wedge \underset{2}{L_1} \right\rangle P_1$$

Remember that the angle bracket expressions are scalar quantities. $\langle Z \rangle$ is to be interpreted as the scalar coefficient of the *n*-element Z when expressed in the current basis. See section 3.6.

We can expand a complete formula in a similar manner to get an alternative expression for the conic section. For example, the first formula is

$$\star\mathcal{C}\left[\mathbf{P} \wedge \mathbf{P_1} \wedge \left(\underset{2}{\mathbf{L_1}} \vee \left(\mathbf{P_0} \wedge \left(\underset{2}{\mathbf{L_2}} \vee (\mathbf{P} \wedge \mathbf{P_2})\right)\right)\right)\right] == 0$$

$$\left(\left\langle P \wedge \underset{2}{L_2} \right\rangle \left\langle P_2 \wedge \underset{2}{L_1} \right\rangle - \left\langle P \wedge \underset{2}{L_1} \right\rangle \left\langle P_2 \wedge \underset{2}{L_2} \right\rangle\right) P \wedge P_0 \wedge P_1 + \left\langle P \wedge \underset{2}{L_2} \right\rangle \left\langle P_0 \wedge \underset{2}{L_1} \right\rangle P \wedge P_1 \wedge P_2 == 0$$

$$\left(\left\langle \mathbf{P} \wedge \underset{2}{\mathbf{L_2}} \right\rangle \left\langle \mathbf{P_2} \wedge \underset{2}{\mathbf{L_1}} \right\rangle - \left\langle \mathbf{P} \wedge \underset{2}{\mathbf{L_1}} \right\rangle \left\langle \mathbf{P_2} \wedge \underset{2}{\mathbf{L_2}} \right\rangle\right) \mathbf{P} \wedge \mathbf{P_0} \wedge \mathbf{P_1} + \left\langle \mathbf{P} \wedge \underset{2}{\mathbf{L_2}} \right\rangle \left\langle \mathbf{P_0} \wedge \underset{2}{\mathbf{L_1}} \right\rangle \mathbf{P} \wedge \mathbf{P_1} \wedge \mathbf{P_2} == 0 \qquad 4.29$$

This is just one of a number of alternative forms. Expanding the second formula gives a different expression. In order to show that they are in fact equal we need to note that in the formulae there are four points involved, of which only three are independent in the plane. Hence we can arbitrarily express one of them (say P) in terms of the other three. (We use the symbol wP to emphasize that the expression we substitute is actually a weighted point).

```
wP = a P0 + b P1 + c P2;
```

In this form, the P_i can be visualized as representing a point basis for the plane, and a, b and c as variable coefficients of the weighted point wP in this basis. The formulae for comparison then become

$$\mathcal{S}_1 = \mathbf{wP} \wedge \mathbf{P_1} \wedge \left(\underset{2}{\mathbf{L_1}} \vee \left(\mathbf{P_0} \wedge \left(\underset{2}{\mathbf{L_2}} \vee (\mathbf{wP} \wedge \mathbf{P_2})\right)\right)\right);$$

$$\mathcal{S}_1 = \mathbf{GrassmannCollect[Expand[}\star\mathcal{C}\mathbf{[}\mathcal{S}_1\mathbf{]]]}$$

$$\left(a^2 \left\langle P_0 \wedge \underset{2}{L_1}\right\rangle \left\langle P_0 \wedge \underset{2}{L_2}\right\rangle + a\,b \left\langle P_0 \wedge \underset{2}{L_1}\right\rangle \left\langle P_1 \wedge \underset{2}{L_2}\right\rangle + a\,c \left\langle P_0 \wedge \underset{2}{L_2}\right\rangle \left\langle P_2 \wedge \underset{2}{L_1}\right\rangle + b\,c \left\langle P_1 \wedge \underset{2}{L_2}\right\rangle \left\langle P_2 \wedge \underset{2}{L_1}\right\rangle - b\,c \left\langle P_1 \wedge \underset{2}{L_1}\right\rangle \left\langle P_2 \wedge \underset{2}{L_2}\right\rangle\right) P_0 \wedge P_1 \wedge P_2$$

$$\mathcal{S}_2 = \mathbf{P_0} \wedge \left(\underset{2}{\mathbf{L_1}} \vee (\mathbf{wP} \wedge \mathbf{P_1})\right) \wedge \left(\underset{2}{\mathbf{L_2}} \vee (\mathbf{wP} \wedge \mathbf{P_2})\right);$$

$$\mathcal{S}_2 = \mathbf{GrassmannCollect[Expand[}\star\mathcal{C}\mathbf{[}\mathcal{S}_2\mathbf{]]]}$$

$$\left(a^2 \left\langle P_0 \wedge \underset{2}{L_1}\right\rangle \left\langle P_0 \wedge \underset{2}{L_2}\right\rangle + a\,b \left\langle P_0 \wedge \underset{2}{L_1}\right\rangle \left\langle P_1 \wedge \underset{2}{L_2}\right\rangle + a\,c \left\langle P_0 \wedge \underset{2}{L_2}\right\rangle \left\langle P_2 \wedge \underset{2}{L_1}\right\rangle + b\,c \left\langle P_1 \wedge \underset{2}{L_2}\right\rangle \left\langle P_2 \wedge \underset{2}{L_1}\right\rangle - b\,c \left\langle P_1 \wedge \underset{2}{L_1}\right\rangle \left\langle P_2 \wedge \underset{2}{L_2}\right\rangle\right) P_0 \wedge P_1 \wedge P_2$$

Clearly these two expressions are equal.

$$\mathcal{S}_1 == \mathcal{S}_2$$

```
True
```

◆ Proof by the Common Factor Axiom (Allan Cortzen)

The following *very* simple proof of the identity of the two formulae $\mathcal{S}_1$ and $\mathcal{S}_2$ which shows that it is simply an application of the Common Factor Axiom, is due to Allan Cortzen (personal communication, February 2010).

Applying formula [3.34] (a dual version of the Common Factor Axiom) to the 2-elements A, B, and C in the plane we have

$$(\mathbf{A} \vee \mathbf{B}) \wedge \mathbf{C} == \mathbf{A} \wedge (\mathbf{B} \vee \mathbf{C})$$

We retrieve the identity to be proved (and thus prove the identity) by putting

$$\mathbf{A} == \mathbf{P} \wedge \mathbf{P_1}$$
$$\mathbf{B} == \mathbf{L_1}$$
$$\mathbf{C} == \mathbf{P_0} \wedge (\mathbf{L_2} \vee (\mathbf{P} \wedge \mathbf{P_2}))$$

Conic sections through five points

Since it is well known that a unique conic section may be constructed through any five points in the plane, we ought to be able to convert the formula above to one involving only these points, rather than involving any lines. Already we can see directly from the formula that P must pass through both P_1 and P_2. Since we have supposed the lines to be non-parallel, they will intersect in some point, O say. We can then pick two more points, one on each line, R_1 and R_2 say, with which to define the lines.

$$\mathbf{L_i} == \mathbf{O} \wedge \mathbf{R_i}$$

But since any points on the lines (except O) will be satisfactory, we can choose them advantageously as the intersections with the lines $P_0 \wedge P_i$.

$$R_1 \equiv L_1 \vee (P_0 \wedge P_2)$$
$$R_2 \equiv L_2 \vee (P_0 \wedge P_1)$$

The point P_0 is then the intersection of the lines $P_1 \wedge R_2$ and $P_2 \wedge R_1$.

$$P_0 \equiv (P_1 \wedge R_2) \vee (P_2 \wedge R_1)$$

On making these substitutions the equation becomes

$$P_0 \wedge (L_1 \vee (P \wedge P_1)) \wedge (L_2 \vee (P \wedge P_2)) == 0$$

$$((P_1 \wedge R_2) \vee (P_2 \wedge R_1)) \wedge ((O \wedge R_1) \vee (P \wedge P_1)) \wedge ((O \wedge R_2) \vee (P \wedge P_2)) == 0 \qquad 4.30$$

Note the symmetry. There are five fixed points, O, P_1, P_2, R_1, R_2, and the variable point P. Each point occurs twice. Swapping P_1 and P_2 while at the same time swapping R_1 and R_2 leaves the equation unchanged.

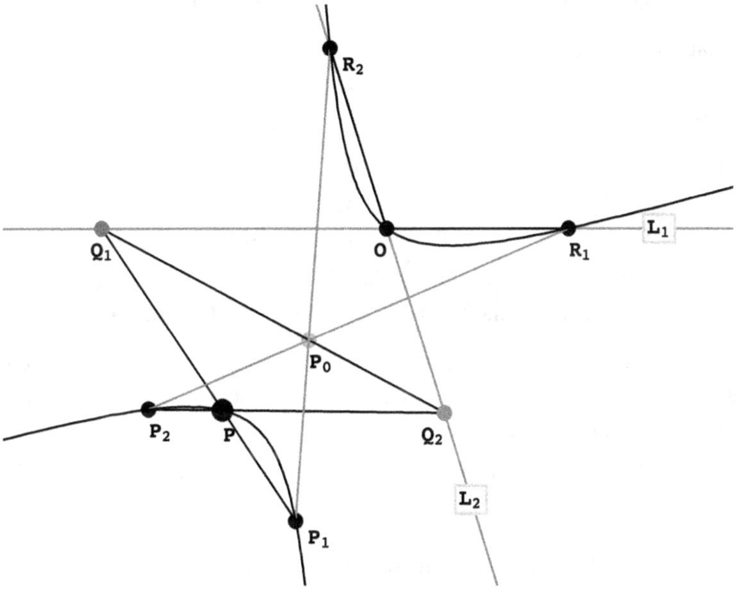

Constructing an hyperbola through 5 points

We can show that these five points satisfy the equation, and thus all lie on the conic section. But since a conic section constructed through five points is unique, this equation represents the general conic section, or general quadric curve in the plane.

We show this as follows. First, we can immediately read off from the equation that it is satisfied when P coincides with either P_1 or P_2. To show it for O, R_1 and R_2 we can expand the regressive products by the Common Factor Theorem (see section 3.6). Here we use `CongruenceSimplify` ($\star C$) as we did in the previous section to obtain the different view of the equation.

```
DeclareExtraVectorSymbols[{P, O, P₁, P₂, R₁, R₂}];
C₁ = ★C[((P₁ ∧ R₂) ∨ (P₂ ∧ R₁)) ∧ ((O ∧ R₁) ∨ (P ∧ P₁)) ∧ ((O ∧ R₂) ∨ (P ∧ P₂))] == 0
```

$$\langle O \wedge P \wedge P_1\rangle \langle P \wedge P_2 \wedge R_2\rangle \langle P_2 \wedge R_1 \wedge R_2\rangle O \wedge P_1 \wedge R_1 - \langle O \wedge P \wedge P_2\rangle \langle P \wedge P_1 \wedge R_1\rangle \langle P_2 \wedge R_1 \wedge R_2\rangle O \wedge P_1 \wedge R_2 + \langle O \wedge P \wedge P_1\rangle \langle P \wedge P_2 \wedge R_2\rangle \langle P_1 \wedge P_2 \wedge R_1\rangle O \wedge R_1 \wedge R_2 - \langle O \wedge P \wedge P_1\rangle \langle O \wedge P \wedge P_2\rangle \langle P_2 \wedge R_1 \wedge R_2\rangle P_1 \wedge R_1 \wedge R_2 == 0$$

It is clear from the first factor in each of the terms that the equation is satisfied when P is equal to O. But it is not yet clear that the equation is satisfied when P is equal to R_1 or R_2. To show this we try applying `CongruenceSimplify` to a rearranged form where the factors of the regressive products are reversed.

```
C₂ = ★C[((P₂ ∧ R₁) ∨ (P₁ ∧ R₂)) ∧ ((P ∧ P₁) ∨ (O ∧ R₁)) ∧ ((P ∧ P₂) ∨ (O ∧ R₂))] == 0
```

$$\langle O \wedge P \wedge R_1\rangle \langle O \wedge P_2 \wedge R_2\rangle \langle P_1 \wedge R_1 \wedge R_2\rangle P \wedge P_1 \wedge P_2 - \langle O \wedge P \wedge R_1\rangle \langle O \wedge P_2 \wedge R_2\rangle \langle P_1 \wedge P_2 \wedge R_2\rangle P \wedge P_1 \wedge R_1 + \langle O \wedge P \wedge R_2\rangle \langle O \wedge P_1 \wedge R_1\rangle \langle P_1 \wedge P_2 \wedge R_2\rangle P \wedge P_2 \wedge R_1 - \langle O \wedge P \wedge R_1\rangle \langle O \wedge P \wedge R_2\rangle \langle P_1 \wedge P_2 \wedge R_2\rangle P_1 \wedge P_2 \wedge R_1 == 0$$

Now we see that not only is the equation satisfied when P is equal to O, but also when P is equal to R_1. We try again with the factors of only the second and third regressive products reversed.

```
C₃ = ★C[((P₁ ∧ R₂) ∨ (P₂ ∧ R₁)) ∧ ((P ∧ P₁) ∨ (O ∧ R₁)) ∧ ((P ∧ P₂) ∨ (O ∧ R₂))] == 0
```

$$-\langle O \wedge P \wedge R_2\rangle \langle O \wedge P_1 \wedge R_1\rangle \langle P_2 \wedge R_1 \wedge R_2\rangle P \wedge P_1 \wedge P_2 + \langle O \wedge P \wedge R_1\rangle \langle O \wedge P_2 \wedge R_2\rangle \langle P_1 \wedge P_2 \wedge R_1\rangle P \wedge P_1 \wedge R_2 - \langle O \wedge P \wedge R_2\rangle \langle O \wedge P_1 \wedge R_1\rangle \langle P_1 \wedge P_2 \wedge R_1\rangle P \wedge P_2 \wedge R_2 + \langle O \wedge P \wedge R_1\rangle \langle O \wedge P \wedge R_2\rangle \langle P_1 \wedge P_2 \wedge R_1\rangle P_1 \wedge P_2 \wedge R_2 == 0$$

This time we see immediately that the equation is satisfied for P equal to R_2.

Of course it would have been more direct to show these sorts of results by replacing the point by P in the equation, and then simplifying. For example, replacing R_2 in the above equation, then simplifying, gives zero as expected.

```
★C[((P₁ ∧ P) ∨ (P₂ ∧ R₁)) ∧ ((P ∧ P₁) ∨ (O ∧ R₁)) ∧ ((P ∧ P₂) ∨ (O ∧ P))]
```

0

We have thus shown that the five fixed points, O, P_1, P_2, R_1, R_2, all lie on the conic section.

- As a second example of a conic plotted from formula [4.30], here is a circle. The only difference from the hyperbola above is the relative location of the five given points.

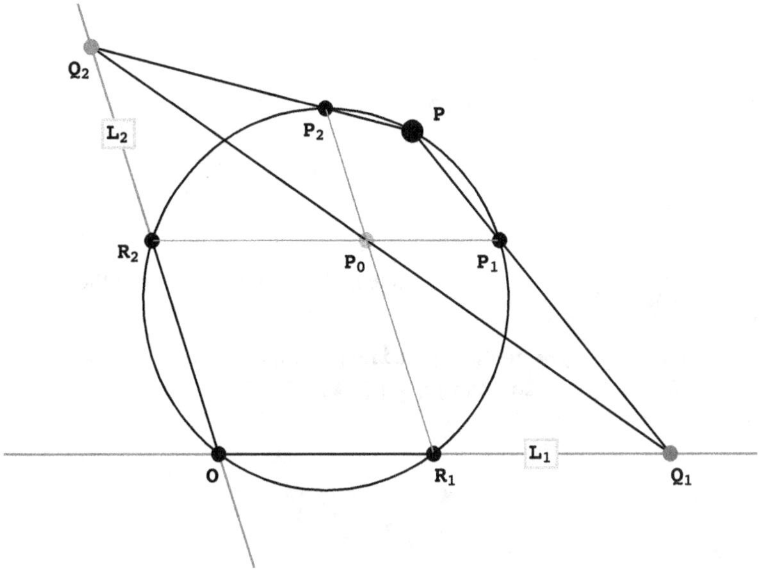

Constructing a circle through 5 points

Dual constructions

The Principle of Duality discussed in section 3.2 says that to every formula such as

$$P_0 \wedge (L_1 \vee (P \wedge P_1)) \wedge (L_2 \vee (P \wedge P_2)) == 0$$

there corresponds a dual formula in which exterior and regressive products are interchanged, and m-elements are interchanged with $(n-m)$-elements. In this planar case, points are interchanged with lines. Thus we can write another formula for a conic section in the plane as

$$\star\mathcal{P}_2;\ \mathcal{D}_1 = L_0 \vee (P_1 \wedge (\mathcal{L} \vee L_1)) \vee (P_2 \wedge (\mathcal{L} \vee L_2));$$

where the P_i are fixed points, the L_i are fixed lines, *and* $\mathcal{L}$ *is a variable line.*

This formula says that three lines intersect at one point. One line L_0 is fixed. The other two are of the form

$$\mathbb{L}_i == P_1 \wedge (\mathcal{L} \vee L_1)$$

To see most directly how such a formula translates into an algebraic equation, we take a numerical example with fixed points and lines defined:

$$P_1 = \star\mathbb{O} + 2\,e_1 + e_2;$$
$$P_2 = \star\mathbb{O} + e_1 + 2\,e_2;$$
$$L_0 = (\star\mathbb{O} + 5\,e_1) \wedge (\star\mathbb{O} + 4\,e_2);$$
$$L_1 = \star\mathbb{O} \wedge (\star\mathbb{O} + e_1);$$
$$L_2 = (\star\mathbb{O} - e_1) \wedge (\star\mathbb{O} + 3\,e_2);$$

We can define the variable line as the product of its intersections with the coordinate axes:

$$\star\star S[\xi, \psi];\ \mathcal{L} = (\star\mathbb{O} + \xi\,e_1) \wedge (\star\mathbb{O} + \psi\,e_2);$$

Substituting these values into the geometric equation and simplifying gives

```
Expand[★C[𝒟1]] == 0
```

$198\,\xi\,\psi - 90\,\xi^2\,\psi - 21\,\psi^2 - 80\,\xi\,\psi^2 + 37\,\xi^2\,\psi^2 == 0$

We can solve for ψ in terms of ξ, but remember that these are line coordinates, not point coordinates.

$$\psi = \frac{18\,(5\,\xi^2 - 11\,\xi)}{37\,\xi^2 - 80\,\xi - 21};$$

By plotting a line for each pair of line coordinates, we see that they form the envelope of a conic section as expected.

```
Graphics[Table[{GrayLevel[.3], Line[{{2 ξ, -ψ}, {-ξ, 2 ψ}}]},
  {ξ, -20, 20, .05}], PlotRange → {{-1, 3}, {-.5, 3}},
 ImageSize → {4 × 72, 4 × 72}]
```

Constructing conics in space

To construct a conic in a (bound vector) 3-space we need only take a 2-space formula such as

$P \wedge P_1 \wedge (L_1 \vee (P_0 \wedge (L_2 \vee (P \wedge P_2)))) == 0$

and replace the 2-space hyperplanes (the lines) with 3-space hyperplanes (planes). Thus we can rewrite the formula as

$$P \wedge P_1 \wedge (\Pi_1 \vee (P_0 \wedge (\Pi_2 \vee (P \wedge L)))) == 0 \qquad 4.31$$

Note that the point P_2 becomes the line L in order to make the innermost term into a plane. The construction is then as follows:

1. Construct the fixed points, lines and planes: P_0, P_1, L, Π_1, and Π_2.
2. Construct a plane Π_c through L to intersect plane Π_2 in line L_2.

```
L2 ≡ Π2 ∨ Πc
Πc == P ∧ L
```

3. Construct a plane Π_b through line L_2 and point P_0 to intersect plane Π_1 in line L_1.

```
L1 ≡ Π1 ∨ Πb
Πb == P0 ∧ L2
```

4. Construct a plane Π_a through line L_1 and point P_1 to intersect line Π_c in point P.

```
0 == P ∧ Πa
Πa == P1 ∧ L1
```

5. Repeat the construction by drawing a new plane Π_c through L.

We now derive the algebraic equation in the same manner as we have previously for the planar cases.

First, define the variable point P and the fixed entities.

```
★A; ★𝒫3; ★★S[ξ, ψ, ζ]; P = ★O + ξ e1 + ψ e2 + ζ e3;
★★S[{a, b, c, i, j, k, p, q, r, s, u, v, w, x, y, z}];
P0 = ★O + a e1 + b e2 + c e2;
P1 = ★O + i e1 + j e2 + k e3;
L = (★O + p e1 + q e2) ∧ (★O + r e2 + s e3);
Π1 = (★O + u e1) ∧ (★O + v e2) ∧ (★O + w e3);
Π2 = (★O + x e1) ∧ (★O + y e2) ∧ (★O + z e3);
```

Next, compute the algebraic formula.

```
ExpandAndCollect[★C[P ∧ P1 ∧ (Π1 ∨ (P0 ∧ (Π2 ∨ (P ∧ L))))] / (★O ∧ e1 ∧ e2 ∧ e3),
  {ξ, ψ, ζ}, 3]
```

```
-b k p s u v w x y - c k p s u v w x y + a k q s u v w x y - b k p r u v w x z - c k p r u v w x z - b i q s u v w x z -
 c i q s u v w x z + a j q s u v w x z - a k p r u v w y z - b i p s u v w y z - c i p s u v w y z + a j p s u v w y z +
 b k p s u v x y z + c k p s u v x y z - a k q s u v x y z + b j p s u w x y z + c j p s u w x y z - a j q s u w x y z +
 b i p s v w x y z + c i p s v w x y z - a i q s v w x y z + k p r u v w x y z - j p s u v w x y z + i q s u v w x y z +
 (-b j p s u w x y - c j p s u w x y + a j q s u w x y - b i p s v w x y - c i p s v w x y + a i q s v w x y +
   b k p u v w x y + c k p u v w x y - a k q u v w x y + a k r u v w x y - k p r u v w x y + b p s u v w x y +
   c p s u v w x y - a q s u v w x y + b k p r u v x z + c k p r u v x z + b i q s u v x z + c i q s u v x z -
   a j q s u v x z + b i q u v w x z + c i q u v w x z - a j q u v w x z - b i r u v w x z - c i r u v w x z +
   a j r u v w x z + b p r u v w x z + c p r u v w x z - j p r u v w x z + a k p r u v y z + b i p s u v y z +
   c i p s u v y z - a j p s u v y z + b i p u v w y z + c i p u v w y z - a j p u v w y z + a p r u v w y z -
   i p r u v w y z - b k p u v x y z - c k p u v x y z + a k q u v x y z - a k r u v x y z - b p s u v x y z -
   c p s u v x y z + j p s u v x y z + a q s u v x y z - i q s u v x y z - b j p u w x y z - c j p u w x y z +
   a j q u w x y z - a j r u w x y z + j p r u w x y z - b i p v w x y z - c i p v w x y z + a i q v w x y z -
   a i r v w x y z + i p r v w x y z + j p u v w x y z - i q u v w x y z + i r u v w x y z - p r u v w x y z) ζ +
 (b j p u w x y + c j p u w x y - a j q u w x y + a j r u w x y - j p r u w x y + b i p v w x y +
   c i p v w x y - a i q v w x y + a i r v w x y - i p r v w x y - b p u v w x y - c p u v w x y +
   a q u v w x y - a r u v w x y + p r u v w x y - b i q u v x z - c i q u v x z + a j q u v x z +
   b i r u v x z + c i r u v x z - a j r u v x z - b p r u v x z - c p r u v x z + j p r u v x z -
   b i p u v y z - c i p u v y z + a j p u v y z - a p r u v y z + i p r u v y z + b p u v x y z +
   c p u v x y z - j p u v x y z - a q u v x y z + i q u v x y z + a r u v x y z - i r u v x y z) ζ² +
 (b k p s v w x y + c k p s v w x y - a k q s v w x y + b k s u v w x y + c k s u v w x y - k q s u v w x y +
   b k p r v w x z + c k p r v w x z + b i q s v w x z + c i q s v w x z - a j q s v w x z - b k q u v w x z -
   c k q u v w x z + b k r u v w x z + c k r u v w x z + b q s u v w x z + c q s u v w x z - j q s u v w x z -
   b k p s u v y z - c k p s u v y z + a k q s u v y z - b j p s u w y z - c j p s u w y z + a j q s u w y z +
   a k p r v w y z - a j p s v w y z + a i q s v w y z - a k q u v w y z + a k r u v w y z + b i s u v w y z +
```

$$
\begin{aligned}
&c i s u v w y z - a j s u v w y z + b p s u v w y z + c p s u v w y z - i q s u v w y z - b k s u v x y z - \\
&c k s u v x y z + k q s u v x y z - b j s u w x y z - c j s u w x y z + j q s u w x y z - \\
&k p r v w x y z - b i s v w x y z - c i s v w x y z - b p s v w x y z - c p s v w x y z + j p s v w x y z + \\
&a q s v w x y z + k q u v w x y z - k r u v w x y z + j s u v w x y z - q s u v w x y z)\ \xi + \\
&(b j s u w x y + c j s u w x y - j q s u w x y - b k p v w x y - c k p v w x y + a k q v w x y - \\
&a k r v w x y + k p r v w x y + b i s v w x y + c i s v w x y - i q s v w x y - b s u v w x y - \\
&c s u v w x y + q s u v w x y + b k q u v x z + c k q u v x z - b k r u v x z - c k r u v x z - b q s u v x z - \\
&c q s u v x z + j q s u v x z - b i q v w x z - c i q v w x z + a j q v w x z + b i r v w x z + c i r v w x z - \\
&a j r v w x z - b p r v w x z - c p r v w x z + j p r v w x z + b k p u v y z + c k p u v y z - \\
&k p r u v y z - b i s u v y z - c i s u v y z + a j s u v y z - a q s u v y z + i q s u v y z + b j p u w y z + \\
&c j p u w y z - a j q u w y z + a j r u w y z - j p r u w y z + a j p v w y z - a i q v w y z + \\
&a i r v w y z - a p r v w y z - b p u v w y z - c p u v w y z + a q u v w y z - a r u v w y z + \\
&p r u v w y z - k q u v x y z + k r u v x y z + b s u v x y z + c s u v x y z - j s u v x y z + b p v w x y z + \\
&c p v w x y z - j p v w x y z - a q v w x y z + i q v w x y z + a r v w x y z - i r v w x y z)\ \zeta\, \xi + \\
&(- b k s v w x y - c k s v w x y + k q s v w x y + b k q v w x z + c k q v w x z - b k r v w x z - c k r v w x z - \\
&b q s v w x z - c q s v w x z + j q s v w x z + b k s u v y z + c k s u v y z - k q s u v y z + b j s u w y z + \\
&c j s u w y z - j q s u w y z + a k q v w y z - a k r v w y z + a j s v w y z - a q s v w y z - b s u v w y z - \\
&c s u v w y z + q s u v w y z - k q v w x y z + k r v w x y z + b s v w x y z + c s v w x y z - j s v w x y z) \\
&\xi^2 + (b k p s u w x y + c k p s u w x y - a k q s u w x y - a k s u v w x y + k p s u v w x y - \\
&b k p s u v x z - c k p s u v x z + a k q s u v x z + b k p r u w x z + c k p r u w x z - b j p s u w x z - \\
&c j p s u w x z + b i q s u w x z + c i q s u w x z - b i p s v w x z - c i p s v w x z + a i q s v w x z + \\
&b k p u v w x z + c k p u v w x z + b i s u v w x z + c i s u v w x z - a j s u v w x z + j p s u v w x z - \\
&a q s u v w x z + a k p r u w y z + b i p s u w y z + c i p s u w y z - a j p s u w y z + \\
&a k p u v w y z - a p s u v w y z + i p s u v w y z + a k s u v x y z - k p s u v x y z - k p r u w x y z + \\
&a j s u w x y z - b p s u w x y z - c p s u w x y z + a q s u w x y z - i q s u w x y z + \\
&a i s v w x y z - i p s v w x y z - k p u v w x y z - i s u v w x y z + p s u v w x y z)\ \psi + \\
&(- b k p u w x y - c k p u w x y + a k q u w x y - a k r u w x y + k p r u w x y - a j s u w x y + \\
&j p s u w x y - a i s v w x y + i p s v w x y + a s u v w x y - p s u v w x y - a k q u v x z + \\
&a k r u v x z - k p r u v x z - b i s u v x z - c i s u v x z + a j s u v x z + b p s u v x z + c p s u v x z - \\
&j p s u v x z + b j p u w x z + c j p u w x z - b i q u w x z - c i q u w x z + b i r u w x z + \\
&c i r u w x z - b p r u w x z - c p r u w x z + b i p v w x z + c i p v w x z - a i q v w x z + a i r v w x z - \\
&i p r v w x z - b p u v w x z - c p u v w x z + a q u v w x z - a r u v w x z + p r u v w x z - \\
&a k p u v y z + a p s u v y z - i p s u v y z - b i p u w y z - c i p u w y z + a j p u w y z - \\
&a p r u w y z + i p r u w y z + k p u v x y z - a s u v x y z + i s u v x y z + b p u w x y z + \\
&c p u w x y z - j p u w x y z - a q u w x y z + i q u w x y z + a r u w x y z - i r u w x y z)\ \zeta\, \psi + \\
&(- b k s u w x y - c k s u w x y + k q s u w x y + a k s v w x y - k p s v w x y + b k s u v x z + c k s u v x z - \\
&k q s u v x z + b k q u w x z + c k q u w x z - b k r u w x z - c k r u w x z + b j s u w x z + c j s u w x z - \\
&b q s u w x z - c q s u w x z - b k p v w x z - c k p v w x z + a j s v w x z + b p s v w x z + c p s v w x z - \\
&j p s v w x z - i q s v w x z - b s u v w x z - c s u v w x z + q s u v w x z - a k s u v y z + k p s u v y z + \\
&a k q u w y z - a k r u w y z - b i s u w y z - c i s u w y z + j p s u w y z - a q s u w y z + i q s u w y z - \\
&a k p v w y z - a i s v w y z + a p s v w y z + a s u v w y z - p s u v w y z - k q u w x y z + k r u w x y z + \\
&b s u w x y z + c s u w x y z - j s u w x y z + k p v w x y z - a s v w x y z + i s v w x y z)\ \xi\, \psi + \\
&(a k s u w x y - k p s u w x y - a k s u v x z + k p s u v x z - b k p u w x z - c k p u w x z - b i s u w x z - \\
&c i s u w x z + b p s u w x z + c p s u w x z - a i s v w x z + i p s v w x z + a s u v w x z - p s u v w x z - \\
&a k p u w y z + a p s u w y z - i p s u w y z + k p u w x y z - a s u w x y z + i s u w x y z)\ \psi^2
\end{aligned}
$$

Substituting some random values for the coordinates of the fixed points, line and planes gives us the equation:

$$
\begin{aligned}
&4.753 - 0.900375\, \zeta + 2.06412\, \zeta^2 - 12.6175\, \xi + 2.27084\, \zeta\, \xi + \\
&\quad 3.12987\, \xi^2 + 2.82363\, \psi + 2.80525\, \zeta\, \psi + 2.39794\, \xi\, \psi - 0.486937\, \psi^2 == 0
\end{aligned}
$$

The graphic below depicts this conic together with its defining points, line and planes. The construction lines are not shown as they are difficult to portray unambiguously in one view.

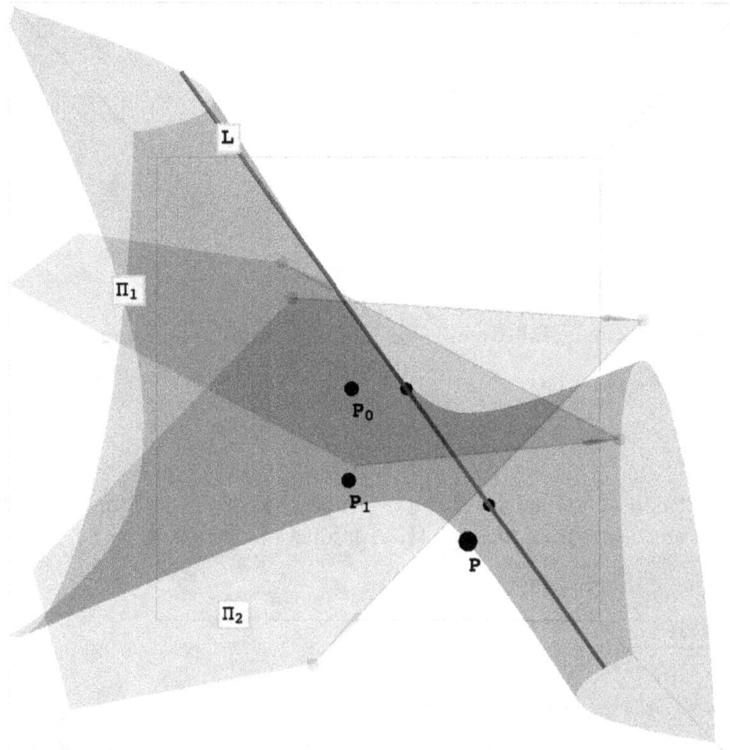

Constructing a conic from two points, a line and two planes

Note that the line L and the point P_1 are on the surface. This is clear from the geometric equation.

A geometric equation for a cubic in the plane

To this point we have explored just second degree equations in the plane and space. In this section we will explore a geometric formula for a third degree equation (cubic) in the plane. As for the quadratic equation, we will assemble the formula from the condition that three points are collinear, but this time, all three points will depend non-trivially on the variable point P. The point P itself will, instead of being at the intersection of two lines, be at the intersection of three lines. We will write down the proposed formula, and then show that it leads to a cubic equation for the coordinates of the point P. Note carefully that *this cubic formula is not a prescription to construct*, as it is not the simple intersection of two elements.

$$\mathcal{D}_1 = (\mathbf{L}_1 \vee (\mathbf{P} \wedge \mathbf{P}_1)) \wedge (\mathbf{L}_2 \vee (\mathbf{P} \wedge \mathbf{P}_2)) \wedge (\mathbf{L}_3 \vee (\mathbf{P} \wedge \mathbf{P}_3));$$

As before we define the points and lines in terms of coordinates.

```
★𝒫2; ★★S[ξ, ψ]; ★★S[{a, b, c, d, j, k, p, q, r, s, u, v}];
P = ★O + ξ e1 + ψ e2;
P1 = ★O + a e1 + b e2;
P2 = ★O + c e1 + d e2;
P3 = ★O + j e1 + k e2;
K1 = ★O + p e1 + q e2;
K2 = ★O + r e1 + s e2;
K3 = ★O + u e1 + v e2;
L1 = K2 ∧ K3;
L2 = K1 ∧ K3;
L3 = K2 ∧ K1;
```

A typical factor of the formula, for example the first, simplifies to the weighted point:

```
Q1 = ★C[L1 ∨ (P ∧ P1)]

★O ((s - v) (a - ξ) + (-r + u) (b - ψ)) + ((s u - r v) (a - ξ) + (-r + u) (b ξ - a ψ)) e1 +
  ((s u - r v) (b - ψ) + (-s + v) (b ξ - a ψ)) e2
```

Remember that ★C is a shorthand for `CongruenceSimplify`. It automatically eliminates the congruence factor ★c arising from the regressive product in the current basis.

The complete formula simplifies to

```
        ★C[𝒟1]
𝒟2 = ――――――――――― == 0
      ★O ∧ e1 ∧ e2

((s u - r v) (b - ψ) + (-s + v) (b ξ - a ψ))
    (((q - s) (j - ξ) - (p - r) (k - ψ)) ((q u - p v) (c - ξ) + (-p + u) (d ξ - c ψ)) +
      ((q - v) (c - ξ) + (-p + u) (d - ψ)) ((q r - p s) (-j + ξ) + (p - r) (k ξ - j ψ))) +
  (-(s u - r v) (a - ξ) - (-r + u) (b ξ - a ψ))
    (((q - s) (j - ξ) - (p - r) (k - ψ)) ((q u - p v) (d - ψ) + (-q + v) (d ξ - c ψ)) +
      ((q - v) (c - ξ) + (-p + u) (d - ψ)) ((-q r + p s) (k - ψ) + (q - s) (k ξ - j ψ))) +
  ((s - v) (a - ξ) + (-r + u) (b - ψ)) ((-(q u - p v) (d - ψ) + (q - v) (d ξ - c ψ))
      ((q r - p s) (-j + ξ) + (p - r) (k ξ - j ψ)) + ((q u - p v) (c - ξ) +
        (-p + u) (d ξ - c ψ)) ((-q r + p s) (k - ψ) + (q - s) (k ξ - j ψ))) == 0
```

It is easy to check that the expression is indeed a cubic, either by inspection, or application of `ExpandAndCollect` as in the previous section. (We do not show the expression in its expanded form as it has over 2000 terms.)

It is instructive perhaps to see a graphical depiction of how the formula works. In the graphic below we plot a typical cubic in the plane defined by the three points P_1, P_2, P_3, and the three lines L_1, L_2, and L_3. The point P is located by the condition that the three points Q_1, Q_2, and Q_3 are concurrent, where each of the points Q_i is defined as the intersection of the line $P \wedge P_i$ with the line L_i.

```
Qi == Li ∨ (P ∧ Pi)
```

As can be seen from the graphic, the cubic curve passes through nine special points: the three points P_1, P_2, P_3, the three intersections of the three lines L_1, L_2, and L_3, and the intersections R_i (unnamed on the graphic) of these three lines with the lines through the points P_i taken two at a time.

```
𝒟1 == (L1 ∨ (P ∧ P1)) ∧ (L2 ∨ (P ∧ P2)) ∧ (L3 ∨ (P ∧ P3))
```

```
R₁ == L₁ v (P₂ ʌ P₃)
R₂ == L₂ v (P₃ ʌ P₁)
R₃ == L₃ v (P₁ ʌ P₂)
```

These special points are displayed as large semi-opaque points.

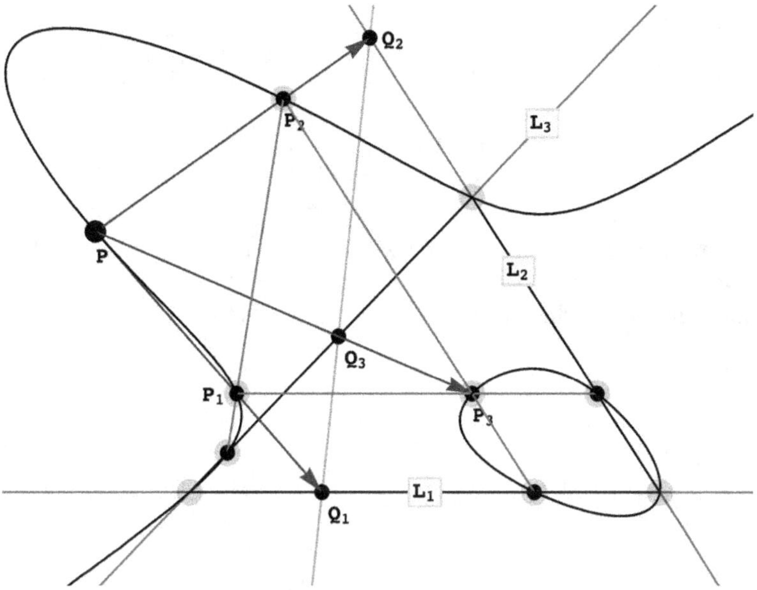

A cubic equation defined by three points and three lines

Pascal's Theorem

Our discussion up to this point leads us to Pascal's Theorem which may be stated: *If a hexagon is inscribed in a conic, then opposite sides intersect in collinear points.*

In the previous sections we have shown that the geometric equation below, involving five fixed points and a variable point P, leads to the algebraic equation of a conic section:

$$((P_1 \wedge R_2) \vee (P_2 \wedge R_1)) \wedge ((O \wedge R_1) \vee (P \wedge P_1)) \wedge ((O \wedge R_2) \vee (P \wedge P_2)) == 0$$

This is therefore the converse to Pascal's Theorem: *If opposite sides of a hexagon intersect in three collinear points, then the hexagon may be inscribed in a conic.*

To explore Pascal's Theorem further we find it conceptually advantageous to start with a fresh notation. We denote the six points by P_1, P_2, P_3, P_4, P_5, and P_6, and suppose that they are placed on the conic in arbitrary (but of course separate) positions. Let us consider the hexagon formed by these points in order, so that the sides S_1 to S_6 may be written, starting with P_1 as

```
S₁ = P₁ ʌ P₂;  S₂ = P₂ ʌ P₃;  S₃ = P₃ ʌ P₄;  S₄ = P₄ ʌ P₅;  S₅ = P₅ ʌ P₆;
S₆ = P₆ ʌ P₁;
```

The pairs of opposite sides can then be intersected to give the collinear points Z_1, Z_2 and Z_3.

```
Z₁ = S₁ v S₄;  Z₂ = S₂ v S₅;  Z₃ = S₃ v S₆;
```

We will call these points *Pascal points*. The *Pascal line* $\mathbb{L}$ on which these points lie can be written alternatively as

$$\mathbb{L} \equiv Z_1 \wedge Z_2 \equiv Z_1 \wedge Z_3 \equiv Z_2 \wedge Z_3$$

The formula is then constructed from the condition that the points Z_1, Z_2, and Z_3 all lie on $\mathbb{L}$, that is, their exterior product is zero.

$$Z_1 \wedge Z_2 \wedge Z_3 == 0$$

$$((P_1 \wedge P_2) \vee (P_4 \wedge P_5)) \wedge ((P_2 \wedge P_3) \vee (P_5 \wedge P_6)) \wedge ((P_3 \wedge P_4) \vee (P_6 \wedge P_1)) == 0$$

$$((P_1 \wedge P_2) \vee (P_4 \wedge P_5)) \wedge ((P_2 \wedge P_3) \vee (P_5 \wedge P_6)) \wedge ((P_3 \wedge P_4) \vee (P_6 \wedge P_1)) == 0 \qquad 4.32$$

The graphic below depicts these entities for six points P_i on a circle, which to show the generality of the theorem, are not placed in order. The extensions to the sides of the hexagon S_i are shown as thinner grey lines, while the Pascal line is shown as the thickest line.

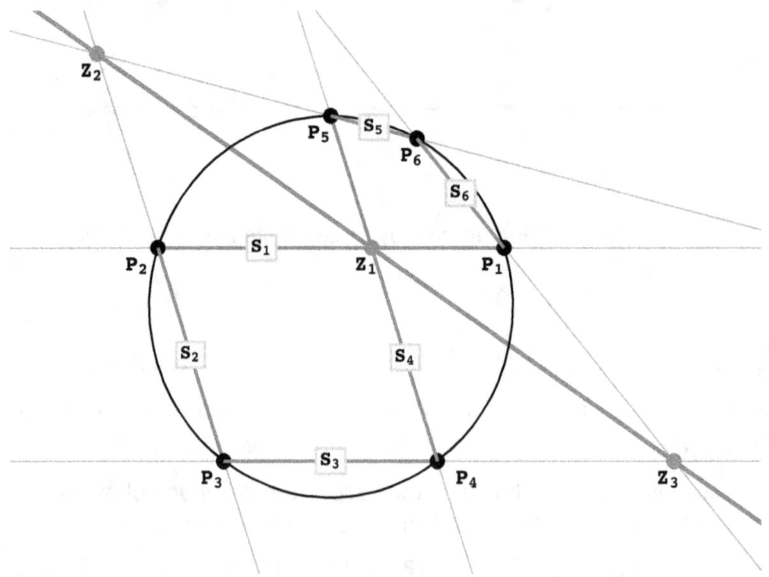

Pascal's theorem

As may be observed, this is essentially the same graphic as *Constructing a circle through 5 points* in the section above on *Conic sections through five points*, except that the points have been renamed as follows:

$P_1 \to P_1,\ P_2 \to P_5,\ R_1 \to P_4,\ R_2 \to P_2,\ O \to P_3,\ P \to P_6,\ Q_1 \to Z_3,\ Q_2 \to Z_2,$
$P_0 \to Z_1$

◆ Demonstration of Pascal's Theorem in an ellipse

The diagram above demonstrates Pascal's Theorem in a conic in which two opposite sides cross within the conic, hence making the Pascal line intersect it. The following example demonstrates the theorem for an ellipse in which all the opposite sides intersect outside the conic. We take the ellipse

$$\left(\frac{x}{4}\right)^2 + \left(\frac{y}{3}\right)^2 == 1$$

and choose some symmetrically placed points around it:

```
EllipsePoints =
```

$$\left\{P_1 \to \star O + 3\,e_2,\ P_2 \to \star O + 2\,e_1 - \frac{3\sqrt{3}}{2}\,e_2,\ P_3 \to \star O + 4\,e_1,\ P_4 \to \star O - 3\,e_2,\ P_5 \to \star O - 4\,e_1,\ P_6 \to \star O - 2\,e_1 - \frac{3\sqrt{3}}{2}\,e_2\right\};$$

We defined the sides S_i above by

$$S_1 = P_1 \wedge P_2;\ S_2 = P_2 \wedge P_3;\ S_3 = P_3 \wedge P_4;\ S_4 = P_4 \wedge P_5;\ S_5 = P_5 \wedge P_6;$$
$$S_6 = P_6 \wedge P_1;$$

To compute the three Pascal points we take the regressive product of opposite sides, substitute in the values `EllipsePoints`, and use `CongruenceSimplify` (in its short form $\star C$) and `ToPointForm` to do the simplifications.

$$\star\mathcal{P}_2;$$
$$\{Z_1, Z_2, Z_3\} = \text{ToPointForm}[\star C[\{S_1 \vee S_4, S_2 \vee S_5, S_3 \vee S_6\} \ /.\ \text{EllipsePoints}]]$$

$$\left\{\star O + 4\left(-1+\sqrt{3}\right) e_1 - 3\sqrt{3}\,e_2,\ \star O - 3\sqrt{3}\,e_2,\ \star O + \left(4-4\sqrt{3}\right) e_1 - 3\sqrt{3}\,e_2\right\}$$

Pascal's Theorem says that these three points are collinear, hence their exterior product should be zero. This is easy to confirm.

$$\star\mathcal{G}[Z_1 \wedge Z_2 \wedge Z_3]$$

0

The Pascal line may be obtained as the exterior product of any pair of the points.

$$\{\star\mathcal{G}[Z_1 \wedge Z_2],\ \star\mathcal{G}[Z_1 \wedge Z_3],\ \star\mathcal{G}[Z_2 \wedge Z_3]\}$$

$$\left\{\left(4-4\sqrt{3}\right) \star O \wedge e_1 + 12\left(-3+\sqrt{3}\right) e_1 \wedge e_2,\right.$$
$$\left(8-8\sqrt{3}\right) \star O \wedge e_1 + 24\left(-3+\sqrt{3}\right) e_1 \wedge e_2,$$
$$\left.\left(4-4\sqrt{3}\right) \star O \wedge e_1 + 12\left(-3+\sqrt{3}\right) e_1 \wedge e_2\right\}$$

Since the weight is not important, we can divide through by it and simplify to obtain the Pascal line in the form

$$\mathbb{L} \equiv \left(\star O - 3\sqrt{3}\,e_2\right) \wedge e_1$$

The figure below shows the line below the ellipse.

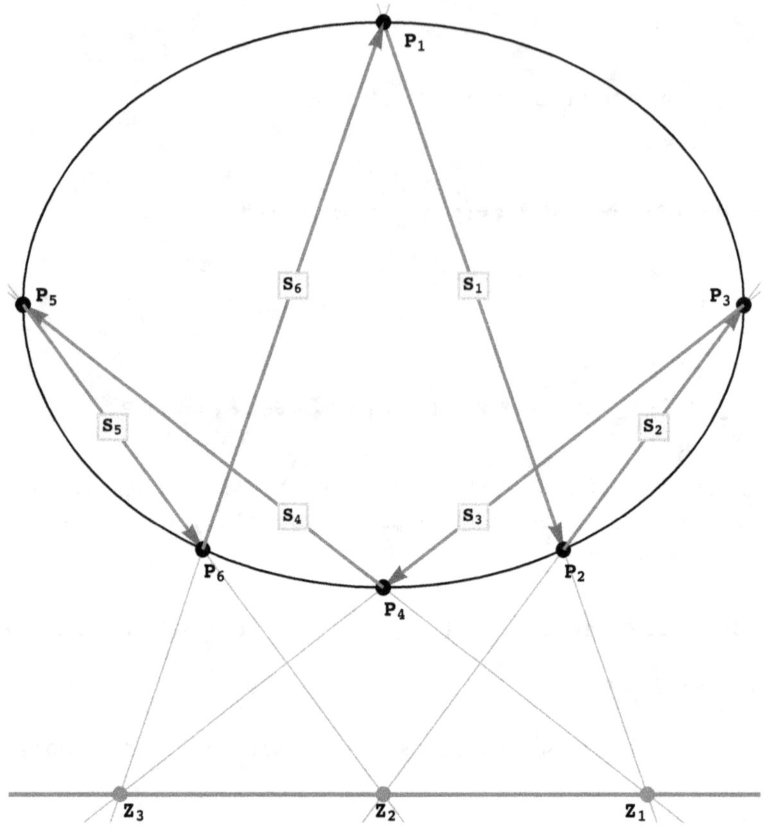

Pascal's theorem in an ellipse

◆ Demonstration of Pascal's Theorem for a regular hexagon

An interesting case occurs when we choose the conic to be a circle and the hexagon to be a regular hexagon, since the pairs of opposite sides are parallel, *and hence do not intersect*. We begin by generating a list of rules for replacing the P_i by the points on a regular hexagon.

```
RegularHexagonPoints =
 Table[Pi → ★O + Cos[i π / 3] e1 + Sin[i π / 3] e2, {i, 1, 6}]
```

$$\left\{P_1 \to \star\mathbb{O} + \frac{e_1}{2} + \frac{\sqrt{3}\ e_2}{2},\ P_2 \to \star\mathbb{O} - \frac{e_1}{2} + \frac{\sqrt{3}\ e_2}{2},\ P_3 \to \star\mathbb{O} - e_1,\ P_4 \to \star\mathbb{O} - \frac{e_1}{2} - \frac{\sqrt{3}\ e_2}{2},\ P_5 \to \star\mathbb{O} + \frac{e_1}{2} - \frac{\sqrt{3}\ e_2}{2},\ P_6 \to \star\mathbb{O} + e_1\right\}$$

The intersections of the opposite sides of the regular hexagon gives

```
{Z1, Z2, Z3} = ★C[{S1 ∨ S4, S2 ∨ S5, S3 ∨ S6} /. RegularHexagonPoints]
```

$$\left\{-\sqrt{3}\ e_1,\ -\frac{1}{2}\sqrt{3}\ e_1 - \frac{3\ e_2}{2},\ \frac{\sqrt{3}\ e_1}{2} - \frac{3\ e_2}{2}\right\}$$

Note that these are all *vectors*, not points as expected, because the opposite sides are all parallel.

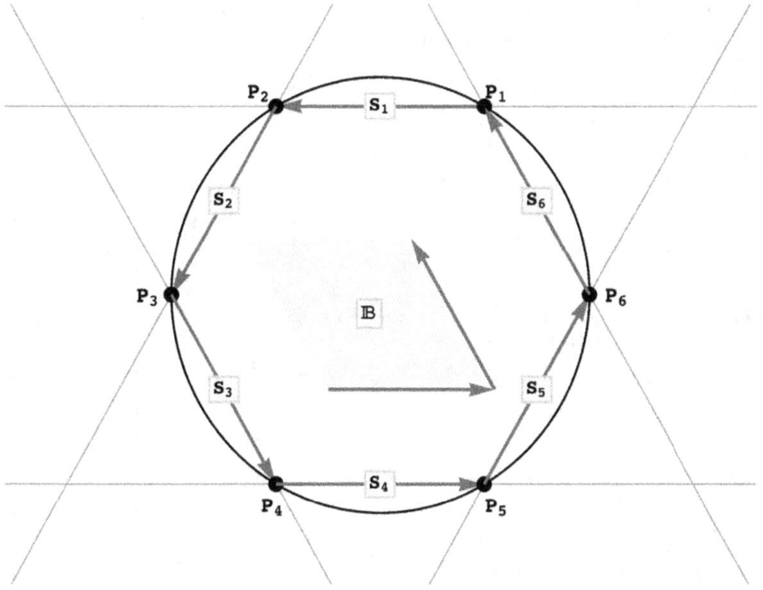

Pascal's theorem for a regular hexagon
Opposite sides intersect at points at infinity lying in the same bivector $\mathbb{B}$

These vectors are dependent (as of course they must be in the plane); and consequently, their exterior product is zero.

$\star\mathcal{G}[\mathbf{Z}_1 \wedge \mathbf{Z}_2 \wedge \mathbf{Z}_3]$

0

Thus Pascal's Theorem is demonstrated in this example to hold for pairs of sides intersecting in *points at infinity* (vectors). The three points at infinity lie on the same *line at infinity*. That is, *the three vectors lie in the same bivector*. (Of course, this must be the case for any three vectors in the plane.)

Hexagons in a conic

We have seen how Pascal's Theorem establishes a line associated with any hexagon in a conic. Suppose we have fixed the positions of the six points anywhere we wish on the conic, and named them in any order we wish. How many essentially different hexagons (whose sides may of course cross each other) can we form by joining these six points in some order?

As before, we name these points P_1, P_2, P_3, P_4, P_5 and P_6. Without loss of generality we can start with P_1 and join it in any of five ways to the other five points. From that point we then have four ways of forming the next side; then three ways; then two ways. This makes 5×4×3×2 = 120 hexagons. However, for each hexagon $P_1\ P_a\ P_b\ P_c\ P_d\ P_e$ there corresponds the same hexagon traversed in the reverse order $P_1\ P_e\ P_d\ P_c\ P_b\ P_a$. Thus there are just 60 essentially different hexagons. We use *Mathematica's* `Permutations` function to list them and `Union` to eliminate the ones traversed in reverse order.

```
hexagons =
  Join[{P1}, #] & /@Union[Permutations[{P2, P3, P4, P5, P6}],
    SameTest → (#1 === Reverse[#2] &)]
```

```
{{P1, P2, P3, P4, P5, P6}, {P1, P2, P3, P4, P6, P5}, {P1, P2, P3, P5, P4, P6},
 {P1, P2, P3, P5, P6, P4}, {P1, P2, P3, P6, P4, P5}, {P1, P2, P3, P6, P5, P4},
 {P1, P2, P4, P3, P5, P6}, {P1, P2, P4, P3, P6, P5}, {P1, P2, P4, P5, P3, P6},
 {P1, P2, P4, P5, P6, P3}, {P1, P2, P4, P6, P3, P5}, {P1, P2, P4, P6, P5, P3},
 {P1, P2, P5, P3, P4, P6}, {P1, P2, P5, P3, P6, P4}, {P1, P2, P5, P4, P3, P6},
 {P1, P2, P5, P4, P6, P3}, {P1, P2, P5, P6, P3, P4}, {P1, P2, P5, P6, P4, P3},
 {P1, P2, P6, P3, P4, P5}, {P1, P2, P6, P3, P5, P4}, {P1, P2, P6, P4, P3, P5},
 {P1, P2, P6, P4, P5, P3}, {P1, P2, P6, P5, P3, P4}, {P1, P2, P6, P5, P4, P3},
 {P1, P3, P2, P4, P5, P6}, {P1, P3, P2, P4, P6, P5}, {P1, P3, P2, P5, P4, P6},
 {P1, P3, P2, P5, P6, P4}, {P1, P3, P2, P6, P4, P5}, {P1, P3, P2, P6, P5, P4},
 {P1, P3, P4, P2, P5, P6}, {P1, P3, P4, P2, P6, P5}, {P1, P3, P4, P5, P2, P6},
 {P1, P3, P4, P6, P2, P5}, {P1, P3, P5, P2, P4, P6}, {P1, P3, P5, P2, P6, P4},
 {P1, P3, P5, P4, P2, P6}, {P1, P3, P5, P6, P2, P4}, {P1, P3, P6, P2, P4, P5},
 {P1, P3, P6, P2, P5, P4}, {P1, P3, P6, P4, P2, P5}, {P1, P3, P6, P5, P2, P4},
 {P1, P4, P2, P3, P5, P6}, {P1, P4, P2, P3, P6, P5}, {P1, P4, P2, P5, P3, P6},
 {P1, P4, P2, P6, P3, P5}, {P1, P4, P3, P2, P5, P6}, {P1, P4, P3, P2, P6, P5},
 {P1, P4, P3, P5, P2, P6}, {P1, P4, P3, P6, P2, P5}, {P1, P4, P5, P2, P3, P6},
 {P1, P4, P5, P3, P2, P6}, {P1, P4, P6, P2, P3, P5}, {P1, P4, P6, P3, P2, P5},
 {P1, P5, P2, P3, P4, P6}, {P1, P5, P2, P4, P3, P6}, {P1, P5, P3, P2, P4, P6},
 {P1, P5, P3, P4, P2, P6}, {P1, P5, P4, P2, P3, P6}, {P1, P5, P4, P3, P2, P6}}
```

We take the example of the ellipse and its points from the previous section and display their hexagons.

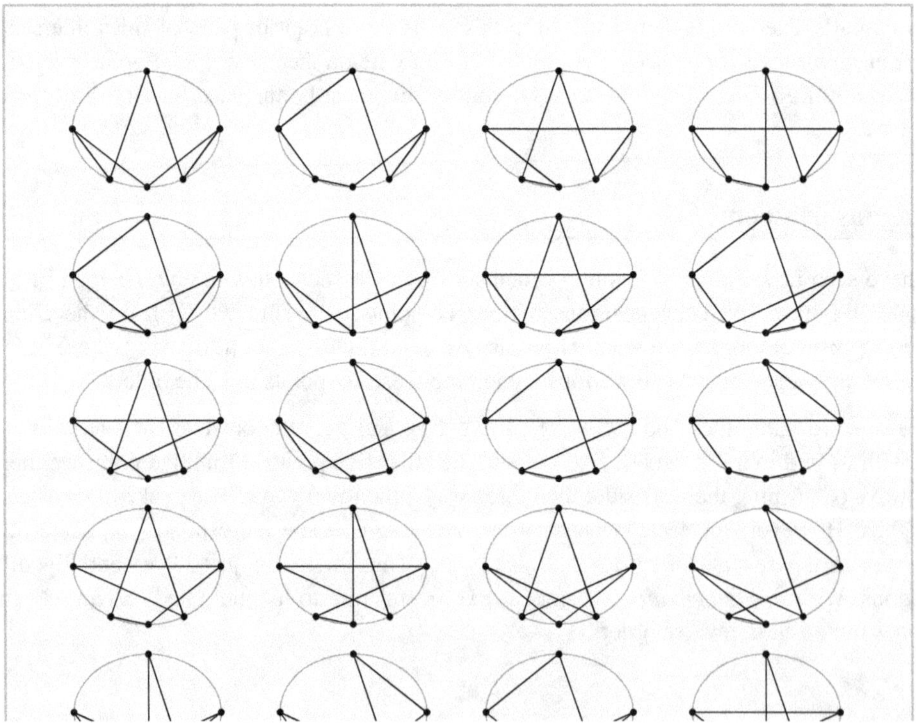

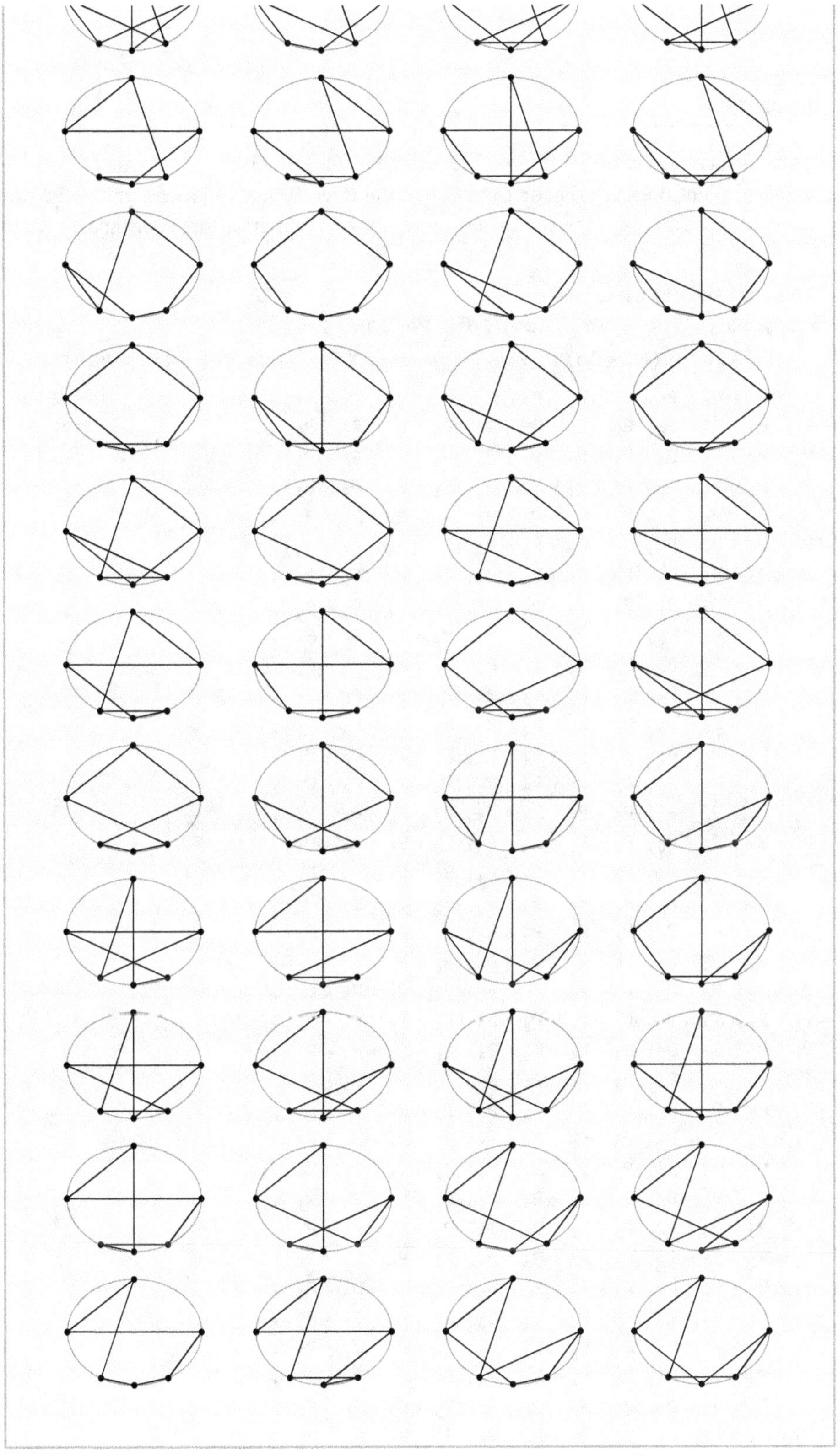

The 60 hexagons formed from six points on an ellipse

Pascal points

When two opposite sides of a hexagon are extended they intersect in a point which we have called a Pascal point. Pascal's Theorem says that the three Pascal points lie on the one line. We can convert each hexagon in the list of hexagons above to a list of three formulae, one formula for each point.

```
allPascalPointFormulae =
  hexagons /. {Pa_, Pb_, Pc_, Pd_, Pe_, Pf_} :→
    { (Pa ∧ Pb) ∨ (Pd ∧ Pe), (Pb ∧ Pc) ∨ (Pe ∧ Pf), (Pc ∧ Pd) ∨ (Pf ∧ Pa) }
```

$\{\{P_1 \wedge P_2 \vee P_4 \wedge P_5,\ P_2 \wedge P_3 \vee P_5 \wedge P_6,\ P_3 \wedge P_4 \vee P_6 \wedge P_1\},$
$\{P_1 \wedge P_2 \vee P_4 \wedge P_6,\ P_2 \wedge P_3 \vee P_6 \wedge P_5,\ P_3 \wedge P_4 \vee P_5 \wedge P_1\},$
$\{P_1 \wedge P_2 \vee P_5 \wedge P_4,\ P_2 \wedge P_3 \vee P_4 \wedge P_6,\ P_3 \wedge P_5 \vee P_6 \wedge P_1\},$
$\{P_1 \wedge P_2 \vee P_5 \wedge P_6,\ P_2 \wedge P_3 \vee P_6 \wedge P_4,\ P_3 \wedge P_5 \vee P_4 \wedge P_1\},$
$\{P_1 \wedge P_2 \vee P_6 \wedge P_4,\ P_2 \wedge P_3 \vee P_4 \wedge P_5,\ P_3 \wedge P_6 \vee P_5 \wedge P_1\},$
$\{P_1 \wedge P_2 \vee P_6 \wedge P_5,\ P_2 \wedge P_3 \vee P_5 \wedge P_4,\ P_3 \wedge P_6 \vee P_4 \wedge P_1\},$
$\{P_1 \wedge P_2 \vee P_3 \wedge P_5,\ P_2 \wedge P_4 \vee P_5 \wedge P_6,\ P_4 \wedge P_3 \vee P_6 \wedge P_1\},$
$\{P_1 \wedge P_2 \vee P_3 \wedge P_6,\ P_2 \wedge P_4 \vee P_6 \wedge P_5,\ P_4 \wedge P_3 \vee P_5 \wedge P_1\},$
$\{P_1 \wedge P_2 \vee P_5 \wedge P_3,\ P_2 \wedge P_4 \vee P_3 \wedge P_6,\ P_4 \wedge P_5 \vee P_6 \wedge P_1\},$
$\{P_1 \wedge P_2 \vee P_5 \wedge P_6,\ P_2 \wedge P_4 \vee P_6 \wedge P_3,\ P_4 \wedge P_5 \vee P_3 \wedge P_1\},$
$\{P_1 \wedge P_2 \vee P_6 \wedge P_3,\ P_2 \wedge P_4 \vee P_3 \wedge P_5,\ P_4 \wedge P_6 \vee P_5 \wedge P_1\},$
$\{P_1 \wedge P_2 \vee P_6 \wedge P_5,\ P_2 \wedge P_4 \vee P_5 \wedge P_3,\ P_4 \wedge P_6 \vee P_3 \wedge P_1\},$
$\{P_1 \wedge P_2 \vee P_3 \wedge P_4,\ P_2 \wedge P_5 \vee P_4 \wedge P_6,\ P_5 \wedge P_3 \vee P_6 \wedge P_1\},$
$\{P_1 \wedge P_2 \vee P_3 \wedge P_6,\ P_2 \wedge P_5 \vee P_6 \wedge P_4,\ P_5 \wedge P_3 \vee P_4 \wedge P_1\},$
$\{P_1 \wedge P_2 \vee P_4 \wedge P_3,\ P_2 \wedge P_5 \vee P_3 \wedge P_6,\ P_5 \wedge P_4 \vee P_6 \wedge P_1\},$
$\{P_1 \wedge P_2 \vee P_4 \wedge P_6,\ P_2 \wedge P_5 \vee P_6 \wedge P_3,\ P_5 \wedge P_4 \vee P_3 \wedge P_1\},$
$\{P_1 \wedge P_2 \vee P_6 \wedge P_3,\ P_2 \wedge P_5 \vee P_3 \wedge P_4,\ P_5 \wedge P_6 \vee P_4 \wedge P_1\},$
$\{P_1 \wedge P_2 \vee P_6 \wedge P_4,\ P_2 \wedge P_5 \vee P_4 \wedge P_3,\ P_5 \wedge P_6 \vee P_3 \wedge P_1\},$
$\{P_1 \wedge P_2 \vee P_3 \wedge P_4,\ P_2 \wedge P_6 \vee P_4 \wedge P_5,\ P_6 \wedge P_3 \vee P_5 \wedge P_1\},$
$\{P_1 \wedge P_2 \vee P_3 \wedge P_5,\ P_2 \wedge P_6 \vee P_5 \wedge P_4,\ P_6 \wedge P_3 \vee P_4 \wedge P_1\},$
$\{P_1 \wedge P_2 \vee P_4 \wedge P_3,\ P_2 \wedge P_6 \vee P_3 \wedge P_5,\ P_6 \wedge P_4 \vee P_5 \wedge P_1\},$
$\{P_1 \wedge P_2 \vee P_4 \wedge P_5,\ P_2 \wedge P_6 \vee P_5 \wedge P_3,\ P_6 \wedge P_4 \vee P_3 \wedge P_1\},$
$\{P_1 \wedge P_2 \vee P_5 \wedge P_3,\ P_2 \wedge P_6 \vee P_3 \wedge P_4,\ P_6 \wedge P_5 \vee P_4 \wedge P_1\},$
$\{P_1 \wedge P_2 \vee P_5 \wedge P_4,\ P_2 \wedge P_6 \vee P_4 \wedge P_3,\ P_6 \wedge P_5 \vee P_3 \wedge P_1\},$
$\{P_1 \wedge P_3 \vee P_4 \wedge P_5,\ P_3 \wedge P_2 \vee P_5 \wedge P_6,\ P_2 \wedge P_4 \vee P_6 \wedge P_1\},$
$\{P_1 \wedge P_3 \vee P_4 \wedge P_6,\ P_3 \wedge P_2 \vee P_6 \wedge P_5,\ P_2 \wedge P_4 \vee P_5 \wedge P_1\},$
$\{P_1 \wedge P_3 \vee P_5 \wedge P_4,\ P_3 \wedge P_2 \vee P_4 \wedge P_6,\ P_2 \wedge P_5 \vee P_6 \wedge P_1\},$
$\{P_1 \wedge P_3 \vee P_5 \wedge P_6,\ P_3 \wedge P_2 \vee P_6 \wedge P_4,\ P_2 \wedge P_5 \vee P_4 \wedge P_1\},$
$\{P_1 \wedge P_3 \vee P_6 \wedge P_4,\ P_3 \wedge P_2 \vee P_4 \wedge P_5,\ P_2 \wedge P_6 \vee P_5 \wedge P_1\},$
$\{P_1 \wedge P_3 \vee P_6 \wedge P_5,\ P_3 \wedge P_2 \vee P_5 \wedge P_4,\ P_2 \wedge P_6 \vee P_4 \wedge P_1\},$
$\{P_1 \wedge P_3 \vee P_2 \wedge P_5,\ P_3 \wedge P_4 \vee P_5 \wedge P_6,\ P_4 \wedge P_2 \vee P_6 \wedge P_1\},$
$\{P_1 \wedge P_3 \vee P_2 \wedge P_6,\ P_3 \wedge P_4 \vee P_6 \wedge P_5,\ P_4 \wedge P_2 \vee P_5 \wedge P_1\},$
$\{P_1 \wedge P_3 \vee P_5 \wedge P_2,\ P_3 \wedge P_4 \vee P_2 \wedge P_6,\ P_4 \wedge P_5 \vee P_6 \wedge P_1\},$
$\{P_1 \wedge P_3 \vee P_6 \wedge P_2,\ P_3 \wedge P_4 \vee P_2 \wedge P_5,\ P_4 \wedge P_6 \vee P_5 \wedge P_1\},$
$\{P_1 \wedge P_3 \vee P_2 \wedge P_4,\ P_3 \wedge P_5 \vee P_4 \wedge P_6,\ P_5 \wedge P_2 \vee P_6 \wedge P_1\},$
$\{P_1 \wedge P_3 \vee P_2 \wedge P_6,\ P_3 \wedge P_5 \vee P_6 \wedge P_4,\ P_5 \wedge P_2 \vee P_4 \wedge P_1\},$
$\{P_1 \wedge P_3 \vee P_4 \wedge P_2,\ P_3 \wedge P_5 \vee P_2 \wedge P_6,\ P_5 \wedge P_4 \vee P_6 \wedge P_1\},$

$$\begin{array}{l}
\{P_1\wedge P_3\vee P_6\wedge P_2,\ P_3\wedge P_5\vee P_2\wedge P_4,\ P_5\wedge P_6\vee P_4\wedge P_1\},\\
\{P_1\wedge P_3\vee P_2\wedge P_4,\ P_3\wedge P_6\vee P_4\wedge P_5,\ P_6\wedge P_2\vee P_5\wedge P_1\},\\
\{P_1\wedge P_3\vee P_2\wedge P_5,\ P_3\wedge P_6\vee P_5\wedge P_4,\ P_6\wedge P_2\vee P_4\wedge P_1\},\\
\{P_1\wedge P_3\vee P_4\wedge P_2,\ P_3\wedge P_6\vee P_2\wedge P_5,\ P_6\wedge P_4\vee P_5\wedge P_1\},\\
\{P_1\wedge P_3\vee P_5\wedge P_2,\ P_3\wedge P_6\vee P_2\wedge P_4,\ P_6\wedge P_5\vee P_4\wedge P_1\},\\
\{P_1\wedge P_4\vee P_3\wedge P_5,\ P_4\wedge P_2\vee P_5\wedge P_6,\ P_2\wedge P_3\vee P_6\wedge P_1\},\\
\{P_1\wedge P_4\vee P_3\wedge P_6,\ P_4\wedge P_2\vee P_6\wedge P_5,\ P_2\wedge P_3\vee P_5\wedge P_1\},\\
\{P_1\wedge P_4\vee P_5\wedge P_3,\ P_4\wedge P_2\vee P_3\wedge P_6,\ P_2\wedge P_5\vee P_6\wedge P_1\},\\
\{P_1\wedge P_4\vee P_6\wedge P_3,\ P_4\wedge P_2\vee P_3\wedge P_5,\ P_2\wedge P_6\vee P_5\wedge P_1\},\\
\{P_1\wedge P_4\vee P_2\wedge P_5,\ P_4\wedge P_3\vee P_5\wedge P_6,\ P_3\wedge P_2\vee P_6\wedge P_1\},\\
\{P_1\wedge P_4\vee P_2\wedge P_6,\ P_4\wedge P_3\vee P_6\wedge P_5,\ P_3\wedge P_2\vee P_5\wedge P_1\},\\
\{P_1\wedge P_4\vee P_5\wedge P_2,\ P_4\wedge P_3\vee P_2\wedge P_6,\ P_3\wedge P_5\vee P_6\wedge P_1\},\\
\{P_1\wedge P_4\vee P_6\wedge P_2,\ P_4\wedge P_3\vee P_2\wedge P_5,\ P_3\wedge P_6\vee P_5\wedge P_1\},\\
\{P_1\wedge P_4\vee P_2\wedge P_3,\ P_4\wedge P_5\vee P_3\wedge P_6,\ P_5\wedge P_2\vee P_6\wedge P_1\},\\
\{P_1\wedge P_4\vee P_3\wedge P_2,\ P_4\wedge P_5\vee P_2\wedge P_6,\ P_5\wedge P_3\vee P_6\wedge P_1\},\\
\{P_1\wedge P_4\vee P_2\wedge P_3,\ P_4\wedge P_6\vee P_3\wedge P_5,\ P_6\wedge P_2\vee P_5\wedge P_1\},\\
\{P_1\wedge P_4\vee P_3\wedge P_2,\ P_4\wedge P_6\vee P_2\wedge P_5,\ P_6\wedge P_3\vee P_5\wedge P_1\},\\
\{P_1\wedge P_5\vee P_3\wedge P_4,\ P_5\wedge P_2\vee P_4\wedge P_6,\ P_2\wedge P_3\vee P_6\wedge P_1\},\\
\{P_1\wedge P_5\vee P_4\wedge P_3,\ P_5\wedge P_2\vee P_3\wedge P_6,\ P_2\wedge P_4\vee P_6\wedge P_1\},\\
\{P_1\wedge P_5\vee P_2\wedge P_4,\ P_5\wedge P_3\vee P_4\wedge P_6,\ P_3\wedge P_2\vee P_6\wedge P_1\},\\
\{P_1\wedge P_5\vee P_4\wedge P_2,\ P_5\wedge P_3\vee P_2\wedge P_6,\ P_3\wedge P_4\vee P_6\wedge P_1\},\\
\{P_1\wedge P_5\vee P_2\wedge P_3,\ P_5\wedge P_4\vee P_3\wedge P_6,\ P_4\wedge P_2\vee P_6\wedge P_1\},\\
\{P_1\wedge P_5\vee P_3\wedge P_2,\ P_5\wedge P_4\vee P_2\wedge P_6,\ P_4\wedge P_3\vee P_6\wedge P_1\}\}
\end{array}$$

This list is comprised of 3×60 formulae. But some of them define the same point because we can interchange the factors in the regressive product, or interchange the factors in each exterior product, without affecting the result. To reduce these formulae to a list of unique formulae, we can apply `GrassmannSimplify` to put the products into canonical order, resulting in the duplicates only differing by a sign (which is unimportant).

```
★𝒫2; pascalPointFormulae =
 Union[★S[Flatten[allPascalPointFormulae]] /. -p_ :→ p]
```

$$\begin{array}{l}
\{P_1\wedge P_2\vee P_3\wedge P_4,\ P_1\wedge P_2\vee P_3\wedge P_5,\ P_1\wedge P_2\vee P_3\wedge P_6,\\
\ P_1\wedge P_2\vee P_4\wedge P_5,\ P_1\wedge P_2\vee P_4\wedge P_6,\ P_1\wedge P_2\vee P_5\wedge P_6,\\
\ P_1\wedge P_3\vee P_2\wedge P_4,\ P_1\wedge P_3\vee P_2\wedge P_5,\ P_1\wedge P_3\vee P_2\wedge P_6,\\
\ P_1\wedge P_3\vee P_4\wedge P_5,\ P_1\wedge P_3\vee P_4\wedge P_6,\ P_1\wedge P_3\vee P_5\wedge P_6,\ P_1\wedge P_4\vee P_2\wedge P_3,\\
\ P_1\wedge P_4\vee P_2\wedge P_5,\ P_1\wedge P_4\vee P_2\wedge P_6,\ P_1\wedge P_4\vee P_3\wedge P_5,\ P_1\wedge P_4\vee P_3\wedge P_6,\\
\ P_1\wedge P_4\vee P_5\wedge P_6,\ P_1\wedge P_5\vee P_2\wedge P_3,\ P_1\wedge P_5\vee P_2\wedge P_4,\ P_1\wedge P_5\vee P_2\wedge P_6,\\
\ P_1\wedge P_5\vee P_3\wedge P_4,\ P_1\wedge P_5\vee P_3\wedge P_6,\ P_1\wedge P_5\vee P_4\wedge P_6,\ P_1\wedge P_6\vee P_2\wedge P_3,\\
\ P_1\wedge P_6\vee P_2\wedge P_4,\ P_1\wedge P_6\vee P_2\wedge P_5,\ P_1\wedge P_6\vee P_3\wedge P_4,\ P_1\wedge P_6\vee P_3\wedge P_5,\\
\ P_1\wedge P_6\vee P_4\wedge P_5,\ P_2\wedge P_3\vee P_4\wedge P_5,\ P_2\wedge P_3\vee P_4\wedge P_6,\ P_2\wedge P_3\vee P_5\wedge P_6,\\
\ P_2\wedge P_4\vee P_3\wedge P_5,\ P_2\wedge P_4\vee P_3\wedge P_6,\ P_2\wedge P_4\vee P_5\wedge P_6,\ P_2\wedge P_5\vee P_3\wedge P_4,\\
\ P_2\wedge P_5\vee P_3\wedge P_6,\ P_2\wedge P_5\vee P_4\wedge P_6,\ P_2\wedge P_6\vee P_3\wedge P_4,\ P_2\wedge P_6\vee P_3\wedge P_5,\\
\ P_2\wedge P_6\vee P_4\wedge P_5,\ P_3\wedge P_4\vee P_5\wedge P_6,\ P_3\wedge P_5\vee P_4\wedge P_6,\ P_3\wedge P_6\vee P_4\wedge P_5\}
\end{array}$$

There are thus in general 45 *distinct* Pascal point formulae for the hexagons of a conic. Of course, since at this stage we have yet to discuss Pascal lines, these formulae are also valid for the 60 hexagons formed from *any* six points - not necessarily those constrained to lie on a conic.

It should be noted that these are distinct *formulae*. The number of *actual distinct points* resulting from their application to any six given points may be less. Some may be duplicates, for example, when the hexagon has some symmetry, and some may reduce to vectors when the hexagon sides are parallel.

For the ellipse and hexagon points discussed above, we can apply the formulae to get the actual points.

```
pascalPoints = ToPointForm[★C[pascalPointFormulae //. EllipsePoints]]
```

$$\left\{\star O + \left(4 - \frac{4}{\sqrt{3}}\right) e_1 - \sqrt{3}\, e_2,\ \star O + \left(8 - 4\sqrt{3}\right) e_1,\ \star O + 2\left(-1 + \sqrt{3}\right) e_1 - \frac{3}{2}\left(-1 + \sqrt{3}\right) e_2,\right.$$

$$\star O + 4\left(-1 + \sqrt{3}\right) e_1 - 3\sqrt{3}\, e_2,\ \star O + \frac{4 e_1}{\sqrt{3}} - 2\sqrt{3}\, e_2,\ \star O + 2\left(1 + \sqrt{3}\right) e_1 - \frac{9}{2}\left(1 + \sqrt{3}\right) e_2,$$

$$\star O + \left(4 + \frac{4}{\sqrt{3}}\right) e_1 - \sqrt{3}\, e_2,\ \star O + 4\left(2 + \sqrt{3}\right) e_1 - 3\left(1 + \sqrt{3}\right) e_2,\ \star O + 2\left(2 + \sqrt{3}\right) e_1 - \frac{3\sqrt{3}\, e_2}{2},$$

$$-96 e_1 + 72 e_2,\ \star O + 4\left(1 + \sqrt{3}\right) e_1 - 3\sqrt{3}\, e_2,\ \star O - 4\left(2 + \sqrt{3}\right) e_1 + 3\left(3 + \sqrt{3}\right) e_2,$$

$$\star O - 3\sqrt{3}\, e_2,\ \star O - \sqrt{3}\, e_2,\ \star O - \frac{3\sqrt{3}\, e_2}{2},\ \star O,\ \star O - \sqrt{3}\, e_2,\ \star O - 3\sqrt{3}\, e_2,$$

$$\star O + 4\left(2 + \sqrt{3}\right) e_1 + 3\left(3 + \sqrt{3}\right) e_2,\ \star O - 4\left(1 + \sqrt{3}\right) e_1 - 3\sqrt{3}\, e_2,$$

$$\star O - 2\left(2 + \sqrt{3}\right) e_1 - \frac{3\sqrt{3}\, e_2}{2},\ 96 e_1 + 72 e_2,\ \star O - 4\left(2 + \sqrt{3}\right) e_1 - 3\left(1 + \sqrt{3}\right) e_2,$$

$$\star O - \frac{4}{3}\left(3 + \sqrt{3}\right) e_1 - \sqrt{3}\, e_2,\ \star O - 2\left(1 + \sqrt{3}\right) e_1 - \frac{9}{2}\left(1 + \sqrt{3}\right) e_2,\ \star O - \frac{4 e_1}{\sqrt{3}} - 2\sqrt{3}\, e_2,$$

$$\star O + \left(2 - 2\sqrt{3}\right) e_1 - \frac{3}{2}\left(-1 + \sqrt{3}\right) e_2,\ \star O + \left(4 - 4\sqrt{3}\right) e_1 - 3\sqrt{3}\, e_2,$$

$$\star O + 4\left(-2 + \sqrt{3}\right) e_1,\ \star O + \left(-4 + \frac{4}{\sqrt{3}}\right) e_1 - \sqrt{3}\, e_2,\ \star O + \left(8 - 4\sqrt{3}\right) e_1 + 3\left(-3 + \sqrt{3}\right) e_2,$$

$$\star O + 2\left(-1 + \sqrt{3}\right) e_1 - \frac{9}{2}\left(-1 + \sqrt{3}\right) e_2,\ \star O - 3\sqrt{3}\, e_2,\ \star O + 4\left(2 + \sqrt{3}\right) e_1,$$

$$\star O - 2\left(1 + \sqrt{3}\right) e_1 - \frac{3}{2}\left(1 + \sqrt{3}\right) e_2,\ \star O + \left(2 - 2\sqrt{3}\right) e_1 - \frac{9}{2}\left(-1 + \sqrt{3}\right) e_2,$$

$$\star O + \left(8 - 4\sqrt{3}\right) e_1 + \left(3 - 3\sqrt{3}\right) e_2,\ \star O - \sqrt{3}\, e_2,\ \star O + 2\left(1 + \sqrt{3}\right) e_1 - \frac{3}{2}\left(1 + \sqrt{3}\right) e_2,$$

$$\star O + \left(4 - 2\sqrt{3}\right) e_1 - \frac{3\sqrt{3}\, e_2}{2},\ -48\sqrt{3}\, e_1,\ \star O + 2\left(-2 + \sqrt{3}\right) e_1 - \frac{3\sqrt{3}\, e_2}{2},$$

$$\left.\star O + 4\left(-2 + \sqrt{3}\right) e_1 + 3\left(-3 + \sqrt{3}\right) e_2,\ \star O - 4\left(2 + \sqrt{3}\right) e_1,\ \star O + 4\left(-2 + \sqrt{3}\right) e_1 + \left(3 - 3\sqrt{3}\right) e_2\right\}$$

Two of these points occur in triplicate $\star O - \sqrt{3}\, e_2$ and $\star O - 3\sqrt{3}\, e_2$; and three of the entities are vectors (points at infinity).

```
vectors = Select[pascalPoints, VectorFormQ]
```

$$\left\{-96 e_1 + 72 e_2,\ 96 e_1 + 72 e_2,\ -48\sqrt{3}\, e_1\right\}$$

Thus, when we plot these points we should be able to count 38 points other than those on the ellipse, and 3 vectors (shown as grey lines).

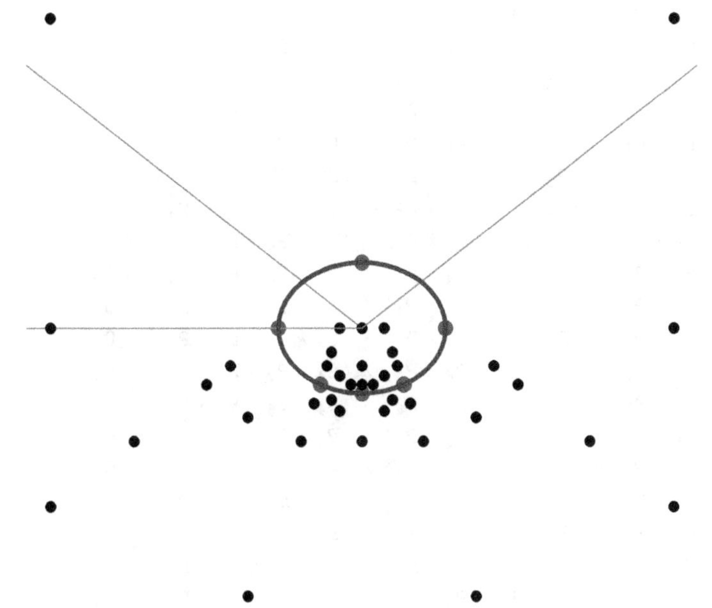

Six hexagon vertices, 38 distinct Pascal points and three vectors

Pascal lines

Since there are 60 essentially different hexagons one can form from 6 given points, there are thus 60 corresponding *Pascal lines*. We can display the equations for these lines.

```
PascalLineEquations =
  allPascalPointFormulae /. {Q1_, Q2_, Q3_} :→ Q1 ∧ Q2 ∧ Q3 == 0

{(P1 ∧ P2 ∨ P4 ∧ P5) ∧ (P2 ∧ P3 ∨ P5 ∧ P6) ∧ (P3 ∧ P4 ∨ P6 ∧ P1) == 0,
 (P1 ∧ P2 ∨ P4 ∧ P6) ∧ (P2 ∧ P3 ∨ P6 ∧ P5) ∧ (P3 ∧ P4 ∨ P5 ∧ P1) == 0,
 (P1 ∧ P2 ∨ P5 ∧ P4) ∧ (P2 ∧ P3 ∨ P4 ∧ P6) ∧ (P3 ∧ P5 ∨ P6 ∧ P1) == 0,
 (P1 ∧ P2 ∨ P5 ∧ P6) ∧ (P2 ∧ P3 ∨ P6 ∧ P4) ∧ (P3 ∧ P5 ∨ P4 ∧ P1) == 0,
 (P1 ∧ P2 ∨ P6 ∧ P4) ∧ (P2 ∧ P3 ∨ P4 ∧ P5) ∧ (P3 ∧ P6 ∨ P5 ∧ P1) == 0,
 (P1 ∧ P2 ∨ P6 ∧ P5) ∧ (P2 ∧ P3 ∨ P5 ∧ P4) ∧ (P3 ∧ P6 ∨ P4 ∧ P1) == 0,
 (P1 ∧ P2 ∨ P3 ∧ P5) ∧ (P2 ∧ P4 ∨ P5 ∧ P6) ∧ (P4 ∧ P3 ∨ P6 ∧ P1) == 0,
 (P1 ∧ P2 ∨ P3 ∧ P6) ∧ (P2 ∧ P4 ∨ P6 ∧ P5) ∧ (P4 ∧ P3 ∨ P5 ∧ P1) == 0,
 (P1 ∧ P2 ∨ P5 ∧ P3) ∧ (P2 ∧ P4 ∨ P3 ∧ P6) ∧ (P4 ∧ P5 ∨ P6 ∧ P1) == 0,
 (P1 ∧ P2 ∨ P5 ∧ P6) ∧ (P2 ∧ P4 ∨ P6 ∧ P3) ∧ (P4 ∧ P5 ∨ P3 ∧ P1) == 0,
 (P1 ∧ P2 ∨ P6 ∧ P3) ∧ (P2 ∧ P4 ∨ P3 ∧ P5) ∧ (P4 ∧ P6 ∨ P5 ∧ P1) == 0,
 (P1 ∧ P2 ∨ P6 ∧ P5) ∧ (P2 ∧ P4 ∨ P5 ∧ P3) ∧ (P4 ∧ P6 ∨ P3 ∧ P1) == 0,
 (P1 ∧ P2 ∨ P3 ∧ P4) ∧ (P2 ∧ P5 ∨ P4 ∧ P6) ∧ (P5 ∧ P3 ∨ P6 ∧ P1) == 0,
 (P1 ∧ P2 ∨ P3 ∧ P6) ∧ (P2 ∧ P5 ∨ P6 ∧ P4) ∧ (P5 ∧ P3 ∨ P4 ∧ P1) == 0,
 (P1 ∧ P2 ∨ P4 ∧ P3) ∧ (P2 ∧ P5 ∨ P3 ∧ P6) ∧ (P5 ∧ P4 ∨ P6 ∧ P1) == 0,
 (P1 ∧ P2 ∨ P4 ∧ P6) ∧ (P2 ∧ P5 ∨ P6 ∧ P3) ∧ (P5 ∧ P4 ∨ P3 ∧ P1) == 0,
 (P1 ∧ P2 ∨ P6 ∧ P3) ∧ (P2 ∧ P5 ∨ P3 ∧ P4) ∧ (P5 ∧ P6 ∨ P4 ∧ P1) == 0,
 (P1 ∧ P2 ∨ P6 ∧ P4) ∧ (P2 ∧ P5 ∨ P4 ∧ P3) ∧ (P5 ∧ P6 ∨ P3 ∧ P1) == 0,
```

$(P_1 \wedge P_2 \vee P_3 \wedge P_4) \wedge (P_2 \wedge P_6 \vee P_4 \wedge P_5) \wedge (P_6 \wedge P_3 \vee P_5 \wedge P_1) == 0,$
$(P_1 \wedge P_2 \vee P_3 \wedge P_5) \wedge (P_2 \wedge P_6 \vee P_5 \wedge P_4) \wedge (P_6 \wedge P_3 \vee P_4 \wedge P_1) == 0,$
$(P_1 \wedge P_2 \vee P_4 \wedge P_3) \wedge (P_2 \wedge P_6 \vee P_3 \wedge P_5) \wedge (P_6 \wedge P_4 \vee P_5 \wedge P_1) == 0,$
$(P_1 \wedge P_2 \vee P_4 \wedge P_5) \wedge (P_2 \wedge P_6 \vee P_5 \wedge P_3) \wedge (P_6 \wedge P_4 \vee P_3 \wedge P_1) == 0,$
$(P_1 \wedge P_2 \vee P_5 \wedge P_3) \wedge (P_2 \wedge P_6 \vee P_3 \wedge P_4) \wedge (P_6 \wedge P_5 \vee P_4 \wedge P_1) == 0,$
$(P_1 \wedge P_2 \vee P_5 \wedge P_4) \wedge (P_2 \wedge P_6 \vee P_4 \wedge P_3) \wedge (P_6 \wedge P_5 \vee P_3 \wedge P_1) == 0,$
$(P_1 \wedge P_3 \vee P_4 \wedge P_5) \wedge (P_3 \wedge P_2 \vee P_5 \wedge P_6) \wedge (P_2 \wedge P_4 \vee P_6 \wedge P_1) == 0,$
$(P_1 \wedge P_3 \vee P_4 \wedge P_6) \wedge (P_3 \wedge P_2 \vee P_6 \wedge P_5) \wedge (P_2 \wedge P_4 \vee P_5 \wedge P_1) == 0,$
$(P_1 \wedge P_3 \vee P_5 \wedge P_4) \wedge (P_3 \wedge P_2 \vee P_4 \wedge P_6) \wedge (P_2 \wedge P_5 \vee P_6 \wedge P_1) == 0,$
$(P_1 \wedge P_3 \vee P_5 \wedge P_6) \wedge (P_3 \wedge P_2 \vee P_6 \wedge P_4) \wedge (P_2 \wedge P_5 \vee P_4 \wedge P_1) == 0,$
$(P_1 \wedge P_3 \vee P_6 \wedge P_4) \wedge (P_3 \wedge P_2 \vee P_4 \wedge P_5) \wedge (P_2 \wedge P_6 \vee P_5 \wedge P_1) == 0,$
$(P_1 \wedge P_3 \vee P_6 \wedge P_5) \wedge (P_3 \wedge P_2 \vee P_5 \wedge P_4) \wedge (P_2 \wedge P_6 \vee P_4 \wedge P_1) == 0,$
$(P_1 \wedge P_3 \vee P_2 \wedge P_5) \wedge (P_3 \wedge P_4 \vee P_5 \wedge P_6) \wedge (P_4 \wedge P_2 \vee P_6 \wedge P_1) == 0,$
$(P_1 \wedge P_3 \vee P_2 \wedge P_6) \wedge (P_3 \wedge P_4 \vee P_6 \wedge P_5) \wedge (P_4 \wedge P_2 \vee P_5 \wedge P_1) == 0,$
$(P_1 \wedge P_3 \vee P_5 \wedge P_2) \wedge (P_3 \wedge P_4 \vee P_2 \wedge P_6) \wedge (P_4 \wedge P_5 \vee P_6 \wedge P_1) == 0,$
$(P_1 \wedge P_3 \vee P_6 \wedge P_2) \wedge (P_3 \wedge P_4 \vee P_2 \wedge P_5) \wedge (P_4 \wedge P_6 \vee P_5 \wedge P_1) == 0,$
$(P_1 \wedge P_3 \vee P_2 \wedge P_4) \wedge (P_3 \wedge P_5 \vee P_4 \wedge P_6) \wedge (P_5 \wedge P_2 \vee P_6 \wedge P_1) == 0,$
$(P_1 \wedge P_3 \vee P_2 \wedge P_6) \wedge (P_3 \wedge P_5 \vee P_6 \wedge P_4) \wedge (P_5 \wedge P_2 \vee P_4 \wedge P_1) == 0,$
$(P_1 \wedge P_3 \vee P_4 \wedge P_2) \wedge (P_3 \wedge P_5 \vee P_2 \wedge P_6) \wedge (P_5 \wedge P_4 \vee P_6 \wedge P_1) == 0,$
$(P_1 \wedge P_3 \vee P_6 \wedge P_2) \wedge (P_3 \wedge P_5 \vee P_2 \wedge P_4) \wedge (P_5 \wedge P_6 \vee P_4 \wedge P_1) == 0,$
$(P_1 \wedge P_3 \vee P_2 \wedge P_4) \wedge (P_3 \wedge P_6 \vee P_4 \wedge P_5) \wedge (P_6 \wedge P_2 \vee P_5 \wedge P_1) == 0,$
$(P_1 \wedge P_3 \vee P_2 \wedge P_5) \wedge (P_3 \wedge P_6 \vee P_5 \wedge P_4) \wedge (P_6 \wedge P_2 \vee P_4 \wedge P_1) == 0,$
$(P_1 \wedge P_3 \vee P_4 \wedge P_2) \wedge (P_3 \wedge P_6 \vee P_2 \wedge P_5) \wedge (P_6 \wedge P_4 \vee P_5 \wedge P_1) == 0,$
$(P_1 \wedge P_3 \vee P_5 \wedge P_2) \wedge (P_3 \wedge P_6 \vee P_2 \wedge P_4) \wedge (P_6 \wedge P_5 \vee P_4 \wedge P_1) == 0,$
$(P_1 \wedge P_4 \vee P_3 \wedge P_5) \wedge (P_4 \wedge P_2 \vee P_5 \wedge P_6) \wedge (P_2 \wedge P_3 \vee P_6 \wedge P_1) == 0,$
$(P_1 \wedge P_4 \vee P_3 \wedge P_6) \wedge (P_4 \wedge P_2 \vee P_6 \wedge P_5) \wedge (P_2 \wedge P_3 \vee P_5 \wedge P_1) == 0,$
$(P_1 \wedge P_4 \vee P_5 \wedge P_3) \wedge (P_4 \wedge P_2 \vee P_3 \wedge P_6) \wedge (P_2 \wedge P_5 \vee P_6 \wedge P_1) == 0,$
$(P_1 \wedge P_4 \vee P_6 \wedge P_3) \wedge (P_4 \wedge P_2 \vee P_3 \wedge P_5) \wedge (P_2 \wedge P_6 \vee P_5 \wedge P_1) == 0,$
$(P_1 \wedge P_4 \vee P_2 \wedge P_5) \wedge (P_4 \wedge P_3 \vee P_5 \wedge P_6) \wedge (P_3 \wedge P_2 \vee P_6 \wedge P_1) == 0,$
$(P_1 \wedge P_4 \vee P_2 \wedge P_6) \wedge (P_4 \wedge P_3 \vee P_6 \wedge P_5) \wedge (P_3 \wedge P_2 \vee P_5 \wedge P_1) == 0,$
$(P_1 \wedge P_4 \vee P_5 \wedge P_2) \wedge (P_4 \wedge P_3 \vee P_2 \wedge P_6) \wedge (P_3 \wedge P_5 \vee P_6 \wedge P_1) == 0,$
$(P_1 \wedge P_4 \vee P_6 \wedge P_2) \wedge (P_4 \wedge P_3 \vee P_2 \wedge P_5) \wedge (P_3 \wedge P_6 \vee P_5 \wedge P_1) == 0,$
$(P_1 \wedge P_4 \vee P_2 \wedge P_3) \wedge (P_4 \wedge P_5 \vee P_3 \wedge P_6) \wedge (P_5 \wedge P_2 \vee P_6 \wedge P_1) == 0,$
$(P_1 \wedge P_4 \vee P_3 \wedge P_2) \wedge (P_4 \wedge P_5 \vee P_2 \wedge P_6) \wedge (P_5 \wedge P_3 \vee P_6 \wedge P_1) == 0,$
$(P_1 \wedge P_4 \vee P_2 \wedge P_3) \wedge (P_4 \wedge P_6 \vee P_3 \wedge P_5) \wedge (P_6 \wedge P_2 \vee P_5 \wedge P_1) == 0,$
$(P_1 \wedge P_4 \vee P_3 \wedge P_2) \wedge (P_4 \wedge P_6 \vee P_2 \wedge P_5) \wedge (P_6 \wedge P_3 \vee P_5 \wedge P_1) == 0,$
$(P_1 \wedge P_5 \vee P_3 \wedge P_4) \wedge (P_5 \wedge P_2 \vee P_4 \wedge P_6) \wedge (P_2 \wedge P_3 \vee P_6 \wedge P_1) == 0,$
$(P_1 \wedge P_5 \vee P_4 \wedge P_3) \wedge (P_5 \wedge P_2 \vee P_3 \wedge P_6) \wedge (P_2 \wedge P_4 \vee P_6 \wedge P_1) == 0,$
$(P_1 \wedge P_5 \vee P_2 \wedge P_4) \wedge (P_5 \wedge P_3 \vee P_4 \wedge P_6) \wedge (P_3 \wedge P_2 \vee P_6 \wedge P_1) == 0,$
$(P_1 \wedge P_5 \vee P_4 \wedge P_2) \wedge (P_5 \wedge P_3 \vee P_2 \wedge P_6) \wedge (P_3 \wedge P_4 \vee P_6 \wedge P_1) == 0,$
$(P_1 \wedge P_5 \vee P_2 \wedge P_3) \wedge (P_5 \wedge P_4 \vee P_3 \wedge P_6) \wedge (P_4 \wedge P_2 \vee P_6 \wedge P_1) == 0,$
$(P_1 \wedge P_5 \vee P_3 \wedge P_2) \wedge (P_5 \wedge P_4 \vee P_2 \wedge P_6) \wedge (P_4 \wedge P_3 \vee P_6 \wedge P_1) == 0\}$

We can verify these equations for the points we have chosen on the ellipse.

```
Union[Simplify[★C[PascalLineEquations /. EllipsePoints]]]
```

```
{True}
```

To generate formulae for the Pascal lines themselves we only have to choose any two of its Pascal points. Here we choose the first two.

```
PascalLineFormulae = allPascalPointFormulae /. {Q1_, Q2_, Q3_} :→ Q1 ∧ Q2
```

$$\{(P_1\wedge P_2\vee P_4\wedge P_5)\wedge(P_2\wedge P_3\vee P_5\wedge P_6),\ (P_1\wedge P_2\vee P_4\wedge P_6)\wedge(P_2\wedge P_3\vee P_6\wedge P_5),$$
$$(P_1\wedge P_2\vee P_5\wedge P_4)\wedge(P_2\wedge P_3\vee P_4\wedge P_6),\ (P_1\wedge P_2\vee P_5\wedge P_6)\wedge(P_2\wedge P_3\vee P_6\wedge P_4),$$
$$(P_1\wedge P_2\vee P_6\wedge P_4)\wedge(P_2\wedge P_3\vee P_4\wedge P_5),\ (P_1\wedge P_2\vee P_6\wedge P_5)\wedge(P_2\wedge P_3\vee P_5\wedge P_4),$$
$$(P_1\wedge P_2\vee P_3\wedge P_5)\wedge(P_2\wedge P_4\vee P_5\wedge P_6),\ (P_1\wedge P_2\vee P_3\wedge P_6)\wedge(P_2\wedge P_4\vee P_6\wedge P_5),$$
$$(P_1\wedge P_2\vee P_5\wedge P_3)\wedge(P_2\wedge P_4\vee P_3\wedge P_6),\ (P_1\wedge P_2\vee P_5\wedge P_6)\wedge(P_2\wedge P_4\vee P_6\wedge P_3),$$
$$(P_1\wedge P_2\vee P_6\wedge P_3)\wedge(P_2\wedge P_4\vee P_3\wedge P_5),\ (P_1\wedge P_2\vee P_6\wedge P_5)\wedge(P_2\wedge P_4\vee P_5\wedge P_3),$$
$$(P_1\wedge P_2\vee P_3\wedge P_4)\wedge(P_2\wedge P_5\vee P_4\wedge P_6),\ (P_1\wedge P_2\vee P_3\wedge P_6)\wedge(P_2\wedge P_5\vee P_6\wedge P_4),$$
$$(P_1\wedge P_2\vee P_4\wedge P_3)\wedge(P_2\wedge P_5\vee P_3\wedge P_6),\ (P_1\wedge P_2\vee P_4\wedge P_6)\wedge(P_2\wedge P_5\vee P_6\wedge P_3),$$
$$(P_1\wedge P_2\vee P_6\wedge P_3)\wedge(P_2\wedge P_5\vee P_3\wedge P_4),\ (P_1\wedge P_2\vee P_6\wedge P_4)\wedge(P_2\wedge P_5\vee P_4\wedge P_3),$$
$$(P_1\wedge P_2\vee P_3\wedge P_4)\wedge(P_2\wedge P_6\vee P_4\wedge P_5),\ (P_1\wedge P_2\vee P_3\wedge P_5)\wedge(P_2\wedge P_6\vee P_5\wedge P_4),$$
$$(P_1\wedge P_2\vee P_4\wedge P_3)\wedge(P_2\wedge P_6\vee P_3\wedge P_5),\ (P_1\wedge P_2\vee P_4\wedge P_5)\wedge(P_2\wedge P_6\vee P_5\wedge P_3),$$
$$(P_1\wedge P_2\vee P_5\wedge P_3)\wedge(P_2\wedge P_6\vee P_3\wedge P_4),\ (P_1\wedge P_2\vee P_5\wedge P_4)\wedge(P_2\wedge P_6\vee P_4\wedge P_3),$$
$$(P_1\wedge P_3\vee P_4\wedge P_5)\wedge(P_3\wedge P_2\vee P_5\wedge P_6),\ (P_1\wedge P_3\vee P_4\wedge P_6)\wedge(P_3\wedge P_2\vee P_6\wedge P_5),$$
$$(P_1\wedge P_3\vee P_5\wedge P_4)\wedge(P_3\wedge P_2\vee P_4\wedge P_6),\ (P_1\wedge P_3\vee P_5\wedge P_6)\wedge(P_3\wedge P_2\vee P_6\wedge P_4),$$
$$(P_1\wedge P_3\vee P_6\wedge P_4)\wedge(P_3\wedge P_2\vee P_4\wedge P_5),\ (P_1\wedge P_3\vee P_6\wedge P_5)\wedge(P_3\wedge P_2\vee P_5\wedge P_4),$$
$$(P_1\wedge P_3\vee P_2\wedge P_5)\wedge(P_3\wedge P_4\vee P_5\wedge P_6),\ (P_1\wedge P_3\vee P_2\wedge P_6)\wedge(P_3\wedge P_4\vee P_6\wedge P_5),$$
$$(P_1\wedge P_3\vee P_5\wedge P_2)\wedge(P_3\wedge P_4\vee P_2\wedge P_6),\ (P_1\wedge P_3\vee P_6\wedge P_2)\wedge(P_3\wedge P_4\vee P_2\wedge P_5),$$
$$(P_1\wedge P_3\vee P_2\wedge P_4)\wedge(P_3\wedge P_5\vee P_4\wedge P_6),\ (P_1\wedge P_3\vee P_2\wedge P_6)\wedge(P_3\wedge P_5\vee P_6\wedge P_4),$$
$$(P_1\wedge P_3\vee P_4\wedge P_2)\wedge(P_3\wedge P_5\vee P_2\wedge P_6),\ (P_1\wedge P_3\vee P_6\wedge P_2)\wedge(P_3\wedge P_5\vee P_2\wedge P_4),$$
$$(P_1\wedge P_3\vee P_2\wedge P_4)\wedge(P_3\wedge P_6\vee P_4\wedge P_5),\ (P_1\wedge P_3\vee P_2\wedge P_5)\wedge(P_3\wedge P_6\vee P_5\wedge P_4),$$
$$(P_1\wedge P_3\vee P_4\wedge P_2)\wedge(P_3\wedge P_6\vee P_2\wedge P_5),\ (P_1\wedge P_3\vee P_5\wedge P_2)\wedge(P_3\wedge P_6\vee P_2\wedge P_4),$$
$$(P_1\wedge P_4\vee P_3\wedge P_5)\wedge(P_4\wedge P_2\vee P_5\wedge P_6),\ (P_1\wedge P_4\vee P_3\wedge P_6)\wedge(P_4\wedge P_2\vee P_6\wedge P_5),$$
$$(P_1\wedge P_4\vee P_5\wedge P_3)\wedge(P_4\wedge P_2\vee P_3\wedge P_6),\ (P_1\wedge P_4\vee P_6\wedge P_3)\wedge(P_4\wedge P_2\vee P_3\wedge P_5),$$
$$(P_1\wedge P_4\vee P_2\wedge P_5)\wedge(P_4\wedge P_3\vee P_5\wedge P_6),\ (P_1\wedge P_4\vee P_2\wedge P_6)\wedge(P_4\wedge P_3\vee P_6\wedge P_5),$$
$$(P_1\wedge P_4\vee P_5\wedge P_2)\wedge(P_4\wedge P_3\vee P_2\wedge P_6),\ (P_1\wedge P_4\vee P_6\wedge P_2)\wedge(P_4\wedge P_3\vee P_2\wedge P_5),$$
$$(P_1\wedge P_4\vee P_2\wedge P_3)\wedge(P_4\wedge P_5\vee P_3\wedge P_6),\ (P_1\wedge P_4\vee P_3\wedge P_2)\wedge(P_4\wedge P_5\vee P_2\wedge P_6),$$
$$(P_1\wedge P_4\vee P_2\wedge P_3)\wedge(P_4\wedge P_6\vee P_3\wedge P_5),\ (P_1\wedge P_4\vee P_3\wedge P_2)\wedge(P_4\wedge P_6\vee P_2\wedge P_5),$$
$$(P_1\wedge P_5\vee P_3\wedge P_4)\wedge(P_5\wedge P_2\vee P_4\wedge P_6),\ (P_1\wedge P_5\vee P_4\wedge P_3)\wedge(P_5\wedge P_2\vee P_3\wedge P_6),$$
$$(P_1\wedge P_5\vee P_2\wedge P_4)\wedge(P_5\wedge P_3\vee P_4\wedge P_6),\ (P_1\wedge P_5\vee P_4\wedge P_2)\wedge(P_5\wedge P_3\vee P_2\wedge P_6),$$
$$(P_1\wedge P_5\vee P_2\wedge P_3)\wedge(P_5\wedge P_4\vee P_3\wedge P_6),\ (P_1\wedge P_5\vee P_3\wedge P_2)\wedge(P_5\wedge P_4\vee P_2\wedge P_6)\}$$

We can explicitly compute the Pascal lines (bound vectors) for the ellipse by substituting in our chosen points.

```
PascalLines = allPascalPointFormulae /.
  {Q1_, Q2_, Q3_} :→
   Simplify[ToPointForm[★C[Q1 /. EllipsePoints]] ∧
     ToPointForm[★C[Q2 /. EllipsePoints]]]
```

$$\Big\{\left(\star\mathbb{O}+4\left(-1+\sqrt{3}\right)e_1-3\sqrt{3}\,e_2\right)\wedge\left(\star\mathbb{O}-3\sqrt{3}\,e_2\right),\ \left(\star\mathbb{O}+\frac{4\,e_1}{\sqrt{3}}-2\sqrt{3}\,e_2\right)\wedge\left(\star\mathbb{O}-3\sqrt{3}\,e_2\right),$$

$$\left(\star\mathbb{O}+4\left(-1+\sqrt{3}\right)e_1-3\sqrt{3}\,e_2\right)\wedge\left(\star\mathbb{O}+2\left(-1+\sqrt{3}\right)e_1-\frac{9}{2}\left(-1+\sqrt{3}\right)e_2\right),$$

$$\left(\star\mathbb{O}+2\left(1+\sqrt{3}\right)e_1-\frac{9}{2}\left(1+\sqrt{3}\right)e_2\right)\wedge\left(\star\mathbb{O}+2\left(-1+\sqrt{3}\right)e_1-\frac{9}{2}\left(-1+\sqrt{3}\right)e_2\right),$$

$$\left(\star\mathbb{O}+\frac{4\,e_1}{\sqrt{3}}-2\sqrt{3}\,e_2\right)\wedge\left(\star\mathbb{O}-4\left(-2+\sqrt{3}\right)e_1+3\left(-3+\sqrt{3}\right)e_2\right),$$

$$\left(\star\mathbb{O}+2\left(1+\sqrt{3}\right)e_1-\frac{9}{2}\left(1+\sqrt{3}\right)e_2\right)\wedge\left(\star\mathbb{O}-4\left(-2+\sqrt{3}\right)e_1+3\left(-3+\sqrt{3}\right)e_2\right),$$

$$\left(\star O - 4\left(-2+\sqrt{3}\right) e_1\right) \wedge \left(\star O - 2\left(-1+\sqrt{3}\right) e_1 - \frac{9}{2}\left(-1+\sqrt{3}\right) e_2\right),$$

$$\left(\star O + 2\left(-1+\sqrt{3}\right) e_1 - \frac{3}{2}\left(-1+\sqrt{3}\right) e_2\right) \wedge \left(\star O - 2\left(-1+\sqrt{3}\right) e_1 - \frac{9}{2}\left(-1+\sqrt{3}\right) e_2\right),$$

$$\left(\star O - 4\left(-2+\sqrt{3}\right) e_1\right) \wedge \left(\star O - 2\left(1+\sqrt{3}\right) e_1 - \frac{3}{2}\left(1+\sqrt{3}\right) e_2\right),$$

$$\left(\star O + 2\left(1+\sqrt{3}\right) e_1 - \frac{9}{2}\left(1+\sqrt{3}\right) e_2\right) \wedge \left(\star O - 2\left(1+\sqrt{3}\right) e_1 - \frac{3}{2}\left(1+\sqrt{3}\right) e_2\right),$$

$$\left(\star O + 2\left(-1+\sqrt{3}\right) e_1 - \frac{3}{2}\left(-1+\sqrt{3}\right) e_2\right) \wedge \left(\star O + 4\left(2+\sqrt{3}\right) e_1\right),$$

$$\left(\star O + 2\left(1+\sqrt{3}\right) e_1 - \frac{9}{2}\left(1+\sqrt{3}\right) e_2\right) \wedge \left(\star O + 4\left(2+\sqrt{3}\right) e_1\right),$$

$$\left(\star O + \left(4 - \frac{4}{\sqrt{3}}\right) e_1 - \sqrt{3}\, e_2\right) \wedge \left(\star O + 2\left(1+\sqrt{3}\right) e_1 - \frac{3}{2}\left(1+\sqrt{3}\right) e_2\right),$$

$$\left(\star O + 2\left(-1+\sqrt{3}\right) e_1 - \frac{3}{2}\left(-1+\sqrt{3}\right) e_2\right) \wedge \left(\star O + 2\left(1+\sqrt{3}\right) e_1 - \frac{3}{2}\left(1+\sqrt{3}\right) e_2\right),$$

$$\left(\star O + \left(4 - \frac{4}{\sqrt{3}}\right) e_1 - \sqrt{3}\, e_2\right) \wedge \left(\star O - \sqrt{3}\, e_2\right), \; \left(\star O + \frac{4 e_1}{\sqrt{3}} - 2\sqrt{3}\, e_2\right) \wedge \left(\star O - \sqrt{3}\, e_2\right),$$

$$\left(\star O + 2\left(-1+\sqrt{3}\right) e_1 - \frac{3}{2}\left(-1+\sqrt{3}\right) e_2\right) \wedge \left(\star O - 4\left(-2+\sqrt{3}\right) e_1 - 3\left(-1+\sqrt{3}\right) e_2\right),$$

$$\left(\star O + \frac{4 e_1}{\sqrt{3}} - 2\sqrt{3}\, e_2\right) \wedge \left(\star O - 4\left(-2+\sqrt{3}\right) e_1 - 3\left(-1+\sqrt{3}\right) e_2\right),$$

$$\left(\star O + \left(4 - \frac{4}{\sqrt{3}}\right) e_1 - \sqrt{3}\, e_2\right) \wedge \left(\star O + 2\left(-2+\sqrt{3}\right) e_1 - \frac{3\sqrt{3}\, e_2}{2}\right),$$

$$\left(\star O - 4\left(-2+\sqrt{3}\right) e_1\right) \wedge \left(\star O + 2\left(-2+\sqrt{3}\right) e_1 - \frac{3\sqrt{3}\, e_2}{2}\right),$$

$$\left(\star O + \left(4 - \frac{4}{\sqrt{3}}\right) e_1 - \sqrt{3}\, e_2\right) \wedge \left(-48\sqrt{3}\, e_1\right), \; \left(\star O + 4\left(-1+\sqrt{3}\right) e_1 - 3\sqrt{3}\, e_2\right) \wedge \left(48\sqrt{3}\, e_1\right),$$

$$\left(\star O - 4\left(-2+\sqrt{3}\right) e_1\right) \wedge \left(\star O - 2\left(-2+\sqrt{3}\right) e_1 - \frac{3\sqrt{3}\, e_2}{2}\right),$$

$$\left(\star O + 4\left(-1+\sqrt{3}\right) e_1 - 3\sqrt{3}\, e_2\right) \wedge \left(\star O - 2\left(-2+\sqrt{3}\right) e_1 - \frac{3\sqrt{3}\, e_2}{2}\right),$$

$$\left(-96\, e_1 + 72\, e_2\right) \wedge \left(\star O - 3\sqrt{3}\, e_2\right), \; \left(\star O + 4\left(1+\sqrt{3}\right) e_1 - 3\sqrt{3}\, e_2\right) \wedge \left(\star O - 3\sqrt{3}\, e_2\right),$$

$$\left(96\, e_1 - 72\, e_2\right) \wedge \left(\star O + 2\left(-1+\sqrt{3}\right) e_1 - \frac{9}{2}\left(-1+\sqrt{3}\right) e_2\right),$$

$$\left(\star O - 4\left(2+\sqrt{3}\right) e_1 + 3\left(3+\sqrt{3}\right) e_2\right) \wedge \left(\star O + 2\left(-1+\sqrt{3}\right) e_1 - \frac{9}{2}\left(-1+\sqrt{3}\right) e_2\right),$$

$$\left(\star O + 4\left(1+\sqrt{3}\right) e_1 - 3\sqrt{3}\, e_2\right) \wedge \left(\star O - 4\left(-2+\sqrt{3}\right) e_1 + 3\left(-3+\sqrt{3}\right) e_2\right),$$

$$\left(\star O - 4\left(2+\sqrt{3}\right) e_1 + 3\left(3+\sqrt{3}\right) e_2\right) \wedge \left(\star O - 4\left(-2+\sqrt{3}\right) e_1 + 3\left(-3+\sqrt{3}\right) e_2\right),$$

$$\left(\star O + 4\left(2+\sqrt{3}\right) e_1 - 3\left(1+\sqrt{3}\right) e_2\right) \wedge \left(\star O + 4\left(-2+\sqrt{3}\right) e_1 + 3\left(-3+\sqrt{3}\right) e_2\right),$$

$$\left(\star 0 + 2\left(2+\sqrt{3}\right)e_1 - \frac{3\sqrt{3}\,e_2}{2}\right)\wedge\left(\star 0 + 4\left(-2+\sqrt{3}\right)e_1 + 3\left(-3+\sqrt{3}\right)e_2\right),$$

$$\left(\star 0 + 4\left(2+\sqrt{3}\right)e_1 - 3\left(1+\sqrt{3}\right)e_2\right)\wedge\left(\star 0 - 2\left(-2+\sqrt{3}\right)e_1 - \frac{3\sqrt{3}\,e_2}{2}\right),$$

$$\left(\star 0 + 2\left(2+\sqrt{3}\right)e_1 - \frac{3\sqrt{3}\,e_2}{2}\right)\wedge\left(\star 0 - 4\left(-2+\sqrt{3}\right)e_1 - 3\left(-1+\sqrt{3}\right)e_2\right),$$

$$\left(\star 0 + \frac{4}{3}\left(3+\sqrt{3}\right)e_1 - \sqrt{3}\,e_2\right)\wedge\left(\star 0 - 4\left(2+\sqrt{3}\right)e_1\right),$$

$$\left(\star 0 + 2\left(2+\sqrt{3}\right)e_1 - \frac{3\sqrt{3}\,e_2}{2}\right)\wedge\left(\star 0 - 4\left(2+\sqrt{3}\right)e_1\right),$$

$$\left(\star 0 + \frac{4}{3}\left(3+\sqrt{3}\right)e_1 - \sqrt{3}\,e_2\right)\wedge\left(48\sqrt{3}\,e_1\right),$$

$$\left(\star 0 + 2\left(2+\sqrt{3}\right)e_1 - \frac{3\sqrt{3}\,e_2}{2}\right)\wedge\left(\star 0 + 4\left(2+\sqrt{3}\right)e_1\right),$$

$$\left(\star 0 + \frac{4}{3}\left(3+\sqrt{3}\right)e_1 - \sqrt{3}\,e_2\right)\wedge\left(\star 0 + 4\left(-2+\sqrt{3}\right)e_1 - 3\left(-1+\sqrt{3}\right)e_2\right),$$

$$\left(\star 0 + 4\left(2+\sqrt{3}\right)e_1 - 3\left(1+\sqrt{3}\right)e_2\right)\wedge\left(\star 0 + 4\left(-2+\sqrt{3}\right)e_1 - 3\left(-1+\sqrt{3}\right)e_2\right),$$

$$\left(\star 0 + \frac{4}{3}\left(3+\sqrt{3}\right)e_1 - \sqrt{3}\,e_2\right)\wedge\left(\star 0 - \sqrt{3}\,e_2\right),$$

$$\left(\star 0 + 4\left(2+\sqrt{3}\right)e_1 - 3\left(1+\sqrt{3}\right)e_2\right)\wedge\left(\star 0 - 2\left(1+\sqrt{3}\right)e_1 - \frac{3}{2}\left(1+\sqrt{3}\right)e_2\right),$$

$$\star 0\wedge\left(\star 0 - 2\left(-1+\sqrt{3}\right)e_1 - \frac{9}{2}\left(-1+\sqrt{3}\right)e_2\right),$$

$$\left(\star 0 - \sqrt{3}\,e_2\right)\wedge\left(\star 0 - 2\left(-1+\sqrt{3}\right)e_1 - \frac{9}{2}\left(-1+\sqrt{3}\right)e_2\right),$$

$$\star 0\wedge\left(\star 0 - 2\left(1+\sqrt{3}\right)e_1 - \frac{3}{2}\left(1+\sqrt{3}\right)e_2\right),\ \left(\star 0 - \sqrt{3}\,e_2\right)\wedge\left(\star 0 + 4\left(2+\sqrt{3}\right)e_1\right),$$

$$\left(\star 0 - \sqrt{3}\,e_2\right)\wedge\left(\star 0 + 4\left(-2+\sqrt{3}\right)e_1 + 3\left(-3+\sqrt{3}\right)e_2\right),$$

$$\left(\star 0 - \frac{3\sqrt{3}\,e_2}{2}\right)\wedge\left(\star 0 + 4\left(-2+\sqrt{3}\right)e_1 + 3\left(-3+\sqrt{3}\right)e_2\right),$$

$$\left(\star 0 - \sqrt{3}\,e_2\right)\wedge\left(\star 0 - 2\left(-2+\sqrt{3}\right)e_1 - \frac{3\sqrt{3}\,e_2}{2}\right),$$

$$\left(\star 0 - \frac{3\sqrt{3}\,e_2}{2}\right)\wedge\left(\star 0 - 4\left(-2+\sqrt{3}\right)e_1 - 3\left(-1+\sqrt{3}\right)e_2\right),$$

$$\left(\star 0 - 3\sqrt{3}\,e_2\right)\wedge\left(\star 0 + 4\left(-2+\sqrt{3}\right)e_1 - 3\left(-1+\sqrt{3}\right)e_2\right),$$

$$\left(\star 0 - 3\sqrt{3}\,e_2\right)\wedge\left(\star 0 + 2\left(-2+\sqrt{3}\right)e_1 - \frac{3\sqrt{3}\,e_2}{2}\right),\ \left(\star 0 - 3\sqrt{3}\,e_2\right)\wedge\left(\star 0 - 4\left(2+\sqrt{3}\right)e_1\right),$$

$$\left(\star 0 - 3\sqrt{3}\,e_2\right)\wedge\left(\star 0 + 2\left(1+\sqrt{3}\right)e_1 - \frac{3}{2}\left(1+\sqrt{3}\right)e_2\right),$$

$$\left(96\,e_1 + 72\,e_2\right)\wedge\left(\star 0 + 2\left(1+\sqrt{3}\right)e_1 - \frac{3}{2}\left(1+\sqrt{3}\right)e_2\right),$$

$(-96\, e_1 - 72\, e_2) \wedge \left(\star O - \sqrt{3}\, e_2\right),\ \left(\star O - 4\left(1+\sqrt{3}\right) e_1 - 3\sqrt{3}\, e_2\right) \wedge \left(\star O - 4\left(2+\sqrt{3}\right) e_1\right),$
$\left(\star O - 4\left(1+\sqrt{3}\right) e_1 - 3\sqrt{3}\, e_2\right) \wedge \left(-48\sqrt{3}\, e_1\right),$
$\left(\star O + 4\left(2+\sqrt{3}\right) e_1 + 3\left(3+\sqrt{3}\right) e_2\right) \wedge \left(\star O + 4\left(-2+\sqrt{3}\right) e_1 - 3\left(-1+\sqrt{3}\right) e_2\right),$
$\left(\star O + 4\left(2+\sqrt{3}\right) e_1 + 3\left(3+\sqrt{3}\right) e_2\right) \wedge \left(\star O + 2\left(-2+\sqrt{3}\right) e_1 - \frac{3\sqrt{3}\, e_2}{2}\right)\}$

We can plot these 60 Pascal lines on their Pascal points. Lines normally have an infinite extent, but to make it clearer which lines correspond to which points, we have plotted them spanning just their three points, where of course one of the points may be at infinity. Thus if the three Pascal points (or vectors) of a Pascal line were P_a, P_b, and P_c we have plotted the line segments (P_a, P_b), (P_b, P_c), and (P_c, P_a). If this is being read as a *Mathematica* notebook, actual line expressions are discernible as Tooltips. However care should be taken when attempting to correlate the expressions shown with lines, as some lines may overlap.

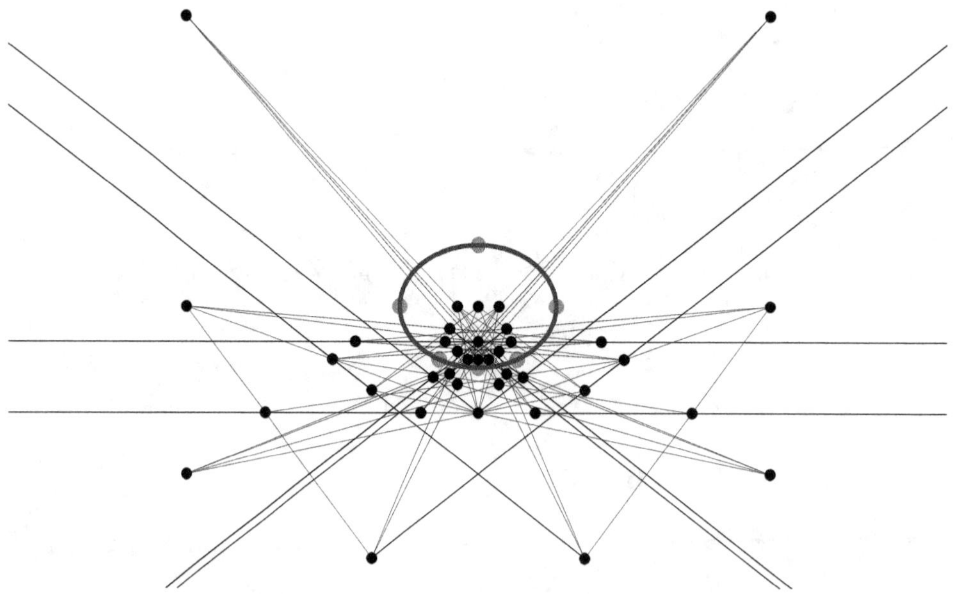

Sixty Pascal lines each joining its three points or points at infinity

Finally, we plot the sixty Pascal lines and bivectors of a regular hexagon. In this case we have shown the arrowheads on the line-bound vectors and on the bivectors. The lack of apparent symmetry is an artifact caused by the vectors and bivectors showing more information (their signs, weights and orientation) than the geometric lines and 2-directions that they are being used to represent.

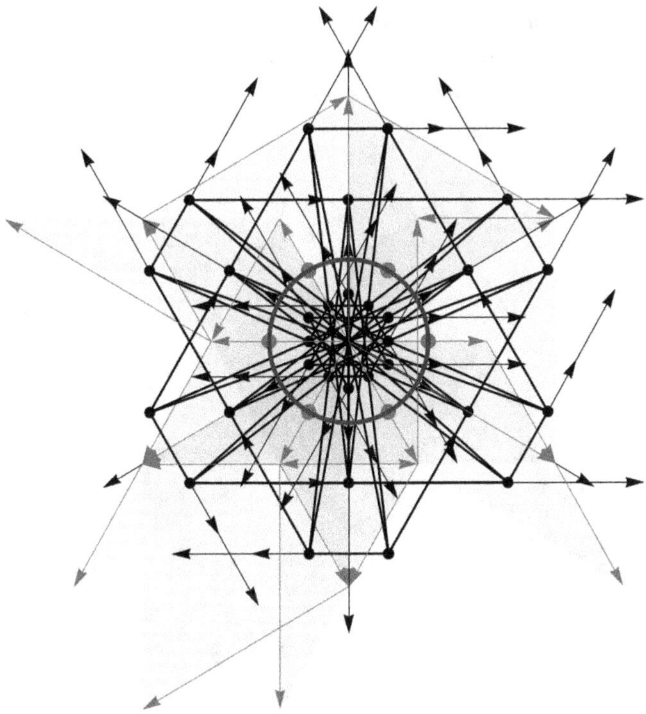

The sixty Pascal lines and bivectors of a regular hexagon

4.15 Summary

In this chapter we have explored the implications of interpreting one of the elements of the underlying linear space as an (origin) *point* (and the rest as vectors). This is a slightly different approach to that adopted by Grassmann and workers in the earlier Grassmannian tradition, of interpreting all the basis elements of the underlying linear space as *points*, and then treating any consequent point differences as vectors.

Although the modern context is to interpret elements of a linear space most commonly as vectors, workers in the early nineteenth century treated the point as the paramount entity of a linear space. Indeed it was Hamilton who coined the term *vector* in 1853 as noted in the Historical Note in section 4.2.

As we hope to have demonstrated in this chapter, the modern context is significantly poorer for ignoring the possibility of interpreting the elements of a linear space such that they can include points. We can probably trace this development back to Gibbs' early work before he became familiar with Grassmann's work, where he unwrapped Hamilton's quaternion operation into the dot and cross product.

Of course, this chapter involves no metric concepts other than the comparison of weights of points (their scalar factors) or the relative lengths of vectors in the same direction. This has enabled us to see what types of geometric theorems and constructions can be done without the concept of a metric. Or, what is equivalent, with an arbitrary metric. The notion of *congruence* has been central.

The next two chapters will introduce the two seminal concepts with which we will explore metric geometry: the *complement*, and the *interior product*. Armed with these, we will also be able to define the notions of length and angle, and more specifically, *orthogonality*. Because the Grassmann algebra extends the vector algebra to higher grade entities, we will find these notions also extend naturally to higher grade entities.

5 The Complement

5.1 Introduction

Up to this point various linear spaces and the dual exterior and regressive product operations have been introduced. The elements of these spaces were incommensurable unless they were congruent, that is, nowhere was there involved the concept of measure or magnitude of an element by which it could be compared with any other element. Thus the subject of the last chapter on Geometric Interpretations was explicitly non-metric geometry; or, to put it another way, it was what geometry is before the ability to compare or measure is added.

The question then arises, how do we associate a measure with the elements of a Grassmann algebra in a consistent way? Of course, we already know the approach that has developed over the last century, that of defining a metric tensor. In this book we will indeed define a metric tensor, but we will take an approach which develops from the concepts of the exterior product and the duality operations which are the foundations of the Grassmann algebra. This will enable us to see how the metric tensor on the underlying linear space generates metric tensors on the exterior linear spaces of higher grade; the implications of the symmetry of the metric tensor; and how to consistently generalize the notion of inner product to elements of arbitrary grade.

One of the consequences of the anti-symmetry of the exterior product of 1-elements is that the exterior linear space of m-elements has the same dimension as the exterior linear space of $(n-m)$-elements. We have already seen this property evidenced in the notions of duality and cobasis elements. And one of the consequences of the notion of duality is that the regressive product of an m-element and an $(n-m)$-element is a *scalar*. Thus there is the opportunity of defining for each m-element a corresponding 'co-m-element' of grade $n-m$ such that the regressive product of these two elements gives a scalar. We will see that this scalar measures the square of the 'magnitude' of the m-element or the $(n-m)$-element, and corresponds to the inner product of either of them with itself. We will also see that the notion of orthogonality is *defined* by the correspondence between m-elements and their 'co-m-elements'. But most importantly, the definition of this inner product as a regressive product of an m-element with a 'co-m-element' is immediately generalizable to elements of arbitrary grade, thus permitting a theory of interior products to be developed which is consistent with the exterior and regressive product axioms and which, via the notion of 'co-m-element', leads to explicit and easily derived formulae between elements of arbitrary (and possibly different) grade.

The foundation of the notion of measure or metric then is the notion of 'co-m-element'. In this book we use the term *complement* rather than 'co-m-element'. In this chapter we will develop the notion of complement in preparation for the development of the notions of interior product and orthogonality in the next chapter.

The complement of an element will be denoted with a horizontal bar over the element. For example, the complements of A, $A + B$, and $A \wedge B$ will be denoted $\overline{A}$, $\overline{A + B}$ and $\overline{A \wedge B}$.

Finally, it should be noted that the term 'complement' may be used either to refer to an operation (the operation of taking the complement of an element), or to the element itself (which is the result of the operation).

◆ *Historical Note*

Grassmann introduced the notion of complement (*Ergänzung*) into the *Ausdehnungslehre* of 1862 [Grassmann 1862]. He denoted the complement of an element x by preceding it with a vertical bar, *viz* |x. For mnemonic reasons, particularly in the derivation of formulae using the Complement Axiom, the notation for the complement used in this book is rather the horizontal bar: $\overline{x}$.

In discussing the complement, Grassmann defines the product of the *n* basis elements (the basis *n*-element) to be unity. That is $[e_1 e_2 \ldots e_n] = 1$ or, in the present notation, $e_1 \wedge e_2 \wedge \ldots \wedge e_n = 1$. Since Grassmann discussed only the *Euclidean* complement (equivalent to imposing a Euclidean metric $g_{ij} = \delta_{ij}$), this statement in the present notation is equivalent to $\overline{1} = 1$. The introduction of such an identity, however, destroys the essential duality between $\underset{m}{\Lambda}$ and $\underset{n-m}{\Lambda}$ which requires rather the identity $\overline{\overline{1}} = 1$. In current terminology, equating $e_1 \wedge e_2 \wedge \ldots \wedge e_n$ to 1 is equivalent to equating *n*-elements (or pseudo-scalars) and scalars. All other writers in the earlier Grassmannian tradition (for example, Hyde, Whitehead and Forder) followed Grassmann's approach. This enabled them to use the same notation for both the progressive and regressive products. While being an attractive approach in a Euclidean system, it is not tenable for general metric spaces. The tenets upon which the *Ausdehnungslehre* are based are so geometrically fundamental however, that it is readily extended to more general metrics.

◆ *On the use of the term 'complement'*

Grassmann's term *Ergänzung* may be translated as either *complement* or *supplement* (as well as other meanings, for example *completion*). Among the earlier workers in the Grassmannian tradition, Whitehead [1898] used supplement, while Hyde [1884-1906] used complement. Kannenberg [2000], in his excellent translation of the *Ausdehnungslehre* of 1862, has used supplement.

Whitehead also gives an interesting historical note on extensions of the *Ausdehnungslehre* to general metric concepts. [Book VI (Theory of Metrics) in *A Treatise on Universal Algebra* (p 369)]

In more modern works the *Ergänzung* for general metrics has become known as the *Hodge Star* operator. But we do not use the term here since our aim is to develop the concept by showing how it can be built straightforwardly on the foundations laid by Grassmann.

In this text we have chosen to use the word complement for its more recent added evocative meanings in set theory and linear algebra (*viz* orthogonal complement) and its mnemonic evocation of the concept of co-*m*-element. But most especially, to slightly differentiate it from Grassmann's original use as may be found in Kannenberg's translation; because our aim is to extend its meaning to include a more general symmetric metric than Grassmann initially envisaged.

5.2 Axioms for the Complement

The grade of a complement

◆ $\overline{1}$: **The complement of an *m*-element is an (*n–m*)-element**

$$\underset{m}{\alpha} \in \underset{m}{\Lambda} \implies \overline{\underset{m}{\alpha}} \in \underset{n-m}{\Lambda} \qquad 5.1$$

The grade of the complement of an element is the complementary grade of the element. (The *complementary grade* of a grade *m* in an algebra with underlying linear space of dimension *n*, is *n–m*.)

The linearity of the complement operation

◆ $\overline{2}$: **The complement operation is linear**

$$\overline{a\,\underset{m}{\alpha} + b\,\underset{k}{\beta}} == \overline{a\,\underset{m}{\alpha}} + \overline{b\,\underset{k}{\beta}} == a\,\overline{\underset{m}{\alpha}} + b\,\overline{\underset{k}{\beta}} \qquad 5.2$$

For scalars a and b, the complement of a sum of elements (perhaps of different grades) is the sum of the complements of the elements. The complement of a scalar multiple of an element is the scalar multiple of the complement of the element.

The complement axiom

◆ $\overline{3}$: **The complement of a product is the dual product of the complements**

The complement of an exterior product of elements is the regressive product of the complements of the elements.

$$\overline{\underset{m}{\alpha} \wedge \underset{k}{\beta}} == \overline{\underset{m}{\alpha}} \vee \overline{\underset{k}{\beta}} \qquad 5.3$$

The complement of a regressive product of elements is the exterior product of the complements of the elements.

$$\overline{\underset{m}{\alpha} \vee \underset{k}{\beta}} == \overline{\underset{m}{\alpha}} \wedge \overline{\underset{k}{\beta}} \qquad 5.4$$

Note that for the terms on each side of the expression [5.3] to be non-zero we require $m+k \leq n$, while in expression [5.4] we require $m+k \geq n$.

Expressions [5.3] and [5.4] are duals of each other. We call these dual expressions the *complement axiom*. Note its enticing similarity to De Morgan's law in Boolean algebra. If we wish, we can confirm this duality by applying the *GrassmannAlgebra* function `Dual`.

$$\mathbf{Dual}\left[\overline{\underset{m}{\alpha}\wedge\underset{k}{\beta}} == \overline{\underset{m}{\alpha}}\vee\overline{\underset{k}{\beta}}\right]$$

$$\overline{\underset{m}{\alpha}\vee\underset{k}{\beta}} == \overline{\underset{m}{\alpha}}\wedge\overline{\underset{k}{\beta}}$$

This axiom is of central importance in the development of the properties of the complement and interior, inner and scalar products, and formulae relating these with exterior and regressive products. In particular, it permits us to be able consistently to generate the complements of basis *m*-elements from the complements of basis 1-elements, and hence via the linearity axiom, the complements of arbitrary elements.

The forms [5.3] and [5.4] may be written for any number of elements. To see this, let $\underset{k}{\beta} = \underset{r}{\gamma}\wedge\underset{s}{\mu}$ and substitute for $\underset{k}{\beta}$ in expression [5.3]:

$$\overline{\underset{m}{\alpha}\wedge\underset{r}{\gamma}\wedge\underset{s}{\mu}} == \overline{\underset{m}{\alpha}}\vee\overline{\underset{r}{\gamma}\wedge\underset{s}{\mu}} == \overline{\underset{m}{\alpha}}\vee\overline{\underset{r}{\gamma}}\vee\overline{\underset{s}{\mu}}$$

In general then, by extending this process, expressions [5.3] and [5.4] may be stated for products of elements A_i of any grade.

$$\overline{A_1\wedge A_2\wedge\cdots\wedge A_z} == \overline{A_1}\vee\overline{A_2}\vee\cdots\vee\overline{A_z} \qquad 5.5$$

$$\overline{A_1\vee A_2\vee\cdots\vee A_z} == \overline{A_1}\wedge\overline{A_2}\wedge\cdots\wedge\overline{A_z} \qquad 5.6$$

In Grassmann's work, this axiom was hidden in his notation. However, since modern notation explicitly distinguishes the progressive and regressive products, this axiom needs to be explicitly stated.

The complement of a complement axiom

◆ $\overline{4}$: The complement of the complement of an element is congruent to itself

Axiom $\overline{1}$ says that the complement of an *m*-element is an (*n*–*m*)-element. Clearly then the complement of an (*n*–*m*)-element is an *m*-element. Thus the complement of the complement of an *m*-element is itself an *m*-element.

Consider an *m*-element A. There are two aspects of the algebra which we would specifically like any axiom defining the *m*-element $\overline{\overline{A}}$ to facilitate:

1. $\overline{\overline{A}}$ is, apart from a possible sign, equal to A.

2. The scalar measure of A is equal the scalar measure of its complement $\overline{A}$.

As we will see in what follows, requirement **1** will constrain the complement mapping from A to $\overline{A}$ to be symmetric - equivalent to the metric tensor being symmetric.

Requirement **2** constrains the possible sign relating $\overline{\overline{A}}$ to A to a specific form. To discover this form, we need briefly to signal how we will define the *measure* of an *m*-element by means of the inner product in Chapter 6. The inner product of an element A with itself (also the square of its measure) will be defined as the regressive product of the element A with its complement $\overline{A}$. If the measure of A is to be equal to the measure of its complement $\overline{A}$ then we have

$$A \vee \overline{A} == \overline{A} \vee \overline{\overline{A}}$$

If now, we invoke requirement **1** by replacing $\overline{\overline{A}}$ with a signed version of A, and then reorder the factors, we have an equality expressing the required sign.

$$A \vee \overline{A} == \overline{A} \vee (-1)^{\phi} A == (-1)^{\phi + m(n-m)} A \vee \overline{A}$$

This may be satisfied by taking ϕ equal to $m(n-m)$, leading finally to the complement of a complement axiom.

$$\overline{\overline{\alpha_m}} == (-1)^{m(n-m)} \alpha_m \qquad 5.7$$

As it turns out, the form of this axiom is fundamental to the Grassmann algebra, underpinning the properties and symmetries which make it such a beautiful structure.

In particular, the complement of the complement of unity is unity, of a scalar is the scalar itself, and of an n-element is the n-element itself.

$$\overline{\overline{1}} == 1 \qquad \overline{\overline{a}} == a \qquad \overline{\overline{\alpha_n}} == \alpha_n \qquad 5.8$$

The complement of the complement of any element in a 3-space is the element itself, since $(-1)^{m(3-m)}$ is unity for m equal to 0, 1, 2, or 3.

Alternatively, we can say that $\overline{\overline{A}}$ is equal to A *except when* A *is of odd degree in an even-dimensional space*.

The simplest case of an element of odd degree in an even-dimensional space is a vector in a vector 2-space.

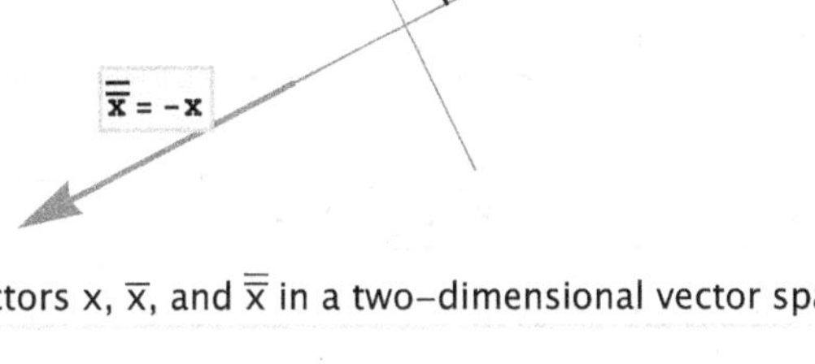

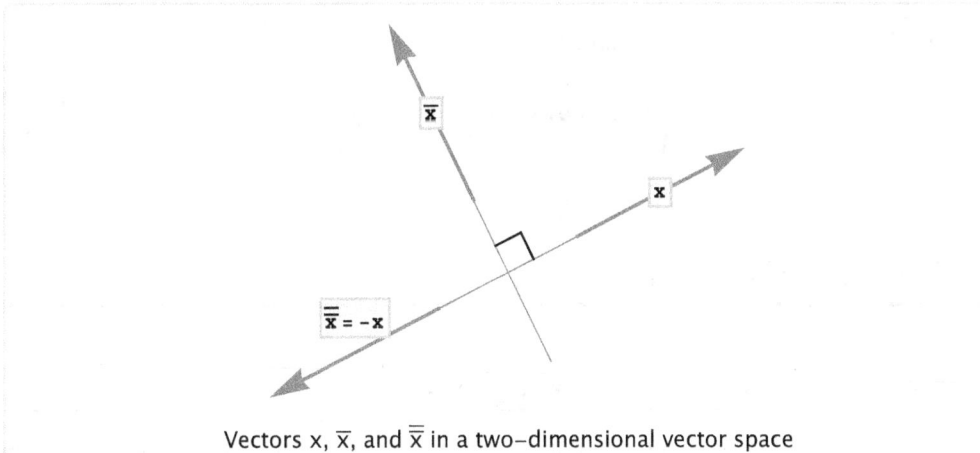

Vectors x, $\overline{x}$, and $\overline{\overline{x}}$ in a two–dimensional vector space

The complement of unity

◆ $\overline{5}$: **The complement of unity is equal to the unit *n*-element**

By the complement of a complement axiom [5.7] we note that the complement of $\overline{1}$ is equal to 1. And from Chapter 3 we also note that the dual of $\underset{n}{1}$ is equal to 1. We now *formally identify* $\overline{1}$ with the unit *n*-element $\underset{n}{1}$.

$$\underset{n}{1} == \overline{1} \qquad 5.9$$

In Chapter 3, equation [3.10] says that the unit *n*-element is congruent to any basis *n*-element, and expressed this result in a form involving an (as yet undetermined) scalar congruence factor ★c.

$$e_1 \wedge e_2 \wedge \ldots \wedge e_n == ★c\,\underset{n}{1}$$

The basis *n*-element can be denoted explicitly by the ordered exterior product of the basis elements, or implicitly by the cobasis of unity in that basis.

$$e_1 \wedge e_2 \wedge \ldots \wedge e_n == \underline{1}$$

For compactness we will often use the cobasis of unity denotation $\underline{1}$ for the basis *n*-element and denote its measure by $|\underline{1}|$. Since the measure of an *n*-element may be interpreted as its (*n*-dimensional) 'volume' (see Chapter 6), the 'boxiness' of $|\underline{1}|$ may (with a large stretch of the imagination) mnemonically suggest a length, area or volume according to the dimension of the space.

Hence we can write the measure of the basis *n*-element in any of the forms:

$$|e_1 \wedge e_2 \wedge \ldots \wedge e_n| == |\underline{1}| == ★c \qquad 5.10$$

And the unit *n*-element in any of the forms:

$$\overline{1} == \underset{n}{1} == \frac{e_1 \wedge e_2 \wedge \ldots \wedge e_n}{|e_1 \wedge e_2 \wedge \ldots \wedge e_n|} == \frac{\underline{1}}{|\underline{1}|} == \hat{\underline{1}} \qquad 5.11$$

5.3 Defining the Complement

The symmetry relation for regressive products

In this section we derive an important result we will need in the development of the complement operation to follow (and in the development of the interior product in the next chapter): if we take two *m*-elements A and B, say, then the (scalar) regressive product of A with the complement of B is equal to the regressive product of B with the complement of A.

To show this we begin with the special case [3.32] of the Common Factor Axiom which relates the exterior and regressive products of an *m*-element and an (*n*–*m*)-element:

$$\underset{m}{\alpha} \wedge \underset{n-m}{\beta} == \left(\underset{m}{\alpha} \vee \underset{n-m}{\beta}\right) \underset{n}{1}$$

If now we put $\underset{n-m}{\beta}$ equal to $\overline{\underset{m}{\beta}}$ and $\underset{n}{1}$ equal to $\overline{1}$ we obtain

$$\underset{m}{\alpha} \wedge \overline{\underset{m}{\beta}} == \left(\underset{m}{\alpha} \vee \overline{\underset{m}{\beta}}\right) \overline{1}$$

Since $\underset{m}{\alpha} \vee \overline{\underset{m}{\beta}}$ is a scalar, and the complement of $\overline{1}$ is 1, taking the complement of this equation gives

$$\overline{\underset{m}{\alpha} \wedge \overline{\underset{m}{\beta}}} == \underset{m}{\alpha} \vee \overline{\underset{m}{\beta}}$$

Using the complement axiom [5.3] and axiom [5.7] for the complement of a complement allows us to transform the left hand side.

$$\overline{\underset{m}{\alpha} \wedge \overline{\underset{m}{\beta}}} == \overline{\underset{m}{\alpha}} \vee \overline{\overline{\underset{m}{\beta}}} == (-1)^{m\,(n-m)}\, \overline{\underset{m}{\alpha}} \vee \underset{m}{\beta} == \underset{m}{\beta} \vee \overline{\underset{m}{\alpha}}$$

Thus we can write:

$$\underset{m}{\alpha} \vee \overline{\underset{m}{\beta}} == \underset{m}{\beta} \vee \overline{\underset{m}{\alpha}} \qquad 5.12$$

We call this formula *the symmetry relation for regressive products*. In section 6.5 we will see that this is also the justification for the symmetry of the inner product and metric tensor.

We should remark that the derivation of this formula relied on three axioms: the Common Factor Axiom [3.32], the Complement Axiom [5.3], and the complement of a complement axiom [5.7].

- For example, if e_i and e_j are two basis elements, then our later justification for the symmetry of their scalar product is:

$$e_i \vee \overline{e_j} == e_j \vee \overline{e_i} \qquad 5.13$$

The complement of an *m*-element

To define the complement of a general *m*-element, A say, we need only define the complement of basis *m*-elements, since by the linearity axiom $\overline{2}$ [5.2], we have that for a general *m*-element expressed as a linear combination of basis *m*-elements, its complement is the corresponding linear combination of the complements of the basis *m*-elements.

$$A == \sum_i a_i \underset{m}{e_i} \iff \overline{A} == \sum_i a_i \overline{\underset{m}{e_i}} \qquad 5.14$$

The complements of the basis *m*-elements however, cannot be defined independently of the complements of basis elements in exterior linear spaces of other grades, since they are related by the complement axioms [5.3] and [5.4]. For example, the complement of a basis 4-element may also be expressed as the regressive product of the complements of two basis 2-elements, or as the regressive product of the complement of a basis 3-element and the complement of a basis 1-

element.

$$\overline{e_1 \wedge e_2 \wedge e_3 \wedge e_4} == \overline{e_1 \wedge e_2} \vee \overline{e_3 \wedge e_4} == \overline{e_1 \wedge e_2 \wedge e_3} \vee \overline{e_4}$$

The complement axiom enables us to define the complement of a basis m-element in terms *only* of the complements of basis 1-elements.

$$\overline{e_1 \wedge e_2 \wedge \ldots \wedge e_m} == \overline{e_1} \vee \overline{e_2} \vee \ldots \vee \overline{e_m} \qquad 5.15$$

Thus in order to define the complement of any element in a Grassmann algebra, we only need to define the complements of the basis 1-elements, that is, the correspondence between basis 1-elements and basis (n–1)-elements.

The defining identity for the complement of an *m*-element

Below we derive an identity for the complement of a general m-element A which will be pivotal in our mechanism for being able *explicitly* to define the complement of any element.

From the complement axiom [5.3] we can write the complement of A as a regressive product with the unit n-element.

$$\overline{A} == \overline{1 \wedge A} == \overline{1} \vee \overline{A}$$

And from [5.11] we can write $\overline{1}$ in terms of the basis n-element as

$$\overline{1} == \frac{e_1 \wedge e_2 \wedge \ldots \wedge e_n}{|e_1 \wedge e_2 \wedge \ldots \wedge e_n|} == \frac{1}{|\underline{1}|} e_1 \wedge e_2 \wedge \ldots \wedge e_n$$

Substituting for $\overline{1}$, we obtain the identity

$$|\underline{1}|\, \overline{A} == (e_1 \wedge e_2 \wedge \ldots \wedge e_n) \vee \overline{A}$$

Applying the special case of the Common Factor Theorem [3.50] allows us now to give an identity for the complement $\overline{A}$ of the m-element A as a linear combination of basis m-elements.

$$|\underline{1}|\, \overline{A} == (e_1 \wedge e_2 \wedge \ldots \wedge e_n) \vee \overline{A}$$

$$== \sum_{i_1 \ldots i_m} \left((e_{i_1} \wedge \ldots \wedge e_{i_m}) \vee \overline{A}\right) (-1)^{K_m} e_1 \wedge \ldots \wedge \Box_{i_1} \wedge \ldots \wedge \Box_{i_m} \wedge \ldots \wedge e_n \qquad 5.16$$

$$K_m == \sum_{\gamma==1}^{m} i_\gamma + \frac{1}{2} m (m+1)$$

The summation is over all the essentially different products $e_{i_1} \wedge \ldots \wedge e_{i_m}$ that maintains their basis order.

This is the identity we will use in what follows to enable the explicit determination of a complement.

Defining the complement of a basis element

Now let us consider the complement of a basis element e_j. In [5.16] above, put A equal to e_j.

$$|\underline{1}|\,\overline{e_j} == (e_1 \wedge e_2 \wedge \ldots \wedge e_n) \vee \overline{e_j}$$

$$== \sum_{i==1}^{n} \left(e_i \vee \overline{e_j}\right) (-1)^{i-1} e_1 \wedge \cdots \wedge \Box_i \wedge \cdots \wedge e_n$$

Suppose now that, *although we do not know a specific value for* $\overline{e_j}$ *yet, we do know the regressive products of* $\overline{e_j}$ *with each of the basis elements* e_i, that is, we know the *scalars* $e_i \vee \overline{e_j}$. Let us denote these scalars by g_{ij} and note that by [5.12], the g_{ij} are symmetric, that is g_{ij} is equal to g_{ji}.

$$\overline{e_j} == \frac{1}{|\underline{1}|} \sum_{i==1}^{n} g_{ij} (-1)^{i-1} e_1 \wedge \cdots \wedge \Box_i \wedge \cdots \wedge e_n$$

This can be written in terms of cobasis elements as:

$$\overline{e_j} == \frac{1}{|\underline{1}|} \sum_{i==1}^{n} g_{ij}\, \underline{e_i} \qquad g_{ij} == g_{ji} \qquad 5.17$$

This is our complement mapping between $\underset{1}{\Lambda}$ and $\underset{n-1}{\Lambda}$. The g_{ij} may be recognized as the elements of the metric tensor. We cannot define the complement of an element without introducing these scalars as additional structure into the algebra. But they are the only scalars we need: we will show in what follows that the complement mapping between $\underset{m}{\Lambda}$ and $\underset{n-m}{\Lambda}$ may also be defined in terms of the g_{ij}.

Defining the complement of the basis *n*-element

Taking the complement of [5.11] and applying [5.8] gives

$$\overline{\overline{1}} == \frac{\overline{e_1 \wedge e_2 \wedge \ldots \wedge e_n}}{|\underline{1}|} == 1$$

Hence we can write the complement of the basis *n*-element as

$$\overline{e_1 \wedge e_2 \wedge \ldots \wedge e_n} == |\underline{1}| \qquad 5.18$$

showing that the complement of a basis *n*-element is also its measure.

Now in [5.16] above, put A equal to $e_1 \wedge e_2 \wedge \ldots \wedge e_n$.

$$|\underline{1}|\,\overline{e_1 \wedge e_2 \wedge \ldots \wedge e_n} == (e_1 \wedge e_2 \wedge \ldots \wedge e_n) \vee \overline{e_1 \wedge e_2 \wedge \ldots \wedge e_n}$$

On the right side apply [5.15] to expand the complement as a regressive product. Then, since the complement of a 1-element is an (*n*–1)-element we can apply [3.51] to get

$$(e_1 \wedge e_2 \wedge \ldots \wedge e_n) \vee \overline{e_1 \wedge e_2 \wedge \ldots \wedge e_n}$$

$$== (e_1 \wedge e_2 \wedge \ldots \wedge e_n) \vee (\overline{e_1} \vee \overline{e_2} \vee \ldots \vee \overline{e_n}) == \mathrm{Det}\left[e_i \vee \overline{e_j}\right] == \mathrm{Det}\left[g_{ij}\right]$$

Substituting back gives

$$|\underline{1}|\,\overline{e_1 \wedge e_2 \wedge \ldots \wedge e_n} == |\underline{1}|^2 == \mathrm{Det}\left[g_{ij}\right]$$

$$|\underline{1}|^2 == \mathrm{Det}[g_{ij}] \qquad 5.19$$

Let G be the the matrix of the elements of the metric tensor g_{ij}, and $|G|$ its determinant. Then

$$|e_1 \wedge \ldots \wedge e_n| == |\underline{1}| == \sqrt{|G|} \qquad 5.20$$

This is an important relationship: the measure of the *n*-element is equal to the square root of the determinant of the metric tensor.

(Note that when applied to a matrix the pair of vertical bars $|\square|$ signifies its determinant. But when applied to an element of the algebra, it signifies its measure: the square root of its inner square.)

Complements of basis elements in matrix form

We can define all the complements of the basis 1-elements in one matrix equation. Let B be the column matrix of basis elements, and G be the matrix $[g_{ij}]$ of the g_{ij}. We have shown that G is symmetric; we suppose it also to be non-singular.

$$B == \begin{pmatrix} e_1 \\ e_2 \\ \ldots \\ e_n \end{pmatrix} \qquad G == [g_{ij}] == \begin{pmatrix} g_{1,1} & g_{1,2} & \cdots & g_{1,n} \\ g_{1,2} & g_{2,2} & \cdots & g_{2,n} \\ \ldots & \ldots & \ldots & \ldots \\ g_{1,n} & g_{2,n} & \cdots & g_{n,n} \end{pmatrix}$$

Substituting for $|\underline{1}|$ from [5.20] into formula [5.17] gives

$$\overline{e_j} == \frac{1}{\sqrt{|G|}} \sum_{i=1}^{n} g_{ij}\, \underline{e_i} \qquad g_{ij} == g_{ji}$$

Thes equations can be collected in the forms

$$\begin{pmatrix} \overline{e_1} \\ \overline{e_2} \\ \ldots \\ \overline{e_n} \end{pmatrix} == \frac{1}{\sqrt{|G|}} \begin{pmatrix} g_{1,1} & g_{1,2} & \cdots & g_{1,n} \\ g_{1,2} & g_{2,2} & \cdots & g_{2,n} \\ \ldots & \ldots & \ldots & \ldots \\ g_{1,n} & g_{2,n} & \cdots & g_{n,n} \end{pmatrix} \begin{pmatrix} \underline{e_1} \\ \underline{e_2} \\ \ldots \\ \underline{e_n} \end{pmatrix} \qquad g_{ij} == g_{ji} \qquad 5.21$$

$$\overline{B} == \frac{1}{\sqrt{|G|}} G\, \underline{B} \qquad G == G^T \qquad |G| == \mathrm{Det}[G] \qquad 5.22$$

Complements of cobasis elements in matrix form

From this matrix formulation it is easy to compute the complements of the cobasis elements. Taking the complement of both sides of formula [5.22] above gives:

$$\overline{B} == \frac{1}{\sqrt{|G|}}\, G\, \underline{B}$$

Note that the symbol $\overline{\underline{B}}$ must be read as *the complement of the cobasis elements* since the 'cobasis of a complement' is undefined.

A formula for the complements of the cobasis elements is obtained by inverting the matrix `G` and applying the complement of a complement axiom [5.7].

$$\overline{\underline{B}} == (-1)^{n-1} \sqrt{|G|}\; G^{-1}\, B \qquad 5.23$$

Alternatively, since the adjoint of `G` can be written

$$\mathrm{Adj}[G] == |G|\, G^{-1}$$

formula [5.23] becomes

$$\overline{\underline{B}} == (-1)^{n-1} \frac{1}{\sqrt{|G|}}\, \mathrm{Adj}[G]\, B == \frac{1}{\sqrt{|G|}}\, \mathrm{Adj}[G]\, \underline{\underline{B}} \qquad 5.24$$

Comparing this result to [5.22] shows that they are of the same form if `B` is replaced by $\underline{B}$ and the matrix factor `G` is replaced by `Adj[G]`.

The metric on the cobasis elements is thus the adjoint of the metric on the basis elements.

◆ Example: The metric on the cobasis elements in a 2-space

As an example, let us look at the metric on the cobasis elements in a 2-space. The cobasis palette is:

```
★𝓑2; DeclareMetric[g]; CobasisPalette
```

Cobasis Palette

Basis	Cobasis
1	$e_1 \wedge e_2$
e_1	e_2
e_2	$-e_1$
$e_1 \wedge e_2$	1

We can write equations [5.22] and [5.24] from the palette.

$$\overline{B} == \frac{1}{\sqrt{|G|}}\, G\, \underline{B} \;\Leftrightarrow\; \begin{pmatrix} \overline{e_1} \\ \overline{e_2} \end{pmatrix} == \frac{1}{\sqrt{|G|}} \begin{pmatrix} g_{1,1} & g_{1,2} \\ g_{1,2} & g_{2,2} \end{pmatrix} \begin{pmatrix} e_2 \\ -e_1 \end{pmatrix}$$

$$\overline{\underline{B}} == \frac{1}{\sqrt{|G|}}\, \mathrm{Adj}[G]\, \underline{\underline{B}} \;\Leftrightarrow\; \begin{pmatrix} \overline{e_2} \\ \overline{-e_1} \end{pmatrix} == \frac{1}{\sqrt{|G|}} \begin{pmatrix} g_{2,2} & -g_{1,2} \\ -g_{1,2} & g_{1,1} \end{pmatrix} \begin{pmatrix} -e_1 \\ -e_2 \end{pmatrix}$$

$$|G| == \mathrm{Det}\left[\begin{pmatrix} g_{1,1} & g_{1,2} \\ g_{1,2} & g_{2,2} \end{pmatrix}\right]$$

We see that they lead to same results.

$$\overline{e_1} == \frac{1}{\sqrt{|G|}} (g_{1,1}\, e_2 - g_{1,2}\, e_1) \qquad \overline{e_2} == \frac{1}{\sqrt{|G|}} (g_{1,2}\, e_2 - g_{2,2}\, e_1)$$

Adjoints

Because the adjoint of a metric matrix plays an important role in the relationship between the metrics of a space, we take the opportunity in this section briefly to summarize some of its properties.

◆ Definition

Consider a *square* matrix M with elements m_{ij} in the ith row and jth column.

The *adjoint* of M is the transpose of the matrix of the cofactors of the elements m_{ij}.

The *cofactor* of the element m_{ij} is $(-1)^{i+j}$ times the minor of m_{ij}.

The *minor* of an element m_{ij} is the determinant of the sub-matrix formed by deleting the row and column containing m_{ij}.

The adjoint of a matrix M will be denoted AdjM or Adj[M], its determinant $|M|$, and its order p.

◆ Inverse

$$\mathrm{Adj}[M] == |M|\, M^{-1} \qquad \mathrm{Adj}\left[M^{-1}\right] == (\mathrm{Adj}[M])^{-1} == \frac{1}{|M|} M \tag{5.25}$$

◆ Scalar multiples

$$\mathrm{Adj}[a\, M] == a^{p-1}\, \mathrm{Adj}[M] \tag{5.26}$$

◆ Adjoint of an adjoint

$$\mathrm{Adj}[\mathrm{Adj}[M]] == |M|^{p-2}\, M \tag{5.27}$$

5.4 Metrics

Complements of basis *m*-elements

◆ Complements of basis 2-elements

Now let us consider the complement of a basis 2-element $e_{j_1} \wedge e_{j_2}$. Putting A equal to $e_{j_1} \wedge e_{j_2}$ in our complement defining identity [5.16] gives

$$\underline{|1|}\;\overline{e_{j_1} \wedge e_{j_2}} == (e_1 \wedge e_2 \wedge \ldots \wedge e_n) \vee \overline{e_{j_1} \wedge e_{j_2}}$$
$$== \sum_{i_1, i_2} \left((e_{i_1} \wedge e_{i_2}) \vee \overline{e_{j_1} \wedge e_{j_2}}\right) (-1)^{i_1 + i_2 - 1} e_1 \wedge \ldots \wedge \square_{i_1} \wedge \ldots \wedge \square_{i_2} \wedge \ldots \wedge e_n$$

Here, the summation is over the $\frac{1}{2}n(n-1)$ essentially different exterior products of two basis elements.

The scalar coefficients in this sum are the metric elements induced by g_{ij} onto $\underset{2}{\Lambda}$. So by analogy with the basis 1-element case, let us denote them by:

$$g_{\{i_1, i_2\}, \{j_1, j_2\}} == (e_{i_1} \wedge e_{i_2}) \vee \overline{e_{j_1} \wedge e_{j_2}} \qquad 5.28$$

Now the complement can be written as before as a linear combination of all the cobasis elements of $\underset{2}{\Lambda}$.

$$\overline{e_{j_1} \wedge e_{j_2}} == \frac{1}{\sqrt{|G|}} \sum_{i_1, i_2} g_{\{i_1, i_2\}, \{j_1, j_2\}} \underline{e_{i_1} \wedge e_{i_2}} \qquad 5.29$$

◆ **Complements of basis *m*-elements**

It is clear from this process that the complement of a basis *m*-element can be formulated as:

$$\overline{e_{j_1} \wedge \ldots \wedge e_{j_m}} == \frac{1}{\sqrt{|G|}} \sum g_{\{i_1, \ldots, i_m\}, \{j_1, \ldots, j_m\}} \underline{e_{i_1} \wedge \ldots \wedge e_{i_m}} \qquad 5.30$$

Here, the summation is a linear combination of all the cobasis elements of $\underset{m}{\Lambda}$.

The metric on $\underset{m}{\Lambda}$

The *metric tensor* on $\underset{m}{\Lambda}$ is defined by the components $g_{\{i_1, \ldots, i_m\}, \{j_1, \ldots, j_m\}}$ induced on $\underset{m}{\Lambda}$ by the metric tensor g_{ij} of $\underset{1}{\Lambda}$.

$$g_{\{i_1, \ldots, i_m\}, \{j_1, \ldots, j_m\}} == (e_{i_1} \wedge \ldots \wedge e_{i_m}) \vee \overline{e_{j_1} \wedge \ldots \wedge e_{j_m}} \qquad 5.31$$

We call such a metric on $\underset{m}{\Lambda}$ an *m*-metric, and denote it or its matrix components, by G_m. For simplicity, the metric on the underlying linear space $\underset{1}{\Lambda}$ may be denoted simply by G.

$$G_m == \left[g_{\{i_1, \ldots, i_m\}, \{j_1, \ldots, j_m\}}\right] \qquad G == \left[g_{ij}\right] \qquad 5.32$$

The metric on $\underset{n-m}{\Lambda}$ will be denoted G_{n-m}, and will be called the *cometric* of G_m.

From the symmetry relation for regressive products [5.12] above, we note that [5.31] shows the induced metric to be also, like g_{ij}, symmetric.

$$g_{\{i_1,\ldots,i_m\},\{j_1,\ldots,j_m\}} == g_{\{j_1,\ldots,j_m\},\{i_1,\ldots,i_m\}} \qquad g_{ij} == g_{ji} \qquad 5.33$$

We will show in the next chapter that, as might have been expected, the components of the metric tensors on any exterior product space are simply the inner products of the basis elements.

◆ The metric on $\underset{0}{\Lambda}$

The unit for measuring *in any dimension* on $\underset{0}{\Lambda}$ is just the unit of the field, which we have taken to be 1 (unity). Putting *m* equal to 0 in [5.31] gives a metric element (and thus a metric tensor) of unity consistent with this expectation.

$$g_{\{\},\{\}} == 1 \vee \overline{1} == 1$$

Hence the matrix of metric elements G_0 is, in any dimension, the matrix containing the unit 1 as its single element.

$$G_0 == (\ 1\ ) \qquad 5.34$$

◆ The metric on $\underset{n}{\Lambda}$

The metric on $\underset{n}{\Lambda}$ may be obtained from [5.31], [5.11]and [5.20] as

$$g_{\{1,2,\ldots,n\},\{1,2,\ldots,n\}} == (e_1 \wedge \ldots \wedge e_n) \vee \overline{e_1 \wedge \ldots \wedge e_n}$$

$$== \left(|\underline{1}|\ \overline{1}\right) \vee \left(|\underline{1}|\ \overline{\overline{1}}\right) == |\underline{1}|^2\ \overline{1} \vee 1 == |\underline{1}|^2 == |G|$$

Thus the metric tensor on $\underset{n}{\Lambda}$ contains only one element.

$$g_{\{1,2,\ldots,n\},\{1,2,\ldots,n\}} == |G| \qquad 5.35$$

So the metric matrix on $\underset{n}{\Lambda}$ also only contains one element.

$$G_n == (\ |G|\ ) \qquad 5.36$$

Hence the matrix of metric elements G_n is, in any dimension, the matrix whose single element is the determinant of the matrix G of the metric tensor of $\underset{1}{\Lambda}$.

To complete the ways of expressing the determinant of the metric tensor, remember from [5.19] that it is also equal to the square of the measure $|\underline{1}|$ of the basis *n*-element.

$$|G| == |e_1 \wedge \ldots \wedge e_n|^2 == |\underline{1}|^2 == |\overline{e_1 \wedge \ldots \wedge e_n}|^2 \qquad 5.37$$

Metrics in a 3-space

◆ The metric on $\underset{1}{\Lambda}$

To fix ideas let us look at the specific case of a vector 3-space. The metric on $\underset{1}{\Lambda}$ may be given by the matrix of 6 different scalars g_{ij}, reduced from 9 by the symmetry condition. These scalars may be either imposed on the space or derived from some auxiliary conditions, for example, those defining the coordinate system.

$$G == [g_{ij}] == \begin{pmatrix} g_{1,1} & g_{1,2} & g_{1,3} \\ g_{1,2} & g_{2,2} & g_{2,3} \\ g_{1,3} & g_{2,3} & g_{3,3} \end{pmatrix}$$

Remember that in Chapter 6 we will be showing that the element $g_{i,j}$ is simply the scalar product of basis element e_i with basis element e_j.

◆ The metric on $\underset{2}{\Lambda}$

From formula [5.31] we have the metric elements on $\underset{2}{\Lambda}$ defined by

$$g_{\{i_1,i_2\},\{j_1,j_2\}} == (e_{i_1} \wedge e_{i_2}) \vee \overline{e_{j_1} \wedge e_{j_2}}$$

And from [5.15] and [3.51] we can write the right hand side as

$$(e_{i_1} \wedge e_{i_2}) \vee (\overline{e_{j_1}} \vee \overline{e_{j_2}}) == \mathrm{Det}\left[\begin{pmatrix} e_{i_1} \vee \overline{e_{j_1}} & e_{i_1} \vee \overline{e_{j_2}} \\ e_{i_2} \vee \overline{e_{j_1}} & e_{i_2} \vee \overline{e_{j_2}} \end{pmatrix}\right]$$

Hence the metric elements on $\underset{2}{\Lambda}$ in a space of any number of dimensions are of the form

$$g_{\{i_1,i_2\},\{j_1,j_2\}} == \begin{vmatrix} g_{i_1,j_1} & g_{i_1,j_2} \\ g_{i_2,j_1} & g_{i_2,j_2} \end{vmatrix} \qquad 5.38$$

For the 3-dimensional case, the exterior products in the definition of the metric elements show that the only composite indices $\{i_1, i_2\}$ and $\{j_1, j_2\}$ leading to non-zero elements are $\{1,2\}, \{1,3\}, \{2,3\}$.

Hence from formula [5.29] the complete metric G_2 on $\underset{2}{\Lambda}$ in 3-space is given by the matrix of minors.

$$G_2 == [g_{\{i_1,i_2\},\{j_1,j_2\}}] == \begin{pmatrix} g_{\{1,2\},\{1,2\}} & g_{\{1,3\},\{1,2\}} & g_{\{2,3\},\{1,2\}} \\ g_{\{1,2\},\{1,3\}} & g_{\{1,3\},\{1,3\}} & g_{\{2,3\},\{1,3\}} \\ g_{\{1,2\},\{2,3\}} & g_{\{1,3\},\{2,3\}} & g_{\{2,3\},\{2,3\}} \end{pmatrix}$$

Writing out these elements according to [5.38] gives

$$G_2 == \begin{pmatrix} \begin{vmatrix} g_{1,1} & g_{1,2} \\ g_{2,1} & g_{2,2} \end{vmatrix} & \begin{vmatrix} g_{1,1} & g_{1,2} \\ g_{3,1} & g_{3,2} \end{vmatrix} & \begin{vmatrix} g_{2,1} & g_{2,2} \\ g_{3,1} & g_{3,2} \end{vmatrix} \\ \begin{vmatrix} g_{1,1} & g_{1,3} \\ g_{2,1} & g_{2,3} \end{vmatrix} & \begin{vmatrix} g_{1,1} & g_{1,3} \\ g_{3,1} & g_{3,3} \end{vmatrix} & \begin{vmatrix} g_{2,1} & g_{2,3} \\ g_{3,1} & g_{3,3} \end{vmatrix} \\ \begin{vmatrix} g_{1,2} & g_{1,3} \\ g_{2,2} & g_{2,3} \end{vmatrix} & \begin{vmatrix} g_{1,2} & g_{1,3} \\ g_{3,2} & g_{3,3} \end{vmatrix} & \begin{vmatrix} g_{2,2} & g_{2,3} \\ g_{3,2} & g_{3,3} \end{vmatrix} \end{pmatrix}$$

Observe how the symmetry of G_2 depends on the symmetry of G.

◆ The metric on $\underset{3}{\Lambda}$

From formula [5.36] the metric on $\underset{3}{\Lambda}$ is given by the matrix containing the single scalar determinant of G.

$$G_3 == (\,|G|\,) == \left(\begin{vmatrix} g_{1,1} & g_{1,2} & g_{1,3} \\ g_{1,2} & g_{2,2} & g_{2,3} \\ g_{1,3} & g_{2,3} & g_{3,3} \end{vmatrix}\right)$$

As shown in [5.37] this single metric element of $\underset{3}{\Lambda}$ is the square of the measure ('volume') of the basis 3-element. It thus 'measures' the basis 3-element.

$$|G| == |e_1 \wedge e_2 \wedge e_3|^2 == |\underline{1}|^2$$

◆ The Metric Palette

Alternatively you can display a palette of all the metric tensors of the space you are working in with *GrassmannAlgebra's* `MetricPalette`. The elements of the metric tensors are generated in a form explicitly displaying their symmetry.

```
★𝓑3; DeclareMetric[g]; MetricPalette
```

Metric Palette

$\underset{0}{\Lambda}$	(1)
$\underset{1}{\Lambda}$	$\begin{pmatrix} g_{1,1} & g_{1,2} & g_{1,3} \\ g_{1,2} & g_{2,2} & g_{2,3} \\ g_{1,3} & g_{2,3} & g_{3,3} \end{pmatrix}$
$\underset{2}{\Lambda}$	$\begin{pmatrix} -g_{1,2}^2 + g_{1,1}\,g_{2,2} & -g_{1,2}\,g_{1,3} + g_{1,1}\,g_{2,3} & -g_{1,3}\,g_{2,2} + g_{1,2}\,g_{2,3} \\ -g_{1,2}\,g_{1,3} + g_{1,1}\,g_{2,3} & -g_{1,3}^2 + g_{1,1}\,g_{3,3} & -g_{1,3}\,g_{2,3} + g_{1,2}\,g_{3,3} \\ -g_{1,3}\,g_{2,2} + g_{1,2}\,g_{2,3} & -g_{1,3}\,g_{2,3} + g_{1,2}\,g_{3,3} & -g_{2,3}^2 + g_{2,2}\,g_{3,3} \end{pmatrix}$
$\underset{3}{\Lambda}$	$\left(-g_{1,3}^2\,g_{2,2} + 2\,g_{1,2}\,g_{1,3}\,g_{2,3} - g_{1,1}\,g_{2,3}^2 - g_{1,2}^2\,g_{3,3} + g_{1,1}\,g_{2,2}\,g_{3,3}\right)$

Complements of basis *m*-elements in matrix form

We can immediately extend the results above to the complements of basis *m*-elements. Let $\underset{m}{\mathrm{B}}$ be the column matrix of basis *m*-elements, and G_m be the matrix of metric elements $g_{\{i_1,\ldots,i_m\},\{j_1,\ldots,j_m\}}$. G_m is symmetric and non-singular.

Then the formulae

$$\overline{e_{j_1} \wedge \ldots \wedge e_{j_m}} == \frac{1}{\sqrt{|G|}} \sum g_{\{i_1,\ldots,i_m\},\{j_1,\ldots,j_m\}}\, \underline{e_{i_1} \wedge \ldots \wedge e_{i_m}}$$

can be collected in the form below. This form is similar to [5.22] for 1-elements.

$$\overline{\underset{m}{\mathrm{B}}} == \frac{1}{\sqrt{|G|}}\, G_m\, \underline{\underset{m}{\mathrm{B}}} \qquad G_m == G_m^{\mathrm{T}} \tag{5.39}$$

◆ The complements of the cobasis *m*-elements

Taking the complement of both sides of the formula above gives:

$$\overline{\underset{m}{\overline{\mathbf{B}}}} == \frac{1}{\sqrt{|G|}} G_m \underset{\overline{m}}{\overline{\mathbf{B}}} == (-1)^{m\,(n-m)} \underset{m}{\mathbf{B}}$$

Hence the formula for the complements of the cobasis *m*-elements is obtained by inverting the matrix G_m.

$$\overline{\underset{\overline{m}}{\mathbf{B}}} == (-1)^{m\,(n-m)} \sqrt{|G|}\; G_m^{-1} \underset{m}{\mathbf{B}} \qquad 5.40$$

Following the same process *mutatis mutandis* which lead to deriving formula [5.24] we also have:

$$\overline{\underset{\overline{m}}{\mathbf{B}}} == \frac{1}{\sqrt{|G|}} \mathrm{Adj}[G_m] \underset{\overline{\overline{m}}}{\mathbf{B}} \qquad 5.41$$

The metric on the cobasis *m*-elements is thus the adjoint of the metric on the basis *m*-elements.

Composing metrics

As it turns out, the metric G_m on $\underset{m}{\Lambda}$ is simply the *m*th compound matrix of G. It is also a *Gram matrix*.

The *compound matrix* of order *m* of a matrix A is the matrix formed from the all *m*×*m* minors of A arranged with their index sets in lexicographic order.

Hence we can compose the components $g_{\{i_1,\ldots,i_m\},\{j_1,\ldots,j_m\}}$ of G_m by composing the *m*th compound matrix of G.

$$G_m == \left[g_{\{i_1,\ldots,i_m\},\{j_1,\ldots,j_m\}}\right] \qquad 5.42$$

- In *GrassmannAlgebra* you can compose the compound matrix of order m of a matrix M by entering CompoundMatrix[m][M]. For example in 3-space:

$$G = \begin{pmatrix} g_{1,1} & g_{1,2} & g_{1,3} \\ g_{1,2} & g_{2,2} & g_{2,3} \\ g_{1,3} & g_{2,3} & g_{3,3} \end{pmatrix};\ \mathbf{MatrixForm[CompoundMatrix[2][G]]}$$

$$\begin{pmatrix} -g_{1,2}^2 + g_{1,1} g_{2,2} & -g_{1,2} g_{1,3} + g_{1,1} g_{2,3} & -g_{1,3} g_{2,2} + g_{1,2} g_{2,3} \\ -g_{1,2} g_{1,3} + g_{1,1} g_{2,3} & -g_{1,3}^2 + g_{1,1} g_{3,3} & -g_{1,3} g_{2,3} + g_{1,2} g_{3,3} \\ -g_{1,3} g_{2,2} + g_{1,2} g_{2,3} & -g_{1,3} g_{2,3} + g_{1,2} g_{3,3} & -g_{2,3}^2 + g_{2,2} g_{3,3} \end{pmatrix}$$

- You can also declare your metric and then use GradeMetric.

```
DeclareMetric[g]; MatrixForm[GradeMetric[2]]
```

$$\begin{pmatrix} -g_{1,2}^2 + g_{1,1} g_{2,2} & -g_{1,2} g_{1,3} + g_{1,1} g_{2,3} & -g_{1,3} g_{2,2} + g_{1,2} g_{2,3} \\ -g_{1,2} g_{1,3} + g_{1,1} g_{2,3} & -g_{1,3}^2 + g_{1,1} g_{3,3} & -g_{1,3} g_{2,3} + g_{1,2} g_{3,3} \\ -g_{1,3} g_{2,2} + g_{1,2} g_{2,3} & -g_{1,3} g_{2,3} + g_{1,2} g_{3,3} & -g_{2,3}^2 + g_{2,2} g_{3,3} \end{pmatrix}$$

◆ Composing G_m

We can define G and G_m so that they are composed automatically if entered, providing the basis and metric of the space have already been declared.

For example, consider a 3-space with diagonal metric. First declare the basis and the metric.

```
★𝓑3; DeclareMetric[{{g11, 0, 0}, {0, g22, 0}, {0, 0, g33}}];
Gm_ := Expand[GradeMetric[m]]
```

Then the metrics for the space can be composed by entering:

```
MatrixForm /@ {G0, G1, G2, G3}
```

$$\left\{(1),\ \begin{pmatrix} g_{11} & 0 & 0 \\ 0 & g_{22} & 0 \\ 0 & 0 & g_{33} \end{pmatrix},\ \begin{pmatrix} g_{11}\,g_{22} & 0 & 0 \\ 0 & g_{11}\,g_{33} & 0 \\ 0 & 0 & g_{22}\,g_{33} \end{pmatrix},\ (g_{11}\,g_{22}\,g_{33})\right\}$$

The determinant of an m-metric

As we noted above, we call a metric on $\underset{m}{\Lambda}$ an m-metric, and denote the matrix of its components by G_m. The determinant of the matrix G_m will be denoted $|G_m|$. This follows conventional notation and should not conflict with the meaning of the bracketing bar notation used elsewhere to signify the measure of an element, since G_m will always be an array of scalars, whilst a measure will always be of a single m-element.

◆ Determinants of scalars and n-elements

The only possibly conflicting cases are when the notation is applied to signify the determinant of an array containing a single scalar, a say, or to signify the measure of a scalar 0-element a. However, in both cases the result is the same.

$$|\{\{a\}\}| == a == |a|$$

Hence for G_0 and G_n we have

$$|G_0| == |\{\{1\}\}| == 1 == |1|$$

$$|G_n| == |\{\{\mathrm{Det}[G]\}\}| == |\{\{|G|\}\}| == |G|$$

◆ Determinants of metrics in 3-space

In the case of a 3-space, the only m-metric whose determinant we have not yet determined is G_2.

```
★𝓑3; DeclareMetric[g]; Gm_ := Expand[GradeMetric[m]];
MatrixForm[G2]
```

$$\begin{pmatrix} -g_{1,2}^2 + g_{1,1}\,g_{2,2} & -g_{1,2}\,g_{1,3} + g_{1,1}\,g_{2,3} & -g_{1,3}\,g_{2,2} + g_{1,2}\,g_{2,3} \\ -g_{1,2}\,g_{1,3} + g_{1,1}\,g_{2,3} & -g_{1,3}^2 + g_{1,1}\,g_{3,3} & -g_{1,3}\,g_{2,3} + g_{1,2}\,g_{3,3} \\ -g_{1,3}\,g_{2,2} + g_{1,2}\,g_{2,3} & -g_{1,3}\,g_{2,3} + g_{1,2}\,g_{3,3} & -g_{2,3}^2 + g_{2,2}\,g_{3,3} \end{pmatrix}$$

Since the coefficients of this determinant are of the 6th order, we guess that the determinant might be equal to the square of $|G|$. And indeed it is.

```
Expand[Det[G₂] == Det[G]²]
```

```
True
```

◆ **Determinants of *m*-metrics in an *n*-space.**

Since the determinant of the 2-metric in a 3-space can be expressed as the square of the determinant of G, let us explore if this result might generalize.

The number of components in G_m in an *n*-space is equal to the dimension of $\underset{m}{\Lambda}$, that is, $\binom{n}{m}$.

The order of the component polynomials of G_m is *m*.

The order of the component polynomials in the determinant of G_m is therefore $m\binom{n}{m}$.

The number of components in G, and hence the order of the polynomial Det[G] is *n*.

The expression for the power of Det[G] that we are seeking is therefore $\frac{m}{n}\binom{n}{m}$.

This expression simplifies to $\binom{n-1}{m-1}$.

Below, we test this out for all metrics in spaces up to and including dimension 4.

$$\mathrm{Det}[G_m] == \mathrm{Det}[G]^{\binom{n-1}{m-1}} \tag{5.43}$$

The formula also holds for square matrices that are not symmetric.

```
Table[★𝓑n; G = DeclareMetric[g]; Gm_ := Expand[GradeMetric[m]];
  Expand[Det[Gm] == Det[G]^Binomial[n-1,m-1]], {n, 1, 4}, {m, 0, n}]
```

```
{{True, True}, {True, True, True},
  {True, True, True, True}, {True, True, True, True, True}}
```

5.5 Cometrics

The relation between a metric and its cometric in 3-space

The elements of the basis $\underset{m}{B}$ of $\underset{m}{\Lambda}$ are (apart from their ordering and a possible sign) the same as the *cobasis* elements $\underline{\underset{n-m}{B}}$ of the basis $\underset{n-m}{B}$ of $\underset{n-m}{\Lambda}$. Concomitantly, the elements of the metric G_m are (apart from their ordering and a possible sign) the same as the elements of the *adjoint* of G_{n-m}.

In this section we look for the transformation which will convert between the metric G_m and its *cometric* G_{n-m}. We begin by exploring the relationship between $\underset{m}{B}$ and $\underline{\underset{n-m}{B}}$.

But to fix ideas, let us first take the case of a 3-space.

◆ Example in a 3-space

We can display the bases and cobases of a 3-space by entering `CobasisPalette`.

```
★𝓑3; CobasisPalette
```

Cobasis Palette

Basis	Cobasis
1	$e_1 \wedge e_2 \wedge e_3$
e_1	$e_2 \wedge e_3$
e_2	$-(e_1 \wedge e_3)$
e_3	$e_1 \wedge e_2$
$e_1 \wedge e_2$	e_3
$e_1 \wedge e_3$	$-e_2$
$e_2 \wedge e_3$	e_1
$e_1 \wedge e_2 \wedge e_3$	1

We want a matrix Q, say, which will transform the basis B of $\underset{1}{\Lambda}$ to the cobasis $\underset{\underline{2}}{\text{B}}$ of $\underset{2}{\Lambda}$.

$$\underset{\underline{2}}{\mathbf{B}} == \mathbf{Q\,B}$$

We can write Q by inspection from the palette entries.

$$\begin{pmatrix} e_3 \\ -e_2 \\ e_1 \end{pmatrix} == \begin{pmatrix} 0 & 0 & 1 \\ 0 & -1 & 0 \\ 1 & 0 & 0 \end{pmatrix} \begin{pmatrix} e_1 \\ e_2 \\ e_3 \end{pmatrix} \qquad Q = \begin{pmatrix} 0 & 0 & 1 \\ 0 & -1 & 0 \\ 1 & 0 & 0 \end{pmatrix}$$

Now let us take the metric G and its cometric G_2.

```
★𝓑3; DeclareMetric[g]; G = Metric; MatrixForm[G]
```

$$\begin{pmatrix} g_{1,1} & g_{1,2} & g_{1,3} \\ g_{1,2} & g_{2,2} & g_{2,3} \\ g_{1,3} & g_{2,3} & g_{3,3} \end{pmatrix}$$

```
G2 = GradeMetric[2]; MatrixForm[G2]
```

$$\begin{pmatrix} -g_{1,2}^2 + g_{1,1}\,g_{2,2} & -g_{1,2}\,g_{1,3} + g_{1,1}\,g_{2,3} & -g_{1,3}\,g_{2,2} + g_{1,2}\,g_{2,3} \\ -g_{1,2}\,g_{1,3} + g_{1,1}\,g_{2,3} & -g_{1,3}^2 + g_{1,1}\,g_{3,3} & -g_{1,3}\,g_{2,3} + g_{1,2}\,g_{3,3} \\ -g_{1,3}\,g_{2,2} + g_{1,2}\,g_{2,3} & -g_{1,3}\,g_{2,3} + g_{1,2}\,g_{3,3} & -g_{2,3}^2 + g_{2,2}\,g_{3,3} \end{pmatrix}$$

Transforming G_2 by Q (note that it is symmetric and thus equal to its own transpose) gives us the adjoint of G.

```
AdjG = Q.G2.Q; MatrixForm[AdjG]
```

$$\begin{pmatrix} -g_{2,3}^2 + g_{2,2}\,g_{3,3} & g_{1,3}\,g_{2,3} - g_{1,2}\,g_{3,3} & -g_{1,3}\,g_{2,2} + g_{1,2}\,g_{2,3} \\ g_{1,3}\,g_{2,3} - g_{1,2}\,g_{3,3} & -g_{1,3}^2 + g_{1,1}\,g_{3,3} & g_{1,2}\,g_{1,3} - g_{1,1}\,g_{2,3} \\ -g_{1,3}\,g_{2,2} + g_{1,2}\,g_{2,3} & g_{1,2}\,g_{1,3} - g_{1,1}\,g_{2,3} & -g_{1,2}^2 + g_{1,1}\,g_{2,2} \end{pmatrix}$$

It is straightforward to see that this matrix `AdjG` is indeed the adjoint of G.

```
Expand[G.AdjG == Det[G] IdentityMatrix[3]]
True
```

Thus *in a 3-space* we have that $Q.G_2.Q$ is the adjoint of G. That is:

$$G.Q.G_2.Q \;==\; |G|\; I$$

Since the square of Q is the identity matrix, pre- and post-multiplying this equation by Q and then simplifying gives the alternate result that $Q.G.Q$ is the adjoint of G_2.

$$Q.G.Q.G_2 \;==\; |G|\; I$$

Hence *in a 3-space only* we have that:

$$G.Q.G_2.Q \;==\; Q.G.Q.G_2 \;==\; |G|\; I \qquad 5.44$$

$$\text{Adj } G \;==\; Q.G_2.Q \qquad\qquad \text{Adj } G_2 \;==\; Q.G.Q \qquad 5.45$$

We will explore the general case below.

The general relation between a metric and its cometric

We are looking for a relationship between a metric G_m, and the adjoint of its cometric G_{n-m}. We begin by supposing that we know the relationship, defined by a matrix Q_m say, between a basis $\underset{m}{B}$, of $\underset{m}{\Lambda}$ and the cobasis $\underline{\underset{n-m}{B}}$ of $\underset{n-m}{\Lambda}$.

$$\underline{\underset{n-m}{B}} \;==\; Q_m \underset{m}{B}$$

But this relationship is also valid by replacing *m* by *n–m*.

$$\underline{\underset{m}{B}} \;==\; Q_{n-m} \underset{n-m}{B}$$

From formula [5.22] for the complements of basis *m*-elements we can also write the formula for the complements of basis (*n–m*)-elements.

$$\overline{\underset{m}{B}} \;==\; \frac{1}{\sqrt{|G|}}\, G_m\, \underline{\underset{m}{B}} \qquad\qquad \overline{\underset{n-m}{B}} \;==\; \frac{1}{\sqrt{|G|}}\, G_{n-m}\, \underline{\underset{n-m}{B}}$$

Now use the definitions involving Q_m and Q_{n-m} above to eliminate the cobases $\underline{\underset{m}{B}}$ and $\underline{\underset{n-m}{B}}$.

$$\overline{\underset{m}{B}} \;==\; \frac{1}{\sqrt{|G|}}\, G_m\, Q_{n-m}\, \underset{n-m}{B} \qquad\qquad \overline{\underset{n-m}{B}} \;==\; \frac{1}{\sqrt{|G|}}\, G_{n-m}\, Q_m\, \underset{m}{B}$$

Take the complement of the first of these equations,

$$\overline{\overline{\underset{m}{B}}} \;==\; \frac{1}{\sqrt{|G|}}\, G_m\, Q_{n-m}\, \overline{\underset{n-m}{B}} \;==\; (-1)^{m\,(n-m)}\, \underset{m}{B}$$

and then substitute for $\overline{\underset{n-m}{B}}$ from the second equation.

$$\frac{1}{\sqrt{|G|}} G_m \mathfrak{Q}_{n-m} \frac{1}{\sqrt{|G|}} G_{n-m} \mathfrak{Q}_m \underset{m}{B} == (-1)^{m(n-m)} \underset{m}{B}$$

$$G_m \mathfrak{Q}_{n-m} G_{n-m} \mathfrak{Q}_m == (-1)^{m(n-m)} |G| I \qquad 5.46$$

Premultiply both sides of this equation by G_m^{-1} and replace it by $\frac{1}{|G|}$ Adj[G_m].

$$\mathfrak{Q}_{n-m} G_{n-m} \mathfrak{Q}_m == (-1)^{m(n-m)} \text{Adj}[G_m] \qquad 5.47$$

Substituting *n–m* for *m* gives us the two alternative forms.

$$G_{n-m} \mathfrak{Q}_m G_m \mathfrak{Q}_{n-m} == (-1)^{m(n-m)} |G| I \qquad 5.48$$

$$\mathfrak{Q}_m G_m \mathfrak{Q}_{n-m} == (-1)^{m(n-m)} \text{Adj}[G_{n-m}] \qquad 5.49$$

◆ Example in a 2-space

We declare a general metric in a 2-space. Here G, G_1 and G_{2-1} are the same matrices.

```
★𝓑2; DeclareMetric[g]; G = Metric;
```

In the section to follow we will discover that $\mathfrak{Q}_1$, equal to $\mathfrak{Q}_{2-1}$ in a 2-space is:

$$\mathfrak{Q}_1 = \begin{pmatrix} 0 & 1 \\ -1 & 0 \end{pmatrix}$$

Composing equation [5.49] for these matrices gives:

```
Simplify[𝔔1 . G . 𝔔1 == - Det[G] Inverse[G]]
```

```
True
```

Computing a Q matrix

We want the matrix $\mathfrak{Q}_m$ such that

$$\underset{\underline{n-m}}{B} == \mathfrak{Q}_m \underset{m}{B}$$

Let us take the example of basis 2-elements in a 5-space.

```
★𝓑5; B = GradeBasis[2]
     2
```

$\{e_1 \wedge e_2, e_1 \wedge e_3, e_1 \wedge e_4, e_1 \wedge e_5,$
$e_2 \wedge e_3, e_2 \wedge e_4, e_2 \wedge e_5, e_3 \wedge e_4, e_3 \wedge e_5, e_4 \wedge e_5\}$

Then $\underset{\underline{n-m}}{B}$ becomes $\underset{\underline{3}}{B}$

```
B = GradeCobasis[3]
3
```

$\{e_4 \wedge e_5, -(e_3 \wedge e_5), e_3 \wedge e_4, e_2 \wedge e_5, -(e_2 \wedge e_4),$
$e_2 \wedge e_3, -(e_1 \wedge e_5), e_1 \wedge e_4, -(e_1 \wedge e_3), e_1 \wedge e_2\}$

If we reverse $\underset{3}{B}$, we get the basis elements of $\underset{2}{B}$ with the signs attached that we want to extract.

```
Reverse[B_3]
```

```
{e1 ∧ e2, -(e1 ∧ e3), e1 ∧ e4, -(e1 ∧ e5),
 e2 ∧ e3, -(e2 ∧ e4), e2 ∧ e5, e3 ∧ e4, -(e3 ∧ e5), e4 ∧ e5}
```

Mathematica interprets the quotient of two lists of equal length as the list of the quotients of their elements, so we can extract the list of signs as

$$S = \frac{\text{Reverse}[\underset{3}{B}]}{\underset{2}{B}}$$

```
{1, -1, 1, -1, 1, -1, 1, 1, -1, 1}
```

The transformation matrix Q_2 that we want can now be composed as the *reverse* of the diagonal matrix with these signs.

$$\begin{pmatrix} e_4 \wedge e_5 \\ -(e_3 \wedge e_5) \\ e_3 \wedge e_4 \\ e_2 \wedge e_5 \\ -(e_2 \wedge e_4) \\ e_2 \wedge e_3 \\ -(e_1 \wedge e_5) \\ e_1 \wedge e_4 \\ -(e_1 \wedge e_3) \\ e_1 \wedge e_2 \end{pmatrix} == \begin{pmatrix} 0 & 0 & 0 & 0 & 0 & 0 & 0 & 0 & 0 & 1 \\ 0 & 0 & 0 & 0 & 0 & 0 & 0 & 0 & -1 & 0 \\ 0 & 0 & 0 & 0 & 0 & 0 & 0 & 1 & 0 & 0 \\ 0 & 0 & 0 & 0 & 0 & 0 & 1 & 0 & 0 & 0 \\ 0 & 0 & 0 & 0 & 0 & -1 & 0 & 0 & 0 & 0 \\ 0 & 0 & 0 & 0 & 1 & 0 & 0 & 0 & 0 & 0 \\ 0 & 0 & 0 & -1 & 0 & 0 & 0 & 0 & 0 & 0 \\ 0 & 0 & 1 & 0 & 0 & 0 & 0 & 0 & 0 & 0 \\ 0 & -1 & 0 & 0 & 0 & 0 & 0 & 0 & 0 & 0 \\ 1 & 0 & 0 & 0 & 0 & 0 & 0 & 0 & 0 & 0 \end{pmatrix} \begin{pmatrix} e_1 \wedge e_2 \\ e_1 \wedge e_3 \\ e_1 \wedge e_4 \\ e_1 \wedge e_5 \\ e_2 \wedge e_3 \\ e_2 \wedge e_4 \\ e_2 \wedge e_5 \\ e_3 \wedge e_4 \\ e_3 \wedge e_5 \\ e_4 \wedge e_5 \end{pmatrix}$$

◆ Computing Q_m

These steps are straightforward to collect into a computational formula for the Q_m matrix of $\underset{m}{\Lambda}$ in the currently declared space. ★D is the dimension of the space.

```
Q_m_ :=
  Reverse[DiagonalMatrix[Reverse[GradeCobasis[★D - m]] / GradeBasis[m]]]
```

We can easily test out this formula. Below we make sure it works up to 10 dimensions.

```
And @@ Flatten[Table[★𝓑_n; Q_m.GradeBasis[m] == GradeCobasis[n - m],
    {n, 1, 10}, {m, 0, n}]]
```

```
True
```

◆ Q matrices for $\underset{0}{\Lambda}$ and $\underset{n}{\Lambda}$

Independent of the dimension of the space, the Q_0 and Q_n are simply 1×1 identity matrices.

```
Table[(★𝓑_n; MatrixForm /@ {Q_0, Q_n}), {n, 1, 3}] // Column
```

```
{(1), (1)}
{(1), (1)}
{(1), (1)}
```

◆ Example: The Q matrices for a 2-space

The only non-trivial Q matrix in a 2-space is Q_1.

```
★𝓑₂; Q₁ // MatrixForm
```

$$\begin{pmatrix} 0 & 1 \\ -1 & 0 \end{pmatrix}$$

Tabulating Q matrices

For future reference we tabulate the Q matrices in spaces of dimension 1 to 5. From this we can see that Q_m is equal to Q_{n-m} in all spaces up to and including 4-space. However in 5-space, Q_2 and Q_3 are not equal, but rather are transposes of each other.

◆ Q matrices in 1-space

```
★𝓑₁; MatrixForm /@ Table[Qₘ, {m, 0, 1}]
```

$$\{(1), (1)\}$$

◆ Q matrices in 2-space

```
★𝓑₂; MatrixForm /@ Table[Qₘ, {m, 0, 2}]
```

$$\left\{(1), \begin{pmatrix} 0 & 1 \\ -1 & 0 \end{pmatrix}, (1)\right\}$$

◆ Q matrices in 3-space

```
★𝓑₃; MatrixForm /@ Table[Qₘ, {m, 0, 3}]
```

$$\left\{(1), \begin{pmatrix} 0 & 0 & 1 \\ 0 & -1 & 0 \\ 1 & 0 & 0 \end{pmatrix}, \begin{pmatrix} 0 & 0 & 1 \\ 0 & -1 & 0 \\ 1 & 0 & 0 \end{pmatrix}, (1)\right\}$$

◆ Q matrices in 4-space

```
★𝓑₄; MatrixForm /@ Table[Qₘ, {m, 0, 4}]
```

$$\left\{(1), \begin{pmatrix} 0 & 0 & 0 & 1 \\ 0 & 0 & -1 & 0 \\ 0 & 1 & 0 & 0 \\ -1 & 0 & 0 & 0 \end{pmatrix}, \begin{pmatrix} 0 & 0 & 0 & 0 & 0 & 1 \\ 0 & 0 & 0 & 0 & -1 & 0 \\ 0 & 0 & 0 & 1 & 0 & 0 \\ 0 & 0 & 1 & 0 & 0 & 0 \\ 0 & -1 & 0 & 0 & 0 & 0 \\ 1 & 0 & 0 & 0 & 0 & 0 \end{pmatrix}, \begin{pmatrix} 0 & 0 & 0 & 1 \\ 0 & 0 & -1 & 0 \\ 0 & 1 & 0 & 0 \\ -1 & 0 & 0 & 0 \end{pmatrix}, (1)\right\}$$

◆ Q matrices in 5-space

```
★ℬ5; MatrixForm /@ Table[Qm, {m, 0, 2}]
```

$$\left\{(1),\ \begin{pmatrix} 0&0&0&0&1\\ 0&0&0&-1&0\\ 0&0&1&0&0\\ 0&-1&0&0&0\\ 1&0&0&0&0 \end{pmatrix},\ \begin{pmatrix} 0&0&0&0&0&0&0&0&0&1\\ 0&0&0&0&0&0&0&0&-1&0\\ 0&0&0&0&0&0&0&1&0&0\\ 0&0&0&0&0&0&1&0&0&0\\ 0&0&0&0&0&-1&0&0&0&0\\ 0&0&0&0&1&0&0&0&0&0\\ 0&0&0&-1&0&0&0&0&0&0\\ 0&0&1&0&0&0&0&0&0&0\\ 0&-1&0&0&0&0&0&0&0&0\\ 1&0&0&0&0&0&0&0&0&0 \end{pmatrix}\right\}$$

```
MatrixForm /@ Table[Qm, {m, 3, 5}]
```

$$\left\{\begin{pmatrix} 0&0&0&0&0&0&0&0&0&1\\ 0&0&0&0&0&0&0&0&-1&0\\ 0&0&0&0&0&0&0&1&0&0\\ 0&0&0&0&0&0&-1&0&0&0\\ 0&0&0&0&0&1&0&0&0&0\\ 0&0&0&0&-1&0&0&0&0&0\\ 0&0&0&1&0&0&0&0&0&0\\ 0&0&1&0&0&0&0&0&0&0\\ 0&-1&0&0&0&0&0&0&0&0\\ 1&0&0&0&0&0&0&0&0&0 \end{pmatrix},\ \begin{pmatrix} 0&0&0&0&1\\ 0&0&0&-1&0\\ 0&0&1&0&0\\ 0&-1&0&0&0\\ 1&0&0&0&0 \end{pmatrix},\ (1)\right\}$$

Properties of Q matrices

◆ Inverse

Because the Q matrices are dependent only on the ordering of the bases, and not on the metric, equations [5.46] to [5.49] will still be valid for any metric if G_m and G_{n-m} are replaced by the Euclidean metrics, that is, by identity matrices, and the determinant of G by unity.

$$\mathfrak{Q}_{n-m}\ \mathfrak{Q}_m == \mathfrak{Q}_m\ \mathfrak{Q}_{n-m} == (-1)^{m\,(n-m)}\ I \qquad 5.50$$

Thus we see that Q_m and Q_{n-m} are (possibly signed) inverses of each other.

$$\mathfrak{Q}_m^{-1} == (-1)^{m\,(n-m)}\ \mathfrak{Q}_{n-m} \qquad \mathfrak{Q}_{n-m}^{-1} == (-1)^{m\,(n-m)}\ \mathfrak{Q}_m \qquad 5.51$$

This is straightforward to test. Below we verify the result in spaces up to 10 dimensions.

```
And @@ Flatten[Table[★ℬn; (-1)^(m (n-m)) Qm.Qn-m
    == IdentityMatrix[Binomial[n, m]], {n, 1, 10}, {m, 0, n}]]
True
```

◆ Transpose

Because the entries in the Q matrices are just either +1 or -1, transposing reverses these numbers on the diagonal, and it is straightforward to see that the product of a Q matrix with its transpose is the identity.

Thus the transpose of a Q matrix is also its inverse.

$$\mathfrak{Q}_m^{-1} == \mathfrak{Q}_m^{T} == (-1)^{m\,(n-m)}\,\mathfrak{Q}_{n-m} \qquad 5.52$$

$$\mathfrak{Q}_{n-m}^{-1} == \mathfrak{Q}_{n-m}^{T} == (-1)^{m\,(n-m)}\,\mathfrak{Q}_m \qquad 5.53$$

```
And @@ Flatten[Table[★𝓑n; Qm.Transpose[Qm]
    == IdentityMatrix[Binomial[n, m]], {n, 1, 10}, {m, 0, n}]]
```

```
True
```

◆ Determinant

The determinants of the $\mathfrak{Q}_m$ and $\mathfrak{Q}_{n-m}$ may be either +1 or -1, but they are always of the same sign. We can show this by taking the determinant of [5.50] and noting that the determinant of a scalar multiple of an identity matrix is the multiple to the power of the order of the matrix.

$$|\mathfrak{Q}_m|\ |\mathfrak{Q}_{n-m}| == (-1)^{m\,(n-m)\binom{n}{m}}$$

We can show that the right hand side is always unity if we can show that the power is always even. The factor $m(n-m)$ is always even unless n is even and m is odd - so we only need to show that in this case $\binom{n}{m}$ is even. To show this we draw on a well-known relation for binomial coefficients (see for example http://mathworld.wolfram.com/BinomialCoefficient.html).

$$\binom{n}{m} == \frac{m+1}{n-m}\binom{n}{m+1}$$

We can see immediately that $\binom{n}{m}$ is even because the first factor is the quotient of an even integer by an odd integer - hence it must be even. This means the right hand side of the formula for the product of the two determinants above is always unity, and thus also that the determinants always have the same sign.

$$|\mathfrak{Q}_m|\ |\mathfrak{Q}_{n-m}| == 1 \qquad 5.54$$

Here are the actual determinants for spaces up to 12 dimensions. The left hand column is the dimension of the space.

```
Grid[Table[(★𝓑n; {n, Table[Det[𝔔m], {m, 0, n}]}), {n, 1, 12}]]
```

```
1                   {1, 1}
2                 {1, 1, 1}
3               {1, 1, 1, 1}
4             {1, 1, -1, 1, 1}
5           {1, 1, -1, -1, 1, 1}
6         {1, 1, -1, 1, -1, 1, 1}
7       {1, 1, -1, -1, -1, -1, 1, 1}
8        {1, 1, 1, 1, -1, 1, 1, 1, 1}
9      {1, 1, 1, 1, -1, -1, 1, 1, 1, 1}
10    {1, 1, 1, 1, -1, 1, -1, 1, 1, 1, 1}
11  {1, 1, 1, 1, -1, -1, -1, -1, 1, 1, 1, 1}
12 {1, 1, -1, 1, -1, 1, 1, 1, -1, 1, -1, 1, 1}
```

◆ Symmetry

A transformed symmetric matrix such as $\mathfrak{Q}_m\, G_m\, \mathfrak{Q}_{n-m}$ is itself symmetric since it reduces to its own transpose.

$$(\mathfrak{Q}_m\, G_m\, \mathfrak{Q}_{n-m})^T == \mathfrak{Q}_{n-m}{}^T\, G_m{}^T\, \mathfrak{Q}_m{}^T$$
$$== \left((-1)^{m\,(n-m)}\, \mathfrak{Q}_m\right) G_m \left((-1)^{m\,(n-m)}\, \mathfrak{Q}_{n-m}\right) == \mathfrak{Q}_m\, G_m\, \mathfrak{Q}_{n-m}$$

$$(\mathfrak{Q}_m\, G_m\, \mathfrak{Q}_{n-m})^T == \mathfrak{Q}_m\, G_m\, \mathfrak{Q}_{n-m} \qquad 5.55$$

Testing the formulae

We can use *Mathematica* to test out these formulae for example [5.48]. Below we test out the cases for all grades *m* in spaces of dimensions *n* up to and including 5. An output of `True` means all the cases calculated are true.

To make the verification process clearer, we first define a function which computes one of the equations for a given *n* and *m*.

```
GQD[n_, m_] := (★𝓑n; DeclareMetric[g]; Expand[Gn-m . 𝔔m . Gm . 𝔔n-m] ==
    Expand[(-1)^(m (n-m)) Det[Metric]] IdentityMatrix[Binomial[n, m]])
```

Now we use this function to calculate the 20 cases.

```
And @@ Flatten[Table[GQD[n, m], {n, 1, 5}, {m, 0, n}]]
True
```

5.6 The Euclidean Complement

Tabulating Euclidean complements of basis elements

The *Euclidean complement* of a basis *m*-element may be defined as its cobasis element, equivalent to the metric matrices G_m being identity matrices, and the measure ('volume') $|\underline{1}|$ of the basis n-element (also equal to the square root of the determinant of the metric G) equaling unity. Conceptually this is the simplest correspondence we can define between basis m-elements and basis $(n-m)$-elements.

Below we tabulate the basis-complement pairs for spaces of two, three and four dimensions by using the *GrassmannAlgebra* function `ComplementPalette`.

◆ Basis elements and their Euclidean complements in 2-space

```
★𝓑₂; ComplementPalette
```

Complement Palette

Basis	Complement
1	$e_1 \wedge e_2$
e_1	e_2
e_2	$-e_1$
$e_1 \wedge e_2$	1

◆ Basis elements and their Euclidean complements in 3-space

```
★𝓑₃; ComplementPalette
```

Complement Palette

Basis	Complement
1	$e_1 \wedge e_2 \wedge e_3$
e_1	$e_2 \wedge e_3$
e_2	$-(e_1 \wedge e_3)$
e_3	$e_1 \wedge e_2$
$e_1 \wedge e_2$	e_3
$e_1 \wedge e_3$	$-e_2$
$e_2 \wedge e_3$	e_1
$e_1 \wedge e_2 \wedge e_3$	1

◆ Basis elements and their Euclidean complements in 4-space

```
⋆𝓑4; ComplementPalette
```

Complement Palette

Basis	Complement
1	$e_1 \wedge e_2 \wedge e_3 \wedge e_4$
e_1	$e_2 \wedge e_3 \wedge e_4$
e_2	$-(e_1 \wedge e_3 \wedge e_4)$
e_3	$e_1 \wedge e_2 \wedge e_4$
e_4	$-(e_1 \wedge e_2 \wedge e_3)$
$e_1 \wedge e_2$	$e_3 \wedge e_4$
$e_1 \wedge e_3$	$-(e_2 \wedge e_4)$
$e_1 \wedge e_4$	$e_2 \wedge e_3$
$e_2 \wedge e_3$	$e_1 \wedge e_4$
$e_2 \wedge e_4$	$-(e_1 \wedge e_3)$
$e_3 \wedge e_4$	$e_1 \wedge e_2$
$e_1 \wedge e_2 \wedge e_3$	e_4
$e_1 \wedge e_2 \wedge e_4$	$-e_3$
$e_1 \wedge e_3 \wedge e_4$	e_2
$e_2 \wedge e_3 \wedge e_4$	$-e_1$
$e_1 \wedge e_2 \wedge e_3 \wedge e_4$	1

Formulae for the Euclidean complement of basis elements

In this section we summarize some of the formulae involving Euclidean complements of basis elements.

The Euclidean complement of a basis 1-element is its cobasis (*n*–1)-element. For a basis 1-element e_i then we have:

$$\overline{e_i} == \underline{e_i} == (-1)^{i-1}\, e_1 \wedge \cdots \wedge \square_i \wedge \cdots \wedge e_n \qquad 5.56$$

This simple relationship extends naturally to complements of basis elements of any grade.

$$\overline{\underset{m}{e_i}} == \underline{\underset{m}{e_i}} \qquad 5.57$$

This may also be written:

$$\overline{e_{i_1} \wedge \cdots \wedge e_{i_m}} == (-1)^{K_m}\, e_1 \wedge \cdots \wedge \square_{i_1} \wedge \cdots \wedge \square_{i_m} \wedge \cdots \wedge e_n$$
$$K_m == \sum_{\gamma=1}^{m} i_\gamma + \frac{1}{2}\, m\,(m+1) \qquad 5.58$$

where the symbol $\square_i$ means the corresponding element is missing from the product.

In particular, the unit *n*-element is now the basis *n*-element:

$$\overline{1} == \underline{1} == e_1 \wedge e_2 \wedge \cdots \wedge e_n \qquad 5.59$$

and the complement of the basis *n*-element is just unity.

$$1 == \overline{e_1 \wedge e_2 \wedge \cdots \wedge e_n} \qquad 5.60$$

The measure of the basis *n*-element is unity.

$$\left|\underline{1}\right| == \left|e_1 \wedge e_2 \wedge \cdots \wedge e_n\right| == 1 \qquad 5.61$$

The matrices G_m of all the metric tensors are just identity matrices I of order $\binom{n}{m}$.

$$g_{ij} == \delta_{ij} \qquad G_m == I \qquad 5.62$$

Finally, we can see that exterior and regressive products of basis elements with complements take on particularly simple forms.

$$\underset{m}{e_i} \wedge \overline{\underset{m}{e_j}} == \delta_{ij}\,\overline{1} \qquad 5.63$$

$$\underset{m}{e_i} \vee \overline{\underset{m}{e_j}} == \delta_{ij} \qquad 5.64$$

These forms will be the basis for the definition of the Euclidean inner product in the next chapter.

We note that the concept of cobasis which we introduced in section 2.6, despite its formal similarity to the Euclidean complement, is only a notational convenience. We do not define it for linear combinations of elements as we do for the complement operation.

Products leading to a scalar or *n*-element

In this section we collect together some simple *Euclidean* results which we will find useful in the rest of the book.

From the basis element formulae above we can immediately derive formulae for the exterior and regressive products of two *m*-elements A and B, say.

$$A == \sum_i a_i \underset{m}{e_i} \qquad B == \sum_i b_i \underset{m}{e_i}$$

$$\overline{A} == \sum_i a_i \overline{\underset{m}{e_i}} \qquad \overline{B} == \sum_i b_i \overline{\underset{m}{e_i}}$$

$$A \wedge \overline{B} == \left(\sum_i a_i \underset{m}{e_i}\right) \wedge \left(\sum_k b_k \overline{\underset{m}{e_k}}\right) == \left(\sum_i a_i b_i\right) \overline{1}$$

$$A \wedge \overline{B} == \left(\sum_i a_i b_i\right) \overline{1} \qquad 5.65$$

Taking the complement of this formula, or else doing a derivation *mutatis mutandis*, leads to the regressive product form.

$$A \vee \overline{B} == \sum_i a_i b_i \qquad 5.66$$

Hence formula [5.65] can be written

$$A \wedge \overline{B} == \left(A \vee \overline{B}\right) \overline{1} \qquad 5.67$$

A common case is for the 1-element.

$$x == \sum_i a_i e_i \qquad y == \sum_i b_i e_i$$

$$x \wedge \overline{y} == \left(\sum_i a_i b_i\right) \overline{1} \qquad x \vee \overline{y} == \sum_i a_i b_i \qquad 5.68$$

In the case A is equal to B, we obtain

$$A \wedge \overline{A} == \left(\sum_i a_i^2\right) \overline{1} \qquad A \vee \overline{A} == \sum_i a_i^2 \qquad 5.69$$

5.7 Exploring Complements

Alternative forms for complements

As a consequence of the complement axioms [5.3] and [5.4] and the fact that multiplication by a scalar is equivalent to exterior multiplication (see section 2.4) we can write the complement of a scalar, an *m*-element, and a scalar multiple of an *m*-element in several alternative forms.

◆ Alternative forms for the complement of a scalar

From the linearity axiom $\overline{2}$ [5.2] we have $\overline{a\,A} = a\,\overline{A}$, hence:

$$\overline{a} == \overline{a\,1} == a\,\overline{1} == a \wedge \overline{1}$$

The complement of a scalar can then be expressed in any of the following forms:

$$\overline{a} == \overline{1 \wedge a} == \overline{1} \vee \overline{a} == 1 \wedge \overline{a} == 1\,\overline{a} == a \wedge \overline{1} == a\,\overline{1} \qquad 5.70$$

◆ **Alternative forms for the complement of an *m*-element**

$$\overline{A} == \overline{1 \wedge A} == \overline{1} \vee \overline{A} == 1 \wedge \overline{A} == 1\, \overline{A} \qquad 5.71$$

◆ **Alternative forms for the complement of a scalar multiple of an *m*-element**

$$\overline{a\, A} == \overline{a \wedge A} == \overline{a} \vee \overline{A} == a \wedge \overline{A} == a\, \overline{A} \qquad 5.72$$

Orthogonality

The specific complement mapping that we impose on a Grassmann algebra will *define* the notion of orthogonality for that algebra. A simple element and its complement (or scalar multiples of them) will be referred to as being *orthogonal* to each other. In standard linear space terminology, the *space* of a simple element and the *space* of its complement are said to be *orthogonal complements* of each other.

This orthogonality is *total*. That is, every 1-element in a given simple *m*-element is orthogonal to every 1-element in the complement of the *m*-element.

In the special case of a 2-dimensional vector space, the complement of a vector is itself a vector, and we *represent* this orthogonality geometrically by depicting the vectors at right angles to each other.

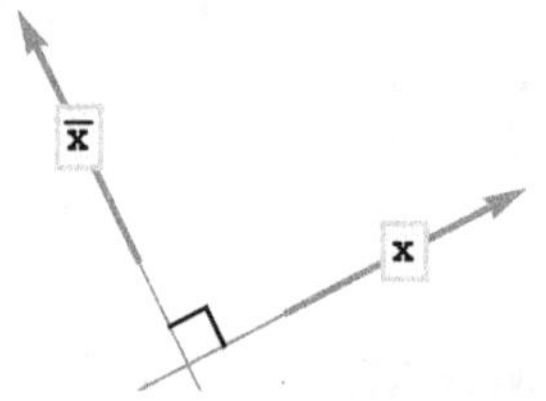

A vector and its complement in a two-dimensional vector space

In a vector three-space, the complement of a vector is a bivector. And of course the complement of a bivector is a vector.

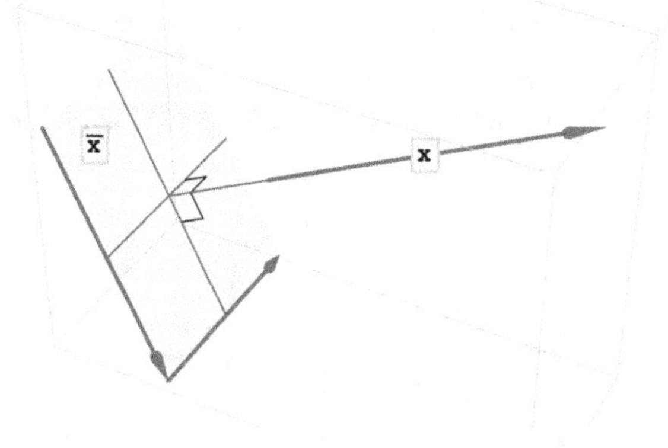

A vector and its bivector complement in a three-dimensional vector space

More generally, we define two simple elements A and B of the same grade to be orthogonal if the exterior product of one with the complement of the other is zero.

$$\overline{A} \wedge B == 0 \qquad A \wedge \overline{B} == 0 \qquad 5.73$$

Note carefully that this definition only requires $\overline{A}$ and B (or $\overline{B}$ and A) to have just *one* 1-element in common for their exterior product to be zero, and hence for them to be deemed orthogonal. If one of the factors in the exterior product totally contains the other we say that A and B are *totally orthogonal*.

By taking the complement of these equations, and applying both the complement axiom [5.3], and the complement of a complement axiom [5.7], we see that we can also say that two simple elements of the same grade A and B, are orthogonal if the regressive product of one with the complement of the other is zero.

$$\overline{A} \vee B == 0 \qquad A \vee \overline{B} == 0 \qquad 5.74$$

We will develop these results in more depth in Chapter 6.

◆ Examples

- In a vector 2-space, a vector x has a vector complement $\overline{x}$ which, by definition, is said to be orthogonal to it. A second vector y such that $\overline{x} \wedge y = 0$ must be congruent to (that is, a scalar multiple of) $\overline{x}$, and hence by definition, orthogonal to x. Similarly for $x \wedge \overline{y} = 0$.

- In a vector 3-space, a vector x has a bivector complement $\overline{x}$ which, by definition, is said to be orthogonal to it. A second vector y such that $\overline{x} \wedge y = 0$ must belong to the bivector $\overline{x}$, and hence by definition, be orthogonal to x.

- In a vector 3-space, a bivector A has a vector complement $\overline{A}$ which, by definition, is said to be orthogonal to it. A second bivector B such that $\overline{A} \wedge B = 0$ must contain the vector $\overline{A}$, and hence by definition, be orthogonal to A.

Visualizing the complement axiom

We can use this notion of orthogonality to visualize the complement axiom geometrically.

Consider the bivector $x \wedge y$. Then $\overline{x \wedge y}$ is orthogonal to $x \wedge y$. But since $\overline{x \wedge y} = \overline{x} \vee \overline{y}$ this also means that the intersection of the two $(n-1)$-spaces defined by $\overline{x}$ and $\overline{y}$ is orthogonal to $x \wedge y$. We can depict this in 3-space as follows:

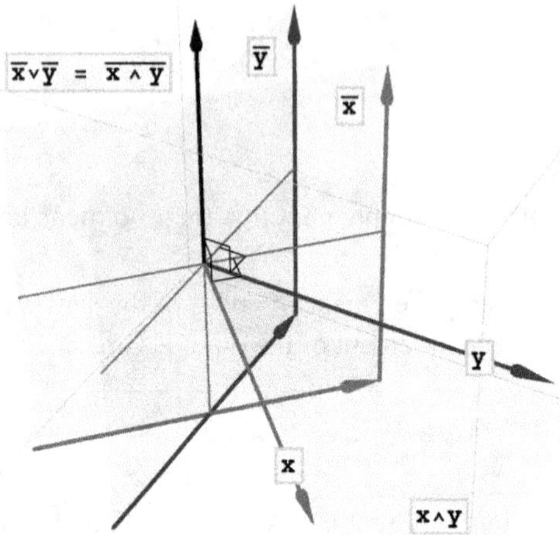

Visualizing the complement axiom in vector three-space

The regressive product in terms of complements

All the operations in the Grassmann algebra can be expressed in terms *only* of the exterior product and the complement operations. It is this fact that makes the complement so important for an understanding of the algebra.

In particular the regressive product (discussed in Chapter 3) and the interior product (to be discussed in the next chapter) have simple representations in terms of the exterior and complement operations.

We have already introduced the complement axiom $\overline{3}$ [5.4] as part of the definition of the complement.

$$\overline{\underset{m}{\alpha} \vee \underset{k}{\beta}} == \overline{\underset{m}{\alpha}} \wedge \overline{\underset{k}{\beta}}$$

Taking the complement of both sides of this axiom, noting that the grade of $\underset{m}{\alpha} \vee \underset{k}{\beta}$ is $m+k-n$, and using the complement of a complement axiom $\overline{4}$ [5.7] gives:

$$\overline{\overline{\underset{m}{\alpha} \vee \underset{k}{\beta}}} == \overline{\overline{\underset{m}{\alpha}} \wedge \overline{\underset{k}{\beta}}} == (-1)^{(m+k)\,(m+k-n)}\, \underset{m}{\alpha} \vee \underset{k}{\beta}$$

Hence, any regressive product can be written as the complement of an exterior product of complements.

$$\underset{m}{\alpha} \vee \underset{k}{\beta} == (-1)^{(m+k)\,(m+k-n)}\,\overline{\overline{\underset{m}{\alpha}} \wedge \overline{\underset{k}{\beta}}} \qquad 5.75$$

The expression of a regressive product in terms of an exterior product and a *double* application of the complement operation poses an interesting conundrum. On the left hand side, there is a product which is defined in a completely non-metric way, and is thus *independent of any metric* imposed on the space. On the right hand side we have a double application of the complement operation, *each of which requires a metric*. This leads us to the notion that *the complement operation applied twice in this way is independent of any metric used to define it*. A second application of the complement operation effectively 'cancels out' the metric elements introduced during the first application.

The complement of a simple element is simple

The complement of a simple element is simple.

Consider an simple element A expressed in factored form as $\alpha_1 \wedge \alpha_2 \wedge \ldots$ where the α_i are 1-elements. Take the complement of A and using the complement axiom [5.5] express it as the regressive product of complements of its factors.

$$\overline{A} == \overline{\alpha_1 \wedge \alpha_2 \wedge \ldots} == \overline{\alpha_1} \vee \overline{\alpha_2} \vee \ldots$$

Since each of the (n–1)-elements $\overline{\alpha_i}$ is simple (see section 2.10), their regressive product (see section 3.7) is simple, and the assertion is proven.

Glimpses of the inner product

In this section we summarize some of the comments we have made to this point regarding the relationship of the complement operation to the inner product and measure operations that we will explore in the next chapter.

The exterior product of an *m*-element with the complement of another *m*-element is either zero, or an *n*-element. If it is zero, we say that the elements are orthogonal. Orthogonality was discussed in a previous section.

If on the other hand the product is an *n*-element, it must be a scalar multiple (a, say) of the unit *n*-element $\overline{1}$. Suppose A and B are such *m*-elements.

$$A \wedge \overline{B} == a\,\overline{1}$$

Taking the complement of this equation gives the scalar a.

$$\overline{A \wedge \overline{B}} == a\,\overline{\overline{1}} == a$$

But we can also rewrite the left hand side as a regressive product:

$$\overline{A \wedge \overline{B}} == \overline{A} \vee \overline{\overline{B}} == (-1)^{m\,(n-m)}\,\overline{A} \vee B == B \vee \overline{A}$$

$$A \wedge \overline{B} == \left(B \vee \overline{A}\right)\overline{1} \qquad 5.76$$

In the case that the two *m*-elements are identical we have the result that the exterior product of an *m*-element with its complement is a scalar multiple of the unit *n*-element, and that this multiple is the regressive product of the *m*-element with its complement. For an *m*-element A, this scalar can be used to define its *measure*, denoted $|A|$.

$$A \wedge \overline{A} == \left(A \vee \overline{A}\right) \overline{1} == |A|^2 \overline{1} \qquad 5.77$$

We will see in the following chapter how these formulae are the foundation of the definition of the inner product.

Idempotent complements

In the way we have defined the complement, no *simple* element (except zero) can be equal to its own complement. This is because the exterior product of a simple element with its complement is congruent to the unit *n*-element (which is not zero).

However, this is not necessarily true for *non-simple* elements. Let S be the sum of a simple *m*-element ($\underset{m}{\alpha}$, say) and its complement.

$$S == \underset{m}{\alpha} + \overline{\underset{m}{\alpha}}$$

Then the complement of S is equal to

$$\overline{S} == \overline{\underset{m}{\alpha}} + \overline{\overline{\underset{m}{\alpha}}} == \overline{\underset{m}{\alpha}} + (-1)^{m\,(n-m)} \underset{m}{\alpha}$$

Since $(-1)^{m\,(n-m)}$ is equal to $(-1)^{m\,(n-1)}$, we can see that $S = \overline{S}$ if and only if the grade *m* is even or the dimension *n* of the space is odd.

$$\overline{\underset{m}{\alpha} + \overline{\underset{m}{\alpha}}} == \underset{m}{\alpha} + \overline{\underset{m}{\alpha}} \qquad (-1)^{m\,(n-m)} == 1 \qquad 5.78$$

◆ **Example: When the grade of an element is even**

In a space of arbitrary dimension, the complement of the sum of a scalar a and its *n*-element complement is identical to itself.

$$\overline{a + \overline{a}} == a + \overline{a}$$

In particular, for the case that a is unity:

$$\overline{1 + \overline{1}} == 1 + \overline{1}$$

Again in a space of arbitrary dimension, the complement of the sum of a 2-element B and its (*n*–2)-element complement $\overline{B}$ does not change the sum.

$$\overline{B + \overline{B}} == B + \overline{B}$$

◆ **Example: When the dimension of the space is 3**

In a 3-space, the complement of a sum of a 1-element and its 2-element complement is identical to itself.

$$\overline{x + \overline{x}} == x + \overline{x}$$

And the complement of a sum of a 2-element and its 1-element complement is also identical to itself.

$$\overline{B + \overline{B}} == B + \overline{B}$$

5.8 Working with Metrics

Working with metrics

In this section we collect in one place the various ways in which you can use *GrassmannAlgebra* to work with metrics. Some of the techniques discussed have already been used in the previous sections.

In *GrassmannAlgebra*, there are three basic ways in which you might apply a metric.

First, you can transform expressions involving interior, inner, or scalar products to expressions specifically involving scalar products of basis elements. Your results involving these scalar products of basis elements are valid for any metric. If you wish to make a specific metric apply to your results, you can declare that metric, and use `ToMetricElements` to perform the substitution of the currently declared metric elements for the corresponding scalar products of basis elements. We will discuss these transformations after we have introduced the interior product in the next chapter.

Second, if you have elements expressed in terms of basis elements, you can compute their complements by first taking their complement (applying `GrassmannComplement`, or using the `OverBar` in the *GrassmannAlgebra* palette), and then applying `ConvertComplements`. This will use the currently declared metric to convert the complements of basis elements to complementary basis elements. We will discuss these transformations in section 5.9 below.

Third, if you want to display the metrics induced by the various exterior product spaces, you can use `GradeMetric` for an individual space or `MetricPalette` to display all the induced metrics. We discuss these in the sections below.

The default metric

In order to simplify complements and interior products, and any products defined in terms of them, *GrassmannAlgebra* needs to know what metric has been imposed on the underlying linear space $\underset{1}{\Lambda}$.

Grassmann and all those writing in the early tradition of the *Ausdehnungslehre* tacitly assumed a Euclidean metric; that is, one in which $g_{ij} = \delta_{ij}$. This metric is also the one tacitly assumed in beginning presentations of the three-dimensional vector calculus, and is most evident in the definition of the cross product as a vector *normal* to the factors of the product.

The default metric assumed by *GrassmannAlgebra* is the Euclidean metric. In the case that the *GrassmannAlgebra* package has just been loaded, entering `Metric` will show the components of the default Euclidean metric tensor.

```
Metric
{{1, 0, 0}, {0, 1, 0}, {0, 0, 1}}
```

These components are arranged as the elements of a 3×3 matrix because the default basis is 3-dimensional.

```
Basis
```

$\{e_1, e_2, e_3\}$

However, if the dimension of the basis is changed, the default metric changes accordingly.

```
DeclareBasis[{★O, i, j, k}]
```

```
{★O, i, j, k}
```

```
Metric
```

```
{{1, 0, 0, 0}, {0, 1, 0, 0}, {0, 0, 1, 0}, {0, 0, 0, 1}}
```

We will sometimes take the liberty of referring to the currently declared metric as *the* metric.

Declaring a metric

GrassmannAlgebra permits you to declare a metric as a matrix with any numeric or symbolic components. There are four conditions to which a valid matrix must conform in order to be a metric:

1. It must be symmetric (and hence square).
2. Its order must be the same as the dimension of the declared linear space.
3. It must be non-singular.
4. Its components must be scalars.

It is up to the user to ensure the first three conditions. The fourth is handled by *GrassmannAlgebra* automatically assuming all symbols in the matrix to be scalars.

◆ Example: Declaring a metric

We remove all default declarations and return to a vector 3-space by entering `★A` from the palette, create a 3×3 matrix `G`, and then declare it as the metric.

```
★A; G = {{α, 0, ν}, {0, β, 0}, {ν, 0, γ}}; DeclareMetric[G];
MatrixForm[G]
```

$$\begin{pmatrix} \alpha & 0 & \nu \\ 0 & \beta & 0 \\ \nu & 0 & \gamma \end{pmatrix}$$

We can verify that this is indeed the metric by entering `Metric`.

```
Metric
```

$\{\{\alpha, 0, \nu\}, \{0, \beta, 0\}, \{\nu, 0, \gamma\}\}$

We can verify that *GrassmannAlgebra* now considers α, β, and ν to be scalars by using `ScalarQ`.

```
ScalarQ[{α, β, ν}]
```

```
{True, True, True}
```

Declaring a general metric

For theoretical calculations it is sometimes useful to be able quickly to declare a metric of general symbolic elements. We can do this as described in the previous section, or we can use the *GrassmannAlgebra* function `DeclareMetric[g]` where `g` is any symbol.

```
★𝔅3; G = DeclareMetric[g]; MatrixForm[G]
```

$$\begin{pmatrix} g_{1,1} & g_{1,2} & g_{1,3} \\ g_{1,2} & g_{2,2} & g_{2,3} \\ g_{1,3} & g_{2,3} & g_{3,3} \end{pmatrix}$$

Note that the matrix `G` is symmetric.

You can check that the components of `G` are now considered scalars.

```
ScalarQ[G]
```

```
{{True, True, True}, {True, True, True}, {True, True, True}}
```

You can test to see if a symbol is a component of the currently declared metric by using `MetricSymbolQ`.

```
MetricSymbolQ[{g2,3, g23, g3,4}]
```

```
{True, False, False}
```

Calculating induced metrics

The *GrassmannAlgebra* function for calculating the metric induced on $\underset{m}{\Lambda}$ by the metric in $\underset{1}{\Lambda}$ is `GradeMetric[m]`.

◆ Induced general metrics

Suppose you are working in 3-space with the general metric defined above. Then the metrics on $\underset{0}{\Lambda}, \underset{1}{\Lambda}, \underset{2}{\Lambda}, \underset{3}{\Lambda}$ can be calculated as follows:

```
★𝔅3; DeclareMetric[g];
```

```
G0 = GradeMetric[0]; MatrixForm[G0]
```

$$(1)$$

```
G1 = GradeMetric[1]; MatrixForm[G1]
```

$$\begin{pmatrix} g_{1,1} & g_{1,2} & g_{1,3} \\ g_{1,2} & g_{2,2} & g_{2,3} \\ g_{1,3} & g_{2,3} & g_{3,3} \end{pmatrix}$$

```
G2 = GradeMetric[2]; MatrixForm[G2]
```

$$\begin{pmatrix} -g_{1,2}^2 + g_{1,1}\, g_{2,2} & -g_{1,2}\, g_{1,3} + g_{1,1}\, g_{2,3} & -g_{1,3}\, g_{2,2} + g_{1,2}\, g_{2,3} \\ -g_{1,2}\, g_{1,3} + g_{1,1}\, g_{2,3} & -g_{1,3}^2 + g_{1,1}\, g_{3,3} & -g_{1,3}\, g_{2,3} + g_{1,2}\, g_{3,3} \\ -g_{1,3}\, g_{2,2} + g_{1,2}\, g_{2,3} & -g_{1,3}\, g_{2,3} + g_{1,2}\, g_{3,3} & -g_{2,3}^2 + g_{2,2}\, g_{3,3} \end{pmatrix}$$

```
G3 = GradeMetric[3]; MatrixForm[G3]
```

$$\left(-g_{1,3}^2\, g_{2,2} + 2\, g_{1,2}\, g_{1,3}\, g_{2,3} - g_{1,1}\, g_{2,3}^2 - g_{1,2}^2\, g_{3,3} + g_{1,1}\, g_{2,2}\, g_{3,3} \right)$$

Note that the metrics are either scalar, or symmetric. The metric in $\underset{3}{\Lambda}$ is of course just the determinant of the metric in $\underset{1}{\Lambda}$.

◆ Example: Induced specific metrics

We return to the metric discussed in a previous example.

```
★𝓑₃; G = {{α, 0, ν}, {0, β, 0}, {ν, 0, γ}}; DeclareMetric[G];
MatrixForm[G]
```

$$\begin{pmatrix} \alpha & 0 & \nu \\ 0 & \beta & 0 \\ \nu & 0 & \gamma \end{pmatrix}$$

The induced metrics are

```
MatrixForm[GradeMetric[#]] & /@ {0, 1, 2, 3}
```

$$\left\{ (1), \begin{pmatrix} \alpha & 0 & \nu \\ 0 & \beta & 0 \\ \nu & 0 & \gamma \end{pmatrix}, \begin{pmatrix} \alpha\beta & 0 & -\beta\nu \\ 0 & \alpha\gamma - \nu^2 & 0 \\ -\beta\nu & 0 & \beta\gamma \end{pmatrix}, (\alpha\beta\gamma - \beta\nu^2) \right\}$$

◆ Verifying the symmetry of the induced metrics

It is easy to verify the symmetry of the induced metrics in any particular case by inspection. However to automate this we need only ask *Mathematica* to compare the metric to its transpose. For example we can verify the symmetry of the metric induced on $\underset{3}{\Lambda}$ by a general metric in 4-space.

```
★𝓑₄; DeclareMetric[g]; G₃ = GradeMetric[3];
G₃ == Transpose[G₃]
True
```

Creating palettes of induced metrics

You can create a palette of all the induced metrics by declaring a metric and then entering `MetricPalette`.

The induced metrics of a Euclidean vector 4-space are all Euclidean.

```
★A; ★𝓑4; MetricPalette
```

Metric Palette

$\underset{0}{\Lambda}$	(1)
$\underset{1}{\Lambda}$	$\begin{pmatrix} 1 & 0 & 0 & 0 \\ 0 & 1 & 0 & 0 \\ 0 & 0 & 1 & 0 \\ 0 & 0 & 0 & 1 \end{pmatrix}$
$\underset{2}{\Lambda}$	$\begin{pmatrix} 1 & 0 & 0 & 0 & 0 & 0 \\ 0 & 1 & 0 & 0 & 0 & 0 \\ 0 & 0 & 1 & 0 & 0 & 0 \\ 0 & 0 & 0 & 1 & 0 & 0 \\ 0 & 0 & 0 & 0 & 1 & 0 \\ 0 & 0 & 0 & 0 & 0 & 1 \end{pmatrix}$
$\underset{3}{\Lambda}$	$\begin{pmatrix} 1 & 0 & 0 & 0 \\ 0 & 1 & 0 & 0 \\ 0 & 0 & 1 & 0 \\ 0 & 0 & 0 & 1 \end{pmatrix}$
$\underset{4}{\Lambda}$	(1)

Here are the metrics for a non-Euclidean vector 3-space.

```
★𝓑3; G = {{α, 0, ν}, {0, β, 0}, {ν, 0, γ}}; DeclareMetric[G];
MetricPalette
```

Metric Palette

$\underset{0}{\Lambda}$	(1)
$\underset{1}{\Lambda}$	$\begin{pmatrix} \alpha & 0 & \nu \\ 0 & \beta & 0 \\ \nu & 0 & \gamma \end{pmatrix}$
$\underset{2}{\Lambda}$	$\begin{pmatrix} \alpha\beta & 0 & -\beta\nu \\ 0 & \alpha\gamma - \nu^2 & 0 \\ -\beta\nu & 0 & \beta\gamma \end{pmatrix}$
$\underset{3}{\Lambda}$	$(\alpha\beta\gamma - \beta\nu^2)$

Here are general metrics for a vector 3-space.

```
⋆𝔅3; DeclareMetric[g]; MetricPalette
```

Metric Palette

$\underset{0}{\Lambda}$	(1)
$\underset{1}{\Lambda}$	$\begin{pmatrix} g_{1,1} & g_{1,2} & g_{1,3} \\ g_{1,2} & g_{2,2} & g_{2,3} \\ g_{1,3} & g_{2,3} & g_{3,3} \end{pmatrix}$
$\underset{2}{\Lambda}$	$\begin{pmatrix} -g_{1,2}^2 + g_{1,1} g_{2,2} & -g_{1,2} g_{1,3} + g_{1,1} g_{2,3} & -g_{1,3} g_{2,2} + g_{1,2} g_{2,3} \\ -g_{1,2} g_{1,3} + g_{1,1} g_{2,3} & -g_{1,3}^2 + g_{1,1} g_{3,3} & -g_{1,3} g_{2,3} + g_{1,2} g_{3,3} \\ -g_{1,3} g_{2,2} + g_{1,2} g_{2,3} & -g_{1,3} g_{2,3} + g_{1,2} g_{3,3} & -g_{2,3}^2 + g_{2,2} g_{3,3} \end{pmatrix}$
$\underset{3}{\Lambda}$	$(-g_{1,3}^2 g_{2,2} + 2 g_{1,2} g_{1,3} g_{2,3} - g_{1,1} g_{2,3}^2 - g_{1,2}^2 g_{3,3} + g_{1,1} g_{2,2} g_{3,3})$

You can paste any of these metric matrices into your notebook by clicking on the palette.

The determinant of the metric tensor

The determinant of the metric tensor is an important scalar of any metric space. Earlier in this chapter we denoted the determinant of a metric tensor G as $|G|$, and showed in section 5.3 that this determinant is the square of the measure of the basis *n*-element of the space [5.20].

$$|G| == |e_1 \wedge \ldots \wedge e_n|^2 == |\underline{1}|^2$$

In order to compute with metrics we need a symbol which denotes the determinant of the currently declared metric, and in *GrassmannAlgebra* this symbol is ⋆g. You can evaluate ⋆g with `ToMetricElements`. For example, if you are working in a 3-space and declare a metric G you could then enter `ToMetricElements[⋆g]` to evaluate the determinant.

```
⋆𝔅3; DeclareMetric[α]; ToMetricElements[⋆g]
```

$$-\alpha_{1,3}^2 \alpha_{2,2} + 2 \alpha_{1,2} \alpha_{1,3} \alpha_{2,3} - \alpha_{1,1} \alpha_{2,3}^2 - \alpha_{1,2}^2 \alpha_{3,3} + \alpha_{1,1} \alpha_{2,2} \alpha_{3,3}$$

Hence we extend the expressions for this important scalar with a computationally-oriented one.

$$|G| == |e_1 \wedge \ldots \wedge e_n|^2 == |\underline{1}|^2 == \star g$$

$$|G| == \mathrm{Det}[G] == \star g \qquad 5.79$$

Most often this scalar is used to "normalize" or "unitize" an expression in some way. So that whereas a formula might display $|\underline{1}|$, an expression resulting from a computation will display $\sqrt{\star g}$.

5.9 Calculating Complements

Entering complements

To enter the expression for the complement of a Grassmann expression X in *GrassmannAlgebra*, enter `GrassmannComplement[X]`. For example to enter the expression for the complement of

$4\, e_1 \wedge e_2$, enter:

```
GrassmannComplement[4 e1 ∧ e2]
```

$\overline{4\, e_1 \wedge e_2}$

Or, you can simply select the expression and click the ■ button on the *GrassmannAlgebra* palette.

Note that an expression with a bar over it symbolizes another expression: the complement of the original expression - not a command to effect a transformation. Hence no transformations will occur by entering an overbarred expression.

`GrassmannComplement` will also work for lists, matrices, or tensors of elements. For example, here is a matrix:

```
M = {{b e1 ∧ e2, d e2}, {-d e2, f e3 ∧ e1}}; MatrixForm[M]
```

$$\begin{pmatrix} b\, e_1 \wedge e_2 & d\, e_2 \\ -d\, e_2 & f\, e_3 \wedge e_1 \end{pmatrix}$$

And here is its complement:

```
MatrixForm[M̄]
```

$$\begin{pmatrix} \overline{b\, e_1 \wedge e_2} & \overline{d\, e_2} \\ \overline{-d\, e_2} & \overline{f\, e_3 \wedge e_1} \end{pmatrix}$$

Creating palettes of complements of basis elements

GrassmannAlgebra has a function `ComplementPalette` for tabulating basis elements and their complements in the currently declared basis and the currently declared metric. Once the palette is generated you can paste from it by clicking on any of the buttons.

◆ Euclidean metric

Here we repeat the construction of the Euclidean complement palette of section 5.6 in a 2-space.

```
★𝓑2; ComplementPalette
```

Complement Palette

Basis	Complement
1	$e_1 \wedge e_2$
e_1	e_2
e_2	$-e_1$
$e_1 \wedge e_2$	1

Palettes of complements for other bases are obtained by declaring the bases, then entering the command `ComplementPalette`.

```
DeclareBasis[{j, k}]; ComplementPalette
```

Complement Palette

Basis	Complement
1	$j \wedge k$
j	k
k	$-j$
$j \wedge k$	1

◆ Non-Euclidean metric

◇ 2-space

Here is the complement palette in a 2-space with general metric.

```
⋆ℬ2; DeclareMetric[g]; ComplementPalette
```

Complement Palette

Basis	Complement
1	$\frac{e_1 \wedge e_2}{\sqrt{\star g}}$
e_1	$\frac{e_2\, g_{1,1}}{\sqrt{\star g}} - \frac{e_1\, g_{1,2}}{\sqrt{\star g}}$
e_2	$\frac{e_2\, g_{1,2}}{\sqrt{\star g}} - \frac{e_1\, g_{2,2}}{\sqrt{\star g}}$
$e_1 \wedge e_2$	$\sqrt{\star g}$

- The symbol $\star g$ represents the determinant of the metric tensor. It (or expressions involving it) can be evaluated in any space with any declared metric by using `ToMetricElements`.

```
ToMetricElements[⋆g]
```

$-g_{1,2}^2 + g_{1,1}\, g_{2,2}$

The palette is arranged to 'collapse' the full determinant expression under the symbol $\star g$ since it is often large. However if you wish, you can include the full form of $\star g$ in the palette by substituting for it.

```
ComplementPalette /. ★g → ToMetricElements[★g]
```

Complement Palette

Basis	Complement
1	$\frac{e_1 \wedge e_2}{\sqrt{-g_{1,2}^2+g_{1,1} g_{2,2}}}$
e_1	$\frac{e_2 g_{1,1}}{\sqrt{-g_{1,2}^2+g_{1,1} g_{2,2}}} - \frac{e_1 g_{1,2}}{\sqrt{-g_{1,2}^2+g_{1,1} g_{2,2}}}$
e_2	$\frac{e_2 g_{1,2}}{\sqrt{-g_{1,2}^2+g_{1,1} g_{2,2}}} - \frac{e_1 g_{2,2}}{\sqrt{-g_{1,2}^2+g_{1,1} g_{2,2}}}$
$e_1 \wedge e_2$	$\sqrt{-g_{1,2}^2 + g_{1,1} g_{2,2}}$

◇ 3-space

Here is the corresponding palette in 3-space.

```
★A; DeclareMetric[g]; ComplementPalette
```

Complement Palette

Basis	Complement
1	$\frac{e_1 \wedge e_2 \wedge e_3}{\sqrt{\star g}}$
e_1	$\frac{g_{1,3}\, e_1 \wedge e_2}{\sqrt{\star g}} - \frac{g_{1,2}\, e_1 \wedge e_3}{\sqrt{\star g}} + \frac{g_{1,1}\, e_2 \wedge e_3}{\sqrt{\star g}}$
e_2	$\frac{g_{2,3}\, e_1 \wedge e_2}{\sqrt{\star g}} - \frac{g_{2,2}\, e_1 \wedge e_3}{\sqrt{\star g}} + \frac{g_{1,2}\, e_2 \wedge e_3}{\sqrt{\star g}}$
e_3	$\frac{g_{3,3}\, e_1 \wedge e_2}{\sqrt{\star g}} - \frac{g_{2,3}\, e_1 \wedge e_3}{\sqrt{\star g}} + \frac{g_{1,3}\, e_2 \wedge e_3}{\sqrt{\star g}}$
$e_1 \wedge e_2$	$\frac{e_3 \left(-g_{1,2}^2+g_{1,1} g_{2,2}\right)}{\sqrt{\star g}} + \frac{e_2 \left(g_{1,2} g_{1,3}-g_{1,1} g_{2,3}\right)}{\sqrt{\star g}} + \frac{e_1 \left(-g_{1,3} g_{2,2}+g_{1,2} g_{2,3}\right)}{\sqrt{\star g}}$
$e_1 \wedge e_3$	$\frac{e_3 \left(-g_{1,2} g_{1,3}+g_{1,1} g_{2,3}\right)}{\sqrt{\star g}} + \frac{e_2 \left(g_{1,3}^2-g_{1,1} g_{3,3}\right)}{\sqrt{\star g}} + \frac{e_1 \left(-g_{1,3} g_{2,3}+g_{1,2} g_{3,3}\right)}{\sqrt{\star g}}$
$e_2 \wedge e_3$	$\frac{e_3 \left(-g_{1,3} g_{2,2}+g_{1,2} g_{2,3}\right)}{\sqrt{\star g}} + \frac{e_2 \left(g_{1,3} g_{2,3}-g_{1,2} g_{3,3}\right)}{\sqrt{\star g}} + \frac{e_1 \left(-g_{2,3}^2+g_{2,2} g_{3,3}\right)}{\sqrt{\star g}}$
$e_1 \wedge e_2 \wedge e_3$	$\sqrt{\star g}$

The determinant of this metric tensor is

```
ToMetricElements[★g]
```

$$-g_{1,3}^2\, g_{2,2} + 2\, g_{1,2}\, g_{1,3}\, g_{2,3} - g_{1,1}\, g_{2,3}^2 - g_{1,2}^2\, g_{3,3} + g_{1,1}\, g_{2,2}\, g_{3,3}$$

Converting complements of basis elements

The *GrassmannAlgebra* function for converting complements of basis elements to other basis elements is `ConvertComplements`. If the declared metric is non-Euclidean, the result will include elements of the metric tensor.

◆ Example: Converting complements of individual basis elements in a Euclidean space

Here is the palette of complements for a Euclidean 3-space.

```
★A; ComplementPalette
```

Complement Palette

Basis	Complement
1	$e_1 \wedge e_2 \wedge e_3$
e_1	$e_2 \wedge e_3$
e_2	$-(e_1 \wedge e_3)$
e_3	$e_1 \wedge e_2$
$e_1 \wedge e_2$	e_3
$e_1 \wedge e_3$	$-e_2$
$e_2 \wedge e_3$	e_1
$e_1 \wedge e_2 \wedge e_3$	1

We could have obtained these complements in three steps:
1. Generate the basis elements (using `GrassmannBases`)
2. Take their complements (using `GrassmannComplement`)
3. Convert the complements to other basis elements (using `ConvertComplements`).

```
GrassmannBases
```

$\{\{1\}, \{e_1, e_2, e_3\}, \{e_1 \wedge e_2, e_1 \wedge e_3, e_2 \wedge e_3\}, \{e_1 \wedge e_2 \wedge e_3\}\}$

```
GrassmannComplement[GrassmannBases]
```

$\{\{\overline{1}\}, \{\overline{e_1}, \overline{e_2}, \overline{e_3}\}, \{\overline{e_1 \wedge e_2}, \overline{e_1 \wedge e_3}, \overline{e_2 \wedge e_3}\}, \{\overline{e_1 \wedge e_2 \wedge e_3}\}\}$

```
ConvertComplements[GrassmannComplement[GrassmannBases]]
```

$\{\{e_1 \wedge e_2 \wedge e_3\}, \{e_2 \wedge e_3, -(e_1 \wedge e_3), e_1 \wedge e_2\}, \{e_3, -e_2, e_1\}, \{1\}\}$

◆ Example: Converting complements of individual basis elements in a non-Euclidean space

To convert complements of basis elements in a non-Euclidean space, follow the same three steps as in the previous section.

```
★𝓑3; DeclareMetric[{{α, 0, ν}, {0, β, 0}, {ν, 0, γ}}] // MatrixForm
```

$$\begin{pmatrix} \alpha & 0 & \nu \\ 0 & \beta & 0 \\ \nu & 0 & \gamma \end{pmatrix}$$

Here is the palette of complements for this metric.

```
ComplementPalette
```

Complement Palette

Basis	Complement
1	$\frac{e_1\wedge e_2\wedge e_3}{\sqrt{\star g}}$
e_1	$\frac{\nu\, e_1\wedge e_2}{\sqrt{\star g}}+\frac{\alpha\, e_2\wedge e_3}{\sqrt{\star g}}$
e_2	$-\frac{\beta\, e_1\wedge e_3}{\sqrt{\star g}}$
e_3	$\frac{\gamma\, e_1\wedge e_2}{\sqrt{\star g}}+\frac{\nu\, e_2\wedge e_3}{\sqrt{\star g}}$
$e_1\wedge e_2$	$-\frac{\beta\,\nu\, e_1}{\sqrt{\star g}}+\frac{\alpha\,\beta\, e_3}{\sqrt{\star g}}$
$e_1\wedge e_3$	$\frac{(-\alpha\,\gamma+\nu^2)\, e_2}{\sqrt{\star g}}$
$e_2\wedge e_3$	$\frac{\beta\,\gamma\, e_1}{\sqrt{\star g}}-\frac{\beta\,\nu\, e_3}{\sqrt{\star g}}$
$e_1\wedge e_2\wedge e_3$	$\sqrt{\star g}$

We could have obtained these complements in three steps as in the previous example.

```
GrassmannBases
```

$\{\{1\},\{e_1,e_2,e_3\},\{e_1\wedge e_2,e_1\wedge e_3,e_2\wedge e_3\},\{e_1\wedge e_2\wedge e_3\}\}$

```
GrassmannComplement[GrassmannBases]
```

$\{\{\overline{1}\},\{\overline{e_1},\overline{e_2},\overline{e_3}\},\{\overline{e_1\wedge e_2},\overline{e_1\wedge e_3},\overline{e_2\wedge e_3}\},\{\overline{e_1\wedge e_2\wedge e_3}\}\}$

```
ConvertComplements[GrassmannComplement[GrassmannBases]]
```

$$\left\{\left\{\frac{e_1\wedge e_2\wedge e_3}{\sqrt{\star g}}\right\},\left\{\frac{\nu\, e_1\wedge e_2}{\sqrt{\star g}}+\frac{\alpha\, e_2\wedge e_3}{\sqrt{\star g}},-\frac{\beta\, e_1\wedge e_3}{\sqrt{\star g}},\frac{\gamma\, e_1\wedge e_2}{\sqrt{\star g}}+\frac{\nu\, e_2\wedge e_3}{\sqrt{\star g}}\right\},\right.$$
$$\left.\left\{-\frac{\beta\,\nu\, e_1}{\sqrt{\star g}}+\frac{\alpha\,\beta\, e_3}{\sqrt{\star g}},\frac{(-\alpha\,\gamma+\nu^2)\, e_2}{\sqrt{\star g}},\frac{\beta\,\gamma\, e_1}{\sqrt{\star g}}-\frac{\beta\,\nu\, e_3}{\sqrt{\star g}}\right\},\left\{\sqrt{\star g}\right\}\right\}$$

The determinant of the metric tensor is

```
ToMetricElements[★g]
```

$\alpha\,\beta\,\gamma-\beta\,\nu^2$

- As described earlier, you can include the full form of ★g in the palette by using a substitution rule.

```
ComplementPalette /. ★g → ToMetricElements[★g]
```

◆ Example: Verifying the complement of a complement axiom for basis elements

The complement of a complement of any element in 3-space is the element itself. We can verify this for the basis elements by applying the complement conversion procedure twice. This example verifies the result for a general metric in 3-space.

```
DeclareMetric[g]
```

$\{\{g_{1,1},g_{1,2},g_{1,3}\},\{g_{1,2},g_{2,2},g_{2,3}\},\{g_{1,3},g_{2,3},g_{3,3}\}\}$

```
G = ConvertComplements[GrassmannComplement[GrassmannBases]]
```

$$\left\{\left\{\frac{e_1\wedge e_2\wedge e_3}{\sqrt{\star g}}\right\},\left\{\frac{g_{1,3}\, e_1\wedge e_2}{\sqrt{\star g}}-\frac{g_{1,2}\, e_1\wedge e_3}{\sqrt{\star g}}+\frac{g_{1,1}\, e_2\wedge e_3}{\sqrt{\star g}},\right.\right.$$
$$\frac{g_{2,3}\, e_1\wedge e_2}{\sqrt{\star g}}-\frac{g_{2,2}\, e_1\wedge e_3}{\sqrt{\star g}}+\frac{g_{1,2}\, e_2\wedge e_3}{\sqrt{\star g}},$$
$$\left.\frac{g_{3,3}\, e_1\wedge e_2}{\sqrt{\star g}}-\frac{g_{2,3}\, e_1\wedge e_3}{\sqrt{\star g}}+\frac{g_{1,3}\, e_2\wedge e_3}{\sqrt{\star g}}\right\},$$
$$\left\{\frac{e_3\left(-g_{1,2}^2+g_{1,1}\, g_{2,2}\right)}{\sqrt{\star g}}+\frac{e_2\left(g_{1,2}\, g_{1,3}-g_{1,1}\, g_{2,3}\right)}{\sqrt{\star g}}+\frac{e_1\left(-g_{1,3}\, g_{2,2}+g_{1,2}\, g_{2,3}\right)}{\sqrt{\star g}},\right.$$
$$\frac{e_3\left(-g_{1,2}\, g_{1,3}+g_{1,1}\, g_{2,3}\right)}{\sqrt{\star g}}+\frac{e_2\left(g_{1,3}^2-g_{1,1}\, g_{3,3}\right)}{\sqrt{\star g}}+\frac{e_1\left(-g_{1,3}\, g_{2,3}+g_{1,2}\, g_{3,3}\right)}{\sqrt{\star g}},$$
$$\frac{e_3\left(-g_{1,3}\, g_{2,2}+g_{1,2}\, g_{2,3}\right)}{\sqrt{\star g}}+\frac{e_2\left(g_{1,3}\, g_{2,3}-g_{1,2}\, g_{3,3}\right)}{\sqrt{\star g}}+$$
$$\left.\left.\frac{e_1\left(-g_{2,3}^2+g_{2,2}\, g_{3,3}\right)}{\sqrt{\star g}}\right\},\left\{\sqrt{\star g}\right\}\right\}$$

Taking the complement of these and converting them again returns us to the original list of basis elements.

```
ConvertComplements[GrassmannComplement[G]]
```

$$\{\{1\},\{e_1, e_2, e_3\},\{e_1\wedge e_2, e_1\wedge e_3, e_2\wedge e_3\},\{e_1\wedge e_2\wedge e_3\}\}$$

◆ **Example: Converting complements of basis elements in general expressions**

- `ConvertComplements` will take any Grassmann expression and convert any complements of basis elements in it. Suppose we are in a 3-space with a Euclidean metric.

$$\star\mathbf{A};\ \mathbf{ConvertComplements}\left[2+a\,\overline{\overline{\overline{\overline{b\, e_2}}\wedge e_2}\wedge\overline{\overline{c\, e_3}}}\right]$$

$$2+a\, b\, c\, e_1\wedge e_2$$

If the metric is more general we get the more general result.

$$\mathbf{DeclareMetric[g];\ F = ConvertComplements}\left[2+a\,\overline{\overline{\overline{\overline{b\, e_2}}\wedge e_2}\wedge\overline{\overline{c\, e_3}}}\right]$$

$$2+\frac{a\, b\, c\, g_{2,2}\, g_{3,3}\, e_1\wedge e_2}{\sqrt{\star g}}-\frac{a\, b\, c\, g_{2,2}\, g_{2,3}\, e_1\wedge e_3}{\sqrt{\star g}}+\frac{a\, b\, c\, g_{1,3}\, g_{2,2}\, e_2\wedge e_3}{\sqrt{\star g}}$$

```
Simplify[F]
```

$$\frac{1}{\sqrt{\star g}}\left(2\sqrt{\star g}+a\, b\, c\, g_{2,2}\left(g_{3,3}\, e_1\wedge e_2-g_{2,3}\, e_1\wedge e_3+g_{1,3}\, e_2\wedge e_3\right)\right)$$

- `ConvertComplements` will also apply some simplification rules to expressions involving non-basis elements.

$\star$**A; ConvertComplements**$\left[2 + a\,\overline{\overline{\overline{b\,e_2} \wedge e_2} \wedge \overline{\overline{c\,x}}}\right]$

$2 + a\,b\,c\,\overline{x}$

DeclareMetric[g]; ConvertComplements$\left[2 + a\,\overline{\overline{\overline{b\,e_2} \wedge e_2} \wedge \overline{\overline{c\,x}}}\right]$

$2 + a\,b\,c\,\overline{x}\,g_{2,2}$

Simplifying expressions involving complements

`ConvertComplements` has been specifically designed to convert expressions involving complements of basis elements according to the currently declared metric of the space. Generally the results will differ when the metrics differ.

`GrassmannSimplify` (alias $\star\mathcal{G}$), on the other hand, only includes simplification rules which are independent of the metric. Generally, these rules apply when an expression involving complements can be expressed without the complements, or can be expressed by another form deemed simpler, often involving the interior product (which will be discussed in the next chapter).

- In the example discussed in the previous section we see that using `GrassmannSimplify` gives a result involving the interior product $e_2 \ominus e_2$ (in this case a scalar product) which is valid independent of the metric.

$\star\mathcal{G}\left[2 + a\,\overline{\overline{\overline{b\,e_2} \wedge e_2} \wedge \overline{\overline{c\,x}}}\right]$

$2 + a\,b\,c\,(e_2 \ominus e_2)\,\overline{x}$

In the case of a Euclidean metric, the scalar product $e_2 \ominus e_2$ in the result is equal to `1`. In the case of a general metric it is equal to $g_{2,2}$.

- Here are some more examples in 3-space. We turn on `ShowPrecedence` with its alias $\star$`P`.

$\star$**A;** $\star$**P;** $\star\mathcal{G}\left[\left\{\overline{\overline{x}}, \overline{\overline{\overline{\overline{x}}}}, x \vee \overline{y}, \overline{\overline{x} \wedge y}, (u \wedge v) \vee \overline{y}, \overline{\overline{u \wedge v} \wedge y}, (u \wedge v) \vee \overline{y} \vee \overline{z},\right.\right.$
$\left.\left.\overline{\overline{u \wedge v} \wedge y \wedge z}, (x \wedge \overline{y}) \ominus z, \overline{(x \wedge y) \vee z}\right\}\right]$

$\{x, x, x \ominus y, x \ominus y, (u \wedge v) \ominus y, (u \wedge v) \ominus y,$
$(u \wedge v) \ominus (y \wedge z), (u \wedge v) \ominus (y \wedge z), (x \ominus y)\,\overline{z}, (\overline{x \wedge y}) \wedge \overline{z}\}$

The dimension of the space may change the sign of some expressions, or indeed make them zero. Here are the same examples in 4-space, giving two sign-changes, and one zero.

$\star\mathcal{B}_4$**;** $\star$**P;** $\star\mathcal{G}\left[\left\{\overline{\overline{x}}, \overline{\overline{\overline{\overline{x}}}}, x \vee \overline{y}, \overline{\overline{x} \wedge y}, (u \wedge v) \vee \overline{y}, \overline{\overline{u \wedge v} \wedge y}, (u \wedge v) \vee \overline{y} \vee \overline{z},\right.\right.$
$\left.\left.\overline{\overline{u \wedge v} \wedge y \wedge z}, (x \wedge \overline{y}) \ominus z, \overline{(x \wedge y) \vee z}\right\}\right]$

$\{-x, x, x \ominus y, -(x \ominus y), (u \wedge v) \ominus y, (u \wedge v) \ominus y,$
$(u \wedge v) \ominus (y \wedge z), (u \wedge v) \ominus (y \wedge z), (x \ominus y)\,\overline{z}, 0\}$

- `GrassmannSimplify` will also handle expressions in different dimensions with symbolic grades.

```
Grid[Table[★𝓑n; {n, ★𝒢[{α̿, ᾱ∧β (overbarred), ᾱ∨β (overbarred)}]}, {n, 2, 4}]]
```

$$\mathtt{Grid}\left[\mathtt{Table}\left[\star\mathcal{B}_n;\ \left\{n,\ \star\mathcal{G}\left[\left\{\overline{\overline{\alpha_m}},\ \overline{\overline{\alpha_m}\wedge\beta_k},\ \overline{\overline{\alpha_m}\vee\beta_k}\right\}\right]\right\},\ \{n, 2, 4\}\right]\right]$$

2	$\left\{(-1)^m \alpha_m,\ (-1)^m \left(\alpha_m \ominus \beta_k\right),\ (-1)^{k+m} \alpha_m \vee \beta_k\right\}$
3	$\left\{\alpha_m,\ \alpha_m \ominus \beta_k,\ \alpha_m \vee \beta_k\right\}$
4	$\left\{(-1)^m \alpha_m,\ (-1)^m \left(\alpha_m \ominus \beta_k\right),\ (-1)^{k+m} \alpha_m \vee \beta_k\right\}$

Converting expressions involving complements to specified forms

In the usual way of developing the Grassmann algebra, the most basic product is the exterior product. The regressive product is then developed as a 'dual' product to the exterior product (that is, it has the same axiom structure, but different interpretation on its symbols). The introduction of a complement operation (mapping) then enables the definition of the interior product (and its special cases the inner product and the scalar product). The interior, inner and scalar products are discussed in the next chapter.

The exterior and regressive products are taken as basic. If a complement operation (or equivalently, a metric) is introduced (defined) on the space, then the exterior and regressive products can be expressed in terms of each other and the complement. The interior product operation can also be expressed in terms either of the exterior product and the complement, or else the regressive product and the complement. To complete the suite of conversion possibilities, the exterior and regressive products can be expressed in terms of the interior product and the complement, and the complement of an element can be expressed as its interior product with the unit n-element.

In this section, we discuss only the conversions between the exterior product, the regressive product, and the complement. The examples below show the effects of these conversions on a single expression in both 3 and 4 dimensional spaces. In a 3-space, the complement of the complement of an element of any grade is the element itself, hence these conversions in a 3-space show no extra signs. In spaces of other dimension however, the results may show signs dependent on the grades of the elements. This is exemplified by the 4-dimensional cases.

◆ Example: Conversions in 3 and 4 dimensional spaces

As an example expression we take the regressive product of two exterior products.

$$\star\mathbf{A};\ \mathbf{X} = \left(\alpha_m \wedge \beta_k\right) \vee \left(\gamma_p \wedge \delta_q\right);$$

Converting the exterior products to regressive products and complements yields a possible sign change in 4-space depending on the grades of the elements.

```
{★𝓑3; ExteriorToRegressive[X], ★𝓑4; ExteriorToRegressive[X]}
```

$$\left\{\overline{\overline{\alpha_m}\vee\overline{\beta_k}}\vee\overline{\overline{\gamma_p}\vee\overline{\delta_q}},\ \left((-1)^{k+m}\,\overline{\overline{\alpha_m}\vee\overline{\beta_k}}\right)\vee\left((-1)^{p+q}\,\overline{\overline{\gamma_p}\vee\overline{\delta_q}}\right)\right\}$$

Converting the regressive products to exterior products and complements also yields a possible sign change.

```
{★𝓑3; RegressiveToExterior[X], ★𝓑4; RegressiveToExterior[X]}
```

$$\left\{\overline{\underset{m}{\alpha}\wedge\underset{k}{\beta}\wedge\overline{\underset{p}{\gamma}\wedge\underset{q}{\delta}}},\ (-1)^{k+m+p+q}\,\overline{\underset{m}{\alpha}\wedge\underset{k}{\beta}\wedge\overline{\underset{p}{\gamma}\wedge\underset{q}{\delta}}}\right\}$$

Converting regressive products of basis elements in a metric space

In a metric space we can compute the regressive product of basis elements in terms of exterior products of basis elements.

◆ Example: The regressive product of two bivectors in Euclidean space

Here is the regressive product of two bivectors

```
★A; Z = ★𝔹a ∨ ★𝔹b
```

$$(a_1\, e_1 \wedge e_2 + a_2\, e_1 \wedge e_3 + a_3\, e_2 \wedge e_3) \vee (b_1\, e_1 \wedge e_2 + b_2\, e_1 \wedge e_3 + b_3\, e_2 \wedge e_3)$$

Expanding and simplifying the product gives

```
Zs = ★𝒢[Z]
```

$$(-a_2\, b_1 + a_1\, b_2)\ (e_1 \wedge e_2) \vee (e_1 \wedge e_3) + (-a_3\, b_1 + a_1\, b_3)\ (e_1 \wedge e_2) \vee (e_2 \wedge e_3) + (-a_3\, b_2 + a_2\, b_3)\ (e_1 \wedge e_3) \vee (e_2 \wedge e_3)$$

Now convert the regressive products to exterior products and complements.

```
Ze = RegressiveToExterior[Z]
```

$$\overline{\overline{(a_1\, e_1 \wedge e_2)} \wedge \overline{(b_1\, e_1 \wedge e_2)}} + \overline{\overline{(a_1\, e_1 \wedge e_2)} \wedge \overline{(b_2\, e_1 \wedge e_3)}} + \overline{\overline{(a_1\, e_1 \wedge e_2)} \wedge \overline{(b_3\, e_2 \wedge e_3)}} + \overline{\overline{(a_2\, e_1 \wedge e_3)} \wedge \overline{(b_1\, e_1 \wedge e_2)}} + \overline{\overline{(a_2\, e_1 \wedge e_3)} \wedge \overline{(b_2\, e_1 \wedge e_3)}} + \overline{\overline{(a_2\, e_1 \wedge e_3)} \wedge \overline{(b_3\, e_2 \wedge e_3)}} + \overline{\overline{(a_3\, e_2 \wedge e_3)} \wedge \overline{(b_1\, e_1 \wedge e_2)}} + \overline{\overline{(a_3\, e_2 \wedge e_3)} \wedge \overline{(b_2\, e_1 \wedge e_3)}} + \overline{\overline{(a_3\, e_2 \wedge e_3)} \wedge \overline{(b_3\, e_2 \wedge e_3)}}$$

Finally, convert the complements of the basis elements.

```
ConvertComplements[Ze]
```

$$(-a_2\, b_1 + a_1\, b_2)\ e_1 + (-a_3\, b_1 + a_1\, b_3)\ e_2 + (-a_3\, b_2 + a_2\, b_3)\ e_3$$

◆ Example: The regressive product of two bivectors in non-Euclidean space

Take the same regressive product of bivectors as in the previous example. Declare a non-Euclidean metric.

```
DeclareMetric[g]
```

$$\{\{g_{1,1}, g_{1,2}, g_{1,3}\}, \{g_{1,2}, g_{2,2}, g_{2,3}\}, \{g_{1,3}, g_{2,3}, g_{3,3}\}\}$$

Then convert the complements of the basis elements in Z_e again.

```
ConvertComplements[Ze]
```

$$\sqrt{\star g}\ (-a_2\, b_1 + a_1\, b_2)\ e_1 + \sqrt{\star g}\ (-a_3\, b_1 + a_1\, b_3)\ e_2 + \sqrt{\star g}\ (-a_3\, b_2 + a_2\, b_3)\ e_3$$

◆ Example: The regressive product of three elements in Euclidean 3-space

Here is a *pair* of regressive products of three basis elements in a 3-space.

```
★A; X = {(e1 ∧ e2) ∨ (e2 ∧ e3) ∨ (e3 ∧ e1), (e1 ∧ e2 ∧ e3) ∨ (e1 ∧ e2 ∧ e3) ∨ e1};
```

Convert the regressive products to exterior products and complements.

```
Xe = RegressiveToExterior[X]
```

$$\left\{\overline{\overline{e_1 \wedge e_2} \wedge \overline{e_2 \wedge e_3} \wedge \overline{e_3 \wedge e_1}},\ \overline{\overline{e_1 \wedge e_2 \wedge e_3} \wedge \overline{e_1 \wedge e_2 \wedge e_3} \wedge \overline{e_1}}\right\}$$

Convert the complements. ConvertComplements assumes the currently declared metric - in this case the default Euclidean metric.

```
ConvertComplements[Xe]
```

$\{1, e_1\}$

◆ Example: The regressive product of three elements in non-Euclidean 3-space

We now change the metric to a general metric.

```
DeclareMetric[g]
```

$\{\{g_{1,1}, g_{1,2}, g_{1,3}\}, \{g_{1,2}, g_{2,2}, g_{2,3}\}, \{g_{1,3}, g_{2,3}, g_{3,3}\}\}$

Applying ConvertComplements assumes this metric, and shows us that both results now have the determinant of the metric tensor as multipliers.

```
ConvertComplements[Xe]
```

$\{\star g, \star g\, e_1\}$

- Note that because their were *two* regressive product operations, $\sqrt{\star g}$ was generated *twice* (giving $\star g$). In the previous example of the regressive product of bivectors, there was one regressive product operation, so $\sqrt{\star g}$ was generated once.

5.10 Complements in a vector space

The Euclidean complement in a vector 2-space

So far in this chapter we have for the most part discussed elements without an express vectorial or bound-vectorial interpretation. Hence all the results apply to either interpretation. In this section we draw on the conceptual familiarity of the vectorial language and re-explore some aspects of the complement operation specifically in vector spaces.

Consider a vector x in a 2-dimensional Euclidean vector space expressed in terms of basis vectors e_1 and e_2.

$$x == a\, e_1 + b\, e_2$$

Since this is a *Euclidean* vector space, we can depict the basis vectors at right angles to each other. But note that since it is a *vector* space, we do not depict an origin.

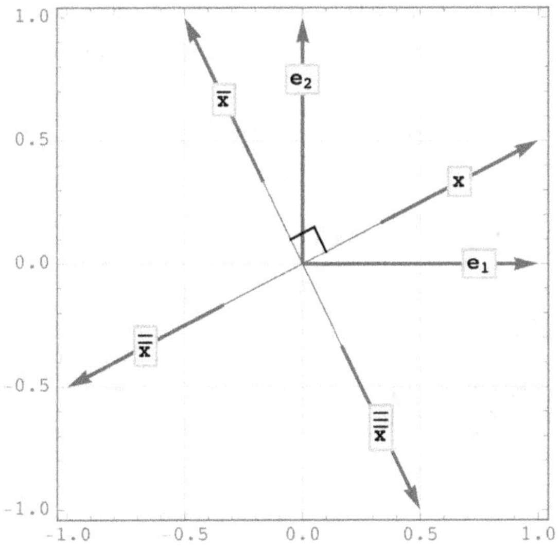

Complementing a two-dimensional vector

The complement of x is given by:

$$\overline{x} == a\,\overline{e_1} + b\,\overline{e_2} == a\,e_2 - b\,e_1$$

Remember, the Euclidean complement of a basis element is formally identical to its cobasis element, and a basis element and its cobasis element are defined by their exterior product being the *unit* basis *n*-element, in this case $e_1 \wedge e_2$.

It is clear from the depiction above that x and $\overline{x}$ are at right angles to each other, thus verifying our geometric interpretation of the algebraic notion of orthogonality: *a simple element and its complement are orthogonal.*

Taking the complement of $\overline{x}$ gives $-x$:

$$\overline{\overline{x}} == a\,\overline{e_2} - b\,\overline{e_1} == -a\,e_1 - b\,e_2 == -x$$

Or, we could have used the complement of a complement axiom:

$$\overline{\overline{x}} == (-1)^{1\,(2-1)}\,x == -x$$

Continuing to take complements we find that we eventually return to the original element.

$$\overline{\overline{\overline{x}}} == -\overline{x}$$

$$\overline{\overline{\overline{\overline{x}}}} == x$$

In a vector 2-space, taking the complement of a vector is thus equivalent to *rotating the vector by one right angle in the direction from* e_1 *to* e_2.

The non-Euclidean complement in a vector 2-space

Suppose now a vector x in a 2-dimensional non-Euclidean vector space expressed similarly in terms of basis vectors e_1 and e_2.

$x == a\,e_1 + b\,e_2$

Since this is a *non*-Euclidean vector space, the basis vectors are generally not at right angles to each other. We declare a general metric, and display the Complement and Metric palettes.

```
★𝓑2; DeclareMetric[g];
Grid[{{ComplementPalette, MetricPalette}}]
```

Complement Palette

Basis	Complement
1	$\frac{e_1 \wedge e_2}{\sqrt{\star g}}$
e_1	$\frac{e_2\, g_{1,1}}{\sqrt{\star g}} - \frac{e_1\, g_{1,2}}{\sqrt{\star g}}$
e_2	$\frac{e_2\, g_{1,2}}{\sqrt{\star g}} - \frac{e_1\, g_{2,2}}{\sqrt{\star g}}$
$e_1 \wedge e_2$	$\sqrt{\star g}$

Metric Palette

$\underset{0}{\Lambda}$	(1)
$\underset{1}{\Lambda}$	$\begin{pmatrix} g_{1,1} & g_{1,2} \\ g_{1,2} & g_{2,2} \end{pmatrix}$
$\underset{2}{\Lambda}$	$(-g_{1,2}^2 + g_{1,1}\, g_{2,2})$

The complement of x is given by:

$\overline{x} == a\,\overline{e_1} + b\,\overline{e_2} ==$ ConvertComplements$[a\,\overline{e_1} + b\,\overline{e_2}]$

$$\overline{x} == a\,\overline{e_1} + b\,\overline{e_2} == \frac{e_2\,(a\,g_{1,1} + b\,g_{1,2})}{\sqrt{\star g}} + \frac{e_1\,(-a\,g_{1,2} - b\,g_{2,2})}{\sqrt{\star g}}$$

Taking the complement of $\overline{x}$ still gives $-x$:

$\overline{\overline{x}} == a\,\overline{\overline{e_1}} + b\,\overline{\overline{e_2}} ==$ ConvertComplements$\left[a\,\overline{\overline{e_1}} + b\,\overline{\overline{e_2}}\right]$

$$\overline{\overline{x}} == a\,\overline{\overline{e_1}} + b\,\overline{\overline{e_2}} == -a\,e_1 - b\,e_2$$

And taking the complement of $\overline{\overline{x}}$ still gives $-\overline{x}$:

$\overline{\overline{\overline{x}}} == a\,\overline{\overline{\overline{e_1}}} + b\,\overline{\overline{\overline{e_2}}} ==$ ConvertComplements$\left[a\,\overline{\overline{\overline{e_1}}} + b\,\overline{\overline{\overline{e_2}}}\right]$

$$\overline{\overline{\overline{x}}} == a\,\overline{\overline{\overline{e_1}}} + b\,\overline{\overline{\overline{e_2}}} == \frac{e_2\,(-a\,g_{1,1} - b\,g_{1,2})}{\sqrt{\star g}} + \frac{e_1\,(a\,g_{1,2} + b\,g_{2,2})}{\sqrt{\star g}}$$

◆ Example

Suppose we take a simple metric and calculate $x = \frac{1}{2}(e_1 + e_2)$ and its complements.

```
★𝓑2; M = DeclareMetric[{{2, 1}, {1, 1}}];
Grid[{{ComplementPalette, MetricPalette}}]
```

Complement Palette

Basis	Complement
1	$e_1 \wedge e_2$
e_1	$-e_1 + 2\,e_2$
e_2	$-e_1 + e_2$
$e_1 \wedge e_2$	1

Metric Palette

$\underset{0}{\Lambda}$	(1)
$\underset{1}{\Lambda}$	$\begin{pmatrix} 2 & 1 \\ 1 & 1 \end{pmatrix}$
$\underset{2}{\Lambda}$	(1)

$$\mathbf{x} == \frac{\mathbf{e_1 + e_2}}{\mathbf{2}};$$

$$\overline{\mathbf{x}} == \mathbf{ConvertComplements}[\overline{\frac{\mathbf{e_1 + e_2}}{\mathbf{2}}}]$$

$$\overline{x} == -e_1 + \frac{3\, e_2}{2}$$

$$\overline{\overline{\mathbf{x}}} == \mathbf{ConvertComplements}[\overline{\overline{\frac{\mathbf{e_1 + e_2}}{\mathbf{2}}}}]$$

$$\overline{\overline{x}} == -\frac{e_1}{2} - \frac{e_2}{2}$$

$$\overline{\overline{\overline{\mathbf{x}}}} == \mathbf{ConvertComplements}[\overline{\overline{\overline{\frac{\mathbf{e_1 + e_2}}{\mathbf{2}}}}}]$$

$$\overline{\overline{\overline{x}}} == e_1 - \frac{3\, e_2}{2}$$

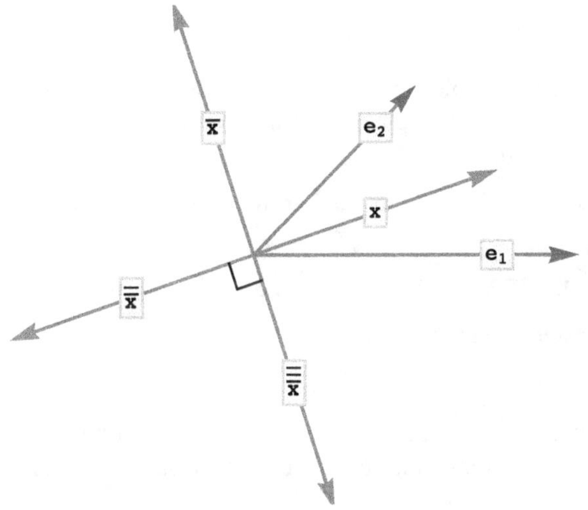

Complementing a two-dimensional vector in a non-Euclidean space

We can see that although e_1 is not a unit vector, and the basis vectors are not orthogonal, that as for the Euclidean case, $\overline{\overline{x}}$ is equal to $-x$, $\overline{\overline{\overline{x}}}$ is equal to $-\overline{x}$, and $\overline{x}$ is orthogonal to x.

The standard conceptual vehicle for exploring the notion of orthogonality is the interior product and its special cases, the inner and scalar products. These we will develop in the next chapter where we will revisit examples of this type, and connect them more closely with familiar concepts.

The Euclidean complement in a vector 3-space

We now explore the Euclidean complement of a vector x in a vector 3-space. This differs somewhat from the vector 2-space case, since the complement of a vector is a bivector, and the complement of a bivector is a vector.

$$x == a\,e_1 + b\,e_2 + c\,e_3$$

For reference we display the complement palette.

```
★𝓑3; ComplementPalette
```

Complement Palette

Basis	Complement
1	$e_1 \wedge e_2 \wedge e_3$
e_1	$e_2 \wedge e_3$
e_2	$-(e_1 \wedge e_3)$
e_3	$e_1 \wedge e_2$
$e_1 \wedge e_2$	e_3
$e_1 \wedge e_3$	$-e_2$
$e_2 \wedge e_3$	e_1
$e_1 \wedge e_2 \wedge e_3$	1

The complement of x is therefore given by:

$$\overline{x} == a\,\overline{e_1} + b\,\overline{e_2} + c\,\overline{e_3} == a\,e_2 \wedge e_3 - b\,e_1 \wedge e_3 + c\,e_1 \wedge e_2$$

Remember, the Euclidean complement of a basis element is formally identical to its cobasis element, and a basis element and its cobasis element are defined by their exterior product being the *unit* basis *n*-element, in this case $e_1 \wedge e_2 \wedge e_3$.

Taking the complement of $\overline{x}$ gives x. Note that this result differs in sign from the two-dimensional case.

$$\overline{\overline{x}} == a\,\overline{e_2 \wedge e_3} - b\,\overline{e_1 \wedge e_3} + c\,\overline{e_1 \wedge e_2} == a\,e_1 + b\,e_2 + c\,e_3 == x$$

Or, we could have used the complement of a complement axiom [5.7].

$$\overline{\overline{x}} == (-1)^{1\,(3-1)}\,x == x$$

Since in a vector 3-space, the complement of the complement of a vector or of a bivector returns us to the original element, continuing to take complements as we did in the two-dimensional case leads us to no further elements.

$$\overline{\overline{\overline{\overline{x}}}} == \overline{\overline{x}} == x \qquad \overline{\overline{\overline{x}}} == \overline{x}$$

The bivector $\overline{x}$ is a sum of components, each in one of the coordinate bivectors. We have already seen in section 2.10 that a bivector in a 3-space is simple, and hence can be factored into the exterior product of two vectors. It is in this form that the bivector will be most easily interpreted as a geometric entity. There is an infinity of vectors that are orthogonal to the vector x, but they are all contained in the bivector $\overline{x}$. You can express the bivector $\overline{x}$ as the exterior product of two vectors again in an infinity of ways. The *GrassmannAlgebra* function `ExteriorFactorize` finds one of them for you. Others may be obtained by adding vectors from the first factor to the second factor, or *vice-versa*.

```
ExteriorFactorize[a e2 ∧ e3 - b e1 ∧ e3 + c e1 ∧ e2]
```

$$c\left(e_1 - \frac{a\,e_3}{c}\right) \wedge \left(e_2 - \frac{b\,e_3}{c}\right)$$

The vector x is orthogonal to each of these vectors since the exterior product of $\overline{x}$ with each of them is zero.

```
★G[{(e1 - a e3/c) ∧ (a e2 ∧ e3 - b e1 ∧ e3 + c e1 ∧ e2),
  (e2 - b e3/c) ∧ (a e2 ∧ e3 - b e1 ∧ e3 + c e1 ∧ e2)}]
{0, 0}
```

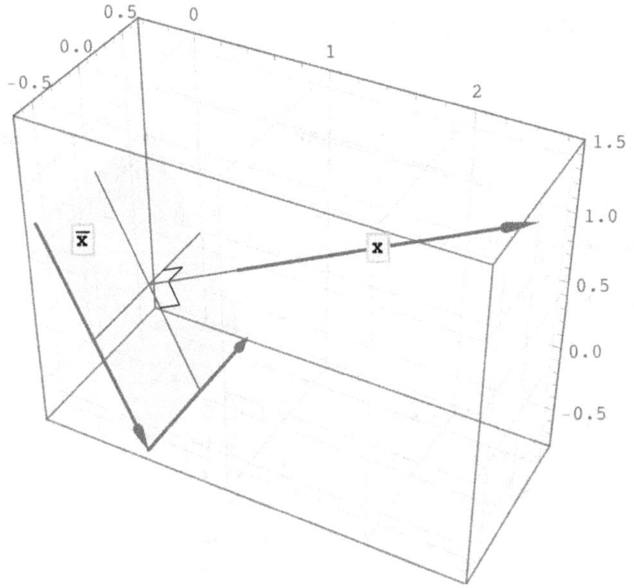

A vector and its bivector as complements in a vector 3-space

The complements of each of these vectors will be bivectors which contain the original vector x. We can verify this easily with `ConvertComplements` and `GrassmannSimplify`.

```
★G[ConvertComplements[
  {(\overline{c e1 - a e3}) ∧ (a e1 + b e2 + c e3), (\overline{c e2 - b e3}) ∧ (a e1 + b e2 + c e3)}]]
{0, 0}
```

The non-Euclidean complement in a vector 3-space

Suppose now a vector x in a 3-dimensional non-Euclidean vector space.

```
x == a e1 + b e2 + c e3
★B3; DeclareMetric[g];
```

The complement of x is given by:

$\overline{\mathbf{x}} == \mathbf{a}\,\overline{\mathbf{e_1}} + \mathbf{b}\,\overline{\mathbf{e_2}} + \mathbf{c}\,\overline{\mathbf{e_3}}$ == **ConvertComplements**$[\mathbf{a}\,\overline{\mathbf{e_1}} + \mathbf{b}\,\overline{\mathbf{e_2}} + \mathbf{c}\,\overline{\mathbf{e_3}}]$

$$\overline{x} == a\,\overline{e_1} + b\,\overline{e_2} + c\,\overline{e_3} == \frac{1}{\sqrt{\star g}}\,(a\,g_{1,3} + b\,g_{2,3} + c\,g_{3,3})\;e_1 \wedge e_2 + \frac{1}{\sqrt{\star g}}\,(-a\,g_{1,2} - b\,g_{2,2} - c\,g_{2,3})\;e_1 \wedge e_3 + \frac{1}{\sqrt{\star g}}\,(a\,g_{1,1} + b\,g_{1,2} + c\,g_{1,3})\;e_2 \wedge e_3$$

Taking the complement of $\overline{x}$ gives x.

$\overline{\overline{\mathbf{x}}} == \mathbf{a}\,\overline{\overline{\mathbf{e_1}}} + \mathbf{b}\,\overline{\overline{\mathbf{e_2}}} + \mathbf{c}\,\overline{\overline{\mathbf{e_3}}}$ == **ConvertComplements**$\left[\mathbf{a}\,\overline{\overline{\mathbf{e_1}}} + \mathbf{b}\,\overline{\overline{\mathbf{e_2}}} + \mathbf{c}\,\overline{\overline{\mathbf{e_3}}}\right]$

$$\overline{\overline{x}} == a\,\overline{\overline{e_1}} + b\,\overline{\overline{e_2}} + c\,\overline{\overline{e_3}} == a\,e_1 + b\,e_2 + c\,e_3$$

◆ Example

Suppose we take the metric used in the example above on the non-Euclidean complement in a vector 2-space and extend it orthogonally to the third dimension.

```
★𝓑3; M = DeclareMetric[{{2, 1, 0}, {1, 1, 0}, {0, 0, 1}}];
Grid[{{ComplementPalette, MetricPalette}}]
```

Complement Palette

Basis	Complement
1	$e_1 \wedge e_2 \wedge e_3$
e_1	$-(e_1 \wedge e_3) + 2\,e_2 \wedge e_3$
e_2	$-(e_1 \wedge e_3) + e_2 \wedge e_3$
e_3	$e_1 \wedge e_2$
$e_1 \wedge e_2$	e_3
$e_1 \wedge e_3$	$e_1 - 2\,e_2$
$e_2 \wedge e_3$	$e_1 - e_2$
$e_1 \wedge e_2 \wedge e_3$	1

Metric Palette

$\underset{0}{\Lambda}$	(1)
$\underset{1}{\Lambda}$	$\begin{pmatrix} 2 & 1 & 0 \\ 1 & 1 & 0 \\ 0 & 0 & 1 \end{pmatrix}$
$\underset{2}{\Lambda}$	$\begin{pmatrix} 1 & 0 & 0 \\ 0 & 2 & 1 \\ 0 & 1 & 1 \end{pmatrix}$
$\underset{3}{\Lambda}$	(1)

We now calculate the complement of the same vector $x = \frac{1}{2}\,(e_1 + e_2)$ and verify that $\overline{\overline{x}}$ is equal to x.

$$\mathbf{x} == \frac{\mathbf{e_1} + \mathbf{e_2}}{2};$$

$\overline{\mathbf{x}}$ == **ConvertComplements**$\left[\overline{\frac{\mathbf{e_1} + \mathbf{e_2}}{2}}\right]$

$$\overline{x} == -(e_1 \wedge e_3) + \frac{3\,e_2 \wedge e_3}{2}$$

$\overline{\overline{\mathbf{x}}}$ == **ConvertComplements**$\left[\overline{\overline{\frac{\mathbf{e_1} + \mathbf{e_2}}{2}}}\right]$

$$\overline{\overline{x}} == \frac{e_1}{2} + \frac{e_2}{2}$$

If we factorize the bivector $\overline{x}$ using `ExteriorFactorize`, we obtain

$$\textbf{ExteriorFactorize}\left[-(e_1 \wedge e_3) + \frac{3\, e_2 \wedge e_3}{2}\right]$$

$$-\left(\left(e_1 - \frac{3\, e_2}{2}\right) \wedge e_3\right)$$

and observe again that the vector x is orthogonal to each of these factors, since clearly the exterior product of the complement of x with each of these factors is zero.

$$\star\mathcal{G}\left[\left\{\left(e_1 - \frac{3\, e_2}{2}\right) \wedge \left(-(e_1 \wedge e_3) + \frac{3\, e_2 \wedge e_3}{2}\right),\ e_3 \wedge \left(-(e_1 \wedge e_3) + \frac{3\, e_2 \wedge e_3}{2}\right)\right\}\right]$$

{0, 0}

5.11 Complements in a bound space

Metrics in a bound space

If we want to interpret one element of a linear space as an origin point, and yet retain the definitions of complement and metric developed so far, we need to consider which complement mappings or forms of metric make sense, or are useful in some degree.

In traditional approaches to the algebraic representation of geometry, the notion of applying a metric to points has been eschewed as largely devoid of use - causing writers in the field, perhaps starting with Gibbs, to concentrate on describing the physical world in vectorial terms. However, it is not 'use' that is the important consideration, but rather 'consistency'. We have introduced a bound vector space as an *interpretation only* on the underlying linear space of the algebra, and since we have shown the notion of metric is a consistent addition to a general linear space, we must also allow it onto the bound space. Viewed from this angle, any bound space can have a metric - all we need to do is to give it an interpretation when used with bound elements. When used with elements interpreted as 'free' - like vectors and bivectors - the metric has its standard 'free' interpretation as a way of measuring and comparing vectorially interpreted elements.

Our objective then is to explore the implications of a metric as a way of measuring and comparing 'bound' elements. We will be constrained by the algebraic properties of the complement operation and metric that we have developed so far. This could mean of course that we may not find the results 'useful' when they are interpreted in a bound space.

More general metrics make sense in vector spaces, because the entities all have the same interpretation. But keeping the interpretation of points and vectors distinct leads us to consider hybrid metrics in bound spaces in which the vector subspace has a general metric but the origin is orthogonal to all vectors. We can therefore adopt a Euclidean metric for the origin relative to the vectors, and a more general metric for the vector subspace.

This hybrid metric is of course just one among many, depending on the applications in which one is interested. Nevertheless it retains a generality acceptable for a wide range of applications, and in the interests of simplicity will be the only one adopted for bound spaces in this book.

The hybrid metric

We define a *hybrid metric* on a bound vector space with basis of the form $\{\star\mathbb{O}, e_1, e_2, \ldots, e_n\}$ to be a metric G of the form:

$$G == \begin{pmatrix} 1 & 0 & \cdots & 0 \\ 0 & g_{1,1} & \cdots & g_{1,n} \\ \vdots & \vdots & \cdots & \vdots \\ 0 & g_{1,n} & \cdots & g_{n,n} \end{pmatrix} \qquad 5.80$$

Here, the g_{ij} have a determinant denoted $|g_{ij}|$ with the properties of a standard (symmetric, non-singular) metric as developed earlier in this chapter.

In a bound vector *n*-space, all the formulae for complements are still valid because a bound vector *n*-space and an (*n*+1)-space whose elements involve no interpretation do not algebraically differ. But because the vector subspace (of dimension *n*) of the bound vector *n*-space is so important, we will find it useful to define a complement operation in this vector subspace that pretends that the subspace is the whole space and obeys the axioms and results for a complement operation in an *n*-space.

We will denote the complement operation in the vector subspace by using an *overvector* $\vec{\square}$ (instead of an overbar $\overline{\square}$), and call it a *vector space complement*. The vector space complement (overvector) operation can be entered from the Basic Operations palette by using the button under the complement (overbar) button.

The orthogonality of the origin to *m*-vectors

Note that the introduction of this hybrid metric onto a space means that, because of the zeros in the metric, we are *defining* the origin $\star\mathbb{O}$ to be orthogonal to all vectors α which are linear combinations of the e_i. In the next chapter we will develop the notion of orthogonality in its more usual form in terms of the interior and inner products. But for this chapter we will use the more fundamental definitions in terms of the regressive and exterior products.

In [5.73] and [5.74] we have introduced the notion of orthogonality between two elements of the same grade. Thus whenever we use the hybrid metric, we will always have that

$$\alpha \wedge \overline{\star\mathbb{O}} == \overline{\alpha} \wedge \star\mathbb{O} == 0 \qquad 5.81$$

And, consequently, by taking the complement:

$$\alpha \vee \overline{\star\mathbb{O}} == \overline{\alpha} \vee \star\mathbb{O} == 0 \qquad 5.82$$

Now consider the regressive product of a simple *m*-vector with the complement of the origin. Let the *m*-vector be of the form $A \wedge \alpha$. Now apply the Product Formula [3.68].

$$(A \wedge \alpha) \vee \overline{\star\mathbb{O}} == \left(A \vee \overline{\star\mathbb{O}}\right) \wedge \alpha + (-1)^m A \wedge \left(\alpha \vee \overline{\star\mathbb{O}}\right)$$

The second term on the right side is zero because $\alpha \vee \overline{\star\mathbb{O}}$ is zero. And the first term on the right will be zero if $A \vee \overline{\star\mathbb{O}}$ is zero. If we then apply the Product Formula recursively we will find that this is indeed the case. Hence the origin is orthogonal to any simple *m*-vector. And by the linearity of the regressive product it is thus also orthogonal to *any* linear combination of vectorial elements.

$$\underset{m}{A} \vee \overline{\star\mathbb{O}} == \overline{\underset{m}{A}} \wedge \star\mathbb{O} == 0 \qquad 5.83$$

This formula is also true of any 1-element x which is orthogonal to all the 1-element factors of a simple *m*-element A.

$$\underset{m}{A} \vee \overline{x} == \overline{\underset{m}{A}} \wedge x == 0 \qquad \underset{m}{A} == A_1 \wedge \ldots \wedge A_m \qquad A_i \vee \overline{x} == 0 \qquad 5.84$$

Unit elements in a bound vector space

In a bound vector space with the hybrid metric tensor G [5.80], the determinant of the metric tensor is the same whether it be for the hybrid metric G or the metric g_{ij} of the vector subspace.

$$|G| == |g_{ij}| == |\star g| == |\underline{1}|^2 \qquad 5.85$$

Consider a bound vector space with basis $\{\star\mathbb{O}, e_1, e_2, \ldots, e_n\}$. As before, the *unit element of the space* is denoted by $\overline{1}$ and is congruent to the product of the basis elements.

$$\overline{1} == \frac{1}{|\underline{1}|} \star\mathbb{O} \wedge e_1 \wedge e_2 \wedge \cdots \wedge e_n \qquad 5.86$$

The unit element of the vector subspace, the *unit n-vector*, is denoted with an ‘overvector’ as $\vec{1}$ and is congruent to the product *only* of the basis vectors.

$$\vec{1} == \frac{1}{|\underline{1}|} e_1 \wedge e_2 \wedge \cdots \wedge e_n \qquad 5.87$$

Hence the unit element of the bound vector space is the unit *n*-vector bound through the origin.

$$\overline{1} == \star\mathbb{O} \wedge \vec{1} \qquad 5.88$$

Complement equivalences in a bound vector 2-space

Consider a bound vector 2-space with basis $\{\star\mathbb{O}, e_1, e_2\}$.

First, let us declare a basis $\{e_1, e_2\}$ and metric for its vector subspace, and then compose the palette of vector space complements of the subspace basis elements.

```
★𝓑2; DeclareMetric[{{g11, g12}, {g12, g22}}];
MatrixForm[Metric]
```

$$\begin{pmatrix} g_{11} & g_{12} \\ g_{12} & g_{22} \end{pmatrix}$$

```
ComplementPalette
```

Complement Palette

Basis	Complement
1	$\frac{e_1 \wedge e_2}{\sqrt{\star g}}$
e_1	$\frac{e_2\, g_{11}}{\sqrt{\star g}} - \frac{e_1\, g_{12}}{\sqrt{\star g}}$
e_2	$\frac{e_2\, g_{12}}{\sqrt{\star g}} - \frac{e_1\, g_{22}}{\sqrt{\star g}}$
$e_1 \wedge e_2$	$\sqrt{\star g}$

If we were to be in a bound vector space, these would be vector space complements, designated by an 'overvector', so the palette correspondences would be written as:

$$\overrightarrow{1} == \frac{e_1 \wedge e_2}{\sqrt{\star g}}$$

$$\overrightarrow{e_1} == \frac{e_2\, g_{11}}{\sqrt{\star g}} - \frac{e_1\, g_{12}}{\sqrt{\star g}}$$

$$\overrightarrow{e_2} == \frac{e_2\, g_{12}}{\sqrt{\star g}} - \frac{e_1\, g_{22}}{\sqrt{\star g}}$$

$$\overrightarrow{e_1 \wedge e_2} == \sqrt{\star g}$$

Now let us declare a basis $\{\star O, e_1, e_2\}$ and hybrid metric for the bound vector 2-space, and then compose the palette of complements of the basis elements.

```
★𝒫2; DeclareMetric[{{1, 0, 0}, {0, g11, g12}, {0, g12, g22}}];
MatrixForm[Metric]
```

$$\begin{pmatrix} 1 & 0 & 0 \\ 0 & g_{11} & g_{12} \\ 0 & g_{12} & g_{22} \end{pmatrix}$$

```
ComplementPalette
```

Complement Palette

Basis	Complement
1	$\frac{\star\mathbb{O}\wedge e_1\wedge e_2}{\sqrt{\star g}}$
$\star\mathbb{O}$	$\frac{e_1\wedge e_2}{\sqrt{\star g}}$
e_1	$\star\mathbb{O}\wedge\left(-\frac{e_2\, g_{11}}{\sqrt{\star g}}+\frac{e_1\, g_{12}}{\sqrt{\star g}}\right)$
e_2	$\star\mathbb{O}\wedge\left(-\frac{e_2\, g_{12}}{\sqrt{\star g}}+\frac{e_1\, g_{22}}{\sqrt{\star g}}\right)$
$\star\mathbb{O}\wedge e_1$	$\frac{e_2\, g_{11}}{\sqrt{\star g}}-\frac{e_1\, g_{12}}{\sqrt{\star g}}$
$\star\mathbb{O}\wedge e_2$	$\frac{e_2\, g_{12}}{\sqrt{\star g}}-\frac{e_1\, g_{22}}{\sqrt{\star g}}$
$e_1\wedge e_2$	$\sqrt{\star g}\,\star\mathbb{O}$
$\star\mathbb{O}\wedge e_1\wedge e_2$	$\sqrt{\star g}$

These are bound vector space complements, designated by an 'overbar'. But the palette correspondences can also be written in terms of the vector space complements determined from the first palette.

$$\overline{1} == \frac{\star\mathbb{O}\wedge e_1\wedge e_2}{\sqrt{\star g}} == \star\mathbb{O}\wedge\overrightarrow{1}$$

$$\overline{\star\mathbb{O}} == \frac{e_1\wedge e_2}{\sqrt{\star g}} == \overrightarrow{1}$$

$$\overline{e_1} == \star\mathbb{O}\wedge\left(-\frac{e_2\, g_{11}}{\sqrt{\star g}}+\frac{e_1\, g_{12}}{\sqrt{\star g}}\right) == -\star\mathbb{O}\wedge\overrightarrow{e_1}$$

$$\overline{e_2} == \star\mathbb{O}\wedge\left(-\frac{e_2\, g_{12}}{\sqrt{\star g}}+\frac{e_1\, g_{22}}{\sqrt{\star g}}\right) == -\star\mathbb{O}\wedge\overrightarrow{e_2}$$

$$\overline{\star\mathbb{O}\wedge e_1} == \frac{e_2\, g_{11}}{\sqrt{\star g}}-\frac{e_1\, g_{12}}{\sqrt{\star g}} == \overrightarrow{e_1}$$

$$\overline{\star\mathbb{O}\wedge e_2} == \frac{e_2\, g_{12}}{\sqrt{\star g}}-\frac{e_1\, g_{22}}{\sqrt{\star g}} == \overrightarrow{e_2}$$

$$\overline{e_1\wedge e_2} == \sqrt{\star g}\,\star\mathbb{O} == \overrightarrow{e_1\wedge e_2}\,\star\mathbb{O} == \star\mathbb{O}\wedge\overrightarrow{e_1\wedge e_2}$$

$$\overline{\star\mathbb{O}\wedge e_1\wedge e_2} == \sqrt{\star g} == \overrightarrow{e_1\wedge e_2}$$

In the next section we will look at the equivalent correspondences in a bound vector 3-space, and following that, summarize the results in some more general formulae.

Complement equivalences in a bound vector 3-space

To confirm the pattern of correspondences, we repeat the previous example (in condensed form) for a bound vector 3-space with basis $\{\star\mathbb{O}, e_1, e_2, e_3\}$.

First, let us declare a basis $\{e_1, e_2, e_3\}$ and metric for its vector subspace, and then compose the palette of vector space complements of the subspace basis elements.

```
★𝓑3; DeclareMetric[{{g11, g12, g13}, {g12, g22, g23}, {g13, g23, g33}}];
MatrixForm[Metric]
```

$$\begin{pmatrix} g_{11} & g_{12} & g_{13} \\ g_{12} & g_{22} & g_{23} \\ g_{13} & g_{23} & g_{33} \end{pmatrix}$$

```
ComplementPalette
```

Complement Palette

Basis	Complement
1	$\frac{e_1\wedge e_2\wedge e_3}{\sqrt{\star g}}$
e_1	$\frac{g_{13}\, e_1\wedge e_2}{\sqrt{\star g}} - \frac{g_{12}\, e_1\wedge e_3}{\sqrt{\star g}} + \frac{g_{11}\, e_2\wedge e_3}{\sqrt{\star g}}$
e_2	$\frac{g_{23}\, e_1\wedge e_2}{\sqrt{\star g}} - \frac{g_{22}\, e_1\wedge e_3}{\sqrt{\star g}} + \frac{g_{12}\, e_2\wedge e_3}{\sqrt{\star g}}$
e_3	$\frac{g_{33}\, e_1\wedge e_2}{\sqrt{\star g}} - \frac{g_{23}\, e_1\wedge e_3}{\sqrt{\star g}} + \frac{g_{13}\, e_2\wedge e_3}{\sqrt{\star g}}$
$e_1\wedge e_2$	$\frac{e_3\,(-g_{12}^2+g_{11}\, g_{22})}{\sqrt{\star g}} + \frac{e_2\,(g_{12}\, g_{13}-g_{11}\, g_{23})}{\sqrt{\star g}} + \frac{e_1\,(-g_{13}\, g_{22}+g_{12}\, g_{23})}{\sqrt{\star g}}$
$e_1\wedge e_3$	$\frac{e_3\,(-g_{12}\, g_{13}+g_{11}\, g_{23})}{\sqrt{\star g}} + \frac{e_2\,(g_{13}^2-g_{11}\, g_{33})}{\sqrt{\star g}} + \frac{e_1\,(-g_{13}\, g_{23}+g_{12}\, g_{33})}{\sqrt{\star g}}$
$e_2\wedge e_3$	$\frac{e_3\,(-g_{13}\, g_{22}+g_{12}\, g_{23})}{\sqrt{\star g}} + \frac{e_2\,(g_{13}\, g_{23}-g_{12}\, g_{33})}{\sqrt{\star g}} + \frac{e_1\,(-g_{23}^2+g_{22}\, g_{33})}{\sqrt{\star g}}$
$e_1\wedge e_2\wedge e_3$	$\sqrt{\star g}$

Again, if we were to be in a bound vector space, these would be vector space complements, designated by an 'overvector', so typical palette correspondences would be written as:

$$\vec{1} == \frac{e_1\wedge e_2\wedge e_3}{\sqrt{\star g}}$$

$$\overrightarrow{e_1} == \frac{g_{13}\, e_1\wedge e_2}{\sqrt{\star g}} - \frac{g_{12}\, e_1\wedge e_3}{\sqrt{\star g}} + \frac{g_{11}\, e_2\wedge e_3}{\sqrt{\star g}}$$

$$\overrightarrow{e_1\wedge e_2} == \frac{e_3\,(g_{11}\, g_{22} - g_{12}^2)}{\sqrt{\star g}} + \frac{e_2\,(g_{12}\, g_{13} - g_{11}\, g_{23})}{\sqrt{\star g}} + \frac{e_1\,(g_{12}\, g_{23} - g_{13}\, g_{22})}{\sqrt{\star g}}$$

$$\overrightarrow{e_1\wedge e_2\wedge e_3} == \sqrt{\star g}$$

Now let us declare a basis $\{\star\mathbb{O}, e_1, e_2, e_3\}$ and metric for the bound vector 3-space, and then compose the palette of complements of the basis elements.

```
★𝒫3;
DeclareMetric[{{1, 0, 0, 0}, {0, g11, g12, g13}, {0, g12, g22, g23},
  {0, g13, g23, g33}}];
MatrixForm[Metric]
```

$$\begin{pmatrix} 1 & 0 & 0 & 0 \\ 0 & g_{11} & g_{12} & g_{13} \\ 0 & g_{12} & g_{22} & g_{23} \\ 0 & g_{13} & g_{23} & g_{33} \end{pmatrix}$$

```
ComplementPalette
```

Complement Palette

Basis	Complement
1	$\frac{\star\mathbb{O}\wedge e_1\wedge e_2\wedge e_3}{\sqrt{\star g}}$
$\star\mathbb{O}$	$\frac{e_1\wedge e_2\wedge e_3}{\sqrt{\star g}}$
e_1	$\star\mathbb{O}\wedge\left(-\frac{g_{13}\,e_1\wedge e_2}{\sqrt{\star g}}+\frac{g_{12}\,e_1\wedge e_3}{\sqrt{\star g}}-\frac{g_{11}\,e_2\wedge e_3}{\sqrt{\star g}}\right)$
e_2	$\star\mathbb{O}\wedge\left(-\frac{g_{23}\,e_1\wedge e_2}{\sqrt{\star g}}+\frac{g_{22}\,e_1\wedge e_3}{\sqrt{\star g}}-\frac{g_{12}\,e_2\wedge e_3}{\sqrt{\star g}}\right)$
e_3	$\star\mathbb{O}\wedge\left(-\frac{g_{33}\,e_1\wedge e_2}{\sqrt{\star g}}+\frac{g_{23}\,e_1\wedge e_3}{\sqrt{\star g}}-\frac{g_{13}\,e_2\wedge e_3}{\sqrt{\star g}}\right)$
$\star\mathbb{O}\wedge e_1$	$\frac{g_{13}\,e_1\wedge e_2}{\sqrt{\star g}}-\frac{g_{12}\,e_1\wedge e_3}{\sqrt{\star g}}+\frac{g_{11}\,e_2\wedge e_3}{\sqrt{\star g}}$
$\star\mathbb{O}\wedge e_2$	$\frac{g_{23}\,e_1\wedge e_2}{\sqrt{\star g}}-\frac{g_{22}\,e_1\wedge e_3}{\sqrt{\star g}}+\frac{g_{12}\,e_2\wedge e_3}{\sqrt{\star g}}$
$\star\mathbb{O}\wedge e_3$	$\frac{g_{33}\,e_1\wedge e_2}{\sqrt{\star g}}-\frac{g_{23}\,e_1\wedge e_3}{\sqrt{\star g}}+\frac{g_{13}\,e_2\wedge e_3}{\sqrt{\star g}}$
$e_1\wedge e_2$	$\star\mathbb{O}\wedge\left(\frac{e_3\,(-g_{12}^2+g_{11}\,g_{22})}{\sqrt{\star g}}+\frac{e_2\,(g_{12}\,g_{13}-g_{11}\,g_{23})}{\sqrt{\star g}}+\frac{e_1\,(-g_{13}\,g_{22}+g_{12}\,g_{23})}{\sqrt{\star g}}\right)$
$e_1\wedge e_3$	$\star\mathbb{O}\wedge\left(\frac{e_3\,(-g_{12}\,g_{13}+g_{11}\,g_{23})}{\sqrt{\star g}}+\frac{e_2\,(g_{13}^2-g_{11}\,g_{33})}{\sqrt{\star g}}+\frac{e_1\,(-g_{13}\,g_{23}+g_{12}\,g_{33})}{\sqrt{\star g}}\right)$
$e_2\wedge e_3$	$\star\mathbb{O}\wedge\left(\frac{e_3\,(-g_{13}\,g_{22}+g_{12}\,g_{23})}{\sqrt{\star g}}+\frac{e_2\,(g_{13}\,g_{23}-g_{12}\,g_{33})}{\sqrt{\star g}}+\frac{e_1\,(-g_{23}^2+g_{22}\,g_{33})}{\sqrt{\star g}}\right)$
$\star\mathbb{O}\wedge e_1\wedge e_2$	$\frac{e_3\,(-g_{12}^2+g_{11}\,g_{22})}{\sqrt{\star g}}+\frac{e_2\,(g_{12}\,g_{13}-g_{11}\,g_{23})}{\sqrt{\star g}}+\frac{e_1\,(-g_{13}\,g_{22}+g_{12}\,g_{23})}{\sqrt{\star g}}$
$\star\mathbb{O}\wedge e_1\wedge e_3$	$\frac{e_3\,(-g_{12}\,g_{13}+g_{11}\,g_{23})}{\sqrt{\star g}}+\frac{e_2\,(g_{13}^2-g_{11}\,g_{33})}{\sqrt{\star g}}+\frac{e_1\,(-g_{13}\,g_{23}+g_{12}\,g_{33})}{\sqrt{\star g}}$
$\star\mathbb{O}\wedge e_2\wedge e_3$	$\frac{e_3\,(-g_{13}\,g_{22}+g_{12}\,g_{23})}{\sqrt{\star g}}+\frac{e_2\,(g_{13}\,g_{23}-g_{12}\,g_{33})}{\sqrt{\star g}}+\frac{e_1\,(-g_{23}^2+g_{22}\,g_{33})}{\sqrt{\star g}}$
$e_1\wedge e_2\wedge e_3$	$-\sqrt{\star g}\,\star\mathbb{O}$
$\star\mathbb{O}\wedge e_1\wedge e_2\wedge e_3$	$\sqrt{\star g}$

These are bound vector space complements, designated by an 'overbar', and as in the previous section, the palette correspondences can also be written in terms of the vector space complements determined from the first palette.

$$\overline{1} == \frac{\star\mathbb{O}\wedge e_1\wedge e_2\wedge e_3}{\sqrt{\star g}} == \star\mathbb{O}\wedge\vec{1}$$

$$\overline{\star\mathbb{O}} == \frac{e_1 \wedge e_2 \wedge e_3}{\sqrt{\star g}} == \overrightarrow{1}$$

$$\overline{e_1} == \star\mathbb{O} \wedge \left(-\frac{g_{13}\, e_1 \wedge e_2}{\sqrt{\star g}} + \frac{g_{12}\, e_1 \wedge e_3}{\sqrt{\star g}} - \frac{g_{11}\, e_2 \wedge e_3}{\sqrt{\star g}}\right) == -\star\mathbb{O} \wedge \overrightarrow{e_1}$$

$$\overline{\star\mathbb{O} \wedge e_1} == \frac{g_{13}\, e_1 \wedge e_2}{\sqrt{\star g}} - \frac{g_{12}\, e_1 \wedge e_3}{\sqrt{\star g}} + \frac{g_{11}\, e_2 \wedge e_3}{\sqrt{\star g}} == \overrightarrow{e_1}$$

$$\overline{e_1 \wedge e_2} == \star\mathbb{O} \wedge \left(\frac{e_3\left(g_{11}\, g_{22} - g_{12}{}^2\right)}{\sqrt{\star g}} + \frac{e_2\,(g_{12}\, g_{13} - g_{11}\, g_{23})}{\sqrt{\star g}} + \frac{e_1\,(g_{12}\, g_{23} - g_{13}\, g_{22})}{\sqrt{\star g}}\right)$$

$$== \star\mathbb{O} \wedge \overrightarrow{e_1 \wedge e_2}$$

$$\overline{\star\mathbb{O} \wedge e_1 \wedge e_2} == \frac{e_3\left(g_{11}\, g_{22} - g_{12}^2\right)}{\sqrt{\star g}} + \frac{e_2\,(g_{12}\, g_{13} - g_{11}\, g_{23})}{\sqrt{\star g}} + \frac{e_1\,(g_{12}\, g_{23} - g_{13}\, g_{22})}{\sqrt{\star g}}$$

$$== \overrightarrow{e_1 \wedge e_2}$$

$$\overline{e_1 \wedge e_2 \wedge e_3} == -\sqrt{\star g}\ \star\mathbb{O} == -\overrightarrow{e_1 \wedge e_2 \wedge e_3}\ \star\mathbb{O} == -\star\mathbb{O} \wedge \overrightarrow{e_1 \wedge e_2 \wedge e_3}$$

$$\overline{\star\mathbb{O} \wedge e_1 \wedge e_2 \wedge e_3} == \sqrt{\star g} == \overrightarrow{e_1 \wedge e_2 \wedge e_3}$$

Complement equivalences in a bound vector *n*-space

Collected below are the complement equivalences obtained from the 2- and 3-dimensional cases. Following this collection, we summarize the formulae for the arbitrary dimensional case. In the interests of minimizing complexity we do not pursue their general proof since the formulae are straightforward to computationally verify in any specific case.

◇ Summary: Complement equivalences in a bound vector 2-space

$$\overline{1} == \star\mathbb{O} \wedge \overrightarrow{1}$$

$$\overline{\star\mathbb{O}} == \overrightarrow{1}$$

$$\overline{e_i} == -\star\mathbb{O} \wedge \overrightarrow{e_i}$$

$$\overline{\star\mathbb{O} \wedge e_i} == \overrightarrow{e_i}$$

$$\overline{e_1 \wedge e_2} == \star\mathbb{O} \wedge \overrightarrow{e_1 \wedge e_2}$$

$$\overline{\star\mathbb{O} \wedge e_1 \wedge e_2} == \overrightarrow{e_1 \wedge e_2}$$

◇ Summary: Complement equivalences in a bound vector 3-space

$$\overline{1} == \star\mathbb{O} \wedge \overrightarrow{1}$$

$$\overline{\star\mathbb{O}} == \overrightarrow{1}$$

$$\overline{e_i} == -\star\mathbb{O} \wedge \overrightarrow{e_i}$$

$$\overline{\star\mathbb{O} \wedge e_i} == \overrightarrow{e_i}$$

$$\overline{e_i \wedge e_j} == \star\mathbb{O} \wedge \overrightarrow{e_i \wedge e_j}$$

$$\overline{\star\mathbb{O} \wedge e_i \wedge e_j} == \overrightarrow{e_i \wedge e_j}$$

$$\overline{e_1 \wedge e_2 \wedge e_3} == - \star\mathbb{O} \wedge \overrightarrow{e_1 \wedge e_2 \wedge e_3}$$

$$\overline{\star\mathbb{O} \wedge e_1 \wedge e_2 \wedge e_3} == \overrightarrow{e_1 \wedge e_2 \wedge e_3}$$

◇ Summary: Complement equivalences in a bound vector *n*-space

From the two cases above, we see that the equivalences do not depend on the dimension of the space.

$$\overline{1} == \star\mathbb{O} \wedge \overrightarrow{1} \qquad \overline{\star\mathbb{O}} == \overrightarrow{1} \tag{5.89}$$

$$\overline{e_i} == - \star\mathbb{O} \wedge \overrightarrow{e_i} \qquad \overline{\star\mathbb{O} \wedge e_i} == \overrightarrow{e_i} \tag{5.90}$$

$$\overline{e_i \wedge e_j} == \star\mathbb{O} \wedge \overrightarrow{e_i \wedge e_j} \qquad \overline{\star\mathbb{O} \wedge e_i \wedge e_j} == \overrightarrow{e_i \wedge e_j} \tag{5.91}$$

$$\overline{e_i \wedge e_j \wedge e_k} == - \star\mathbb{O} \wedge \overrightarrow{e_i \wedge e_j \wedge e_k} \qquad \overline{\star\mathbb{O} \wedge e_i \wedge e_j \wedge e_k} == \overrightarrow{e_i \wedge e_j \wedge e_k} \tag{5.92}$$

However, with the inclusion of a sign $(-1)^m$ these four sets of formulae can be seen to be examples of the more general case:

$$\overline{\underset{m}{e_i}} == (-1)^m \star\mathbb{O} \wedge \overrightarrow{\underset{m}{e_i}} \qquad \overline{\star\mathbb{O} \wedge \underset{m}{e_i}} == \overrightarrow{\underset{m}{e_i}} \tag{5.93}$$

And using [5.14] we can write the corresponding formulae for general *m*-vectors $\underset{m}{\alpha}$.

$$\overline{\underset{m}{\alpha}} == (-1)^m \star\mathbb{O} \wedge \overrightarrow{\underset{m}{\alpha}} \qquad \overline{\star\mathbb{O} \wedge \underset{m}{\alpha}} == \overrightarrow{\underset{m}{\alpha}} \tag{5.94}$$

- The vector space complement of any exterior product involving the origin $\star\mathbb{O}$ is undefined.

The complement of the complement of an *m*-vector in a bound space

Consider a bound *n*-space. Since the dimension of the space is *n*+1, the complement of the complement of an *m*-vector $\underset{m}{\alpha}$ is given by:

$$\overline{\overline{\underset{m}{\alpha}}} == (-1)^{m\,(n+1-m)} \underset{m}{\alpha}$$

The vector space complement of the vector space complement of $\underset{m}{\alpha}$ in the vector subspace (of dimension *n*) is

$$\overrightarrow{\overrightarrow{\underset{m}{\alpha}}} == (-1)^{m\,(n-m)} \underset{m}{\alpha}$$

Hence since both double complements are congruent to $\underset{m}{\alpha}$ we can write:

$$\overline{\overline{\underset{m}{\alpha}}} == (-1)^{m}\, \overrightarrow{\overrightarrow{\underset{m}{\alpha}}} \qquad 5.95$$

We can also verify this using the formulae [5.94].

$$\overline{\underset{m}{\alpha}} == (-1)^{m} \star\mathbb{O} \wedge \overrightarrow{\underset{m}{\alpha}}$$

$$\overline{\overline{\underset{m}{\alpha}}} == (-1)^{m}\, \overline{\star\mathbb{O} \wedge \overrightarrow{\underset{m}{\alpha}}} == (-1)^{m}\, \overrightarrow{\overrightarrow{\underset{m}{\alpha}}}$$

Calculating with vector space complements

◆ Entering a vector space complement

To enter a vector space complement of a Grassmann expression `X` in *GrassmannAlgebra* you can either use the *GrassmannAlgebra* Basic Operations palette by selecting the expression `X` and clicking the button, or simply enter `OverVector[X]` directly.

`OverVector[`$e_1 \wedge e_2$`]`

$\overrightarrow{e_1 \wedge e_2}$

◆ Simplifying a vector space complement

If the basis of your currently declared space does not contain the origin $\star\mathbb{O}$, then the vector space complement (`OverVector`) operation is equivalent to the normal (`OverBar`) operation, and `ConvertComplements` will treat them as the same.

$\star\mathcal{B}_3$; `ConvertComplements`$\left[\left\{\overrightarrow{e_1}, \overrightarrow{A}, \overline{e_1}, \overline{A}\right\}\right]$

$\left\{e_2 \wedge e_3, \overline{A}, e_2 \wedge e_3, \overline{A}\right\}$

If, on the other hand, the basis of your currently declared space does contain the origin $\star\mathbb{O}$, then `ConvertComplements` will convert any expressions containing `OverVector` complements to their equivalent `OverBar` forms.

$\star\mathcal{P}_3$; `ConvertComplements`$\left[\left\{\overrightarrow{e_1}, \overrightarrow{A}, \overline{e_1}, \overline{A}\right\}\right]$

$\left\{e_2 \wedge e_3, \overline{\star\mathbb{O} \wedge A}, -(\star\mathbb{O} \wedge e_2 \wedge e_3), \overline{A}\right\}$

◆ The vector space complement of the origin

Note that the vector space complement of the origin, or any exterior product of elements involving the origin, is undefined, and will be left unevaluated by `ConvertComplements`.

$\star\mathcal{P}_3$; `ConvertComplements`$\left[\left\{\overrightarrow{\star\mathbb{O}}, \overrightarrow{\star\mathbb{O} \wedge e_1}, \overrightarrow{\star\mathbb{O} + x}, \overrightarrow{\star\mathbb{O} + e_1}\right\}\right]$

$\left\{\overrightarrow{\star\mathbb{O}}, \overrightarrow{\star\mathbb{O} \wedge e_1}, \overrightarrow{\star\mathbb{O} \wedge x} + \overrightarrow{\star\mathbb{O}}, \overrightarrow{\star\mathbb{O}} + e_2 \wedge e_3\right\}$

5.12 Complements of bound elements

The Euclidean complement of a point in the plane

In order to fix the geometrical ideas involved in complementing bound elements, we will first look at the two simplest cases in detail: the Euclidean complement of a point in the plane and of a point in space. Then we will look at more general formulations.

Suppose we are working in the Euclidean plane with a basis of an origin point $\star\mathbb{O}$ and two basis vectors e_1 and e_2. Let us declare this basis then call for a palette of basis elements and their complements.

```
★A; ★𝒫2; ComplementPalette
```

Complement Palette

Basis	Complement
1	$\star\mathbb{O} \wedge e_1 \wedge e_2$
$\star\mathbb{O}$	$e_1 \wedge e_2$
e_1	$-(\star\mathbb{O} \wedge e_2)$
e_2	$\star\mathbb{O} \wedge e_1$
$\star\mathbb{O} \wedge e_1$	e_2
$\star\mathbb{O} \wedge e_2$	$-e_1$
$e_1 \wedge e_2$	$\star\mathbb{O}$
$\star\mathbb{O} \wedge e_1 \wedge e_2$	1

The complement palette tells us that each basis vector is orthogonal to the *axis* involving the other basis vector. For example, e_2 is orthogonal to the axis $\star\mathbb{O} \wedge e_1$.

Suppose now we take a general vector $x = a\, e_1 + b\, e_2$. The complement of this vector is:

$$\overline{x} == a\,\overline{e_1} + b\,\overline{e_2} == -a\,\star\mathbb{O} \wedge e_2 + b\,\star\mathbb{O} \wedge e_1 == \star\mathbb{O} \wedge (b\, e_1 - a\, e_2)$$

Again, this is an axis (bound vector) through the origin orthogonal to the vector x.

Now let us take a point $P = \star\mathbb{O} + x$ and explore what element is orthogonal to this point.

$$\overline{P} == \overline{\star\mathbb{O}} + \overline{x} == e_1 \wedge e_2 + \star\mathbb{O} \wedge (b\, e_1 - a\, e_2)$$

The effect of adding the bivector $e_1 \wedge e_2$ to $\overline{x}$ is to shift the bound vector $\star\mathbb{O} \wedge (b\, e_1 - a\, e_2)$ parallel to itself. We can factor $e_1 \wedge e_2$ into the exterior product of two vectors, one parallel to $b\, e_1 - a\, e_2$.

$$e_1 \wedge e_2 == z \wedge (b\, e_1 - a\, e_2)$$

Now we can write $\overline{P}$ as:

$$\overline{P} == (\star\mathbb{O} + z) \wedge (b\, e_1 - a\, e_2)$$

The vector z is the position vector of *any* point on the line defined by the bound vector $\overline{P}$. A particular point of interest is the point on the line closest to the point P or the origin. The position vector z would then be a vector orthogonal to the direction of the line. Thus we can write $e_1 \wedge e_2$ as:

$$e_1 \wedge e_2 == -\frac{(a\, e_1 + b\, e_2) \wedge (b\, e_1 - a\, e_2)}{a^2 + b^2}$$

The final expression for the complement of the point P can then be written as a bound vector in a direction orthogonal to the position vector of P.

$$P == \star\mathbb{O} + a\, e_1 + b\, e_2 \qquad \overline{P} == \left(\star\mathbb{O} - \frac{a\, e_1 + b\, e_2}{a^2 + b^2}\right) \wedge (b\, e_1 - a\, e_2) \tag{5.96}$$

This formula turns out to be a special case of a more general formula which we will derive later in the section, and for which we now propose some convenient notation.

$$P == \star\mathbb{O} + x \qquad P^* == \star\mathbb{O} + x^* \qquad x^* == -\frac{x}{|x|^2} \tag{5.97}$$

We call the point P^* the *inverse point* to P. Inverse points are situated on the same line through the origin, and on opposite sides of it. The product of their distances from the origin is unity.

Remembering that the vector space complement of the position vector x is denoted $\vec{x}$, we can now rewrite our derived formula for the complement of P in a more condensed form.

$$\overline{P} == P^* \wedge (-\vec{x}) \tag{5.98}$$

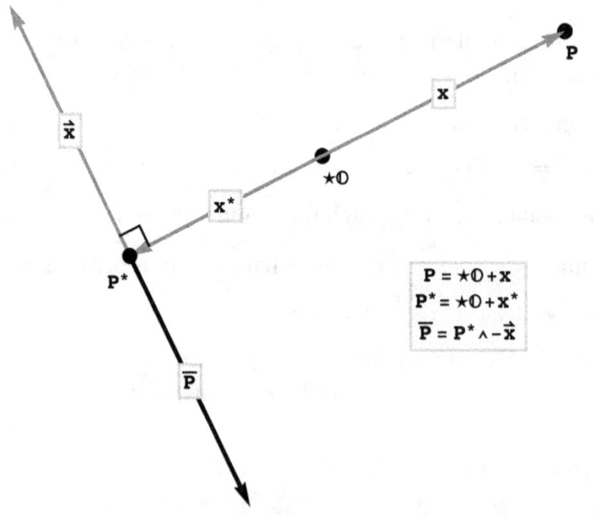

The Euclidean complement of a point in the plane

Thus, the complement of a point in the plane is a bound vector through its inverse point in a direction orthogonal to its position vector.

Of course, since the bound space of a plane is a three-dimensional linear space, the complement of the complement of a point is the point itself, just like the complement of a complement of a vector in a vector 3-space is the vector itself.

The Euclidean complement of a point in a bound 3-space

The Euclidean complement of a point in a bound 2-space (the plane) is a bound vector (line) orthogonal to it. In a bound 3-space (space) it is a bound bivector (plane) orthogonal to it. Determining this bound bivector follows *mutatis mutandis* what we determined for the bound 2-space, so we will only outline the process here.

Let the basis for the bound 3-space be $\{\star\mathbb{O}, e_1, e_2, e_3\}$. Each basis vector is orthogonal to the *coordinate plane* involving the other basis vectors. For example, e_2 is orthogonal to the coordinate plane $\star\mathbb{O} \wedge e_1 \wedge e_3$.

Suppose now we take a general vector $x = a\, e_1 + b\, e_2 + c\, e_3$. The complement of this vector is:

$$\overline{x} == a\,\overline{e_1} + b\,\overline{e_2} + c\,\overline{e_3} == \star\mathbb{O} \wedge (-a\, e_2 \wedge e_3 + b\, e_1 \wedge e_3 - c\, e_1 \wedge e_2)$$

This is a bound bivector (plane) through the origin orthogonal to the vector (direction) x.

The complement of the point $P = \star\mathbb{O} + x$ is

$$\overline{P} == \overline{\star\mathbb{O}} + \overline{x} == e_1 \wedge e_2 \wedge e_3 + \star\mathbb{O} \wedge (-a\, e_2 \wedge e_3 + b\, e_1 \wedge e_3 - c\, e_1 \wedge e_2)$$

The effect of adding the trivector $e_1 \wedge e_2 \wedge e_3$ to the bound bivector $\overline{x}$ is to shift it parallel to itself. Suppose we factor the trivector as:

$$e_1 \wedge e_2 \wedge e_3 == z \wedge (-a\, e_2 \wedge e_3 + b\, e_1 \wedge e_3 - c\, e_1 \wedge e_2)$$

Now we can write $\overline{P}$ as:

$$\overline{P} == (\star\mathbb{O} + z) \wedge (-a\, e_2 \wedge e_3 + b\, e_1 \wedge e_3 - c\, e_1 \wedge e_2)$$

The vector z is the position vector of *any* point on the plane defined by the bound bivector $\overline{P}$. A particular point of interest is the point on the plane closest to the point P or the origin. The position vector z would then be a vector orthogonal to the plane. Thus we can write z as:

$$z == -\frac{a\, e_1 + b\, e_2 + c\, e_3}{a^2 + b^2 + c^2} == x^*$$

The final expression for the complement of the point P can then be written in the same form as for the plane in the previous section.

$$\overline{P} == \left(\star\mathbb{O} - \frac{(a\, e_1 + b\, e_2 + c\, e_3)}{a^2 + b^2 + c^2}\right) \wedge (-a\, e_2 \wedge e_3 + b\, e_1 \wedge e_3 - c\, e_1 \wedge e_2)$$

$$\overline{P} == P^* \wedge \left(-\overline{\vec{x}}\right) == (\star\mathbb{O} + x^*) \wedge \left(-\overline{\vec{x}}\right) \qquad 5.99$$

$$P == \star\mathbb{O} + x == \star\mathbb{O} + a\, e_1 + b\, e_2 + c\, e_3$$

$$P^* == \star\mathbb{O} + x^* == \star\mathbb{O} - \frac{x}{|x|^2} == \star\mathbb{O} - \frac{a\, e_1 + b\, e_2 + c\, e_3}{a^2 + b^2 + c^2} \qquad 5.100$$

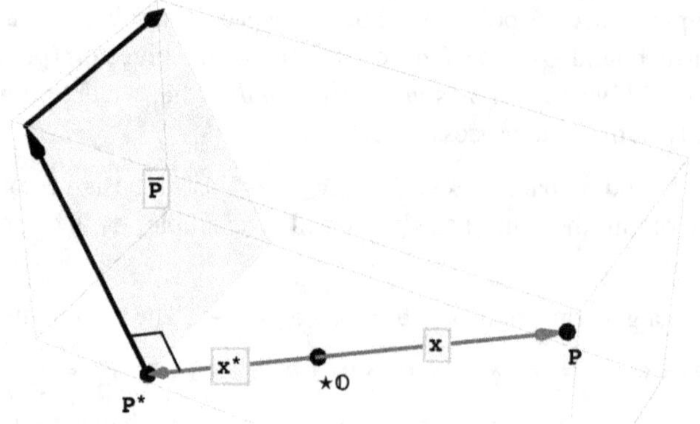

The Euclidean complement of a point in a bound 3-space

The complement of a point in a bound *n*-space

In the previous two sections we have discussed the geometry of obtaining the Euclidean complement of a point in bound 2- and 3-spaces. It is however simpler to derive the general equations directly from [5.94].

$$\overline{P} == \overline{\star\mathbb{O}} + \overline{x} == \vec{1} - \star\mathbb{O} \wedge \vec{\overline{x}}$$

From [5.77] we have

$$x \wedge \vec{\overline{x}} == \left(x \vee \vec{\overline{x}}\right) \vec{1} == |x|^2 \, \vec{1}$$

Substituting for $\vec{1}$ in the expression for $\overline{P}$:

$$\overline{P} == \frac{x \wedge \vec{\overline{x}}}{|x|^2} - \star\mathbb{O} \wedge \vec{\overline{x}} == \left(\star\mathbb{O} - \frac{x}{|x|^2}\right) \wedge \left(-\vec{\overline{x}}\right)$$

$$\overline{P} == P^* \wedge \left(-\vec{\overline{x}}\right) \qquad P^* == \star\mathbb{O} - \frac{x}{|x|^2} \qquad 5.101$$

This formula applies to the complement of a point in a space of any dimension and vector space metric.

◆ The inverse point of an inverse point

The point P^* has been called the *inverse point* to P.

$$P == \star\mathbb{O} + x \qquad P^* == \star\mathbb{O} - \frac{x}{|x|^2}$$

The inverse point of an inverse point is the point itself. Or, what is equivalent: P is also the inverse point to P^*.

$$P^{**} == \star\mathbb{O} - \frac{-\frac{x}{|x|^2}}{\left|-\frac{x}{|x|^2}\right|^2} == P$$

$$P^{**} == P \qquad 5.102$$

◆ **A point and its inverse point are orthogonal**

From equation [5.73], the points P and P^* are orthogonal since

$$P^* \wedge \overline{P} == P^* \wedge P^* \wedge \left(-\overrightarrow{\overline{x}}\right) == 0$$

The complement of a bound element

We have expressed the complement of a point as (the negative of) the vector space complement of its position vector bound through its inverse point. We would like to see to what extent this form of expression also applies to the complements of more general bound elements.

Consider an *m*-vector $\underset{m}{\alpha}$ bound through a point P with position vector x. By taking its complement and then transforming the resulting terms by using equations [5.94] we get that the complement of a bound *m*-vector may be written as the sum of a bound and a free term.

$$\overline{P \wedge \underset{m}{\alpha}} == \overline{\star\mathbb{O} \wedge \underset{m}{\alpha}} + \overline{x \wedge \underset{m}{\alpha}} == \overrightarrow{\underset{m}{\overline{\alpha}}} - \star\mathbb{O} \wedge \overrightarrow{\overline{\underset{m}{\alpha} \wedge x}}$$

$$\overline{P \wedge \underset{m}{\alpha}} == \star\mathbb{O} \wedge \left(-\overrightarrow{\overline{\underset{m}{\alpha} \wedge x}}\right) + \overrightarrow{\underset{m}{\overline{\alpha}}} \qquad 5.103$$

To get this into the form we want, we need to factor $\overrightarrow{\underset{m}{\overline{\alpha}}}$ so that one of its factors is equal to $\overline{\underset{m}{\alpha} \wedge x}$. We will also find that the resulting formula is simpler and considerably more intuitive if we choose (without loss of generality) the point on the bound *m*-vector to be orthogonal to the *m*-vector. Because of our hybrid metric in which the origin is orthogonal to all vectors, this is equivalent to requiring that the position vector x of P is orthogonal to $\underset{m}{\alpha}$.

A direct way to ensure these conditions is to use the Product Formula [3.68] with $\underset{k}{\beta}$ equal to x and $\underset{n-1}{x}$ equal to $\overline{x}$.

$$\left(\underset{m}{\alpha} \wedge x\right) \vee \overline{x} == \left(\underset{m}{\alpha} \vee \overline{x}\right) \wedge x + (-1)^m \underset{m}{\alpha} \wedge (x \vee \overline{x})$$

Now take the complement of each of these terms. The first and last terms become

$$\overline{\left(\underset{m}{\alpha} \wedge x\right) \vee \overline{x}} == \overline{\underset{m}{\alpha} \wedge x} \wedge \overline{\overline{x}} == (-1)^n \overline{\underset{m}{\alpha} \wedge x} \wedge x == (-1)^n (-1)^{m+1} \star\mathbb{O} \wedge \overrightarrow{\overline{\underset{m}{\alpha} \wedge x}} \wedge x$$

$$(-1)^m \overline{\underset{m}{\alpha} \wedge (x \vee \overline{x})} == (-1)^m |x|^2 \underset{m}{\overline{\alpha}} == |x|^2 \star\mathbb{O} \wedge \overrightarrow{\underset{m}{\overline{\alpha}}}$$

The first term on the right hand side will be zero because we are positing that P is chosen so that x is orthogonal to $\overline{\underset{m}{\alpha}}$, or what is equivalent, x belongs to the complement of $\overline{\underset{m}{\alpha}}$.

$$\left(\underset{m}{\alpha} \vee \overline{x}\right) \wedge x == 0 \quad \Leftrightarrow \quad \overline{\underset{m}{\alpha}} \wedge x == 0 \quad \Leftrightarrow \quad \underset{m}{\vec{\alpha}} \wedge x == 0$$

By factoring out the origin from the first and last terms we can now express $\underset{m}{\vec{\alpha}}$ in the form we want.

$$\underset{m}{\vec{\alpha}} == (-1)^{n-m+1}\, \overrightarrow{\underset{m}{\alpha} \wedge x} \wedge \frac{x}{|x|^2} == \frac{x}{|x|^2} \wedge \overrightarrow{\underset{m}{\alpha} \wedge x}$$

Substituting back into [5.103] gives us the final result.

$$\overline{P \wedge \underset{m}{\alpha}} == \star\mathbb{O} \wedge \left(-\overrightarrow{\underset{m}{\alpha} \wedge x}\right) + \frac{x}{|x|^2} \wedge \overrightarrow{\underset{m}{\alpha} \wedge x} == \left(\star\mathbb{O} - \frac{x}{|x|^2}\right) \wedge \left(-\overrightarrow{\underset{m}{\alpha} \wedge x}\right)$$

$$P == \star\mathbb{O} + x \qquad P^* == \star\mathbb{O} - \frac{x}{|x|^2}$$

$$\overline{P \wedge \underset{m}{\alpha}} == P^* \wedge \left(-\overrightarrow{\underset{m}{\alpha} \wedge x}\right) \qquad \underset{m}{\vec{\alpha}} \wedge x == 0 \qquad 5.104$$

The complement of the complement of a bound element

We already know from the complement of a complement axiom [5.7] that

$$\overline{\overline{P \wedge \underset{m}{\alpha}}} == (-1)^{(m+1)\,(n+1-(m+1))}\, P \wedge \underset{m}{\alpha} == (-1)^{n\,(m+1)}\, P \wedge \underset{m}{\alpha}$$

In [5.104] we have shown the results that the complement of a bound *m*-vector whose position vector is orthogonal to the *m*-vector, is also a bound element of the same form through its inverse point. Therefore we ought to be able to recover this congruence by taking the complement of the new bound element through the inverse point, again using [5.104]. This is not a particularly straightforward piece of manipulation, but it does serve as an exercise to confirm the results of [5.104]. Note that we will need to use [5.83] and [5.84], since the formula is only valid under the orthogonality conditions. The steps below are only an outline.

$$\overline{P \wedge \underset{m}{\alpha}} == P^* \wedge \left(-\overrightarrow{\underset{m}{\alpha} \wedge x}\right) == P^* \wedge A$$

$$\overline{\overline{P \wedge \underset{m}{\alpha}}} == \overline{P^* \wedge A} == P^{**} \wedge \left(\overline{-A \wedge \left(-\frac{x}{|x|^2}\right)}\right) == P \wedge \left(\frac{1}{|x|^2}\, \overrightarrow{A \wedge x}\right)$$

$$\overrightarrow{A \wedge x} == \overline{\star\mathbb{O} \wedge A \wedge x} == -\overline{\star\mathbb{O} \wedge \overline{\star\mathbb{O} \wedge \underset{m}{\alpha} \wedge x} \wedge x}$$

$$== -(-1)^{(m+2)\,(n+1-(m+2))}\,(-1)^{(n+1-(m+2))} \left(\star\mathbb{O} \wedge \underset{m}{\alpha} \wedge x\right) \vee \overline{\star\mathbb{O}} \vee \overline{x}$$

$$== (-1)^{n\,(m+1)+m} \left(\left(\star\mathbb{O} \wedge \left(\underset{m}{\alpha} \wedge x\right)\right) \vee \overline{\star\mathbb{O}}\right) \vee \overline{x}$$

By the Product Formula [3.68] we can expand the first factor of this expression and note that $\star\mathbb{O} \vee \overline{\star\mathbb{O}}$ is unity due the hybrid metric, and the last term is zero because the origin is orthogonal to all vectors.

$$\left(\star\mathbb{O} \wedge \left(\underset{m}{\alpha} \wedge x\right)\right) \vee \overline{\star\mathbb{O}} == \left(\star\mathbb{O} \vee \overline{\star\mathbb{O}}\right) \wedge \left(\underset{m}{\alpha} \wedge x\right) - \star\mathbb{O} \wedge \left(\left(\underset{m}{\alpha} \wedge x\right) \vee \overline{\star\mathbb{O}}\right)$$

$$\overrightarrow{\overline{A \wedge x}} == (-1)^{n\,(m+1)+m} \left(\underset{m}{\alpha} \wedge x\right) \vee \overline{x}$$

Applying the Product Formula [3.68] again to the right hand side we find that since we are requiring that the position vector of P be orthogonal to the *m*-vector, the first term is zero. The second term is congruent to the *m*-vector.

$$\left(\underset{m}{\alpha} \wedge x\right) \vee \overline{x} == \left(\underset{m}{\alpha} \vee \overline{x}\right) \wedge x + (-1)^{m} \underset{m}{\alpha} \wedge (x \vee \overline{x}) == (-1)^{m} \underset{m}{\alpha} \,|x|^{2}$$

Thus we have finally that the expected complement of a complement formula is recovered.

$$\overline{\overline{P \wedge \underset{m}{\alpha}}} == P \wedge \left(\frac{1}{|x|^{2}} \overrightarrow{\overline{A \wedge x}}\right) == (-1)^{n\,(m+1)} P \wedge \left(\frac{1}{|x|^{2}} \underset{m}{\alpha} \,|x|^{2}\right) == (-1)^{n\,(m+1)} P \wedge \underset{m}{\alpha}$$

Entities which are of the same type as their complements

The grade of a bound *m*-vector is *m*+1. If the dimension of the vector subspace is *n* then the grade of the complement of the bound *m*-vector is *n*+1–(*m*+1). These two grades are equal when *n* is equal to 2*m*+1. Thus an entity can be congruent to its complement whenever *n* is equal to 2*m*+1.

For *m* = 0, the complement of a point is a point in a bound 1-space (the line). For *m* = 1, the complement of a bound vector (or line) is a bound vector (or line) in a bound 3-space (space). We explore these two special cases below.

◆ The complement of a point in a line

As expected, if $\underset{m}{\alpha}$ is unity [5.104] reduces to the formula for the complement of a point. Note that the orthogonality condition $\vec{\underset{m}{\alpha}} \wedge x = 0$ of the formula is always satisfied because the vector x will always belong to the unit *n*-vector $\vec{1}$.

$$\overline{P \wedge 1} == P^{*} \wedge \left(-\,\overrightarrow{\overline{1 \wedge x}}\right) \qquad \vec{1} \wedge x == 0$$

$$P == \star\mathbb{O} + x \qquad P^{*} == \star\mathbb{O} - \frac{x}{|x|^{2}} \qquad \overline{P} == P^{*} \wedge \left(-\,\vec{\overline{x}}\right) \qquad 5.105$$

■ In a bound 1-space with basis $\{\star\mathbb{O},\, e_1\}$ say, any vector x is congruent to e_1.

$$\vec{\overline{x}} == a\,\overrightarrow{\overline{e_1}} == a\sqrt{\star g} == a\sqrt{g_{11}} == a\sqrt{|e_1|^{2}} == a\,|e_1| == |x|$$

$$P == \star\mathbb{O} + x \qquad P^{*} == \star\mathbb{O} - \frac{x}{|x|^{2}} \qquad \overline{P} == -\,|x|\,P^{*} \qquad 5.106$$

Thus the complement of a point in a line is its inverse point weighted by the (negative) magnitude of its position vector.

◆ The complement of a line in space

The complement of a line $P \wedge \alpha$ with P (or its position vector) orthogonal to α, is again given by [5.104].

$$\overline{P \wedge \alpha} == P^* \wedge \left(-\overline{\alpha \wedge \vec{x}}\right) == P^* \wedge (\overline{x \wedge \vec{\alpha}}) \qquad \vec{\alpha} \wedge x == 0$$

- In a bound 3-space $\overline{x \wedge \vec{\alpha}}$ is just the cross product of x and α. See Chapter 6, section 6.10 for a discussion of the cross product and its generalizations. Hence we can write

$$\overline{P \wedge \alpha} == P^* \wedge (x \times \alpha)$$

Hence in this bound 3-space, the complement of a line L in a direction α through point P orthogonal to α is another line $\overline{L}$, where $\overline{L}$ is in a direction orthogonal to L and passes through the inverse point to P.

Complementing either of these lines gives the other line.

$$L == P \wedge \alpha \qquad \overline{L} == P^* \wedge (x \times \alpha) \qquad \overline{\overline{L}} == L \tag{5.107}$$

Thus the complement of a line in space is a line through its inverse point in a direction orthogonal to the plane it makes with the origin.

Euclidean complements of bound elements

We began this section with a discussion from first principles of the Euclidean complement of a point in bound 2- and 3-space (the plane and space). In this section we briefly look at how the general formula [5.104] applies to lines and planes in the plane and space.

◆ The Euclidean complement of a bound vector

From [5.104] we have

$$\overline{P \wedge \alpha} == \left(\star O - \frac{x}{|x|^2}\right) \wedge (\overline{x \wedge \vec{\alpha}})$$

As a specific example, let x be e_1, and α be e_2.

- In the plane, the complement of this bound vector becomes a point.

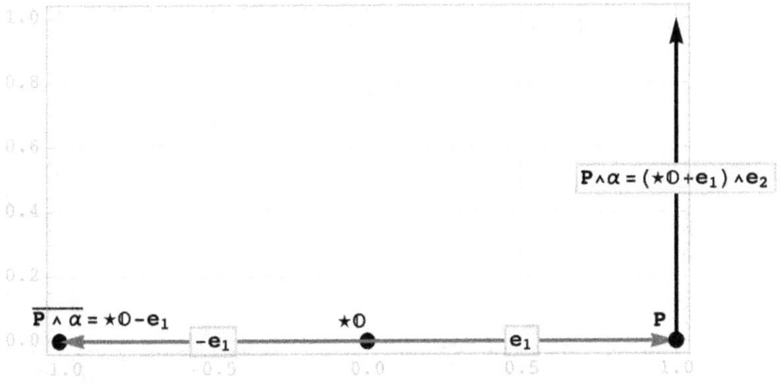

The Euclidean complement of a bound vector in the plane

$$\overline{P \wedge e_2} == (\star 0 - e_1) \wedge (\overline{e_1 \wedge e_2}) == (\star 0 - e_1) \wedge (1) == \star 0 - e_1$$

- In space, the complement of this bound vector becomes a second bound vector (orthogonal to the first one).

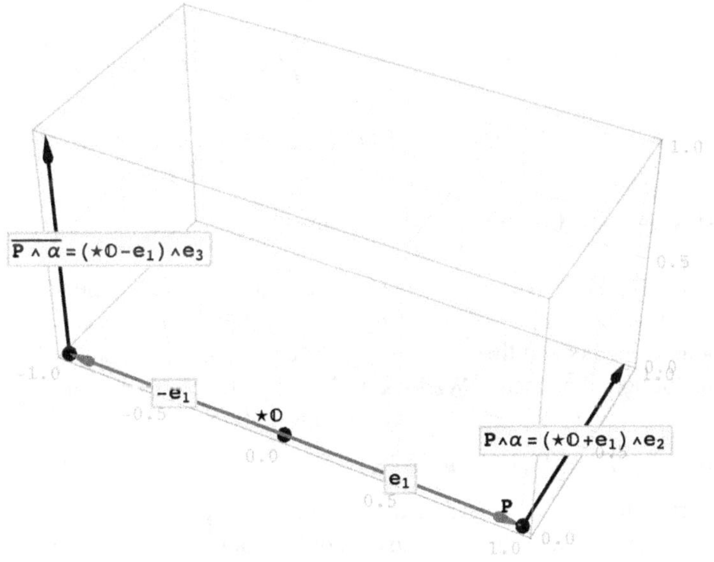

The Euclidean complement of a bound vector in space

$$\overline{P \wedge e_2} == (\star 0 - e_1) \wedge (\overline{e_1 \wedge e_2}) == (\star 0 - e_1) \wedge e_3$$

◆ **The Euclidean complement of a bound bivector**

Formula [5.104] gives

$$\overline{P \wedge \alpha \wedge \beta} == \left(\star 0 - \frac{x}{|x|^2}\right) \wedge \left(-\overline{x \wedge \alpha \wedge \beta}\right)$$

Again, let $\mathbf{x}$ be $\mathbf{e}_1$, and α be $\mathbf{e}_2$, and put β equal to $\mathbf{e}_3$.

- In space, the complement of this bound bivector becomes a (weighted) point.

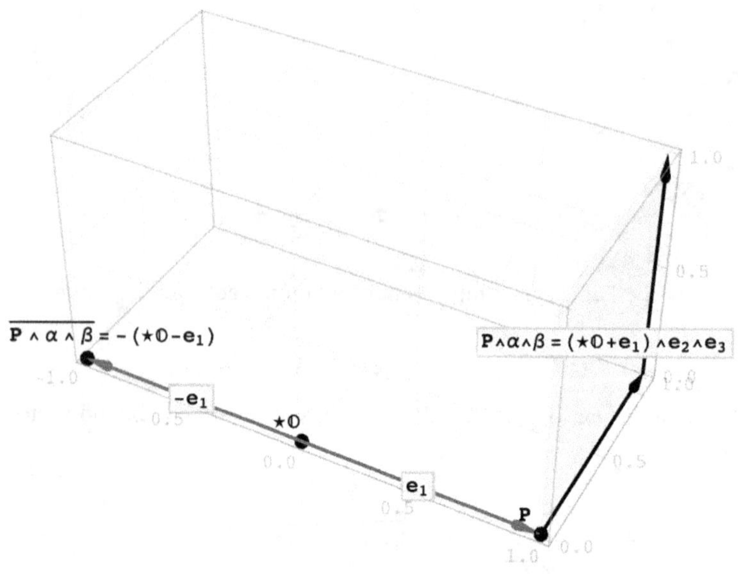

The Euclidean complement of a bound bivector in space

$$\overline{P \wedge e_2 \wedge e_3} == (\star\mathbb{O} - e_1) \wedge \left(-\overline{e_1 \wedge e_2 \wedge e_3}\right) == (\star\mathbb{O} - e_1) \wedge (-1) == -(\star\mathbb{O} - e_1)$$

The regressive product of point complements

In the sections above, we have explored the simplest cases of complements of bound elements. There is another way to compose these results if we express the bound element as an exterior product of bound elements and then apply the complement axiom to obtain a regressive product of complements of bound elements. We look at the simplest case first: that of the complement of a bound vector expressed as the product of two points. Let

$$\mathbf{L} == \mathbf{P} \wedge \mathbf{Q} \qquad \overline{\mathbf{L}} == \overline{\mathbf{P} \wedge \mathbf{Q}} == \overline{\mathbf{P}} \vee \overline{\mathbf{Q}} \equiv \mathbf{R}$$

In geometric terms, this can be interpreted as saying that the complement of a line defined by two points is the intersection of the complements of the points.

- The simplest case is in the plane, where the complement of the points are themselves lines, and whose intersection is a point.
  The relationships between the triplets of points $\{P, Q, R\}$, their inverse points $\{P^*, Q^*, R^*\}$, and the lines $\{\overline{P}, \overline{Q}, \overline{R}\}$ are shown on the diagram below.

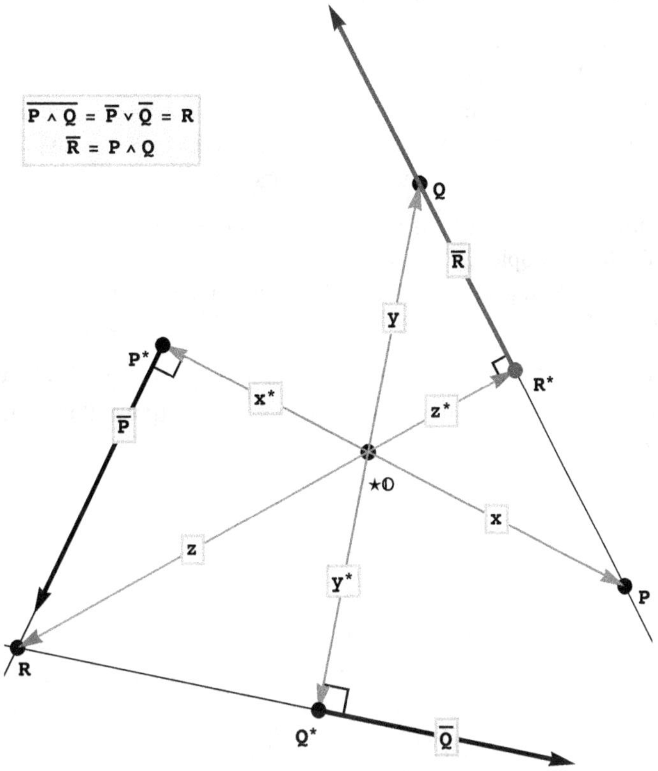

The regressive product of point complements

- In a bound 3-space, the same relations hold, but their interpretations are different. The complement $\overline{P}$ of a point is a *bound bivector* as we have depicted earlier. The complements of two different points P and Q yield two distinct bound bivectors which intersect in a *line* R, say. And the complement of this line R is the line passing through the points P and Q.
- The complement of a bound bivector can be composed as the regressive product of the complements of three points. In a bound 3-space, this results in the intersection of three bound bivectors yielding a point. The complement of this point is a bound bivector passing through the original three points.

The simple examples explored in the sections above are straightforwardly extended *mutatis mutandis* to higher dimensional bound vector spaces, but of course they are more challenging to depict!

5.13 Reciprocal Bases

Reciprocal bases

In this section we introduce the concept of a basis reciprocal to a declared basis, summarizing formulae relating these two sets of elements, their cobases and complements, and enabling us to write them in a particularly symmetric form. However, since an understanding of this section is not necessary for an understanding of the rest of the book, it may be skipped without detriment.

GrassmannAlgebra is set up to allow only one basis at a time actually active, which we have referred to throughout this book as *the declared basis*. So any 'basis' reciprocal to this declared basis *is not a declared basis* but simply a set of 1-elements in a special reciprocal-like relation to it. (Though to make things terminologically simpler, we will abuse this distinction somewhat by using the terms 'basis elements', 'cobasis elements' and 'complement' for both.)

To align our notation with current usage we will suppose the declared basis is denoted with subscripted indices, for example e_1. A basis reciprocal to the declared basis has the same number of elements as the declared basis, and following standard practice we denote them with superscripted indices, for example e^1.

In the underlying linear space, the metric tensor g_{ij} or its matrix G forms the relationship between the declared basis and its reciprocal basis. We can express this relationship, using the Einstein summation convention or matrices, as:

$$e_i == g_{ij}\, e^j \qquad \underset{1}{B} == \overset{1}{B} == G\, \overset{1}{B} \tag{5.108}$$

This relationship induces a metric tensor $\overset{m}{g}_{ij}$ on $\underset{m}{\Lambda}$.

$$\underset{m}{e_i} == \overset{m}{g}_{ij}\, \underset{m}{e^j} \qquad \underset{m}{B} == G_m\, \overset{m}{B} \tag{5.109}$$

The complement of a basis element

The complement of a basis 1-element of the declared basis has already been defined in [5.17] by:

$$\overline{e_j} == \frac{1}{|\underline{1}|} \sum_{i=1}^{n} g_{ij}\, \underline{e_i} \qquad g_{ij} == g_{ji}$$

Interchanging *i* and *j*, and adopting the summation convention gives:

$$\overline{e_i} == \frac{1}{|\underline{1}|} g_{ij}\, \underline{e_j} \qquad \overline{B} == \frac{1}{|\underline{1}|} G\, \underline{B} \tag{5.110}$$

Remember that $|\underline{1}|$ is the measure of the *n*-element of the *declared* basis, and whose square is equal to the determinant $|g_{ij}|$ or $|G|$ of the metric tensor defined on the declared basis.

$$G == (g_{ij}) \qquad \star g == |g_{ij}| == \mathrm{Det}[G] \qquad |\underline{1}| == \sqrt{\star g}$$

Taking the complement of formula [5.108] and substituting for $\overline{e_i}$ in formula [5.110] gives:

$$\overline{e^{i}} == \frac{1}{\lvert\underline{1}\rvert}\,\underline{e_{i}} \qquad \overline{\overset{1}{B}} == \frac{1}{\lvert\underline{1}\rvert}\,\underline{B} \tag{5.111}$$

The reciprocal relation to this is:

$$\overline{e_{i}} == \lvert\underline{1}\rvert\,\underline{e^{i}} \qquad \overline{B} == \lvert\underline{1}\rvert\,\underline{\overset{1}{B}} \tag{5.112}$$

These formulae can be extended to basis elements of $\underset{m}{\Lambda}$.

$$\overline{\underset{m}{e_{i}}} == \lvert\underline{1}\rvert\,\underline{\underset{m}{e^{i}}} \qquad \overline{\underset{m}{B}} == \lvert\underline{1}\rvert\,\underline{\overset{m}{B}} \tag{5.113}$$

$$\overline{\underset{m}{e^{i}}} == \frac{1}{\lvert\underline{1}\rvert}\,\underline{\underset{m}{e_{i}}} \qquad \overline{\overset{m}{B}} == \frac{1}{\lvert\underline{1}\rvert}\,\underline{\underset{m}{B}} \tag{5.114}$$

Particular cases of these formulae are:

$$\frac{1}{\lvert\underline{1}\rvert}\,e_{1}\wedge e_{2}\wedge\cdots\wedge e_{n} == \overline{1} == \lvert\underline{1}\rvert\,e^{1}\wedge e^{2}\wedge\cdots\wedge e^{n} \tag{5.115}$$

$$\overline{e_{1}\wedge e_{2}\wedge\cdots\wedge e_{n}} == \lvert\underline{1}\rvert \qquad \overline{\underset{n}{B}} == \lvert\underline{1}\rvert \tag{5.116}$$

$$\overline{e^{1}\wedge e^{2}\wedge\cdots\wedge e^{n}} == \frac{1}{\lvert\underline{1}\rvert} \qquad \overline{\overset{n}{B}} == \frac{1}{\lvert\underline{1}\rvert} \tag{5.117}$$

$$\overline{e_{i}} == \lvert\underline{1}\rvert\,(-1)^{i-1}\,e^{1}\wedge\cdots\wedge\square^{i}\wedge\cdots\wedge e^{n} \tag{5.118}$$

$$\overline{e^{i}} == \frac{1}{\lvert\underline{1}\rvert}\,(-1)^{i-1}\,e_{1}\wedge\cdots\wedge\square_{i}\wedge\cdots\wedge e_{n} \tag{5.119}$$

$$\overline{e_{i_1}\wedge\cdots\wedge e_{i_m}} == \lvert\underline{1}\rvert\,(-1)^{K_m}\,e^{1}\wedge\cdots\wedge\square^{i_1}\wedge\cdots\wedge\square^{i_m}\wedge\cdots\wedge e^{n} \tag{5.120}$$

$$\overline{e^{i_1}\wedge\cdots\wedge e^{i_m}} == \frac{1}{\lvert\underline{1}\rvert}\,(-1)^{K_m}\,e_{1}\wedge\cdots\wedge\square_{i_1}\wedge\cdots\wedge\square_{i_m}\wedge\cdots\wedge e_{n} \tag{5.121}$$

$$K_m = \sum_{\gamma=1}^{m} i_\gamma + \frac{1}{2} m (m + 1) \qquad 5.122$$

Here, the occurrence of the symbol □ means the corresponding element is missing from the product.

The complement of a cobasis element

The complements of declared basis and reciprocal basis elements are given in terms of their respective cobasis elements by formulae [5.113] and [5.114] as:

$$\overline{\underset{m}{B}} = \lvert\underline{1}\rvert \, \underline{\overset{m}{B}} \qquad \overline{\overset{m}{B}} = \frac{1}{\lvert\underline{1}\rvert} \underline{\underset{m}{B}}$$

By taking the complements of these formulae it is straightforward to obtain formulae for the complements of cobasis elements in both bases.

$$\overline{\overline{\underset{m}{B}}} = \lvert\underline{1}\rvert \, \overline{\underline{\overset{m}{B}}} = (-1)^{m (n-m)} \underset{m}{B} \qquad \overline{\overline{\overset{m}{B}}} = \frac{1}{\lvert\underline{1}\rvert} \overline{\underline{\underset{m}{B}}} = (-1)^{m (n-m)} \overset{m}{B}$$

$$\overline{\underline{\underset{m}{e_i}}} = \lvert\underline{1}\rvert \, (-1)^{m (n-m)} \underset{m}{e^i} \qquad \overline{\underline{\underset{m}{B}}} = \lvert\underline{1}\rvert \, (-1)^{m (n-m)} \overset{m}{B} \qquad 5.123$$

$$\overline{\underline{\underset{m}{e^i}}} = \frac{1}{\lvert\underline{1}\rvert} (-1)^{m (n-m)} \underset{m}{e_i} \qquad \overline{\underline{\overset{m}{B}}} = \frac{1}{\lvert\underline{1}\rvert} (-1)^{m (n-m)} \underset{m}{B} \qquad 5.124$$

Products of basis elements

For reference we list below the formulae for the exterior and regressive products of elements and their complements in both the declared basis and its reciprocal basis.

$$\underset{m}{e_i} \wedge \overline{\underset{m}{e^j}} = \delta_i^j \, \overline{1} \qquad \underset{m}{e_i} \vee \overline{\underset{m}{e^j}} = \delta_i^j \qquad 5.125$$

$$\underset{m}{e^i} \wedge \overline{\underset{m}{e_j}} = \delta_j^i \, \overline{1} \qquad \underset{m}{e^i} \vee \overline{\underset{m}{e_j}} = \delta_j^i \qquad 5.126$$

$$\underset{m}{e_i} \wedge \overline{\underset{m}{e_j}} = \overset{m}{g}_{ij} \, \overline{1} \qquad \underset{m}{e_i} \vee \overline{\underset{m}{e_j}} = \overset{m}{g}_{ij} \qquad 5.127$$

$$\underset{m}{e^i} \wedge \overline{\underset{m}{e^j}} == g^{m\,ij}\; \overline{1} \qquad \underset{m}{e^i} \vee \overline{\underset{m}{e^j}} == g^{m\,ij} \tag{5.128}$$

5.14 Summary

In this chapter we began by discussing how we could 'measure' and compare m-elements by establishing a correspondence between m-elements and $(n-m)$-elements. We discovered that provided we set a few requirements (or axioms) on the form of the correspondence we required, we only needed to establish this correspondence between basis 1-elements and basis $(n-1)$-elements, because this also induced the ability to measure on all the other spaces.

The correspondence was defined by introducing a *complement* operation. The complement of an m-element was its corresponding $(n-m)$-element and was denoted by placing a bar over it.

This complement operation had to obey a *linearity axiom*, a *complement axiom*, and a *complement of a complement axiom*. The complement axiom relates the complements of products, while the complement of a complement axiom relates the complements of a single element.

$$\overline{\underset{m}{\alpha} \wedge \underset{k}{\beta}} == \overline{\underset{m}{\alpha}} \vee \overline{\underset{k}{\beta}} \qquad \overline{\underset{m}{\alpha} \vee \underset{k}{\beta}} == \overline{\underset{m}{\alpha}} \wedge \overline{\underset{k}{\beta}} \qquad \overline{\overline{\underset{m}{\alpha}}} == (-1)^{m\,(n-m)}\, \underset{m}{\alpha}$$

The explicit formula for the complement of a basis element was a consequence of the Common Factor Theorem developed in Chapter 3, and is able most neatly to be written as a linear combination of cobasis elements. The scalar multiples in the linear combination turn out to be the elements of the (symmetric) metric tensor divided by the square root of its determinant, which is also the measure $|\underline{1}|$ of the basis n-element of the space.

$$\overline{e_j} == \frac{1}{|\underline{1}|} \sum_{i=1}^{n} g_{ij}\, \underline{e_i} \qquad g_{ij} == g_{ji} \qquad G == (g_{ij})$$

$$\overline{B} == \frac{1}{|\underline{1}|}\, G\, \underline{B} \qquad G == G^T \qquad |\underline{1}| == \sqrt{\star g} \qquad \star g == \mathrm{Det}[G]$$

These results apply to spaces whose basis elements are endowed with any interpretation. In the case of bound vector spaces in which one of the basis elements is interpreted geometrically as an origin point, we introduced the notion of hybrid metric to ensure that while the vectors can enjoy any metric, the origin is always orthogonal to them. This leads to developing two complement operations - one on the bound space (the complement), and one on the vector subspace of the bound space (the vector space complement). Using these two complements operations we were able to find geometrically intuitive formulae for the complements of bound elements.

To this point the relationship of these concepts to the notions of interior, inner or scalar products was only very briefly mentioned. This important relationship will be addressed in the next chapter.

6 The Interior Product

6.1 Introduction

To this point we have defined three important operations in the Grassmann algebra: the *exterior product*, the *regressive product*, and the *complement*.

In this chapter we introduce the *interior product*, the fourth operation of fundamental importance to the algebra. The interior product of two elements is defined as the regressive product of one element with the complement of the other. Whilst the exterior product of an *m*-element and a *k*-element generates an (*m*+*k*)-element, the interior product of an *m*-element and a *k*-element ($m \geq k$) generates an (*m*–*k*)-element. This means that the interior product of two elements of equal grade is a 0-element (or scalar). The interior product of an element with itself is a scalar, and it is this scalar that is used to define the *measure* of the element.

The interior product of two 1-elements corresponds to the usual notion of *inner*, *scalar*, or *dot product*. But we will see that the notion of measure is not restricted to 1-elements. Just as one may associate the measure of a vector with a *length*, the measure of a bivector may be associated with an *area*, and the measure of a trivector with a *volume*.

If the exterior product of an *m*-element and a 1-element is zero, then it is known that the 1-element is contained in the *m*-element. If the interior product of an *m*-element and a 1-element is zero, then this means that the 1-element is contained in the *complement* of the *m*-element. In this case it may be said that the 1-element is *orthogonal* to the *m*-element.

The basing of the notion of interior product on the notions of regressive product and complement follows here the Grassmannian tradition rather than that of the current literature which often introduces the inner product onto a linear space as an arbitrary extra definition. We do this in the belief that it is the most straightforward way to obtain consistency within the algebra and to see and exploit the relationships between the notions of exterior product, regressive product, complement and interior product, and to discover and prove formulae relating them.

We use the term 'interior' in addition to 'inner' to signal that the products are not quite the same. In traditional usage the inner product has resulted in a scalar. The interior product is however more general, being able also to operate on two elements of different grades. We reserve the term *inner product* for the interior product of two elements of the same grade. An inner product of two elements of grade 1 is called a *scalar product*. Inner products are symmetric.

Thus, an interior product may or may not be scalar. If it is scalar, it may also be called an inner product. Inner products of 1-elements may also be called scalar products. An interior product is scalar if and only if the grades of its factors are the same.

We denote the interior, inner and scalar products by a circle with a 'bar' through it, thus ⊖. This is to signify that it has a more extended meaning than the inner product. Thus the interior product of A with B is denoted $A \ominus B$. This product is zero if the grade of A is less than the grade of B. Thus the order is important if the grades are different: for a non-zero product the element of higher grade should be on the left. The interior product has the same left associativity as the negation operator, or minus sign. It is possible to define both left and right interior products, but in practice the added complexity is not rewarded by an increase in utility.

In Volume 2 we will introduce the generalized Grassmann product. This is a product that subsumes and generalizes the exterior and interior products; and we will see that the interior product of two elements can be expressed as a certain generalized product, which is independent of the order of its factors.

The beauty of the regressive product is that it is faithfully dual to the exterior product and is critical to simplifying projective geometry. However, results involving it often depend on the dimension of the current space: for example, interchanging the order of two factors. On the other hand, the interior product, like the exterior product, does not inherently depend on the dimension of the current space. The exterior product is a grade-increasing operation, while the interior product is a grade-reducing operation. One is antisymmetric, the other symmetric. Together they form a powerful and complementary duo. In Volume 2 we will see how they unite to define hypercomplex and Clifford products.

6.2 Defining the Interior Product

Definition of the inner product

Grassmann *defined* the complement $\overline{A}$ of a non-zero element A such that the regressive product of A by $\overline{A}$ *in that order*, was always a *positive* scalar. That is $A \vee \overline{A}$ was always positive for any element of any grade in a space of any dimension. As is well known, this positivity is a naturally accepted property of Euclidean inner products.

This immediately suggests a definition for the inner product of an element with itself as $A \vee \overline{A}$, and, by extension, a formula for the inner product of *any* two elements A and B of the same grade as $A \vee \overline{B}$.

The *inner product* of two elements A and B *of the same grade* is denoted $A \ominus B$ and is defined by the *scalar* $A \vee \overline{B}$.

$$A \ominus B == A \vee \overline{B} \qquad 6.1$$

The result of taking the regressive product of an *m*-element A and its complement $\overline{A}$ in the reverse order introduces a potential change of sign.

$$\overline{A} \vee A == (-1)^{m\,(n-m)} A \vee \overline{A}$$

Because of the form of Grassmann's definition of the complement, the element $A \vee \overline{A}$ is always a positive scalar. However, it is clear from the above that the element $\overline{A} \vee A$ may well not be. And furthermore, it may depend on the dimension of the space. Hence in the definition of the inner product it is important that the complemented element be the second factor.

A *scalar product* is the inner product of two 1-elements.

Forms of the inner product

Before beginning our exploration of the interior product, we look at some simple forms of the inner product which we will use as reference. More detail on the inner product may be found in section 6.5 below.

One of the special cases [3.28] of the Common Factor Axiom [3.22] is

$$\left\{\underset{m}{\alpha} \vee \underset{k}{\beta} == \left(\underset{m}{\alpha} \wedge \underset{k}{\beta}\right) \vee 1 \in \underset{0}{\Lambda},\ m + k - n == 0\right\}$$

Putting $\underset{k}{\beta}$ equal to $\underset{m}{\overline{\beta}}$ then gives immediately

$$\underset{m}{\alpha} \vee \underset{m}{\overline{\beta}} == \underset{m}{\alpha} \ominus \underset{m}{\beta} == \left(\underset{m}{\alpha} \wedge \underset{m}{\overline{\beta}}\right) \vee 1$$

$$\underset{m}{\alpha} \ominus \underset{m}{\beta} == \left(\underset{m}{\alpha} \wedge \underset{m}{\overline{\beta}}\right) \vee 1 \qquad 6.2$$

The dual [3.32] to this special case of the Common Factor Axiom says that an exterior product of two elements whose grades sum to the dimension of the space is equal to the (scalar) regressive product of the two elements multiplied by the unit n-element.

$$\left\{\underset{m}{\alpha} \wedge \underset{k}{\beta} == \left(\underset{m}{\alpha} \vee \underset{k}{\beta}\right) \underset{n}{1} \in \underset{n}{\Lambda},\ m + k - n == 0\right\}$$

Now that we have defined the complement operation, the unit n-element $\underset{n}{1}$ can finally be identified with $\overline{1}$. We can then rewrite this dual axiom as

$$\underset{m}{\alpha} \wedge \underset{m}{\overline{\beta}} == \left(\underset{m}{\alpha} \vee \underset{m}{\overline{\beta}}\right) \overline{1} == \left(\underset{m}{\alpha} \ominus \underset{m}{\beta}\right) \overline{1}$$

$$\underset{m}{\alpha} \wedge \underset{m}{\overline{\beta}} == \left(\underset{m}{\alpha} \ominus \underset{m}{\beta}\right) \overline{1} \qquad 6.3$$

Taking the complement of this equation gives another form for the inner product.

$$\underset{m}{\alpha} \ominus \underset{m}{\beta} == \overline{\underset{m}{\alpha} \wedge \underset{m}{\overline{\beta}}} \qquad 6.4$$

Definition of the interior product

The fundamentals of Grassmann's algebra (based on the two products and the complement operation) are however strong enough for the expression $A \vee \overline{B}$ to be meaningful for elements of *any* grade, leading to the definition of the *interior product*.

The *interior product* of $\underset{m}{\alpha}$ and $\underset{k}{\beta}$ is denoted $\underset{m}{\alpha} \ominus \underset{k}{\beta}$ and is defined by the (m–k)-element $\underset{m}{\alpha} \vee \underset{k}{\overline{\beta}}$.

$$\underset{m}{\alpha} \ominus \underset{k}{\beta} == \underset{m}{\alpha} \vee \underset{k}{\overline{\beta}} \quad \in \underset{m-k}{\Lambda} \qquad m \geq k \qquad 6.5$$

Thus, *the interior product does not depend on the dimension of the space*, since the dependence on the dimension of the regressive product and the complement operations 'cancel' each other out. This makes it, like the exterior product, of special importance in the algebra.

If $m < k$, then $\underset{m}{\alpha} \vee \underset{k}{\overline{\beta}}$ is necessarily zero (otherwise its grade would be negative), hence:

$$\underset{m}{\alpha} \ominus \underset{k}{\beta} == 0 \qquad m < k \tag{6.6}$$

◆ An important convention

In order to avoid unnecessarily distracting caveats on every formula involving interior products, in the rest of this book we will suppose that *the grade of the first factor is always greater than or equal to the grade of the second factor*. The formulae will remain true even if this is not the case, but they will usually be trivially so by virtue of their terms reducing to zero.

◆ *Historical Note*

Grassmann and workers in the Grassmannian tradition define the interior product of two elements as the product of one with the complement of the other, the product being either exterior or regressive depending on which interpretation produces a non-zero result. Furthermore, when the grades of the elements are equal, it is defined either way. This definition involves the confusion between scalars and n-elements discussed in section 5.1 (equivalent to assuming a Euclidean metric and identifying scalars with pseudo-scalars). It is to obviate this inconsistency and restriction on generality that the approach adopted here bases its definition of the interior product *explicitly* on the *regressive* exterior product.

Left and right interior products

To deal with the apparent asymmetry of the interior product, some authors (for example E. Tonti *On the Formal Structure of Physical Theories*, page 68, 1975) have introduced the notion of left and right interior products equivalent to the definitions:

$$\begin{array}{llll} \underset{m}{\alpha} \vdash \underset{k}{\beta} == \underset{m}{\alpha} \vee \overline{\underset{k}{\beta}} & \in \underset{m-k}{\Lambda} & m \geq k \\ \underset{m}{\alpha} \dashv \underset{k}{\beta} == \overline{\underset{m}{\alpha}} \vee \underset{k}{\beta} & \in \underset{k-m}{\Lambda} & k \geq m \end{array} \tag{6.7}$$

Here, $\underset{m}{\alpha} \vdash \underset{k}{\beta}$ is the *right interior product*, and $\underset{m}{\alpha} \dashv \underset{k}{\beta}$ is the *left interior product*.

In this book we have adopted the right interior product as the only one considered, and called it *the* interior product. This follows Grassmann's approach in having the complemented element be the second factor. For the algebraic explorations in this book this single interior product suffices.

$$\underset{m}{\alpha} \vdash \underset{k}{\beta} == \underset{m}{\alpha} \vee \overline{\underset{k}{\beta}} == \underset{m}{\alpha} \ominus \underset{k}{\beta} \in \underset{m-k}{\Lambda} \qquad m \geq k \tag{6.8}$$

By interchanging the order of the factors in one of the regressive products, and then interchanging the m-element with the k-element, we can see that they are related by the formula:

$$\underset{m}{\alpha} \vdash \underset{k}{\beta} == (-1)^{k\,(n-m)}\, \underset{k}{\beta} \dashv \underset{m}{\alpha} \qquad m \geq k \tag{6.9}$$

Note that this relationship depends not only on the grade of the elements, but also on the dimension of the space.

One reason for using the left interior product might be when it is visually desirable that a 1-element act from the left, as, for example, in the traditional application of a differential operator.

In Volume 2 we will discover a new product (the generalized Grassmann product) which leads to the same interior product independent of the order of its factors, thus retrieving, in this wider system, the sought-after symmetry.

Implications of the regressive product axioms

By expressing one or more elements as a complement, the relations of the regressive product axiom set may be rewritten in terms of the interior product, thus yielding some of its more fundamental properties.

◆ ⊖ 6: The interior product of an *m*-element and a *k*-element is an (*m*–*k*)-element

The grade of the interior product of two elements is the difference of their grades.

$$\underset{m}{\alpha} \in \underset{m}{\Lambda},\ \underset{k}{\beta} \in \underset{k}{\Lambda} \quad \Longrightarrow \quad \underset{m}{\alpha} \ominus \underset{k}{\beta} \ \in \underset{m-k}{\Lambda} \tag{6.10}$$

Thus, in contradistinction to the regressive product, the grade of an interior product does not depend on the dimension of the underlying linear space.

If the grade of the first factor is less than that of the second, the interior product is zero.

$$\underset{m}{\alpha} \ominus \underset{k}{\beta} == 0 \qquad m < k \tag{6.11}$$

◆ ⊖ 7: The interior product is not associative

The following formulae are derived directly from the associativity of the regressive product, and show that the interior product is *not* associative.

$$\left(\underset{m}{\alpha} \ominus \underset{k}{\beta}\right) \ominus \underset{r}{\gamma} == \underset{m}{\alpha} \ominus \left(\underset{k}{\beta} \wedge \underset{r}{\gamma}\right) \tag{6.12}$$

$$\left(\underset{m}{\alpha} \vee \underset{k}{\beta}\right) \ominus \underset{r}{\gamma} == \underset{m}{\alpha} \vee \left(\underset{k}{\beta} \ominus \underset{r}{\gamma}\right) \tag{6.13}$$

◆ ⊖ 8: The unit scalar 1 is the identity for the interior product

The interior product of an element with the unit scalar 1 does not change it.

$$\underset{m}{\alpha} \ominus 1 == \underset{m}{\alpha} \tag{6.14}$$

The interior product of an element with a scalar is equivalent to ordinary scalar multiplication.

$$\underset{m}{\alpha} \ominus a == \underset{m}{\alpha} \wedge a == a \underset{m}{\alpha} \qquad 6.15$$

The interior product of scalars is equivalent to ordinary scalar multiplication.

$$a \ominus b == a\, b \qquad 6.16$$

The unit scalar 1 is the identity for the interior product.

$$1 \ominus 1 == 1 \qquad 6.17$$

The complement operation may be viewed as the interior product with the unit *n*-element $\overline{1}$.

$$\overline{\underset{m}{\alpha}} == \overline{1} \ominus \underset{m}{\alpha} \qquad 6.18$$

◆ **⊖ 9: The inverse of a scalar with respect to the interior product is its complement**

The interior product of the complement of a scalar and the reciprocal of the scalar is the unit *n*-element.

$$\overline{1} == \overline{a} \ominus \frac{1}{a} \qquad 6.19$$

◆ **⊖ 10: The interior product of two elements is congruent to the interior product of their complements in reverse order**

The interior product of two elements is equal (apart from a possible sign) to the interior product of their complements in reverse order.

$$\underset{m}{\alpha} \ominus \underset{k}{\beta} == (-1)^{(n-m)\,(m-k)}\, \overline{\underset{k}{\beta}} \ominus \overline{\underset{m}{\alpha}} \qquad 6.20$$

If the elements are of the same grade, the interior product of two elements is equal to the interior product of their complements. (It will be shown later that, because of the symmetry of the metric tensor, the interior product of two elements of the same grade is symmetric, hence the order of the factors on either side of the equation may be reversed.)

$$\underset{m}{\alpha} \ominus \underset{m}{\beta} == \overline{\underset{m}{\beta}} \ominus \overline{\underset{m}{\alpha}} \qquad 6.21$$

◆ **⊖ 11: An interior product with zero is zero**

Every exterior linear space has a zero element whose interior product with any other element is zero.

$$\underset{m}{\alpha} \ominus 0 == 0 == 0 \ominus \underset{m}{\alpha} \qquad 6.22$$

◆ ⊖ 12: The interior product is distributive over addition

The interior product is both left and right distributive over addition.

$$\left(\underset{m}{\alpha} + \underset{m}{\beta}\right) \ominus \underset{r}{\gamma} == \underset{m}{\alpha} \ominus \underset{r}{\gamma} + \underset{m}{\beta} \ominus \underset{r}{\gamma} \qquad 6.23$$

$$\underset{m}{\alpha} \ominus \left(\underset{r}{\beta} + \underset{r}{\gamma}\right) == \underset{m}{\alpha} \ominus \underset{r}{\beta} + \underset{m}{\alpha} \ominus \underset{r}{\gamma} \qquad 6.24$$

Orthogonality

Orthogonality is a concept generated by the complement operation but often most simply expressed in terms of the interior product.

If x is a 1-element and A is a simple m-element, then x and (the 1-element factors of) A may be said to be *linearly dependent* if and only if $A \wedge x = 0$, that is, x is contained in (the space of) A.

Similarly, x and A are said to be *orthogonal* if and only if $\overline{A} \wedge x = 0$, that is, x is contained in $\overline{A}$.

By taking the complement of $\overline{A} \wedge x = 0$, we can see that $\overline{A} \wedge x = 0$ if and only if $A \vee \overline{x} = 0$. Thus it may also be said that a 1-element x is orthogonal to a simple element A if and only if their interior product is zero, that is, $A \ominus x = 0$.

If $X = x_1 \wedge x_2 \wedge \cdots \wedge x_k$ then A and X are said to be *totally orthogonal* if and only if $A \ominus x_i = 0$ for all x_i contained in X.

However, for $A \ominus (x_1 \wedge x_2 \wedge \cdots \wedge x_k)$ to be zero it is only necessary that *one* of the x_i be orthogonal to A. To show this, suppose it to be (without loss of generality) x_1. Then by formula **⊖7** [6.12] we can write $A \ominus (x_1 \wedge x_2 \wedge \cdots \wedge x_k)$ as $(A \ominus x_1) \ominus (x_2 \wedge \cdots \wedge x_k)$, whence it becomes immediately clear that if $A \ominus x_1$ is zero then so is the product $A \ominus X$.

Just as the vanishing of the exterior product of two *simple* elements A and X implies only that they have some 1-element in common, so the vanishing of their interior product implies only that $\overline{A}$ and X (where the grade of A is not less than the grade of X) have some 1-element in common, and conversely. That is, there is a 1-element x such that the following implications hold.

$$A \wedge X == 0 \quad \Leftrightarrow \quad \{A \wedge x == 0,\; X \wedge x == 0\} \qquad 6.25$$

$$A \ominus X == 0 \quad \Leftrightarrow \quad \{\overline{A} \wedge x == 0,\; X \wedge x == 0\} \qquad 6.26$$

$$A \ominus X == 0 \quad \Leftrightarrow \quad \{A \ominus x == 0,\; X \wedge x == 0\} \qquad 6.27$$

Example: The interior product of a simple bivector with a vector

In this section we preview the simplest case of an interior product of elements which is not an inner product - that of a bivector with a vector. At this stage we just sketch the computation sufficiently to be able to depict it geometrically. We will revisit the general case and justify the concepts and formulae used in more detail later in the chapter.

Let x be a vector and $B = x_1 \wedge x_2$ be a simple bivector in a space of *any* number of dimensions. The interior product of the bivector B with the vector x is the vector $B \ominus x$, which is orthogonal to both x and its orthogonal projection $x^{\llcorner}$ onto B.

The vector $B \ominus x$ can be expanded by the Interior Common Factor Theorem (which we will discuss later in section 6.4) to give

$$B \ominus x == (x_1 \wedge x_2) \ominus x == (x \ominus x_1)\, x_2 - (x \ominus x_2)\, x_1$$

Since $B \ominus x$ is expressed as a linear combination of x_1 and x_2 it is clearly *contained in* the bivector B. Thus

$$B \wedge (B \ominus x) == 0$$

The resulting vector $B \ominus x$ is also *orthogonal* to x. We can show this by taking its scalar product with x and using formula [6.12].

$$(B \ominus x) \ominus x == B \ominus (x \wedge x) == 0$$

The unit bivector $\hat{B}$ is defined by

$$\hat{B} == \frac{B}{\sqrt{B \ominus B}}$$

The orthogonal projection $x^{\#}$ of x onto B is given by

$$x^{\#} == -\hat{B} \ominus \left(\hat{B} \ominus x\right)$$

And the component $x^{\llcorner}$ of x orthogonal to B is given by

$$x^{\llcorner} == \left(\hat{B} \wedge x\right) \ominus \hat{B}$$

◆ The interior product of a bivector and a vector in a vector 3-space

The sketch above indicates that in a vector *n*-space, $B \ominus x$ lies in B and is orthogonal to x. It is also orthogonal to the components of x *in* B ($x^{\#}$) and *orthogonal to* B ($x^{\llcorner}$). We can depict this in a vector 3-space as follows

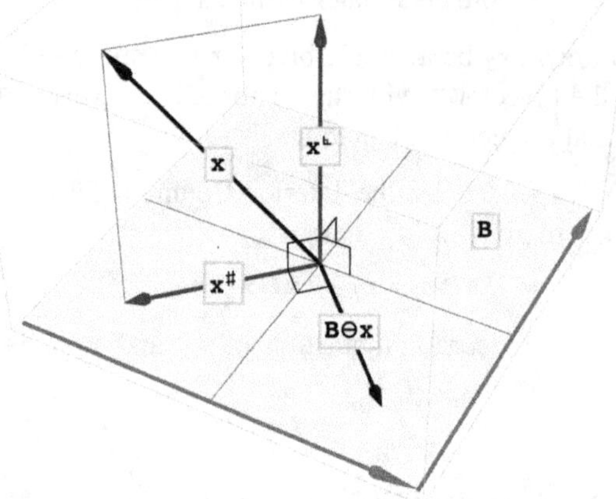

The interior product of a simple bivector and a vector

◆ The interior product of a bivector and a vector in a vector 2-space

In a vector 2-space, the same analysis is valid. However, $x^{\perp}$ is zero because it involves the exterior product of three vectors in a 2-space; and $x^{\#}$ is identical to x because the unit bivector $\hat{B}$ must be equal to the unit 2-element of the space $\overline{1}$ (see formulae [5.11]).

$$x^{\perp} == \left(\hat{B} \wedge x\right) \ominus \hat{B} == 0$$

$$x^{\#} == -\hat{B} \ominus \left(\hat{B} \ominus x\right) == -\overline{1} \ominus \left(\overline{1} \ominus x\right) == -\overline{1} \ominus \overline{x} == -\left(\overline{\overline{x}}\right) == x$$

Hence the taking the interior product of a vector with a bivector in a vector 2-space may be seen as rotating it by $\frac{\pi}{2}$.

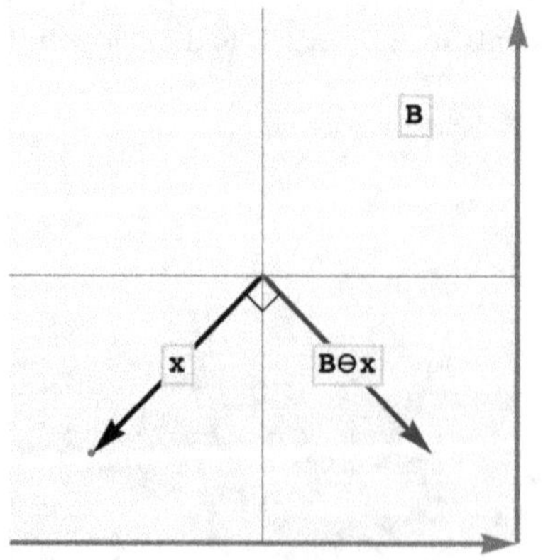

The interior product of a bivector with a vector in a vector 2-space

6.3 Properties of the Interior Product

Implications of the complement axioms

In addition to the definitions and formulae we have already derived for the interior product there are many more that can be derived by successive application of just three groups of axioms - the anti-commutativity, complement, and complement of a complement axioms. We list these axioms here again for reference in the derivations to follow.

Anticommutativity axioms

$$\underset{m}{\alpha} \wedge \underset{k}{\beta} == (-1)^{m k} \underset{k}{\beta} \wedge \underset{m}{\alpha} \qquad \underset{m}{\alpha} \vee \underset{k}{\beta} == (-1)^{(n-k)\,(n-m)} \underset{k}{\beta} \vee \underset{m}{\alpha} \tag{6.28}$$

Complement axioms

$$\overline{\underset{m}{\alpha} \wedge \underset{k}{\beta}} == \overline{\underset{m}{\alpha}} \vee \overline{\underset{k}{\beta}} \qquad \overline{\underset{m}{\alpha} \vee \underset{k}{\beta}} == \overline{\underset{m}{\alpha}} \wedge \overline{\underset{k}{\beta}} \tag{6.29}$$

Complement of a complement axiom

$$\overline{\overline{\underset{m}{\alpha}}} == (-1)^{m\,(n-m)} \underset{m}{\alpha} \tag{6.30}$$

Our aim in this section is to show some of the various ways in which the interior product of two elements, or complements of elements, can be expressed in terms of either exterior or regressive products, together (perhaps) with the complement operation. Derivations are made without comment but use formulae [6.10] to [6.24] derived from the regressive product axioms, together with the product and complement axioms above.

◇ Formula 1

$$\underset{m}{\alpha} \ominus \underset{k}{\beta} == \underset{m}{\alpha} \vee \overline{\underset{k}{\beta}} == (-1)^{k\,(n-m)} \overline{\underset{k}{\beta}} \vee \underset{m}{\alpha}$$

$$\underset{m}{\alpha} \ominus \underset{k}{\beta} == \underset{m}{\alpha} \vee \overline{\underset{k}{\beta}} == (-1)^{k\,(n-m)} \overline{\underset{k}{\beta}} \vee \underset{m}{\alpha} \tag{6.31}$$

◇ Formula 2

$$\underset{m}{\alpha} \ominus \underset{k}{\beta} == \underset{m}{\alpha} \vee \overline{\underset{k}{\beta}} == (-1)^{m\,(n-m)} \overline{\overline{\underset{m}{\alpha}}} \vee \overline{\underset{k}{\beta}} == (-1)^{m\,(n-m)} \overline{\overline{\underset{m}{\alpha}} \wedge \underset{k}{\beta}}$$

$$\underset{m}{\alpha} \ominus \underset{k}{\beta} == (-1)^{m\,(n-m)} \overline{\overline{\underset{m}{\alpha}} \wedge \underset{k}{\beta}} == (-1)^{(m+k)\,(n-m)} \overline{\underset{k}{\beta} \wedge \overline{\underset{m}{\alpha}}} \tag{6.32}$$

◇ Formula 3

$$\overline{\underset{m}{\alpha}}\ominus\underset{k}{\beta} == \overline{\underset{m}{\alpha}}\vee\overline{\underset{k}{\beta}} == \overline{\underset{m}{\alpha}\wedge\underset{k}{\beta}} == (-1)^{m\,k}\,\overline{\underset{k}{\beta}}\ominus\underset{m}{\alpha}$$

$$\overline{\underset{m}{\alpha}}\ominus\underset{k}{\beta} == \overline{\underset{m}{\alpha}}\vee\overline{\underset{k}{\beta}} == \overline{\underset{m}{\alpha}\wedge\underset{k}{\beta}} == (-1)^{m\,k}\,\overline{\underset{k}{\beta}}\ominus\underset{m}{\alpha} \qquad 6.33$$

◇ Formula 4

$$\underset{m}{\alpha}\ominus\overline{\underset{k}{\beta}} == \underset{m}{\alpha}\vee\overline{\overline{\underset{k}{\beta}}} == (-1)^{k\,(n-k)}\,\underset{m}{\alpha}\vee\underset{k}{\beta}$$

$$\underset{m}{\alpha}\ominus\overline{\underset{k}{\beta}} == (-1)^{k\,(n-k)}\,\underset{m}{\alpha}\vee\underset{k}{\beta} \qquad 6.34$$

◇ Formula 5

$$\overline{\underset{m}{\alpha}}\ominus\overline{\underset{k}{\beta}} == (-1)^{k\,(n-k)}\,\overline{\underset{m}{\alpha}}\vee\underset{k}{\beta} == (-1)^{k\,(n-k)+m\,(n-k)}\,\underset{k}{\beta}\vee\overline{\underset{m}{\alpha}}$$

$$\overline{\underset{m}{\alpha}}\ominus\overline{\underset{k}{\beta}} == (-1)^{(m+k)\,(n-k)}\,\underset{k}{\beta}\ominus\underset{m}{\alpha} \qquad 6.35$$

◇ Formula 6

$$\overline{\underset{m}{\alpha}\ominus\underset{k}{\beta}} == \overline{\underset{m}{\alpha}\vee\overline{\underset{k}{\beta}}} == (-1)^{k\,(n-k)}\,\overline{\underset{m}{\alpha}}\wedge\underset{k}{\beta} == (-1)^{k\,(n-k)+k\,(n-m)}\,\underset{k}{\beta}\wedge\overline{\underset{m}{\alpha}}$$

$$\overline{\underset{m}{\alpha}\ominus\underset{k}{\beta}} == (-1)^{k\,(m+k)}\,\underset{k}{\beta}\wedge\overline{\underset{m}{\alpha}} \qquad 6.36$$

◇ Formula 7

$$\overline{\overline{\underset{m}{\alpha}}\ominus\underset{k}{\beta}} == \overline{\overline{\underset{m}{\alpha}}\vee\overline{\underset{k}{\beta}}} == \overline{\overline{\underset{m}{\alpha}}}\wedge\overline{\overline{\underset{k}{\beta}}} == (-1)^{m\,(n-m)+k\,(n-k)}\,\underset{m}{\alpha}\wedge\underset{k}{\beta}$$

$$\overline{\overline{\underset{m}{\alpha}}\ominus\underset{k}{\beta}} == \overline{\overline{\underset{m}{\alpha}\wedge\underset{k}{\beta}}} == (-1)^{(m+k)\,(n-(m+k))}\,\underset{m}{\alpha}\wedge\underset{k}{\beta}$$

$$\overline{\overline{\underset{m}{\alpha}}\ominus\underset{k}{\beta}} == (-1)^{(m+k)\,(n-(m+k))}\,\underset{m}{\alpha}\wedge\underset{k}{\beta} == (-1)^{m\,k}\,\overline{\overline{\underset{k}{\beta}}\ominus\underset{m}{\alpha}} \qquad 6.37$$

◇ Formula 8

$$\overline{\underset{m}{\alpha}\ominus\overline{\underset{k}{\beta}}} == (-1)^{k\,(n-k)}\,\overline{\underset{m}{\alpha}\vee\underset{k}{\beta}} == (-1)^{k\,(n-k)}\,\overline{\underset{m}{\alpha}}\wedge\overline{\underset{k}{\beta}}$$

$$\overline{\underset{m}{\alpha} \ominus \underset{k}{\beta}} == (-1)^{k\,(n-k)}\, \overline{\underset{m}{\alpha}} \wedge \overline{\underset{k}{\beta}} \qquad 6.38$$

◇ **Formula 9**

$$\overline{\underset{m}{\alpha}} \ominus \overline{\underset{k}{\beta}} == (-1)^{k\,(n-k)}\, \overline{\underset{m}{\alpha} \vee \underset{k}{\beta}} == (-1)^{k\,(n-k)+m\,(n-m)}\, \underset{m}{\alpha} \wedge \overline{\underset{k}{\beta}}$$

$$\overline{\underset{m}{\alpha}} \ominus \overline{\underset{k}{\beta}} == (-1)^{k\,(n-k)+m\,(n-m)}\, \underset{m}{\alpha} \wedge \overline{\underset{k}{\beta}} \qquad 6.39$$

Extended interior products

The interior product of an element with the exterior product of several elements (of any grades) may be shown directly from its definition to be a type of 'extended' interior product.

$$A \ominus (B \wedge C \wedge \ldots \wedge Z) == A \vee (\overline{B \wedge C \wedge \ldots \wedge Z}) == A \vee \left(\overline{B} \vee \overline{C} \vee \ldots \vee \overline{Z}\right)$$
$$== \left(\left(\left(A \vee \overline{B}\right) \vee \overline{C}\right) \vee \ldots\right) \vee \overline{Z} == (((A \ominus B) \ominus C) \ominus \ldots) \ominus Z$$

$$A \ominus (B \wedge C \wedge \ldots \wedge Z) == (((A \ominus B) \ominus C) \ominus \ldots) \ominus Z \qquad 6.40$$

This formula is one of the most valuably useful in the algebra.

If the interior product of *any* of the elements B, ..., Z with A is zero, then the complete interior product is zero. We can see this by rearranging the factors in the exterior product to bring that factor into first position.

By interchanging the order of the factors in the exterior product we can derive alternative expressions. For the case of two factors we have

$$A \ominus \left(\underset{k}{B} \wedge \underset{r}{C}\right) == (-1)^{k\,r}\, A \ominus \left(\underset{r}{C} \wedge \underset{k}{B}\right) \quad \Longrightarrow \quad \left(A \ominus \underset{k}{B}\right) \ominus \underset{r}{C} == (-1)^{k\,r} \left(A \ominus \underset{r}{C}\right) \ominus \underset{k}{B}$$

$$\left(A \ominus \underset{k}{B}\right) \ominus \underset{r}{C} == (-1)^{k\,r} \left(A \ominus \underset{r}{C}\right) \ominus \underset{k}{B} \qquad 6.41$$

The precedence of the interior product

When an algebra has multiple operations, expressions generally need parentheses to make their meaning clear. But we can also set rules by which to interpret a default meaning for expressions without parentheses. These rules introduce a *precedence* onto the set of operations. In *Mathematica* every operation has a natural inbuilt precedence which *GrassmannAlgebra* accepts.

```
Precedence /@ {InteriorProduct, RegressiveProduct, ExteriorProduct}
```

```
{330., 430., 440.}
```

Thus, of these three operations, the interior product takes the least precedence. *Grassmann-Algebra* knows this, so that when you enter an expression involving several product operations without parentheses, it will assume its inbuilt precedence. And in fact, if you enter an expression with parentheses which conforms to the inbuilt precedences, *Mathematica* will strip your parentheses in the output - a feature which may be highly confusing if the expression is complex. For example, the following inputs, since they all conform to the inbuilt precedence are all output with the parentheses stripped.

```
{w ∧ x ∨ y ⊖ z, (w ∧ x ∨ y) ⊖ z, (w ∧ x) ∨ y ⊖ z}
{w ∧ x ∨ y ⊖ z, w ∧ x ∨ y ⊖ z, w ∧ x ∨ y ⊖ z}
```

The interior product operator has the same precedence as the minus operator (hence the choice of `CircleMinus` (⊖) for it). So again, if you parenthesize an expression according to this precedence and enter it, your parentheses will be stripped.

```
(((w ⊖ x) ⊖ y) ⊖ z ) ⊖ …
w ⊖ x ⊖ y ⊖ z ⊖ …
```

In order to show all inbuilt precedences in an expression involving Grassmann products, *GrassmannAlgebra* has a command `ShowPrecedence` that you can enter at any stage in your work. If you want to turn this facility off you can enter `HidePrecedence`. You can also turn these on and off from the 'Show product precedences' checkbox on the *GrassmannAlgebra* palette. This has the advantage that you can see at any time whether the facility is turned on or off.

If we now enter `ShowPrecedence` and then enter the previous inputs, we get the inbuilt precedences displayed.

```
ShowPrecedence
{w ∧ x ∨ y ⊖ z, (w ∧ x ∨ y) ⊖ z, (w ∧ x) ∨ y ⊖ z, (((w ⊖ x) ⊖ y) ⊖ z ) ⊖ …}
{((w ∧ x) ∨ y) ⊖ z, ((w ∧ x) ∨ y) ⊖ z, ((w ∧ x) ∨ y) ⊖ z, (((w ⊖ x) ⊖ y) ⊖ z) ⊖ (…)}
```

Of course you can override any of the inbuilt precedences with your own parentheses. If `ShowPrecedence` is on, these overrides will be retained in the output.

```
{w ∧ x ∨ (y ⊖ z), w ∧ (x ∨ y ⊖ z), w ∧ (x ∨ y) ⊖ z, x ⊖ (y ⊖ (z ⊖ (w ⊖ …)))}
{(w ∧ x) ∨ (y ⊖ z), w ∧ ((x ∨ y) ⊖ z), (w ∧ (x ∨ y)) ⊖ z, x ⊖ (y ⊖ (z ⊖ (w ⊖ (…))))}
```

- For `ShowPrecedence` to parenthesize expressions as you would expect, it needs them to be valid Grassmann expressions. In particular it needs the symbols to be declared scalar or vector symbols, or be underscripted symbols. Other symbols not recognized by *GrassmannAlgebra* will be flagged with an extra pair of parentheses around them. Here, `W`, `X`, and `Z` have not been declared as either scalar or vector symbols, and hence are flagged with extra parentheses.

$$\{w \wedge x \vee y \ominus z,\ W \wedge X \vee \underset{m}{Y} \ominus Z\}$$

$$\{((w \wedge x) \vee y) \ominus z,\ (((W) \wedge (X)) \vee \underset{m}{Y}) \ominus (Z)\}$$

- Because of the intricacy of the code to override *Mathematica* and display the precedence parentheses, parts of the resulting expressions may not be editable. If you intend to edit an output expression by hand, you may need to restore *Mathematica's* free reign first by entering `HidePrecedence` before generating the output. For example, you cannot edit the preceding output.

- The code for the `ShowPrecedence` and `HidePrecedence` commands and palette button was written for *GrassmannAlgebra* by David Park [Park, 2010]. It is a critically important and gratefully acknowledged facility for understanding the output from Grassmann algebra computations.

Converting interior products to exterior and regressive products

GrassmannAlgebra has a number of functions for converting between exterior, regressive, and interior products. These conversions will usually involve the complement operation.

We give some examples below. Note carefully however, that the conversion is carried out *in the currently declared space*, that is, *GrassmannAlgebra* will use the current dimension of the space for its conversion so that the formula for any attached sign may be simplified. All of the functions are effectively listable.

To make the outputs a little easier to read, we turn on `ShowPrecedence`.

`ShowPrecedence`

◆ Converting interior products to regressive products

Converting interior products to regressive products is independent of the dimension of the space since the definitional formula is used.

$$\mathbf{A} = \left\{\left(\underset{m}{\alpha} \ominus \underset{k}{\beta}\right) \ominus \underset{j}{\gamma},\ \underset{m}{\alpha} \ominus \left(\underset{k}{\beta} \ominus \underset{j}{\gamma}\right)\right\};$$

`★𝓑3; InteriorToRegressive[A]`

$$\left\{\underset{m}{\alpha} \vee \overline{\underset{k}{\beta}} \vee \overline{\underset{j}{\gamma}},\ \underset{m}{\alpha} \vee \overline{\left(\underset{k}{\beta} \vee \overline{\underset{j}{\gamma}}\right)}\right\}$$

◆ Converting interior products to exterior products

In a 3-space, the complement of a complement of an element is the element itself, hence there are no extra signs resulting from the conversion.

`★𝓑3; InteriorToExterior[A]`

$$\left\{\overline{\left(\overline{\overline{\overline{\underset{m}{\alpha}} \wedge \underset{k}{\beta}}}\right) \wedge \underset{j}{\gamma}},\ \overline{\overline{\underset{m}{\alpha}} \wedge \left(\overline{\overline{\underset{k}{\beta}} \wedge \underset{j}{\gamma}}\right)}\right\}$$

In a 4-space however, this is not so, so there may be extra signs resulting from the conversion.

`★𝓑4; InteriorToExterior[A]`

$$\left\{(-1)^{k+m}\,\overline{\overline{\left((-1)^{m}\,\overline{\overline{\overline{\underset{m}{\alpha}} \wedge \underset{k}{\beta}}}\right)} \wedge \underset{j}{\gamma}},\ (-1)^{m}\,\overline{\overline{\underset{m}{\alpha}} \wedge \left((-1)^{k}\,\overline{\overline{\underset{k}{\beta}} \wedge \underset{j}{\gamma}}\right)}\right\}$$

- You can aggregate and simplify the signs by applying `AggregateSigns`.

`InteriorToExterior[A] // AggregateSigns`

$$\left\{(-1)^{k}\,\overline{\left(\overline{\overline{\overline{\underset{m}{\alpha}} \wedge \underset{k}{\beta}}}\right) \wedge \underset{j}{\gamma}},\ (-1)^{k+m}\,\overline{\overline{\underset{m}{\alpha}} \wedge \left(\overline{\overline{\underset{k}{\beta}} \wedge \underset{j}{\gamma}}\right)}\right\}$$

◆ Converting exterior and regressive products to interior products

Converting exterior and regressive products to interior products may utilize the conversion in which the complement of an element is replaced by its interior product with the unit n-element $\overline{1}$.

$$\mathbf{B} = \left\{\underset{m}{\alpha}\wedge\underset{k}{\beta}\wedge\underset{j}{\gamma},\ \left(\underset{m}{\alpha}\wedge\underset{k}{\beta}\right)\vee\underset{j}{\gamma}\right\};$$

In a 3-space, conversion to interior products gives

```
★𝓑3; ToInteriorForm[B]
```

$$\left\{(-1)^{j\,m+k\,m}\left(\overline{1}\ominus\left(\left(\overline{1}\ominus\underset{k}{\beta}\right)\ominus\left(\underset{j}{\gamma}\wedge\underset{m}{\alpha}\right)\right)\right),\ (-1)^{1+j+k+j\,k+m+j\,m}\left(\underset{j}{\gamma}\ominus\left(\left(\overline{1}\ominus\underset{m}{\alpha}\right)\ominus\underset{k}{\beta}\right)\right)\right\}$$

But in an even-dimensional 4-space, the signs may be different.

```
★𝓑4; ToInteriorForm[B]
```

$$\left\{(-1)^{j+k+m+j\,m+k\,m}\left(\overline{1}\ominus\left(\left(\overline{1}\ominus\underset{k}{\beta}\right)\ominus\left(\underset{j}{\gamma}\wedge\underset{m}{\alpha}\right)\right)\right),\ (-1)^{k+j\,k+m+j\,m}\left(\underset{j}{\gamma}\ominus\left(\left(\overline{1}\ominus\underset{m}{\alpha}\right)\ominus\underset{k}{\beta}\right)\right)\right\}$$

And in a 5-space, the signs return to those of the (also odd) 3-space.

```
★𝓑5; ToInteriorForm[B]
```

$$\left\{(-1)^{j\,m+k\,m}\left(\overline{1}\ominus\left(\left(\overline{1}\ominus\underset{k}{\beta}\right)\ominus\left(\underset{j}{\gamma}\wedge\underset{m}{\alpha}\right)\right)\right),\ (-1)^{1+j+k+j\,k+m+j\,m}\left(\underset{j}{\gamma}\ominus\left(\left(\overline{1}\ominus\underset{m}{\alpha}\right)\ominus\underset{k}{\beta}\right)\right)\right\}$$

The complement form of interior products

Formula [6.20] says that an interior product of elements can always be expressed as the interior product of their complements in reverse order.

$$\underset{m}{\alpha}\ominus\underset{k}{\beta} == (-1)^{(m-k)\,(n-m)}\,\overline{\underset{k}{\beta}}\ominus\overline{\underset{m}{\alpha}}$$

Apart from this interior product, there are two other commonly occurring forms. The first is

$$\left(\underset{m}{\alpha}\wedge\underset{j}{\xi}\right)\ominus\underset{k}{\beta} == (-1)^{(m+j-k)\,(n-m-j)}\,\overline{\underset{k}{\beta}}\ominus\overline{\underset{m}{\alpha}\wedge\underset{j}{\xi}} == (-1)^{(m+j-k)\,(n-m-j)}\,\overline{\underset{k}{\beta}}\ominus\left(\overline{\underset{m}{\alpha}}\ominus\underset{j}{\xi}\right)$$

$$\left(\underset{m}{\alpha}\wedge\underset{j}{\xi}\right)\ominus\underset{k}{\beta} == (-1)^{(m+j-k)\,(n-m-j)}\,\overline{\underset{k}{\beta}}\ominus\left(\overline{\underset{m}{\alpha}}\ominus\underset{j}{\xi}\right) \qquad 6.42$$

The second form is

$$\underset{m}{\alpha}\ominus\left(\underset{k}{\beta}\ominus\underset{j}{\xi}\right) == (-1)^{(m-k+j)\,(n-m)}\,\overline{\underset{k}{\beta}\ominus\underset{j}{\xi}}\ominus\overline{\underset{m}{\alpha}} == (-1)^{(m-k+j)\,(n-m)+j\,(n-j)}\left(\overline{\underset{k}{\beta}}\wedge\underset{j}{\xi}\right)\ominus\overline{\underset{m}{\alpha}}$$

$$\underset{m}{\alpha}\ominus\left(\underset{k}{\beta}\ominus\underset{j}{\xi}\right) == (-1)^{(m-k+j)\,(n-m)+j\,(n-j)}\left(\overline{\underset{k}{\beta}}\wedge\underset{j}{\xi}\right)\ominus\overline{\underset{m}{\alpha}} \qquad 6.43$$

These two forms may (apart from a possible sign) be converted to the other form by replacing $\underset{m}{\alpha}$ by $\underset{k}{\overline{\beta}}$ and $\underset{k}{\beta}$ by $\underset{m}{\overline{\alpha}}$. Ignoring the signs gives us the congruences below.

$$\left(\underset{m}{\alpha}\wedge\underset{j}{\xi}\right)\ominus\underset{k}{\beta} \equiv \underset{k}{\overline{\beta}}\ominus\left(\underset{m}{\overline{\alpha}}\ominus\underset{j}{\xi}\right) \qquad \underset{m}{\alpha}\ominus\left(\underset{k}{\beta}\ominus\underset{j}{\xi}\right)\equiv\left(\underset{k}{\overline{\beta}}\wedge\underset{j}{\xi}\right)\ominus\underset{m}{\overline{\alpha}} \tag{6.44}$$

The forms of these expressions become even closer when $\underset{k}{\beta}$ is congruent to $\underset{m}{\alpha}$.

$$\left(\underset{m}{\alpha}\wedge\underset{j}{\xi}\right)\ominus\underset{m}{\alpha} \equiv \underset{m}{\overline{\alpha}}\ominus\left(\underset{m}{\overline{\alpha}}\ominus\underset{j}{\xi}\right) \qquad \underset{m}{\alpha}\ominus\left(\underset{m}{\alpha}\ominus\underset{j}{\xi}\right)\equiv\left(\underset{m}{\overline{\alpha}}\wedge\underset{j}{\xi}\right)\ominus\underset{m}{\overline{\alpha}} \tag{6.45}$$

In the common case where $\underset{j}{\xi}$ is a 1-element, x, say, these two formulae reduce to:

$$\left(\underset{m}{\alpha}\wedge x\right)\ominus\underset{k}{\beta} == (-1)^{(m-k+1)\,(n-m+1)}\,\underset{k}{\overline{\beta}}\ominus\left(\underset{m}{\overline{\alpha}}\ominus x\right) \tag{6.46}$$

$$\underset{m}{\alpha}\ominus\left(\underset{k}{\beta}\ominus x\right) == (-1)^{(m-k)\,(n-m)+m+1}\left(\underset{k}{\overline{\beta}}\wedge x\right)\ominus\underset{m}{\overline{\alpha}} \tag{6.47}$$

6.4 The Interior Common Factor Theorem

The Interior Common Factor Formula

The dual Common Factor Axiom [3.31] introduced in Chapter 3 is:

$$\left\{\left(\underset{m}{\alpha}\vee\underset{j}{\mu}\right)\wedge\left(\underset{k}{\beta}\vee\underset{j}{\mu}\right) == \left(\underset{m}{\alpha}\vee\underset{k}{\beta}\vee\underset{j}{\mu}\right)\wedge\underset{j}{\mu}\in\underset{j}{\Lambda},\quad m+k+j-2n==0\right\}$$

Interchanging the order of the factors and replacing $\underset{m}{\alpha}$ and $\underset{k}{\beta}$ by their complements gives

$$\left\{\left(\underset{j}{\mu}\vee\underset{m}{\overline{\alpha}}\right)\wedge\left(\underset{j}{\mu}\vee\underset{k}{\overline{\beta}}\right) == \left(\underset{j}{\mu}\vee\underset{m}{\overline{\alpha}}\vee\underset{k}{\overline{\beta}}\right)\wedge\underset{j}{\mu}\in\underset{j}{\Lambda},\quad j==m+k\right\}$$

Finally, by replacing $\underset{m}{\overline{\alpha}}\vee\underset{k}{\overline{\beta}}$ with $\overline{\underset{m}{\alpha}\wedge\underset{k}{\beta}}$ we get the Interior Common Factor Formula.

$$\left(\underset{j}{\mu}\ominus\underset{m}{\alpha}\right)\wedge\left(\underset{j}{\mu}\ominus\underset{k}{\beta}\right) == \left(\underset{j}{\mu}\ominus\left(\underset{m}{\alpha}\wedge\underset{k}{\beta}\right)\right)\underset{j}{\mu} \qquad j==m+k \tag{6.48}$$

In this case $\underset{j}{\mu}$ is simple, the expression is a *j*-element, and $j = m+k$.

The Interior Common Factor Theorem

The Common Factor Theorem introduced in section 3.6 enabled an explicit expression for a regressive product to be derived. We now derive the interior product version called the Interior Common Factor Theorem. This theorem is a source of many useful relations in Grassmann algebra.

We start with the Common Factor Theorem in the A form [3.42].

$$\underset{m}{\alpha} \vee \underset{s}{\beta} == \sum_{i==1}^{\nu} \left(\underset{m-j}{\alpha_i} \wedge \underset{s}{\beta} \right) \vee \underset{j}{\alpha_i}$$

$$\underset{m}{\alpha} == \underset{m-j}{\alpha_1} \wedge \underset{j}{\alpha_1} == \underset{m-j}{\alpha_2} \wedge \underset{j}{\alpha_2} == \ldots == \underset{m-j}{\alpha_\nu} \wedge \underset{j}{\alpha_\nu}$$

where $\underset{m}{\alpha}$ is simple, $j = m+s-n$, and $\nu = \binom{m}{j}$.

Suppose now that $\underset{s}{\beta} = \overline{\underset{k}{\beta}}$, $k = n-s = m-j$. Substituting for $\underset{s}{\beta}$ and using formula [6.3] allows us to write the bracketed term in the sum as:

$$\underset{m-j}{\alpha_i} \wedge \underset{s}{\beta} == \underset{k}{\alpha_i} \wedge \overline{\underset{k}{\beta}} == \left(\underset{k}{\alpha_i} \ominus \underset{k}{\beta} \right) \overline{1}$$

Thus the Common Factor Theorem is now written in terms of the interior product rather than the regressive product. In this form it will be called the Interior Common Factor Theorem.

$$\underset{m}{\alpha} \ominus \underset{k}{\beta} == \sum_{i==1}^{\nu} \left(\underset{k}{\alpha_i} \ominus \underset{k}{\beta} \right) \underset{m-k}{\alpha_i}$$

$$\underset{m}{\alpha} == \underset{k}{\alpha_1} \wedge \underset{m-k}{\alpha_1} == \underset{k}{\alpha_2} \wedge \underset{m-k}{\alpha_2} == \ldots == \underset{k}{\alpha_\nu} \wedge \underset{m-k}{\alpha_\nu} \qquad 6.49$$

where $k \leq m$, $\nu = \binom{m}{k}$, and $\underset{m}{\alpha}$ is simple.

The formula indicates that an interior product of a simple element with another, not necessarily simple, element of equal or lower grade, may be expressed in terms of the factors of the simple element of higher grade.

When $\underset{m}{\alpha}$ is not simple, it may always be expressed as the sum of simple components.

$$\underset{m}{\alpha} == \underset{m}{\alpha_1} + \underset{m}{\alpha_2} + \underset{m}{\alpha_3} + \ldots$$

From the linearity of the interior product, the Interior Common Factor Theorem may then be applied to each of the simple terms.

$$\underset{m}{\alpha} \ominus \underset{k}{\beta} == \left(\underset{m}{\alpha_1} + \underset{m}{\alpha_2} + \underset{m}{\alpha_3} + \ldots \right) \ominus \underset{k}{\beta} == \underset{m}{\alpha_1} \ominus \underset{k}{\beta} + \underset{m}{\alpha_2} \ominus \underset{k}{\beta} + \underset{m}{\alpha_3} \ominus \underset{k}{\beta} + \ldots$$

Thus, the Interior Common Factor Theorem may be applied to the interior product of any two elements, simple or non-simple. Indeed, the application to components in the preceding expansion may be applied to the components of an element even if it is simple.

A more explicit form of the Interior Common Factor Theorem may be obtained by writing the simple element in factored form.

$$(\alpha_1 \wedge \alpha_2 \wedge \ldots \wedge \alpha_m) \ominus \underset{k}{\beta} = \sum_{i_1 \ldots i_m} \left((\alpha_{i_1} \wedge \ldots \wedge \alpha_{i_k}) \ominus \underset{k}{\beta} \right) (-1)^{K_k} \alpha_1 \wedge \ldots \wedge \Box_{i_1} \wedge \ldots \wedge \Box_{i_k} \wedge \ldots \wedge \alpha_m$$
$$K_k = \sum_{\gamma=1}^{k} i_\gamma + \frac{1}{2} k (k+1) \qquad 6.50$$

where $\Box_j$ means α_j is missing from the product.

Examples of the Interior Common Factor Theorem

In this section we take the formula just developed and give examples for specific values of *m* and *k*. We do this using the *GrassmannAlgebra* function InteriorCommonFactorTheorem which generates the specific case of the theorem when given an argument of the form $\underset{m}{\alpha} \ominus \underset{k}{\beta}$. The grades *m* and *k* must be given as positive integers. The formulae are independent of the dimension of the currently declared space.

◆ **β is a scalar**

$$(\alpha_1 \wedge \alpha_2 \wedge \ldots \wedge \alpha_m) \ominus \underset{0}{\beta} = \underset{0}{\beta} (\alpha_1 \wedge \alpha_2 \wedge \ldots \wedge \alpha_m) \qquad 6.51$$

◆ **β is a 1-element**

$$(\alpha_1 \wedge \alpha_2 \wedge \ldots \wedge \alpha_m) \ominus \beta = \sum_{i=1}^{m} (\alpha_i \ominus \beta) (-1)^{i-1} \alpha_1 \wedge \ldots \wedge \Box_i \wedge \ldots \wedge \alpha_m \qquad 6.52$$

InteriorCommonFactorTheorem$\left[\underset{2}{\alpha} \ominus \underset{1}{\beta}\right]$

$$\underset{2}{\alpha} \ominus \underset{1}{\beta} = \alpha_1 \wedge \alpha_2 \ominus \underset{1}{\beta} = -\left(\alpha_2 \ominus \underset{1}{\beta}\right) \alpha_1 + \left(\alpha_1 \ominus \underset{1}{\beta}\right) \alpha_2$$

InteriorCommonFactorTheorem$\left[\underset{3}{\alpha} \ominus \underset{1}{\beta}\right]$

$$\underset{3}{\alpha} \ominus \underset{1}{\beta} = \alpha_1 \wedge \alpha_2 \wedge \alpha_3 \ominus \underset{1}{\beta} = \left(\alpha_3 \ominus \underset{1}{\beta}\right) \alpha_1 \wedge \alpha_2 - \left(\alpha_2 \ominus \underset{1}{\beta}\right) \alpha_1 \wedge \alpha_3 + \left(\alpha_1 \ominus \underset{1}{\beta}\right) \alpha_2 \wedge \alpha_3$$

InteriorCommonFactorTheorem$\left[\underset{4}{\alpha} \ominus \underset{1}{\beta}\right]$

$$\underset{4}{\alpha} \ominus \underset{1}{\beta} = \alpha_1 \wedge \alpha_2 \wedge \alpha_3 \wedge \alpha_4 \ominus \underset{1}{\beta} = -\left(\alpha_4 \ominus \underset{1}{\beta}\right) \alpha_1 \wedge \alpha_2 \wedge \alpha_3 +$$
$$\left(\alpha_3 \ominus \underset{1}{\beta}\right) \alpha_1 \wedge \alpha_2 \wedge \alpha_4 - \left(\alpha_2 \ominus \underset{1}{\beta}\right) \alpha_1 \wedge \alpha_3 \wedge \alpha_4 + \left(\alpha_1 \ominus \underset{1}{\beta}\right) \alpha_2 \wedge \alpha_3 \wedge \alpha_4$$

◆ β **is a 2-element**

$$(\alpha_1 \wedge \alpha_2 \wedge \ldots \wedge \alpha_m) \ominus \underset{2}{\beta} == \sum_{i,j} \left((\alpha_i \wedge \alpha_j) \ominus \underset{2}{\beta} \right) (-1)^{i+j-1} \alpha_1 \wedge \ldots \wedge \square_i \wedge \ldots \wedge \square_j \wedge \ldots \wedge \alpha_m \quad 6.53$$

```
InteriorCommonFactorTheorem[α ⊖ β]
                            3   2
```

$$\underset{3}{\alpha} \ominus \underset{2}{\beta} == \alpha_1 \wedge \alpha_2 \wedge \alpha_3 \ominus \underset{2}{\beta} == \left(\underset{2}{\beta} \ominus \alpha_2 \wedge \alpha_3\right) \alpha_1 - \left(\underset{2}{\beta} \ominus \alpha_1 \wedge \alpha_3\right) \alpha_2 + \left(\underset{2}{\beta} \ominus \alpha_1 \wedge \alpha_2\right) \alpha_3$$

```
InteriorCommonFactorTheorem[α ⊖ β]
                            4   2
```

$$\underset{4}{\alpha} \ominus \underset{2}{\beta} == \alpha_1 \wedge \alpha_2 \wedge \alpha_3 \wedge \alpha_4 \ominus \underset{2}{\beta} ==$$
$$\left(\underset{2}{\beta} \ominus \alpha_3 \wedge \alpha_4\right) \alpha_1 \wedge \alpha_2 - \left(\underset{2}{\beta} \ominus \alpha_2 \wedge \alpha_4\right) \alpha_1 \wedge \alpha_3 + \left(\underset{2}{\beta} \ominus \alpha_2 \wedge \alpha_3\right) \alpha_1 \wedge \alpha_4 +$$
$$\left(\underset{2}{\beta} \ominus \alpha_1 \wedge \alpha_4\right) \alpha_2 \wedge \alpha_3 - \left(\underset{2}{\beta} \ominus \alpha_1 \wedge \alpha_3\right) \alpha_2 \wedge \alpha_4 + \left(\underset{2}{\beta} \ominus \alpha_1 \wedge \alpha_2\right) \alpha_3 \wedge \alpha_4$$

◆ β **is a 3-element**

```
InteriorCommonFactorTheorem[α ⊖ β]
                            4   3
```

$$\underset{4}{\alpha} \ominus \underset{3}{\beta} == \alpha_1 \wedge \alpha_2 \wedge \alpha_3 \wedge \alpha_4 \ominus \underset{3}{\beta} == -\left(\underset{3}{\beta} \ominus \alpha_2 \wedge \alpha_3 \wedge \alpha_4\right) \alpha_1 +$$
$$\left(\underset{3}{\beta} \ominus \alpha_1 \wedge \alpha_3 \wedge \alpha_4\right) \alpha_2 - \left(\underset{3}{\beta} \ominus \alpha_1 \wedge \alpha_2 \wedge \alpha_4\right) \alpha_3 + \left(\underset{3}{\beta} \ominus \alpha_1 \wedge \alpha_2 \wedge \alpha_3\right) \alpha_4$$

◆ β **is a (m–1)-element**

$$(\alpha_1 \wedge \alpha_2 \wedge \ldots \wedge \alpha_m) \ominus \underset{m-1}{\beta} == \sum_{i=1}^{m} (-1)^{m-i} \left((\alpha_1 \wedge \ldots \wedge \square_i \wedge \ldots \wedge \alpha_m) \ominus \underset{m-1}{\beta} \right) \alpha_i \quad 6.54$$

```
InteriorCommonFactorTheorem[α ⊖ β]
                            3   2
```

$$\underset{3}{\alpha} \ominus \underset{2}{\beta} == \alpha_1 \wedge \alpha_2 \wedge \alpha_3 \ominus \underset{2}{\beta} == \left(\underset{2}{\beta} \ominus \alpha_2 \wedge \alpha_3\right) \alpha_1 - \left(\underset{2}{\beta} \ominus \alpha_1 \wedge \alpha_3\right) \alpha_2 + \left(\underset{2}{\beta} \ominus \alpha_1 \wedge \alpha_2\right) \alpha_3$$

The computational form of the Interior Common Factor Theorem

The Interior Common Factor Theorem rewritten in terms of the symbols we have been using for multilinear forms is

$$\underset{m}{A} \ominus \underset{k}{B} == \sum_{i==1}^{\nu} \left(\underset{k}{A_i} \ominus \underset{k}{B} \right) \underset{m-k}{A_i}$$
$$\underset{m}{A} == \underset{k}{A_1} \wedge \underset{m-k}{A_1} == \underset{k}{A_2} \wedge \underset{m-k}{A_2} == \ldots == \underset{k}{A_\nu} \wedge \underset{m-k}{A_\nu} \qquad 6.55$$

where $k \leq m$, and $\nu = \binom{m}{k}$, and $\underset{m}{A}$ is simple.

Just as we did for the Common Factor Theorem in section 3.6, we can express this in a computational form. If

$$\star\mathbb{S}_k\left[\underset{m}{A}\right] == \left\{\underset{k}{A_1}, \underset{k}{A_2}, \ldots, \underset{k}{A_\nu}\right\} \qquad \star\mathbb{S}^k\left[\underset{m}{A}\right] == \left\{\underset{m-k}{A_1}, \underset{m-k}{A_2}, \ldots, \underset{m-k}{A_\nu}\right\} \qquad 6.56$$

then

$$\underset{m}{A} \ominus \underset{k}{B} == \sum_{i==1}^{\nu} \left(\star\mathbb{S}_k\left[\underset{m}{A}\right]_{[[i]]} \ominus \underset{k}{B} \right) \star\mathbb{S}^k\left[\underset{m}{A}\right]_{[[i]]} \qquad 6.57$$

where $k \leq m$, and $\nu = \binom{m}{k}$, and $\underset{m}{A}$ is simple.

Or, using *Mathematica*'s inbuilt Listability

$$\underset{m}{A} \ominus \underset{k}{B} == \star\Sigma\left[\left(\star\mathbb{S}_k\left[\underset{m}{A}\right] \ominus \underset{k}{B} \right) \star\mathbb{S}^k\left[\underset{m}{A}\right]\right] \qquad 6.58$$

Or again using *Mathematica*'s `Dot` function to do the summation.

$$\underset{m}{A} \ominus \underset{k}{B} == \left(\star\mathbb{S}_k\left[\underset{m}{A}\right] \ominus \underset{k}{B} \right) . \star\mathbb{S}^k\left[\underset{m}{A}\right] \qquad 6.59$$

(The `Dot` function is an operation on lists, and should not be confused with the interior product.)

In the next section we will discuss the inner product (where the grades of the factors are equal), and show that the inner product is symmetric. Using this symmetry property we can write the above formulae in the alternative forms:

$$\underset{m}{A} \ominus \underset{k}{B} == \sum_{i==1}^{\nu} \left(\underset{k}{B} \ominus \star\mathbb{S}_k\left[\underset{m}{A}\right]_{[[i]]} \right) \star\mathbb{S}^k\left[\underset{m}{A}\right]_{[[i]]} \qquad 6.60$$

$$\underset{m}{A} \ominus \underset{k}{B} == \star\Sigma\left[\left(\underset{k}{B} \ominus \star\mathbb{S}_k\left[\underset{m}{A}\right] \right) \star\mathbb{S}^k\left[\underset{m}{A}\right]\right] \qquad 6.61$$

$$\underset{m}{A} \ominus \underset{k}{B} == \left(\underset{k}{B} \ominus \star S_k\left[\underset{m}{A}\right]\right) . \star S^k\left[\underset{m}{A}\right] \tag{6.62}$$

◆ Examples

Here are the three computational forms compared to the result from the `InteriorCommonFactorTheorem` function which implements the initially derived theorem.

```
ShowPrecedence; m = 3; k = 2; ν = 3;
```

$$\underset{m}{A} \ominus \underset{k}{B} == \sum_{i=1}^{\nu} \left(\underset{k}{B} \ominus \star S_k\left[\underset{m}{A}\right]_{[[i]]}\right) \star S^k\left[\underset{m}{A}\right]_{[[i]]}$$

$$\underset{3}{A} \ominus \underset{2}{B} == \left(\underset{2}{B} \ominus (A_2 \wedge A_3)\right) A_1 - \left(\underset{2}{B} \ominus (A_1 \wedge A_3)\right) A_2 + \left(\underset{2}{B} \ominus (A_1 \wedge A_2)\right) A_3$$

$$\underset{m}{A} \ominus \underset{k}{B} == \star\Sigma\left[\left(\underset{k}{B} \ominus \star S_k\left[\underset{m}{A}\right]\right) \star S^k\left[\underset{m}{A}\right]\right]$$

$$\underset{3}{A} \ominus \underset{2}{B} == \left(\underset{2}{B} \ominus (A_2 \wedge A_3)\right) A_1 - \left(\underset{2}{B} \ominus (A_1 \wedge A_3)\right) A_2 + \left(\underset{2}{B} \ominus (A_1 \wedge A_2)\right) A_3$$

$$\underset{m}{A} \ominus \underset{k}{B} == \left(\underset{k}{B} \ominus \star S_k\left[\underset{m}{A}\right]\right) . \star S^k\left[\underset{m}{A}\right]$$

$$\underset{3}{A} \ominus \underset{2}{B} == \left(\underset{2}{B} \ominus (A_2 \wedge A_3)\right) A_1 - \left(\underset{2}{B} \ominus (A_1 \wedge A_3)\right) A_2 + \left(\underset{2}{B} \ominus (A_1 \wedge A_2)\right) A_3$$

$$\text{InteriorCommonFactorTheorem}\left[\underset{3}{A} \ominus \underset{2}{B}\right]$$

$$\underset{3}{A} \ominus \underset{2}{B} == (A_1 \wedge A_2 \wedge A_3) \ominus \underset{2}{B} == \left(\underset{2}{B} \ominus (A_2 \wedge A_3)\right) A_1 - \left(\underset{2}{B} \ominus (A_1 \wedge A_3)\right) A_2 + \left(\underset{2}{B} \ominus (A_1 \wedge A_2)\right) A_3$$

Converting interior products to inner and scalar products

◆ Converting interior products to inner products

To convert the interior products in an expression to inner products you can use the *GrassmannAlgebra* command `ToInnerProducts`.

$$\text{ToInnerProducts}[a\,(e_1 \wedge e_2 \wedge e_3) \ominus (e_1 \wedge e_2) + b\,(e_1 \wedge e_2 \wedge e_3) \ominus (e_2 \wedge e_3)]$$

$$\begin{aligned}&(a\,((e_1 \wedge e_2) \ominus (e_2 \wedge e_3)) + b\,((e_2 \wedge e_3) \ominus (e_2 \wedge e_3)))\,e_1 +\\ &\quad(-a\,((e_1 \wedge e_2) \ominus (e_1 \wedge e_3)) - b\,((e_1 \wedge e_3) \ominus (e_2 \wedge e_3)))\,e_2 +\\ &\quad(a\,((e_1 \wedge e_2) \ominus (e_1 \wedge e_2)) + b\,((e_1 \wedge e_2) \ominus (e_2 \wedge e_3)))\,e_3\end{aligned}$$

◆ Converting interior products to scalar products

To convert the interior or inner products in an expression to scalar products you can use the *GrassmannAlgebra* command `ToScalarProducts`.

$$\text{ToScalarProducts}[a\,(e_1 \wedge e_2 \wedge e_3) \ominus (e_1 \wedge e_2)]$$

$$\begin{aligned}&a\,(-(e_1 \ominus e_3)\,(e_2 \ominus e_2) + (e_1 \ominus e_2)\,(e_2 \ominus e_3))\,e_1 -\\ &\quad a\,(-(e_1 \ominus e_3)\,(e_2 \ominus e_1) + (e_1 \ominus e_1)\,(e_2 \ominus e_3))\,e_2 +\\ &\quad a\,(-(e_1 \ominus e_2)\,(e_2 \ominus e_1) + (e_1 \ominus e_1)\,(e_2 \ominus e_2))\,e_3\end{aligned}$$

◆ Converting interior products to metric elements

To convert the interior, inner or scalar products in an expression to metric elements you can use the *GrassmannAlgebra* command `ToMetricElements`.

For example, if the metric is the default Euclidean metric with orthonormal basis the above expression becomes:

```
ToMetricElements[a (e1 ∧ e2 ∧ e3) ⊖ (e1 ∧ e2)]
```

$a\, e_3$

And for a general metric you will get the original expression with the general metric elements substituted for the scalar products.

```
DeclareMetric[g]; ToMetricElements[a (e1 ∧ e2 ∧ e3) ⊖ (e1 ∧ e2)]
```

$$a\, e_3 \left(-g_{1,2}^2 + g_{1,1}\, g_{2,2}\right) + a\, e_2\, (g_{1,2}\, g_{1,3} - g_{1,1}\, g_{2,3}) + a\, e_1\, (-g_{1,3}\, g_{2,2} + g_{1,2}\, g_{2,3})$$

6.5 The Inner Product

The symmetry of the inner product

It is a normally accepted feature of the inner product that it be symmetric - that is, independent of the order of its factors. To show that our definition does indeed imply this symmetry we only need to do some simple transformations on the right hand side of the equation [6.4]. (We have presaged these in section 5.3.)

$$\underset{m}{\alpha} \ominus \underset{m}{\beta} == \overline{\underset{m}{\alpha} \wedge \overline{\underset{m}{\beta}}} == \overline{\underset{m}{\alpha}} \vee \overline{\overline{\underset{m}{\beta}}} == (-1)^{m\,(n-m)}\, \overline{\underset{m}{\alpha}} \vee \underset{m}{\beta} == \underset{m}{\beta} \vee \overline{\underset{m}{\alpha}} == \underset{m}{\beta} \ominus \underset{m}{\alpha}$$

$$\underset{m}{\alpha} \ominus \underset{m}{\beta} == \underset{m}{\beta} \ominus \underset{m}{\alpha} \qquad 6.63$$

This symmetry, is of course, a property that we would expect of an inner product. However it may be of interest to recall that the major result which permits this symmetry is that the complement operation has been required to satisfy the complement of a complement axiom [5.7], that is, that the complement of the complement of an element should be, apart from a possible sign, equal to the element itself.

The inner product of complements

We have already shown earlier in ⊖10, formula [6.21], that the inner product of two elements is equal to the inner product of their complements in reverse order. This followed from the more general axiom for interior products. However, since we have now shown that the inner product is symmetric, we can immediately see that the inner product of two elements is equal to the inner product of their complements.

$$\underset{m}{\alpha} \ominus \underset{m}{\beta} == \overline{\underset{m}{\beta}} \ominus \overline{\underset{m}{\alpha}} == \overline{\underset{m}{\alpha}} \ominus \overline{\underset{m}{\beta}}$$

$$\underset{m}{\alpha} \ominus \underset{m}{\beta} == \underset{m}{\overline{\alpha}} \ominus \underset{m}{\overline{\beta}} \quad 6.64$$

The inner product as a determinant

The inner product of simple elements can be expressed as a determinant of scalar products. This result is of seminal importance since it generalizes the notions of scalar product, metric, measure and angle to general elements of the algebra.

We begin with the form

$$\underset{m}{\alpha} \ominus \underset{m}{\beta} == (\alpha_1 \wedge \alpha_2 \wedge \ldots \wedge \alpha_m) \ominus (\beta_1 \wedge \beta_2 \wedge \ldots \wedge \beta_m)$$

The expression $(\alpha_1 \wedge \alpha_2 \wedge \ldots \wedge \alpha_m) \ominus (\beta_1 \wedge \beta_2 \wedge \ldots \wedge \beta_m)$ can be written using the formula for extended interior products [6.40] as

$$((\alpha_1 \wedge \alpha_2 \wedge \ldots \wedge \alpha_m) \ominus \beta_1) \ominus (\beta_2 \wedge \ldots \wedge \beta_m)$$

or, by rearranging the β factors to bring β_j to the beginning of the product, written as

$$\left((\alpha_1 \wedge \alpha_2 \wedge \ldots \wedge \alpha_m) \ominus \beta_j\right) \ominus \left((-1)^{j-1} \left(\beta_1 \wedge \beta_2 \wedge \ldots \wedge \square_j \wedge \ldots \wedge \beta_m\right)\right)$$

The Interior Common Factor Theorem [6.52] gives the first factor of this expression as

$$(\alpha_1 \wedge \alpha_2 \wedge \ldots \wedge \alpha_m) \ominus \beta_j == \sum_{i=1}^{m} \left(\alpha_i \ominus \beta_j\right) (-1)^{i-1} \alpha_1 \wedge \alpha_2 \wedge \ldots \wedge \square_i \wedge \ldots \wedge \alpha_m$$

Thus, $(\alpha_1 \wedge \alpha_2 \wedge \ldots \wedge \alpha_m) \ominus (\beta_1 \wedge \beta_2 \wedge \ldots \wedge \beta_m)$ becomes

$$(\alpha_1 \wedge \alpha_2 \wedge \ldots \wedge \alpha_m) \ominus (\beta_1 \wedge \beta_2 \wedge \ldots \wedge \beta_m) == \sum_{i=1}^{m} \left(\alpha_i \ominus \beta_j\right) (-1)^{i+j} (\alpha_1 \wedge \ldots \wedge \square_i \wedge \ldots \wedge \alpha_m) \ominus \left(\beta_1 \wedge \ldots \wedge \square_j \wedge \ldots \wedge \beta_m\right) \quad 6.65$$

If this process is repeated, we find the strikingly simple formula

$$(\alpha_1 \wedge \alpha_2 \wedge \ldots \wedge \alpha_m) \ominus (\beta_1 \wedge \beta_2 \wedge \ldots \wedge \beta_m) == \mathrm{Det}\left[\alpha_i \ominus \beta_j\right] \quad 6.66$$

$$(\alpha_1 \wedge \alpha_2 \wedge \ldots \wedge \alpha_m) \ominus (\beta_1 \wedge \beta_2 \wedge \ldots \wedge \beta_m) == \begin{vmatrix} \alpha_1 \ominus \beta_1 & \alpha_1 \ominus \beta_2 & \alpha_1 \ominus \ldots & \alpha_1 \ominus \beta_m \\ \alpha_2 \ominus \beta_1 & \alpha_2 \ominus \beta_2 & \alpha_2 \ominus \ldots & \alpha_2 \ominus \beta_m \\ \ldots \ominus \beta_1 & \ldots \ominus \beta_2 & \ldots \ominus \ldots & \ldots \ominus \beta_m \\ \alpha_m \ominus \beta_1 & \alpha_m \ominus \beta_2 & \alpha_m \ominus \ldots & \alpha_m \ominus \beta_m \end{vmatrix} \quad 6.67$$

Calculating inner products

The expression for the inner product of two simple elements can be developed in *Grassmann-Algebra* by using either the operation `ComposeScalarProductMatrix` or the operation `ToScalarProducts`.

◆ ComposeScalarProductMatrix

- `ComposeScalarProductMatrix[A⊖B]` develops the inner product of `A` and `B` into the matrix of the scalar products of their factors. You can use *Mathematica*'s inbuilt `Det` operation on this matrix to get its determinant, and hence the inner product.

$$M_1 = \text{ComposeScalarProductMatrix}\left[\underset{3}{\alpha} \ominus \underset{3}{\beta}\right];\ \text{MatrixForm}[M_1]$$

$$\begin{pmatrix} \alpha_1 \ominus \beta_1 & \alpha_1 \ominus \beta_2 & \alpha_1 \ominus \beta_3 \\ \alpha_2 \ominus \beta_1 & \alpha_2 \ominus \beta_2 & \alpha_2 \ominus \beta_3 \\ \alpha_3 \ominus \beta_1 & \alpha_3 \ominus \beta_2 & \alpha_3 \ominus \beta_3 \end{pmatrix}$$

$$\text{Det}[M_1]$$

$$-(\alpha_1 \ominus \beta_3)(\alpha_2 \ominus \beta_2)(\alpha_3 \ominus \beta_1) + (\alpha_1 \ominus \beta_2)(\alpha_2 \ominus \beta_3)(\alpha_3 \ominus \beta_1) + (\alpha_1 \ominus \beta_3)(\alpha_2 \ominus \beta_1)(\alpha_3 \ominus \beta_2) - (\alpha_1 \ominus \beta_1)(\alpha_2 \ominus \beta_3)(\alpha_3 \ominus \beta_2) - (\alpha_1 \ominus \beta_2)(\alpha_2 \ominus \beta_1)(\alpha_3 \ominus \beta_3) + (\alpha_1 \ominus \beta_1)(\alpha_2 \ominus \beta_2)(\alpha_3 \ominus \beta_3)$$

- If you just want to *display* M_1 to look like a determinant, you can use *Mathematica's* `Grid` and `BracketingBar`.

$$\text{BracketingBar}[\text{Grid}[M_1]]$$

$$\begin{vmatrix} \alpha_1 \ominus \beta_1 & \alpha_1 \ominus \beta_2 & \alpha_1 \ominus \beta_3 \\ \alpha_2 \ominus \beta_1 & \alpha_2 \ominus \beta_2 & \alpha_2 \ominus \beta_3 \\ \alpha_3 \ominus \beta_1 & \alpha_3 \ominus \beta_2 & \alpha_3 \ominus \beta_3 \end{vmatrix}$$

- If we reverse the order of the two elements and then take the transpose of the resulting scalar product matrix we get a matrix M_2 which differs from M_1 only by the ordering of its scalar products.

$$M_2 = \text{ComposeScalarProductMatrix}\left[\underset{3}{\beta} \ominus \underset{3}{\alpha}\right];$$

$$\text{MatrixForm}[\text{Transpose}[M_2]]$$

$$\begin{pmatrix} \beta_1 \ominus \alpha_1 & \beta_2 \ominus \alpha_1 & \beta_3 \ominus \alpha_1 \\ \beta_1 \ominus \alpha_2 & \beta_2 \ominus \alpha_2 & \beta_3 \ominus \alpha_2 \\ \beta_1 \ominus \alpha_3 & \beta_2 \ominus \alpha_3 & \beta_3 \ominus \alpha_3 \end{pmatrix}$$

This shows clearly how the symmetry of the inner product depends on the symmetry of the scalar product.

◆ ToScalarProducts

- To obtain the expansion of the inner product more directly, you can use the more general `ToScalarProducts` discussed in section 6.4 above. (Converting an inner or interior product to scalar products is one of the most common operations in the algebra).

$$\text{ToScalarProducts}\left[\underset{3}{\alpha} \ominus \underset{3}{\beta}\right]$$

$$-(\alpha_1 \ominus \beta_3)(\alpha_2 \ominus \beta_2)(\alpha_3 \ominus \beta_1) + (\alpha_1 \ominus \beta_2)(\alpha_2 \ominus \beta_3)(\alpha_3 \ominus \beta_1) + (\alpha_1 \ominus \beta_3)(\alpha_2 \ominus \beta_1)(\alpha_3 \ominus \beta_2) - (\alpha_1 \ominus \beta_1)(\alpha_2 \ominus \beta_3)(\alpha_3 \ominus \beta_2) - (\alpha_1 \ominus \beta_2)(\alpha_2 \ominus \beta_1)(\alpha_3 \ominus \beta_3) + (\alpha_1 \ominus \beta_1)(\alpha_2 \ominus \beta_2)(\alpha_3 \ominus \beta_3)$$

Inner products of basis elements

The following formulae collect together various results for the inner product of two basis m-elements. They are given to complete the discussion of reciprocal bases in section 5.13 and are not needed for an understaning of the rest of the material.

Here, $\underset{m}{e_a}$ and $\underset{m}{e_b}$ refer to any basis elements.

$$\underset{m}{e_a} \ominus \underset{m}{e_b} == \underset{m}{e_b} \ominus \underset{m}{e_a} == \overline{\underset{m}{e_a}} \ominus \overline{\underset{m}{e_b}} == \overline{\underset{m}{e_b}} \ominus \overline{\underset{m}{e_a}} \qquad 6.68$$

$$\underset{m}{e_i} \ominus \underset{m}{e^j} == \delta_i^j \qquad \underset{m}{e^i} \ominus \underset{m}{e_j} == \delta_j^i \qquad 6.69$$

$$\underset{m}{e_i} \ominus \underset{m}{e_j} == \overline{\underset{m}{e_i}} \ominus \overline{\underset{m}{e_j}} == \left|g_{ij}\right| \underline{\underset{m}{e^i}} \ominus \underline{\underset{m}{e^j}} == \overset{m}{g}_{ij} \qquad 6.70$$

$$\underset{m}{e^i} \ominus \underset{m}{e^j} == \overline{\underset{m}{e^i}} \ominus \overline{\underset{m}{e^j}} == \frac{1}{\left|g_{ij}\right|} \underline{\underset{m}{e_i}} \ominus \underline{\underset{m}{e_j}} == \overset{m}{g}{}^{ij} \qquad 6.71$$

Here, $\left|g_{ij}\right|$ is the determinant of the metric tensor.

6.6 The Measure of an *m*-element

The definition of measure

The inner product $A \ominus A$ of an m-element A with itself may be called its *inner square*. By definition of the interior product this is a scalar quantity. And since A is 'squared', if it is non-zero, it is always positive.

The *measure of an m-element* A is denoted $|A|$ and defined as the positive square root of its inner square.

$$|A| == \sqrt{A \ominus A} == |-A| \qquad 6.72$$

Since by [6.64] the inner square of an element is equal to the inner square of its complement, the measure of an element is equal to the measure of its complement. See also section 5.2.

$$|A| == |\overline{A}| == |-\overline{A}| \qquad 6.73$$

The *measure of a scalar* or the measure of its complement is equal to its absolute value.

$$|a| == |\overline{a}| == \sqrt{a^2} \qquad 6.74$$

Thus the *measure of unity* or the *measure of the unit n-element* (or their negatives) are, as would be expected, equal to unity.

$$|1| == |-1| == |\overline{1}| == \left|-\overline{1}\right| == 1 \qquad 6.75$$

The measure of a scalar multiple a of an element A is the product of their measures.

$$|a\,A| == |a|\;|A| \qquad 6.76$$

Since the basis *n*-element is also the cobasis of 1, the measure of the basis *n*-element can always be written as the measure of the cobasis of 1. (We have already used this designation for convenience in the previous chapter.)

$$|e_1 \wedge e_2 \wedge \ldots \wedge e_n| == |\underline{1}| \qquad 6.77$$

By equation [6.67], the *measure of a simple m-element* A is equal to the square root of the determinant of the matrix of the scalar products of its factors.

$$\begin{array}{c} A == \alpha_1 \wedge \alpha_2 \wedge \ldots \wedge \alpha_m \\ |A|^2 == (\alpha_1 \wedge \alpha_2 \wedge \ldots \wedge \alpha_m) \ominus (\alpha_1 \wedge \alpha_2 \wedge \ldots \wedge \alpha_m) == \mathrm{Det}\left[\alpha_i \ominus \alpha_j\right] \end{array} \qquad 6.78$$

In a *Euclidean metric*, the measure of an element is given by the familiar 'root sum of squares' formula.

$$A == \sum_i a_i \underset{m}{e_i} \quad \Longrightarrow \quad |A| == \sqrt{\sum a_i^2} \qquad 6.79$$

Unit elements

The *unit m-element* associated with the element A is denoted $\hat{A}$ and may now be defined as the element divided by its measure.

$$\hat{A} == \frac{A}{|A|} \qquad 6.80$$

The *unit scalar* $\hat{a}$ corresponding to the scalar a, is therefore equal to +1 if a is positive and is equal to -1 if a is negative.

$$\hat{a} == \frac{a}{|a|} == \begin{cases} 1 & a > 0 \\ -1 & a < 0 \end{cases} \qquad 6.81$$

$$\hat{1} == 1 \qquad \widehat{-1} == -1 \qquad 6.82$$

A *unit n-element* $\hat{\underset{n}{A}}$ corresponding to the *n*-element $\underset{n}{A} = a\, e_1 \wedge \ldots \wedge e_n$ is equal to the unit *n*-element $\overline{1}$ if a is positive, and equal to $-\overline{1}$ if a is negative.

$$\hat{\underset{n}{A}} == \frac{\underset{n}{A}}{\left|\underset{n}{A}\right|} == \frac{a\, e_1 \wedge \ldots \wedge e_n}{|a|\; |\underline{1}|} == \begin{cases} \overline{1} & a > 0 \\ -\overline{1} & a < 0 \end{cases} \qquad 6.83$$

$$\hat{\overline{1}} == \overline{1} \qquad \widehat{-\overline{1}} == -\overline{1} \qquad 6.84$$

The *measure of a unit m-element* $\hat{A}$ is therefore always unity.

$$|\hat{A}| == \sqrt{\hat{A} \ominus \hat{A}} == \sqrt{\frac{A}{|A|} \ominus \frac{A}{|A|}} == 1$$

$$|\hat{A}| == 1 \qquad 6.85$$

Calculating measures

To calculate the measure of any simple element you can use the *GrassmannAlgebra* function `Measure`.

```
Measure[x ∧ y]
```

$$\sqrt{-(x \ominus y)^2 + (x \ominus x)\,(y \ominus y)}$$

`Measure` will compute the measure of a simple element using the currently declared metric.

For example, with the default Euclidean metric, the measure of this 3-element is 5.

```
Measure[(2 e1 - 3 e2) ∧ (e1 + e2 + e3) ∧ (5 e2 + e3)]
```

5

With a general metric in 3-space

```
DeclareMetric[g]
```

$\{\{g_{1,1}, g_{1,2}, g_{1,3}\}, \{g_{1,2}, g_{2,2}, g_{2,3}\}, \{g_{1,3}, g_{2,3}, g_{3,3}\}\}$

```
Measure[(2 e1 - 3 e2) ∧ (e1 + e2 + e3) ∧ (5 e2 + e3)]
```

$$\sqrt{\left(50\, g_{1,2}\, g_{1,3}\, g_{2,3} + 25\, g_{1,1}\, g_{2,2}\, g_{3,3} - 25\left(g_{1,3}^2\, g_{2,2} + g_{1,1}\, g_{2,3}^2 + g_{1,2}^2\, g_{3,3}\right)\right)}$$

`Measure` will automatically apply any simplifications due to the symmetry of the scalar product.

The measure of vectorial elements

The measure of a vectorial *m*-element has a well-defined geometric significance. The results below apply in spaces of any dimension.

◆ Scalars: m = 0

$$|a|^2 == a^2$$

The measure of a scalar is its absolute value.

```
Measure[a]
```

$$\sqrt{a^2}$$

◆ Vectors: m = 1

$$|x|^2 == x \ominus x$$

If x is interpreted as a *vector*, then $|x|$ is called the *length* of **x**.

```
Measure[x]
```

$$\sqrt{x \ominus x}$$

◆ Bivectors: m = 2

$$|x \wedge y|^2 == (x \wedge y) \ominus (x \wedge y) == \begin{vmatrix} x \ominus x & x \ominus y \\ y \ominus x & y \ominus y \end{vmatrix}$$

If $x \wedge y$ is interpreted as a *bivector*, then $|x \wedge y|$ is called the *area* of $x \wedge y$. The scalar $|x \wedge y|$ is in fact the area of the parallelogram formed by the vectors x and y.

```
Measure[x ∧ y]
```

$$\sqrt{-(x \ominus y)^2 + (x \ominus x)\ (y \ominus y)}$$

Because of the nilpotent properties of the exterior product, *the measure of the bivector is independent of the way in which it is expressed in terms of its vector factors*.

```
Measure[x ∧ (a x + y)]
```

$$\sqrt{-(x \ominus y)^2 + (x \ominus x)\ (y \ominus y)}$$

This is equivalent to the well-known fact that parallelograms on the same base and between the same parallels, have equal area.

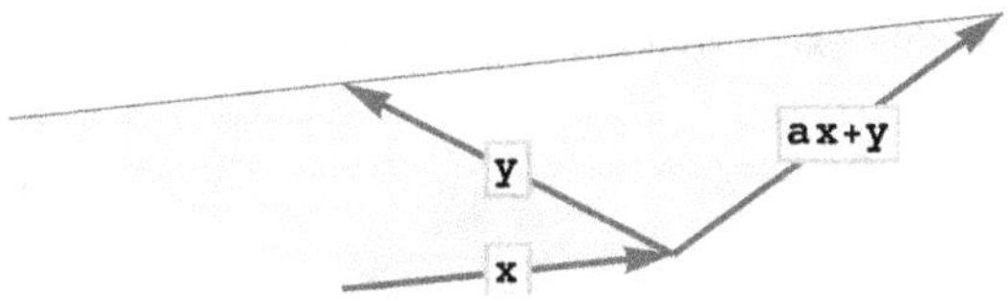

The measure of a bivector is independent of its shape

◆ **Trivectors: m = 3**

$$|x\wedge y\wedge z|^2 == (x\wedge y\wedge z)\ominus(x\wedge y\wedge z) == \begin{vmatrix} x\ominus x & x\ominus y & x\ominus z \\ y\ominus x & y\ominus y & y\ominus z \\ z\ominus x & z\ominus y & z\ominus z \end{vmatrix}$$

If $x\wedge y\wedge z$ is interpreted as a *trivector*, then $|x\wedge y\wedge z|$ is called the *volume* of $x\wedge y\wedge z$. The scalar $|x\wedge y\wedge z|$ is in fact the volume of the parallelepiped formed by the vectors x, y and z.

```
Measure[x ∧ y ∧ z]
```

$$\sqrt{\left(-(x\ominus z)^2 (y\ominus y) + 2 (x\ominus y) (x\ominus z) (y\ominus z) - (x\ominus x) (y\ominus z)^2 - (x\ominus y)^2 (z\ominus z) + (x\ominus x) (y\ominus y) (z\ominus z)\right)}$$

Again, the measure of trivector is independent of the way in which it is expressed in terms of its vector factors. For example

```
Measure[x ∧ (y + a x) ∧ (z + b x + c y)]
```

$$\sqrt{\left(-(x\ominus z)^2 (y\ominus y) + 2 (x\ominus y) (x\ominus z) (y\ominus z) - (x\ominus x) (y\ominus z)^2 - (x\ominus y)^2 (z\ominus z) + (x\ominus x) (y\ominus y) (z\ominus z)\right)}$$

The measure of orthogonal elements

Consider the inner square of a simple *m*-element $\alpha_1\wedge\alpha_2\wedge\ldots\wedge\alpha_m$, then by formula [6.67] we can write it in determinant form as

$$(\alpha_1\wedge\alpha_2\wedge\ldots\wedge\alpha_m)\ominus(\alpha_1\wedge\alpha_2\wedge\ldots\wedge\alpha_m) == \begin{vmatrix} \alpha_1\ominus\alpha_1 & \alpha_1\ominus\alpha_2 & \alpha_1\ominus\ldots & \alpha_1\ominus\alpha_m \\ \alpha_2\ominus\alpha_1 & \alpha_2\ominus\alpha_2 & \alpha_2\ominus\ldots & \alpha_2\ominus\alpha_m \\ \ldots\ominus\alpha_1 & \ldots\ominus\alpha_2 & \ldots\ominus\ldots & \ldots\ominus\alpha_m \\ \alpha_m\ominus\alpha_1 & \alpha_m\ominus\alpha_2 & \alpha_m\ominus\ldots & \alpha_m\ominus\alpha_m \end{vmatrix}$$

In the case that the α_i are mutually orthogonal, that is $\alpha_i\ominus\alpha_j = 0$, the determinant simplifies to

$$(\alpha_1\wedge\alpha_2\wedge\ldots\wedge\alpha_m)\ominus(\alpha_1\wedge\alpha_2\wedge\ldots\wedge\alpha_m) == \begin{vmatrix} \alpha_1\ominus\alpha_1 & 0 & \ldots & 0 \\ 0 & \alpha_2\ominus\alpha_2 & \ldots & 0 \\ \ldots & \ldots & \ldots & \ldots \\ 0 & 0 & \ldots & \alpha_m\ominus\alpha_m \end{vmatrix}$$

$$== (\alpha_1\ominus\alpha_1)(\alpha_2\ominus\alpha_2)\ldots(\alpha_m\ominus\alpha_m) == |\alpha_1|^2 |\alpha_2|^2 \ldots |\alpha_m|^2$$

$$|\alpha_1\wedge\alpha_2\wedge\ldots\wedge\alpha_m| == |\alpha_1| |\alpha_2| \ldots |\alpha_m| \qquad \alpha_i\ominus\alpha_j = 0 \qquad 6.86$$

If the α_i are mutually orthogonal *except* for α_1 and α_2 (that is, $\alpha_1\ominus\alpha_2$ is not necessarily zero) then the determinant becomes

$$\left((\alpha_1\ominus\alpha_1)(\alpha_2\ominus\alpha_2) - (\alpha_1\ominus\alpha_2)^2\right)(\alpha_3\ominus\alpha_3)\ldots(\alpha_m\ominus\alpha_m)$$

Since this first factor is just the inner square of $\alpha_1\wedge\alpha_2$ we can write

$$|\alpha_1\wedge\alpha_2\wedge\ldots\wedge\alpha_m| == |\alpha_1\wedge\alpha_2| |\alpha_3| \ldots |\alpha_m| \qquad \alpha_1\ominus\alpha_2 \neq 0 \bigwedge \alpha_i\ominus\alpha_j = 0 \qquad 6.87$$

Note that in an inner square we can always rearrange the factors in any convenient order without affecting the result.

If an m-element has a number of factors, say the first k, which are mutually orthogonal and the rest which are not necessarily so, we can write

$$\left|\alpha_1 \wedge \ldots \wedge \alpha_k \wedge \alpha_{k+1} \wedge \ldots \wedge \alpha_m\right| == \left|\alpha_1\right| \ldots \left|\alpha_k\right| \left|\alpha_{k+1} \wedge \ldots \wedge \alpha_m\right| \qquad 6.88$$

Here, the α_1 to α_k are mutually orthogonal, and the α_{k+1} to α_m are not necessarily so.

If the α_1 to α_k are *also* unit vectors then they can be removed from the inner square.

$$\left|\hat{\alpha}_1 \wedge \ldots \wedge \hat{\alpha}_k \wedge \alpha_{k+1} \wedge \ldots \wedge \alpha_m\right| == \left|\alpha_{k+1} \wedge \ldots \wedge \alpha_m\right| \qquad 6.89$$

The measure of bound elements

In this section we explore some relationships for measures of bound elements in a bound vector n-space under the hybrid metric introduced in Chapter 5. This metric has the properties that:
1. The scalar product of the origin with itself (and hence its measure) is unity.
2. The scalar product of the origin with any vector or m-vector is zero.
3. The scalar product of any two vectors is given by the metric tensor on the n-dimensional vector subspace of the bound vector n-space.

We summarize these properties as follows. A is an m-vector.

$$\star\mathbb{O} \ominus \star\mathbb{O} == 1 \qquad A \ominus \star\mathbb{O} == 0 \qquad e_i \ominus e_j == g_{ij} \qquad 6.90$$

◆ The measure of a point

Since a point P is defined as the sum of the origin $\star\mathbb{O}$ and the point's position vector x relative to the origin, we can write

$$P \ominus P == (\star\mathbb{O} + x) \ominus (\star\mathbb{O} + x) == \star\mathbb{O} \ominus \star\mathbb{O} + 2\,\star\mathbb{O} \ominus x + x \ominus x == 1 + x \ominus x$$

$$P == \star\mathbb{O} + x \implies |P|^2 == 1 + |x|^2 \qquad 6.91$$

Although this is at first sight a strange result, it is due entirely to the hybrid metric referred to above. Unlike a vector difference of two points whose measure starts off from zero and increases continuously as the two points move further apart, the measure of a single point starts off from unity as it coincides with the origin and increases continuously as it moves further away from it.

The measure of the difference of two points is, as expected, equal to the measure of the resulting vector.

In Volume 2 we will explore dual points. A *dual point* is one whose position vector is multiplied by Clifford's dual unit ω, which has the property that its square is zero. This means that the measure of *any* dual point is unity.

◆ The measure of an m-vector bound through the origin

The measure of an m-vector bound through the origin is equal to the measure of the m-vector.

To show this, suppose first a simple m-vector A. The inner square of A bound through the origin is $(\star\mathbb{O} \wedge A) \ominus (\star\mathbb{O} \wedge A)$. If now we apply the extended interior product formula [6.40] we obtain

$$(\star\mathbb{O} \wedge A) \ominus (\star\mathbb{O} \wedge A) == ((\star\mathbb{O} \wedge A) \ominus \star\mathbb{O}) \ominus A$$

But the factor $(\star\mathbb{O} \wedge A) \ominus \star\mathbb{O}$ can be expanded by example [6.52] of the Interior Common Factor Theorem. Let

$$\star\mathbb{O} \wedge A == \alpha_1 \wedge \alpha_2 \wedge \cdots \wedge \alpha_m \wedge \alpha_{m+1}$$

Then formula [6.52] with α_1 equal to $\star\mathbb{O}$ gives

$$(\alpha_1 \wedge \alpha_2 \wedge \cdots \wedge \alpha_m \wedge \alpha_{m+1}) \ominus \star\mathbb{O} == \sum_{i=1}^{m+1} (\alpha_i \ominus \star\mathbb{O}) (-1)^{i-1} \alpha_1 \wedge \cdots \wedge \square_i \wedge \cdots \wedge \alpha_{m+1}$$

But because of our hybrid metric, the scalar product $\alpha_i \ominus \star\mathbb{O}$ is equal to zero whenever α_i is a vector. That is, in all cases except for α_1 where $\alpha_1 \ominus \star\mathbb{O}$ is equal to $\star\mathbb{O} \ominus \star\mathbb{O}$, and thus equal to unity. Thus we have

$$(\star\mathbb{O} \wedge A) \ominus \star\mathbb{O} == A \qquad 6.92$$

$$(\star\mathbb{O} \wedge A) \ominus (\star\mathbb{O} \wedge A) == ((\star\mathbb{O} \wedge A) \ominus \star\mathbb{O}) \ominus A == A \ominus A$$

$$|\star\mathbb{O} \wedge A| == |A| \qquad 6.93$$

◆ The measure of a bound m-vector

The measure of a bound m-vector bound through a more general point P with position vector x is now straightforward to determine by expanding its inner square.

$$(P \wedge A) \ominus (P \wedge A) == ((\star\mathbb{O} + x) \wedge A) \ominus ((\star\mathbb{O} + x) \wedge A)$$
$$== (\star\mathbb{O} \wedge A) \ominus (\star\mathbb{O} \wedge A) + (\star\mathbb{O} \wedge A) \ominus (x \wedge A) + (x \wedge A) \ominus (\star\mathbb{O} \wedge A) + (x \wedge A) \ominus (x \wedge A)$$

The second and third terms are zero. To see this note first that they are equal due to the symmetry of the inner product. Then apply the extended interior product formula [6.40] again.

$$(x \wedge A) \ominus (\star\mathbb{O} \wedge A) == ((x \wedge A) \ominus \star\mathbb{O}) \ominus A == 0 \ominus A == 0$$

By our previous result [6.93] the first term is equal to the inner square of the m-vector A so we have

$$(P \wedge A) \ominus (P \wedge A) == A \ominus A + (x \wedge A) \ominus (x \wedge A)$$

$$|P \wedge A|^2 == |A|^2 + |x \wedge A|^2 \qquad 6.94$$

Note that in the case x is zero, this reduces to [6.93]. And in the case the m-vector is a unit m-vector we have

$$\left|P \wedge \hat{A}\right|^2 == 1 + \left|x \wedge \hat{A}\right|^2 \qquad \left|\star\mathbb{O} \wedge \hat{A}\right|^2 == 1 \qquad 6.95$$

◆ The measure of a bound m-vector with orthogonal position vector

If the position vector of the point of a bound m-vector is orthogonal to the m-vector, then we can apply the same simplification to $(\bar{x} \wedge A) \ominus (\bar{x} \wedge A)$ as we did to $(\bigstar\mathbb{O} \wedge A) \ominus (\bigstar\mathbb{O} \wedge A)$ using [6.92]. Let us denote the orthogonal position vector by $x^{\llcorner}$. Then

$$(x^{\llcorner} \wedge A) \ominus (x^{\llcorner} \wedge A) == (x^{\llcorner} \ominus x^{\llcorner})\, (A \ominus A)$$

Hence we can rewrite [6.94] as

$$|P \wedge A|^2 == |A|^2 + |x \wedge A|^2 == |A|^2 + |x^{\llcorner}|^2\, |A|^2$$

$$|P \wedge A|^2 == |A|^2 \left(1 + |x^{\llcorner}|^2\right) \qquad A \ominus x^{\llcorner} == 0 \qquad 6.96$$

And if A is a unit m-element we obtain a formula which looks very much like that for a single point.

$$\left|P \wedge \hat{A}\right|^2 == 1 + |x^{\llcorner}|^2 \qquad A \ominus x^{\llcorner} == 0 \qquad 6.97$$

Thus the measure of a bound unit m-vector indicates its minimum distance from the origin.

Determining the *m*-vector of a bound *m*-vector

The hybrid metric that we have chosen to explore has the property that the origin is orthogonal to any m-vector. Hence if we take the interior product of any bound m-vector with the origin the result will be to 'unbind' or 'free' the m-vector.

$$(P \wedge A) \ominus \bigstar\mathbb{O} == ((\bigstar\mathbb{O} + x) \wedge A) \ominus \bigstar\mathbb{O} == (\bigstar\mathbb{O} \wedge A) \ominus \bigstar\mathbb{O} + (x \wedge A) \ominus \bigstar\mathbb{O} == A$$

The second term is zero and by [6.92] the first term is equal to A, so we have

$$(P \wedge A) \ominus \bigstar\mathbb{O} == A \qquad 6.98$$

◆ Example

Suppose we are in a bound 3-space, and we consider the bound bivector expressed as the product of three points: $P \wedge Q \wedge R$. Such a 3-element could, for example, define a plane. We want to find the bivector.

```
★P3; P = ★O + e1; Q = ★O + e2; R = ★O + e3; Π = P ∧ Q ∧ R;
```

Taking the interior product of Π with $\bigstar\mathbb{O}$ and then expanding and simplifying gives four terms.

```
B1 = ★G[Π ⊖ ★O]
```

$$\bigstar\mathbb{O} \wedge e_1 \wedge e_2 \ominus \bigstar\mathbb{O} - \bigstar\mathbb{O} \wedge e_1 \wedge e_3 \ominus \bigstar\mathbb{O} + \bigstar\mathbb{O} \wedge e_2 \wedge e_3 \ominus \bigstar\mathbb{O} + e_1 \wedge e_2 \wedge e_3 \ominus \bigstar\mathbb{O}$$

We can convert these to scalar products using `ToScalarProducts` (which is based on the Interior Common Factor Theorem).

```
B2 = ToScalarProducts[B1]
```

$$\bigstar\mathbb{O} \wedge ((\bigstar\mathbb{O} \ominus e_2 - \bigstar\mathbb{O} \ominus e_3)\, e_1 + (-(\bigstar\mathbb{O} \ominus e_1) + \bigstar\mathbb{O} \ominus e_3)\, e_2 + (\bigstar\mathbb{O} \ominus e_1 - \bigstar\mathbb{O} \ominus e_2)\, e_3) + (\bigstar\mathbb{O} \ominus \bigstar\mathbb{O} + \bigstar\mathbb{O} \ominus e_3)\, e_1 \wedge e_2 + (-(\bigstar\mathbb{O} \ominus \bigstar\mathbb{O}) - \bigstar\mathbb{O} \ominus e_2)\, e_1 \wedge e_3 + (\bigstar\mathbb{O} \ominus \bigstar\mathbb{O} + \bigstar\mathbb{O} \ominus e_1)\, e_2 \wedge e_3$$

Now, when we apply ToMetricElements, all the scalar products of basis vectors with the origin are put to zero, and we get the required bivector.

B_3 = ToMetricElements[B_2]

$e_1 \wedge e_2 - e_1 \wedge e_3 + e_2 \wedge e_3$

This bivector is simple, and if we wish, we can factorize it.

B_4 = ExteriorFactorize[B_3]

$(e_1 - e_3) \wedge (e_2 - e_3)$

In this case we can rewrite this bivector in terms of points just as we would if we were deriving it geometrically from the plane of $P \wedge Q \wedge R$.

$$B_5 == (P - R) \wedge (Q - R)$$

Expanding this bivector gives an important form related to the original simple exterior product $P \wedge Q \wedge R$.

$$B_6 == P \wedge Q - P \wedge R + Q \wedge R$$

This form was discussed in section [4.6] on *m*-Planes: The *m*-vector of a bound *m*-vector.

Exploring a bound measure

As we have seen so far, we can compare the relative magnitudes of elements, bound or free, by comparing their measures. However, a bound element possesses another quality over a free element - its position. In this section we explore whether there might be some sort of measure of a bound element which also captures this relative position. One possibility is to construct a weighted point out of a bound *m*-element $P \wedge A$ by taking its interior product with the unit *m*-element $\hat{A}$.

Let us begin by looking at the interior product of $P \wedge A$ with A.

$$(P \wedge A) \ominus A \quad == \quad ((\star\mathbb{O} + x) \wedge A) \ominus A \quad == \quad (\star\mathbb{O} \wedge A) \ominus A + (x \wedge A) \ominus A$$

In example [6.54] if we replace *m* by *m*+1, put α_1 equal to $\star\mathbb{O}$, and $\underset{m}{\beta}$ and $\alpha_2 \wedge \cdots \wedge \alpha_{m+1}$ equal to A, we get $(\star\mathbb{O} \wedge A) \ominus A$ as the sum of terms

$$(\alpha_1 \wedge \alpha_2 \wedge \cdots \wedge \alpha_{m+1}) \ominus A \quad == \quad \sum_{i=1}^{m+1} (-1)^{m+1-i} \left((\alpha_1 \wedge \cdots \wedge \square_i \wedge \cdots \wedge \alpha_{m+1}) \ominus A \right) \alpha_i$$

But the first term on the right hand side is the only non-zero term because the origin does not figure in the inner product. The terms in which the origin does figure in the inner product are zero because they can be rearranged by the extended interior products formula [6.40] to involve the zero interior product $A \ominus \star\mathbb{O}$ as a factor. Thus we can write

$$(P \wedge A) \ominus A \quad == \quad (-1)^m (A \ominus A) \star\mathbb{O} + (x \wedge A) \ominus A \quad == \quad (A \ominus A) \left((-1)^m \star\mathbb{O} + \left(x \wedge \hat{A}\right) \ominus \hat{A} \right)$$

In section 6.11 below we will see that $\left(x \wedge \hat{A}\right) \ominus \hat{A}$ is $(-1)^m$ times the component of x orthogonal to A so that $(P \wedge A) \ominus A$ now becomes

$$(P \wedge A) \ominus A \quad == \quad (-1)^m (A \ominus A) (\star\mathbb{O} + x^{\perp})$$

Dividing through by the measure of A finally gives us the bound measure of $P \wedge A$ as its interior product with $\hat{A}$.

$$(P \wedge A) \ominus \hat{A} == (-1)^m \,|A|\, P^{\llcorner} \qquad P^{\llcorner} == \star\mathcal{O} + x^{\llcorner} \qquad A \ominus x^{\llcorner} == 0 \tag{6.99}$$

Hence we have defined the *bound measure* of a bound element $P \wedge A$ as the weighted point whose weight is (apart from a possible sign) the measure of A, and whose position is the point $P^{\llcorner}$ on $P \wedge A$ closest to the origin. The sign is positive for A even, and negative for A odd.

This is only an exploration and this concept of bound measure may have no use at all!

6.7 Induced Metric Tensors

The metric tensor as a tensor of inner products

Chapter 5 showed how the notions of metric and induced metric could be defined and explored using the exterior, regressive and complement operations. This enabled us to get an insight into the fundamental mechanisms underpinning the comparison and measurement of elements of a linear space. However, there kept arising expressions of the form $A \vee \overline{B}$, where A and B are of the same grade; and particularly where A and B were basis *m*-elements of the same space. We have since called these expressions inner products, and we will find in what follows that they are the natural mechanism, both conceptually and algebraically, with which to explore metrics.

The elements of the metric tensor of $\underset{m}{\Lambda}$ are simply inner products of basis elements of $\underset{m}{\Lambda}$.

The matrix of elements G_m of the metric tensor of $\underset{m}{\Lambda}$ is the outer square of the list of basis elements of $\underset{m}{\Lambda}$.

We use the term *outer square* of a list to denote the outer product of the list with itself - but using the *interior product* as the product operation between the list elements.

The outer product here is the list equivalent of the tensor outer product, as implemented by *Mathematica's* `Outer` function.

For example, the matrix G_1 (also denoted G) in a 3-space is

```
★𝓑3; MatrixForm[Outer[InteriorProduct, Basis, Basis]]
```

$$\begin{pmatrix} e_1 \ominus e_1 & e_1 \ominus e_2 & e_1 \ominus e_3 \\ e_2 \ominus e_1 & e_2 \ominus e_2 & e_2 \ominus e_3 \\ e_3 \ominus e_1 & e_3 \ominus e_2 & e_3 \ominus e_3 \end{pmatrix}$$

And the matrix G_2 of metric elements induced on $\underset{2}{\Lambda}$ is

```
ShowPrecedence;
MatrixForm[Outer[InteriorProduct, GradeBasis[2], GradeBasis[2]]]
```

$$\begin{pmatrix} (e_1 \wedge e_2) \ominus (e_1 \wedge e_2) & (e_1 \wedge e_2) \ominus (e_1 \wedge e_3) & (e_1 \wedge e_2) \ominus (e_2 \wedge e_3) \\ (e_1 \wedge e_3) \ominus (e_1 \wedge e_2) & (e_1 \wedge e_3) \ominus (e_1 \wedge e_3) & (e_1 \wedge e_3) \ominus (e_2 \wedge e_3) \\ (e_2 \wedge e_3) \ominus (e_1 \wedge e_2) & (e_2 \wedge e_3) \ominus (e_1 \wedge e_3) & (e_2 \wedge e_3) \ominus (e_2 \wedge e_3) \end{pmatrix}$$

Such arrays, and hence induced metric tensors, will always be symmetric because inner products are symmetric.

Induced metric tensors

The metric G_m is said to be *induced* onto $\underset{m}{\Lambda}$ by the metric G on $\underset{1}{\Lambda}$ because G_m can be derived from the elements of G. This is simply because any inner product can always be decomposed into sums of products of scalar products - the determinant of its scalar product matrix.

For example we can write $(e_1 \wedge e_2) \ominus (e_1 \wedge e_3)$ as

$$(e_1 \wedge e_2) \ominus (e_1 \wedge e_3) == \begin{vmatrix} e_1 \ominus e_1 & e_1 \ominus e_3 \\ e_2 \ominus e_1 & e_2 \ominus e_3 \end{vmatrix}$$

In *GrassmannAlgebra* you can use `ComposeScalarProductMatrix` to construct a *matrix* of these scalar products.

```
MatrixForm[ComposeScalarProductMatrix[(e1 ∧ e2) ⊖ (e1 ∧ e3)]]
```

$$\begin{pmatrix} e_1 \ominus e_1 & e_1 \ominus e_3 \\ e_2 \ominus e_1 & e_2 \ominus e_3 \end{pmatrix}$$

To get the *determinant*, apply `Det` to the matrix.

```
Det[ComposeScalarProductMatrix[(e1 ∧ e2) ⊖ (e1 ∧ e3)]]
```

$$-(e_1 \ominus e_3)\ (e_2 \ominus e_1) + (e_1 \ominus e_1)\ (e_2 \ominus e_3)$$

Hence the expansion of the inner product becomes

$$(e_1 \wedge e_2) \ominus (e_1 \wedge e_3) == (e_1 \ominus e_1)\ (e_2 \ominus e_3) - (e_1 \ominus e_3)\ (e_2 \ominus e_1)$$

`ComposeScalarProductMatrix` may also be applied to a complete matrix of inner products. Consider again the matrix G_2 of metric elements induced on $\underset{2}{\Lambda}$.

```
G2 = Outer[InteriorProduct, GradeBasis[2], GradeBasis[2]];
ShowPrecedence; MatrixForm[G2]
```

$$\begin{pmatrix} (e_1 \wedge e_2) \ominus (e_1 \wedge e_2) & (e_1 \wedge e_2) \ominus (e_1 \wedge e_3) & (e_1 \wedge e_2) \ominus (e_2 \wedge e_3) \\ (e_1 \wedge e_3) \ominus (e_1 \wedge e_2) & (e_1 \wedge e_3) \ominus (e_1 \wedge e_3) & (e_1 \wedge e_3) \ominus (e_2 \wedge e_3) \\ (e_2 \wedge e_3) \ominus (e_1 \wedge e_2) & (e_2 \wedge e_3) \ominus (e_1 \wedge e_3) & (e_2 \wedge e_3) \ominus (e_2 \wedge e_3) \end{pmatrix}$$

```
sG2 = ComposeScalarProductMatrix[G2]; MatrixForm[sG2]
```

$$\begin{pmatrix} \begin{pmatrix} e_1 \ominus e_1 & e_1 \ominus e_2 \\ e_2 \ominus e_1 & e_2 \ominus e_2 \end{pmatrix} & \begin{pmatrix} e_1 \ominus e_1 & e_1 \ominus e_3 \\ e_2 \ominus e_1 & e_2 \ominus e_3 \end{pmatrix} & \begin{pmatrix} e_1 \ominus e_2 & e_1 \ominus e_3 \\ e_2 \ominus e_2 & e_2 \ominus e_3 \end{pmatrix} \\ \begin{pmatrix} e_1 \ominus e_1 & e_1 \ominus e_2 \\ e_3 \ominus e_1 & e_3 \ominus e_2 \end{pmatrix} & \begin{pmatrix} e_1 \ominus e_1 & e_1 \ominus e_3 \\ e_3 \ominus e_1 & e_3 \ominus e_3 \end{pmatrix} & \begin{pmatrix} e_1 \ominus e_2 & e_1 \ominus e_3 \\ e_3 \ominus e_2 & e_3 \ominus e_3 \end{pmatrix} \\ \begin{pmatrix} e_2 \ominus e_1 & e_2 \ominus e_2 \\ e_3 \ominus e_1 & e_3 \ominus e_2 \end{pmatrix} & \begin{pmatrix} e_2 \ominus e_1 & e_2 \ominus e_3 \\ e_3 \ominus e_1 & e_3 \ominus e_3 \end{pmatrix} & \begin{pmatrix} e_2 \ominus e_2 & e_2 \ominus e_3 \\ e_3 \ominus e_2 & e_3 \ominus e_3 \end{pmatrix} \end{pmatrix}$$

The induced metric is the matrix of the *determinants* of the submatrices. We can display this specifically as

```
M = MatrixForm[Map[BracketingBar[Grid[#]] &, sG2, {2}]]
```

$$\begin{pmatrix} \begin{vmatrix} e_1 \ominus e_1 & e_1 \ominus e_2 \\ e_2 \ominus e_1 & e_2 \ominus e_2 \end{vmatrix} & \begin{vmatrix} e_1 \ominus e_1 & e_1 \ominus e_3 \\ e_2 \ominus e_1 & e_2 \ominus e_3 \end{vmatrix} & \begin{vmatrix} e_1 \ominus e_2 & e_1 \ominus e_3 \\ e_2 \ominus e_2 & e_2 \ominus e_3 \end{vmatrix} \\ \begin{vmatrix} e_1 \ominus e_1 & e_1 \ominus e_2 \\ e_3 \ominus e_1 & e_3 \ominus e_2 \end{vmatrix} & \begin{vmatrix} e_1 \ominus e_1 & e_1 \ominus e_3 \\ e_3 \ominus e_1 & e_3 \ominus e_3 \end{vmatrix} & \begin{vmatrix} e_1 \ominus e_2 & e_1 \ominus e_3 \\ e_3 \ominus e_2 & e_3 \ominus e_3 \end{vmatrix} \\ \begin{vmatrix} e_2 \ominus e_1 & e_2 \ominus e_2 \\ e_3 \ominus e_1 & e_3 \ominus e_2 \end{vmatrix} & \begin{vmatrix} e_2 \ominus e_1 & e_2 \ominus e_3 \\ e_3 \ominus e_1 & e_3 \ominus e_3 \end{vmatrix} & \begin{vmatrix} e_2 \ominus e_2 & e_2 \ominus e_3 \\ e_3 \ominus e_2 & e_3 \ominus e_3 \end{vmatrix} \end{pmatrix}$$

Converting to metric elements

The standard notation for the element $e_i \ominus e_j$ of the metric tensor of $\underset{1}{\Lambda}$ is g_{ij}. In this book we will also use $g_{i,j}$ so that *Mathematica* can distinguish the indices. To convert from the basis notation to the standard notation you can use the *GrassmannAlgebra* operation `ToMetricElements`.

For example to convert the matrix `M` of the section above to one in standard notation just apply `ToMetricElements` to it.

```
DeclareMetric[g]; ToMetricElements[M]
```

$$\begin{pmatrix} \begin{vmatrix} g_{1,1} & g_{1,2} \\ g_{1,2} & g_{2,2} \end{vmatrix} & \begin{vmatrix} g_{1,1} & g_{1,3} \\ g_{1,2} & g_{2,3} \end{vmatrix} & \begin{vmatrix} g_{1,2} & g_{1,3} \\ g_{2,2} & g_{2,3} \end{vmatrix} \\ \begin{vmatrix} g_{1,1} & g_{1,2} \\ g_{1,3} & g_{2,3} \end{vmatrix} & \begin{vmatrix} g_{1,1} & g_{1,3} \\ g_{1,3} & g_{3,3} \end{vmatrix} & \begin{vmatrix} g_{1,2} & g_{1,3} \\ g_{2,3} & g_{3,3} \end{vmatrix} \\ \begin{vmatrix} g_{1,2} & g_{2,2} \\ g_{1,3} & g_{2,3} \end{vmatrix} & \begin{vmatrix} g_{1,2} & g_{2,3} \\ g_{1,3} & g_{3,3} \end{vmatrix} & \begin{vmatrix} g_{2,2} & g_{2,3} \\ g_{2,3} & g_{3,3} \end{vmatrix} \end{pmatrix}$$

This standard notation is generally used in theoretical calculations, and often when an unspecified metric is needed. In *GrassmannAlgebra*, applying `DeclareMetric` to an undefined symbol as we have done here will generate such an unspecified metric.

However, you can also declare a *specific* metric in *GrassmannAlgebra*, so that whenever you apply `ToMetricElements` to an expression involving inner or scalar products of basis elements it substitutes the specific scalars (or scalar functions) you have declared. For example

```
{MatrixForm[DeclareMetric[{{a, 0, 1}, {0, b, 0}, {1, 0, c}}]],
 ToMetricElements[M]}
```

$$\left\{\begin{pmatrix} a & 0 & 1 \\ 0 & b & 0 \\ 1 & 0 & c \end{pmatrix}, \begin{pmatrix} \begin{vmatrix} a & 0 \\ 0 & b \end{vmatrix} & \begin{vmatrix} a & 1 \\ 0 & 0 \end{vmatrix} & \begin{vmatrix} 0 & 1 \\ b & 0 \end{vmatrix} \\ \begin{vmatrix} a & 0 \\ 1 & 0 \end{vmatrix} & \begin{vmatrix} a & 1 \\ 1 & c \end{vmatrix} & \begin{vmatrix} 0 & 1 \\ 0 & c \end{vmatrix} \\ \begin{vmatrix} 0 & b \\ 1 & 0 \end{vmatrix} & \begin{vmatrix} 0 & 0 \\ 1 & c \end{vmatrix} & \begin{vmatrix} b & 0 \\ 0 & c \end{vmatrix} \end{pmatrix}\right\}$$

- It should be noted that until we specify scalars for the $e_i \ominus e_j$, or constrain the $e_i \ominus e_j$ amongst themselves in some way, then *we have not specified a metric for the space*.

Displaying induced metric tensors as a matrix of matrices

GrassmannAlgebra has an inbuilt function `MetricMatrix` for displaying an induced metric in matrix form. For example, we can display the induced metric on $\underset{3}{\Lambda}$ in a 4-space.

```
★B4; DeclareMetric[g]
```

$\{\{g_{1,1}, g_{1,2}, g_{1,3}, g_{1,4}\}, \{g_{1,2}, g_{2,2}, g_{2,3}, g_{2,4}\},$
$\{g_{1,3}, g_{2,3}, g_{3,3}, g_{3,4}\}, \{g_{1,4}, g_{2,4}, g_{3,4}, g_{4,4}\}\}$

```
MetricMatrix[3] // MatrixForm
```

$$\begin{pmatrix}
\begin{pmatrix} g_{1,1} & g_{1,2} & g_{1,3} \\ g_{1,2} & g_{2,2} & g_{2,3} \\ g_{1,3} & g_{2,3} & g_{3,3} \end{pmatrix} &
\begin{pmatrix} g_{1,1} & g_{1,2} & g_{1,4} \\ g_{1,2} & g_{2,2} & g_{2,4} \\ g_{1,3} & g_{2,3} & g_{3,4} \end{pmatrix} &
\begin{pmatrix} g_{1,1} & g_{1,3} & g_{1,4} \\ g_{1,2} & g_{2,3} & g_{2,4} \\ g_{1,3} & g_{3,3} & g_{3,4} \end{pmatrix} &
\begin{pmatrix} g_{1,2} & g_{1,3} & g_{1,4} \\ g_{2,2} & g_{2,3} & g_{2,4} \\ g_{2,3} & g_{3,3} & g_{3,4} \end{pmatrix} \\
\begin{pmatrix} g_{1,1} & g_{1,2} & g_{1,3} \\ g_{1,2} & g_{2,2} & g_{2,3} \\ g_{1,4} & g_{2,4} & g_{3,4} \end{pmatrix} &
\begin{pmatrix} g_{1,1} & g_{1,2} & g_{1,4} \\ g_{1,2} & g_{2,2} & g_{2,4} \\ g_{1,4} & g_{2,4} & g_{4,4} \end{pmatrix} &
\begin{pmatrix} g_{1,1} & g_{1,3} & g_{1,4} \\ g_{1,2} & g_{2,3} & g_{2,4} \\ g_{1,4} & g_{3,4} & g_{4,4} \end{pmatrix} &
\begin{pmatrix} g_{1,2} & g_{1,3} & g_{1,4} \\ g_{2,2} & g_{2,3} & g_{2,4} \\ g_{2,4} & g_{3,4} & g_{4,4} \end{pmatrix} \\
\begin{pmatrix} g_{1,1} & g_{1,2} & g_{1,3} \\ g_{1,3} & g_{2,3} & g_{3,3} \\ g_{1,4} & g_{2,4} & g_{3,4} \end{pmatrix} &
\begin{pmatrix} g_{1,1} & g_{1,2} & g_{1,4} \\ g_{1,3} & g_{2,3} & g_{3,4} \\ g_{1,4} & g_{2,4} & g_{4,4} \end{pmatrix} &
\begin{pmatrix} g_{1,1} & g_{1,3} & g_{1,4} \\ g_{1,3} & g_{3,3} & g_{3,4} \\ g_{1,4} & g_{3,4} & g_{4,4} \end{pmatrix} &
\begin{pmatrix} g_{1,2} & g_{1,3} & g_{1,4} \\ g_{2,3} & g_{3,3} & g_{3,4} \\ g_{2,4} & g_{3,4} & g_{4,4} \end{pmatrix} \\
\begin{pmatrix} g_{1,2} & g_{2,2} & g_{2,3} \\ g_{1,3} & g_{2,3} & g_{3,3} \\ g_{1,4} & g_{2,4} & g_{3,4} \end{pmatrix} &
\begin{pmatrix} g_{1,2} & g_{2,2} & g_{2,4} \\ g_{1,3} & g_{2,3} & g_{3,4} \\ g_{1,4} & g_{2,4} & g_{4,4} \end{pmatrix} &
\begin{pmatrix} g_{1,2} & g_{2,3} & g_{2,4} \\ g_{1,3} & g_{3,3} & g_{3,4} \\ g_{1,4} & g_{3,4} & g_{4,4} \end{pmatrix} &
\begin{pmatrix} g_{2,2} & g_{2,3} & g_{2,4} \\ g_{2,3} & g_{3,3} & g_{3,4} \\ g_{2,4} & g_{3,4} & g_{4,4} \end{pmatrix}
\end{pmatrix}$$

Of course, we are only *displaying* the metric tensor in this form. Its elements are correctly the *determinants* of the sub-matrices displayed, not the matrices themselves.

Calculating induced metric tensors

In the last chapter we saw how we could use the *GrassmannAlgebra* function `GradeMetric[m]` to calculate the matrix of metric tensor elements induced on $\underset{m}{\Lambda}$ by the metric tensor declared on $\underset{1}{\Lambda}$. For example, you can declare a general metric in a 3-space:

```
★B3; DeclareMetric[g]
```

$$\{\{g_{1,1}, g_{1,2}, g_{1,3}\}, \{g_{1,2}, g_{2,2}, g_{2,3}\}, \{g_{1,3}, g_{2,3}, g_{3,3}\}\}$$

and then ask for the metric induced on $\underset{2}{\Lambda}$.

```
GradeMetric[2]
```

$$\{\{-g_{1,2}^2 + g_{1,1}\, g_{2,2},\ -g_{1,2}\, g_{1,3} + g_{1,1}\, g_{2,3},\ -g_{1,3}\, g_{2,2} + g_{1,2}\, g_{2,3}\},$$
$$\{-g_{1,2}\, g_{1,3} + g_{1,1}\, g_{2,3},\ -g_{1,3}^2 + g_{1,1}\, g_{3,3},\ -g_{1,3}\, g_{2,3} + g_{1,2}\, g_{3,3}\},$$
$$\{-g_{1,3}\, g_{2,2} + g_{1,2}\, g_{2,3},\ -g_{1,3}\, g_{2,3} + g_{1,2}\, g_{3,3},\ -g_{2,3}^2 + g_{2,2}\, g_{3,3}\}\}$$

You can also get a palette of induced metrics by entering `MetricPalette`. For example in a 2-space

```
★B2; DeclareMetric[g]; MetricPalette
```

Metric Palette

$\underset{0}{\Lambda}$	(1)
$\underset{1}{\Lambda}$	$\begin{pmatrix} g_{1,1} & g_{1,2} \\ g_{1,2} & g_{2,2} \end{pmatrix}$
$\underset{2}{\Lambda}$	$(-g_{1,2}^2 + g_{1,1}\, g_{2,2})$

But you can also get the metric palette in basis form by doing a simple substitution.

```
MetricPalette /. gi_,j_ :→ Basis[[i]] ⊖ Basis[[j]]
```

Metric Palette

$\overset{0}{\Lambda}$	(1)
$\overset{1}{\Lambda}$	$\begin{pmatrix} e_1 \ominus e_1 & e_1 \ominus e_2 \\ e_1 \ominus e_2 & e_2 \ominus e_2 \end{pmatrix}$
$\overset{2}{\Lambda}$	$(-(e_1 \ominus e_2)^2 + (e_1 \ominus e_1)\,(e_2 \ominus e_2))$

6.8 Product Formulae for Interior Products

The basic Interior Product Formula

In discussing product formulae for the interior product we refer to section 3.10 of Chapter 3 on Product Formulae for Regressive Products. In transforming the results of that section to results involving interior products, we may often simply transform expressions or their duals involving regressive products with $(n-m)$-elements $\square \vee \underset{n-m}{x}$ into expressions involving interior products with m-elements $\square \ominus \underset{m}{x}$.

$$\square \vee \underset{n-m}{x} \;\Longrightarrow\; \square \vee \overline{\underset{m}{x}} \;\Longrightarrow\; \square \ominus \underset{m}{x}$$

The Interior Common Factor Theorem theorem plays the same role in this development that the Common Factor Theorem did in Chapter 3.

◆ The basic Interior Product Formula

The basic Interior Product Formula is a transformation of the first regressive Product Formula [3.68] in Chapter 3. Beginning with this formula

$$\left(\underset{m}{\alpha} \wedge \underset{k}{\beta}\right) \vee \underset{n-1}{x} == \left(\underset{m}{\alpha} \vee \underset{n-1}{x}\right) \wedge \underset{k}{\beta} + (-1)^m \underset{m}{\alpha} \wedge \left(\underset{k}{\beta} \vee \underset{n-1}{x}\right)$$

replace $\square \vee \underset{n-1}{x}$ with $\square \ominus x$ to get

$$\left(\underset{m}{\alpha} \wedge \underset{k}{\beta}\right) \ominus x == \left(\underset{m}{\alpha} \ominus x\right) \wedge \underset{k}{\beta} + (-1)^m \underset{m}{\alpha} \wedge \left(\underset{k}{\beta} \ominus x\right) \qquad 6.100$$

◆ Derivation using the Interior Common Factor Theorem

The basic Interior Product Formula may also be shown (for simple elements) using the Interior Common Factor Theorem in the form [6.52].

$$(\alpha_1 \wedge \alpha_2 \wedge \cdots \wedge \alpha_m) \ominus \beta == \sum_{i=1}^{m} (\alpha_i \ominus \beta)\,(-1)^{i-1}\,\alpha_1 \wedge \cdots \wedge \square_i \wedge \cdots \wedge \alpha_m$$

In this formula put β equal to x, and $\alpha_1 \wedge \alpha_2 \wedge \cdots \wedge \alpha_m$ equal to $(\alpha_1 \wedge \cdots \wedge \alpha_m) \wedge (\beta_1 \wedge \cdots \wedge \beta_k)$.

$$\left(\underset{m}{\alpha} \wedge \underset{k}{\beta}\right) \ominus x == ((\alpha_1 \wedge \cdots \wedge \alpha_m) \wedge (\beta_1 \wedge \cdots \wedge \beta_k)) \ominus x$$

$$== \sum_{i==1}^{m} (-1)^{i-1} (\alpha_i \ominus x) (\alpha_1 \wedge \cdots \wedge \square_i \wedge \cdots \wedge \alpha_m) \wedge (\beta_1 \wedge \cdots \wedge \beta_k)$$

$$+ \sum_{j==1}^{k} (-1)^{m+j-1} (\beta_j \ominus x) (\alpha_1 \wedge \cdots \wedge \alpha_m) \wedge (\beta_1 \wedge \cdots \wedge \square_j \wedge \cdots \wedge \beta_k)$$

$$== \left(\underset{m}{\alpha} \ominus x\right) \wedge (\beta_1 \wedge \cdots \wedge \beta_k) + (-1)^m (\alpha_1 \wedge \cdots \wedge \alpha_m) \wedge \left(\underset{k}{\beta} \ominus x\right)$$

$$== \left(\underset{m}{\alpha} \ominus x\right) \wedge \underset{k}{\beta} + (-1)^m \underset{m}{\alpha} \wedge \left(\underset{k}{\beta} \ominus x\right)$$

Deriving Interior Product Formulae

◆ Deriving Interior Product Formulae manually

If x is of a grade higher than 1, then similar relations hold, but with extra terms on the right-hand side. For example, if we replace x by $x_1 \wedge x_2$ and note that

$$\left(\underset{m}{\alpha} \wedge \underset{k}{\beta}\right) \ominus (x_1 \wedge x_2) == \left(\left(\underset{m}{\alpha} \wedge \underset{k}{\beta}\right) \ominus x_1\right) \ominus x_2$$

then the right-hand side may be expanded by applying the basic Interior Product Formula [6.100] twice. The first application is

$$\left(\underset{m}{\alpha} \wedge \underset{k}{\beta}\right) \ominus (x_1 \wedge x_2) == \left(\left(\underset{m}{\alpha} \ominus x_1\right) \wedge \underset{k}{\beta} + (-1)^m \underset{m}{\alpha} \wedge \left(\underset{k}{\beta} \ominus x_1\right)\right) \ominus x_2$$

The second application is to each term on the right hand side.

$$\left(\left(\underset{m}{\alpha} \ominus x_1\right) \wedge \underset{k}{\beta}\right) \ominus x_2 == \left(\left(\underset{m}{\alpha} \ominus x_1\right) \ominus x_2\right) \wedge \underset{k}{\beta} + (-1)^{m-1} \left(\underset{m}{\alpha} \ominus x_1\right) \wedge \left(\underset{k}{\beta} \ominus x_2\right)$$

$$(-1)^m \left(\underset{m}{\alpha} \wedge \left(\underset{k}{\beta} \ominus x_1\right)\right) \ominus x_2 == (-1)^m \left(\underset{m}{\alpha} \ominus x_2\right) \wedge \left(\underset{k}{\beta} \ominus x_1\right) + \underset{m}{\alpha} \wedge \left(\left(\underset{k}{\beta} \ominus x_1\right) \ominus x_2\right)$$

Adding these equations and remembering that $\square \ominus (x_1 \wedge x_2) = (\square \ominus x_1) \ominus x_2$ gives

$$\left(\underset{m}{\alpha} \wedge \underset{k}{\beta}\right) \ominus (x_1 \wedge x_2) == \left(\underset{m}{\alpha} \ominus (x_1 \wedge x_2)\right) \wedge \underset{k}{\beta} + \underset{m}{\alpha} \wedge \left(\underset{k}{\beta} \ominus (x_1 \wedge x_2)\right)$$
$$- (-1)^m \left(\left(\underset{m}{\alpha} \ominus x_1\right) \wedge \left(\underset{k}{\beta} \ominus x_2\right) - \left(\underset{m}{\alpha} \ominus x_2\right) \wedge \left(\underset{k}{\beta} \ominus x_1\right)\right) \qquad 6.101$$

As with the product formula for regressive products, successive application doubles the number of terms. We started with two terms on the right hand side of the basic Interior Product Formula. By applying it again we obtain a Product Formula with four terms on the right hand side. The next application would give us eight terms.

Thus the Product Formula for $\left(\underset{m}{\alpha} \wedge \underset{k}{\beta}\right) \ominus (x_1 \wedge \ldots \wedge x_j)$ would lead to 2^j terms.

◆ Deriving Interior Product Formulae from the dual regressive Product Formulae

A second derivation may be obtained directly by first computing the `Dual` of a regressive Product Formula, and then directly using the substitution

$$\square \vee \underset{n-m}{x} \Longrightarrow \square \vee \underset{m}{\overline{x}} \Longrightarrow \square \ominus \underset{m}{x}$$

For example, here is a regressive Product Formula.

```
ShowPrecedence; F1 = DeriveProductFormula[(α_m ∨ β_k) ∧ x_2]
```

$$(\underset{m}{\alpha} \vee \underset{k}{\beta}) \wedge x_1 \wedge x_2 == \underset{m}{\alpha} \vee (\underset{k}{\beta} \wedge x_1 \wedge x_2) +$$
$$(-1)^{\star n-m}\left(-\left((\underset{m}{\alpha} \wedge x_1) \vee (\underset{k}{\beta} \wedge x_2)\right) + (\underset{m}{\alpha} \wedge x_2) \vee (\underset{k}{\beta} \wedge x_1)\right) + (\underset{m}{\alpha} \wedge x_1 \wedge x_2) \vee \underset{k}{\beta}$$

Compute its dual.

```
F2 = Dual[F1]
```

$$(\underset{m}{\alpha} \wedge \underset{k}{\beta}) \vee \underset{-1+\star n}{x_1} \vee \underset{-1+\star n}{x_2} == \underset{m}{\alpha} \wedge (\underset{k}{\beta} \vee \underset{-1+\star n}{x_1} \vee \underset{-1+\star n}{x_2}) + (-1)^m$$
$$\left(-\left((\underset{m}{\alpha} \vee \underset{-1+\star n}{x_1}) \wedge (\underset{k}{\beta} \vee \underset{-1+\star n}{x_2})\right) + (\underset{m}{\alpha} \vee \underset{-1+\star n}{x_2}) \wedge (\underset{k}{\beta} \vee \underset{-1+\star n}{x_1})\right) + (\underset{m}{\alpha} \vee \underset{-1+\star n}{x_1} \vee \underset{-1+\star n}{x_2}) \wedge \underset{k}{\beta}$$

Convert the regressive products into interior products.

```
F3 = F2 //. A_ ∨ x_⋆n-1 :→ A ⊖ x
```

$$((\underset{m}{\alpha} \wedge \underset{k}{\beta}) \ominus x_1) \ominus x_2 == ((\underset{m}{\alpha} \ominus x_1) \ominus x_2) \wedge \underset{k}{\beta} +$$
$$(-1)^m\left(-\left((\underset{m}{\alpha} \ominus x_1) \wedge (\underset{k}{\beta} \ominus x_2)\right) + (\underset{m}{\alpha} \ominus x_2) \wedge (\underset{k}{\beta} \ominus x_1)\right) + \underset{m}{\alpha} \wedge ((\underset{k}{\beta} \ominus x_1) \ominus x_2)$$

Use the extended interior products formula [6.40] to put the result in the final form.

```
F3 /. (Z_ ⊖ x_) ⊖ y_ :→ Z ⊖ (x ∧ y)
```

$$(\underset{m}{\alpha} \wedge \underset{k}{\beta}) \ominus (x_1 \wedge x_2) == (-1)^m\left(-\left((\underset{m}{\alpha} \ominus x_1) \wedge (\underset{k}{\beta} \ominus x_2)\right) + (\underset{m}{\alpha} \ominus x_2) \wedge (\underset{k}{\beta} \ominus x_1)\right) +$$
$$(\underset{m}{\alpha} \ominus (x_1 \wedge x_2)) \wedge \underset{k}{\beta} + \underset{m}{\alpha} \wedge (\underset{k}{\beta} \ominus (x_1 \wedge x_2))$$

◆ Deriving Interior Product Formulae sequentially

We can automate this process by encoding the two rules used in the derivation above and having *Mathematica* apply them.

The first rule is the basic Interior Product Formula [6.100].

$$\left(\underset{m}{\alpha} \wedge \underset{k}{\beta}\right) \ominus x == \left(\underset{m}{\alpha} \ominus x\right) \wedge \underset{k}{\beta} + (-1)^m \underset{m}{\alpha} \wedge \left(\underset{k}{\beta} \ominus x\right)$$

which we encode as R_1.

```
R1 := (s_. (A_ ∧ B_) ⊖ x_ :→ s (A ⊖ x) ∧ B + s (-1)^⋆G[A] A ∧ (B ⊖ x))
```

The second rule is the extended interior product formula [6.40]

$$(\mathbf{Z} \ominus \mathbf{x}) \ominus \mathbf{y} \; == \; \mathbf{Z} \ominus (\mathbf{x} \wedge \mathbf{y})$$

which we encode as R_2.

$$\mathbf{R_2} := ((\mathbf{Z\_} \ominus \mathbf{x\_}) \ominus \mathbf{y\_} :\to \mathbf{Z} \ominus (\mathbf{x} \wedge \mathbf{y}))$$

We will also need to declare the x_i as extra vector symbols.

$$\star\star\mathbf{V}[\mathbf{x\_}];$$

To apply these rules, start with an exterior product. Call it F_0.

$$\mathbf{F_0} = \left(\underset{m}{\alpha} \wedge \underset{k}{\beta}\right);$$

Take the interior product of F_0 with x_1, and apply both rules as many times as they are applicable to get F_1.

$$\mathbf{F_1} = \mathbf{F_0} \ominus \mathbf{x_1} \; //. \; \{\mathbf{R_1}, \mathbf{R_2}\}$$

$$(\underset{m}{\alpha} \ominus x_1) \wedge \underset{k}{\beta} + (-1)^m \underset{m}{\alpha} \wedge (\underset{k}{\beta} \ominus x_1)$$

Repeat by taking the interior product of F_1 with x_2, expanding, and applying the rules to get F_2.

$$\mathbf{F_2} = \star\mathcal{E}[\mathbf{F_1} \ominus \mathbf{x_2}] \; //. \; \{\mathbf{R_1}, \mathbf{R_2}\}$$

$$(-1)^{-1+m} (\underset{m}{\alpha} \ominus x_1) \wedge (\underset{k}{\beta} \ominus x_2) + (-1)^m (\underset{m}{\alpha} \ominus x_2) \wedge (\underset{k}{\beta} \ominus x_1) +$$
$$(\underset{m}{\alpha} \ominus (x_1 \wedge x_2)) \wedge \underset{k}{\beta} + (-1)^{2m} \underset{m}{\alpha} \wedge (\underset{k}{\beta} \ominus (x_1 \wedge x_2))$$

We can simplify the signs by using *GrassmannAlgebra's* `SimplifyPowerSigns`.

$$\mathbf{F_2} = \star\mathcal{E}[\mathbf{F_1} \ominus \mathbf{x_2}] \; //. \; \{\mathbf{R_1}, \mathbf{R_2}\} \; // \; \mathbf{SimplifyPowerSigns}$$

$$(-1)^{1+m} (\underset{m}{\alpha} \ominus x_1) \wedge (\underset{k}{\beta} \ominus x_2) +$$
$$(-1)^m (\underset{m}{\alpha} \ominus x_2) \wedge (\underset{k}{\beta} \ominus x_1) + (\underset{m}{\alpha} \ominus (x_1 \wedge x_2)) \wedge \underset{k}{\beta} + \underset{m}{\alpha} \wedge (\underset{k}{\beta} \ominus (x_1 \wedge x_2))$$

To get succeeding formulae, we simply repeat this process.

$$\mathbf{F_3} = \star\mathcal{E}[\mathbf{F_2} \ominus \mathbf{x_3}] \; //. \; \{\mathbf{R_1}, \mathbf{R_2}\} \; // \; \mathbf{SimplifyPowerSigns}$$

$$(\underset{m}{\alpha} \ominus x_1) \wedge (\underset{k}{\beta} \ominus (x_2 \wedge x_3)) - (\underset{m}{\alpha} \ominus x_2) \wedge (\underset{k}{\beta} \ominus (x_1 \wedge x_3)) +$$
$$(\underset{m}{\alpha} \ominus x_3) \wedge (\underset{k}{\beta} \ominus (x_1 \wedge x_2)) + (-1)^m (\underset{m}{\alpha} \ominus (x_1 \wedge x_2)) \wedge (\underset{k}{\beta} \ominus x_3) +$$
$$(-1)^{1+m} (\underset{m}{\alpha} \ominus (x_1 \wedge x_3)) \wedge (\underset{k}{\beta} \ominus x_2) + (-1)^m (\underset{m}{\alpha} \ominus (x_2 \wedge x_3)) \wedge (\underset{k}{\beta} \ominus x_1) +$$
$$(\underset{m}{\alpha} \ominus (x_1 \wedge x_2 \wedge x_3)) \wedge \underset{k}{\beta} + (-1)^m \underset{m}{\alpha} \wedge (\underset{k}{\beta} \ominus (x_1 \wedge x_2 \wedge x_3))$$

Collecting the terms with the same signs gives

$$
\begin{aligned}
&\left(\underset{m}{\alpha} \wedge \underset{k}{\beta}\right) \ominus (x_1 \wedge x_2 \wedge x_3) == \\
&\left(\underset{m}{\alpha} \ominus x_1\right) \wedge \left(\underset{k}{\beta} \ominus (x_2 \wedge x_3)\right) - \\
&\left(\underset{m}{\alpha} \ominus x_2\right) \wedge \left(\underset{k}{\beta} \ominus (x_1 \wedge x_3)\right) + \left(\underset{m}{\alpha} \ominus x_3\right) \wedge \left(\underset{k}{\beta} \ominus (x_1 \wedge x_2)\right) + \\
&(-1)^m \left(\left(\underset{m}{\alpha} \ominus (x_1 \wedge x_2)\right) \wedge \left(\underset{k}{\beta} \ominus x_3\right) - \left(\underset{m}{\alpha} \ominus (x_1 \wedge x_3)\right) \wedge \left(\underset{k}{\beta} \ominus x_2\right) + \right. \\
&\left. \left(\underset{m}{\alpha} \ominus (x_2 \wedge x_3)\right) \wedge \left(\underset{k}{\beta} \ominus x_1\right) + \underset{m}{\alpha} \wedge \left(\underset{k}{\beta} \ominus (x_1 \wedge x_2 \wedge x_3)\right)\right) + \\
&\left(\underset{m}{\alpha} \ominus (x_1 \wedge x_2 \wedge x_3)\right) \wedge \underset{k}{\beta}
\end{aligned}
\qquad 6.102
$$

Deriving Interior Product Formulae automatically

Just as we did in Chapter 3, it is straightforward to encapsulate this sequential derivation in a function for automatic application. It is not necessary to enter these functions since the principle one `DeriveInteriorProductFormula` is built in to *GrassmannAlgebra*.

◆ 1. Define the rules to be applied at each step

```
DeriveInteriorProductFormulaOnce[F_, x_] :=
  GrassmannExpand[F ⊖ x] //. {s_. A_ ∧ B_ ⊖ z_ :→
    s (A ⊖ z) ∧ B + s (-1)^RawGrade[A] A ∧ (B ⊖ z), Z_ ⊖ z_ ⊖ y_ :→ Z ⊖ z ∧ y};
```

◆ 2. Apply the rules as many times as required

```
DeriveInteriorProductFormula[(A_ ∧ B_) ⊖ X_] :=
  (A ∧ B) ⊖ ComposeSimpleForm[X, 1] ==
   SimplifyPowerSigns[Fold[DeriveInteriorProductFormulaOnce,
     A ∧ B, {ComposeSimpleForm[X, 1]} /. Wedge → Sequence]];
```

◆ Check the results

$\star\star V[x\_]$; H_1 = DeriveInteriorProductFormula$\left[\left(\underset{m}{\alpha} \wedge \underset{k}{\beta}\right) \ominus x_1\right]$

$$(\underset{m}{\alpha} \wedge \underset{k}{\beta}) \ominus x_1 == (\underset{m}{\alpha} \ominus x_1) \wedge \underset{k}{\beta} + (-1)^m \underset{m}{\alpha} \wedge (\underset{k}{\beta} \ominus x_1)$$

H_2 = DeriveInteriorProductFormula$\left[\left(\underset{m}{\alpha} \wedge \underset{k}{\beta}\right) \ominus \underset{2}{x}\right]$

$$
\begin{aligned}
&(\underset{m}{\alpha} \wedge \underset{k}{\beta}) \ominus (x_1 \wedge x_2) == (-1)^{1+m} (\underset{m}{\alpha} \ominus x_1) \wedge (\underset{k}{\beta} \ominus x_2) + \\
&\quad (-1)^m (\underset{m}{\alpha} \ominus x_2) \wedge (\underset{k}{\beta} \ominus x_1) + (\underset{m}{\alpha} \ominus (x_1 \wedge x_2)) \wedge \underset{k}{\beta} + \underset{m}{\alpha} \wedge (\underset{k}{\beta} \ominus (x_1 \wedge x_2))
\end{aligned}
$$

$$H_3 = \text{DeriveInteriorProductFormula}\left[\left(\underset{m}{\alpha} \wedge \underset{k}{\beta}\right) \ominus \underset{3}{x}\right]$$

$$\left(\underset{m}{\alpha} \wedge \underset{k}{\beta}\right) \ominus (x_1 \wedge x_2 \wedge x_3) == \left(\underset{m}{\alpha} \ominus x_1\right) \wedge \left(\underset{k}{\beta} \ominus (x_2 \wedge x_3)\right) - \left(\underset{m}{\alpha} \ominus x_2\right) \wedge \left(\underset{k}{\beta} \ominus (x_1 \wedge x_3)\right) +$$
$$\left(\underset{m}{\alpha} \ominus x_3\right) \wedge \left(\underset{k}{\beta} \ominus (x_1 \wedge x_2)\right) + (-1)^m \left(\underset{m}{\alpha} \ominus (x_1 \wedge x_2)\right) \wedge \left(\underset{k}{\beta} \ominus x_3\right) +$$
$$(-1)^{1+m} \left(\underset{m}{\alpha} \ominus (x_1 \wedge x_3)\right) \wedge \left(\underset{k}{\beta} \ominus x_2\right) + (-1)^m \left(\underset{m}{\alpha} \ominus (x_2 \wedge x_3)\right) \wedge \left(\underset{k}{\beta} \ominus x_1\right) +$$
$$\left(\underset{m}{\alpha} \ominus (x_1 \wedge x_2 \wedge x_3)\right) \wedge \underset{k}{\beta} + (-1)^m \underset{m}{\alpha} \wedge \left(\underset{k}{\beta} \ominus (x_1 \wedge x_2 \wedge x_3)\right)$$

These give the same results as the previously manually derived formulae.

$$\left\{\left(\underset{m}{\alpha} \wedge \underset{k}{\beta} \ominus x_1 == F_1\right) === H_1,\right.$$
$$\left(\underset{m}{\alpha} \wedge \underset{k}{\beta} \ominus x_1 \wedge x_2 == F_2\right) === H_2,$$
$$\left.\left(\underset{m}{\alpha} \wedge \underset{k}{\beta} \ominus x_1 \wedge x_2 \wedge x_3 == F_3\right) === H_3\right\}$$

{True, True, True}

Exploring the computable form of the Interior Product Formula

If we inspect the different cases of the Interior Product Formulae for example [6.101] and [6.102]. we can see that the terms are the products generated by

$$\left(\underset{m}{\alpha} \ominus \star S^\star\left[\underset{j}{x}\right]\right) \wedge \left(\underset{k}{\beta} \ominus \star S_\star\left[\underset{j}{x}\right]\right)$$

where $\star S_\star\left[\underset{j}{x}\right]$ and $\star S^\star\left[\underset{j}{x}\right]$ are the complete span and complete cospan of $\underset{j}{x}$ (see section 2.12).

For example if we add (and simplify) the terms for *j* equal to 2 we get

$$\star S\left[\star\Sigma\left[\left(\underset{m}{\alpha} \ominus \star S^\star\left[\underset{2}{x}\right]\right) \wedge \left(\underset{k}{\beta} \ominus \star S_\star\left[\underset{2}{x}\right]\right)\right]\right]$$

$$-\left(\left(\underset{m}{\alpha} \ominus x_1\right) \wedge \left(\underset{k}{\beta} \ominus x_2\right)\right) + \left(\underset{m}{\alpha} \ominus x_2\right) \wedge \left(\underset{k}{\beta} \ominus x_1\right) + \left(\underset{m}{\alpha} \ominus (x_1 \wedge x_2)\right) \wedge \underset{k}{\beta} + \underset{m}{\alpha} \wedge \left(\underset{k}{\beta} \ominus (x_1 \wedge x_2)\right)$$

The right hand side of [6.101] is

$$-(-1)^m \left(\left(\underset{m}{\alpha} \ominus x_1\right) \wedge \left(\underset{k}{\beta} \ominus x_2\right) - \left(\underset{m}{\alpha} \ominus x_2\right) \wedge \left(\underset{k}{\beta} \ominus x_1\right)\right) +$$
$$\left(\underset{m}{\alpha} \ominus (x_1 \wedge x_2)\right) \wedge \underset{k}{\beta} + \underset{m}{\alpha} \wedge \left(\underset{k}{\beta} \ominus (x_1 \wedge x_2)\right)$$

Clearly although the terms are the same, there needs to be an adjustment for sign. We need an alternating list of 1s and $(-1)^m$s of length *j*+1 which we can get by including the sign $(-1)^{\star r[j]\, m}$ ($\star r[j]$ was introduced in section 2.12). For example, for *j* equal to 3 we get

$$(-1)^{\star r[3]\, m}$$

$$\{1, (-1)^m, 1, (-1)^m\}$$

$$\star S\left[\star\Sigma\left[(-1)^{\star r[2]\,m}\left(\underset{m}{\alpha}\ominus\star\mathbb{S}^{\star}\left[\underset{2}{\mathbf{x}}\right]\right)\wedge\left(\underset{k}{\beta}\ominus\star\mathbb{S}_{\star}\left[\underset{2}{\mathbf{x}}\right]\right)\right]\right]$$

$$-(-1)^{m}\left(\underset{m}{\alpha}\ominus x_1\right)\wedge\left(\underset{k}{\beta}\ominus x_2\right)+(-1)^{m}\left(\underset{m}{\alpha}\ominus x_2\right)\wedge\left(\underset{k}{\beta}\ominus x_1\right)+$$

$$\left(\underset{m}{\alpha}\ominus(x_1\wedge x_2)\right)\wedge\underset{k}{\beta}+\underset{m}{\alpha}\wedge\left(\underset{k}{\beta}\ominus(x_1\wedge x_2)\right)$$

This formula also gives the correct result for other values of *j*.

$$\left(\underset{m}{\alpha}\wedge\underset{k}{\beta}\right)\ominus\underset{j}{\mathbf{x}}==\star\Sigma\left[(-1)^{\star r[j]\,m}\left(\underset{m}{\alpha}\ominus\star\mathbb{S}^{\star}\left[\underset{j}{\mathbf{x}}\right]\right)\wedge\left(\underset{k}{\beta}\ominus\star\mathbb{S}_{\star}\left[\underset{j}{\mathbf{x}}\right]\right)\right]$$

We prove it in the next section.

The computable form of the Interior Product Formula

To derive the Interior Product Formula in its computable form we can take the Product Formula [3.72] and apply the complement operation. The Product Formula is:

$$\left(\underset{m}{\alpha}\vee\underset{k}{\beta}\right)\wedge\underset{j}{\mathbf{x}}==\star\Sigma\left[(-1)^{\star r[j]\,(\star n-m)}\left(\underset{m}{\alpha}\wedge\star\mathbb{S}^{\star}\left[\underset{j}{\mathbf{x}}\right]\right)\vee\left(\underset{k}{\beta}\wedge\star\mathbb{S}_{\star}\left[\underset{j}{\mathbf{x}}\right]\right)\right]$$

We want to transform the formula so that the left hand side $\left(\underset{m}{\alpha}\vee\underset{k}{\beta}\right)\wedge\underset{j}{\mathbf{x}}$ reads $\left(\underset{m}{\alpha}\wedge\underset{k}{\beta}\right)\ominus\underset{j}{\mathbf{x}}$. To do this we begin by taking its complement.

$$\overline{\left(\underset{m}{\alpha}\vee\underset{k}{\beta}\right)\wedge\underset{j}{\mathbf{x}}}==\left(\underset{m}{\overline{\alpha}}\wedge\underset{k}{\overline{\beta}}\right)\vee\underset{j}{\overline{\mathbf{x}}}==\left(\underset{m}{\overline{\alpha}}\wedge\underset{k}{\overline{\beta}}\right)\ominus\underset{j}{\mathbf{x}}$$

This means we need to take the complement of the right hand side and express it so that it contains only $\underset{m}{\overline{\alpha}}$ and $\underset{k}{\overline{\beta}}$, rather than $\underset{m}{\alpha}$ and $\underset{k}{\beta}$. Taking the complement of a typical term gives:

$$\overline{\left(\underset{m}{\alpha}\wedge\underset{j-s}{\mathbf{s}}\right)\vee\left(\underset{k}{\beta}\wedge\underset{s}{\mathbf{s}}\right)}==\left(\underset{m}{\overline{\alpha}}\vee\underset{j-s}{\overline{\mathbf{s}}}\right)\wedge\left(\underset{k}{\overline{\beta}}\vee\underset{s}{\overline{\mathbf{s}}}\right)==\left(\underset{m}{\overline{\alpha}}\ominus\underset{j-s}{\mathbf{s}}\right)\wedge\left(\underset{k}{\overline{\beta}}\ominus\underset{s}{\mathbf{s}}\right)$$

We can now write our transformed Product Formula as:

$$\left(\underset{m}{\overline{\alpha}}\wedge\underset{k}{\overline{\beta}}\right)\ominus\underset{j}{\mathbf{x}}==\star\Sigma\left[(-1)^{\star r[j]\,(\star n-m)}\left(\underset{m}{\overline{\alpha}}\ominus\star\mathbb{S}^{\star}\left[\underset{j}{\mathbf{x}}\right]\right)\wedge\left(\underset{k}{\overline{\beta}}\ominus\star\mathbb{S}_{\star}\left[\underset{j}{\mathbf{x}}\right]\right)\right]$$

Since $\underset{m}{\alpha}$ and $\underset{k}{\beta}$ are general elements, we can replace $\underset{m}{\overline{\alpha}}$ by $\underset{m}{\alpha}$ and $\underset{k}{\overline{\beta}}$ by $\underset{k}{\beta}$ provided we also replace $\star n-m$ by m in the power sign. Hence we obtain the computable form of the Interior Product Formula.

$$\left(\underset{m}{\alpha}\wedge\underset{k}{\beta}\right)\ominus\underset{j}{\mathbf{x}}==\star\Sigma\left[(-1)^{\star r[j]\,m}\left(\underset{m}{\alpha}\ominus\star\mathbb{S}^{\star}\left[\underset{j}{\mathbf{x}}\right]\right)\wedge\left(\underset{k}{\beta}\ominus\star\mathbb{S}_{\star}\left[\underset{j}{\mathbf{x}}\right]\right)\right] \qquad \textbf{6.103}$$

We should particularly take note that this interior product form of the Product Formula no longer explicitly displays the dimension $\star n$ of the space. It is therefore *independent* of the dimension of the space. This independence is indeed a characteristic of any formula in which the product operations are only exterior products and/or interior products The exterior product sums grades, the interior product subtracts them, and neither operation specifically refers to the dimension of the space in which the operations are occurring.

This general Interior Product Formula is one of the most important formulae in the Grassmann algebra.

◆ Examples

◇ j = 1

$$\left(\underset{m}{\alpha}\wedge\underset{k}{\beta}\right)\ominus\underset{1}{x} == \star\mathcal{S}\left[\star\Sigma\left[(-1)^{\star r[1]\,m}\left(\underset{m}{\alpha}\ominus\star\mathbb{S}^{\star}\left[\underset{1}{x}\right]\right)\wedge\left(\underset{k}{\beta}\ominus\star\mathbb{S}_{\star}\left[\underset{1}{x}\right]\right)\right]\right]$$

$$\left(\underset{m}{\alpha}\wedge\underset{k}{\beta}\right)\ominus\underset{1}{x} == \left(\underset{m}{\alpha}\ominus\underset{1}{x}\right)\wedge\underset{k}{\beta}+(-1)^{m}\underset{m}{\alpha}\wedge\left(\underset{k}{\beta}\ominus\underset{1}{x}\right)$$

◇ j = 2

$$\left(\underset{m}{\alpha}\wedge\underset{k}{\beta}\right)\ominus\underset{2}{x} == \star\mathcal{S}\left[\star\Sigma\left[(-1)^{\star r[2]\,m}\left(\underset{m}{\alpha}\ominus\star\mathbb{S}^{\star}\left[\underset{2}{x}\right]\right)\wedge\left(\underset{k}{\beta}\ominus\star\mathbb{S}_{\star}\left[\underset{2}{x}\right]\right)\right]\right]$$

$$\left(\underset{m}{\alpha}\wedge\underset{k}{\beta}\right)\ominus\underset{2}{x} == -(-1)^{m}\left(\underset{m}{\alpha}\ominus x_1\right)\wedge\left(\underset{k}{\beta}\ominus x_2\right)+ (-1)^{m}\left(\underset{m}{\alpha}\ominus x_2\right)\wedge\left(\underset{k}{\beta}\ominus x_1\right)+\left(\underset{m}{\alpha}\ominus(x_1\wedge x_2)\right)\wedge\underset{k}{\beta}+\underset{m}{\alpha}\wedge\left(\underset{k}{\beta}\ominus(x_1\wedge x_2)\right)$$

◆ When the x_i are all orthogonal to one of the factors

A special case that we will find useful later is when the 1-element factors x_i are *all* orthogonal to one of the general factors, say $\underset{k}{\beta}$. In this case there remains only one non-zero term in the expansion, that in which there are no interior products of $\underset{k}{\beta}$ with any of the x_i.

$$\left(\underset{m}{\alpha}\wedge\underset{k}{\beta}\right)\ominus(x_1\wedge\ldots\wedge x_j) == \left(\underset{m}{\alpha}\ominus(x_1\wedge\ldots\wedge x_j)\right)\wedge\underset{k}{\beta} \qquad \underset{k}{\beta}\ominus x_i == 0 \qquad 6.104$$

More specifically, if *j* and *m* are equal, the result is a scalar multiple of $\underset{k}{\beta}$.

$$\left(\underset{m}{\alpha}\wedge\underset{k}{\beta}\right)\ominus(x_1\wedge\ldots\wedge x_m) == \left(\underset{m}{\alpha}\ominus(x_1\wedge\ldots\wedge x_m)\right)\underset{k}{\beta} \qquad \underset{k}{\beta}\ominus x_i == 0 \qquad 6.105$$

Comparing derivations of the Interior Product Formulae

◆ Encapsulating the computable formula

We can easily encapsulate this computable formula in a function for easy comparison with the results from our automatic derivation above using `DeriveInteriorProductFormula`.

We call this function `ComputeInteriorProductFormula`.

```
ComputeInteriorProductFormula[(A_ ∧ B_) ⊖ X_] :=
  (A ∧ B) ⊖ ComposeSimpleForm[X, 1] == ⋆S[⋆Σ[
     (-1)^(⋆r[RawGrade[X]] RawGrade[A]) (A ⊖ ⋆S^⋆[X]) ∧ (B ⊖ ⋆S_⋆[X])]];
```

◆ Comparing results

We can compare the results obtained by our two formulations.

```
★𝓑₈; Table[ComputeInteriorProductFormula[(α_m ∧ β_k) ⊖ x_j] ===
  (DeriveInteriorProductFormula[(α_m ∧ β_k) ⊖ x_j] /. ((-1)^(m+1) → -(-1)^m)),
 {j, 1, 8}]
```

```
{True, True, True, True, True, True, True, True}
```

The invariance of Interior Product Formula

In Chapter 2 we saw that the m:k-forms based on a simple factorized m-element X were, like the m-element, independent of the factorization.

```
★Σ[(G₁ ⊕ ★S_k[X]) ⊙ (G₂ ⊗ ★S^k[X])]
```

Just as in Chapter 3, we can see that the Interior Product Formula is composed of a number of m:k-forms, and, as expected, shows the same invariance. For example, suppose we generate the formula F_1 for a 3-element X equal to $x \wedge y \wedge z$:

```
F₁ = ComputeInteriorProductFormula[(α_m ∧ β_k) ⊖ (x ∧ y ∧ z)]
```

$$\left(\underset{m}{\alpha} \wedge \underset{k}{\beta}\right) \ominus (x \wedge y \wedge z) == \left(\underset{m}{\alpha} \ominus x\right) \wedge \left(\underset{k}{\beta} \ominus (y \wedge z)\right) - \left(\underset{m}{\alpha} \ominus y\right) \wedge \left(\underset{k}{\beta} \ominus (x \wedge z)\right) + \left(\underset{m}{\alpha} \ominus z\right) \wedge \left(\underset{k}{\beta} \ominus (x \wedge y)\right) +$$
$$(-1)^m \left(\underset{m}{\alpha} \ominus (x \wedge y)\right) \wedge \left(\underset{k}{\beta} \ominus z\right) - (-1)^m \left(\underset{m}{\alpha} \ominus (x \wedge z)\right) \wedge \left(\underset{k}{\beta} \ominus y\right) +$$
$$(-1)^m \left(\underset{m}{\alpha} \ominus (y \wedge z)\right) \wedge \left(\underset{k}{\beta} \ominus x\right) + \left(\underset{m}{\alpha} \ominus (x \wedge y \wedge z)\right) \wedge \underset{k}{\beta} + (-1)^m \underset{m}{\alpha} \wedge \left(\underset{k}{\beta} \ominus (x \wedge y \wedge z)\right)$$

We can express the 3-element as a product of different 1-element factors by adding to any given factor, scalar multiples of the other factors. For example

```
F₂ = ComputeInteriorProductFormula[(α_m ∧ β_k) ⊖ (x ∧ y ∧ (z + a x + b y))]
```

$$\left(\underset{m}{\alpha} \wedge \underset{k}{\beta}\right) \ominus (x \wedge y \wedge (a\,x + b\,y + z)) == \left(\underset{m}{\alpha} \ominus x\right) \wedge \left(\underset{k}{\beta} \ominus (y \wedge (a\,x + b\,y + z))\right) - \left(\underset{m}{\alpha} \ominus y\right) \wedge \left(\underset{k}{\beta} \ominus (x \wedge (a\,x + b\,y + z))\right) +$$
$$\left(\underset{m}{\alpha} \ominus (a\,x + b\,y + z)\right) \wedge \left(\underset{k}{\beta} \ominus (x \wedge y)\right) + (-1)^m \left(\underset{m}{\alpha} \ominus (x \wedge y)\right) \wedge \left(\underset{k}{\beta} \ominus (a\,x + b\,y + z)\right) -$$
$$(-1)^m \left(\underset{m}{\alpha} \ominus (x \wedge (a\,x + b\,y + z))\right) \wedge \left(\underset{k}{\beta} \ominus y\right) +$$
$$(-1)^m \left(\underset{m}{\alpha} \ominus (y \wedge (a\,x + b\,y + z))\right) \wedge \left(\underset{k}{\beta} \ominus x\right) +$$
$$\left(\underset{m}{\alpha} \ominus (x \wedge y \wedge (a\,x + b\,y + z))\right) \wedge \underset{k}{\beta} + (-1)^m \underset{m}{\alpha} \wedge \left(\underset{k}{\beta} \ominus (x \wedge y \wedge (a\,x + b\,y + z))\right)$$

Applying `GrassmannExpandAndSimplify` to these expressions shows that they are equal.

```
★𝒢[F₁ == F₂]
```

True

An alternative form for the Interior Product Formula

Again, as in Chapter 3, we can rearrange the Interior Product Formula by interchanging the span and cospan so that the span elements become associated with $\underset{m}{\alpha}$ and the cospan elements become associated with $\underset{k}{\beta}$ (and changing some signs).

Thus we can write the Interior Product Formula in the alternative form

$$\left(\underset{m}{\alpha} \wedge \underset{k}{\beta}\right) \ominus \underset{j}{x} == \star\Sigma\left[(-1)^{\star r[j]\, m} (-1)^{\star rr[j]} \star R\left[\left(\underset{m}{\alpha} \ominus \star S_{\star}\left[\underset{j}{x}\right]\right) \wedge \left(\underset{k}{\beta} \ominus \star S^{\star}\left[\underset{j}{x}\right]\right)\right]\right] \qquad 6.106$$

◆ Encapsulating the alternative computable formula

Just as we encapsulated the first computable formula, we can similarly encapsulate this alternative one. We call it `ComputeInteriorProductFormulaB`.

```
ComputeInteriorProductFormulaB[(A_ ∧ B_) ⊖ X_] :=
  (A ∧ B) ⊖ ComposeSimpleForm[X, 1] == ★S[★Σ[
     (-1)^(★r[RawGrade[X]] RawGrade[A]) (-1)^(★rr[RawGrade[X]])
      ★R[(A ⊖ ★S★[X]) ∧ (B ⊖ ★S★[X])]]];
```

◆ Comparing results

Comparing the results of `ComputeInteriorProductFormula` with the new function `ComputeInteriorProductFormulaB` verifies their equivalence for values of *j* from 1 to 8.

```
★𝓑₈; Table[★𝒢[ComputeInteriorProductFormula[(α_m ∧ β_k) ⊖ x_j]] ===
   ★𝒢[ComputeInteriorProductFormulaB[(α_m ∧ β_k) ⊖ x_j]], {j, 1, 8}]
```

{True, True, True, True, True, True, True, True}

Interior Product Formula B

There is a second Interior Product Formula which we can derive from the Product Formula [3.72]. Formula [3.72] is

$$\left(\underset{m}{\alpha} \vee \underset{k}{\beta}\right) \wedge \underset{j}{x} == \star\Sigma\left[(-1)^{\star r[j]\,(\star n-m)} \left(\underset{m}{\alpha} \wedge \star S^{\star}\left[\underset{j}{x}\right]\right) \vee \left(\underset{k}{\beta} \wedge \star S_{\star}\left[\underset{j}{x}\right]\right)\right]$$

Replace $\underset{k}{\beta}$ by $\underset{k}{\overline{\beta}}$ to get

$$\left(\underset{m}{\alpha} \vee \underset{k}{\overline{\beta}}\right) \wedge \underset{j}{\mathbf{x}} == \star\Sigma\left[(-1)^{\star r[j]\,(\star n-m)}\left(\underset{m}{\alpha} \wedge \star\mathbb{S}^{\star}\left[\underset{j}{\mathbf{x}}\right]\right) \vee \left(\underset{k}{\overline{\beta}} \wedge \star\mathbb{S}_{\star}\left[\underset{j}{\mathbf{x}}\right]\right)\right]$$

The terms on the right can be rewritten as

$$\left(\underset{m}{\alpha} \wedge \underset{j-s}{\mathbf{s}}\right) \vee \left(\underset{k}{\overline{\beta}} \wedge \underset{s}{\mathbf{s}}\right) == (-1)^{(\star n-k+s)\,(k-s)}\left(\underset{m}{\alpha} \wedge \underset{j-s}{\mathbf{s}}\right) \vee \overline{\overline{\left(\underset{k}{\overline{\beta}} \wedge \underset{s}{\mathbf{s}}\right)}}$$

$$== (-1)^{(\star n-k+s)\,(k-s)}\left(\underset{m}{\alpha} \wedge \underset{j-s}{\mathbf{s}}\right) \ominus \overline{\left(\underset{k}{\overline{\beta}} \wedge \underset{s}{\mathbf{s}}\right)}$$

$$== (-1)^{(\star n-k+s)\,(k-s)}\,(-1)^{k\,(\star n-k)}\left(\underset{m}{\alpha} \wedge \underset{j-s}{\mathbf{s}}\right) \ominus \left(\underset{k}{\beta} \ominus \underset{s}{\mathbf{s}}\right)$$

$$== (-1)^{s\,(\star n-s)}\left(\underset{m}{\alpha} \wedge \underset{j-s}{\mathbf{s}}\right) \ominus \left(\underset{k}{\beta} \ominus \underset{s}{\mathbf{s}}\right)$$

When s is even, $(-1)^{s\,(\star n-s)}$ is unity. When s is odd $(-1)^{s\,(\star n-s)}$ is $(-1)^{\star n+1}$. For j equal to 1, s takes the values $\{0, 1\}$, and so there are two terms in the expansion:

$$\left\{\left(\underset{m}{\alpha} \wedge \underset{1}{\mathbf{s}}\right) \ominus \left(\underset{k}{\beta} \ominus \underset{0}{\mathbf{s}}\right),\ (-1)^{1+\star n}\left(\left(\underset{m}{\alpha} \wedge \underset{0}{\mathbf{s}}\right) \ominus \left(\underset{k}{\beta} \ominus \underset{1}{\mathbf{s}}\right)\right)\right\}$$

When j is equal to 2, s takes the values $\{0, 1, 2\}$, and so there three terms in the expansion:

$$\left\{\left(\underset{m}{\alpha} \wedge \underset{2}{\mathbf{s}}\right) \ominus \left(\underset{k}{\beta} \ominus \underset{0}{\mathbf{s}}\right),\ (-1)^{1+\star n}\left(\left(\underset{m}{\alpha} \wedge \underset{1}{\mathbf{s}}\right) \ominus \left(\underset{k}{\beta} \ominus \underset{1}{\mathbf{s}}\right)\right),\ \left(\underset{m}{\alpha} \wedge \underset{0}{\mathbf{s}}\right) \ominus \left(\underset{k}{\beta} \ominus \underset{2}{\mathbf{s}}\right)\right\}$$

When j is equal to 3, s takes the values $\{0, 1, 2, 3\}$, and so there four terms in the expansion:

$$\left\{\left(\underset{m}{\alpha} \wedge \underset{3}{\mathbf{s}}\right) \ominus \left(\underset{k}{\beta} \ominus \underset{0}{\mathbf{s}}\right),\ (-1)^{1+\star n}\left(\left(\underset{m}{\alpha} \wedge \underset{2}{\mathbf{s}}\right) \ominus \left(\underset{k}{\beta} \ominus \underset{1}{\mathbf{s}}\right)\right),\ \left(\underset{m}{\alpha} \wedge \underset{1}{\mathbf{s}}\right) \ominus \left(\underset{k}{\beta} \ominus \underset{2}{\mathbf{s}}\right),\right.$$
$$\left.(-1)^{1+\star n}\left(\left(\underset{m}{\alpha} \wedge \underset{0}{\mathbf{s}}\right) \ominus \left(\underset{k}{\beta} \ominus \underset{3}{\mathbf{s}}\right)\right)\right\}$$

This pattern clearly repeats, showing that the sign $(-1)^{1+\star n}$ applies to each alternate term in the same way as the sign $(-1)^{\star r[1]\,(\star n-m)}$ of the original expression. For example, for j equal to 2 we have

$$(-1)^{\star r[2]\,(\star n-m)}\left(\left(\underset{m}{\alpha} \wedge \underset{j-s}{\mathbf{s}}\right) \ominus \left(\underset{k}{\beta} \ominus \underset{s}{\mathbf{s}}\right)\right)$$

$$\left\{\left(\underset{m}{\alpha} \wedge \underset{j-s}{\mathbf{s}}\right) \ominus \left(\underset{k}{\beta} \ominus \underset{s}{\mathbf{s}}\right),\ (-1)^{\star n-m}\left(\left(\underset{m}{\alpha} \wedge \underset{j-s}{\mathbf{s}}\right) \ominus \left(\underset{k}{\beta} \ominus \underset{s}{\mathbf{s}}\right)\right),\ \left(\underset{m}{\alpha} \wedge \underset{j-s}{\mathbf{s}}\right) \ominus \left(\underset{k}{\beta} \ominus \underset{s}{\mathbf{s}}\right)\right\}$$

Multiplying these two signs confirms that the new formula no longer shows a dependence on the dimension of the space $\star n$, and the new sign for the new formula becomes $(-1)^{\star r[j]\,(m+1)}$.

$$(-1)^{1+\star n}\,(-1)^{\star n-m} == (-1)^{m+1}$$

Hence the new formula becomes

$$\left(\underset{m}{\alpha} \ominus \underset{k}{\beta}\right) \wedge \underset{j}{\mathbf{x}} == \star\Sigma\left[(-1)^{\star r[j]\,(m+1)}\left(\underset{m}{\alpha} \wedge \star\mathbb{S}^{\star}\left[\underset{j}{\mathbf{x}}\right]\right) \ominus \left(\underset{k}{\beta} \ominus \star\mathbb{S}_{\star}\left[\underset{j}{\mathbf{x}}\right]\right)\right] \qquad 6.107$$

We will call it the Interior Product Formula B.

◆ Examples

◇ j = 1

$$\left(\underset{m}{\alpha}\ominus\underset{k}{\beta}\right)\wedge\underset{1}{x} == \star\mathcal{S}\left[\star\Sigma\left[(-1)^{\star r[1]\,(m+1)}\left(\underset{m}{\alpha}\wedge\star\mathbb{S}^{\star}\left[\underset{1}{x}\right]\right)\ominus\left(\underset{k}{\beta}\ominus\star\mathbb{S}_{\star}\left[\underset{1}{x}\right]\right)\right]\right]$$

$$\left(\underset{m}{\alpha}\ominus\underset{k}{\beta}\right)\wedge\underset{1}{x} == (-1)^{1+m}\left(\underset{m}{\alpha}\ominus\left(\underset{k}{\beta}\ominus\underset{1}{x}\right)\right)+\left(\underset{m}{\alpha}\wedge\underset{1}{x}\right)\ominus\underset{k}{\beta}$$

$$\left(\underset{m}{\alpha}\ominus\underset{k}{\beta}\right)\wedge x == \left(\underset{m}{\alpha}\wedge x\right)\ominus\underset{k}{\beta}+(-1)^{m+1}\underset{m}{\alpha}\ominus\left(\underset{k}{\beta}\ominus x\right) \qquad 6.108$$

◇ j = 2

$$\left(\underset{m}{\alpha}\ominus\underset{k}{\beta}\right)\wedge\underset{2}{x} == \star\mathcal{S}\left[\star\Sigma\left[(-1)^{\star r[2]\,(m+1)}\left(\underset{m}{\alpha}\wedge\star\mathbb{S}^{\star}\left[\underset{2}{x}\right]\right)\ominus\left(\underset{k}{\beta}\ominus\star\mathbb{S}_{\star}\left[\underset{2}{x}\right]\right)\right]\right]$$

$$\left(\underset{m}{\alpha}\ominus\underset{k}{\beta}\right)\wedge\underset{2}{x} == \underset{m}{\alpha}\ominus\left(\underset{k}{\beta}\ominus(x_1\wedge x_2)\right)+(-1)^{m}\left(\left(\underset{m}{\alpha}\wedge x_1\right)\ominus\left(\underset{k}{\beta}\ominus x_2\right)\right)+$$
$$(-1)^{1+m}\left(\left(\underset{m}{\alpha}\wedge x_2\right)\ominus\left(\underset{k}{\beta}\ominus x_1\right)\right)+\left(\underset{m}{\alpha}\wedge x_1\wedge x_2\right)\ominus\underset{k}{\beta}$$

$$\left(\underset{m}{\alpha}\ominus\underset{k}{\beta}\right)\wedge(x_1\wedge x_1) == \left(\underset{m}{\alpha}\wedge x_1\wedge x_1\right)\ominus\underset{k}{\beta}+$$
$$(-1)^{m}\left(\left(\underset{m}{\alpha}\wedge x_1\right)\ominus\left(\underset{k}{\beta}\ominus x_2\right)-\left(\underset{m}{\alpha}\wedge x_2\right)\ominus\left(\underset{k}{\beta}\ominus x_1\right)\right)+\underset{m}{\alpha}\ominus\left(\underset{k}{\beta}\ominus(x_1\wedge x_1)\right) \qquad 6.109$$

◆ Confirming the formula

It is straightforward to confirm that formula behaves as expected. Here we test out the formula for all 48 combinations of values of *m* and *k* from 0 to 3, and *j* from 1 to 3.

```
★𝓑12; Table[X = ComposeSimpleForm[x_j]; A = ComposeSimpleForm[α_m];
  B = ComposeSimpleForm[β_k]; ★𝒢[ToScalarProducts[★𝒢[
     (A⊖B) ∧ X == ★Σ[(-1)^(★r[j] (m+1)) (A ∧ ★S*[X]) ⊖ (B⊖★S⋆[X])]]]],
  {j, 1, 3}, {m, 0, 3}, {k, 0, 3}] // (And @@ Flatten[#] &)
```

```
True
```

The Orthogonal Decomposition Formula

The Interior Product Formula B [6.107] has a special property when the interior product on the left hand side is an inner product. When this is the case, the inner product can be factored out as a scalar, and the formula becomes one which expresses the decomposition of a simple *m*-element in terms of the factors of the inner product. Let us begin by recalling the Interior Product Formula B.

$$\left(\underset{m}{\alpha}\ominus\underset{k}{\beta}\right)\wedge\underset{j}{x} == \star\Sigma\left[(-1)^{\star r[j]\,(m+1)}\left(\underset{m}{\alpha}\wedge\star\mathbb{S}^{\star}\left[\underset{j}{x}\right]\right)\ominus\left(\underset{k}{\beta}\ominus\star\mathbb{S}_{\star}\left[\underset{j}{x}\right]\right)\right]$$

When k is equal to m we can divide through by the inner product.

$$\underset{j}{x} == \frac{1}{\underset{m}{\alpha}\ominus\underset{m}{\beta}}\star\Sigma\left[(-1)^{\star r[j]\,(m+1)}\left(\underset{m}{\alpha}\wedge\star\mathbb{S}^{\star}\left[\underset{j}{x}\right]\right)\ominus\left(\underset{m}{\beta}\ominus\star\mathbb{S}_{\star}\left[\underset{j}{x}\right]\right)\right] \qquad 6.110$$

We can also express the result in terms of unit m-elements:

$$\underset{j}{x} == \frac{1}{\underset{m}{\hat{\alpha}}\ominus\underset{m}{\hat{\beta}}}\star\Sigma\left[(-1)^{\star r[j]\,(m+1)}\left(\underset{m}{\hat{\alpha}}\wedge\star\mathbb{S}^{\star}\left[\underset{j}{x}\right]\right)\ominus\left(\underset{m}{\hat{\beta}}\ominus\star\mathbb{S}_{\star}\left[\underset{j}{x}\right]\right)\right]$$

A particularly important case of this formula is when $\underset{m}{\beta}$ is equal to $\underset{m}{\alpha}$.

$$\underset{j}{x} == \star\Sigma\left[(-1)^{\star r[j]\,(m+1)}\left(\underset{m}{\hat{\alpha}}\wedge\star\mathbb{S}^{\star}\left[\underset{j}{x}\right]\right)\ominus\left(\underset{m}{\hat{\alpha}}\ominus\star\mathbb{S}_{\star}\left[\underset{j}{x}\right]\right)\right] \qquad 6.111$$

We call this the Orthogonal Decomposition Formula. It expresses the decomposition of the simple j-element $\underset{j}{x}$ in terms of a unit simple m-element $\underset{m}{\hat{\alpha}}$. We shall explore this formula in much more detail in what follows.

◆ Examples

◇ j = 1

$$x == \star S\left[\star\Sigma\left[(-1)^{\star r[1]\,(m+1)}\left(\underset{m}{\hat{\alpha}}\wedge\star\mathbb{S}^{\star}[x]\right)\ominus\left(\underset{m}{\hat{\alpha}}\ominus\star\mathbb{S}_{\star}[x]\right)\right]\right]$$

$$x == (-1)^{1+m}\left(\underset{m}{\hat{\alpha}}\ominus\left(\underset{m}{\hat{\alpha}}\ominus x\right)\right)+\left(\underset{m}{\hat{\alpha}}\wedge x\right)\ominus\underset{m}{\hat{\alpha}}$$

$$x == \left(\underset{m}{\hat{\alpha}}\wedge x\right)\ominus\underset{m}{\hat{\alpha}}+(-1)^{m+1}\,\underset{m}{\hat{\alpha}}\ominus\left(\underset{m}{\hat{\alpha}}\ominus x\right) \qquad 6.112$$

◇ j = 2

$$\underset{2}{x} == \star S\left[\star\Sigma\left[(-1)^{\star r[2]\,(m+1)}\left(\underset{m}{\hat{\alpha}}\wedge\star\mathbb{S}^{\star}\left[\underset{2}{x}\right]\right)\ominus\left(\underset{m}{\hat{\alpha}}\ominus\star\mathbb{S}_{\star}\left[\underset{2}{x}\right]\right)\right]\right]$$

$$\underset{2}{x} == \underset{m}{\hat{\alpha}}\ominus\left(\underset{m}{\hat{\alpha}}\ominus(x_1\wedge x_2)\right)+(-1)^{m}\left(\left(\underset{m}{\hat{\alpha}}\wedge x_1\right)\ominus\left(\underset{m}{\hat{\alpha}}\ominus x_2\right)\right)+$$
$$(-1)^{1+m}\left(\left(\underset{m}{\hat{\alpha}}\wedge x_2\right)\ominus\left(\underset{m}{\hat{\alpha}}\ominus x_1\right)\right)+\left(\underset{m}{\hat{\alpha}}\wedge x_1\wedge x_2\right)\ominus\underset{m}{\hat{\alpha}}$$

$$x_1 \wedge x_1 == \left(\underset{m}{\hat{\alpha}} \wedge x_1 \wedge x_1\right) \ominus \underset{m}{\hat{\alpha}} + (-1)^m \left(\left(\underset{m}{\hat{\alpha}} \wedge x_1\right) \ominus \left(\underset{m}{\hat{\alpha}} \ominus x_2\right) - \left(\underset{m}{\hat{\alpha}} \wedge x_2\right) \ominus \left(\underset{m}{\hat{\alpha}} \ominus x_1\right)\right) + \underset{m}{\hat{\alpha}} \ominus \left(\underset{m}{\hat{\alpha}} \ominus (x_1 \wedge x_1)\right) \quad 6.113$$

Orthogonal Decomposition Formula B

The Interior Product Formula B has been given in [6.107] as:

$$\left(\underset{m}{\alpha} \ominus \underset{k}{\beta}\right) \wedge \underset{j}{x} == \star\Sigma\left[(-1)^{\star r[j]\,(m+1)} \left(\underset{m}{\alpha} \wedge \star S^{\star}\left[\underset{j}{x}\right]\right) \ominus \left(\underset{k}{\beta} \ominus \star S_{\star}\left[\underset{j}{x}\right]\right)\right]$$

The Orthogonal Decomposition Formula [6.111] discussed in the previous section was derived from it by putting $\underset{k}{\beta}$ equal to $\underset{m}{\alpha}$. Suppose now instead, we put $\underset{k}{\beta}$ equal to $\underset{j}{x}$.

$$\left(\underset{m}{\alpha} \ominus \underset{j}{x}\right) \wedge \underset{j}{x} == \star\Sigma\left[(-1)^{\star r[j]\,(m+1)} \left(\underset{m}{\alpha} \wedge \star S^{\star}\left[\underset{j}{x}\right]\right) \ominus \left(\underset{j}{x} \ominus \star S_{\star}\left[\underset{j}{x}\right]\right)\right]$$

Before proceeding any further, let us see what the expansion of this formula looks like for small *j*.

◆ Examples

◇ j = 1

$$\left(\underset{m}{\alpha} \ominus x\right) \wedge x == \star\mathcal{S}\left[\star\Sigma\left[(-1)^{\star r[1]\,(m+1)} \left(\underset{m}{\alpha} \wedge \star S^{\star}[x]\right) \ominus (x \ominus \star S_{\star}[x])\right]\right]$$

$$\left(\underset{m}{\alpha} \ominus x\right) \wedge x == \left(\underset{m}{\alpha} \wedge x\right) \ominus x - (-1)^m (x \ominus x)\, \underset{m}{\alpha}$$

$$\left(\underset{m}{\alpha} \ominus x\right) \wedge x == \left(\underset{m}{\alpha} \wedge x\right) \ominus x - (-1)^m (x \ominus x)\, \underset{m}{\alpha} \quad 6.114$$

◇ j = 2

$$\left(\underset{m}{\alpha} \ominus \underset{2}{x}\right) \wedge \underset{2}{x} == \star\mathcal{S}\left[\star\Sigma\left[(-1)^{\star r[2]\,(m+1)} \left(\underset{m}{\alpha} \wedge \star S^{\star}\left[\underset{2}{x}\right]\right) \ominus \left(\underset{2}{x} \ominus \star S_{\star}\left[\underset{2}{x}\right]\right)\right]\right]$$

$$\left(\underset{m}{\alpha} \ominus \underset{2}{x}\right) \wedge \underset{2}{x} == (-1)^m \left(\left(\underset{m}{\alpha} \wedge x_1\right) \ominus \left(\underset{2}{x} \ominus x_2\right)\right) + (-1)^{1+m} \left(\left(\underset{m}{\alpha} \wedge x_2\right) \ominus \left(\underset{2}{x} \ominus x_1\right)\right) + \left(\underset{m}{\alpha} \wedge x_1 \wedge x_2\right) \ominus \underset{2}{x} + \left(\underset{2}{x} \ominus (x_1 \wedge x_2)\right) \underset{m}{\alpha}$$

$$\left(\underset{m}{\alpha} \ominus (x_1 \wedge x_2)\right) \wedge (x_1 \wedge x_2) == \left(\underset{m}{\alpha} \wedge x_1 \wedge x_2\right) \ominus (x_1 \wedge x_2) + ((x_1 \wedge x_2) \ominus (x_1 \wedge x_2))\, \underset{m}{\alpha} + (-1)^m \left(\left(\underset{m}{\alpha} \wedge x_1\right) \ominus ((x_1 \wedge x_2) \ominus x_2) - \left(\underset{m}{\alpha} \wedge x_2\right) \ominus ((x_1 \wedge x_2) \ominus x_1)\right) \quad 6.115$$

◇ $j = 3$

$$\left(\underset{m}{\alpha}\ominus\underset{3}{x}\right)\wedge\underset{3}{x} == \star\mathcal{S}\left[\star\Sigma\left[(-1)^{\star r[3]\,(m+1)}\left(\underset{m}{\alpha}\wedge\star\mathbb{S}^{\star}\left[\underset{3}{x}\right]\right)\ominus\left(\underset{3}{x}\ominus\star\mathbb{S}_{\star}\left[\underset{3}{x}\right]\right)\right]\right]$$

$$\left(\underset{m}{\alpha}\ominus\underset{3}{x}\right)\wedge\underset{3}{x} == \left(\underset{m}{\alpha}\wedge x_1\right)\ominus\left(\underset{3}{x}\ominus(x_2\wedge x_3)\right) - \left(\underset{m}{\alpha}\wedge x_2\right)\ominus\left(\underset{3}{x}\ominus(x_1\wedge x_3)\right) +$$
$$\left(\underset{m}{\alpha}\wedge x_3\right)\ominus\left(\underset{3}{x}\ominus(x_1\wedge x_2)\right) + (-1)^{1+m}\left(\left(\underset{m}{\alpha}\wedge x_1\wedge x_2\right)\ominus\left(\underset{3}{x}\ominus x_3\right)\right) +$$
$$(-1)^{m}\left(\left(\underset{m}{\alpha}\wedge x_1\wedge x_3\right)\ominus\left(\underset{3}{x}\ominus x_2\right)\right) + (-1)^{1+m}\left(\left(\underset{m}{\alpha}\wedge x_2\wedge x_3\right)\ominus\left(\underset{3}{x}\ominus x_1\right)\right) +$$
$$\left(\underset{m}{\alpha}\wedge x_1\wedge x_2\wedge x_3\right)\ominus\underset{3}{x} - (-1)^{m}\left(\underset{3}{x}\ominus(x_1\wedge x_2\wedge x_3)\right)\underset{m}{\alpha}$$

- We can observe from these expansions that they are all of the form

$$\left(\underset{m}{\alpha}\ominus\underset{j}{x}\right)\wedge\underset{j}{x} == \left(\underset{m}{\alpha}\wedge\underset{j}{x}\right)\ominus\underset{j}{x} + \underset{m}{M} + (-1)^{j\,(m+1)}\left(\underset{j}{x}\ominus\underset{j}{x}\right)\underset{m}{\alpha}$$

Here, the first term is due to the zeroth element of the span of $\underset{j}{x}$, the third term is due to the *j*th element of the span, and the middle term $\underset{m}{M}$ is due to the other elements of the span. Thus, we can write a decomposition for the element $\underset{m}{\alpha}$ in terms of $\underset{j}{x}$ as:

$$\underset{m}{\alpha} == (-1)^{j\,(m+1)}\left(\left(\underset{m}{\alpha}\ominus\underset{j}{\hat{x}}\right)\wedge\underset{j}{\hat{x}} - \frac{1}{\underset{j}{x}\ominus\underset{j}{x}}\underset{m}{M} - \left(\underset{m}{\alpha}\wedge\underset{j}{\hat{x}}\right)\ominus\underset{j}{\hat{x}}\right) \qquad 6.116$$

In the case j is equal to 1, the middle terms are zero and we can write

$$\underset{m}{\alpha} == (-1)^{m}\left(\left(\underset{m}{\alpha}\wedge\hat{x}\right)\ominus\hat{x} - \left(\underset{m}{\alpha}\ominus\hat{x}\right)\wedge\hat{x}\right) \qquad 6.117$$

Interior Product Formulae as double sums

For completeness we express the three major Interior Product Formulae developed above as double sums. As discussed in section 3.10, this formulation allows some ambiguity of interpretation of the underscripted symbols when the formula is unwrapped for a given value of *j*.

◆ The Interior Product Formula B as a double sum

Consider the double sum regressive General Product Formula [3.77].

$$\left(\underset{m}{\alpha}\vee\underset{k}{\beta}\right)\wedge\underset{p}{x} == \sum_{r==0}^{p}(-1)^{r\,(n-m)}\sum_{i==1}^{\nu}\left(\underset{m}{\alpha}\wedge\underset{p-r}{x_i}\right)\vee\left(\underset{k}{\beta}\wedge\underset{r}{x_i}\right)$$

$$\underset{p}{x} == \underset{r}{x_1}\wedge\underset{p-r}{x_1} == \underset{r}{x_2}\wedge\underset{p-r}{x_2} == \ldots == \underset{r}{x_\nu}\wedge\underset{p-r}{x_\nu} \qquad \nu == \binom{p}{r}$$

Now make the following replacements.

$$\underset{k}{\beta} \longrightarrow \underset{n-k}{\overline{\beta}} \qquad \underset{n-k}{\overline{\beta}} \wedge \underset{r}{x_i} \longrightarrow (-1)^{r\,(m-r)}\, \overline{\underset{n-k}{\beta} \vee \underset{r}{\overline{x_i}}} \qquad n-k \longrightarrow k$$

These enable the formula to be written

$$\left(\underset{m}{\alpha} \ominus \underset{k}{\beta}\right) \wedge \underset{p}{x} = \sum_{r=0}^{p} (-1)^{r\,(m-r)} \sum_{i=1}^{\nu} \left(\underset{m}{\alpha} \wedge \underset{p-r}{x_i}\right) \ominus \left(\underset{k}{\beta} \ominus \underset{r}{x_i}\right)$$
$$\underset{p}{x} = \underset{r}{x_1} \wedge \underset{p-r}{x_1} = \underset{r}{x_2} \wedge \underset{p-r}{x_2} = \ldots = \underset{r}{x_\nu} \wedge \underset{p-r}{x_\nu}, \qquad \nu = \binom{p}{r} \qquad 6.118$$

◆ **The Interior Product Formula as a double sum**

By a similar process we obtain

$$\left(\underset{m}{\alpha} \wedge \underset{k}{\beta}\right) \ominus \underset{p}{x} = \sum_{r=0}^{p} (-1)^{r\,m} \sum_{i=1}^{\nu} \left(\underset{m}{\alpha} \ominus \underset{p-r}{x_i}\right) \wedge \left(\underset{k}{\beta} \ominus \underset{r}{x_i}\right)$$
$$\underset{p}{x} = \underset{r}{x_1} \wedge \underset{p-r}{x_1} = \underset{r}{x_2} \wedge \underset{p-r}{x_2} = \ldots = \underset{r}{x_\nu} \wedge \underset{p-r}{x_\nu}, \qquad \nu = \binom{p}{r} \qquad 6.119$$

◆ **The Orthogonal Decomposition Formula as a double sum**

Putting $\underset{k}{\beta}$ equal to $\underset{m}{\alpha}$ in Interior Product Formula B [6.118] expresses a simple *p*-element in terms of products with a unit *m*-element.

$$\underset{p}{x} = \sum_{r=0}^{p} (-1)^{r\,(m-r)} \sum_{i=1}^{\nu} \left(\underset{m}{\hat{\alpha}} \wedge \underset{p-r}{x_i}\right) \ominus \left(\underset{m}{\hat{\alpha}} \ominus \underset{r}{x_i}\right)$$
$$\underset{p}{x} = \underset{r}{x_1} \wedge \underset{p-r}{x_1} = \underset{r}{x_2} \wedge \underset{p-r}{x_2} = \ldots = \underset{r}{x_\nu} \wedge \underset{p-r}{x_\nu}, \qquad \nu = \binom{p}{r} \qquad 6.120$$

Extended Interior Product Formulae

Consider the two Interior Product Formulae [6.103] and [6.107] developed above. We refer now to the original form [6.103] as the A form:

$$\left(\underset{m}{\alpha} \wedge \underset{k}{\beta}\right) \ominus \underset{j}{x} = \star\Sigma\left[(-1)^{\star r[j]\, m} \left(\underset{m}{\alpha} \ominus \star S^{\star}\left[\underset{j}{x}\right]\right) \wedge \left(\underset{k}{\beta} \ominus \star S_{\star}\left[\underset{j}{x}\right]\right)\right]$$

The second form is the B form:

$$\left(\underset{m}{\alpha} \ominus \underset{k}{\beta}\right) \wedge \underset{j}{x} = \star\Sigma\left[(-1)^{\star r[j]\,(m+1)} \left(\underset{m}{\alpha} \wedge \star S^{\star}\left[\underset{j}{x}\right]\right) \ominus \left(\underset{k}{\beta} \ominus \star S_{\star}\left[\underset{j}{x}\right]\right)\right]$$

In this section we will look at how these formulae may be extended by taking their product with a further simple element $\underset{s}{z}$, say. We have four principal possibilities. For the A form we have:

$$\left(\left(\underset{m}{\alpha}\wedge\underset{k}{\beta}\right)\ominus\underset{j}{x}\right)\wedge\underset{s}{z} \qquad \left(\left(\underset{m}{\alpha}\wedge\underset{k}{\beta}\right)\ominus\underset{j}{x}\right)\ominus\underset{s}{z}$$

And for the B form we have:

$$\left(\left(\underset{m}{\alpha}\ominus\underset{k}{\beta}\right)\wedge\underset{j}{x}\right)\wedge\underset{s}{z} \qquad \left(\left(\underset{m}{\alpha}\ominus\underset{k}{\beta}\right)\wedge\underset{j}{x}\right)\ominus\underset{s}{z}$$

Let us consider each of these in turn.

◆ $\left(\left(\underset{m}{\alpha}\wedge\underset{k}{\beta}\right)\ominus\underset{j}{x}\right)\wedge\underset{s}{z}$ ◆

This can be expressed by taking the exterior product of $\underset{s}{z}$ with an A form.

$$\left(\left(\underset{m}{\alpha}\wedge\underset{k}{\beta}\right)\ominus\underset{j}{x}\right)\wedge\underset{s}{z} == \star\Sigma\left[(-1)^{\star r[j]\,m}\left(\underset{m}{\alpha}\ominus\star S^{\star}\left[\underset{j}{x}\right]\right)\wedge\left(\underset{k}{\beta}\ominus\star S_{\star}\left[\underset{j}{x}\right]\right)\wedge\underset{s}{z}\right]$$

Or by recognizing that the whole expression is a B form.

$$\left(\left(\underset{m}{\alpha}\wedge\underset{k}{\beta}\right)\ominus\underset{j}{x}\right)\wedge\underset{s}{z} == \star\Sigma\left[(-1)^{\star r[s]\,(m+k+1)}\left(\left(\underset{m}{\alpha}\wedge\underset{k}{\beta}\right)\wedge\star S^{\star}\left[\underset{s}{z}\right]\right)\ominus\left(\underset{j}{x}\ominus\star S_{\star}\left[\underset{s}{z}\right]\right)\right]$$

The equality of these two expressions thus gives rise to an identity between the two forms.

◇ Example 1

Consider the simplest example where all the elements are 1-elements.

$$((\alpha\wedge\beta)\ominus x)\wedge z == \star\mathcal{S}\left[\star\Sigma\left[(-1)^{\star r[1]\,1}\left(\alpha\ominus\star S^{\star}[x]\right)\wedge(\beta\ominus\star S_{\star}[x])\wedge z\right]\right]$$

$$((\alpha\wedge\beta)\ominus x)\wedge z == (x\ominus\beta)\, z\wedge\alpha-(x\ominus\alpha)\, z\wedge\beta$$

$$((\alpha\wedge\beta)\ominus x)\wedge z == \star\mathcal{S}\left[\star\Sigma\left[(-1)^{\star r[1]\,(3)}\left((\alpha\wedge\beta)\wedge\star S^{\star}[z]\right)\ominus(x\ominus\star S_{\star}[z])\right]\right]$$

$$((\alpha\wedge\beta)\ominus x)\wedge z == (z\wedge\alpha\wedge\beta)\ominus x-(x\ominus z)\,\alpha\wedge\beta$$

To show that these are the same result we can apply `ToScalarProducts`.

```
ToScalarProducts[(z ∧ α ∧ β) ⊖ x - (x ⊖ z) α ∧ β]
```

$$(x\ominus\beta)\, z\wedge\alpha-(x\ominus\alpha)\, z\wedge\beta$$

◆ $\left(\left(\underset{m}{\alpha}\wedge\underset{k}{\beta}\right)\ominus\underset{j}{x}\right)\ominus\underset{s}{z}$ ◆

This form can also be expressed using the extended interior product formula [6.40], leading again to two expressions, but this time, both from the A form.

$$\left(\left(\underset{m}{\alpha}\wedge\underset{k}{\beta}\right)\ominus\underset{j}{x}\right)\ominus\underset{s}{z} == \left(\underset{m}{\alpha}\wedge\underset{k}{\beta}\right)\ominus\left(\underset{j}{x}\wedge\underset{s}{z}\right)$$

$$\left(\left(\underset{m}{\alpha}\wedge\underset{k}{\beta}\right)\ominus\underset{j}{x}\right)\ominus\underset{s}{z} == \star\Sigma\left[(-1)^{\star r[j]\,m}\left(\left(\underset{m}{\alpha}\ominus\star S^{\star}\left[\underset{j}{x}\right]\right)\wedge\left(\underset{k}{\beta}\ominus\star S_{\star}\left[\underset{j}{x}\right]\right)\right)\ominus\underset{s}{z}\right]$$

$$\left(\underset{m}{\alpha}\wedge\underset{k}{\beta}\right)\ominus\left(\underset{j}{x}\wedge\underset{s}{z}\right) == \star\Sigma\left[(-1)^{\star r[j+s]\,m}\left(\underset{m}{\alpha}\ominus\star S^{\star}\left[\underset{j}{x}\wedge\underset{s}{z}\right]\right)\wedge\left(\underset{k}{\beta}\ominus\star S_{\star}\left[\underset{j}{x}\wedge\underset{s}{z}\right]\right)\right]$$

◇ **Example 2**

$$\left(\left(\underset{m}{\alpha}\wedge\underset{k}{\beta}\right)\ominus \mathbf{x}\right)\ominus \mathbf{z} == \star\mathcal{S}\left[\star\Sigma\left[(-1)^{\star r[1]\,m}\left(\left(\underset{m}{\alpha}\ominus\star\mathbb{S}^{\star}[\mathbf{x}]\right)\wedge\left(\underset{k}{\beta}\ominus\star\mathbb{S}_{\star}[\mathbf{x}]\right)\right)\ominus \mathbf{z}\right]\right]$$

$$((\underset{m}{\alpha}\wedge\underset{k}{\beta})\ominus \mathbf{x})\ominus \mathbf{z} == ((\underset{m}{\alpha}\ominus \mathbf{x})\wedge\underset{k}{\beta})\ominus \mathbf{z} + (-1)^{m}\left((\underset{m}{\alpha}\wedge(\underset{k}{\beta}\ominus \mathbf{x}))\ominus \mathbf{z}\right)$$

$$\left(\underset{m}{\alpha}\wedge\underset{k}{\beta}\right)\ominus(\mathbf{x}\wedge \mathbf{z}) == \star\mathcal{S}\left[\star\Sigma\left[(-1)^{\star r[2]\,m}\left(\underset{m}{\alpha}\ominus\star\mathbb{S}^{\star}[\mathbf{x}\wedge \mathbf{z}]\right)\wedge\left(\underset{k}{\beta}\ominus\star\mathbb{S}_{\star}[\mathbf{x}\wedge \mathbf{z}]\right)\right]\right]$$

$$(\underset{m}{\alpha}\wedge\underset{k}{\beta})\ominus(\mathbf{x}\wedge \mathbf{z}) == -(-1)^{m}(\underset{m}{\alpha}\ominus \mathbf{x})\wedge(\underset{k}{\beta}\ominus \mathbf{z}) + (-1)^{m}(\underset{m}{\alpha}\ominus \mathbf{z})\wedge(\underset{k}{\beta}\ominus \mathbf{x}) + (\underset{m}{\alpha}\ominus(\mathbf{x}\wedge \mathbf{z}))\wedge\underset{k}{\beta} + \underset{m}{\alpha}\wedge(\underset{k}{\beta}\ominus(\mathbf{x}\wedge \mathbf{z}))$$

This form is an expansion of the first one. We can see this by applying the basic Interior Product Formula [6.100] to each of its two terms.

◆ $(\underset{m}{\alpha}\ominus\underset{k}{\beta})\wedge\underset{j}{\mathbf{x}}\wedge\underset{s}{\mathbf{z}}$ ◆

This form leads to two expressions from the B form. But we can also get a third expression by interchanging the order of the factors in the exterior product.

$$\left(\underset{m}{\alpha}\ominus\underset{k}{\beta}\right)\wedge\underset{j}{\mathbf{x}}\wedge\underset{s}{\mathbf{z}} == (-1)^{j\,s}\left(\underset{m}{\alpha}\ominus\underset{k}{\beta}\right)\wedge\underset{s}{\mathbf{z}}\wedge\underset{j}{\mathbf{x}}$$

$$\left(\left(\underset{m}{\alpha}\ominus\underset{k}{\beta}\right)\wedge\underset{j}{\mathbf{x}}\right)\wedge\underset{s}{\mathbf{z}} == \star\Sigma\left[(-1)^{\star r[j]\,(m+1)}\left(\left(\underset{m}{\alpha}\wedge\star\mathbb{S}^{\star}\left[\underset{j}{\mathbf{x}}\right]\right)\ominus\left(\underset{k}{\beta}\ominus\star\mathbb{S}_{\star}\left[\underset{j}{\mathbf{x}}\right]\right)\right)\wedge\underset{s}{\mathbf{z}}\right]$$

$$\left(\left(\underset{m}{\alpha}\ominus\underset{k}{\beta}\right)\wedge\underset{j}{\mathbf{x}}\right)\wedge\underset{s}{\mathbf{z}} == (-1)^{j\,s}\star\Sigma\left[(-1)^{\star r[s]\,(m+1)}\left(\left(\underset{m}{\alpha}\wedge\star\mathbb{S}^{\star}\left[\underset{s}{\mathbf{z}}\right]\right)\ominus\left(\underset{k}{\beta}\ominus\star\mathbb{S}_{\star}\left[\underset{s}{\mathbf{z}}\right]\right)\right)\wedge\underset{j}{\mathbf{x}}\right]$$

$$\left(\underset{m}{\alpha}\ominus\underset{k}{\beta}\right)\wedge\left(\underset{j}{\mathbf{x}}\wedge\underset{s}{\mathbf{z}}\right) == \star\Sigma\left[(-1)^{\star r[j+s]\,(m+1)}\left(\underset{m}{\alpha}\wedge\star\mathbb{S}^{\star}\left[\underset{j}{\mathbf{x}}\wedge\underset{s}{\mathbf{z}}\right]\right)\ominus\left(\underset{k}{\beta}\ominus\star\mathbb{S}_{\star}\left[\underset{j}{\mathbf{x}}\wedge\underset{s}{\mathbf{z}}\right]\right)\right]$$

◇ **Example 3**

$$\left(\left(\underset{m}{\alpha}\ominus\underset{k}{\beta}\right)\wedge \mathbf{x}\right)\wedge \mathbf{z} == \star\mathcal{S}\left[\star\Sigma\left[(-1)^{\star r[1]\,(m+1)}\left(\left(\underset{m}{\alpha}\wedge\star\mathbb{S}^{\star}[\mathbf{x}]\right)\ominus\left(\underset{k}{\beta}\ominus\star\mathbb{S}_{\star}[\mathbf{x}]\right)\right)\wedge \mathbf{z}\right]\right]$$

$$(\underset{m}{\alpha}\ominus\underset{k}{\beta})\wedge \mathbf{x}\wedge \mathbf{z} == (-1)^{1+m}(\underset{m}{\alpha}\ominus(\underset{k}{\beta}\ominus \mathbf{x}))\wedge \mathbf{z} + ((\underset{m}{\alpha}\wedge \mathbf{x})\ominus\underset{k}{\beta})\wedge \mathbf{z}$$

$$\left(\left(\underset{m}{\alpha}\ominus\underset{k}{\beta}\right)\wedge \mathbf{x}\right)\wedge \mathbf{z} == -\star\mathcal{S}\left[\star\Sigma\left[(-1)^{\star r[1]\,(m+1)}\left(\left(\underset{m}{\alpha}\wedge\star\mathbb{S}^{\star}[\mathbf{z}]\right)\ominus\left(\underset{k}{\beta}\ominus\star\mathbb{S}_{\star}[\mathbf{z}]\right)\right)\wedge \mathbf{x}\right]\right]$$

$$(\underset{m}{\alpha}\ominus\underset{k}{\beta})\wedge \mathbf{x}\wedge \mathbf{z} == -(-1)^{1+m}(\underset{m}{\alpha}\ominus(\underset{k}{\beta}\ominus \mathbf{z}))\wedge \mathbf{x} - ((\underset{m}{\alpha}\wedge \mathbf{z})\ominus\underset{k}{\beta})\wedge \mathbf{x}$$

$$\left(\underset{m}{\alpha}\ominus\underset{k}{\beta}\right)\wedge(\mathbf{x}\wedge \mathbf{z}) == \star\mathcal{S}\left[\star\Sigma\left[(-1)^{\star r[2]\,(m+1)}\left(\underset{m}{\alpha}\wedge\star\mathbb{S}^{\star}[\mathbf{x}\wedge \mathbf{z}]\right)\ominus\left(\underset{k}{\beta}\ominus\star\mathbb{S}_{\star}[\mathbf{x}\wedge \mathbf{z}]\right)\right]\right]$$

$$(\underset{m}{\alpha}\ominus\underset{k}{\beta})\wedge \mathbf{x}\wedge \mathbf{z} == \underset{m}{\alpha}\ominus(\underset{k}{\beta}\ominus(\mathbf{x}\wedge \mathbf{z})) + (-1)^{m}\left((\underset{m}{\alpha}\wedge \mathbf{x})\ominus(\underset{k}{\beta}\ominus \mathbf{z})\right) + (-1)^{1+m}\left((\underset{m}{\alpha}\wedge \mathbf{z})\ominus(\underset{k}{\beta}\ominus \mathbf{x})\right) + (\underset{m}{\alpha}\wedge \mathbf{x}\wedge \mathbf{z})\ominus\underset{k}{\beta}$$

As in the previous example, this formula is an expansion of the first or second formulae if the basic Interior Product Formula [6.100] is again applied to them.

◆ $((\underset{m}{\alpha} \ominus \underset{k}{\beta}) \wedge \underset{j}{x}) \ominus \underset{s}{z}$ ◆

The A form gives

$$\left(\left(\underset{m}{\alpha} \ominus \underset{k}{\beta}\right) \wedge \underset{j}{x}\right) \ominus \underset{s}{z} == \star\Sigma\left[(-1)^{\star r[s]\,(m-k)} \left(\left(\underset{m}{\alpha} \ominus \underset{k}{\beta}\right) \ominus \star S^{\star}\left[\underset{s}{z}\right]\right) \wedge \left(\underset{j}{x} \ominus \star S_{\star}\left[\underset{s}{z}\right]\right)\right]$$

The B form gives

$$\left(\left(\underset{m}{\alpha} \ominus \underset{k}{\beta}\right) \wedge \underset{j}{x}\right) \ominus \underset{s}{z} == \star\Sigma\left[(-1)^{\star r[j]\,(m+1)} \left(\left(\underset{m}{\alpha} \wedge \star S^{\star}\left[\underset{j}{x}\right]\right) \ominus \left(\underset{k}{\beta} \ominus \star S_{\star}\left[\underset{j}{x}\right]\right)\right) \ominus \underset{s}{z}\right]$$

◇ **Example 4**

The forms of the following results are different.

$$(((\alpha_1 \wedge \alpha_2) \ominus \beta) \wedge x) \ominus z ==$$
$$\star S\left[\star\Sigma\left[(-1)^{\star r[1]\,(1)} \left(((\alpha_1 \wedge \alpha_2) \ominus \beta) \ominus \star S^{\star}[z]\right) \wedge (x \ominus \star S_{\star}[z])\right]\right]$$

$$(((\alpha_1 \wedge \alpha_2) \ominus \beta) \wedge x) \ominus z == -x\,((z \wedge \beta) \ominus (\alpha_1 \wedge \alpha_2)) - (x \ominus z)\,((\alpha_1 \wedge \alpha_2) \ominus \beta)$$

$$(((\alpha_1 \wedge \alpha_2) \ominus \beta) \wedge x) \ominus z ==$$
$$\star S\left[\star\Sigma\left[(-1)^{\star r[1]\,(3)} \left(((\alpha_1 \wedge \alpha_2) \wedge \star S^{\star}[x]) \ominus (\beta \ominus \star S_{\star}[x])\right) \ominus z\right]\right]$$

$$(((\alpha_1 \wedge \alpha_2) \ominus \beta) \wedge x) \ominus z == -(x \ominus \beta)\,((\alpha_1 \wedge \alpha_2) \ominus z) - (x \wedge \alpha_1 \wedge \alpha_2) \ominus (z \wedge \beta)$$

But it is straightforward to show they are identical by converting the terms to scalar products.

```
★★V[α1, α2];
★S[ToScalarProducts[-x ((z ∧ β) ⊖ (α1 ∧ α2)) - (x ⊖ z) ((α1 ∧ α2) ⊖ β)]]
```

$$x\,((z \ominus \alpha_2)\,(\beta \ominus \alpha_1) - (z \ominus \alpha_1)\,(\beta \ominus \alpha_2)) + (x \ominus z)\,(\beta \ominus \alpha_2)\,\alpha_1 - (x \ominus z)\,(\beta \ominus \alpha_1)\,\alpha_2$$

```
★S[ToScalarProducts[-(x ⊖ β) ((α1 ∧ α2) ⊖ z) - (x ∧ α1 ∧ α2) ⊖ (z ∧ β)]]
```

$$x\,((z \ominus \alpha_2)\,(\beta \ominus \alpha_1) - (z \ominus \alpha_1)\,(\beta \ominus \alpha_2)) + (x \ominus z)\,(\beta \ominus \alpha_2)\,\alpha_1 - (x \ominus z)\,(\beta \ominus \alpha_1)\,\alpha_2$$

Complementary forms for Decomposition Formulae

In section 6.3 we saw that there were several ways in which we could relate the interior product of two elements and their complements. If we look at Interior Product Formula B for a 1-element x [6.108] we notice that the right hand side is composed of two terms, each of which corresponds to one of the forms discussed in section 6.3. Formula [6.108] is

$$\left(\underset{m}{\alpha} \ominus \underset{k}{\beta}\right) \wedge x == \left(\underset{m}{\alpha} \wedge x\right) \ominus \underset{k}{\beta} + (-1)^{m+1}\, \underset{m}{\alpha} \ominus \left(\underset{k}{\beta} \ominus x\right)$$

By replacing the second term on the right by its complement form [6.47] we obtain after simplifying the signs:

$$\left(\underset{m}{\alpha} \ominus \underset{k}{\beta}\right) \wedge x == \left(\underset{m}{\alpha} \wedge x\right) \ominus \underset{k}{\beta} + (-1)^{(m-k)\,(n-m)} \left(\overline{\underset{k}{\beta}} \wedge x\right) \ominus \overline{\underset{m}{\alpha}} \qquad 6.121$$

By replacing the first term on the right from complement form [6.46] we obtain

$$\left(\underset{m}{\alpha}\ominus\underset{k}{\beta}\right)\wedge x == (-1)^{(m-k+1)\ (n-m+1)}\ \overline{\underset{k}{\beta}}\ominus\left(\overline{\underset{m}{\alpha}}\ominus x\right) + (-1)^{m+1}\ \underset{m}{\alpha}\ominus\left(\underset{k}{\beta}\ominus x\right) \qquad 6.122$$

The point to note here is that in both cases, both terms on the right-hand side now have the same form.

This is particularly suggestive in the case where $\underset{m}{\alpha}$ is equal to $\underset{k}{\beta}$ enabling us to write the decomposition of x in either of two ways:

$$x == \left(\underset{m}{\hat{\alpha}}\wedge x\right)\ominus\underset{m}{\hat{\alpha}} + \left(\overline{\underset{m}{\hat{\alpha}}}\wedge x\right)\ominus\overline{\underset{m}{\hat{\alpha}}} \qquad 6.123$$

$$x == (-1)^{(n-m+1)}\ \overline{\underset{m}{\hat{\alpha}}}\ominus\left(\overline{\underset{m}{\hat{\alpha}}}\ominus x\right) + (-1)^{m+1}\ \underset{m}{\hat{\alpha}}\ominus\left(\underset{m}{\hat{\alpha}}\ominus x\right) \qquad 6.124$$

For example, the decomposition formula in this form shows that the 1-element x has two components constructed *identically* (apart from a possible sign) from the unit *m*-element $\underset{m}{\hat{\alpha}}$ and its complement. And that these components can be expressed in either of two distinct ways.

Complementary forms for Interior Product Formulae

We can also develop complementary forms for the Interior Product Formulae. For example, let us consider Interior Product Formula B [6.107].

$$\left(\underset{m}{\alpha}\ominus\underset{k}{\beta}\right)\wedge\underset{j}{x} == \star\Sigma\left[(-1)^{\star r[j]\ (m+1)}\left(\underset{m}{\alpha}\wedge\star S^{\star}\left[\underset{j}{x}\right]\right)\ominus\left(\underset{k}{\beta}\ominus\star S_{\star}\left[\underset{j}{x}\right]\right)\right]$$

If we replace $\underset{m}{\alpha}$ by the complement of $\underset{k}{\beta}$, and $\underset{k}{\beta}$ by the complement of $\underset{m}{\alpha}$, we get

$$\left(\overline{\underset{k}{\beta}}\ominus\overline{\underset{m}{\alpha}}\right)\wedge\underset{j}{x} == \star\Sigma\left[(-1)^{\star r[j]\ (\star n-k+1)}\left(\overline{\underset{k}{\beta}}\wedge\star S^{\star}\left[\underset{j}{x}\right]\right)\ominus\left(\overline{\underset{m}{\alpha}}\ominus\star S_{\star}\left[\underset{j}{x}\right]\right)\right]$$

However, by [6.20] an interior product of elements can always be expressed as the interior product of their complements in reverse order.

$$\underset{m}{\alpha}\ominus\underset{k}{\beta} == (-1)^{(m-k)\ (\star n-m)}\ \overline{\underset{k}{\beta}}\ominus\overline{\underset{m}{\alpha}}$$

Substituting back in to the previous formula enables us to return to the original left hand side of the equation.

$$\left(\underset{m}{\alpha}\ominus\underset{k}{\beta}\right)\wedge\underset{j}{x} ==$$
$$(-1)^{(m-k)\ (\star n-m)}\ \star\Sigma\left[(-1)^{\star r[j]\ (\star n-k+1)}\left(\overline{\underset{k}{\beta}}\wedge\star S^{\star}\left[\underset{j}{x}\right]\right)\ominus\left(\overline{\underset{m}{\alpha}}\ominus\star S_{\star}\left[\underset{j}{x}\right]\right)\right] \qquad 6.125$$

We can call this the complement form of the Interior Product Formula B because it expands the *same* expression as the Interior Product Formula B, but in the expansion $\underset{m}{\alpha}$ and $\underset{k}{\beta}$ are complemented, and their positions exchanged within each term. There may also be a change in sign of the terms.

◆ Displaying the structure of the computation

The stages in the construction of this formula may be displayed as follows. Begin by replacing j by 2, and enclosing anything we expect to be a list with column matrix brackets |□|.

- **1.** Begin by entering

```
|X_List| := MatrixForm[X]
```

$$(-1)^{(m-k)\,(\star n-m)} \left| \left| (-1)^{\star r[2]\,(\star n-k+1)} \right| \left(\underset{k}{\overline{\beta}} \wedge \left| \star S^{\star}\left[\underset{2}{x}\right] \right| \right) \ominus \left(\underset{m}{\overline{\alpha}} \ominus \left| \star S_{\star}\left[\underset{2}{x}\right] \right| \right) \right|$$

$$(-1)^{(\star n-m)\,(-k+m)} \left| \begin{pmatrix} 1 \\ (-1)^{1+\star n-k} \\ 1 \end{pmatrix} \underset{k}{\overline{\beta}} \wedge \begin{pmatrix} \{x_1 \wedge x_2\} \\ \{x_2,\, -x_1\} \\ \{1\} \end{pmatrix} \ominus \left(\underset{m}{\overline{\alpha}} \ominus \begin{pmatrix} \{1\} \\ \{x_1,\, x_2\} \\ \{x_1 \wedge x_2\} \end{pmatrix} \right) \right|$$

- **2.** Enter this output again.

$$(-1)^{(\star n-m)\,(-k+m)} \left| \begin{pmatrix} 1 \\ (-1)^{1+\star n-k} \\ 1 \end{pmatrix} \underset{k}{\overline{\beta}} \wedge \begin{pmatrix} \{x_1 \wedge x_2\} \\ \{x_2,\, -x_1\} \\ \{1\} \end{pmatrix} \ominus \left(\underset{m}{\overline{\alpha}} \ominus \begin{pmatrix} \{1\} \\ \{x_1,\, x_2\} \\ \{x_1 \wedge x_2\} \end{pmatrix} \right) \right|$$

$$(-1)^{(\star n-m)\,(-k+m)} \begin{pmatrix} \left\{\underset{k}{\overline{\beta}} \wedge x_1 \wedge x_2 \ominus \left(\underset{m}{\overline{\alpha}} \ominus 1\right)\right\} \\ \left\{(-1)^{1+\star n-k}\, \underset{k}{\overline{\beta}} \wedge x_2 \ominus \left(\underset{m}{\overline{\alpha}} \ominus x_1\right),\; (-1)^{1+\star n-k}\, \underset{k}{\overline{\beta}} \wedge -x_1 \ominus \left(\underset{m}{\overline{\alpha}} \ominus x_2\right)\right\} \\ \left\{\underset{k}{\overline{\beta}} \wedge 1 \ominus \left(\underset{m}{\overline{\alpha}} \ominus x_1 \wedge x_2\right)\right\} \end{pmatrix}$$

- **3.** Sum this column of lists.

$$(-1)^{(\star n-m)\,(-k+m)}$$

$$\star\Sigma\left[\begin{pmatrix} \left\{\underset{k}{\overline{\beta}} \wedge x_1 \wedge x_2 \ominus \left(\underset{m}{\overline{\alpha}} \ominus 1\right)\right\} \\ \left\{(-1)^{1+\star n-k}\, \underset{k}{\overline{\beta}} \wedge x_2 \ominus \left(\underset{m}{\overline{\alpha}} \ominus x_1\right),\; (-1)^{1+\star n-k}\, \underset{k}{\overline{\beta}} \wedge -x_1 \ominus \left(\underset{m}{\overline{\alpha}} \ominus x_2\right)\right\} \\ \left\{\underset{k}{\overline{\beta}} \wedge 1 \ominus \left(\underset{m}{\overline{\alpha}} \ominus x_1 \wedge x_2\right)\right\} \end{pmatrix} \right]$$

$$(-1)^{(\star n-m)\,(-k+m)} \left(\underset{k}{\overline{\beta}} \wedge 1 \ominus \left(\underset{m}{\overline{\alpha}} \ominus x_1 \wedge x_2\right) + (-1)^{1+\star n-k}\, \underset{k}{\overline{\beta}} \wedge -x_1 \ominus \left(\underset{m}{\overline{\alpha}} \ominus x_2\right) + (-1)^{1+\star n-k}\, \underset{k}{\overline{\beta}} \wedge x_2 \ominus \left(\underset{m}{\overline{\alpha}} \ominus x_1\right) + \underset{k}{\overline{\beta}} \wedge x_1 \wedge x_2 \ominus \left(\underset{m}{\overline{\alpha}} \ominus 1\right) \right)$$

To get this directly we could of course simply have entered formula [6.125] with j set equal to 2.

- **4.** Simplify if we wish.

```
★S[%]
```

$$(-1)^{\star n\,k+m+\star n\,m+k\,m}\left(\overline{\underset{k}{\beta}}\ominus\left(\overline{\underset{m}{\alpha}}\ominus x_1\wedge x_2\right)\right)+$$

$$(-1)^{\star n+k}\left(\overline{\underset{k}{\beta}}\wedge x_1\ominus\left(\overline{\underset{m}{\alpha}}\ominus x_2\right)\right)+(-1)^{1+\star n+k}\left(\overline{\underset{k}{\beta}}\wedge x_2\ominus\left(\overline{\underset{m}{\alpha}}\ominus x_1\right)\right)+\overline{\underset{k}{\beta}}\wedge x_1\wedge x_2\ominus\overline{\underset{m}{\alpha}}$$

◆ Validation example

Because this formula explicitly involves the complements of elements, validation of the formula requires both sides of it to be evaluated using the same explicit metric. This means that we must apply the *GrassmannAlgebra* function `ToMetricElements` to convert any scalar products which arise.

As an example we take the simplest non-trivial case of 1-elements in a 2-space with a general metric.

```
★A; ★B2; {n, m, k, j} = {2, 1, 1, 1}; DeclareMetric[g]
```

$$\{\{g_{1,1}, g_{1,2}\}, \{g_{1,2}, g_{2,2}\}\}$$

$$f_1 = \left(\underset{m}{\alpha}\ominus\underset{k}{\beta}\right)\wedge\underset{j}{x} == (-1)^{(m-k)\,(n-m)}\star\Sigma\left[(-1)^{\star r[j]\,(n-k+1)}\left(\overline{\underset{k}{\beta}}\wedge\star S^{\star}\left[\underset{j}{x}\right]\right)\ominus\left(\overline{\underset{m}{\alpha}}\ominus\star S_{\star}\left[\underset{j}{x}\right]\right)\right]$$

$$\left(\underset{1}{\alpha}\ominus\underset{1}{\beta}\right)\wedge\underset{1}{x} == \overline{\underset{1}{\beta}}\ominus\left(\overline{\underset{1}{\alpha}}\ominus\underset{1}{x}\right)+\overline{\underset{1}{\beta}}\wedge\underset{1}{x}\ominus\overline{\underset{1}{\alpha}}$$

The first step is to express each of the 1-elements in the current basis.

```
f2 = ★S[ComposeShortBasisForm /@ f1]
```

$$(e_1\,x_1+e_2\,x_2)\wedge((e_1\,\alpha_1+e_2\,\alpha_2)\ominus(e_1\,\beta_1+e_2\,\beta_2)) ==$$
$$(\overline{e_1}\,\beta_1+\overline{e_2}\,\beta_2)\ominus((\overline{e_1}\,\alpha_1+\overline{e_2}\,\alpha_2)\ominus(e_1\,x_1+e_2\,x_2))-$$
$$(e_1\,x_1+e_2\,x_2)\wedge(\overline{e_1}\,\beta_1+\overline{e_2}\,\beta_2)\ominus(\overline{e_1}\,\alpha_1+\overline{e_2}\,\alpha_2)$$

Then the complements are converted using the currently declared metric. This may be a long expression, so here we have only shown part of it.

```
f₃ = ★𝒢[ConvertComplements[f₂]]; Short[f₃, 20]
```

$$e_1\,((e_1\ominus e_1)\,x_1\,\alpha_1\,\beta_1+(e_2\ominus e_2)\,x_1\,\alpha_2\,\beta_2+(e_1\ominus e_2)\,(x_1\,\alpha_2\,\beta_1+x_1\,\alpha_1\,\beta_2))+$$
$$e_2\,((e_1\ominus e_1)\,x_2\,\alpha_1\,\beta_1+(e_2\ominus e_2)\,x_2\,\alpha_2\,\beta_2+(e_1\ominus e_2)\,(x_2\,\alpha_2\,\beta_1+x_2\,\alpha_1\,\beta_2))==$$
$$(e_1\wedge e_2\ominus e_2)$$
$$\left(-\frac{x_1\,\alpha_1\,\beta_1\,g_{1,1}^2}{\star g}-\frac{x_2\,\alpha_1\,\beta_1\,g_{1,1}\,g_{1,2}}{\star g}-\frac{x_1\,\alpha_2\,\beta_1\,g_{1,1}\,g_{1,2}}{\star g}-\frac{x_1\,\alpha_1\,\beta_2\,g_{1,1}\,g_{1,2}}{\star g}-\right.$$
$$\left.\frac{x_2\,\alpha_2\,\beta_1\,g_{1,2}^2}{\star g}-\frac{x_1\,\alpha_2\,\beta_2\,g_{1,2}^2}{\star g}-\frac{x_2\,\alpha_1\,\beta_2\,g_{1,1}\,g_{2,2}}{\star g}-\frac{x_2\,\alpha_2\,\beta_2\,g_{1,2}\,g_{2,2}}{\star g}\right)+$$
$$\ll 2\gg+e_1\left((e_2\ominus e_2)\left(-\frac{x_2\,\alpha_1\,\beta_1\,g_{1,1}\,g_{1,2}}{\star g}-\frac{x_2\,\alpha_2\,\beta_1\,g_{1,2}^2}{\star g}-\right.\right.$$
$$\left.\frac{x_2\,\alpha_1\,\beta_2\,g_{1,1}\,g_{2,2}}{\star g}-\frac{x_2\,\alpha_2\,\beta_2\,g_{1,2}\,g_{2,2}}{\star g}\right)+\ll 1\gg+$$
$$(e_1\ominus e_2)\left(-\frac{x_1\,\alpha_1\,\beta_1\,g_{1,1}\,g_{1,2}}{\star g}+\frac{x_2\,\alpha_1\,\beta_1\,g_{\ll 1\gg}^2}{\star g}-\frac{\ll 1\gg}{\star g}-\ll 1\gg+\right.$$
$$\left.\left.\ll 1\gg+\frac{\ll 1\gg}{\ll 2\gg}-\frac{x_1\ll 3\gg g_{\ll 1\gg}}{\star g}+\frac{x_2\,\alpha_2\,\beta_2\,g_{2,2}^2}{\star g}\right)\right)$$

Replacing the scalar products by their metric elements shows the identity for a general metric, and hence this simple example is true for any metric.

```
ToMetricElements[f₃]
True
```

Interior Product Formulae for 1-elements

As a final summary, we collect here some examples of the results of the previous general formulae for the specific and most common case where x is of grade 1.

◆ The Interior Product Formula

$$\left(\underset{m}{\alpha}\wedge\underset{k}{\beta}\right)\ominus x == \left(\underset{m}{\alpha}\ominus x\right)\wedge\underset{k}{\beta}+(-1)^m\,\underset{m}{\alpha}\wedge\left(\underset{k}{\beta}\ominus x\right) \qquad 6.126$$

◆ The Interior Product Formula B

$$\left(\underset{m}{\alpha}\ominus\underset{k}{\beta}\right)\wedge x == \left(\underset{m}{\alpha}\wedge x\right)\ominus\underset{k}{\beta}+(-1)^{m+1}\,\underset{m}{\alpha}\ominus\left(\underset{k}{\beta}\ominus x\right) \qquad 6.127$$

◆ The Orthogonal Decomposition Formula

$$x == \left(\underset{m}{\hat{\alpha}} \wedge x\right) \ominus \underset{m}{\hat{\alpha}} + (-1)^{m+1} \left(\underset{m}{\hat{\alpha}} \ominus \left(\underset{m}{\hat{\alpha}} \ominus x\right)\right) \qquad 6.128$$

◆ The complement form of the Orthogonal Decomposition Formula

$$x == \left(\underset{m}{\hat{\alpha}} \wedge x\right) \ominus \underset{m}{\hat{\alpha}} + \left(\overline{\underset{m}{\hat{\alpha}}} \wedge x\right) \ominus \overline{\underset{m}{\hat{\alpha}}} \qquad 6.129$$

◆ The Orthogonal Decomposition Formula B

$$\underset{m}{\alpha} == (-1)^{m} \left(\left(\underset{m}{\alpha} \wedge \hat{x}\right) \ominus \hat{x} - \left(\underset{m}{\alpha} \ominus \hat{x}\right) \wedge \hat{x}\right) \qquad 6.130$$

◆ Augmented Interior Product Formula 1

Taking the exterior product of the Interior Product Formula with z, and rearranging the exterior products gives:

$$\left(\left(\underset{m}{\alpha} \wedge \underset{k}{\beta}\right) \ominus x\right) \wedge z == \left(\left(\underset{m}{\alpha} \ominus x\right) \wedge \underset{k}{\beta}\right) \wedge z + (-1)^{m} \left(\underset{m}{\alpha} \wedge \left(\underset{k}{\beta} \ominus x\right)\right) \wedge z$$

$$== \left(\underset{m}{\alpha} \ominus x\right) \wedge \left(\underset{k}{\beta} \wedge z\right) + (-1)^{m} (-1)^{k-1} \left(\underset{m}{\alpha} \wedge z\right) \wedge \left(\underset{k}{\beta} \ominus x\right)$$

$$\left(\left(\underset{m}{\alpha} \wedge \underset{k}{\beta}\right) \ominus x\right) \wedge z == \left(\underset{m}{\alpha} \ominus x\right) \wedge \left(\underset{k}{\beta} \wedge z\right) + (-1)^{m+k-1} \left(\underset{m}{\alpha} \wedge z\right) \wedge \left(\underset{k}{\beta} \ominus x\right) \qquad 6.131$$

◆ Augmented Interior Product Formula 2

An analogous formula is obtained by taking the interior product of the Interior Product Formula B with z, and applying the extended interior product formula [6.40].

$$\left(\left(\underset{m}{\alpha} \ominus \underset{k}{\beta}\right) \wedge x\right) \ominus z == \left(\left(\underset{m}{\alpha} \wedge x\right) \ominus \underset{k}{\beta}\right) \ominus z + (-1)^{m+1} \left(\underset{m}{\alpha} \ominus \left(\underset{k}{\beta} \ominus x\right)\right) \ominus z$$

$$== \left(\underset{m}{\alpha} \wedge x\right) \ominus \left(\underset{k}{\beta} \wedge z\right) + (-1)^{m+1} (-1)^{k-1} \left(\underset{m}{\alpha} \ominus z\right) \ominus \left(\underset{k}{\beta} \ominus x\right)$$

$$\left(\left(\underset{m}{\alpha} \ominus \underset{k}{\beta}\right) \wedge x\right) \ominus z == \left(\underset{m}{\alpha} \wedge x\right) \ominus \left(\underset{k}{\beta} \wedge z\right) + (-1)^{m+k} \left(\underset{m}{\alpha} \ominus z\right) \ominus \left(\underset{k}{\beta} \ominus x\right) \qquad 6.132$$

6.9 Interior Products of Interpreted Elements

Introduction

In this section we consider interior products of interpreted elements, that is, points and vectors and their higher grade products. We have already discussed complements of bound elements in sections 5.11 and 5.12, and the measure of bound elements in section 6.6.

Several of the results of this section have already been derived in section 6.6. We revisit them here for completeness.

As we did there, we will assume a hybrid metric in which the origin $\star\mathbb{O}$ is orthogonal to all vectors and has measure unity. The metric on the vector subspace remains arbitrary. For a discussion of this hybrid metric see section 5.11 and formula [5.80].

With this hybrid metric it is easy to see that the measure $|P|$ of a point P with position vector $\star\mathbb{O} + p$ is given by

$$|P|^2 == P \ominus P == (\star\mathbb{O} + p) \ominus (\star\mathbb{O} + p) == \star\mathbb{O} \ominus \star\mathbb{O} + p \ominus p == 1 + |p|^2$$

$$|\star\mathbb{O}| == 1 \qquad |P|^2 == 1 + |p|^2 \qquad P == \star\mathbb{O} + p \tag{6.133}$$

Thus we conclude that with such a hybrid metric the measure of a point P is greater than the measure of its position vector. This may at first encounter appear strange. Nevertheless it enables us to bring the whole of metric geometry to bear consistently onto spaces in which we have distinguished points and vectors by interpreting one of the basis elements as an origin. And indeed it turns out to lead to some interesting and intuitively acceptable results.

In particular, it is interesting that the *measure of a point is never zero*. The measure of a point starts at 1 at the origin, and then increases as it moves further way from the origin. The only *unit point* is the origin.

However, there are cases in which we would like the measure of *all* points to be unity, for example, in the dual number approach to mechanics (see Volume 2); or in the definition of translation operators in a hypercomplex algebra (see also Volume 2). In these cases we can again retain all the formulae we have developed so far while at the same time accommodating this new requirement by defining a new variant of the point: the *dual point*. A dual point has its position vector multiplied by the *dual unit* ω which has the *formal* property that ω^2 is zero. This symbol was introduced by Clifford in 1873.

$$|P|^2 == P \ominus P == (\star\mathbb{O} + \omega\, p) \ominus (\star\mathbb{O} + \omega\, p) == 1 + \omega^2\, p \ominus p == 1$$

It should be remarked that here too, the applications involving dual points are predicated on the origin having unit measure.

The following formulae involve the interior product of two interpreted elements. In considering these formulae and their derivations it should be remembered that the interior product is zero whenever the right hand factor is of higher grade than that of the left hand factor.

Interior products involving the origin

◆ The measure of the origin

The hybrid metric [5.80] *defines* the interior product of the origin with itself to be unity.

$$|\star\mathbb{O}|^2 == \star\mathbb{O} \ominus \star\mathbb{O} == 1 \qquad |\star\mathbb{O}| == 1 \tag{6.134}$$

◆ The interior product of an *m*-vector with the origin

The hybrid metric also defines the scalar product of the origin with any vector to be zero.

The interior product of the origin with an *m*-vector can be expanded into a sum whose coefficients are the scalar products of the origin with the factors of the *m*-vector. Hence the interior product of the origin with any *m*-vector is zero. If α is a vector, and $\underset{m}{\alpha}$ is a (perhaps non-simple) *m*-vector, then

$$\alpha \ominus \star\mathbb{O} == 0 \qquad \underset{m}{\alpha} \ominus \star\mathbb{O} == 0 \tag{6.135}$$

◆ The interior product of an origin-bound *m*-vector with a *k*-vector

The interior product of an origin-bound *m*-vector with a *k*-vector is an origin-bound (*m*–*k*)-vector. To show this begin with the basic Interior Product Formula [6.100]. Note that *k* must be less than or equal to *m* for this to be a non-trivial relationship, and the *k*-vector must be the second factor in the interior product. Formula [6.100] is

$$\left(\underset{m}{\alpha} \ominus \underset{k}{\beta}\right) \wedge x == \left(\underset{m}{\alpha} \wedge x\right) \ominus \underset{k}{\beta} + (-1)^{m-1} \underset{m}{\alpha} \ominus \left(\underset{k}{\beta} \ominus x\right)$$

Putting x equal to the origin, and noting that the last term becomes zero because $\underset{k}{\beta} \ominus \star\mathbb{O}$ is zero by [6.135] we can rearrange the remaining two terms to give

$$\left(\star\mathbb{O} \wedge \underset{m}{\alpha}\right) \ominus \underset{k}{\beta} == (-1)^k \star\mathbb{O} \wedge \left(\underset{m}{\alpha} \ominus \underset{k}{\beta}\right) \tag{6.136}$$

If the *m*-vector is the *first* factor in the interior product, the result is zero.

$$\underset{m}{\alpha} \ominus \left(\star\mathbb{O} \wedge \underset{k}{\beta}\right) == \left(\underset{m}{\alpha} \ominus \star\mathbb{O}\right) \ominus \underset{k}{\beta} == 0$$

$$\underset{m}{\alpha} \ominus \left(\star\mathbb{O} \wedge \underset{k}{\beta}\right) == 0 \tag{6.137}$$

◆ The interior product of an origin-bound *m*-vector with the origin

The interior product of an origin-bound *m*-vector with the origin is an *m*-vector.

To show this put both x and $\underset{k}{\beta}$ equal to the origin in [6.100]; then the left hand side of the equation becomes zero and we have, after some rearrangement

$$\left(\star\mathbb{O}\wedge\underset{m}{\alpha}\right)\ominus\star\mathbb{O} == \underset{m}{\alpha}\ominus(\star\mathbb{O}\ominus\star\mathbb{O}) == \underset{m}{\alpha}$$

$$\left(\star\mathbb{O}\wedge\underset{m}{\alpha}\right)\ominus\star\mathbb{O} == \underset{m}{\alpha} \qquad 6.138$$

Thus, taking the interior product of an *m*-vector bound through the origin with the origin, frees the *m*-vector. (This is also true more generally for the interior product of *any* bound *m*-vector with the origin. See also [6.98] and [6.149]).

◆ The interior product of two origin-bound multivectors

The interior product of two origin-bound multivectors is a multivector.

By extending the interior product using formula [6.40], and then applying the formulae [6.136] and [6.137] that we have just derived we get

$$\left(\star\mathbb{O}\wedge\underset{m}{\alpha}\right)\ominus\left(\star\mathbb{O}\wedge\underset{k}{\beta}\right)$$

$$== (-1)^{k}\left(\left(\left(\star\mathbb{O}\wedge\underset{m}{\alpha}\right)\ominus\underset{k}{\beta}\right)\ominus\star\mathbb{O}\right) == \left(\star\mathbb{O}\wedge\left(\underset{m}{\alpha}\ominus\underset{k}{\beta}\right)\right)\ominus\star\mathbb{O} == \underset{m}{\alpha}\ominus\underset{k}{\beta}$$

$$\left(\star\mathbb{O}\wedge\underset{m}{\alpha}\right)\ominus\left(\star\mathbb{O}\wedge\underset{k}{\beta}\right) == \underset{m}{\alpha}\ominus\underset{k}{\beta} \qquad 6.139$$

Interior products involving points

In this section we repeat the format of the previous section, but develop analogous formulae for a general point instead of the origin. In all the formulae P is equal to $\star\mathbb{O} + p$.

Note carefully that the formulae we develop here are only valid for the symbols having their expected meaning as either bound or free elements. This is in distinction to the general formulae derived at the beginning of the chapter (for example the basic Interior Product Formula [6.100]) in which the symbols may represent any entities. Remember also that for an interior product to be non-zero, the grade of the first factor must be greater than or equal to the grade of the second.

The derivations use formulae [6.136], [6.137], [6.138] and [6.139].

◆ The measure of a point

The squared measure of a point is the squared measure of its position vector + 1.

$$|P|^2 == |\star\mathbb{O}+p|^2 == 1+|p|^2 \qquad 6.140$$

◆ The scalar product of two points

The scalar product of two points is the scalar product of their position vectors + 1.

$$\mathbf{P} \ominus \mathbf{Q} == (\star\mathbb{O} + \mathbf{p}) \ominus (\star\mathbb{O} + \mathbf{q}) == 1 + \mathbf{p} \ominus \mathbf{q} \qquad 6.141$$

◆ The interior product of an m-vector with a point

The interior product of a point with an *m*-vector is the interior product of its position vector with the *m*-vector.

$$\underset{m}{\alpha} \ominus \mathbf{P} == \underset{m}{\alpha} \ominus (\star\mathbb{O} + \mathbf{p}) == \underset{m}{\alpha} \ominus \mathbf{p} \qquad 6.142$$

◆ The interior product of a bound *m*-vector with a *k*-vector

$$\left(\mathbf{P} \wedge \underset{m}{\alpha}\right) \ominus \underset{k}{\beta} == \left((\star\mathbb{O} + \mathbf{p}) \wedge \underset{m}{\alpha}\right) \ominus \underset{k}{\beta} == (-1)^{k} \star\mathbb{O} \wedge \left(\underset{m}{\alpha} \ominus \underset{k}{\beta}\right) + \left(\mathbf{p} \wedge \underset{m}{\alpha}\right) \ominus \underset{k}{\beta}$$

$$\left(\mathbf{P} \wedge \underset{m}{\alpha}\right) \ominus \underset{k}{\beta} == (-1)^{k} \star\mathbb{O} \wedge \left(\underset{m}{\alpha} \ominus \underset{k}{\beta}\right) + \left(\mathbf{p} \wedge \underset{m}{\alpha}\right) \ominus \underset{k}{\beta} \qquad 6.143$$

◆ The interior product of an *m*-vector with a bound *k*-vector

$$\underset{m}{\alpha} \ominus \left(\mathbf{Q} \wedge \underset{k}{\beta}\right) == \underset{m}{\alpha} \ominus \left(\star\mathbb{O} \wedge \underset{k}{\beta}\right) + \underset{m}{\alpha} \ominus \left(\mathbf{q} \wedge \underset{k}{\beta}\right)$$

$$== \left(\left(\underset{m}{\alpha} \ominus \star\mathbb{O}\right) \ominus \underset{k}{\beta}\right) + \underset{m}{\alpha} \ominus \left(\mathbf{q} \wedge \underset{k}{\beta}\right) == \underset{m}{\alpha} \ominus \left(\mathbf{q} \wedge \underset{k}{\beta}\right)$$

$$\underset{m}{\alpha} \ominus \left(\mathbf{Q} \wedge \underset{k}{\beta}\right) == \underset{m}{\alpha} \ominus \left(\mathbf{q} \wedge \underset{k}{\beta}\right) \qquad 6.144$$

◆ The interior product of two bound multivectors

$$\left(\mathbf{P} \wedge \underset{m}{\alpha}\right) \ominus \left(\mathbf{Q} \wedge \underset{k}{\beta}\right) == \left((\star\mathbb{O} + \mathbf{p}) \wedge \underset{m}{\alpha}\right) \ominus \left((\star\mathbb{O} + \mathbf{q}) \wedge \underset{k}{\beta}\right)$$

$$== \left(\star\mathbb{O} \wedge \underset{m}{\alpha}\right) \ominus \left(\star\mathbb{O} \wedge \underset{k}{\beta}\right) + \left(\star\mathbb{O} \wedge \underset{m}{\alpha}\right) \ominus \left(\mathbf{q} \wedge \underset{k}{\beta}\right) + \left(\mathbf{p} \wedge \underset{m}{\alpha}\right) \ominus \left(\star\mathbb{O} \wedge \underset{k}{\beta}\right) + \left(\mathbf{p} \wedge \underset{m}{\alpha}\right) \ominus \left(\mathbf{q} \wedge \underset{k}{\beta}\right)$$

The first term is

$$\left(\star\mathbb{O} \wedge \underset{m}{\alpha}\right) \ominus \left(\star\mathbb{O} \wedge \underset{k}{\beta}\right) == \underset{m}{\alpha} \ominus \underset{k}{\beta}$$

The second term is

$$\left(\star\mathbb{O} \wedge \underset{m}{\alpha}\right) \ominus \left(\mathbf{q} \wedge \underset{k}{\beta}\right) == (-1)^{k+1} \star\mathbb{O} \wedge \left(\underset{m}{\alpha} \ominus \left(\mathbf{q} \wedge \underset{k}{\beta}\right)\right)$$

The third term is zero.

$$\left(\mathbf{p} \wedge \underset{m}{\alpha}\right) \ominus \left(\star\mathbb{O} \wedge \underset{k}{\beta}\right) == \left(\left(\mathbf{p} \wedge \underset{m}{\alpha}\right) \ominus \star\mathbb{O}\right) \ominus \underset{k}{\beta} == 0$$

Collecting these terms together we have

$$\left(P\wedge\underset{m}{\alpha}\right)\ominus\left(Q\wedge\underset{k}{\beta}\right) == (-1)^{k+1}\star\mathbb{O}\wedge\left(\underset{m}{\alpha}\ominus\left(q\wedge\underset{k}{\beta}\right)\right)+\underset{m}{\alpha}\ominus\underset{k}{\beta}+\left(p\wedge\underset{m}{\alpha}\right)\ominus\left(q\wedge\underset{k}{\beta}\right) \qquad 6.145$$

Special cases

From these formulae, particularly [6.145], we can straightforwardly derive many special cases. Below we present some of the more common ones.

◆ The interior product of a bound *m*-vector with a point

$$\left(P\wedge\underset{m}{\alpha}\right)\ominus Q == -\star\mathbb{O}\wedge\left(\underset{m}{\alpha}\ominus q\right)+\underset{m}{\alpha}+\left(p\wedge\underset{m}{\alpha}\right)\ominus q \qquad 6.146$$

◆ The interior product of a bound *m*-vector with a point (alternative)

If we put x and $\underset{k}{\beta}$ equal to points P and Q in the basic Interior Product Formula [6.100] we have

$$\left(\underset{m}{\alpha}\ominus Q\right)\wedge P == \left(\underset{m}{\alpha}\wedge P\right)\ominus Q+(-1)^{m-1}\underset{m}{\alpha}\ominus(Q\ominus P)$$

After some rearrangement, this gives

$$\left(P\wedge\underset{m}{\alpha}\right)\ominus Q == (P\ominus Q)\,\underset{m}{\alpha}-P\wedge\left(\underset{m}{\alpha}\ominus Q\right) \qquad 6.147$$

◆ When the right hand factor is bound through the origin

When the right hand factor is bound through the origin, q is zero so that

$$\left(P\wedge\underset{m}{\alpha}\right)\ominus\left(\star\mathbb{O}\wedge\underset{k}{\beta}\right) == \underset{m}{\alpha}\ominus\underset{k}{\beta} \qquad 6.148$$

Thus, the interior product of any bound *m*-vector with a *k*-vector bound through the origin results in the (free) interior product of the *m*-vector and the *k*-vector.

◆ When the right hand factor is the origin

When $\underset{k}{\beta}$ is unity [6.148] becomes

$$\left(P\wedge\underset{m}{\alpha}\right)\ominus\star\mathbb{O} == \underset{m}{\alpha} \qquad 6.149$$

We have already seen this result in formula [6.98].

◆ **When the left hand factor is bound through the origin**

When the left hand m-vector is bound through the origin, p is zero but there is an extra term on the right hand side.

$$\left(\star\mathbb{O}\wedge\underset{m}{\alpha}\right)\ominus\left(\mathbf{Q}\wedge\underset{k}{\beta}\right) == (-1)^{k+1}\star\mathbb{O}\wedge\left(\underset{m}{\alpha}\ominus\left(\mathbf{q}\wedge\underset{k}{\beta}\right)\right)+\underset{m}{\alpha}\ominus\underset{k}{\beta} \qquad 6.150$$

◆ **When the bound multivectors are of the same grade**

When the bound multivectors are of the same grade the first term on the right hand side of the above formula is zero since the grade of the second factor of the interior product is greater than the first.

$$\left(\mathbf{P}\wedge\underset{m}{\alpha}\right)\ominus\left(\mathbf{Q}\wedge\underset{m}{\beta}\right) == \underset{m}{\alpha}\ominus\underset{m}{\beta}+\left(\mathbf{p}\wedge\underset{m}{\alpha}\right)\ominus\left(\mathbf{q}\wedge\underset{m}{\beta}\right) \qquad 6.151$$

◆ **When the bound multivectors are equal**

Putting Q equal to P and $\underset{k}{\beta}$ equal to $\underset{m}{\alpha}$ in the previous formula allows us to write the squared measure of $\mathbf{P}\wedge\underset{m}{\alpha}$ as

$$\left(\mathbf{P}\wedge\underset{m}{\alpha}\right)\ominus\left(\mathbf{P}\wedge\underset{m}{\alpha}\right) == \underset{m}{\alpha}\ominus\underset{m}{\alpha}+\left(\mathbf{p}\wedge\underset{m}{\alpha}\right)\ominus\left(\mathbf{p}\wedge\underset{m}{\alpha}\right) \qquad 6.152$$

$$\left|\mathbf{P}\wedge\underset{m}{\alpha}\right|^2 == \left|\underset{m}{\alpha}\right|^2+\left|\mathbf{p}\wedge\underset{m}{\alpha}\right|^2 \qquad 6.153$$

Formula summary for points and bound vectors

◆ **The interior product of an origin-bound vector with a vector**

$$(\star\mathbb{O}\wedge\mathbf{x})\ominus\mathbf{y} == -\star\mathbb{O}\wedge(\mathbf{x}\ominus\mathbf{y}) \qquad 6.154$$

◆ **The interior product of an origin-bound vector with the origin**

$$(\star\mathbb{O}\wedge\mathbf{x})\ominus\star\mathbb{O} == \mathbf{x} \qquad 6.155$$

- **The interior product of two origin-bound vectors**

$$(\star\mathbb{O} \wedge x) \ominus (\star\mathbb{O} \wedge y) == x \ominus y \qquad 6.156$$

- **The interior product of a bound vector with a vector**

$$(P \wedge x) \ominus y == -\star\mathbb{O} \wedge (x \ominus y) + (p \wedge x) \ominus y \qquad 6.157$$

- **The interior product of a vector with a bound vector**

$$x \ominus (Q \wedge y) == x \ominus (q \wedge y) \qquad 6.158$$

- **The interior product of two bound vectors**

$$(P \wedge x) \ominus (Q \wedge y) == x \ominus y + (p \wedge x) \ominus (q \wedge y) \qquad 6.159$$

- **The interior product of a bound vector with a point**

$$(P \wedge x) \ominus Q == -\star\mathbb{O} \wedge (x \ominus q) + x + (p \wedge x) \ominus q \qquad 6.160$$

- **The interior product of a bound vector with a point (alternative)**

$$(P \wedge x) \ominus Q == (P \ominus Q)\, x - P \wedge (x \ominus Q) \qquad 6.161$$

- **When the right hand factor is bound through the origin**

$$(P \wedge x) \ominus (\star\mathbb{O} \wedge y) == x \ominus y \qquad 6.162$$

- **When the right hand factor is the origin**

$$(P \wedge x) \ominus \star\mathbb{O} == x \qquad 6.163$$

- **When the left hand factor is bound through the origin**

$$(\star\mathbb{O} \wedge x) \ominus (Q \wedge y) == \star\mathbb{O} \wedge \left(\underset{m}{\alpha} \ominus (q \wedge y)\right) + x \ominus y \qquad 6.164$$

◆ **When the bound vectors are equal**

$$(P \wedge x) \ominus (P \wedge x) == x \ominus x + (p \wedge x) \ominus (p \wedge x) \quad 6.165$$

$$|P \wedge x|^2 == |x|^2 + |p \wedge x|^2 \quad 6.166$$

6.10 The Cross Product

Defining a generalized cross product

The cross or vector product of the three-dimensional vector algebra of Gibbs *et al.* [Gibbs 1928] corresponds to two operations in Grassmann's more general algebra. *Taking the cross-product of two vectors in three dimensions corresponds to taking the complement of their exterior product.* However, whilst the usual cross product formulation is valid only for vectors in three dimensions, the exterior product formulation is valid for elements of *any* grade in *any* number of dimensions. Therefore the opportunity exists to generalize the concept.

Because our generalization reduces to the usual definition under the usual circumstances, we take the liberty of continuing to refer to the generalized cross product as, simply, the cross product.

The cross product of elements A and B of any grade is denoted $A \times B$ and is defined as the complement of their exterior product. The cross product of an *m*-element and a *k*-element is thus an (*n*–(*m*+*k*))-element.

$$A \times B == \overline{A \wedge B} \quad 6.167$$

This definition preserves the basic property of the cross product: that the cross product of two elements is an element orthogonal to both, and reduces to the usual notion for vectors in a three dimensional metric vector space.

If A and B are both 1-elements then their generalized cross product $A \times B$ is a scalar in 2-space, a 1-element in 3-space, and a 2-element in 4-space.

Cross products involving 1-elements

For 1-elements x, y, z, w, the definition above has the following consequences, independent of the dimension of the *n* of the space.

◆ **The triple cross product**

The triple cross product is a 1-element in any number of dimensions.

$$(x \times y) \times z == \overline{(\overline{x \wedge y}) \wedge z} == (x \wedge y) \ominus z == (z \ominus x)\, y - (z \ominus y)\, x$$

$$(x \times y) \times z == (x \wedge y) \ominus z == (z \ominus x)\, y - (z \ominus y)\, x \qquad 6.168$$

$$x \times (y \times z) == \overline{x \wedge (\overline{y \wedge z})} == \overline{x} \vee (y \wedge z)$$
$$== (-1)^{(n-2)} (y \wedge z) \ominus x == (-1)^{n} ((x \ominus y)\, z - (x \ominus z)\, y)$$

$$x \times (y \times z) == (-1)^{n} (y \wedge z) \ominus x == (-1)^{n} ((x \ominus y)\, z - (x \ominus z)\, y) \qquad 6.169$$

◆ The box product or triple vector product

The box product, or triple vector product, is an $(n-3)$-element, and therefore a scalar only in three dimensions.

$$(x \times y) \ominus z == (\overline{x \wedge y}) \ominus z == \overline{x \wedge y \wedge z}$$

$$(x \times y) \ominus z == \overline{x \wedge y \wedge z} \qquad 6.170$$

Because Grassmann identified n-elements with their scalar Euclidean complements (see the historical note in the introduction to Chapter 5), he considered $x \wedge y \wedge z$ in a 3-space to be a scalar. His notation for the exterior product of elements was to use square brackets $[x\ y\ z]$, thus originating the 'box' product notation for $(x \times y) \ominus z$ used in the three-dimensional vector algebra in the early decades of the twentieth century.

The 'generalized' box product is thus zero in 2-space, a scalar in 3-space, and a 1-element in 4-space.

◆ The scalar product of two cross products

The scalar product of two cross products is a scalar in any number of dimensions.

$$(x \times y) \ominus (z \times w) == (\overline{x \wedge y}) \ominus (\overline{z \wedge w}) == (x \wedge y) \ominus (z \wedge w)$$

$$(x \times y) \ominus (z \times w) == (x \wedge y) \ominus (z \wedge w) \qquad 6.171$$

◆ The cross product of two cross products

The cross product of two cross products is a $(4-n)$-element, and therefore a 1-element only in three dimensions. It corresponds to the regressive product of two exterior products.

$$(x \times y) \times (z \times w) == \overline{(\overline{x \wedge y}) \wedge (\overline{z \wedge w})} == (x \wedge y) \vee (z \wedge w)$$

$$(x \times y) \times (z \times w) == (x \wedge y) \vee (z \wedge w) \qquad 6.172$$

- Note that the cross product of two cross products, unlike the other products above, *does not rely on a metric*, because the expression involves only exterior and regressive products.

Implications of the axioms for the cross product

By expressing one or more elements as a complement, axioms for the exterior and regressive products may be rewritten in terms of the cross product, thus yielding some of its more fundamental properties.

◆ **× 6: The cross product of an *m*-element and a *k*-element is an (*n*–(*m*+*k*))-element**

The grade of the cross product of an *m*-element and a *k*-element is *n*–(*m*+*k*).

$$\underset{m}{\alpha} \in \underset{m}{\Lambda},\ \underset{k}{\beta} \in \underset{k}{\Lambda} \Rightarrow \underset{m}{\alpha} \times \underset{k}{\beta} \in \underset{n-(m+k)}{\Lambda} \tag{6.173}$$

◆ **× 7: The cross product is not associative**

The cross product is not associative. However, it can be expressed in terms of exterior and interior products.

$$\left(\underset{m}{\alpha} \times \underset{k}{\beta}\right) \times \underset{r}{\gamma} == (-1)^{(n-1)\,(m+k)} \left(\underset{m}{\alpha} \wedge \underset{k}{\beta}\right) \ominus \underset{r}{\gamma} \tag{6.174}$$

$$\underset{m}{\alpha} \times \left(\underset{k}{\beta} \times \underset{r}{\gamma}\right) == (-1)^{(n-(k+r))\,(m+k+r)} \left(\underset{k}{\beta} \wedge \underset{r}{\gamma}\right) \ominus \underset{m}{\alpha} \tag{6.175}$$

◆ **× 8: The cross product with unity yields the complement.**

The cross product of an element with the unit scalar 1 yields its complement. Thus the complement operation may be viewed as the cross product with unity.

$$1 \times \underset{m}{\alpha} == \underset{m}{\alpha} \times 1 == \overline{\underset{m}{\alpha}} \tag{6.176}$$

The cross product of an element with a scalar yields that scalar multiple of its complement.

$$a \times \underset{m}{\alpha} == \underset{m}{\alpha} \times a == a\,\overline{\underset{m}{\alpha}} \tag{6.177}$$

◆ **× 9: The cross product of unity with itself is the unit *n*-element.**

The cross product of 1 with itself is the unit *n*-element. The cross product of a scalar and its reciprocal is the unit *n*-element.

$$1 \times 1 == a \times \frac{1}{a} == \overline{1} \tag{6.178}$$

◆ **× 10: The cross product of two 1-elements anti-commutes.**

The cross product of two elements is equal (apart from a possible sign) to their cross product in reverse order. The cross product of two 1-elements is anti-commutative, just as is the exterior product.

$$\underset{m}{\alpha} \times \underset{k}{\beta} == (-1)^{m k} \underset{k}{\beta} \times \underset{m}{\alpha} \qquad 6.179$$

◆ **× 11: The cross product with zero is zero.**

$$0 \times \underset{m}{\alpha} == 0 == \underset{m}{\alpha} \times 0 \qquad 6.180$$

◆ **× 12: The cross product is both left and right distributive under addition.**

$$\left(\underset{m}{\alpha} + \underset{m}{\beta}\right) \times \underset{r}{\gamma} == \underset{m}{\alpha} \times \underset{r}{\gamma} + \underset{m}{\beta} \times \underset{r}{\gamma} \qquad 6.181$$

$$\underset{m}{\alpha} \times \left(\underset{r}{\beta} + \underset{r}{\gamma}\right) == \underset{m}{\alpha} \times \underset{r}{\beta} + \underset{m}{\alpha} \times \underset{r}{\gamma} \qquad 6.182$$

The cross product as a universal product

We have already shown that all products can be expressed in terms of the exterior product and the complement operation. Additionally, **× 8** above shows that the complement operation may be written as the cross product with unity.

$$1 \times \underset{m}{\alpha} == \underset{m}{\alpha} \times 1 == \overline{\underset{m}{\alpha}}$$

We can therefore write the exterior, regressive, and interior products in terms *only* of the cross product.

$$\underset{m}{\alpha} \wedge \underset{k}{\beta} == (-1)^{(m+k)\,(n-(m+k))}\, 1 \times \left(\underset{m}{\alpha} \times \underset{k}{\beta}\right) \qquad 6.183$$

$$\underset{m}{\alpha} \vee \underset{k}{\beta} == (-1)^{m\,(n-m)+k\,(n-k)} \left(1 \times \underset{m}{\alpha}\right) \times \left(1 \times \underset{k}{\beta}\right) \qquad 6.184$$

$$\underset{m}{\alpha} \ominus \underset{k}{\beta} == (-1)^{m\,(n-m)} \left(1 \times \underset{m}{\alpha}\right) \times \underset{k}{\beta} == (-1)^{(m+k)\,(n-m)}\, \underset{k}{\beta} \times \left(1 \times \underset{m}{\alpha}\right) \qquad 6.185$$

Thus for elements A and B *of the same grade* we have

$$\begin{aligned} A \wedge B &== 1 \times (A \times B) \\ A \vee B &== (1 \times A) \times (1 \times B) \\ A \ominus B &== B \times (1 \times A) \end{aligned} \qquad 6.186$$

These formulae show that any result in the Grassmann algebra involving exterior, regressive or interior products, or complements, could be expressed in terms of the generalized cross product alone. This is somewhat reminiscent of the role played by the Scheffer stroke (or Pierce arrow) symbol of Boolean algebra.

Cross product formulae

◆ The complement of a cross product

The complement of a cross product of two elements is, but for a possible sign, the exterior product of the elements.

$$\overline{\underset{m}{\alpha} \times \underset{k}{\beta}} == (-1)^{(m+k)\,(n-(m+k))}\, \underset{m}{\alpha} \wedge \underset{k}{\beta} \qquad 6.187$$

◆ The Common Factor Axiom for cross products

The Common Factor Axiom can be written for $m+k+p = n$.

$$\left(\underset{m}{\alpha} \times \underset{p}{\gamma}\right) \times \left(\underset{k}{\beta} \times \underset{p}{\gamma}\right) == \left(\underset{p}{\gamma} \times \left(\underset{m}{\alpha} \wedge \underset{k}{\beta}\right)\right) \underset{p}{\gamma} == (-1)^{p\,(n-p)} \left(\underset{p}{\gamma} \ominus \left(\underset{m}{\alpha} \times \underset{k}{\beta}\right)\right) \underset{p}{\gamma} \qquad 6.188$$

◆ Product formulae for cross products

All of the product formulae derived previously have counterparts involving cross products. For example product formula [3.68] is

$$\left(\underset{m}{\alpha} \wedge \underset{k}{\beta}\right) \vee \underset{n-1}{x} == \left(\underset{m}{\alpha} \vee \underset{n-1}{x}\right) \wedge \underset{k}{\beta} + (-1)^m \underset{m}{\alpha} \wedge \left(\underset{k}{\beta} \vee \underset{n-1}{x}\right)$$

Taking its complement gives

$$\left(\underset{m}{\alpha} \times \underset{k}{\beta}\right) \wedge x == (-1)^{n-1} \left(\left(\underset{m}{\alpha} \ominus x\right) \times \underset{k}{\beta} + (-1)^m \underset{m}{\alpha} \times \left(\underset{k}{\beta} \ominus x\right)\right) \qquad 6.189$$

◆ The measure of a cross product

The measure of the cross product of two elements is equal to the measure of their exterior product.

$$\left|\underset{m}{\alpha} \times \underset{k}{\beta}\right| == \left|\underset{m}{\alpha} \wedge \underset{k}{\beta}\right| \qquad 6.190$$

6.11 The Triangle Formulae

The triangle components

In this section several formulae will be developed which are generalizations of the fundamental Pythagorean relationship $a^2 = b^2 + c^2$ for the right-angled triangle and the trigonometric identity $\cos^2(\theta) + \sin^2(\theta) = 1$. These formulae will be found useful in determining projections, shortest distances between multiplanes, and a variety of other geometric results.

The results of this section have their basis in the formulae discussed in section 6.8 on product formulae. However, in order to make this important section more coherent we may repeat some of the results obtained previously.

The starting point is the Interior Product Formula B in the form [6.127].

$$\left(\underset{m}{\alpha} \ominus \underset{k}{\beta}\right) \wedge x == \left(\underset{m}{\alpha} \wedge x\right) \ominus \underset{k}{\beta} + (-1)^{m-1} \underset{m}{\alpha} \ominus \left(\underset{k}{\beta} \ominus x\right)$$

If k is equal to m the formula becomes a decomposition formula for the 1-element x into two components.

$$\left(\underset{m}{\alpha} \ominus \underset{m}{\beta}\right) x == \left(\underset{m}{\alpha} \wedge x\right) \ominus \underset{m}{\beta} + (-1)^{m-1} \underset{m}{\alpha} \ominus \left(\underset{m}{\beta} \ominus x\right) \qquad 6.191$$

If, as well, $\underset{m}{\alpha}$ is equal to $\underset{m}{\beta}$ we get a decomposition formula for x which turns out to be a generalization of the Pythagorean relationship where the vector x (the hypotenuse) is decomposed into the two orthogonal legs of the right-angled triangle - one parallel to (contained in) the m-vector $\underset{m}{\alpha}$, and the other orthogonal to it.

$$\left(\underset{m}{\alpha} \ominus \underset{m}{\alpha}\right) x == \left(\underset{m}{\alpha} \wedge x\right) \ominus \underset{m}{\alpha} + (-1)^{m-1} \underset{m}{\alpha} \ominus \left(\underset{m}{\alpha} \ominus x\right) \qquad 6.192$$

$$x == \frac{\left(\underset{m}{\alpha} \wedge x\right) \ominus \underset{m}{\alpha}}{\underset{m}{\alpha} \ominus \underset{m}{\alpha}} + (-1)^{m-1} \frac{\underset{m}{\alpha} \ominus \left(\underset{m}{\alpha} \ominus x\right)}{\underset{m}{\alpha} \ominus \underset{m}{\alpha}} \qquad 6.193$$

However, since our focus is most often on x and its components, most of our exploration will involve the unit form of the formula in which $\underset{m}{\hat{\alpha}}$ is a unit m-element.

$$x == \left(\underset{m}{\hat{\alpha}} \wedge x\right) \ominus \underset{m}{\hat{\alpha}} + (-1)^{m-1} \underset{m}{\hat{\alpha}} \ominus \left(\underset{m}{\hat{\alpha}} \ominus x\right) \qquad 6.194$$

The first component on the right hand side is orthogonal to $\underset{m}{\alpha}$ since by the extended interior products formula [6.40] it is orthogonal to each of the 1-element factors of $\underset{m}{\hat{\alpha}}$.

$$\left(\left(\underset{m}{\hat{\alpha}} \wedge x\right) \ominus \underset{m}{\hat{\alpha}}\right) \ominus \alpha_i == \left(\underset{m}{\hat{\alpha}} \wedge x\right) \ominus \left(\underset{m}{\hat{\alpha}} \wedge \alpha_i\right) == 0$$

We thus call it the *orthogonal component* of x (to $\underset{m}{\hat{\alpha}}$) and denote it by $x^{\llcorner}$.

Since $\underset{m}{\hat{\alpha}} \ominus x$ can be expanded by the interior common factor theorem in the form [6.52] as a linear combination of (*m*–1)-element factors of $\underset{m}{\hat{\alpha}}$, so $\underset{m}{\hat{\alpha}} \ominus \left(\underset{m}{\hat{\alpha}} \ominus x\right)$ can be expanded by the interior common factor theorem in the form [6.54] as a linear combination of 1-element factors of $\underset{m}{\hat{\alpha}}$. Hence the second component of x is contained in (parallel to) $\underset{m}{\hat{\alpha}}$. We thus call it the *parallel component* of x (to $\underset{m}{\hat{\alpha}}$) and denote it by $x^{\#}$.

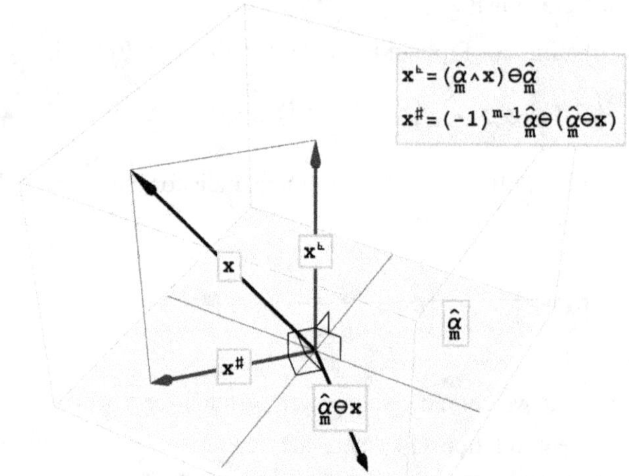

The triangle components

$$x == x^{\llcorner} + x^{\#} \qquad x^{\llcorner} == \left(\underset{m}{\hat{\alpha}} \wedge x\right) \ominus \underset{m}{\hat{\alpha}} \qquad x^{\#} == (-1)^{m-1}\, \underset{m}{\hat{\alpha}} \ominus \left(\underset{m}{\hat{\alpha}} \ominus x\right) \qquad 6.195$$

To show that these two components are themselves orthogonal we form their scalar product, apply the formula for extended interior products [6.40] and note that since $\underset{m}{\hat{\alpha}} \ominus \left(\underset{m}{\hat{\alpha}} \ominus x\right)$ belongs to $\underset{m}{\hat{\alpha}}$, its exterior product with $\underset{m}{\hat{\alpha}}$ is zero.

$$x^{\llcorner} \ominus x^{\#} == \left(\left(\underset{m}{\hat{\alpha}} \wedge x\right) \ominus \underset{m}{\hat{\alpha}}\right) \ominus \left(\underset{m}{\hat{\alpha}} \ominus \left(\underset{m}{\hat{\alpha}} \ominus x\right)\right) == \left(\underset{m}{\hat{\alpha}} \wedge x\right) \ominus \left(\underset{m}{\hat{\alpha}} \wedge \left(\underset{m}{\hat{\alpha}} \ominus \left(\underset{m}{\hat{\alpha}} \ominus x\right)\right)\right) == 0$$

$$x^{\llcorner} \ominus x^{\#} == 0 \qquad 6.196$$

Note that this derivation of component orthogonality is also valid for a given $\underset{m}{\hat{\alpha}}$ if the components belong to two distinct 1-elements (x and y, say). That is, $x^{\llcorner} \ominus y^{\#}$ and $y^{\llcorner} \ominus x^{\#}$ are also zero allowing us to straightforwardly decompose a scalar product.

$$x \ominus y == \left(x^{\perp} + x^{\#}\right) \ominus \left(y^{\perp} + y^{\#}\right) == x^{\perp} \ominus y^{\perp} + x^{\#} \ominus y^{\#} \qquad 6.197$$

However, no such simplification exists for the components of an exterior product. For example

$$x \wedge y == \left(x^{\perp} + x^{\#}\right) \wedge \left(y^{\perp} + y^{\#}\right) == x^{\perp} \wedge y^{\perp} + x^{\perp} \wedge y^{\#} + x^{\#} \wedge y^{\perp} + x^{\#} \wedge y^{\#}$$

Nevertheless, the purely orthogonal and parallel components of an exterior product (in this example, $x^{\perp} \wedge y^{\perp}$ and $x^{\#} \wedge y^{\#}$) will still find application. See for example section 6.15 below.

The measure of the triangular components

The scalar products of each of the orthogonal components with itself are

$$x^{\perp} \ominus x^{\perp} == \left(\left(\hat{\underset{m}{\alpha}} \wedge x\right) \ominus \hat{\underset{m}{\alpha}}\right) \ominus \left(\left(\hat{\underset{m}{\alpha}} \wedge x\right) \ominus \hat{\underset{m}{\alpha}}\right)$$

$$x^{\#} \ominus x^{\#} == \left(\hat{\underset{m}{\alpha}} \ominus \left(\hat{\underset{m}{\alpha}} \ominus x\right)\right) \ominus \left(\hat{\underset{m}{\alpha}} \ominus \left(\hat{\underset{m}{\alpha}} \ominus x\right)\right)$$

These expressions may be simplified fairly directly as follows.

$$x^{\perp} \ominus x^{\perp} == x^{\perp} \ominus \left(x - x^{\#}\right) == x^{\perp} \ominus x == \left(\left(\hat{\underset{m}{\alpha}} \wedge x\right) \ominus \hat{\underset{m}{\alpha}}\right) \ominus x == \left(\hat{\underset{m}{\alpha}} \wedge x\right) \ominus \left(\hat{\underset{m}{\alpha}} \wedge x\right)$$

$$x^{\#} \ominus x^{\#} == x^{\#} \ominus (x - x^{\perp}) == x^{\#} \ominus x == (-1)^{(m-1)} \left(\hat{\underset{m}{\alpha}} \ominus \left(\hat{\underset{m}{\alpha}} \ominus x\right)\right) \ominus x$$

$$== (-1)^{(m-1)} \hat{\underset{m}{\alpha}} \vee \overline{\hat{\underset{m}{\alpha}} \ominus x} \vee \overline{x} == \left(\hat{\underset{m}{\alpha}} \vee \overline{x}\right) \vee \overline{\hat{\underset{m}{\alpha}} \ominus x} == \left(\hat{\underset{m}{\alpha}} \ominus x\right) \ominus \left(\hat{\underset{m}{\alpha}} \ominus x\right)$$

$$x^{\perp} \ominus x^{\perp} == \left(\hat{\underset{m}{\alpha}} \wedge x\right) \ominus \left(\hat{\underset{m}{\alpha}} \wedge x\right) \qquad x^{\#} \ominus x^{\#} == \left(\hat{\underset{m}{\alpha}} \ominus x\right) \ominus \left(\hat{\underset{m}{\alpha}} \ominus x\right) \qquad 6.198$$

The magnitudes of the orthogonal components then become

$$\left|x^{\perp}\right| == \left|\left(\hat{\underset{m}{\alpha}} \wedge x\right) \ominus \hat{\underset{m}{\alpha}}\right| == \left|\hat{\underset{m}{\alpha}} \wedge x\right| \qquad \left|x^{\#}\right| == \left|\hat{\underset{m}{\alpha}} \ominus \left(\hat{\underset{m}{\alpha}} \ominus x\right)\right| == \left|\hat{\underset{m}{\alpha}} \ominus x\right| \qquad 6.199$$

and the Pythagorean relationships become clearly apparent.

$$\left|x^{\perp}\right|^2 + \left|x^{\#}\right|^2 == |x|^2 \qquad \left|\hat{\underset{m}{\alpha}} \wedge \hat{x}\right|^2 + \left|\hat{\underset{m}{\alpha}} \ominus \hat{x}\right|^2 == 1 \qquad 6.200$$

◆ The scalar product of independent parallel components

The formulae for the scalar products of two *independent* 1-elements parallel and orthogonal to the same *m*-element may be derived *mutatis mutandis* to give forms similar to those of [6.198].

$$x^{\perp} \ominus y^{\perp} == \left(\hat{\underset{m}{\alpha}} \wedge x\right) \ominus \left(\hat{\underset{m}{\alpha}} \wedge y\right) \qquad x^{\#} \ominus y^{\#} == \left(\hat{\underset{m}{\alpha}} \ominus x\right) \ominus \left(\hat{\underset{m}{\alpha}} \ominus y\right) \qquad 6.201$$

Complementary forms for the triangle components

If we take the triangle formula [6.194] and replace $\underset{m}{\hat{\alpha}}$ by its complement $\overline{\underset{m}{\hat{\alpha}}}$ we obtain another decomposition for x.

$$x = \left(\overline{\underset{m}{\hat{\alpha}}} \wedge x\right) \ominus \overline{\underset{m}{\hat{\alpha}}} + (-1)^{(n-m)-1}\, \overline{\underset{m}{\hat{\alpha}}} \ominus \left(\overline{\underset{m}{\hat{\alpha}}} \ominus x\right)$$

But formulae [6.46] and [6.47] show that if in each of them we substitute $\underset{m}{\hat{\alpha}}$ for $\underset{m}{\alpha}$ and $\underset{k}{\beta}$ we obtain

$$\left(\underset{m}{\hat{\alpha}} \wedge x\right) \ominus \underset{m}{\hat{\alpha}} = (-1)^{(n-m)-1}\, \overline{\underset{m}{\hat{\alpha}}} \ominus \left(\overline{\underset{m}{\hat{\alpha}}} \ominus x\right)$$

$$(-1)^{m-1}\, \underset{m}{\hat{\alpha}} \ominus \left(\underset{m}{\hat{\alpha}} \ominus x\right) = \left(\overline{\underset{m}{\hat{\alpha}}} \wedge x\right) \ominus \overline{\underset{m}{\hat{\alpha}}}$$

Hence we confirm that $x^{\perp}$ is not only the component of x orthogonal to $\underset{m}{\hat{\alpha}}$, but also the component of x parallel to (contained in) $\overline{\underset{m}{\hat{\alpha}}}$.

$$x^{\perp} = \left(\underset{m}{\hat{\alpha}} \wedge x\right) \ominus \underset{m}{\hat{\alpha}} = (-1)^{(n-m)-1}\, \overline{\underset{m}{\hat{\alpha}}} \ominus \left(\overline{\underset{m}{\hat{\alpha}}} \ominus x\right) \qquad 6.202$$

Similarly we confirm that $x^{\#}$ is not only the component of x parallel to $\underset{m}{\hat{\alpha}}$, but also the component of x orthogonal to $\overline{\underset{m}{\hat{\alpha}}}$.

$$x^{\#} = (-1)^{m-1}\, \underset{m}{\hat{\alpha}} \ominus \left(\underset{m}{\hat{\alpha}} \ominus x\right) = \left(\overline{\underset{m}{\hat{\alpha}}} \wedge x\right) \ominus \overline{\underset{m}{\hat{\alpha}}} \qquad 6.203$$

These results are a simple consequence of the notion of complement. The complement $\overline{\underset{m}{\hat{\alpha}}}$ of an element $\underset{m}{\hat{\alpha}}$ is *by definition* orthogonal to $\underset{m}{\hat{\alpha}}$. Hence if a 1-element is orthogonal to $\underset{m}{\hat{\alpha}}$ it must lie in $\overline{\underset{m}{\hat{\alpha}}}$ - and *vice versa*.

Expanding the components

We have seen earlier in this section that the triangle formula can be written in the form [6.192]:

$$\left(\underset{m}{\alpha} \ominus \underset{m}{\alpha}\right) x = \left(\underset{m}{\alpha} \wedge x\right) \ominus \underset{m}{\alpha} + (-1)^{m-1}\, \underset{m}{\alpha} \ominus \left(\underset{m}{\alpha} \ominus x\right) \qquad \underset{m}{\alpha} = \alpha_1 \wedge \alpha_2 \wedge \ldots \wedge \alpha_m$$

In this formula the terms on the right hand side are scalar multiples of the components of x orthogonal and parallel to the simple *m*-element $\underset{m}{\alpha}$.

$$\left(\underset{m}{\alpha} \ominus \underset{m}{\alpha}\right) x^{\perp} = \left(\underset{m}{\alpha} \wedge x\right) \ominus \underset{m}{\alpha} \qquad \left(\underset{m}{\alpha} \ominus \underset{m}{\alpha}\right) x^{\#} = (-1)^{m-1}\, \underset{m}{\alpha} \ominus \left(\underset{m}{\alpha} \ominus x\right)$$

In this section we develop these components as linear combinations of the 1-element factors α_i of $\underset{m}{\alpha}$. And in the following section we will show how the coefficients of the components simplify when the α_i are orthogonal or orthonormal. Since $x = x^{\perp} + x^{\#}$, we only need to consider one of the terms, since the other can be obtained by subtraction.

Let us consider then the parallel component. To expand $\underset{m}{\alpha} \ominus \left(\underset{m}{\alpha} \ominus x\right)$ we can use the Interior Common Factor Theorem [6.54]:

$$(\alpha_1 \wedge \alpha_2 \wedge \ldots \wedge \alpha_m) \ominus \underset{m-1}{\beta} == \sum_{i=1}^{m} (-1)^{m-i} \left((\alpha_1 \wedge \ldots \wedge \square_i \wedge \ldots \wedge \alpha_m) \ominus \underset{m-1}{\beta} \right) \alpha_i$$

Substituting $\underset{m}{\alpha} \ominus x$ for $\underset{m-1}{\beta}$ and multiplying by $(-1)^{m-1}$ then gives us the expansion

$$\left(\underset{m}{\alpha} \ominus \underset{m}{\alpha}\right) x^{\#} == \sum_{i=1}^{m} (-1)^{i-1} \left((\alpha_1 \wedge \ldots \wedge \square_i \wedge \ldots \wedge \alpha_m) \ominus \left(\underset{m}{\alpha} \ominus x\right) \right) \alpha_i == \sum_{i=1}^{m} a_i \, \alpha_i$$

This expression is a linear combination of the factors α_i. Let us now look at a typical coefficient a_i.

$$a_i == (-1)^{i-1} (\alpha_1 \wedge \ldots \wedge \square_i \wedge \ldots \wedge \alpha_m) \ominus \left(\underset{m}{\alpha} \ominus x\right)$$

Substituting x for β in the Interior Common Factor Theorem in its form [6.52] gives

$$(\alpha_1 \wedge \alpha_2 \wedge \ldots \wedge \alpha_m) \ominus x == \sum_{j=1}^{m} (\alpha_j \ominus x) \; (-1)^{j-1} \alpha_1 \wedge \ldots \wedge \square_j \wedge \ldots \wedge \alpha_m$$

so that a_i is given by a sum of products of inner and scalar products. (Remember that $\square_i$ means that the *i*th factor is missing.)

$$a_i == \sum_{j=1}^{m} (-1)^{i+j} (\alpha_1 \wedge \ldots \wedge \square_i \wedge \ldots \wedge \alpha_m) \ominus (\alpha_1 \wedge \ldots \wedge \square_j \wedge \ldots \wedge \alpha_m) \; (\alpha_j \ominus x)$$

Now let us concentrate on a typical inner product which we will denote by c_{ij}.

$$c_{ij} == (-1)^{i+j} (\alpha_1 \wedge \ldots \wedge \square_i \wedge \ldots \wedge \alpha_m) \ominus (\alpha_1 \wedge \ldots \wedge \square_j \wedge \ldots \wedge \alpha_m)$$

Formula [6.67] tells us that we can write any inner product as a determinant of scalar products:

$$(\alpha_1 \wedge \alpha_2 \wedge \ldots \wedge \alpha_m) \ominus (\alpha_1 \wedge \alpha_2 \wedge \ldots \wedge \alpha_m) == \begin{vmatrix} \alpha_1 \ominus \alpha_1 & \alpha_1 \ominus \alpha_2 & \alpha_1 \ominus \ldots & \alpha_1 \ominus \alpha_m \\ \alpha_2 \ominus \alpha_1 & \alpha_2 \ominus \alpha_2 & \alpha_2 \ominus \ldots & \alpha_2 \ominus \alpha_m \\ \ldots \ominus \alpha_1 & \ldots \ominus \alpha_2 & \ldots \ominus \ldots & \ldots \ominus \alpha_m \\ \alpha_m \ominus \alpha_1 & \alpha_m \ominus \alpha_2 & \alpha_m \ominus \ldots & \alpha_m \ominus \alpha_m \end{vmatrix}$$

In particular, for c_{ij} this becomes

$$c_{ij} == (-1)^{i+j} \begin{vmatrix} \alpha_1 \ominus \alpha_1 & \ldots & \alpha_1 \ominus \square_j & \ldots & \alpha_1 \ominus \alpha_m \\ \ldots & \ldots & \ldots & \ldots & \ldots \\ \square_i \ominus \alpha_1 & \ldots & \square_i \ominus \square_j & \ldots & \square_i \ominus \alpha_m \\ \ldots & \ldots & \ldots & \ldots & \ldots \\ \alpha_m \ominus \alpha_1 & \ldots & \alpha_m \ominus \square_j & \ldots & \alpha_m \ominus \alpha_m \end{vmatrix}$$

Thus c_{ij} is the *cofactor* of the element $\alpha_i \ominus \alpha_j$ in the *matrix*, A say, whose elements are the scalar products of the α_i.

$$A == \begin{pmatrix} \alpha_1 \ominus \alpha_1 & \alpha_1 \ominus \alpha_2 & \alpha_1 \ominus \ldots & \alpha_1 \ominus \alpha_m \\ \alpha_2 \ominus \alpha_1 & \alpha_2 \ominus \alpha_2 & \alpha_2 \ominus \ldots & \alpha_2 \ominus \alpha_m \\ \ldots \ominus \alpha_1 & \ldots \ominus \alpha_2 & \ldots \ominus \ldots & \ldots \ominus \alpha_m \\ \alpha_m \ominus \alpha_1 & \alpha_m \ominus \alpha_2 & \alpha_m \ominus \ldots & \alpha_m \ominus \alpha_m \end{pmatrix} \qquad D == \mathrm{Det}[A]$$

Since A is symmetric, the inverse A^{-1} of A is the matrix of cofactors of A divided by its determinant D. Let C_{ij} be the components of A^{-1}. Then

$$C_{ij} == \frac{c_{ij}}{D} == (-1)^{i+j} \frac{1}{D} \left((\alpha_1 \wedge \ldots \wedge \square_i \wedge \ldots \wedge \alpha_m) \ominus (\alpha_1 \wedge \ldots \wedge \square_j \wedge \ldots \wedge \alpha_m) \right)$$

$$D == (\alpha_1 \wedge \alpha_2 \wedge \ldots \wedge \alpha_m) \ominus (\alpha_1 \wedge \alpha_2 \wedge \ldots \wedge \alpha_m)$$

Our final form for the component of x contained in $\alpha_1 \wedge \alpha_2 \wedge \ldots \wedge \alpha_m$ then becomes

$$x^\# == \sum_{i==1}^{m} \sum_{j==1}^{m} (\alpha_j \ominus x)\, C_{ij}\, \alpha_i \qquad 6.204$$

$$C_{ij} == (-1)^{i+j} \frac{1}{D} \left((\alpha_1 \wedge \ldots \wedge \square_i \wedge \ldots \wedge \alpha_m) \ominus (\alpha_1 \wedge \ldots \wedge \square_j \wedge \ldots \wedge \alpha_m) \right)$$
$$D == (\alpha_1 \wedge \alpha_2 \wedge \ldots \wedge \alpha_m) \ominus (\alpha_1 \wedge \alpha_2 \wedge \ldots \wedge \alpha_m) \qquad 6.205$$

The expression for $x^\#$ can also be written as a product of matrices.

$$x^\# == \begin{pmatrix} \alpha_1 \ominus x & \alpha_2 \ominus x & \ldots & \alpha_m \ominus x \end{pmatrix} \begin{pmatrix} C_{11} & C_{12} & \ldots & C_{1\,m} \\ C_{21} & C_{22} & \ldots & C_{2\,m} \\ \ldots & \ldots & \ldots & \ldots \\ C_{m\,1} & C_{m\,2} & \ldots & C_{m\,m} \end{pmatrix} \begin{pmatrix} \alpha_1 \\ \alpha_2 \\ \ldots \\ \alpha_m \end{pmatrix} \qquad 6.206$$

◆ **Example**

In the case of a 3-element $\alpha_1 \wedge \alpha_2 \wedge \alpha_3$, the component of x parallel to it (contained in it) is given by

$$(\alpha_1 \wedge \alpha_2 \wedge \alpha_3) \ominus (\alpha_1 \wedge \alpha_2 \wedge \alpha_3)\, x^\# ==$$
$$\begin{pmatrix} \alpha_1 \ominus x & \alpha_2 \ominus x & \alpha_3 \ominus x \end{pmatrix}$$
$$\begin{pmatrix} (\alpha_2 \wedge \alpha_3) \ominus (\alpha_2 \wedge \alpha_3) & -(\alpha_2 \wedge \alpha_3) \ominus (\alpha_1 \wedge \alpha_3) & (\alpha_2 \wedge \alpha_3) \ominus (\alpha_1 \wedge \alpha_2) \\ -(\alpha_1 \wedge \alpha_3) \ominus (\alpha_2 \wedge \alpha_3) & (\alpha_1 \wedge \alpha_3) \ominus (\alpha_1 \wedge \alpha_3) & -(\alpha_1 \wedge \alpha_3) \ominus (\alpha_1 \wedge \alpha_2) \\ (\alpha_1 \wedge \alpha_2) \ominus (\alpha_2 \wedge \alpha_3) & -(\alpha_1 \wedge \alpha_2) \ominus (\alpha_1 \wedge \alpha_3) & (\alpha_1 \wedge \alpha_2) \ominus (\alpha_1 \wedge \alpha_2) \end{pmatrix} \begin{pmatrix} \alpha_1 \\ \alpha_2 \\ \alpha_3 \end{pmatrix}$$

Expanding the components in orthogonal factors

Following on from the previous section: if the factors α_i are mutually orthogonal, then the matrices A and A^{-1} become diagonal, and the formula reduces to the standard formula for $x^\#$ as the sum of the projections of x onto the α_i.

$$A == \begin{pmatrix} |\alpha_1|^2 & 0 & \ldots & 0 \\ 0 & |\alpha_2|^2 & \ldots & 0 \\ \ldots & \ldots & \ldots & \ldots \\ 0 & 0 & \ldots & |\alpha_m|^2 \end{pmatrix} \qquad A^{-1} == \begin{pmatrix} |\alpha_1|^{-2} & 0 & \ldots & 0 \\ 0 & |\alpha_2|^{-2} & \ldots & 0 \\ \ldots & \ldots & \ldots & \ldots \\ 0 & 0 & \ldots & |\alpha_m|^{-2} \end{pmatrix}$$

Written as a product of matrices the expression for $x^\#$ becomes

$$x^{\#} == (\alpha_1 \ominus x \ \ \alpha_2 \ominus x \ \ \ldots \ \ \alpha_m \ominus x) \begin{pmatrix} |\alpha_1|^{-2} & 0 & \ldots & 0 \\ 0 & |\alpha_2|^{-2} & \ldots & 0 \\ \ldots & \ldots & \ldots & \ldots \\ 0 & 0 & \ldots & |\alpha_m|^{-2} \end{pmatrix} \begin{pmatrix} \alpha_1 \\ \alpha_2 \\ \ldots \\ \alpha_m \end{pmatrix}$$

$$== \left(\frac{\alpha_1 \ominus x}{|\alpha_1|^2} \ \ \frac{\alpha_2 \ominus x}{|\alpha_2|^2} \ \ \ldots \ \ \frac{\alpha_1 \ominus x}{|\alpha_1|^2} \right) \begin{pmatrix} \alpha_1 \\ \alpha_2 \\ \ldots \\ \alpha_3 \end{pmatrix} == (\hat{\alpha}_1 \ominus x)\,\hat{\alpha}_1 + (\hat{\alpha}_2 \ominus x)\,\hat{\alpha}_2 + \ldots + (\hat{\alpha}_m \ominus x)\,\hat{\alpha}_m$$

The components $x^{\#}$ and $x^{\perp}$ then become particularly simple linear combinations of the α_i: $x^{\#}$ becomes the sum of the projections of x onto the unit 1-elements $\hat{\alpha}_i$ (and $x^{\perp}$ is given by $x^{\perp} = x - x^{\#}$).

$$x^{\#} == \sum_{i==1}^{m} \frac{(\alpha_i \ominus x)\,\alpha_i}{|\alpha_i|^2} == \sum_{i==1}^{m} (\hat{\alpha}_i \ominus x)\,\hat{\alpha}_i \qquad 6.207$$

$$x^{\perp} == x - \sum_{i==1}^{m} \frac{(\alpha_i \ominus x)\,\alpha_i}{|\alpha_i|^2} == x - \sum_{i==1}^{m} (\hat{\alpha}_i \ominus x)\,\hat{\alpha}_i \qquad 6.208$$

◆ The Gram-Schmidt process

Formula [6.208] is the essential basis for the well-known Gram-Schmidt process for orthogonalizing a sequence of 1-elements x_i into a sequence of mutually orthogonal 1-elements α_i.

$$\alpha_i == x_i - \sum_{k==1}^{i-1} \frac{(\alpha_k \ominus x_i)\,\alpha_k}{|\alpha_k|^2} == x_i - \sum_{k==1}^{i-1} (\hat{\alpha}_k \ominus x_i)\,\hat{\alpha}_k \qquad \alpha_1 == x_1 \qquad 6.209$$

◆ Example: Orthonormal factors

The example of the previous section was

$$(\alpha_1 \wedge \alpha_2 \wedge \alpha_3) \ominus (\alpha_1 \wedge \alpha_2 \wedge \alpha_3)\, x^{\#} ==$$
$$(\alpha_1 \ominus x \ \ \alpha_2 \ominus x \ \ \alpha_3 \ominus x)$$
$$\begin{pmatrix} (\alpha_2 \wedge \alpha_3) \ominus (\alpha_2 \wedge \alpha_3) & -(\alpha_2 \wedge \alpha_3) \ominus (\alpha_1 \wedge \alpha_3) & (\alpha_2 \wedge \alpha_3) \ominus (\alpha_1 \wedge \alpha_2) \\ -(\alpha_1 \wedge \alpha_3) \ominus (\alpha_2 \wedge \alpha_3) & (\alpha_1 \wedge \alpha_3) \ominus (\alpha_1 \wedge \alpha_3) & -(\alpha_1 \wedge \alpha_3) \ominus (\alpha_1 \wedge \alpha_2) \\ (\alpha_1 \wedge \alpha_2) \ominus (\alpha_2 \wedge \alpha_3) & -(\alpha_1 \wedge \alpha_2) \ominus (\alpha_1 \wedge \alpha_3) & (\alpha_1 \wedge \alpha_2) \ominus (\alpha_1 \wedge \alpha_2) \end{pmatrix} \begin{pmatrix} \alpha_1 \\ \alpha_2 \\ \alpha_3 \end{pmatrix}$$

To confirm how this simplifies when the α_i are *orthonormal* we can declare them to be basis elements, and adopt the default Euclidean metric which assumes their orthonormality.

```
★𝓑3; DeclareBasis[{α1, α2, α3}];
```

The orthonormal basis vectors form a unit 3-element.

```
ToMetricElements[ToScalarProducts[(α1 ∧ α2 ∧ α3) ⊖ (α1 ∧ α2 ∧ α3)]]
```

1

The matrix of cofactors simplifies to the identity matrix.

$$M = \begin{pmatrix} (\alpha_2 \wedge \alpha_3) \ominus (\alpha_2 \wedge \alpha_3) & -(\alpha_2 \wedge \alpha_3) \ominus (\alpha_1 \wedge \alpha_3) & (\alpha_2 \wedge \alpha_3) \ominus (\alpha_1 \wedge \alpha_2) \\ -(\alpha_1 \wedge \alpha_3) \ominus (\alpha_2 \wedge \alpha_3) & (\alpha_1 \wedge \alpha_3) \ominus (\alpha_1 \wedge \alpha_3) & -(\alpha_1 \wedge \alpha_3) \ominus (\alpha_1 \wedge \alpha_2) \\ (\alpha_1 \wedge \alpha_2) \ominus (\alpha_2 \wedge \alpha_3) & -(\alpha_1 \wedge \alpha_2) \ominus (\alpha_1 \wedge \alpha_3) & (\alpha_1 \wedge \alpha_2) \ominus (\alpha_1 \wedge \alpha_2) \end{pmatrix};$$

```
MatrixForm[ToMetricElements[ToScalarProducts[M]]]
```

$$\begin{pmatrix} 1 & 0 & 0 \\ 0 & 1 & 0 \\ 0 & 0 & 1 \end{pmatrix}$$

The component $x^\#$ is then obtained as the linear combination of the orthonormal α_i that we expected.

$$x^\# == \star\Sigma\left[(\alpha_1 \ominus x \;\; \alpha_2 \ominus x \;\; \alpha_3 \ominus x) . \begin{pmatrix} \alpha_1 \\ \alpha_2 \\ \alpha_3 \end{pmatrix}\right]$$

$$x^\# == (\alpha_1 \ominus x)\, \alpha_1 + (\alpha_2 \ominus x)\, \alpha_2 + (\alpha_3 \ominus x)\, \alpha_3$$

◆ Example: Orthogonal factors

To confirm how the example of the previous section gives simplifies when the α_i are *orthogonal* we can still declare them to be basis elements, but adopt a metric which only reflects their orthogonality, that is, which still allows each to have an arbitrary magnitude.

$$\mathbf{DeclareMetric}[\{\{\alpha_1 \ominus \alpha_1, 0, 0\}, \{0, \alpha_2 \ominus \alpha_2, 0\}, \{0, 0, \alpha_3 \ominus \alpha_3\}\}];$$

$$V = \mathbf{ToMetricElements}[\mathbf{ToScalarProducts}[(\alpha_1 \wedge \alpha_2 \wedge \alpha_3) \ominus (\alpha_1 \wedge \alpha_2 \wedge \alpha_3)]]$$

$$(\alpha_1 \ominus \alpha_1)\, (\alpha_2 \ominus \alpha_2)\, (\alpha_3 \ominus \alpha_3)$$

The matrix of cofactors simplifies to a diagonal matrix.

$$M = \begin{pmatrix} (\alpha_2 \wedge \alpha_3) \ominus (\alpha_2 \wedge \alpha_3) & -(\alpha_2 \wedge \alpha_3) \ominus (\alpha_1 \wedge \alpha_3) & (\alpha_2 \wedge \alpha_3) \ominus (\alpha_1 \wedge \alpha_2) \\ -(\alpha_1 \wedge \alpha_3) \ominus (\alpha_2 \wedge \alpha_3) & (\alpha_1 \wedge \alpha_3) \ominus (\alpha_1 \wedge \alpha_3) & -(\alpha_1 \wedge \alpha_3) \ominus (\alpha_1 \wedge \alpha_2) \\ (\alpha_1 \wedge \alpha_2) \ominus (\alpha_2 \wedge \alpha_3) & -(\alpha_1 \wedge \alpha_2) \ominus (\alpha_1 \wedge \alpha_3) & (\alpha_1 \wedge \alpha_2) \ominus (\alpha_1 \wedge \alpha_2) \end{pmatrix};$$

```
M = ToMetricElements[ToScalarProducts[M]];
MatrixForm[M]
```

$$\begin{pmatrix} (\alpha_2 \ominus \alpha_2)\, (\alpha_3 \ominus \alpha_3) & 0 & 0 \\ 0 & (\alpha_1 \ominus \alpha_1)\, (\alpha_3 \ominus \alpha_3) & 0 \\ 0 & 0 & (\alpha_1 \ominus \alpha_1)\, (\alpha_2 \ominus \alpha_2) \end{pmatrix}$$

The component $x^\#$ is then obtained as the linear combination of the orthogonal α_i that we expected. In terms of normalized $\hat{\alpha}_i$, this is the same result as in the previous example.

$$x^\# == \frac{1}{V} \star\Sigma\left[(\alpha_1 \ominus x \;\; \alpha_2 \ominus x \;\; \alpha_3 \ominus x) . M . \begin{pmatrix} \alpha_1 \\ \alpha_2 \\ \alpha_3 \end{pmatrix}\right] \; // \; \mathbf{Simplify}$$

$$\frac{(\alpha_1 \ominus x)\, \alpha_1}{\alpha_1 \ominus \alpha_1} + \frac{(\alpha_2 \ominus x)\, \alpha_2}{\alpha_2 \ominus \alpha_2} + \frac{(\alpha_3 \ominus x)\, \alpha_3}{\alpha_3 \ominus \alpha_3} == x^\#$$

The triangle formulae for a bivector

As an example to help consolidate the concepts we have discussed above, this section develops in summary form the triangle formulae for a vector decomposed orthogonally to a bivector. This is also an important case with its own applications (for example for computing the common normal to two lines). We begin with formula [6.194]

$$z == \left(\hat{\underset{m}{\alpha}} \wedge z\right) \ominus \hat{\underset{m}{\alpha}} + (-1)^{m-1}\, \hat{\underset{m}{\alpha}} \ominus \left(\hat{\underset{m}{\alpha}} \ominus z\right)$$

and then substitute $\hat{B}$ for $\hat{\underset{m}{\alpha}}$.

$$z == \left(z \wedge \hat{B}\right) \ominus \hat{B} - \hat{B} \ominus \left(\hat{B} \ominus z\right)$$

The 1-element z is thus decomposed into the sum of two components which we denote $z^{\llcorner}$ and $z^{\#}$.

$$z == z^{\llcorner} + z^{\#} \qquad z^{\llcorner} == \left(z \wedge \hat{B}\right) \ominus \hat{B} \qquad z^{\#} == -\hat{B} \ominus \left(\hat{B} \ominus z\right)$$

The component $z^{\llcorner}$ is orthogonal to $\hat{B}$ while the component $z^{\#}$ is parallel to $\hat{B}$. We call $z^{\llcorner}$ the *orthogonal component* of z and $z^{\#}$ the *parallel component* of z. These two components are therefore also orthogonal to each other.

$$\hat{B} \ominus z^{\llcorner} == 0 \qquad \hat{B} \wedge z^{\#} == 0 \qquad z^{\llcorner} \ominus z^{\#} == 0$$

Because of the mutual orthogonality of the components we can write the square of the magnitude of z as the sum of the squares of the magnitudes of its components.

$$|z|^2 == |z^{\llcorner}|^2 + |z^{\#}|^2$$

These squared magnitudes may also be written in a form somewhat simpler than directly from their definitions.

$$|z^{\llcorner}|^2 == \left(\left(z \wedge \hat{B}\right) \ominus \hat{B}\right) \ominus \left(\left(z \wedge \hat{B}\right) \ominus \hat{B}\right) == \left(z \wedge \hat{B}\right) \ominus \left(z \wedge \hat{B}\right)$$

$$|z^{\#}|^2 == \left(-\hat{B} \ominus \left(\hat{B} \ominus z\right)\right) \ominus \left(-\hat{B} \ominus \left(\hat{B} \ominus z\right)\right) == \left(\hat{B} \ominus z\right) \ominus \left(\hat{B} \ominus z\right)$$

$$|z|^2 == \left|z \wedge \hat{B}\right|^2 + \left|\hat{B} \ominus z\right|^2$$

$$\left|\hat{z} \wedge \hat{B}\right|^2 + \left|\hat{B} \ominus \hat{z}\right|^2 == 1$$

- Now let us turn our attention to determining $z^{\llcorner}$ and $z^{\#}$ as functions of z and the 1-element factors which define B. Let $B = x \wedge y$. The square of the measure of B is given by

$$|B|^2 == (x \wedge y) \ominus (x \wedge y) == \mathrm{Det}\left[\begin{pmatrix} x \ominus x & x \ominus y \\ x \ominus y & y \ominus y \end{pmatrix}\right] == (x \ominus x)\,(y \ominus y) - (x \ominus y)^2$$

The parallel component $z^{\#}$ can be expressed in matrix terms as

$$|B|^2\, z^{\#} == \begin{pmatrix} x \ominus z & y \ominus z \end{pmatrix} \begin{pmatrix} y \ominus y & -x \ominus y \\ -x \ominus y & x \ominus x \end{pmatrix} \begin{pmatrix} x \\ y \end{pmatrix}$$

while the orthogonal component $z^{\llcorner}$ is most straightforwardly found as $z - z^{\#}$.

- If we choose to express the bivector in terms of *normal* factors $\hat{x}$ and $\hat{y}$ at an angle θ to each other, the expressions for the components of z simplify a little.

$$|B|^2 == \left(\hat{x} \wedge \hat{y}\right) \ominus \left(\hat{x} \wedge \hat{y}\right) == \mathrm{Det}\left[\begin{pmatrix} 1 & \mathrm{Cos}[\theta] \\ \mathrm{Cos}[\theta] & 1 \end{pmatrix}\right] == \mathrm{Sin}[\theta]^2$$

$$|B|^2\, z^{\#} == \begin{pmatrix} \hat{x} \ominus z & \hat{y} \ominus z \end{pmatrix} \begin{pmatrix} 1 & -\mathrm{Cos}[\theta] \\ -\mathrm{Cos}[\theta] & 1 \end{pmatrix} \begin{pmatrix} \hat{x} \\ \hat{y} \end{pmatrix}$$

$$== \begin{pmatrix} \hat{x} \ominus z & \hat{y} \ominus z \end{pmatrix} \begin{pmatrix} \hat{x} - \mathrm{Cos}[\theta]\, \hat{y} \\ \hat{y} - \mathrm{Cos}[\theta]\, \hat{x} \end{pmatrix}$$

- If we choose to express the bivector in terms of *orthonormal* factors $\hat{x}$ and $\hat{y}$ the expressions for the components of z simplify somewhat further.

$$|B|^2 == (\hat{x}\wedge\hat{y}) \ominus (\hat{x}\wedge\hat{y}) == \mathrm{Det}\left[\begin{pmatrix}1 & 0\\ 0 & 1\end{pmatrix}\right] == 1$$

$$z^{\#} == (\hat{x}\ominus z \;\; \hat{y}\ominus z)\begin{pmatrix}1 & 0\\ 0 & 1\end{pmatrix}\begin{pmatrix}\hat{x}\\ \hat{y}\end{pmatrix} == (\hat{x}\ominus z \;\; \hat{y}\ominus z)\begin{pmatrix}\hat{x}\\ \hat{y}\end{pmatrix} == (\hat{x}\ominus z)\,\hat{x} + (\hat{y}\ominus z)\,\hat{y}$$

$$z^{\llcorner} == z - (\hat{x}\ominus z)\,\hat{x} - (\hat{y}\ominus z)\,\hat{y}$$

◆ Formula summary

We can now summarize the formulae we have developed.

$$z == z^{\llcorner} + z^{\#} \qquad z^{\llcorner} == (z\wedge\hat{B})\ominus\hat{B} \qquad z^{\#} == -\hat{B}\ominus(\hat{B}\ominus z) \tag{6.210}$$

$$\hat{B}\ominus z^{\llcorner} == 0 \qquad \hat{B}\wedge z^{\#} == 0 \qquad z^{\llcorner}\ominus z^{\#} == 0 \tag{6.211}$$

$$|z|^2 == |z^{\llcorner}|^2 + |z^{\#}|^2 == |z\wedge\hat{B}|^2 + |\hat{B}\ominus z|^2 \tag{6.212}$$

If $B = x\wedge y$:

$$|B|^2\, z^{\#} == (x\ominus z \;\; y\ominus z)\begin{pmatrix} y\ominus y & -x\ominus y\\ -x\ominus y & x\ominus x\end{pmatrix}\begin{pmatrix}x\\ y\end{pmatrix}$$

$$|B|^2 == (x\wedge y)\ominus(x\wedge y) == (x\ominus x)\,(y\ominus y) - (x\ominus y)^2 \tag{6.213}$$

If $B = \hat{x}\wedge\hat{y}$ and $\hat{x}\ominus\hat{y} = \mathrm{Cos}[\theta]$:

$$z^{\#} == \frac{1}{\mathrm{Sin}[\theta]^2}(\hat{x}\ominus z \;\; \hat{y}\ominus z)\begin{pmatrix}\hat{x} - \mathrm{Cos}[\theta]\,\hat{y}\\ \hat{y} - \mathrm{Cos}[\theta]\,\hat{x}\end{pmatrix} \tag{6.214}$$

If $B = \hat{x}\wedge\hat{y}$ and $\hat{x}\ominus\hat{y} = 0$:

$$z^{\#} == (\hat{x}\ominus z)\,\hat{x} + (\hat{y}\ominus z)\,\hat{y} \tag{6.215}$$

The triangle formulae for a point and an *m*-vector

In this section we explore the interpretation of the triangle formulae when the 1-element is a point P with position vector p. We begin with the decomposition for P in terms of a unit *m*-vector (formula [6.194]), denoting the first term by $P^{\llcorner}$ and the second by $P^{\#}$. We assume the hybrid metric for interpreted elements [5.80] where the origin is orthogonal to all vectors.

$$\mathbf{P} == \left(\underset{m}{\hat{\alpha}} \wedge \mathbf{P}\right) \ominus \underset{m}{\hat{\alpha}} + (-1)^{m-1} \underset{m}{\hat{\alpha}} \ominus \left(\underset{m}{\hat{\alpha}} \ominus \mathbf{P}\right)$$

$$\mathbf{P}^{\llcorner} == \left(\underset{m}{\hat{\alpha}} \wedge \mathbf{P}\right) \ominus \underset{m}{\hat{\alpha}} \qquad \mathbf{P}^{\#} == (-1)^{m-1} \underset{m}{\hat{\alpha}} \ominus \left(\underset{m}{\hat{\alpha}} \ominus \mathbf{P}\right)$$

If we replace P with $\star\mathbb{O} + p$, the expression for the orthogonal component $P^{\llcorner}$ expands to

$$\mathbf{P}^{\llcorner} == \left(\underset{m}{\hat{\alpha}} \wedge \mathbf{P}\right) \ominus \underset{m}{\hat{\alpha}} == \left(\underset{m}{\hat{\alpha}} \wedge (\star\mathbb{O} + \mathbf{p})\right) \ominus \underset{m}{\hat{\alpha}} == \left(\underset{m}{\hat{\alpha}} \wedge \star\mathbb{O}\right) \ominus \underset{m}{\hat{\alpha}} + \left(\underset{m}{\hat{\alpha}} \wedge \mathbf{p}\right) \ominus \underset{m}{\hat{\alpha}}$$

It is easy to show that the first term on the right hand side of this equation for $P^{\llcorner}$ reduces simply to $\star\mathbb{O}$. We do this by applying the triangle formula to $\star\mathbb{O}$ (instead of P) as follows:

$$\star\mathbb{O} == \left(\underset{m}{\hat{\alpha}} \wedge \star\mathbb{O}\right) \ominus \underset{m}{\hat{\alpha}} + (-1)^{m-1} \underset{m}{\hat{\alpha}} \ominus \left(\underset{m}{\hat{\alpha}} \ominus \star\mathbb{O}\right) == \left(\underset{m}{\hat{\alpha}} \wedge \star\mathbb{O}\right) \ominus \underset{m}{\hat{\alpha}}$$

The second term in the expression for $P^{\llcorner}$ is simply the orthogonal component $p^{\llcorner}$ of the position vector p of P.

$$\mathbf{p}^{\llcorner} == \left(\underset{m}{\hat{\alpha}} \wedge \mathbf{p}\right) \ominus \underset{m}{\hat{\alpha}}$$

Hence $P^{\llcorner}$ is the *point* orthogonal to $\underset{m}{\hat{\alpha}}$.

$$\mathbf{P}^{\llcorner} == \star\mathbb{O} + \mathbf{p}^{\llcorner} == \star\mathbb{O} + \left(\underset{m}{\hat{\alpha}} \wedge \mathbf{p}\right) \ominus \underset{m}{\hat{\alpha}} \qquad 6.216$$

The expression for the parallel component $P^{\#}$ of the point P expands to

$$\mathbf{P}^{\#} == (-1)^{m-1} \underset{m}{\hat{\alpha}} \ominus \left(\underset{m}{\hat{\alpha}} \ominus \mathbf{P}\right) == (-1)^{m-1} \underset{m}{\hat{\alpha}} \ominus \left(\underset{m}{\hat{\alpha}} \ominus (\star\mathbb{O} + \mathbf{p})\right)$$

$$== (-1)^{m-1} \left(\underset{m}{\hat{\alpha}} \ominus \left(\underset{m}{\hat{\alpha}} \ominus \star\mathbb{O}\right) + \underset{m}{\hat{\alpha}} \ominus \left(\underset{m}{\hat{\alpha}} \ominus \mathbf{p}\right)\right)$$

Again the first term is zero leaving just a vector, so that $P^{\#}$ is the *vector* lying in (parallel to) $\underset{m}{\hat{\alpha}}$.

$$\mathbf{P}^{\#} == \mathbf{p}^{\#} == (-1)^{m-1} \underset{m}{\hat{\alpha}} \ominus \left(\underset{m}{\hat{\alpha}} \ominus \mathbf{p}\right) \qquad 6.217$$

In general then, the orthogonal component $P^{\llcorner}$ of a point P is a point, but the parallel component $P^{\#}$ of P is a *vector*.

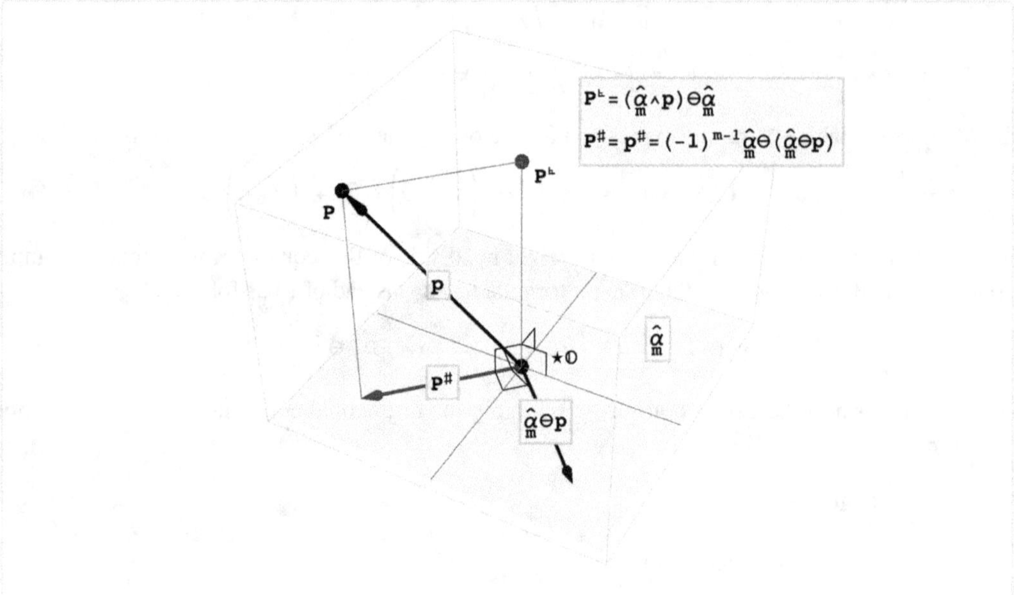

The triangle components for a point

Special cases of the triangle formulae for a point and an *m*-vector

In this section we explore formulae [6.216] and [6.217] for various grades of the *m*-vector.

◆ If *m* is 0

If m is 0, then the unit m-vector is unity, and the formulae show that the component of a point parallel to a scalar is zero, while its component orthogonal to the scalar is the point itself.

$$P^{\#} == p^{\#} == 0 \qquad P^{\llcorner} == \star\mathbb{O} + p^{\llcorner} == \star\mathbb{O} + p == P \qquad \{m == 0\} \tag{6.218}$$

◆ If *m* is 1

If m is 1, then the unit m-vector is a unit vector ($\hat{\alpha}$, say), and the formulae show that the component of a point parallel to $\hat{\alpha}$ is the projection of the position vector of the point onto $\hat{\alpha}$, while its component orthogonal to $\hat{\alpha}$ is the point whose position vector is the orthogonal component of the projection.

$$P^{\#} == p^{\#} == (\hat{\alpha} \ominus p)\, \hat{\alpha}$$

$$P^{\llcorner} == \star\mathbb{O} + p^{\llcorner} == \star\mathbb{O} + (\hat{\alpha} \wedge p) \ominus \hat{\alpha} == \star\mathbb{O} + p - (\hat{\alpha} \ominus p)\, \hat{\alpha} == P - (\hat{\alpha} \ominus p)\, \hat{\alpha}$$

$$P^{\#} == p^{\#} == (\hat{\alpha} \ominus p)\, \hat{\alpha} \qquad P^{\llcorner} == P - (\hat{\alpha} \ominus p)\, \hat{\alpha} \qquad \{m == 1\} \tag{6.219}$$

Clearly, $P^{\#}$ (a vector) is parallel to $\hat{\alpha}$, and the point $P^{\llcorner}$ (or its position vector) is orthogonal to $\hat{\alpha}$.

$$P^{\#} \wedge \hat{\alpha} == p^{\#} \wedge \hat{\alpha} == (\hat{\alpha} \ominus p)\, \hat{\alpha} \wedge \hat{\alpha} == 0$$

$$\mathbf{P}^{\llcorner} \ominus \hat{\alpha} == \mathbf{P} \ominus \hat{\alpha} - (\hat{\alpha} \ominus \mathbf{p})\, \hat{\alpha} \ominus \hat{\alpha} == \mathbf{p} \ominus \hat{\alpha} - (\hat{\alpha} \ominus \mathbf{p}) == 0$$

◆ **If *m* is 2**

If m is 2, then the unit m-vector is a unit bivector ($\hat{B}$, say), and the formulae show that the component of a point parallel to $\hat{B}$ is the projection of the position vector of the point onto $\hat{\alpha}$, while its component orthogonal to $\hat{\alpha}$ is the point whose position vector is the orthogonal component of the projection.

$$\mathbf{P}^{\#} == \mathbf{p}^{\#} == -\hat{\mathbf{B}} \ominus (\hat{\mathbf{B}} \ominus \mathbf{p})$$

$$\mathbf{P}^{\llcorner} == \star\mathbb{O} + \mathbf{p}^{\llcorner} == \star\mathbb{O} + (\hat{\mathbf{B}} \wedge \mathbf{p}) \ominus \hat{\mathbf{B}} == \star\mathbb{O} + \mathbf{p} + \hat{\mathbf{B}} \ominus (\hat{\mathbf{B}} \ominus \mathbf{p}) == \mathbf{P} + \hat{\mathbf{B}} \ominus (\hat{\mathbf{B}} \ominus \mathbf{p})$$

$$\mathbf{P}^{\#} == \mathbf{p}^{\#} == -\hat{\mathbf{B}} \ominus (\hat{\mathbf{B}} \ominus \mathbf{p}) \qquad \mathbf{P}^{\llcorner} == \mathbf{P} + \hat{\mathbf{B}} \ominus (\hat{\mathbf{B}} \ominus \mathbf{p}) \qquad \{m == 2\} \tag{6.220}$$

As before, the vector, $P^{\#}$ is parallel to $\hat{B}$; and the point $P^{\llcorner}$ (or its position vector) is orthogonal to $\hat{B}$.

The orthogonal decomposition of an *m*-element by a 1-element

To this stage we have been concentrating on the decomposition of a 1-element parallel to and orthogonal to an m-element. In this and the following sections we show that, *mutatis mutandis*, we are able to decompose an m-element parallel and orthogonal to a 1-element.

The basic Interior Product Formula [6.100] is

$$\left(\underset{m}{\alpha} \wedge \underset{k}{\beta}\right) \ominus \mathbf{x} == \left(\underset{m}{\alpha} \ominus \mathbf{x}\right) \wedge \underset{k}{\beta} + (-1)^{m}\, \underset{m}{\alpha} \wedge \left(\underset{k}{\beta} \ominus \mathbf{x}\right)$$

By putting $\underset{k}{\beta}$ equal to x we get

$$\left(\underset{m}{\alpha} \wedge \mathbf{x}\right) \ominus \mathbf{x} == \left(\underset{m}{\alpha} \ominus \mathbf{x}\right) \wedge \mathbf{x} + (-1)^{m}\, \underset{m}{\alpha} \wedge (\mathbf{x} \ominus \mathbf{x})$$

Rearranging terms expresses a 'decomposition' of the element $\underset{m}{\alpha}$ in terms of the unit 1-element $\hat{x}$:

$$\underset{m}{\alpha} == \left(\hat{\mathbf{x}} \wedge \underset{m}{\alpha}\right) \ominus \hat{\mathbf{x}} + \hat{\mathbf{x}} \wedge \left(\underset{m}{\alpha} \ominus \hat{\mathbf{x}}\right) \tag{6.221}$$

First we show that the two components of $\underset{m}{\alpha}$ are orthogonal.

$$\left(\left(\hat{\mathbf{x}} \wedge \underset{m}{\alpha}\right) \ominus \hat{\mathbf{x}}\right) \ominus \left(\hat{\mathbf{x}} \wedge \left(\underset{m}{\alpha} \ominus \hat{\mathbf{x}}\right)\right)$$
$$== \left(\hat{\mathbf{x}} \wedge \underset{m}{\alpha}\right) \ominus \left(\hat{\mathbf{x}} \wedge \hat{\mathbf{x}} \wedge \left(\underset{m}{\alpha} \ominus \hat{\mathbf{x}}\right)\right) == \left(\hat{\mathbf{x}} \wedge \underset{m}{\alpha}\right) \ominus 0 == 0$$

Next we show that $\hat{x}$ is orthogonal to the first component, and is contained in (parallel to) the second.

$$\underset{m}{\alpha}^{\llcorner} \ominus \hat{\mathbf{x}} == \left(\left(\hat{\mathbf{x}} \wedge \underset{m}{\alpha}\right) \ominus \hat{\mathbf{x}}\right) \ominus \hat{\mathbf{x}} == \left(\hat{\mathbf{x}} \wedge \underset{m}{\alpha}\right) \ominus (\hat{\mathbf{x}} \wedge \hat{\mathbf{x}}) == 0$$

$$\underset{m}{\alpha}^{\#}\wedge\hat{x} == \hat{x}\wedge\left(\underset{m}{\alpha}\ominus\hat{x}\right)\wedge\hat{x} == 0$$

Now we can characterize these two components of $\underset{m}{\alpha}$ as 'orthogonal' and 'parallel' (to x).

$$\underset{m}{\alpha} == \underset{m}{\alpha}^{\perp} + \underset{m}{\alpha}^{\#} \qquad \underset{m}{\alpha}^{\perp} == \left(\hat{x}\wedge\underset{m}{\alpha}\right)\ominus\hat{x} \qquad \underset{m}{\alpha}^{\#} == \hat{x}\wedge\left(\underset{m}{\alpha}\ominus\hat{x}\right) \tag{6.222}$$

$$\underset{m}{\alpha}^{\perp}\ominus\hat{x} == 0 \qquad \underset{m}{\alpha}^{\#}\wedge\hat{x} == 0 \qquad \underset{m}{\alpha}^{\perp}\ominus\underset{m}{\alpha}^{\#} == 0 \tag{6.223}$$

Note that this derivation of component orthogonality is also valid for a given x if the components belong to two distinct m-elements ($\underset{m}{\alpha}$ and $\underset{m}{\beta}$, say).

$$\left(\left(\hat{x}\wedge\underset{m}{\alpha}\right)\ominus\hat{x}\right)\ominus\left(\left(\underset{m}{\beta}\ominus\hat{x}\right)\wedge\hat{x}\right)$$
$$== \left(\hat{x}\wedge\underset{m}{\alpha}\right)\ominus\left(\hat{x}\wedge\left(\left(\underset{m}{\beta}\ominus\hat{x}\right)\wedge\hat{x}\right)\right) == \left(\hat{x}\wedge\underset{m}{\alpha}\right)\ominus 0 == 0$$

That is, the requisite orthogonality relationships still hold allowing us straightforwardly to decompose an inner product.

$$\underset{m}{\alpha}^{\perp}\ominus\underset{m}{\beta}^{\#} == 0 \qquad \underset{m}{\beta}^{\perp}\ominus\underset{m}{\alpha}^{\#} == 0 \tag{6.224}$$

$$\underset{m}{\alpha}\ominus\underset{m}{\beta} == \left(\underset{m}{\alpha}^{\perp}+\underset{m}{\alpha}^{\#}\right)\ominus\left(\underset{m}{\beta}^{\perp}+\underset{m}{\beta}^{\#}\right) == \underset{m}{\alpha}^{\perp}\ominus\underset{m}{\beta}^{\perp} + \underset{m}{\alpha}^{\#}\ominus\underset{m}{\beta}^{\#} \tag{6.225}$$

The measure of the triangular *m*-element components

To compute the measure of the orthogonal components $\underset{m}{\alpha}^{\perp}$ and $\underset{m}{\alpha}^{\#}$, we can take the inner product square of each component with itself.

$$\underset{m}{\alpha}^{\perp}\ominus\underset{m}{\alpha}^{\perp} == \left(\left(\hat{x}\wedge\underset{m}{\alpha}\right)\ominus\hat{x}\right)\ominus\left(\left(\hat{x}\wedge\underset{m}{\alpha}\right)\ominus\hat{x}\right)$$
$$\underset{m}{\alpha}^{\#}\ominus\underset{m}{\alpha}^{\#} == \left(\left(\underset{m}{\alpha}\ominus\hat{x}\right)\wedge\hat{x}\right)\ominus\left(\left(\underset{m}{\alpha}\ominus\hat{x}\right)\wedge\hat{x}\right)$$

These expressions may be simplified fairly directly as follows.

$$\underset{m}{\alpha}^{\perp}\ominus\underset{m}{\alpha}^{\perp} == \underset{m}{\alpha}^{\perp}\ominus\left(\underset{m}{\alpha}-\underset{m}{\alpha}^{\#}\right) == \underset{m}{\alpha}^{\perp}\ominus\underset{m}{\alpha} == \left(\left(\hat{x}\wedge\underset{m}{\alpha}\right)\ominus\hat{x}\right)\ominus\underset{m}{\alpha} == \left(\hat{x}\wedge\underset{m}{\alpha}\right)\ominus\left(\hat{x}\wedge\underset{m}{\alpha}\right)$$
$$\underset{m}{\alpha}^{\#}\ominus\underset{m}{\alpha}^{\#} == \left(\underset{m}{\alpha}-\underset{m}{\alpha}^{\perp}\right)\ominus\underset{m}{\alpha}^{\#} == \underset{m}{\alpha}\ominus\underset{m}{\alpha}^{\#} == (-1)^{m-1}\underset{m}{\alpha}\ominus\left(\left(\underset{m}{\alpha}\ominus\hat{x}\right)\wedge\hat{x}\right)$$
$$== \underset{m}{\alpha}\ominus\left(\hat{x}\wedge\left(\underset{m}{\alpha}\ominus\hat{x}\right)\right) == \left(\underset{m}{\alpha}\ominus\hat{x}\right)\ominus\left(\underset{m}{\alpha}\ominus\hat{x}\right)$$

$$\underset{m}{\alpha}^{\perp} \ominus \underset{m}{\alpha}^{\perp} = \left(\hat{\mathbf{x}} \wedge \underset{m}{\alpha}\right) \ominus \left(\hat{\mathbf{x}} \wedge \underset{m}{\alpha}\right) \qquad \underset{m}{\alpha}^{\#} \ominus \underset{m}{\alpha}^{\#} = \left(\underset{m}{\alpha} \ominus \hat{\mathbf{x}}\right) \ominus \left(\underset{m}{\alpha} \ominus \hat{\mathbf{x}}\right) \tag{6.226}$$

The magnitudes of the orthogonal components then become

$$\left|\underset{m}{\alpha}^{\perp}\right| = \left|\left(\hat{\mathbf{x}} \wedge \underset{m}{\alpha}\right) \ominus \hat{\mathbf{x}}\right| = \left|\hat{\mathbf{x}} \wedge \underset{m}{\alpha}\right|$$
$$\left|\underset{m}{\alpha}^{\#}\right| = \left|\left(\underset{m}{\alpha} \ominus \hat{\mathbf{x}}\right) \wedge \hat{\mathbf{x}}\right| = \left|\underset{m}{\alpha} \ominus \hat{\mathbf{x}}\right| \tag{6.227}$$

and the Pythagorean relationships become clearly apparent.

$$\left|\underset{m}{\alpha}^{\perp}\right|^2 + \left|\underset{m}{\alpha}^{\#}\right|^2 = \left|\underset{m}{\alpha}\right|^2 \qquad \left|\underset{m}{\hat{\alpha}} \wedge \hat{\mathbf{x}}\right|^2 + \left|\underset{m}{\hat{\alpha}} \ominus \hat{\mathbf{x}}\right|^2 = 1 \tag{6.228}$$

Note that the second formula here which represents the Pythagorean formula in the form $\cos^2(\theta) + \sin^2(\theta) = 1$ is the *same* formula as obtained in [6.200].

The triangle formulae for a vector and a bound m-vector

Suppose now that the m-element we wish to decompose is a bound m-vector. In this section we will explore its orthogonal and parallel components and put them into a form with a more immediate geometric significance then appears from their initial definitions. Denote the bound m-vector by $\mathbb{A}$. Then its decomposition orthogonal and parallel to vector x may be written:

$$\mathbb{A} = \mathbf{P} \wedge \underset{m}{\alpha} = \mathbb{A}^{\perp} + \mathbb{A}^{\#} \qquad \mathbb{A}^{\perp} = \left(\hat{\mathbf{x}} \wedge \mathbb{A}\right) \ominus \hat{\mathbf{x}} \qquad \mathbb{A}^{\#} = \hat{\mathbf{x}} \wedge \left(\mathbb{A} \ominus \hat{\mathbf{x}}\right) \tag{6.229}$$

We will also find it useful to recall the definitions [6.222] of the orthogonal and parallel components of the m-vector of $\mathbb{A}$ and the basic Interior Product Formula [6.100].

$$\underset{m}{\alpha}^{\perp} = \left(\hat{\mathbf{x}} \wedge \underset{m}{\alpha}\right) \ominus \hat{\mathbf{x}} \qquad \underset{m}{\alpha}^{\#} = \hat{\mathbf{x}} \wedge \left(\underset{m}{\alpha} \ominus \hat{\mathbf{x}}\right)$$

$$\left(\underset{m}{\alpha} \wedge \underset{k}{\beta}\right) \ominus \mathbf{x} = \left(\underset{m}{\alpha} \ominus \mathbf{x}\right) \wedge \underset{k}{\beta} + (-1)^m \underset{m}{\alpha} \wedge \left(\underset{k}{\beta} \ominus \mathbf{x}\right)$$

Applying the Interior Product Formula to $\mathbb{A}$ (seen as a product of its point and its m-vector) gives

$$\left(\mathbf{P} \wedge \underset{m}{\alpha}\right) \ominus \hat{\mathbf{x}} = \left(\mathbf{P} \ominus \hat{\mathbf{x}}\right) \underset{m}{\alpha} - \mathbf{P} \wedge \left(\underset{m}{\alpha} \ominus \hat{\mathbf{x}}\right)$$

This is a bound $(m-1)$-vector. In order to cast it into the form of a product of a point and an m-vector we need to determine a vector z, say, such that

$$-\mathbf{P} \wedge \left(\underset{m}{\alpha} \ominus \hat{\mathbf{x}}\right) + \left(\mathbf{P} \ominus \hat{\mathbf{x}}\right) \underset{m}{\alpha} = -(\mathbf{P} + \mathbf{z}) \wedge \left(\underset{m}{\alpha} \ominus \hat{\mathbf{x}}\right)$$

That is, the vector z must satisfy

$$z \wedge \left(\underset{m}{\alpha} \ominus \hat{x}\right) == -\left(P \ominus \hat{x}\right) \underset{m}{\alpha}$$

If $\underset{m}{\alpha}$ is written as the exterior product of 1-element factors we can expand $\underset{m}{\alpha} \ominus \hat{x}$ by the Interior Common Factor Theorem in the form [6.52].

$$(\alpha_1 \wedge \alpha_2 \wedge \cdots \wedge \alpha_m) \ominus \hat{x} == \sum_{i=1}^{m} \left(\alpha_i \ominus \hat{x}\right) (-1)^{i-1} \alpha_1 \wedge \cdots \wedge \square_i \wedge \cdots \wedge \alpha_m$$

If we define z to be the specific scalar multiple of the vector α_j in $\underset{m}{\alpha}$,

$$z == -\frac{P \ominus \hat{x}}{\alpha_j \ominus \hat{x}} \alpha_j$$

and then multiply the previous expansion through by it, we see that this value of z is a solution since the right hand side is zero for all *i* not equal to *j*, and equal to $-\left(P \ominus \hat{x}\right) \underset{m}{\alpha}$ for *i* equal to *j*.

$$z \wedge \left((\alpha_1 \wedge \alpha_2 \wedge \cdots \wedge \alpha_m) \ominus \hat{x}\right)$$

$$== -\frac{P \ominus \hat{x}}{\alpha_j \ominus \hat{x}} \sum_{i=1}^{m} \left(\alpha_i \ominus \hat{x}\right) (-1)^{i-1} \alpha_j \wedge (\alpha_1 \wedge \cdots \wedge \square_i \wedge \cdots \wedge \alpha_m) == -\left(P \ominus \hat{x}\right) \underset{m}{\alpha}$$

We will call this new point $P^{\dagger}$. Note that it, and therefore its position vector, are orthogonal to x.

$$P^{\dagger} == P + z == P - \frac{P \ominus \hat{x}}{\alpha_j \ominus \hat{x}} \alpha_j \qquad P^{\dagger} \ominus \hat{x} == P \ominus \hat{x} - \frac{P \ominus \hat{x}}{\alpha_j \ominus \hat{x}} \alpha_j \ominus \hat{x} == 0$$

Thus we can now write $\left(P \wedge \underset{m}{\alpha}\right) \ominus \hat{x}$, and $\mathbb{A}^{\#}$ in the more geometrically intuitive form we were looking for.

$$\left(P \wedge \underset{m}{\alpha}\right) \ominus \hat{x} == -(P + z) \wedge \left(\underset{m}{\alpha} \ominus \hat{x}\right) == -P^{\dagger} \wedge \left(\underset{m}{\alpha} \ominus \hat{x}\right)$$

$$\mathbb{A}^{\#} == -\hat{x} \wedge P^{\dagger} \wedge \left(\underset{m}{\alpha} \ominus \hat{x}\right) == P^{\dagger} \wedge \hat{x} \wedge \left(\underset{m}{\alpha} \ominus \hat{x}\right) == P^{\dagger} \wedge \underset{m}{\alpha}^{\#}$$

It is now straightforward to develop the expression for $\mathbb{A}^{\llcorner}$ because since the point $P^{\dagger}$ is on $\mathbb{A}$, $\mathbb{A}$ may also be written in terms of the point $P^{\dagger}$.

$$\mathbb{A}^{\llcorner} == \mathbb{A} - \mathbb{A}^{\#} == P \wedge \underset{m}{\alpha} - P^{\dagger} \wedge \underset{m}{\alpha}^{\#}$$

$$== P^{\dagger} \wedge \underset{m}{\alpha} - P^{\dagger} \wedge \underset{m}{\alpha}^{\#} == P^{\dagger} \wedge \left(\underset{m}{\alpha} - \underset{m}{\alpha}^{\#}\right) == P^{\dagger} \wedge \underset{m}{\alpha}^{\llcorner}$$

$$\mathbb{A} == P \wedge \underset{m}{\alpha} == \mathbb{A}^{\llcorner} + \mathbb{A}^{\#} \qquad \mathbb{A}^{\llcorner} == P^{\dagger} \wedge \underset{m}{\alpha}^{\llcorner} \qquad \mathbb{A}^{\#} == P^{\dagger} \wedge \underset{m}{\alpha}^{\#} \tag{6.230}$$

$$P^{\dagger} == P - \frac{P \ominus \hat{x}}{\alpha_j \ominus \hat{x}} \alpha_j \qquad \alpha_j \wedge \underset{m}{\alpha} == 0 \tag{6.231}$$

$$\underset{m}{\alpha} == \underset{m}{\alpha}^{\llcorner} + \underset{m}{\alpha}^{\#} \qquad \underset{m}{\alpha}^{\llcorner} == \left(\hat{x} \wedge \underset{m}{\alpha}\right) \ominus \hat{x} \qquad \underset{m}{\alpha}^{\#} == \hat{x} \wedge \left(\underset{m}{\alpha} \ominus \hat{x}\right) \tag{6.232}$$

Clearly, the bound m-vector $\mathbb{A}$ and its two components all intersect in the common point $P^{\dagger}$. Although there is clearly an infinity of such points because α_j can be any vector belonging to $\underset{m}{\alpha}$, they all have the common property of being orthogonal to x.

Examples of the triangle formulae for a bound m-vector

In this section we explore some lower grade examples of formulae [6.230], [6.231] and [6.232] for the decomposition of a bound m-vector orthogonal to and parallel to a vector.

◆ Decomposition of a point

The special case in which the bound m-vector is simply a point (m equal to 0) may serve to give the simplest example of the way the decomposition works. However, in this case we explore it from the basic definitions, not from the special formulae we derived in the previous section - since these formulae do not apply if there is no vectorial factor in the bound element.

Putting $\underset{m}{\alpha}$ equal to the scalar a shows that with respect to x the orthogonal component $a^{\perp}$ of a is a, but the parallel component $a^{\#}$ is 0.

$$\underset{m}{\alpha}^{\perp} == a^{\perp} == (\hat{x} \wedge a) \ominus \hat{x} == a \qquad \underset{m}{\alpha}^{\#} == a^{\#} == \hat{x} \wedge (a \ominus \hat{x}) == 0$$

The parallel component $P^{\#}$ of P is a *vector*, and is the parallel component $p^{\#}$ of its position vector p. This parallel component is, as might be expected, simply the orthogonal projection of the position vector p onto x.

$$P^{\#} == \hat{x} \wedge (P \ominus \hat{x}) == (p \ominus \hat{x})\, \hat{x} == p^{\#}$$

The orthogonal component $P^{\perp}$ of P is a *point*, whose position vector $p^{\perp}$ is the orthogonal component of the position vector of P. This orthogonal component is, as might be expected, simply the difference between the point P and its parallel component $P^{\#}$.

$$P^{\#} == p^{\#} == (p \ominus \hat{x})\, \hat{x} \qquad P^{\perp} == P - P^{\#} \tag{6.233}$$

◆ Decomposition of a line

Consider now a line $\mathbb{L}$ represented by a bound vector $P \wedge \alpha$, and a direction represented by a unit vector $\hat{x}$. Then we can write the line's components orthogonal and parallel to $\hat{x}$ as:

$$\mathbb{L}^{\perp} == P^{\dagger} \wedge \alpha^{\perp} \qquad \mathbb{L}^{\#} == P^{\dagger} \wedge \alpha^{\#} \qquad \mathbb{L} == P \wedge \alpha$$

$$P^{\dagger} == P - \frac{P \ominus \hat{x}}{\alpha \ominus \hat{x}}\, \alpha \qquad \alpha^{\#} == (\alpha \ominus \hat{x})\, \hat{x} \qquad \alpha^{\perp} == \alpha - \alpha^{\#} == (\hat{x} \wedge \alpha) \ominus \hat{x}$$

$$\mathbb{L}^{\#} == P^{\dagger} \wedge ((\alpha \ominus \hat{x})\, \hat{x}) \qquad \mathbb{L}^{\perp} == P^{\dagger} \wedge (\alpha - (\alpha \ominus \hat{x})\, \hat{x}) \qquad P^{\dagger} == P - \frac{P \ominus \hat{x}}{\alpha \ominus \hat{x}}\, \alpha \tag{6.234}$$

All three lines intersect in the common point $P^{\dagger}$ which is orthogonal to x. The vectors of all three lines and the vector x all lie in the same bivector.

In sum: the vector x has "chosen" a point $P^{\dagger}$ on the original line through which to draw a line parallel to itself. It adjusts the length of the vector of this line so that when it subtracts this new line from the original line it gets a second new line orthogonal to the first new one through the same point. This could have been done in an infinity of ways, but **x** pins down the uniqueness of the operation by choosing the precise point $\mathrm{P}^{\dagger}$ to which it also is orthogonal.

◆ Decomposition of a plane

Consider now a plane $\mathbb{P}$ represented by a bound vector $\mathrm{P}\wedge\alpha\wedge\beta$, and a direction represented by a unit vector $\hat{\mathbf{x}}$. Then we can write its orthogonal and parallel components as:

$$\mathbb{P}^{\perp} == \mathrm{P}^{\dagger}\wedge(\alpha\wedge\beta)^{\perp} \qquad \mathbb{P}^{\#} == \mathrm{P}^{\dagger}\wedge(\alpha\wedge\beta)^{\#}$$

$$\mathbb{P} == \mathrm{P}\wedge\alpha\wedge\beta \qquad \mathrm{P}^{\dagger} == \mathrm{P} - \frac{\mathrm{P}\ominus\hat{\mathbf{x}}}{\alpha\ominus\hat{\mathbf{x}}}\,\alpha$$

$$(\alpha\wedge\beta)^{\#} == \hat{\mathbf{x}}\wedge\left((\alpha\wedge\beta)\ominus\hat{\mathbf{x}}\right) == \hat{\mathbf{x}}\wedge\left(\alpha\ominus\hat{\mathbf{x}}\right)\beta - \hat{\mathbf{x}}\wedge\left(\beta\ominus\hat{\mathbf{x}}\right)\alpha$$

$$(\alpha\wedge\beta)^{\perp} == \left(\hat{\mathbf{x}}\wedge\alpha\wedge\beta\right)\ominus\hat{\mathbf{x}} == \alpha\wedge\beta - (\alpha\wedge\beta)^{\#}$$

$$\mathbb{P}^{\#} == \mathrm{P}^{\dagger}\wedge\hat{\mathbf{x}}\wedge\left((\alpha\wedge\beta)\ominus\hat{\mathbf{x}}\right)$$

$$\mathbb{P}^{\perp} == \mathrm{P}^{\dagger}\wedge\left(\alpha\wedge\beta - \hat{\mathbf{x}}\wedge\left((\alpha\wedge\beta)\ominus\hat{\mathbf{x}}\right)\right) \qquad \mathrm{P}^{\dagger} == \mathrm{P} - \frac{\mathrm{P}\ominus\hat{\mathbf{x}}}{\alpha\ominus\hat{\mathbf{x}}}\,\alpha \qquad 6.235$$

All three planes intersect in the common point $\mathrm{P}^{\dagger}$ which is orthogonal to **x**. The bivectors of all three planes and the vector **x** all lie in the same trivector $\hat{\mathbf{x}}\wedge\alpha\wedge\beta$.

In sum: the vector **x** has "chosen" a point $\mathrm{P}^{\dagger}$ on the original plane through which to draw a plane incorporating its own direction. It adjusts the bivector of this plane so that when it subtracts this new plane from the original plane it gets a second new plane orthogonal to the first new one through the same point. As in the previous example, this could have been done in an infinity of ways, but **x** pins down the uniqueness of the operation by choosing the precise point $\mathrm{P}^{\dagger}$ to which it also is orthogonal.

6.12 Angle

In this section we briefly introduce some notions of *angle* between elements. In the simple case where one of the elements is a vector, we can straightforwardly define the angle between it and a simple *m*-vector as the angle between it and its orthogonal projection on to the *m*-vector by using the Triangle Formulae developed in the previous section.

However, if neither element is a vector, the notion of angle between two elements needs a more complex consideration. For example in a vector 3-space, the angle between two intersecting bivectors may either be defined in the traditional manner as the angle between their vector complements; or by extension of the Triangle notion above as the *maximum* angle obtained by any vector in one bivector with the other bivector. The *minimum* angle so obtained would be zero.

The angle between 1-elements

The interior product enables us to *define*, as is standard practice, the angle between two 1-elements.

$$x \ominus y == |x| \, |y| \, \text{Cos}[\theta]$$

$$x \ominus y == |x| \, |y| \, \text{Cos}[\theta] \qquad 6.236$$

The square of the magnitude of the bivector $x \wedge y$ can be expanded to give

$$|x \wedge y|^2 == (x \wedge y) \ominus (x \wedge y) == (x \ominus x) \, (y \ominus y) - (x \ominus y)^2$$
$$== |x|^2 \, |y|^2 \left(1 - \text{Cos}[\theta]^2\right) == |x|^2 \, |y|^2 \, \text{Sin}[\theta]^2$$

$$|x \wedge y| == |x| \, |y| \, \text{Sin}[\theta] \qquad 6.237$$

Hence we have the fundamental relationships relating the angle between two 1-elements and the magnitudes of their exterior and interior products

$$\left|\hat{x} \wedge \hat{y}\right| == \text{Sin}[\theta] \qquad \hat{x} \ominus \hat{y} == \text{Cos}[\theta] \qquad 6.238$$

$$|x \wedge y|^2 + |x \ominus y|^2 == |x|^2 \, |y|^2 \qquad 6.239$$

$$\left|\hat{x} \wedge \hat{y}\right|^2 + \left|\hat{x} \ominus \hat{y}\right|^2 == 1 \qquad 6.240$$

The angle between a 1-element and a simple *m*-element

We define the angle between a 1-element x and a simple *m*-element $\underset{m}{\alpha}$ as the angle between x and the *component* $x^{\#}$ of x *parallel* to $\underset{m}{\alpha}$ in the triangle relationship.

Recall from formula [6.195] that $x^{\#}$ may be written as

$$x^{\#} == (-1)^{m-1} \, \underset{m}{\hat{\alpha}} \ominus \left(\underset{m}{\hat{\alpha}} \ominus x\right)$$

Then its scalar product with x is given variously as

$$x \ominus x^{\#} == \left(x^{\#} + x^{\perp}\right) \ominus x^{\#} == x^{\#} \ominus x^{\#}$$

$$== \left|x^{\#}\right|^2 == \left|\underset{m}{\hat{\alpha}} \ominus x\right|^2 == |x| \, \left|x^{\#}\right| \, \text{Cos}[\theta]$$

From this we see that the angle between $\underset{m}{\alpha}$ and x is given by the magnitude of the interior product of $\underset{m}{\hat{\alpha}}$ and $\hat{x}$.

$$\text{Cos}[\theta] == \frac{\left|x^{\#}\right|}{|x|} == \left|\underset{m}{\hat{\alpha}} \ominus \hat{x}\right| \qquad 6.241$$

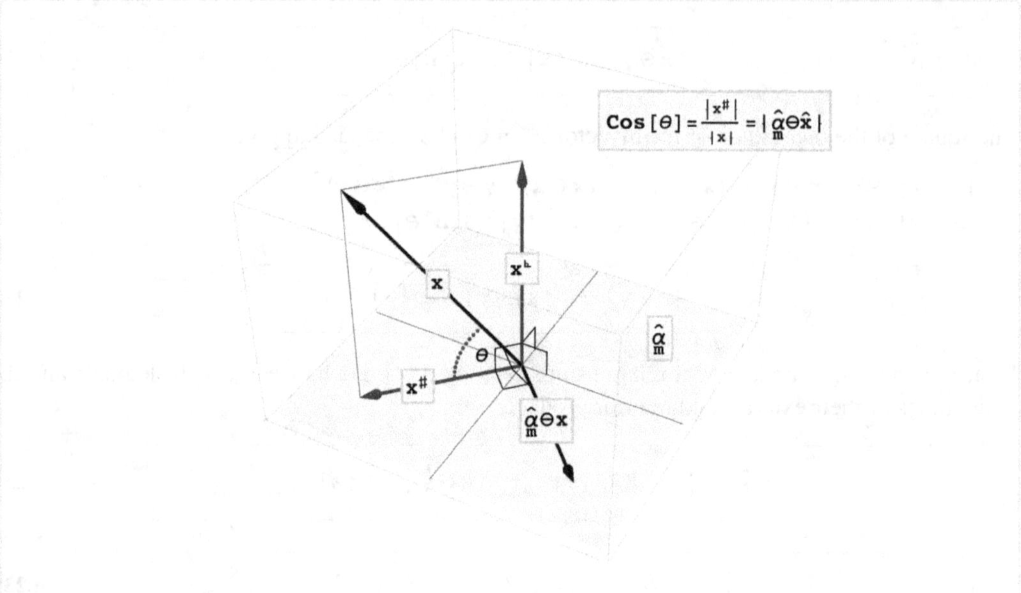

The angle between a vector and a simple unit m–vector

Furthermore, by extracting out the magnitudes of $\underset{m}{\hat{\alpha}}$ and $\hat{x}$ from this formula, we can write the interior product of an m-element with a 1-element in the enticingly familiar form

$$\left|\underset{m}{\alpha} \ominus x\right| == \left|\underset{m}{\alpha}\right| |x| \operatorname{Cos}[\theta] \qquad 6.242$$

And since in formula [6.200] we have already shown that

$$\left|\underset{m}{\hat{\alpha}} \wedge \hat{x}\right|^2 + \left|\underset{m}{\hat{\alpha}} \ominus \hat{x}\right|^2 == 1$$

we can now summarize the results as

$$\left|\underset{m}{\alpha} \ominus x\right| == \left|\underset{m}{\alpha}\right| |x| \operatorname{Cos}[\theta] \qquad \left|\underset{m}{\alpha} \wedge x\right| == \left|\underset{m}{\alpha}\right| |x| \operatorname{Sin}[\theta] \qquad 6.243$$

$$\left|\underset{m}{\hat{\alpha}} \ominus \hat{x}\right| == \operatorname{Cos}[\theta] \qquad \left|\underset{m}{\hat{\alpha}} \wedge \hat{x}\right| == \operatorname{Sin}[\theta] \qquad 6.244$$

Computing angles

To compute the angle between a vector and a simple m-vector, we revert to the interior product expression for $\operatorname{Cos}[\theta]^2$.

$$\text{Cos}[\theta]^2 == \left|\hat{\underset{m}{\alpha}} \ominus \hat{x}\right|^2 == \frac{\left|\underset{m}{\alpha} \ominus x\right|^2}{\left|\underset{m}{\alpha}\right|^2 |x|^2} == \frac{\left(\underset{m}{\alpha} \ominus x\right) \ominus \left(\underset{m}{\alpha} \ominus x\right)}{\left(\underset{m}{\alpha} \ominus \underset{m}{\alpha}\right) (x \ominus x)} \qquad 6.245$$

The interior products in this expression can be converted into scalar products, and each scalar product in turn converted to the form given by [6.236]. But since it will be convenient to name the angles between vectors from the vectors themselves, we denote the vector x and the 1-element vector factors of $\underset{m}{\alpha}$ by subscripted symbols - each different vector with a different subscript. The angle between a given pair of vectors can then be denoted by their pair of subscripts. For example we would write

$$x_1 \ominus x_2 == |x_1| \, |x_2| \, \text{Cos}[\theta_{1,2}]$$

To make this conversion of a scalar product of vectors *distinguished only by their subscripts*, you can use the *GrassmannAlgebra* function `ToAngleForm`. For example

```
★★V[x_]; ToAngleForm[x₁ ⊖ x₂, θ]
```

$|x_1| \, |x_2| \, \text{Cos}[\theta_{1,2}]$

Here, we have first declared all subscripted `x` symbols as vector symbols, and also given `ToAngleForm` the symbol we want it to use to denote the angles with.

`ToAngleForm` will automatically convert any interior products into scalar products. For example the interior product of a vector with a bivector can be expressed in terms of the angles between the vector and the vector factors of the bivector.

```
ToAngleForm[(x₂ ∧ x₃) ⊖ x₁, θ]
```

$|x_1| \, (-|x_3| \, \text{Cos}[\theta_{1,3}] \, x_2 + |x_2| \, \text{Cos}[\theta_{1,2}] \, x_3)$

Let the cosines $\text{Cos}[\theta_{i,j}]$ between vector pairs of elements be called *component cosines*.

The angle between a vector and a bivector

As a simple example of the results of the previous section, take the case where the vector is the vector x_1 and the *m*-vector is the bivector $x_2 \wedge x_3$.

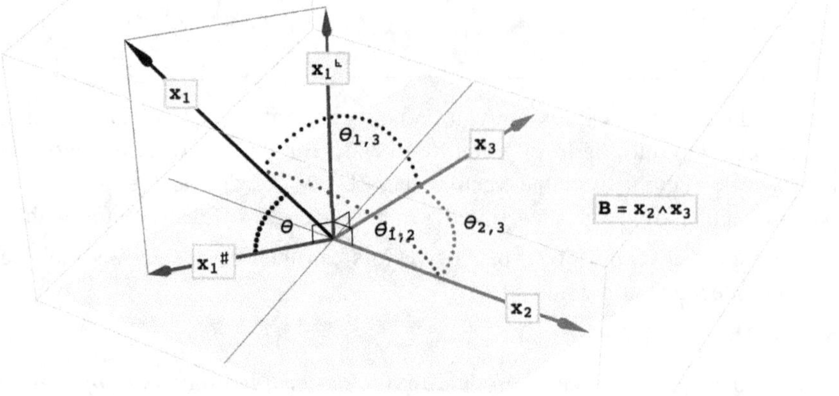

The components of the angle between a vector and a bivector

The cosine of the angle between the vector x_1 and the bivector $x_2 \wedge x_3$ may be obtained from formula [6.245].

$$\text{Cos}[\theta]^2 == \frac{((x_2 \wedge x_3) \ominus x_1) \ominus ((x_2 \wedge x_3) \ominus x_1)}{((x_2 \wedge x_3) \ominus (x_2 \wedge x_3))\ (x_1 \ominus x_1)}$$

We can use `ToAngleForm` to compute $\text{Cos}[\theta]^2$.

$$\star\star\mathbf{V}[x\_];\ \mathbf{ToAngleForm}\left[\frac{((x_2 \wedge x_3) \ominus x_1) \ominus ((x_2 \wedge x_3) \ominus x_1)}{((x_2 \wedge x_3) \ominus (x_2 \wedge x_3))\ (x_1 \ominus x_1)}, \theta\right]$$

$$\left(\text{Cos}[\theta_{1,2}]^2 + \text{Cos}[\theta_{1,3}]^2 - 2\,\text{Cos}[\theta_{1,2}]\ \text{Cos}[\theta_{1,3}]\ \text{Cos}[\theta_{2,3}]\right)\ \text{Csc}[\theta_{2,3}]^2$$

Then it can be seen that the angle between a vector and a bivector can expressed in terms of component cosines according to

$$\text{Cos}[\theta]^2 == \left(\text{Cos}[\theta_{1,2}]^2 + \text{Cos}[\theta_{1,3}]^2 - 2\,\text{Cos}[\theta_{1,2}]\ \text{Cos}[\theta_{1,3}]\ \text{Cos}[\theta_{2,3}]\right) / \left(1 - \text{Cos}[\theta_{2,3}]^2\right) \qquad 6.246$$

This result may be verified by elementary geometry.

Notice that the normalization of the vector and bivector effected by the denominator cancels out the magnitudes of the vectors

The angle between bivectors in a 3-space

Consider two non-parallel bivectors $x_1 \wedge x_2$ and $x_3 \wedge x_4$ in a 3-space.

In a 3-space we can find the vector complements of the bivectors, and then find the angle between these vectors. This is equivalent to the cross product formulation. Let $\ominus$ be the cosine of the angle between the vector complements, then:

$$\text{Cos}[\theta] == \frac{(\overline{x_1 \wedge x_2}) \ominus (\overline{x_3 \wedge x_4})}{|\overline{x_1 \wedge x_2}|\ |\overline{x_3 \wedge x_4}|} == \frac{(x_1 \times x_2) \ominus (x_3 \times x_4)}{|x_1 \times x_2|\ |x_3 \times x_4|}$$

But from formula [6.64] we can remove the complement operations to get the simpler expression:

$$\text{Cos}[\theta] == \frac{(x_1 \wedge x_2) \ominus (x_3 \wedge x_4)}{|x_1 \wedge x_2|\ |x_3 \wedge x_4|} \qquad 6.247$$

Note that, in contradistinction to the 3-space formulation using cross products, this formulation is valid in a space of any number of dimensions.

An actual calculation is most readably expressed by expanding each of the terms separately. We can either represent the x_i in terms of basis elements or deal with them directly. A direct expansion, valid for any metric is:

$$\textbf{ToScalarProducts}[(x_1 \wedge x_2) \ominus (x_3 \wedge x_4)]$$

$$-(x_1 \ominus x_4)\ (x_2 \ominus x_3) + (x_1 \ominus x_3)\ (x_2 \ominus x_4)$$

$$\textbf{Measure}[x_1 \wedge x_2]\ \textbf{Measure}[x_3 \wedge x_4]$$

$$\sqrt{\left(-(x_1 \ominus x_2)^2 + (x_1 \ominus x_1)\ (x_2 \ominus x_2)\right)}\ \sqrt{\left(-(x_3 \ominus x_4)^2 + (x_3 \ominus x_3)\ (x_4 \ominus x_4)\right)}$$

Alternatively, using ToAngleForm we can obtain

$$A = \frac{(x_1 \wedge x_2) \ominus (x_3 \wedge x_4)}{|x_1 \wedge x_2|\ |x_3 \wedge x_4|};\ \text{Cos}[\theta] == \textbf{ToAngleForm}[A, \theta]$$

$$\text{Cos}[\theta] == (|x_1|\ |x_2|\ |x_3|\ |x_4|\ (-\text{Cos}[\theta_{1,4}]\ \text{Cos}[\theta_{2,3}] + \text{Cos}[\theta_{1,3}]\ \text{Cos}[\theta_{2,4}]))\ /\ (|x_1 \wedge x_2|\ |x_3 \wedge x_4|)$$

To eliminate reference to vector and bivector magnitudes in the output, we need to eliminate them in the input.

$$A = \frac{((x_1 \wedge x_2) \ominus (x_3 \wedge x_4))\ ((x_1 \wedge x_2) \ominus (x_3 \wedge x_4))}{((x_1 \wedge x_2) \ominus (x_1 \wedge x_2))\ ((x_3 \wedge x_4) \ominus (x_3 \wedge x_4))};$$

$$\text{Cos}[\theta]^2 == \textbf{ToAngleForm}[A, \theta]$$

$$\text{Cos}[\theta]^2 == (\text{Cos}[\theta_{1,4}]\ \text{Cos}[\theta_{2,3}] - \text{Cos}[\theta_{1,3}]\ \text{Cos}[\theta_{2,4}])^2\ \text{Csc}[\theta_{1,2}]^2\ \text{Csc}[\theta_{3,4}]^2$$

$$\text{Cos}[\theta] == \frac{(x_1 \wedge x_2) \ominus (x_3 \wedge x_4)}{|x_1 \wedge x_2|\ |x_3 \wedge x_4|} == \frac{\text{Cos}[\theta_{1,4}]\ \text{Cos}[\theta_{2,3}] - \text{Cos}[\theta_{1,3}]\ \text{Cos}[\theta_{2,4}]}{\text{Sin}[\theta_{1,2}]\ \text{Sin}[\theta_{3,4}]} \qquad 6.248$$

In the case the two bivectors are each expressed as products of orthogonal vectors, that is $\theta_{1,2}$ and $\theta_{3,4}$ are $\frac{\pi}{2}$, we can write more simply

$$\text{Cos}[\theta] == \text{Cos}[\theta_{1,4}]\ \text{Cos}[\theta_{2,3}] - \text{Cos}[\theta_{1,3}]\ \text{Cos}[\theta_{2,4}] \qquad \theta_{1,2} == \theta_{3,4} == \frac{\pi}{2} \qquad 6.249$$

◆ Example 1

To take the simplest example, suppose take two *unit* bivectors B_1 and B_2 in a 3-space expressed in terms of a set of orthonormal basis vectors with their common vector in the e_2 direction. We have expressed the bivectors in such a way as to expect the angle between them to be ψ.

$$B_1 == e_1 \wedge e_2$$
$$B_2 == (\mathrm{Cos}[\psi]\, e_1 + \mathrm{Sin}[\psi]\, e_3) \wedge e_2$$

$$\mathrm{Cos}[\theta] == B_1 \ominus B_2$$
$$== (e_1 \wedge e_2) \ominus ((\mathrm{Cos}[\psi]\, e_1 + \mathrm{Sin}[\psi]\, e_3) \wedge e_2)$$
$$== (e_1 \ominus (\mathrm{Cos}[\psi]\, e_1 + \mathrm{Sin}[\psi]\, e_3))\, (e_2 \ominus e_2) - (e_1 \ominus e_2)\, (e_2 \ominus (\mathrm{Cos}[\psi]\, e_1 + \mathrm{Sin}[\psi]\, e_3)) == \mathrm{Cos}[\psi]$$

The volume of a parallelepiped

We can calculate the volume of a parallelepiped as the measure of the trivector whose vectors make up the sides of the parallelepiped. *GrassmannAlgebra* provides the function `Measure` for expressing the volume in terms of scalar products:

```
V = Measure[x1 ∧ x2 ∧ x3]
```

$$\sqrt{\left(-(x_1 \ominus x_3)^2\, (x_2 \ominus x_2) + 2\, (x_1 \ominus x_2)\, (x_1 \ominus x_3)\, (x_2 \ominus x_3) - (x_1 \ominus x_1)\, (x_2 \ominus x_3)^2 - (x_1 \ominus x_2)^2\, (x_3 \ominus x_3) + (x_1 \ominus x_1)\, (x_2 \ominus x_2)\, (x_3 \ominus x_3)\right)}$$

Note that this has been simplified somewhat as permitted by the symmetry of the scalar product.

By putting this in its angle form we get the usual expression for the volume of a parallelepiped:

```
ToAngleForm[V, θ]
```

$$\sqrt{\left(-|x_1|^2\, |x_2|^2\, |x_3|^2\, \left(-1 + \mathrm{Cos}[\theta_{1,2}]^2 + \mathrm{Cos}[\theta_{1,3}]^2 - 2\,\mathrm{Cos}[\theta_{1,2}]\,\mathrm{Cos}[\theta_{1,3}]\,\mathrm{Cos}[\theta_{2,3}] + \mathrm{Cos}[\theta_{2,3}]^2\right)\right)}$$

A slight rearrangement gives the volume of the parallelepiped as:

$$V == |x_1|\, |x_2|\, |x_3| \sqrt{\left(1 + 2\,\mathrm{Cos}[\theta_{1,2}]\,\mathrm{Cos}[\theta_{1,3}]\,\mathrm{Cos}[\theta_{2,3}] - \mathrm{Cos}[\theta_{1,2}]^2 - \mathrm{Cos}[\theta_{1,3}]^2 - \mathrm{Cos}[\theta_{2,3}]^2\right)} \qquad 6.250$$

◆ The volume of a 4-dimensional parallelepiped

We can of course use the same approach in any number of dimensions. For example, the 'volume' of a 4-dimensional parallelepiped in terms of the lengths of its sides and the angles between them is:

```
★𝓑4; ToAngleForm[Measure[x1 ∧ x2 ∧ x3 ∧ x4], θ]
```

$$\sqrt{\left(|x_1|^2 |x_2|^2 |x_3|^2 |x_4|^2 \left(1 - \text{Cos}[\theta_{2,3}]^2 - \text{Cos}[\theta_{2,4}]^2 + 2\,\text{Cos}[\theta_{1,3}]\,\text{Cos}[\theta_{1,4}]\,\text{Cos}[\theta_{3,4}] - \text{Cos}[\theta_{3,4}]^2 + 2\,\text{Cos}[\theta_{2,3}]\,\text{Cos}[\theta_{2,4}]\,(-\text{Cos}[\theta_{1,3}]\,\text{Cos}[\theta_{1,4}] + \text{Cos}[\theta_{3,4}]) + 2\,\text{Cos}[\theta_{1,2}]\,(\text{Cos}[\theta_{1,4}]\,(\text{Cos}[\theta_{2,4}] - \text{Cos}[\theta_{2,3}]\,\text{Cos}[\theta_{3,4}]) + \text{Cos}[\theta_{1,3}]\,(\text{Cos}[\theta_{2,3}] - \text{Cos}[\theta_{2,4}]\,\text{Cos}[\theta_{3,4}])) - \text{Cos}[\theta_{1,4}]^2\,\text{Sin}[\theta_{2,3}]^2 - \text{Cos}[\theta_{1,3}]^2\,\text{Sin}[\theta_{2,4}]^2 - \text{Cos}[\theta_{1,2}]^2\,\text{Sin}[\theta_{3,4}]^2\right)\right)}$$

◆ The area of a parallelogram

As a trivial check on the independence of dimension, we can use the same formulation in two dimensions to determine the area of a parallelogram.

```
★𝓑2; ToAngleForm[Measure[x1 ∧ x2], θ]
```

$$\sqrt{|x_1|^2 |x_2|^2 \text{Sin}[\theta_{1,2}]^2}$$

The angle between a point and a simple m-vector

Formula [6.216] shows that the component $P^{\perp}$ of a point P orthogonal to a simple m-vector is the *point* whose position vector $p^{\perp}$ is orthogonal to the m-vector; and formula [6.217] shows that the component $P^{\#}$ of the point P parallel to the m-vector is the *vector* component $p^{\#}$ of the position vector of P parallel to the m-vector.

$$P^{\perp} == \star O + p^{\perp} == \star O + \left(\hat{\underset{m}{\alpha}} \wedge p\right) \ominus \hat{\underset{m}{\alpha}} \qquad P == \star O + p$$

$$P^{\#} == p^{\#} == (-1)^{m-1}\, \hat{\underset{m}{\alpha}} \ominus \left(\hat{\underset{m}{\alpha}} \ominus p\right)$$

The angle between a point P and a simple m-vector is defined according to the definition [6.241] for the angle between any 1-element and an m-vector: as the angle between P and the *component* $P^{\#}$ of P *parallel* to m-vector in the triangle relationship.

$$P \ominus P^{\#} == \left(\star O + p^{\perp} + p^{\#}\right) \ominus p^{\#} == p^{\#} \ominus p^{\#} == \left|p^{\#}\right|^2 == \left|\hat{\underset{m}{\alpha}} \ominus p\right|^2 == |p|\,\left|p^{\#}\right|\,\text{Cos}[\theta]$$

Hence the angle between a point and an m-vector is simply the angle between the position vector of the point and the m-vector.

$$P == \star O + p \qquad P \ominus P^{\#} == \left|\hat{\underset{m}{\alpha}} \ominus p\right|^2 \qquad \text{Cos}[\theta] == \left|\hat{\underset{m}{\alpha}} \ominus \hat{p}\right| \qquad 6.251$$

The angle between simple elements

In section 6.11 we discussed how a vector can be decomposed into two components with respect to a simple unit m-element: one component lying in the m-element, and one component orthogonal to it. The original vector and its two components thus form a right angled triangle, and the ensuing collection of formulae we developed we called the Triangle Formulae.

This right angled triangle was used to define the notion of the angle between the vector and the m-element in the usual trigonometric manner.

In the sections to follow we explore how we might extend this notion to that of the angle between a simple k-vector and a simple m-vector. To fix ideas we first revisit the simplest non-trivial case: the angle between a bivector and a vector.

◆ The angle between a bivector and a vector

This case reduces to the situation we have already explored in the Triangle Formulae in section 6.11. However, let us now adopt the point of view where the vector is the m-vector, and the bivector B makes an angle to it.

Without loss of generality to the notion of angle we can assume the bivector may be expressed as the exterior product of two *orthonormal* vectors x and y. This will make the ensuing results easier to interpret.

To fix ideas, suppose x is e_1, y is e_2 and the m-vector is a general *unit* vector α in a 4-space.

$$\alpha == a\,e_1 + b\,e_2 + c\,e_3 + d\,e_4 \qquad a^2 + b^2 + c^2 + d^2 == 1$$

$$B == x \wedge y == e_1 \wedge e_2$$

Now consider a general unit vector u in the bivector written as a linear combination of x and y.

$$u == \text{Cos}[\psi]\,x + \text{Sin}[\psi]\,y == \text{Cos}[\psi]\,e_1 + \text{Sin}[\psi]\,e_2$$

From the triangle formula [6.200] we *define* the angle between u and α as the angle between u *and its component* $u^{\#}$ parallel to α. The scalar product $u \ominus u^{\#}$ can be written in terms of $\text{Cos}[\theta]$ as

$$u \ominus u^{\#} == |u|\,\left|u^{\#}\right|\,\text{Cos}[\theta] == \left|u^{\#}\right|\,\text{Cos}[\theta]$$

But it can also be written as the square of the magnitude of the parallel component.

$$u \ominus u^{\#} == \left(u^{\llcorner} + u^{\#}\right) \ominus u^{\#} == u^{\#} \ominus u^{\#} == \left|u^{\#}\right|^2$$

Hence $\text{Cos}[\theta]$ becomes simply the magnitude of the parallel component.

$$\text{Cos}[\theta] == \left|u^{\#}\right|$$

Note carefully that this definition of angle requires $\text{Cos}[\theta]$ to be non-negative. Additionally we will also define the angle θ to be positive, restricting it to lie in the range 0 to $\frac{1}{2}\pi$. (This is equivalent to saying the angle between the vector and the bivector is the same as the angle between the bivector and the vector.)

The parallel component of u to α is defined by [6.195] as

$$u^{\#} == \alpha \ominus (\alpha \ominus u) == (\alpha \ominus u)\,\alpha$$

$$\text{Cos}[\theta] == \left|u^{\#}\right| == |\alpha \ominus u|$$

Hence $\text{Cos}[\theta]$ is the positive number $|\alpha \ominus u|$, where θ lies between 0 and $\frac{1}{2}\pi$.

As the angle ψ varies from 0 to 2π, u rotates in the bivector and $\alpha \ominus u$ can be written

$$\alpha \ominus u == (a\,e_1 + b\,e_2 + c\,e_3 + d\,e_4) \ominus (\text{Cos}[\psi]\,e_1 + \text{Sin}[\psi]\,e_2)$$
$$== a\,\text{Cos}[\psi] + b\,\text{Sin}[\psi]$$

Hence the inclination angle θ can be written as a function of the rotation angle ψ as

$$\theta == \text{ArcCos}[\,|\,a\,\text{Cos}[\psi] + b\,\text{Sin}[\psi]\,|\,] \qquad 0 \le \theta \le \frac{1}{2}\pi$$

To visualize what is happening, suppose a is 0.4 and b is 0.5 (remember that the sum of their squares must be less than or equal to 1). We can now plot the inclination angle θ as a function of the rotation parameter ψ.

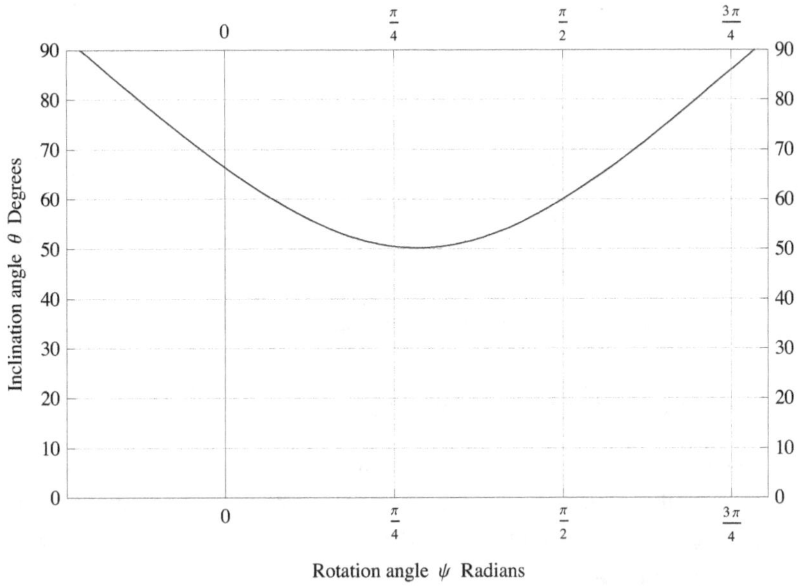

This is as expected. As the vector u rotates in the bivector, it makes a minimum angle θ_m with α (from the graph, 50°) at a rotation angle of ψ_m (from the graph, slightly greater than $\frac{\pi}{4}$). The maximum angle of 90° occurs when u is orthogonal to α at $\psi_m \pm \frac{\pi}{2}$.

Thus there is an infinity of angles between the vector α and the bivector - depending on which vector in the bivector is chosen to form the angle. We may however *define* the angle between the vector α and the bivector B as the *minimum angle* θ_m generated between α and a vector in B. In this example then, we may say that the angle between α and B is 50°.

Remark that although α is required to be a unit vector, it may not be readily apparent in this case that its components other than those in e_1 and e_2 are important. This requirement is important however, since if α *only* had components in e_1 and e_2 it would be a unit vector *lying in the bivector*, and hence the minimum angle would be zero when the rotating vector coincided with it. If α has components other than those in e_1 and e_2 the sum of squares of the coefficients of e_1 and e_2 will not sum to unity, resulting in the angle being greater than zero.

For example suppose now that α is a unit vector lying in the bivector

$$\alpha == 0.8\,e_1 + 0.6\,e_2$$

The angle θ now becomes equal to the angle ψ, and the minimum value is 0 as u coincides with α.

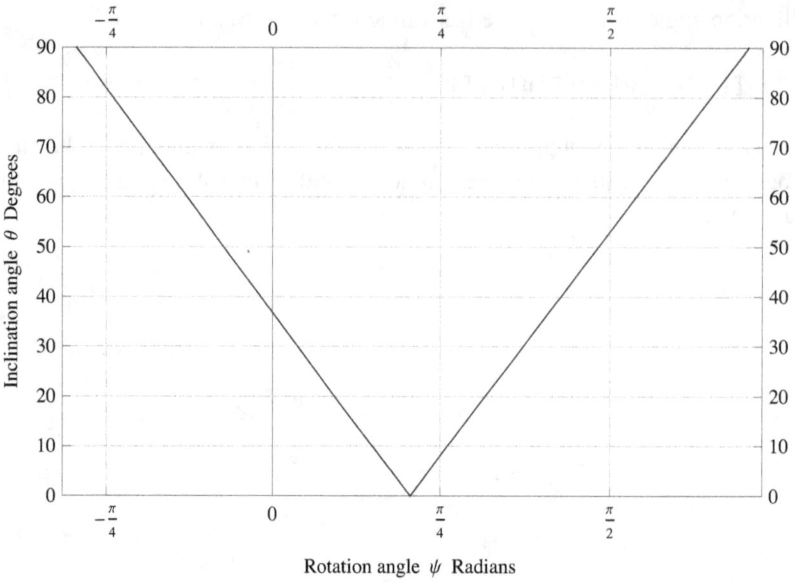

The angle between a bivector and an *m*-vector

Let us now consider the more general case of the angle between a bivector and a *unit* m-vector A. As before, let us express the bivector as the exterior product of two orthonormal vectors x and y, and construct a 'rotating' unit vector u in the bivector as a linear combination of x and y.

$$u == \text{Cos}[\psi]\, x + \text{Sin}[\psi]\, y$$

From triangle formula [6.195] we can write the component of u parallel to (in) A as a linear combination of the parallel components of x and y.

$$u^{\#} == (-1)^{m-1} A \ominus (A \ominus u)$$
$$== \text{Cos}[\psi]\, (-1)^{m-1} A \ominus (A \ominus x) + \text{Sin}[\psi]\, (-1)^{m-1} A \ominus (A \ominus y)$$
$$== \text{Cos}[\psi]\, x^{\#} + \text{Sin}[\psi]\, y^{\#}$$

We want to find the angle θ between u and $u^{\#}$ as a function of ψ. Since u is a unit vector, $\text{Cos}[\theta]$ is simply given by $|u^{\#}|$ as derived in the previous section, or from [6.199] as $|A \ominus u|$.

$$\text{Cos}[\theta] == |u^{\#}| == |A \ominus (A \ominus u)| == |A \ominus u| \qquad 6.252$$

Since $\text{Cos}[\theta]$ is the magnitude of a vector, it is positive. We constrain θ to be positive also.

Clearly if u is orthogonal to A at any point in its rotation, $|u^{\#}|$ will be zero at that point making θ equal to $\frac{1}{2}\pi$. And if u intersects with A at any point in its rotation, $u^{\#}$ will be equal to u, and $|u^{\#}|$ will be unity making θ equal to zero. Thus, as u rotates, we expect θ not to lie outside the bounds of 0 and $\frac{1}{2}\pi$. But we also expect it to vary, perhaps achieving a maximum or a minimum as it does. To look for stationary values, first square the equation for $\text{Cos}[\theta]$, then differentiate with respect to ψ.

$$\text{Cos}[\theta]^2 == |u^{\#}|^2 == u^{\#} \ominus u^{\#}$$

$$2\,\mathrm{Cos}[\theta]\,(-\mathrm{Sin}[\theta])\,\frac{\partial\theta}{\partial\psi} == 2\,u^{\#}\ominus\frac{\partial u^{\#}}{\partial\psi}$$

$$== 2\left(\mathrm{Cos}[\psi]\,x^{\#}+\mathrm{Sin}[\psi]\,y^{\#}\right)\ominus\left(-\,\mathrm{Sin}[\psi]\,x^{\#}+\mathrm{Cos}[\psi]\,y^{\#}\right)$$

$$== 2\left(x^{\#}\ominus y^{\#}\right)\mathrm{Cos}[2\,\psi]+\left(y^{\#}\ominus y^{\#}-x^{\#}\ominus x^{\#}\right)\mathrm{Sin}[2\,\psi]$$

The left hand side is zero (and hence the right hand side is zero) whenever θ is zero (the bivector and the *m*-vector intersect); whenever θ is $\frac{1}{2}\pi$ (the bivector is orthogonal to the *m*-vector); or when $\frac{\partial\theta}{\partial\psi}$ is zero (the angle θ has a stationary value as the angle ψ varies). We can thus determine the value of ψ for these cases from

$$\mathrm{Tan}[2\,\psi] == \frac{2\,(x^{\#}\ominus y^{\#})}{x^{\#}\ominus x^{\#}-y^{\#}\ominus y^{\#}} \qquad 6.253$$

Defining

$$a == 2\left(x^{\#}\ominus y^{\#}\right) \qquad b == x^{\#}\ominus x^{\#}-y^{\#}\ominus y^{\#}$$

then $\mathrm{Sin}[2\psi]$ and $\mathrm{Cos}[2\psi]$ can be written

$$\mathrm{Sin}[2\,\psi] == \frac{a}{\sqrt{a^2+b^2}} \qquad \mathrm{Cos}[2\,\psi] == \frac{b}{\sqrt{a^2+b^2}}$$

Returning to the expression for $\mathrm{Cos}[\theta]^2$ we can substitute these expressions to give a result involving only the scalar products of $x^{\#}$ and $y^{\#}$.

$$\mathrm{Cos}[\theta]^2 == u^{\#}\ominus u^{\#} == \left(\mathrm{Cos}[\psi]\,x^{\#}+\mathrm{Sin}[\psi]\,y^{\#}\right)\ominus\left(\mathrm{Cos}[\psi]\,x^{\#}+\mathrm{Sin}[\psi]\,y^{\#}\right)$$

$$== \frac{1}{2}\left(x^{\#}\ominus x^{\#}+y^{\#}\ominus y^{\#}+\left(x^{\#}\ominus x^{\#}\right)\mathrm{Cos}[2\,\psi]\right.$$
$$\left.-\left(y^{\#}\ominus y^{\#}\right)\mathrm{Cos}[2\,\psi]+2\left(x^{\#}\ominus y^{\#}\right)\mathrm{Sin}[2\,\psi]\right)$$

$$== \frac{1}{2}\left(x^{\#}\ominus x^{\#}+y^{\#}\ominus y^{\#}\right)+\frac{1}{2}\left(\frac{a^2}{\sqrt{a^2+b^2}}+\frac{b^2}{\sqrt{a^2+b^2}}\right)$$

$$== \frac{1}{2}\left(x^{\#}\ominus x^{\#}+y^{\#}\ominus y^{\#}+\sqrt{a^2+b^2}\right)$$

$$\mathrm{Cos}[\theta]^2 == \frac{1}{2}\left(x^{\#}\ominus x^{\#}+y^{\#}\ominus y^{\#}+\sqrt{\left(4\left(x^{\#}\ominus y^{\#}\right)^2+\left(x^{\#}\ominus x^{\#}-y^{\#}\ominus y^{\#}\right)^2\right)}\right) \qquad 6.254$$

$$x^{\#} == (-1)^{m-1}\,A\ominus(A\ominus x) \qquad y^{\#} == (-1)^{m-1}\,A\ominus(A\ominus y) \qquad 6.255$$

$$|A| == |x| == |y| == 1 \qquad x\ominus y == 0 \qquad 6.256$$

◆ An alternative formulation

Formula [6.254] has the hallmarks of being the solution to a quadratic equation in $\mathrm{Cos}[\theta]^2$. Let us retrieve the quadratic.

$$\text{Simplify}\left[\text{Expand}\left[\left(2\,\text{Cos}[\theta]^2 - (x^{\#}\ominus x^{\#} + y^{\#}\ominus y^{\#})\right)^2 - \left(4\,(x^{\#}\ominus y^{\#})^2 + (x^{\#}\ominus x^{\#} - y^{\#}\ominus y^{\#})^2\right) == 0\right]\right]$$

$$(x^{\#}\ominus y^{\#})^2 + (x^{\#}\ominus x^{\#} - \text{Cos}[\theta]^2)\,(-(y^{\#}\ominus y^{\#}) + \text{Cos}[\theta]^2) == 0$$

$$(\text{Cos}[\theta]^2 - x^{\#}\ominus x^{\#})\,(\text{Cos}[\theta]^2 - y^{\#}\ominus y^{\#}) - (x^{\#}\ominus y^{\#})^2 == 0 \qquad 6.257$$

From formulae [6.198] and [6.201] we can write the quadratic for $\text{Cos}[\theta]^2$ in terms of the original *m*-element A and any two orthonormal vectors x and y defining the bivector.

$$\left(\text{Cos}[\theta]^2 - (A\ominus x)\ominus(A\ominus x)\right)\left(\text{Cos}[\theta]^2 - (A\ominus y)\ominus(A\ominus y)\right) - ((A\ominus x)\ominus(A\ominus y))^2 == 0 \qquad 6.258$$

◆ The angle between a bivector and a vector

Let us now apply formula [6.258] to the example of the previous section where A is a *unit* vector α. Direct substitution gives

$$\left(\text{Cos}[\theta]^2 - (\alpha\ominus x)\,(\alpha\ominus x)\right)\left(\text{Cos}[\theta]^2 - (\alpha\ominus y)\,(\alpha\ominus y)\right) - ((\alpha\ominus x)\,(\alpha\ominus y))^2 == 0$$

which simplifies to

$$\text{Cos}[\theta]^2 == (\alpha\ominus x)^2 + (\alpha\ominus y)^2 \qquad 6.259$$

$$|\alpha| == |x| == |y| == 1 \qquad x\ominus y == 0 \qquad 6.260$$

On the other hand, the angle between the vector and the bivector given by triangle formula [6.210] also reduces to the same result.

$$\begin{aligned}\text{Cos}[\theta]^2 &== |-B\ominus(B\ominus\alpha)|^2 == ((x\wedge y)\ominus\alpha)\ominus((x\wedge y)\ominus\alpha)\\ &== ((\alpha\ominus x)\,y - (\alpha\ominus y)\,x)\ominus((\alpha\ominus x)\,y - (\alpha\ominus y)\,x)\\ &== (\alpha\ominus x)^2 + (\alpha\ominus y)^2\end{aligned}$$

■ For the numerical examples of the previous section

$$\text{Cos}[\theta]^2 == (\alpha\ominus x)^2 + (\alpha\ominus y)^2 == (\alpha\ominus e_1)^2 + (\alpha\ominus e_2)^2 == a^2 + b^2$$

When the vector α has components both orthogonal to and parallel to the bivector:

$$\theta == \text{ArcCos}\left[\sqrt{0.4^2 + 0.5^2}\right]\,180/\pi$$

$$\theta == 50.1849$$

When the vector α belongs to the bivector:

```
θ == ArcCos[√(0.8² + 0.6²)] 180 / π
```

```
θ == 0.
```

When the vector α is orthogonal to the bivector:

```
θ == ArcCos[√(0.0² + 0.0²)] 180 / π
```

```
θ == 90.
```

Example: The angle between a bivector and a trivector

Consider a default basis for a 6-space in which the e_i are orthonormal, and construct a unit bivector B as

```
B == x ∧ y == e1 ∧ e2
```

We will construct a random unit simple trivector in the 6-space by first constructing a random simple trivector, simplifying it, and then normalizing it.

```
★ℬ6;
A3 =
  ★𝒢[★Va ∧ ★Vb ∧ ★Vc /. {a_ :→ RandomReal[], b_ :→ RandomReal[],
     c_ :→ RandomReal[]}]
```

```
-0.147982 e1 ∧ e2 ∧ e3 - 0.00742747 e1 ∧ e2 ∧ e4 +
 0.0915464 e1 ∧ e2 ∧ e5 - 0.0686148 e1 ∧ e2 ∧ e6 + 0.0611959 e1 ∧ e3 ∧ e4 +
 0.0564279 e1 ∧ e3 ∧ e5 + 0.145428 e1 ∧ e3 ∧ e6 + 0.0406899 e1 ∧ e4 ∧ e5 -
 0.0210753 e1 ∧ e4 ∧ e6 - 0.11613 e1 ∧ e5 ∧ e6 - 0.169082 e2 ∧ e3 ∧ e4 -
 0.166607 e2 ∧ e3 ∧ e5 + 0.0487217 e2 ∧ e3 ∧ e6 - 0.112962 e2 ∧ e4 ∧ e5 +
 0.0808437 e2 ∧ e4 ∧ e6 + 0.0471098 e2 ∧ e5 ∧ e6 + 0.00442422 e3 ∧ e4 ∧ e5 -
 0.186312 e3 ∧ e4 ∧ e6 - 0.18231 e3 ∧ e5 ∧ e6 - 0.122358 e4 ∧ e5 ∧ e6
```

To speed up the computations, we will find the following combined simplification function useful.

```
F[X_] := Plus @@ ToMetricElements[ToScalarProducts[List @@ ★𝒢[X]]]
```

The unit trivector $\hat{A}$ and components $x^{\#}$ and $y^{\#}$ parallel to it can then be computed as

```
Â = Expand[A3 / √F[A3 ⊖ A3]]
```

```
-0.300889 e1 ∧ e2 ∧ e3 - 0.0151021 e1 ∧ e2 ∧ e4 +
 0.186139 e1 ∧ e2 ∧ e5 - 0.139513 e1 ∧ e2 ∧ e6 + 0.124428 e1 ∧ e3 ∧ e4 +
 0.114734 e1 ∧ e3 ∧ e5 + 0.295696 e1 ∧ e3 ∧ e6 + 0.0827339 e1 ∧ e4 ∧ e5 -
 0.0428521 e1 ∧ e4 ∧ e6 - 0.236125 e1 ∧ e5 ∧ e6 - 0.343792 e2 ∧ e3 ∧ e4 -
 0.338759 e2 ∧ e3 ∧ e5 + 0.0990648 e2 ∧ e3 ∧ e6 - 0.229683 e2 ∧ e4 ∧ e5 +
 0.164378 e2 ∧ e4 ∧ e6 + 0.0957874 e2 ∧ e5 ∧ e6 + 0.00899567 e3 ∧ e4 ∧ e5 -
 0.378825 e3 ∧ e4 ∧ e6 - 0.370687 e3 ∧ e5 ∧ e6 - 0.248787 e4 ∧ e5 ∧ e6
```

```
x# = F[Â ⊖ F[Â ⊖ e1]]
```

```
0.325393 e1 - 0.101016 e2 + 0.176191 e3 + 0.338859 e4 + 0.21883 e5 - 0.12471 e6
```

$$y^{\#} = F[\hat{A} \ominus F[\hat{A} \ominus e_2]]$$

$$-0.101016\, e_1 + 0.476588\, e_2 - 0.12162\, e_3 + 0.0755625\, e_4 + 0.143239\, e_5 + 0.44523\, e_6$$

Using formula [6.254] we compute θ by evaluating

$$\Xi = F[x^{\#} \ominus x^{\#}] + F[y^{\#} \ominus y^{\#}] + \sqrt{\left(4\, F[x^{\#} \ominus y^{\#}]^2 + \left(F[x^{\#} \ominus x^{\#}] - F[y^{\#} \ominus y^{\#}]\right)^2\right)};$$

$$\theta == \text{ArcCos}\left[\sqrt{\frac{1}{2}\,\Xi}\right]$$

$$\theta == 0.758223$$

Thus the minimum angle between any two vectors, one in the bivector, and one in the trivector is about 0.758 radians, or about 43.4 degrees. This then we take (by definition) to be *the* angle between the bivector and the trivector.

6.13 Orthogonal Decomposition

Introduction

In section 3.10 we introduced The Decomposition Formula [3.76] where we showed that it was a special case of the The General Product Formula [3.72] . In Chapter 4 we applied this formula to decompose points, lines and planes with reference to two non-intersecting elements. At this stage we had not yet introduced a metric, and so the decompositions did not require any.

In this chapter, we are equipped with a metric and the notion of orthogonality. So now, instead of decomposing an element with reference to two non-intersecting elements, we can consider the decomposition of an element relative to a single element *and its complement*. This is called *orthogonal decomposition*.

The previous two sections on The Triangle Formulae and on Angle have already introduced the notion of orthogonal decomposition. But the discussion only involved the decomposition of a 1-element. This simple but important case was presented first so that the underlying concepts that we are about to discuss have some familiarity.

The Triangle Formula resulted in just *two* components - a component $x^{\#}$ parallel to $\underset{m}{\hat{\alpha}}$ and a component $x^{\perp}$ orthogonal to $\underset{m}{\hat{\alpha}}$. These were given by the formulae:

$$x == x^{\perp} + x^{\#} \qquad\qquad x^{\perp} \ominus x^{\#} == 0$$

$$x^{\perp} == \left(\underset{m}{\hat{\alpha}} \wedge x\right) \ominus \underset{m}{\hat{\alpha}} \qquad\qquad x^{\#} == (-1)^{m-1}\, \underset{m}{\hat{\alpha}} \ominus \left(\underset{m}{\hat{\alpha}} \ominus x\right)$$

However The Orthogonal Decomposition Formula [6.120] shows how we can decompose a simple k-element $\underset{k}{x}$ in terms of a single unit simple m-element $\underset{m}{\hat{\alpha}}$. We can write this formula either as a double-sum formula:

$$\underset{k}{x} = \sum_{r=0}^{k} (-1)^{r\,(m+1)} \sum_{i=1}^{\nu} \left(\underset{m}{\hat{\alpha}} \wedge \underset{k-r}{x_i}\right) \ominus \left(\underset{m}{\hat{\alpha}} \ominus \underset{r}{x_i}\right)$$
$$\underset{k}{x} = \underset{r}{x_1} \wedge \underset{k-r}{x_1} = \underset{r}{x_2} \wedge \underset{k-r}{x_2} = \ldots = \underset{r}{x_\nu} \wedge \underset{k-r}{x_\nu} \qquad \nu = \binom{k}{r} \tag{6.261}$$

or as a computable one directly translated from it.

$$\underset{k}{x} = \star\Sigma\left[(-1)^{\star r[k]\,(m+1)} \left(\underset{m}{\hat{\alpha}} \wedge \star\mathbb{S}^{\star}\left[\underset{k}{x}\right]\right) \ominus \left(\underset{m}{\hat{\alpha}} \ominus \star\mathbb{S}_{\star}\left[\underset{k}{x}\right]\right)\right] \tag{6.262}$$

For the case $\underset{k}{x}$ is a 1-element we retrieve the formula:

$$x = \left(\underset{m}{\hat{\alpha}} \wedge x\right) \ominus \underset{m}{\hat{\alpha}} + (-1)^{m-1} \underset{m}{\hat{\alpha}} \ominus \left(\underset{m}{\hat{\alpha}} \ominus x\right)$$

For the case $\underset{k}{x}$ is a 2-element, $x \wedge y$ say, we retrieve the formula

$$x \wedge y = \left(\underset{m}{\hat{\alpha}} \wedge x \wedge y\right) \ominus \underset{m}{\hat{\alpha}} +$$
$$(-1)^m \left(\left(\underset{m}{\hat{\alpha}} \wedge x\right) \ominus \left(\underset{m}{\hat{\alpha}} \ominus y\right) - \left(\underset{m}{\hat{\alpha}} \wedge y\right) \ominus \left(\underset{m}{\hat{\alpha}} \ominus x\right)\right) + \underset{m}{\hat{\alpha}} \ominus \left(\underset{m}{\hat{\alpha}} \ominus (x \wedge y)\right)$$

This formula has four components. The first and last appear as if they may indeed be the components of $x \wedge y$ orthogonal and parallel to $\underset{m}{\hat{\alpha}}$. The question arises: what is the geometric significance of the middle two components?

If we write x and y each as the sum of its orthogonal and parallel components we can expand the product $x \wedge y$ in terms of products of the components.

$$x \wedge y = \left(x^{\perp} + x^{\#}\right) \wedge \left(y^{\perp} + y^{\#}\right) = x^{\perp} \wedge y^{\perp} + x^{\perp} \wedge y^{\#} + x^{\#} \wedge y^{\perp} + x^{\#} \wedge y^{\#}$$

In what follows we will see that the terms of these two expressions do indeed correspond, and that this correspondence continues to hold for simple elements $\underset{k}{x}$ of higher grade leading to 2^k terms.

The decomposition formulae we are discussing in this chapter require a metric in order to define the interior product and hence the notion of orthogonality. However, it should be remarked that the basic formula we use (the Interior Product Formula) is a formula directly derived from one we first met in Chapter 3 in its regressive product formulation as the General Product Formula. In Chapter 3 we began to look at the notion of decomposition (without a metric) but left our main discussion to Chapter 4 where we could elucidate the algebra with geometrically interpreted examples.

Following this section on orthogonal decomposition we will look a little more at one of its more important facets: orthogonal projection. This too has its roots in the non-metric geometry of projective geometry. Projection without a metric is discussed in Chapter 4.

Orthogonal decomposition

Consider a simple k-element $\underset{k}{x}$, and the decomposition of each of its factors x_i into two components $x_i^{\perp}$ orthogonal to a simple unit element $\underset{m}{\hat{\alpha}}$, and $x_i^{\#}$ parallel to it. Hence we can write

$$\underset{k}{x} == x_1 \wedge x_2 \wedge \ldots \wedge x_k == \left(x_1^{\perp} + x_1^{\#}\right) \wedge \left(x_2^{\perp} + x_2^{\#}\right) \wedge \ldots \wedge \left(x_k^{\perp} + x_k^{\#}\right)$$

$$== x_1^{\perp} \wedge x_2^{\perp} \wedge \ldots \wedge x_k^{\perp} + x_1^{\perp} \wedge x_2^{\perp} \wedge \ldots \wedge x_k^{\#} +$$
$$\ldots + x_1^{\perp} \wedge x_2^{\#} \wedge \ldots \wedge x_k^{\#} + x_1^{\#} \wedge x_2^{\#} \wedge \ldots \wedge x_k^{\#}$$

The number of terms in this expansion is 2^k - the same number of components as in the span (or cospan) of $\underset{k}{x}$ and hence also the same number of terms as in the decomposition of $\underset{k}{x}$ through its decomposition formulae [6.261] or [6.262].

Let u and v be two arbitrary 1-elements and let their orthogonal and parallel components be given by

$$u^{\perp} == \left(\underset{m}{\hat{\alpha}} \wedge u\right) \ominus \underset{m}{\hat{\alpha}} \qquad v^{\#} == (-1)^{m-1}\, \underset{m}{\hat{\alpha}} \ominus \left(\underset{m}{\hat{\alpha}} \ominus v\right)$$

Then we can show straightforwardly, that $u^{\perp}$ and $v^{\#}$ are orthogonal by applying the extended interior product formula.

$$u^{\perp} \ominus v^{\#} == (-1)^{m-1} \left(\left(\underset{m}{\hat{\alpha}} \wedge u\right) \ominus \underset{m}{\hat{\alpha}}\right) \ominus \left(\underset{m}{\hat{\alpha}} \ominus \left(\underset{m}{\hat{\alpha}} \ominus v\right)\right)$$

$$== (-1)^{m-1} \left(\underset{m}{\hat{\alpha}} \wedge u\right) \ominus \left(\underset{m}{\hat{\alpha}} \wedge \left(\underset{m}{\hat{\alpha}} \ominus \left(\underset{m}{\hat{\alpha}} \ominus v\right)\right)\right) == (-1)^{m-1} \left(\underset{m}{\hat{\alpha}} \wedge u\right) \ominus 0 == 0$$

This also means that if two such 1-elements are each expressed as the sum of their orthogonal and parallel components, their scalar product simplifies to reflect this orthogonality.

$$u \ominus v == \left(u^{\perp} + u^{\#}\right) \ominus \left(v^{\perp} + v^{\#}\right)$$
$$== u^{\perp} \ominus v^{\perp} + u^{\perp} \ominus v^{\#} + u^{\#} \ominus v^{\perp} + u^{\#} \ominus v^{\#} == u^{\perp} \ominus v^{\perp} + u^{\#} \ominus v^{\#}$$

$$u \ominus v == u^{\perp} \ominus v^{\perp} + u^{\#} \ominus v^{\#} \qquad u^{\perp} \ominus v^{\#} == u^{\#} \ominus v^{\perp} == 0 \qquad 6.263$$

In our discussion of the triangle formulae we discovered that the orthogonal component $u^{\perp}$ of a vector u, say, can be expressed as

$$u^{\perp} == \left(\underset{m}{\hat{\alpha}} \wedge u\right) \ominus \underset{m}{\hat{\alpha}}$$

Replacing u by the sum of its two components and then expanding gives $u^{\perp}$ as the sum of two terms.

$$u^{\perp} == \left(\underset{m}{\hat{\alpha}} \wedge \left(u^{\perp} + u^{\#}\right)\right) \ominus \underset{m}{\hat{\alpha}} == \left(\underset{m}{\hat{\alpha}} \wedge u^{\perp}\right) \ominus \underset{m}{\hat{\alpha}} + \left(\underset{m}{\hat{\alpha}} \wedge u^{\#}\right) \ominus \underset{m}{\hat{\alpha}}$$

Since $u^{\perp}$ is orthogonal to $\underset{m}{\hat{\alpha}}$, its exterior product with $\underset{m}{\hat{\alpha}}$ is not zero, and so the first term is not zero. On the other hand, $u^{\#}$ lies in $\underset{m}{\hat{\alpha}}$ and thus its exterior product with $\underset{m}{\hat{\alpha}}$ is zero.

$$u^{\perp} == \left(\underset{m}{\hat{\alpha}} \wedge u^{\perp}\right) \ominus \underset{m}{\hat{\alpha}} \qquad \left(\underset{m}{\hat{\alpha}} \wedge u^{\#}\right) \ominus \underset{m}{\hat{\alpha}} == 0 \qquad 6.264$$

A similar argument applies for $u^{\#}$. Since $u^{\#}$ is parallel to $\underset{m}{\hat{\alpha}}$, its interior product with $\underset{m}{\hat{\alpha}}$ is not zero, and so the second term is not zero. On the other hand, $u^{\perp}$ is orthogonal to $\underset{m}{\hat{\alpha}}$ and thus its interior product with $\underset{m}{\hat{\alpha}}$ is zero.

$$u^{\#} == (-1)^{m-1}\underset{m}{\hat{\alpha}}\ominus\left(\underset{m}{\hat{\alpha}}\ominus\left(u^{\perp}+u^{\#}\right)\right) == (-1)^{m-1}\underset{m}{\hat{\alpha}}\ominus\left(\underset{m}{\hat{\alpha}}\ominus u^{\perp}\right)+(-1)^{m-1}\underset{m}{\hat{\alpha}}\ominus\left(\underset{m}{\hat{\alpha}}\ominus u^{\#}\right)$$

$$u^{\#} == (-1)^{m-1}\underset{m}{\hat{\alpha}}\ominus\left(\underset{m}{\hat{\alpha}}\ominus u^{\#}\right) \qquad \underset{m}{\hat{\alpha}}\ominus\left(\underset{m}{\hat{\alpha}}\ominus u^{\perp}\right) == 0 \qquad 6.265$$

It is easy to see that we can extend the forms of product which are zero to forms in which $u^{\#}$ or $u^{\perp}$ is replaced by an exterior product of factors containing them. The second form is zero because of the extended interior product formula [6.40].

$$\left(\underset{m}{\hat{\alpha}}\wedge\left(\ldots\wedge u^{\#}\wedge\ldots\right)\right)\ominus\underset{m}{\hat{\alpha}} == 0 \qquad \underset{m}{\hat{\alpha}}\ominus\left(\underset{m}{\hat{\alpha}}\ominus\left(\ldots\wedge u^{\perp}\wedge\ldots\right)\right) == 0 \qquad 6.266$$

The decomposition formula applied

To fix ideas, consider a simple 2-element X in which each factor is expressed as the sum of its components orthogonal and parallel to a unit simple m-element $\underset{m}{\hat{\alpha}}$.

$$X = \left(x_1{}^{\perp}+x_1{}^{\#}\right)\wedge\left(x_2{}^{\perp}+x_2{}^{\#}\right);\ \star\star V\left[x_1{}^{\perp},x_1{}^{\#},x_2{}^{\perp},x_2{}^{\#}\right];$$

Decomposing X into components treats each factor as an entity and makes no expansions.

$$X1 = \star\mathcal{S}\left[\star\Sigma\left[(-1)^{\star r[\mathrm{Grade}[X]]\,(m+1)}\left(\underset{m}{\hat{\alpha}}\wedge\star\mathbb{S}^{\star}[X]\right)\ominus\left(\underset{m}{\hat{\alpha}}\ominus\star\mathbb{S}_{\star}[X]\right)\right]\right]$$

$$\underset{m}{\hat{\alpha}}\ominus\left(\underset{m}{\hat{\alpha}}\ominus\left(\left(x_1{}^{\perp}+x_1{}^{\#}\right)\wedge\left(x_2{}^{\perp}+x_2{}^{\#}\right)\right)\right)+$$
$$(-1)^{1+m}\left(\left(\underset{m}{\hat{\alpha}}\wedge\left(-x_1{}^{\perp}-x_1{}^{\#}\right)\right)\ominus\left(\underset{m}{\hat{\alpha}}\ominus\left(x_2{}^{\perp}+x_2{}^{\#}\right)\right)\right)+$$
$$(-1)^{1+m}\left(\left(\underset{m}{\hat{\alpha}}\wedge\left(x_2{}^{\perp}+x_2{}^{\#}\right)\right)\ominus\left(\underset{m}{\hat{\alpha}}\ominus\left(x_1{}^{\perp}+x_1{}^{\#}\right)\right)\right)+\left(\underset{m}{\hat{\alpha}}\wedge\left(x_1{}^{\perp}+x_1{}^{\#}\right)\wedge\left(x_2{}^{\perp}+x_2{}^{\#}\right)\right)\ominus\underset{m}{\hat{\alpha}}$$

Applying `GrassmannExpandAndSimplify` expands the decomposition to give 16 terms.

$$\star P;\ X2 = \star\mathcal{G}[X1]$$

$$\underset{m}{\hat{\alpha}}\ominus\left(\underset{m}{\hat{\alpha}}\ominus\left(x_1{}^{\perp}\wedge x_2{}^{\perp}\right)\right)+\underset{m}{\hat{\alpha}}\ominus\left(\underset{m}{\hat{\alpha}}\ominus\left(x_1{}^{\perp}\wedge x_2{}^{\#}\right)\right)+\underset{m}{\hat{\alpha}}\ominus\left(\underset{m}{\hat{\alpha}}\ominus\left(x_1{}^{\#}\wedge x_2{}^{\perp}\right)\right)+$$
$$\underset{m}{\hat{\alpha}}\ominus\left(\underset{m}{\hat{\alpha}}\ominus\left(x_1{}^{\#}\wedge x_2{}^{\#}\right)\right)+(-1)^{m}\left(\left(\underset{m}{\hat{\alpha}}\wedge x_1{}^{\perp}\right)\ominus\left(\underset{m}{\hat{\alpha}}\ominus x_2{}^{\perp}\right)\right)+$$
$$(-1)^{m}\left(\left(\underset{m}{\hat{\alpha}}\wedge x_1{}^{\perp}\right)\ominus\left(\underset{m}{\hat{\alpha}}\ominus x_2{}^{\#}\right)\right)+(-1)^{m}\left(\left(\underset{m}{\hat{\alpha}}\wedge x_1{}^{\#}\right)\ominus\left(\underset{m}{\hat{\alpha}}\ominus x_2{}^{\perp}\right)\right)+$$
$$(-1)^{m}\left(\left(\underset{m}{\hat{\alpha}}\wedge x_1{}^{\#}\right)\ominus\left(\underset{m}{\hat{\alpha}}\ominus x_2{}^{\#}\right)\right)+(-1)^{1+m}\left(\left(\underset{m}{\hat{\alpha}}\wedge x_2{}^{\perp}\right)\ominus\left(\underset{m}{\hat{\alpha}}\ominus x_1{}^{\perp}\right)\right)+$$
$$(-1)^{1+m}\left(\left(\underset{m}{\hat{\alpha}}\wedge x_2{}^{\perp}\right)\ominus\left(\underset{m}{\hat{\alpha}}\ominus x_1{}^{\#}\right)\right)+(-1)^{1+m}\left(\left(\underset{m}{\hat{\alpha}}\wedge x_2{}^{\#}\right)\ominus\left(\underset{m}{\hat{\alpha}}\ominus x_1{}^{\perp}\right)\right)+$$
$$(-1)^{1+m}\left(\left(\underset{m}{\hat{\alpha}}\wedge x_2{}^{\#}\right)\ominus\left(\underset{m}{\hat{\alpha}}\ominus x_1{}^{\#}\right)\right)+\left(\underset{m}{\hat{\alpha}}\wedge x_1{}^{\perp}\wedge x_2{}^{\perp}\right)\ominus\underset{m}{\hat{\alpha}}+$$
$$\left(\underset{m}{\hat{\alpha}}\wedge x_1{}^{\perp}\wedge x_2{}^{\#}\right)\ominus\underset{m}{\hat{\alpha}}+\left(\underset{m}{\hat{\alpha}}\wedge x_1{}^{\#}\wedge x_2{}^{\perp}\right)\ominus\underset{m}{\hat{\alpha}}+\left(\underset{m}{\hat{\alpha}}\wedge x_1{}^{\#}\wedge x_2{}^{\#}\right)\ominus\underset{m}{\hat{\alpha}}$$

But many of these terms will be zero as observed in the previous section. To identify the zero terms and put them to zero, we can use some rules mimicking the patterns of the zero terms.

```
X3 =
  ★G[X2 /. {α̂ₘ ⊖ (_^⌊) :→ 0, α̂ₘ ∧ (_^#) :→ 0, α̂ₘ ⊖ (___ ∧ _^⌊ ∧ ___) :→ 0,
     α̂ₘ ∧ (___ ∧ _^# ∧ ___) :→ 0}]
```

$$\underset{m}{\hat{\alpha}} \ominus \left(\underset{m}{\hat{\alpha}} \ominus \left(x_1^{\#} \wedge x_2^{\#}\right)\right) + (-1)^m \left(\left(\underset{m}{\hat{\alpha}} \wedge x_1^{\llcorner}\right) \ominus \left(\underset{m}{\hat{\alpha}} \ominus x_2^{\#}\right)\right) +$$
$$(-1)^{1+m} \left(\left(\underset{m}{\hat{\alpha}} \wedge x_2^{\llcorner}\right) \ominus \left(\underset{m}{\hat{\alpha}} \ominus x_1^{\#}\right)\right) + \left(\underset{m}{\hat{\alpha}} \wedge x_1^{\llcorner} \wedge x_2^{\llcorner}\right) \ominus \underset{m}{\hat{\alpha}}$$

Note that there are now just four terms. If we expand the original expression for X we also get four terms.

```
★E[(x1^⌊ + x1^#) ∧ (x2^⌊ + x2^#)]
```

$$x_1^{\llcorner} \wedge x_2^{\llcorner} + x_1^{\llcorner} \wedge x_2^{\#} + x_1^{\#} \wedge x_2^{\llcorner} + x_1^{\#} \wedge x_2^{\#}$$

$$\begin{aligned} &\left(x_1^{\llcorner} + x_1^{\#}\right) \wedge \left(x_2^{\llcorner} + x_2^{\#}\right) \\ &== x_1^{\llcorner} \wedge x_2^{\llcorner} + x_1^{\llcorner} \wedge x_2^{\#} + x_1^{\#} \wedge x_2^{\llcorner} + x_1^{\#} \wedge x_2^{\#} \\ &== \left(\underset{m}{\hat{\alpha}} \wedge \left(x_1^{\llcorner} \wedge x_2^{\llcorner}\right)\right) \ominus \underset{m}{\hat{\alpha}} + (-1)^m \left(\underset{m}{\hat{\alpha}} \wedge x_1^{\llcorner} \ominus \left(\underset{m}{\hat{\alpha}} \ominus x_2^{\#}\right)\right) - \\ &\quad (-1)^m \left(\underset{m}{\hat{\alpha}} \wedge x_2^{\llcorner} \ominus \left(\underset{m}{\hat{\alpha}} \ominus x_1^{\#}\right)\right) + \underset{m}{\hat{\alpha}} \ominus \left(\underset{m}{\hat{\alpha}} \ominus \left(x_1^{\#} \wedge x_2^{\#}\right)\right) \end{aligned} \qquad 6.267$$

In what follows we will show that, as might be suspected, the terms involving the same $x_i^{\llcorner}$ and $x_j^{\#}$ correspond.

OrthogonalDecompose

It is convenient first to define a function to perform the expansion and simplifications above. We will call it OrthogonalDecompose.

```
SetAttributes[OrthogonalDecompose, Listable];
OrthogonalDecompose[X_] :=
  ★G[★G[★Σ[(-1)^(★r[Grade[X]] (m+1)) (α̂ₘ ∧ ★S*[X]) ⊖ (α̂ₘ ⊖ ★S⋆[X])]]] /.
    {α̂ₘ ⊖ _^⌊ :→ 0, α̂ₘ ⊖ ___ ∧ _^⌊ ∧ ___ :→ 0, α̂ₘ ∧ ___ ∧ _^# ∧ ___ :→ 0}]
```

Now, if we apply ShowPrecedence and OrthogonalDecompose to X we get, as expected, the decomposition formula for X.

```
★P; X = x1^⌊ + x1^#; OrthogonalDecompose[X]
```

$$(-1)^{1+m} \left(\underset{m}{\hat{\alpha}} \ominus \left(\underset{m}{\hat{\alpha}} \ominus x_1^{\#}\right)\right) + \left(\underset{m}{\hat{\alpha}} \wedge x_1^{\llcorner}\right) \ominus \underset{m}{\hat{\alpha}}$$

But we can equally well apply OrthogonalDecompose to each component of X to get the same decomposition.

$$\textbf{OrthogonalDecompose}\left[\left\{x_1^{\llcorner},\ x_1^{\#}\right\}\right]$$

$$\left\{\left(\underset{m}{\hat{\alpha}}\wedge x_1^{\llcorner}\right)\ominus\underset{m}{\hat{\alpha}},\ (-1)^{1+m}\left(\underset{m}{\hat{\alpha}}\ominus\left(\underset{m}{\hat{\alpha}}\ominus x_1^{\#}\right)\right)\right\}$$

Our interest is in showing that this behaviour extends to cases where `X` is of grade higher than 1. Suppose `X` is of grade 2. Then, as expected, `OrthogonalDecompose` reproduces the results already obtained above.

$$\mathbf{X} = \left(x_1^{\llcorner}+x_1^{\#}\right)\wedge\left(x_2^{\llcorner}+x_2^{\#}\right);\ \textbf{OrthogonalDecompose}[\mathbf{X}]$$

$$\underset{m}{\hat{\alpha}}\ominus\left(\underset{m}{\hat{\alpha}}\ominus\left(x_1^{\#}\wedge x_2^{\#}\right)\right)+(-1)^{m}\left(\left(\underset{m}{\hat{\alpha}}\wedge x_1^{\llcorner}\right)\ominus\left(\underset{m}{\hat{\alpha}}\ominus x_2^{\#}\right)\right)+$$
$$(-1)^{1+m}\left(\left(\underset{m}{\hat{\alpha}}\wedge x_2^{\llcorner}\right)\ominus\left(\underset{m}{\hat{\alpha}}\ominus x_1^{\#}\right)\right)+\left(\underset{m}{\hat{\alpha}}\wedge x_1^{\llcorner}\wedge x_2^{\llcorner}\right)\ominus\underset{m}{\hat{\alpha}}$$

If however, we apply `OrthogonalDecompose` separately to typical terms in the expansion of `X`, we get the corresponding terms of the decomposition. (The sum of terms above differs in order from the list of the terms below due to *Mathematica's* ordering rules for the terms of a sum. Thus in reading an output comprising a sum, no import should be placed on the ordering of its terms).

$$\textbf{OrthogonalDecompose}\left[\left\{x_1^{\llcorner}\wedge x_2^{\llcorner},\ x_1^{\llcorner}\wedge x_2^{\#},\ x_1^{\#}\wedge x_2^{\llcorner},\ x_1^{\#}\wedge x_2^{\#}\right\}\right]$$

$$\left\{\left(\underset{m}{\hat{\alpha}}\wedge x_1^{\llcorner}\wedge x_2^{\llcorner}\right)\ominus\underset{m}{\hat{\alpha}},\ (-1)^{m}\left(\left(\underset{m}{\hat{\alpha}}\wedge x_1^{\llcorner}\right)\ominus\left(\underset{m}{\hat{\alpha}}\ominus x_2^{\#}\right)\right),\right.$$
$$\left.(-1)^{1+m}\left(\left(\underset{m}{\hat{\alpha}}\wedge x_2^{\llcorner}\right)\ominus\left(\underset{m}{\hat{\alpha}}\ominus x_1^{\#}\right)\right),\ \underset{m}{\hat{\alpha}}\ominus\left(\underset{m}{\hat{\alpha}}\ominus\left(x_1^{\#}\wedge x_2^{\#}\right)\right)\right\}$$

Generalizing this result leads us to surmise that decomposing a product of factors (each the sum of a parallel and an orthogonal component) gives the same result as expanding the factors into exterior products of parallel and orthogonal components, decomposing each exterior product, and then summing the results.

It is easy to confirm that this surmise is likely to be correct by taking some more examples.

$$\mathbf{X} = \left(x_1^{\llcorner}+x_1^{\#}\right)\wedge\left(x_2^{\llcorner}+x_2^{\#}\right)\wedge\left(x_3^{\llcorner}+x_3^{\#}\right);$$

$$\textbf{OrthogonalDecompose}[\mathbf{X}] == \textbf{Plus @@ OrthogonalDecompose}[\textbf{List @@} \star\mathcal{E}[\mathbf{X}]]$$

```
True
```

$$\mathbf{X} = \left(x_1^{\llcorner}+x_1^{\#}\right)\wedge\left(x_2^{\llcorner}+x_2^{\#}\right)\wedge\left(x_3^{\llcorner}+x_3^{\#}\right)\wedge\left(x_4^{\llcorner}+x_4^{\#}\right);$$

$$\textbf{OrthogonalDecompose}[\mathbf{X}] == \textbf{Plus @@ OrthogonalDecompose}[\textbf{List @@} \star\mathcal{E}[\mathbf{X}]]$$

```
True
```

Decomposition components

Let us now try to generalize these results. Suppose for example, that `X` is a 4-element given in terms of parallel and orthogonal components. Expanding `X` and listing its terms gives:

```
X = (x1ᴸ + x1#) ∧ (x2ᴸ + x2#) ∧ (x3ᴸ + x3#) ∧ (x4ᴸ + x4#);
Xe = List @@ ★ℰ[X]
```

$$\{x_1^{\llcorner} \wedge x_2^{\llcorner} \wedge x_3^{\llcorner} \wedge x_4^{\llcorner},\ x_1^{\llcorner} \wedge x_2^{\llcorner} \wedge x_3^{\llcorner} \wedge x_4^{\#},$$
$$x_1^{\llcorner} \wedge x_2^{\llcorner} \wedge x_3^{\#} \wedge x_4^{\llcorner},\ x_1^{\llcorner} \wedge x_2^{\llcorner} \wedge x_3^{\#} \wedge x_4^{\#},$$
$$x_1^{\llcorner} \wedge x_2^{\#} \wedge x_3^{\llcorner} \wedge x_4^{\llcorner},\ x_1^{\llcorner} \wedge x_2^{\#} \wedge x_3^{\llcorner} \wedge x_4^{\#},\ x_1^{\llcorner} \wedge x_2^{\#} \wedge x_3^{\#} \wedge x_4^{\llcorner},$$
$$x_1^{\llcorner} \wedge x_2^{\#} \wedge x_3^{\#} \wedge x_4^{\#},\ x_1^{\#} \wedge x_2^{\llcorner} \wedge x_3^{\llcorner} \wedge x_4^{\llcorner},\ x_1^{\#} \wedge x_2^{\llcorner} \wedge x_3^{\llcorner} \wedge x_4^{\#},$$
$$x_1^{\#} \wedge x_2^{\llcorner} \wedge x_3^{\#} \wedge x_4^{\llcorner},\ x_1^{\#} \wedge x_2^{\llcorner} \wedge x_3^{\#} \wedge x_4^{\#},\ x_1^{\#} \wedge x_2^{\#} \wedge x_3^{\llcorner} \wedge x_4^{\llcorner},$$
$$x_1^{\#} \wedge x_2^{\#} \wedge x_3^{\llcorner} \wedge x_4^{\#},\ x_1^{\#} \wedge x_2^{\#} \wedge x_3^{\#} \wedge x_4^{\llcorner},\ x_1^{\#} \wedge x_2^{\#} \wedge x_3^{\#} \wedge x_4^{\#}\}$$

Applying `OrthogonalDecompose` to each term enables us to tabulate the decompositions.

```
★★V[x_ᴸ, x_#]; ★ℬ4; ShowPrecedence;
Thread[Xe == OrthogonalDecompose[Xe]] // TableForm
```

$$x_1^{\llcorner} \wedge x_2^{\llcorner} \wedge x_3^{\llcorner} \wedge x_4^{\llcorner} == (\underset{m}{\hat{\alpha}} \wedge x_1^{\llcorner} \wedge x_2^{\llcorner} \wedge x_3^{\llcorner} \wedge x_4^{\llcorner}) \ominus \underset{m}{\hat{\alpha}}$$

$$x_1^{\llcorner} \wedge x_2^{\llcorner} \wedge x_3^{\llcorner} \wedge x_4^{\#} == (-1)^{m} \left((\underset{m}{\hat{\alpha}} \wedge x_1^{\llcorner} \wedge x_2^{\llcorner} \wedge x_3^{\llcorner}) \ominus (\underset{m}{\hat{\alpha}} \ominus x_4^{\#})\right)$$

$$x_1^{\llcorner} \wedge x_2^{\llcorner} \wedge x_3^{\#} \wedge x_4^{\llcorner} == (-1)^{1+m} \left((\underset{m}{\hat{\alpha}} \wedge x_1^{\llcorner} \wedge x_2^{\llcorner} \wedge x_4^{\llcorner}) \ominus (\underset{m}{\hat{\alpha}} \ominus x_3^{\#})\right)$$

$$x_1^{\llcorner} \wedge x_2^{\llcorner} \wedge x_3^{\#} \wedge x_4^{\#} == (\underset{m}{\hat{\alpha}} \wedge x_1^{\llcorner} \wedge x_2^{\llcorner}) \ominus (\underset{m}{\hat{\alpha}} \ominus (x_3^{\#} \wedge x_4^{\#}))$$

$$x_1^{\llcorner} \wedge x_2^{\#} \wedge x_3^{\llcorner} \wedge x_4^{\llcorner} == (-1)^{m} \left((\underset{m}{\hat{\alpha}} \wedge x_1^{\llcorner} \wedge x_3^{\llcorner} \wedge x_4^{\llcorner}) \ominus (\underset{m}{\hat{\alpha}} \ominus x_2^{\#})\right)$$

$$x_1^{\llcorner} \wedge x_2^{\#} \wedge x_3^{\llcorner} \wedge x_4^{\#} == -\left((\underset{m}{\hat{\alpha}} \wedge x_1^{\llcorner} \wedge x_3^{\llcorner}) \ominus (\underset{m}{\hat{\alpha}} \ominus (x_2^{\#} \wedge x_4^{\#}))\right)$$

$$x_1^{\llcorner} \wedge x_2^{\#} \wedge x_3^{\#} \wedge x_4^{\llcorner} == (\underset{m}{\hat{\alpha}} \wedge x_1^{\llcorner} \wedge x_4^{\llcorner}) \ominus (\underset{m}{\hat{\alpha}} \ominus (x_2^{\#} \wedge x_3^{\#}))$$

$$x_1^{\llcorner} \wedge x_2^{\#} \wedge x_3^{\#} \wedge x_4^{\#} == (-1)^{m} \left((\underset{m}{\hat{\alpha}} \wedge x_1^{\llcorner}) \ominus (\underset{m}{\hat{\alpha}} \ominus (x_2^{\#} \wedge x_3^{\#} \wedge x_4^{\#}))\right)$$

$$x_1^{\#} \wedge x_2^{\llcorner} \wedge x_3^{\llcorner} \wedge x_4^{\llcorner} == (-1)^{1+m} \left((\underset{m}{\hat{\alpha}} \wedge x_2^{\llcorner} \wedge x_3^{\llcorner} \wedge x_4^{\llcorner}) \ominus (\underset{m}{\hat{\alpha}} \ominus x_1^{\#})\right)$$

$$x_1^{\#} \wedge x_2^{\llcorner} \wedge x_3^{\llcorner} \wedge x_4^{\#} == (\underset{m}{\hat{\alpha}} \wedge x_2^{\llcorner} \wedge x_3^{\llcorner}) \ominus (\underset{m}{\hat{\alpha}} \ominus (x_1^{\#} \wedge x_4^{\#}))$$

$$x_1^{\#} \wedge x_2^{\llcorner} \wedge x_3^{\#} \wedge x_4^{\llcorner} == -\left((\underset{m}{\hat{\alpha}} \wedge x_2^{\llcorner} \wedge x_4^{\llcorner}) \ominus (\underset{m}{\hat{\alpha}} \ominus (x_1^{\#} \wedge x_3^{\#}))\right)$$

$$x_1^{\#} \wedge x_2^{\llcorner} \wedge x_3^{\#} \wedge x_4^{\#} == (-1)^{1+m} \left((\underset{m}{\hat{\alpha}} \wedge x_2^{\llcorner}) \ominus (\underset{m}{\hat{\alpha}} \ominus (x_1^{\#} \wedge x_3^{\#} \wedge x_4^{\#}))\right)$$

$$x_1^{\#} \wedge x_2^{\#} \wedge x_3^{\llcorner} \wedge x_4^{\llcorner} == (\underset{m}{\hat{\alpha}} \wedge x_3^{\llcorner} \wedge x_4^{\llcorner}) \ominus (\underset{m}{\hat{\alpha}} \ominus (x_1^{\#} \wedge x_2^{\#}))$$

$$x_1^{\#} \wedge x_2^{\#} \wedge x_3^{\llcorner} \wedge x_4^{\#} == (-1)^{m} \left((\underset{m}{\hat{\alpha}} \wedge x_3^{\llcorner}) \ominus (\underset{m}{\hat{\alpha}} \ominus (x_1^{\#} \wedge x_2^{\#} \wedge x_4^{\#}))\right)$$

$$x_1^{\#} \wedge x_2^{\#} \wedge x_3^{\#} \wedge x_4^{\llcorner} == (-1)^{1+m} \left((\underset{m}{\hat{\alpha}} \wedge x_4^{\llcorner}) \ominus (\underset{m}{\hat{\alpha}} \ominus (x_1^{\#} \wedge x_2^{\#} \wedge x_3^{\#}))\right)$$

$$x_1^{\#} \wedge x_2^{\#} \wedge x_3^{\#} \wedge x_4^{\#} == \underset{m}{\hat{\alpha}} \ominus (\underset{m}{\hat{\alpha}} \ominus (x_1^{\#} \wedge x_2^{\#} \wedge x_3^{\#} \wedge x_4^{\#}))$$

Let us group the parallel and orthogonal factors in the left hand sides of these formulae. For example, we can write

$$x_1^{\llcorner} \wedge x_2^{\#} \wedge x_3^{\llcorner} \wedge x_4^{\#} \quad == \quad -(x_1^{\llcorner} \wedge x_3^{\llcorner}) \wedge (x_2^{\#} \wedge x_4^{\#})$$

The corresponding right hand side is

$$- \left(\underset{m}{\hat{\alpha}} \wedge (x_1{}^{\perp} \wedge x_3{}^{\perp}) \ominus \left(\underset{m}{\hat{\alpha}} \ominus \left(x_2{}^{\#} \wedge x_4{}^{\#}\right)\right)\right)$$

By inspection we can see that if $\underset{\kappa}{\gamma^{\perp}}$ is the exterior product of all the orthogonal components, and $\underset{k-\kappa}{\gamma^{\#}}$ is the product of all the parallel components the 16 formulae obey the congruence

$$\underset{\kappa}{\gamma^{\perp}} \wedge \underset{k-\kappa}{\gamma^{\#}} \equiv (-1)^{m\kappa} \left(\underset{m}{\hat{\alpha}} \wedge \underset{\kappa}{\gamma^{\perp}}\right) \ominus \left(\underset{m}{\hat{\alpha}} \ominus \underset{k-\kappa}{\gamma^{\#}}\right) \qquad 6.268$$

Note carefully however, that for a given κ there are $\binom{k}{\kappa}$ terms of this form. For example let k equal 4; then for κ equal to 2 there are 6 terms, and for κ equal to 1 and 3, there are four terms. For k equal to 1 these four terms are:

$$x_1{}^{\perp} \wedge x_2{}^{\#} \wedge x_3{}^{\#} \wedge x_4{}^{\#} \qquad x_1{}^{\#} \wedge x_2{}^{\perp} \wedge x_3{}^{\#} \wedge x_4{}^{\#}$$
$$x_1{}^{\#} \wedge x_2{}^{\#} \wedge x_3{}^{\perp} \wedge x_4{}^{\perp} \qquad x_1{}^{\#} \wedge x_2{}^{\#} \wedge x_3{}^{\#} \wedge x_4{}^{\perp}$$

In particular, using [6.195] to adjust the signs for general k, the single purely orthogonal and purely parallel k-elements

$$\underset{k}{x^{\perp}} == \left(\underset{m}{\hat{\alpha}} \wedge \underset{k}{x}\right) \ominus \underset{m}{\hat{\alpha}} == \left(\underset{m}{\hat{\alpha}} \wedge \underset{k}{x^{\perp}}\right) \ominus \underset{m}{\hat{\alpha}} \qquad 6.269$$

$$\underset{k}{x^{\#}} == (-1)^{k\,(m-k)}\, \underset{m}{\hat{\alpha}} \ominus \left(\underset{m}{\hat{\alpha}} \ominus \underset{k}{x}\right) == (-1)^{k\,(m-k)}\, \underset{m}{\hat{\alpha}} \ominus \left(\underset{m}{\hat{\alpha}} \ominus \underset{k}{x^{\#}}\right) \qquad 6.270$$

$$\underset{k}{x} == x_1 \wedge x_2 \wedge \ldots \wedge x_k == \left(x_1{}^{\perp} + x_1{}^{\#}\right) \wedge \left(x_2{}^{\perp} + x_2{}^{\#}\right) \wedge \ldots \wedge \left(x_k{}^{\perp} + x_k{}^{\#}\right) \qquad 6.271$$

$$\underset{k}{x^{\perp}} == x_1{}^{\perp} \wedge x_2{}^{\perp} \wedge \ldots{}^{\perp} \wedge x_k{}^{\perp}$$
$$\underset{k}{x^{\#}} == x_1{}^{\#} \wedge x_2{}^{\#} \wedge \ldots{}^{\#} \wedge x_k{}^{\#} \qquad 6.272$$
$$x_i{}^{\perp} \ominus x_j{}^{\#} == 0$$

In the special case where k is equal to m, these formulae simplify somewhat.

$$\underset{m}{x^{\perp}} == \left(\underset{m}{\hat{\alpha}} \wedge \underset{m}{x}\right) \ominus \underset{m}{\hat{\alpha}} == \left(\underset{m}{\hat{\alpha}} \wedge \underset{m}{x^{\perp}}\right) \ominus \underset{m}{\hat{\alpha}} \qquad 6.273$$

$$\underset{m}{x^{\#}} == \left(\underset{m}{\hat{\alpha}} \ominus \underset{m}{x}\right) \underset{m}{\hat{\alpha}} == \left(\underset{m}{\hat{\alpha}} \ominus \underset{m}{x^{\#}}\right) \underset{m}{\hat{\alpha}} \qquad 6.274$$

Relationships amongst the components

The identities implied by the last formulae of the previous section imply a number of further useful identities based on the fact that the exterior product of any number of groupings of orthogonal (respectively, parallel) factors on the left hand side is equal to the exterior product of the corresponding components on the right hand side. Here are two examples:

$$(x_1{}^{\perp} \wedge \ldots^{\perp} \wedge x_i{}^{\perp}) \wedge (x_{i+1}{}^{\perp} \wedge \ldots^{\perp} \wedge x_k{}^{\perp}) == \left(\left(\underset{m}{\hat{\alpha}} \wedge (x_1{}^{\perp} \wedge \ldots^{\perp} \wedge x_i{}^{\perp})\right) \ominus \underset{m}{\hat{\alpha}}\right) \wedge \left(\left(\underset{m}{\hat{\alpha}} \wedge (x_{i+1}{}^{\perp} \wedge \ldots^{\perp} \wedge x_k{}^{\perp})\right) \ominus \underset{m}{\hat{\alpha}}\right) \qquad 6.275$$

$$\left(x_1{}^{\#} \wedge x_2{}^{\#}\right) \wedge \left(x_3{}^{\#} \wedge x_4{}^{\#}\right) \wedge x_5{}^{\#} == (-1)^{5\,(m-5)} \left(\underset{m}{\hat{\alpha}} \ominus \left(\underset{m}{\hat{\alpha}} \ominus (x_1{}^{\#} \wedge x_2{}^{\#})\right)\right) \wedge \left(\underset{m}{\hat{\alpha}} \ominus \left(\underset{m}{\hat{\alpha}} \ominus (x_3{}^{\#} \wedge x_4{}^{\#})\right)\right) \wedge \left(\underset{m}{\hat{\alpha}} \ominus \left(\underset{m}{\hat{\alpha}} \ominus x_5{}^{\#}\right)\right) \qquad 6.276$$

We can also consider the interior product of components. Let u be any vector in $\underset{m}{\alpha}$. Then $u^{\#}$ is orthogonal to any 1-elements orthogonal to $\underset{m}{\alpha}$, and thus to their exterior product.

$$(x_1{}^{\perp} \wedge \ldots^{\perp} \wedge x_i{}^{\perp}) \ominus u^{\#} == 0$$

In the same way, $u^{\perp}$ is orthogonal to any 1-elements contained in $\underset{m}{\alpha}$, and thus to their exterior product.

$$(x_1{}^{\perp} \wedge \ldots^{\perp} \wedge x_i{}^{\perp}) \ominus u^{\#} == 0 \qquad \left(x_1{}^{\#} \wedge \ldots^{\#} \wedge x_i{}^{\#}\right) \ominus u^{\perp} == 0 \qquad 6.277$$

Clearly, we can take the interior or exterior product of these zeros with any other elements, perhaps apply the extended interior product formula [6.40], and still get zero.

$$\left((x_1{}^{\perp} \wedge \ldots^{\perp} \wedge x_i{}^{\perp}) \ominus u^{\#}\right) \ominus \ldots == 0 \qquad \left(\left(x_1{}^{\#} \wedge \ldots^{\#} \wedge x_i{}^{\#}\right) \ominus u^{\perp}\right) \ominus \ldots == 0 \qquad 6.278$$

$$\left((x_1{}^{\perp} \wedge \ldots^{\perp} \wedge x_i{}^{\perp}) \ominus u^{\#}\right) \wedge \ldots == 0 \qquad \left(\left(x_1{}^{\#} \wedge \ldots^{\#} \wedge x_i{}^{\#}\right) \ominus u^{\perp}\right) \wedge \ldots == 0 \qquad 6.279$$

$$(x_1{}^{\perp} \wedge \ldots^{\perp} \wedge x_i{}^{\perp}) \ominus \left(\ldots \wedge u^{\#} \wedge \ldots\right) == 0 \qquad \left(x_1{}^{\#} \wedge \ldots^{\#} \wedge x_i{}^{\#}\right) \ominus (\ldots \wedge u^{\perp} \wedge \ldots) == 0 \qquad 6.280$$

In particular these formulae show that the totally orthogonal and totally parallel components, $\underset{k}{x}^{\perp}$ and $\underset{k}{x}^{\#}$, are, as may be expected, orthogonal to each other. This can also be straightforwardly established using the Extended Interior Product Formula discussed earlier in this chapter.

$$\underset{k}{x}^{\perp} \ominus \underset{k}{x}^{\#} == (-1)^{k\,(m-k)} \left(\left(\left(\underset{m}{\hat{\alpha}} \wedge \underset{k}{x}\right) \ominus \underset{m}{\hat{\alpha}}\right) \ominus \left(\underset{m}{\hat{\alpha}} \ominus \left(\underset{m}{\hat{\alpha}} \ominus \underset{k}{x}\right)\right)\right)$$
$$== (-1)^{k\,(m-k)} \left(\underset{m}{\hat{\alpha}} \wedge \underset{k}{x}\right) \ominus \left(\underset{m}{\hat{\alpha}} \wedge \left(\underset{m}{\hat{\alpha}} \ominus \left(\underset{m}{\hat{\alpha}} \ominus \underset{k}{x}\right)\right)\right) == (-1)^{k\,(m-k)} \left(\underset{m}{\hat{\alpha}} \wedge \underset{k}{x}\right) \ominus 0 ==$$
$$0$$

Taking the interior products of $\underset{k}{x}^{\perp}$, $\underset{k}{x}^{\#}$ and $\underset{k}{x}$ gives

$$\underset{k}{x}^{\perp} \ominus \underset{k}{x} == \left(\left(\underset{m}{\hat{\alpha}} \wedge \underset{k}{x}\right) \ominus \underset{m}{\hat{\alpha}}\right) \ominus \underset{k}{x} == \left(\underset{m}{\hat{\alpha}} \wedge \underset{k}{x}\right) \ominus \left(\underset{m}{\hat{\alpha}} \wedge \underset{k}{x}\right) == \left|\underset{m}{\hat{\alpha}} \wedge \underset{k}{x}\right|^2$$

$$\underset{k}{x}^{\#} \ominus \underset{k}{x} == (-1)^{k(m-k)} \left(\underset{m}{\hat{\alpha}} \ominus \left(\underset{m}{\hat{\alpha}} \ominus \underset{k}{x}\right)\right) \ominus \underset{k}{x} == (-1)^{k(m-k)} \underset{m}{\hat{\alpha}} \vee \overline{\underset{m}{\hat{\alpha}} \ominus \underset{k}{x}} \vee \overline{\underset{k}{x}}$$

$$== \left(\underset{m}{\hat{\alpha}} \vee \overline{\underset{k}{x}}\right) \vee \overline{\underset{m}{\hat{\alpha}} \ominus \underset{k}{x}} == \left(\underset{m}{\hat{\alpha}} \ominus \underset{k}{x}\right) \ominus \left(\underset{m}{\hat{\alpha}} \ominus \underset{k}{x}\right) == \left|\underset{m}{\hat{\alpha}} \ominus \underset{k}{x}\right|^2$$

$$\underset{k}{x}^{\perp} \ominus \underset{k}{x}^{\perp} == \left(\left(\underset{m}{\hat{\alpha}} \wedge \underset{k}{x}^{\perp}\right) \ominus \underset{m}{\hat{\alpha}}\right) \ominus \underset{k}{x}^{\perp} == \left|\underset{m}{\hat{\alpha}} \wedge \underset{k}{x}^{\perp}\right|^2$$

$$\underset{k}{x}^{\#} \ominus \underset{k}{x}^{\#} == (-1)^{k(m-k)} \left(\underset{m}{\hat{\alpha}} \ominus \left(\underset{m}{\hat{\alpha}} \ominus \underset{k}{x}^{\#}\right)\right) \ominus \underset{k}{x}^{\#} == \left|\underset{m}{\hat{\alpha}} \ominus \underset{k}{x}^{\#}\right|^2$$

$$\underset{k}{x}^{\perp} \ominus \underset{k}{x} == \left|\underset{m}{\hat{\alpha}} \wedge \underset{k}{x}\right|^2 == \underset{k}{x}^{\perp} \ominus \underset{k}{x}^{\perp} == \left|\underset{m}{\hat{\alpha}} \wedge \underset{k}{x}^{\perp}\right|^2 \qquad 6.281$$

$$\underset{k}{x}^{\#} \ominus \underset{k}{x} == \left|\underset{m}{\hat{\alpha}} \ominus \underset{k}{x}\right|^2 == \underset{k}{x}^{\#} \ominus \underset{k}{x}^{\#} == \left|\underset{m}{\hat{\alpha}} \ominus \underset{k}{x}^{\#}\right|^2 \qquad 6.282$$

$$\underset{k}{x}^{\perp} \ominus \underset{k}{x}^{\#} == 0 \qquad 6.283$$

◆ Example

Suppose we have a unit vector $\hat{\alpha}$ and a vector x, then, with little comment, we specialize the above formulae to this simple case.

- The components

$$x^{\perp} == (\hat{\alpha} \wedge x) \ominus \hat{\alpha} == x - (\hat{\alpha} \ominus x)\, \hat{\alpha}$$

$$x^{\#} == \hat{\alpha} \ominus (\hat{\alpha} \ominus x) == (\hat{\alpha} \ominus x)\, \hat{\alpha}$$

$$x^{\perp} \ominus x^{\#} == (x - (\hat{\alpha} \ominus x)\, \hat{\alpha}) \ominus ((\hat{\alpha} \ominus x)\, \hat{\alpha}) == 0$$

- Orthogonal relationships

$$x^{\perp} \ominus x == (x - (\hat{\alpha} \ominus x)\, \hat{\alpha}) \ominus x == x \ominus x - (\hat{\alpha} \ominus x)^2$$

$$|\hat{\alpha} \wedge x|^2 == (\hat{\alpha} \wedge x) \ominus (\hat{\alpha} \wedge x) == x \ominus x - (\hat{\alpha} \ominus x)^2$$

$$x^{\perp} \ominus x^{\perp} == (x - (\hat{\alpha} \ominus x)\, \hat{\alpha}) \ominus (x - (\hat{\alpha} \ominus x)\, \hat{\alpha}) == x \ominus x - (\hat{\alpha} \ominus x)^2$$

$$|\hat{\alpha} \wedge x^{\perp}|^2 == (\hat{\alpha} \wedge x^{\perp}) \ominus (\hat{\alpha} \wedge x^{\perp}) == (\hat{\alpha} \wedge (x - (\hat{\alpha} \ominus x)\, \hat{\alpha})) \ominus (\hat{\alpha} \wedge (x - (\hat{\alpha} \ominus x)\, \hat{\alpha}))$$

$$== (\hat{\alpha} \wedge x) \ominus (\hat{\alpha} \wedge x) == x \ominus x - (\hat{\alpha} \ominus x)^2$$

- Parallel relationships

$$x^{\#} \ominus x == (\hat{\alpha} \ominus x)\, \hat{\alpha} \ominus x == (\hat{\alpha} \ominus x)^2$$

$$|\hat{\alpha} \ominus x|^2 == (\hat{\alpha} \ominus x) \ominus (\hat{\alpha} \ominus x) == (\hat{\alpha} \ominus x)^2$$

$$x^{\#} \ominus x^{\#} == ((\hat{\alpha} \ominus x)\, \hat{\alpha}) \ominus ((\hat{\alpha} \ominus x)\, \hat{\alpha}) == (\hat{\alpha} \ominus x)^2$$

$$\left|\hat{\alpha}\ominus x^{\#}\right|^{2} == \left(\hat{\alpha}\ominus x^{\#}\right)\ominus\left(\hat{\alpha}\ominus x^{\#}\right) == \left(\hat{\alpha}\ominus\left(\left(\hat{\alpha}\ominus x\right)\,\hat{\alpha}\right)\right)\ominus\left(\hat{\alpha}\ominus\left(\left(\hat{\alpha}\ominus x\right)\,\hat{\alpha}\right)\right)$$
$$== \left(\hat{\alpha}\ominus x\right)\ominus\left(\hat{\alpha}\ominus x\right) == \left(\hat{\alpha}\ominus x\right)^{2}$$

Invariance of the decomposition

As we have seen above, the Orthogonal Decomposition Formula [6.262] decomposes a simple element X into a sum of elements whose factors are either orthogonal or parallel to another given simple element.

$$X == \star\Sigma\left[(-1)^{\star r[\text{Grade}[X]]\,(m+1)}\left(\underset{m}{\hat{\alpha}}\wedge\star S^{\star}[X]\right)\ominus\left(\underset{m}{\hat{\alpha}}\ominus\star S_{\star}[X]\right)\right]$$

In this section we will explore the invariance of this operation to the factorization of X. It turns out that although *individual* terms in the decomposition sum may not be invariant to a factorization of X, the complete decomposition sum, the totally orthogonal and totally parallel components, and certain sums of other terms, will be invariant.

Consider the decomposition of a simple 2-element $x\wedge y$. Applying the formula and doing some basic simplification gives the standard result we have obtained previously above.

$$X = x\wedge y;$$
$$A_1 = \star\mathcal{S}\left[\star\Sigma\left[\text{Flatten}\left[(-1)^{\star r[\text{Grade}[X]]\,(m+1)}\left(\underset{m}{\hat{\alpha}}\wedge\star S^{\star}[X]\right)\ominus\left(\underset{m}{\hat{\alpha}}\ominus\star S_{\star}[X]\right)\right]\right]\right]$$

$$\underset{m}{\hat{\alpha}}\ominus\left(\underset{m}{\hat{\alpha}}\ominus x\wedge y\right)+(-1)^{m}\left(\underset{m}{\hat{\alpha}}\wedge x\ominus\left(\underset{m}{\hat{\alpha}}\ominus y\right)\right)+(-1)^{1+m}\left(\underset{m}{\hat{\alpha}}\wedge y\ominus\left(\underset{m}{\hat{\alpha}}\ominus x\right)\right)+\underset{m}{\hat{\alpha}}\wedge x\wedge y\ominus\underset{m}{\hat{\alpha}}$$

Had we represented the bivector $x\wedge y$ differently, say as $x\wedge(y+a\,x)$ we would have obtained:

$$X = x\wedge(y+a\,x);$$
$$A_2 = \star\mathcal{S}\left[\star\Sigma\left[(-1)^{\star r[\text{Grade}[X]]\,(m+1)}\left(\underset{m}{\hat{\alpha}}\wedge\star S^{\star}[X]\right)\ominus\left(\underset{m}{\hat{\alpha}}\ominus\star S_{\star}[X]\right)\right]\right]$$

$$\underset{m}{\hat{\alpha}}\ominus\left(\underset{m}{\hat{\alpha}}\ominus x\wedge(a\,x+y)\right)+(-1)^{m}\left(\underset{m}{\hat{\alpha}}\wedge x\ominus\left(\underset{m}{\hat{\alpha}}\ominus(a\,x+y)\right)\right)+$$
$$(-1)^{1+m}\left(\underset{m}{\hat{\alpha}}\wedge(a\,x+y)\ominus\left(\underset{m}{\hat{\alpha}}\ominus x\right)\right)+\underset{m}{\hat{\alpha}}\wedge x\wedge(a\,x+y)\ominus\underset{m}{\hat{\alpha}}$$

So that we can keep track of the terms, let us put them in to a list.

$$A_3 = \text{List}\;@@\;A_2$$

$$\Big\{\underset{m}{\hat{\alpha}}\ominus\left(\underset{m}{\hat{\alpha}}\ominus x\wedge(a\,x+y)\right),\;(-1)^{m}\left(\underset{m}{\hat{\alpha}}\wedge x\ominus\left(\underset{m}{\hat{\alpha}}\ominus(a\,x+y)\right)\right),$$
$$(-1)^{1+m}\left(\underset{m}{\hat{\alpha}}\wedge(a\,x+y)\ominus\left(\underset{m}{\hat{\alpha}}\ominus x\right)\right),\;\underset{m}{\hat{\alpha}}\wedge x\wedge(a\,x+y)\ominus\underset{m}{\hat{\alpha}}\Big\}$$

Expanding each of the terms gives:

$$\mathbf{A_4} = \star\mathcal{E}[\mathbf{A_3}]$$

$$\Big\{\hat{\underset{m}{\alpha}}\ominus\Big(\hat{\underset{m}{\alpha}}\ominus x\wedge(a\,x)\Big)+\hat{\underset{m}{\alpha}}\ominus\Big(\hat{\underset{m}{\alpha}}\ominus x\wedge y\Big),$$
$$(-1)^m\Big(\hat{\underset{m}{\alpha}}\wedge x\ominus\Big(\hat{\underset{m}{\alpha}}\ominus a\,x\Big)\Big)+(-1)^m\Big(\hat{\underset{m}{\alpha}}\wedge x\ominus\Big(\hat{\underset{m}{\alpha}}\ominus y\Big)\Big),$$
$$(-1)^{1+m}\Big(\hat{\underset{m}{\alpha}}\wedge(a\,x)\ominus\Big(\hat{\underset{m}{\alpha}}\ominus x\Big)\Big)+(-1)^{1+m}\Big(\hat{\underset{m}{\alpha}}\wedge y\ominus\Big(\hat{\underset{m}{\alpha}}\ominus x\Big)\Big),$$
$$\hat{\underset{m}{\alpha}}\wedge x\wedge(a\,x)\ominus\hat{\underset{m}{\alpha}}+\hat{\underset{m}{\alpha}}\wedge x\wedge y\ominus\hat{\underset{m}{\alpha}}\Big\}$$

It is clear that upon simplification the first term of the first and last elements of the list will reduce to zero, while the second and third elements each have an oppositely signed, but otherwise equal added term.

$$\mathbf{A_5} = \star\mathcal{G}[\mathbf{A_4}]$$

$$\Big\{\hat{\underset{m}{\alpha}}\ominus\Big(\hat{\underset{m}{\alpha}}\ominus x\wedge y\Big),\ (-1)^m a\Big(\hat{\underset{m}{\alpha}}\wedge x\ominus\Big(\hat{\underset{m}{\alpha}}\ominus x\Big)\Big)+(-1)^m\Big(\hat{\underset{m}{\alpha}}\wedge x\ominus\Big(\hat{\underset{m}{\alpha}}\ominus y\Big)\Big),$$
$$-(-1)^m a\Big(\hat{\underset{m}{\alpha}}\wedge x\ominus\Big(\hat{\underset{m}{\alpha}}\ominus x\Big)\Big)+(-1)^{1+m}\Big(\hat{\underset{m}{\alpha}}\wedge y\ominus\Big(\hat{\underset{m}{\alpha}}\ominus x\Big)\Big),\ \hat{\underset{m}{\alpha}}\wedge x\wedge y\ominus\hat{\underset{m}{\alpha}}\Big\}$$

By taking a trivial case we can see that this term is not in general zero.

$$(-1)^m a\Big(\hat{\underset{m}{\alpha}}\wedge x\ominus\Big(\hat{\underset{m}{\alpha}}\ominus x\Big)\Big)\ \longrightarrow\ -a\,(\alpha\ominus x)\,\alpha\wedge x$$

Nevertheless, the *sum* of the second and third terms is invariant. If we add all the elements of the list we retrieve the original decomposition of $x\wedge y$. Hence, in general, although the decomposition formula is invariant to factorizations of the input, the *individual* terms in the decomposition sum may not be.

There are however two terms which are invariant to any factorization: the totally orthogonal and totally parallel terms. These are invariant because they involve only the total element to be decomposed.

$$\underset{k}{x}^{\perp} == x_1^{\perp}\wedge x_2^{\perp}\wedge\ldots^{\perp}\wedge x_k^{\perp} == \Big(\hat{\underset{m}{\alpha}}\wedge(x_1^{\perp}\wedge x_2^{\perp}\wedge\ldots^{\perp}\wedge x_k^{\perp})\Big)\ominus\hat{\underset{m}{\alpha}}$$

$$\underset{k}{x}^{\#} == x_1^{\#}\wedge x_2^{\#}\wedge\ldots^{\#}\wedge x_k^{\#} == (-1)^{k\,(m-k)}\,\hat{\underset{m}{\alpha}}\ominus\Big(\hat{\underset{m}{\alpha}}\ominus\big(x_1^{\#}\wedge x_2^{\#}\wedge\ldots^{\#}\wedge x_k^{\#}\big)\Big)$$

6.14 Orthogonalization of Elements

Orthogonalizing the factors of a simple *m*-element

We have already seen in section 2.12 that a multilinear m:k-form based on a simple element A is invariant to the way A is factorized. Thus in order to prove a property of an m:k-form based on A we may always assume a factorization convenient to the proof.

One particularly special factorization is into orthogonal 1-elements, for example by the Gram-Schmidt process. In this section we will explore orthogonalizing a simple element and give a theorem, the Zero Interior Sum Theorem, which uses this orthogonalization in its proof.

Consider a simple m-element A with factors α_i that we would like to replace with mutually orthogonal factors ε_i whose exterior product is congruent to A.

$$A == \alpha_1 \wedge \alpha_2 \wedge \ldots \wedge \alpha_m \equiv \varepsilon_1 \wedge \varepsilon_2 \wedge \ldots \wedge \varepsilon_m$$

This is equivalent to saying we have a set of independent 1-elements $\alpha_1, \alpha_2, \alpha_3, \ldots$, and we wish to create an orthogonal set $\varepsilon_1, \varepsilon_2, \varepsilon_3, \ldots$ spanning the same space.

We begin by choosing one of the elements, α_1 say, arbitrarily and setting this to be the first element ε_1 in the orthogonal set to be created.

$$\varepsilon_1 == \alpha_1$$

To create a second element ε_2 orthogonal to α_1 (and ε_1) within the space concerned, we choose a second element of the space, α_2 say, and form the interior product.

$$\varepsilon_2 == (\alpha_1 \wedge \alpha_2) \ominus \alpha_1 == (\varepsilon_1 \wedge \alpha_2) \ominus \varepsilon_1$$

From our discussion of the triangle formulae in section 6.11, we can see that ε_2 is orthogonal to α_1 (and ε_1). Alternatively we can take their interior product to see (by [6.40]) that it is zero.

$$\varepsilon_2 \ominus \varepsilon_1 == \varepsilon_2 \ominus \alpha_1 == ((\alpha_1 \wedge \alpha_2) \ominus \alpha_1) \ominus \alpha_1 == (\alpha_1 \wedge \alpha_2) \ominus (\alpha_1 \wedge \alpha_1) == 0$$

We can also see that α_2 lies in the 2-element $\varepsilon_1 \wedge \varepsilon_2$ (that is, α_2 is a linear combination of ε_1 and ε_2) by taking their exterior product and expanding it.

$$\varepsilon_1 \wedge ((\varepsilon_1 \wedge \alpha_2) \ominus \varepsilon_1) \wedge \alpha_2$$
$$== \varepsilon_1 \wedge ((\varepsilon_1 \ominus \varepsilon_1)\, \alpha_2 - (\varepsilon_1 \ominus \alpha_2)\, \varepsilon_1) \wedge \alpha_2$$
$$== (\varepsilon_1 \ominus \varepsilon_1)\, \varepsilon_1 \wedge \alpha_2 \wedge \alpha_2 - (\varepsilon_1 \ominus \alpha_2)\, \varepsilon_1 \wedge \varepsilon_1 \wedge \alpha_2 == 0$$

Here we have used the Interior Common Factor Theorem in its form [6.52].

To create a third element ε_3 orthogonal to ε_1 and ε_2 within the space concerned, we choose a third element of the space, α_3 say, and form the interior product.

$$\varepsilon_3 == (\varepsilon_1 \wedge \varepsilon_2 \wedge \alpha_3) \ominus (\varepsilon_1 \wedge \varepsilon_2)$$

Again from the triangle formulae in section 6.11, we have that ε_3 is orthogonal to ε_1 and ε_2. Alternatively we can take their interior products to see that they are zero.

$$\varepsilon_3 \ominus \varepsilon_1 == ((\varepsilon_1 \wedge \varepsilon_2 \wedge \alpha_3) \ominus (\varepsilon_1 \wedge \varepsilon_2)) \ominus \varepsilon_1$$
$$== (\varepsilon_1 \wedge \varepsilon_2 \wedge \alpha_3) \ominus (\varepsilon_1 \wedge \varepsilon_2 \wedge \varepsilon_1) == 0$$

$$\varepsilon_3 \ominus \varepsilon_2 == ((\varepsilon_1 \wedge \varepsilon_2 \wedge \alpha_3) \ominus (\varepsilon_1 \wedge \varepsilon_2)) \ominus e_2$$
$$== (\varepsilon_1 \wedge \varepsilon_2 \wedge \alpha_3) \ominus (\varepsilon_1 \wedge \varepsilon_2 \wedge \varepsilon_2) == 0$$

We can also see that α_3 lies in the 3-element $\varepsilon_1 \wedge \varepsilon_2 \wedge \varepsilon_3$ (that is, α_3 is a linear combination of ε_1, ε_2 and ε_3), or in the general case that α_i lies in the i-element $\varepsilon_1 \wedge \varepsilon_2 \wedge \ldots \wedge \varepsilon_i$, by taking their exterior product and expanding it, or using the Interior Common Factor Theorem in the form [6.54].

We create the rest of the orthogonal elements in a similar manner.

$$\varepsilon_4 == (\varepsilon_1 \wedge \varepsilon_2 \wedge \varepsilon_3 \wedge \alpha_4) \ominus (\varepsilon_1 \wedge \varepsilon_2 \wedge \varepsilon_3)$$
$$\varepsilon_5 == (\varepsilon_1 \wedge \varepsilon_2 \wedge \varepsilon_3 \wedge \varepsilon_4 \wedge \alpha_5) \ominus (\varepsilon_1 \wedge \varepsilon_2 \wedge \varepsilon_3 \wedge \varepsilon_4)$$
$$\ldots$$

In general then, the ith element ε_i of the orthogonal set $\varepsilon_1, \varepsilon_2, \varepsilon_3, \cdots$ is obtained from the previous i–1 elements and the ith element α_i of the original set.

$$\varepsilon_i == (\varepsilon_1 \wedge \varepsilon_2 \wedge \ldots \wedge \varepsilon_{i-1} \wedge \alpha_i) \ominus (\varepsilon_1 \wedge \varepsilon_2 \wedge \ldots \wedge \varepsilon_{i-1}) \qquad 6.284$$

The element ε_i is orthogonal to $\varepsilon_1 \wedge \varepsilon_2 \wedge \ldots \wedge \varepsilon_{i-1}$ and hence to each of the $\varepsilon_1, \varepsilon_2, \ldots, \varepsilon_{i-1}$. The element α_i lies in $\varepsilon_1 \wedge \varepsilon_2 \wedge \ldots \wedge \varepsilon_i$ and hence is a linear combination of $\varepsilon_1, \varepsilon_2, \ldots, \varepsilon_i$.

We have thus determined a factorization $\varepsilon_1 \wedge \varepsilon_2 \wedge \ldots \wedge \varepsilon_m$ *congruent* to A. That is

$$A == \alpha_1 \wedge \alpha_2 \wedge \ldots \wedge \alpha_m == a\, \varepsilon_1 \wedge \varepsilon_2 \wedge \ldots \wedge \varepsilon_m$$

To determine the scalar factor a we can take the measures of the elements.

$$|A| == |\alpha_1 \wedge \alpha_2 \wedge \ldots \wedge \alpha_m| == a\, |\varepsilon_1 \wedge \varepsilon_2 \wedge \ldots \wedge \varepsilon_m|$$

$$A == \alpha_1 \wedge \alpha_2 \wedge \ldots \wedge \alpha_m == |A| \frac{\varepsilon_1 \wedge \varepsilon_2 \wedge \ldots \wedge \varepsilon_m}{|\varepsilon_1 \wedge \varepsilon_2 \wedge \ldots \wedge \varepsilon_m|}$$

Since the ε_i are mutually orthogonal we can then also write

$$A == |A| \frac{\varepsilon_1 \wedge \varepsilon_2 \wedge \ldots \wedge \varepsilon_m}{|\varepsilon_1|\, |\varepsilon_2| \ldots |\varepsilon_m|} \qquad \hat{A} == \hat{\varepsilon}_1 \wedge \hat{\varepsilon}_2 \wedge \ldots \wedge \hat{\varepsilon}_m \qquad 6.285$$

Orthogonalization in terms of the original factors

If, in the development of the previous section, we expand $\varepsilon_1 \wedge \varepsilon_2$ out in terms of α_1 and α_2 by [6.52] we get

$$\varepsilon_1 \wedge \varepsilon_2 == \alpha_1 \wedge ((\alpha_1 \wedge \alpha_2) \ominus \alpha_1)$$
$$== \alpha_1 \wedge ((\alpha_1 \ominus \alpha_1)\, \alpha_2 - (\alpha_1 \ominus \alpha_2)\, \alpha_1) == (\alpha_1 \ominus \alpha_1)\, \alpha_1 \wedge \alpha_2$$

Hence we can also write ε_3 as

$$\varepsilon_3 == (\varepsilon_1 \wedge \varepsilon_2 \wedge \alpha_3) \ominus (\varepsilon_1 \wedge \varepsilon_2) == (\alpha_1 \ominus \alpha_1)^2\, (\alpha_1 \wedge \alpha_2 \wedge \alpha_3) \ominus (\alpha_1 \wedge \alpha_2)$$

Expanding $(\alpha_1 \wedge \alpha_2 \wedge \alpha_3) \ominus (\alpha_1 \wedge \alpha_2)$ out by [6.53] gives

$$(\alpha_1 \wedge \alpha_2 \wedge \alpha_3) \ominus (\alpha_1 \wedge \alpha_2)$$
$$== (\alpha_1 \wedge \alpha_2) \ominus (\alpha_2 \wedge \alpha_3)\, \alpha_1 - (\alpha_1 \wedge \alpha_2) \ominus (\alpha_1 \wedge \alpha_3)\, \alpha_2 + (\alpha_1 \wedge \alpha_2) \ominus (\alpha_1 \wedge \alpha_2)\, \alpha_3$$

Now $\varepsilon_1 \wedge \varepsilon_2 \wedge \varepsilon_3$ becomes

$$\varepsilon_1 \wedge \varepsilon_2 \wedge \varepsilon_3 == ((\alpha_1 \ominus \alpha_1)\, \alpha_1 \wedge \alpha_2) \wedge \left((\alpha_1 \ominus \alpha_1)^2\, (\alpha_1 \wedge \alpha_2 \wedge \alpha_3) \ominus (\alpha_1 \wedge \alpha_2)\right)$$
$$== (\alpha_1 \ominus \alpha_1)^3\, (\alpha_1 \wedge \alpha_2) \ominus (\alpha_1 \wedge \alpha_2)\, \alpha_1 \wedge \alpha_2 \wedge \alpha_3$$

And ε_4 then becomes

$$\varepsilon_4 == (\varepsilon_1 \wedge \varepsilon_2 \wedge \varepsilon_3 \wedge \alpha_4) \ominus (\varepsilon_1 \wedge \varepsilon_2 \wedge \varepsilon_3)$$
$$== (\alpha_1 \ominus \alpha_1)^6\, ((\alpha_1 \wedge \alpha_2) \ominus (\alpha_1 \wedge \alpha_2))^2\, (\alpha_1 \wedge \alpha_2 \wedge \alpha_3 \wedge \alpha_4) \ominus (\alpha_1 \wedge \alpha_2 \wedge \alpha_3)$$

To determine ε_5 in terms of the α_i we can get a little help from `ToInnerProducts`.

```
★P; ToInnerProducts[(α1 ∧ α2 ∧ α3 ∧ α4) ⊖ (α1 ∧ α2 ∧ α3)]
```

$$-(\alpha_1 \wedge \alpha_2 \wedge \alpha_3 \ominus \alpha_2 \wedge \alpha_3 \wedge \alpha_4)\, \alpha_1 + (\alpha_1 \wedge \alpha_2 \wedge \alpha_3 \ominus \alpha_1 \wedge \alpha_3 \wedge \alpha_4)\, \alpha_2 -$$
$$(\alpha_1 \wedge \alpha_2 \wedge \alpha_3 \ominus \alpha_1 \wedge \alpha_2 \wedge \alpha_4)\, \alpha_3 + (\alpha_1 \wedge \alpha_2 \wedge \alpha_3 \ominus \alpha_1 \wedge \alpha_2 \wedge \alpha_3)\, \alpha_4$$

$$\varepsilon_1 \wedge \varepsilon_2 \wedge \varepsilon_3 \wedge \varepsilon_4$$
$$== \left((\alpha_1 \ominus \alpha_1)^3\ (\alpha_1 \wedge \alpha_2 \ominus \alpha_1 \wedge \alpha_2)\ \alpha_1 \wedge \alpha_2 \wedge \alpha_3\right) \wedge (\alpha_1 \ominus \alpha_1)^6\ (\alpha_1 \wedge \alpha_2 \ominus \alpha_1 \wedge \alpha_2)^2\ (\alpha_1 \wedge \alpha_2 \wedge \alpha_3 \ominus \alpha_1 \wedge \alpha_2 \wedge \alpha_3)\ \alpha_4$$
$$== (\alpha_1 \ominus \alpha_1)^9\ (\alpha_1 \wedge \alpha_2 \ominus \alpha_1 \wedge \alpha_2)^3\ (\alpha_1 \wedge \alpha_2 \wedge \alpha_3 \ominus \alpha_1 \wedge \alpha_2 \wedge \alpha_3)\ \alpha_1 \wedge \alpha_2 \wedge \alpha_3 \wedge \alpha_4$$

$$\varepsilon_5 == (\varepsilon_1 \wedge \varepsilon_2 \wedge \varepsilon_3 \wedge \varepsilon_4 \wedge \alpha_5) \ominus (\varepsilon_1 \wedge \varepsilon_2 \wedge \varepsilon_3 \wedge \varepsilon_4)$$
$$== (\alpha_1 \ominus \alpha_1)^{18}\ (\alpha_1 \wedge \alpha_2 \ominus \alpha_1 \wedge \alpha_2)^6\ (\alpha_1 \wedge \alpha_2 \wedge \alpha_3 \ominus \alpha_1 \wedge \alpha_2 \wedge \alpha_3)^2\ (\alpha_1 \wedge \alpha_2 \wedge \alpha_3 \wedge \alpha_4 \wedge \alpha_5) \ominus (\alpha_1 \wedge \alpha_2 \wedge \alpha_3 \wedge \alpha_4)$$

◆ Summary

From these results it can be seen that the magnitudes of the ε_1 are strongly dependent on the magnitudes of the α_i.

$$\varepsilon_1 == \alpha_1$$

$$\varepsilon_2 == (\varepsilon_1 \wedge \alpha_2) \ominus \varepsilon_1 == (\alpha_1 \wedge \alpha_2) \ominus \alpha_1$$

$$\varepsilon_3 == (\varepsilon_1 \wedge \varepsilon_2 \wedge \alpha_3) \ominus (\varepsilon_1 \wedge \varepsilon_2) == (\alpha_1 \ominus \alpha_1)^2\ (\alpha_1 \wedge \alpha_2 \wedge \alpha_3) \ominus (\alpha_1 \wedge \alpha_2)$$

$$\varepsilon_4 == (\varepsilon_1 \wedge \varepsilon_2 \wedge \varepsilon_3 \wedge \alpha_4) \ominus (\varepsilon_1 \wedge \varepsilon_2 \wedge \varepsilon_3)$$
$$== (\alpha_1 \ominus \alpha_1)^6\ ((\alpha_1 \wedge \alpha_2) \ominus (\alpha_1 \wedge \alpha_2))^2\ (\alpha_1 \wedge \alpha_2 \wedge \alpha_3 \wedge \alpha_4) \ominus (\alpha_1 \wedge \alpha_2 \wedge \alpha_3)$$

$$\varepsilon_5 == (\varepsilon_1 \wedge \varepsilon_2 \wedge \varepsilon_3 \wedge \varepsilon_4 \wedge \alpha_5) \ominus (\varepsilon_1 \wedge \varepsilon_2 \wedge \varepsilon_3 \wedge \varepsilon_4)$$
$$== (\alpha_1 \ominus \alpha_1)^{18}\ ((\alpha_1 \wedge \alpha_2) \ominus (\alpha_1 \wedge \alpha_2))^6\ ((\alpha_1 \wedge \alpha_2 \wedge \alpha_3) \ominus (\alpha_1 \wedge \alpha_2 \wedge \alpha_3))^2\ (\alpha_1 \wedge \alpha_2 \wedge \alpha_3 \wedge \alpha_4 \wedge \alpha_5) \ominus (\alpha_1 \wedge \alpha_2 \wedge \alpha_3 \wedge \alpha_4)$$

Depending on the actual orthogonalizing task at hand, it may therefore be advantageous to consider normalizing the α_i (and adjusting `A` accordingly by the product of the magnitudes of the α_i) before orthogonalization using the formula [6.284] for the ε_i.

Example: Orthogonalization of a 3-element

As an example, let us refactor a simple 3-element `A` into the product of three orthogonal 1-elements, where the original factors are given in terms of an orthonormal basis e_i (*GrassmannAlgebra's* default Euclidean basis).

```
★𝓑4;
α1 = 2 e1 + 3 e3 - e4;
α2 = e2 + e3 - e4;
α3 = 5 e2 - 2 e4;
A = α1 ∧ α2 ∧ α3;
```

To do the computations we will need to convert the interior products to scalar products (using `ToScalarProducts`), then convert the scalar products of basis elements to their metric form (using `ToMetricElements`), and do some simplification. Here is a simple piece of code to do this.

```
F[A_⊖B_] := Expand[ToMetricElements[ToScalarProducts[★𝒢[A] ⊖ ★𝒢[B]]]]
```

```
ε1 = 2 e1 + 3 e3 - e4;
ε2 = F[(ε1 ∧ α2) ⊖ ε1]
```

$-8\ e_1 + 14\ e_2 + 2\ e_3 - 10\ e_4$

$\varepsilon_3 = F[(\varepsilon_1 \wedge \varepsilon_2 \wedge \alpha_3) \ominus (\varepsilon_1 \wedge \varepsilon_2)]$

$8624\, e_1 + 7840\, e_2 - 4704\, e_3 + 3136\, e_4$

First, let us check that these elements are mutually orthogonal.

$\{F[\varepsilon_1 \ominus \varepsilon_2], F[\varepsilon_1 \ominus \varepsilon_3], F[\varepsilon_2 \ominus \varepsilon_3]\}$

$\{0, 0, 0\}$

Now normalize them and confirm that they are unit elements.

$$\text{Simplify}\left[\left\{\hat{\varepsilon}_1 = \frac{\varepsilon_1}{\sqrt{F[\varepsilon_1 \ominus \varepsilon_1]}}, \hat{\varepsilon}_2 = \frac{\varepsilon_2}{\sqrt{F[\varepsilon_2 \ominus \varepsilon_2]}}, \hat{\varepsilon}_3 = \frac{\varepsilon_3}{\sqrt{F[\varepsilon_3 \ominus \varepsilon_3]}}\right\}\right]$$

$$\left\{\frac{2\, e_1 + 3\, e_3 - e_4}{\sqrt{14}}, \frac{-4\, e_1 + 7\, e_2 + e_3 - 5\, e_4}{\sqrt{91}}, \frac{11\, e_1 + 10\, e_2 - 6\, e_3 + 4\, e_4}{\sqrt{273}}\right\}$$

$\{F[\hat{\varepsilon}_1 \ominus \hat{\varepsilon}_1], F[\hat{\varepsilon}_2 \ominus \hat{\varepsilon}_2], F[\hat{\varepsilon}_3 \ominus \hat{\varepsilon}_3]\}$

$\{1, 1, 1\}$

The magnitude of A from its original factorization is

$|A| = \sqrt{F[A \ominus A]}$

$2\sqrt{42}$

Hence our new *orthonormal* factorization is

$A == |A|\ \hat{\varepsilon}_1 \wedge \hat{\varepsilon}_2 \wedge \hat{\varepsilon}_3$

$$(2\, e_1 + 3\, e_3 - e_4) \wedge (e_2 + e_3 - e_4) \wedge (5\, e_2 - 2\, e_4) == 2\sqrt{42}\ \frac{2\, e_1 + 3\, e_3 - e_4}{\sqrt{14}} \wedge \frac{-8\, e_1 + 14\, e_2 + 2\, e_3 - 10\, e_4}{2\sqrt{91}} \wedge (8624\, e_1 + 7840\, e_2 - 4704\, e_3 + 3136\, e_4) \Big/ \left(784\sqrt{273}\right)$$

Confirm that this equality holds.

$\star\mathcal{G}\left[A == |A|\ \hat{\varepsilon}_1 \wedge \hat{\varepsilon}_2 \wedge \hat{\varepsilon}_3\right]$

True

The magnitude of the orthogonal factors

Formula [6.199] enables us to simplify the computation of the measure of the new orthogonal 1-elements. Let $E = \varepsilon_1 \wedge \varepsilon_2 \wedge \ldots \wedge \varepsilon_{i-1}$, then [6.284] gives

$$\varepsilon_i == (\varepsilon_1 \wedge \varepsilon_2 \wedge \ldots \wedge \varepsilon_{i-1} \wedge \alpha_i) \ominus (\varepsilon_1 \wedge \varepsilon_2 \wedge \ldots \wedge \varepsilon_{i-1}) == (E \wedge \alpha_i) \ominus E$$

$$\varepsilon_i \ominus \varepsilon_i == ((E \wedge \alpha_i) \ominus E) \ominus ((E \wedge \alpha_i) \ominus E) == (E \ominus E)\ (E \wedge \alpha_i) \ominus (E \wedge \alpha_i)$$

But since the ε_1 to ε_{i-1} are orthogonal we have

$$|E|^2 == |\varepsilon_1 \wedge \varepsilon_2 \wedge \ldots \wedge \varepsilon_{i-1}|^2 == (|\varepsilon_1|\ |\varepsilon_2| \ldots |\varepsilon_{i-1}|)^2 \qquad 6.286$$

$$|\varepsilon_i| == (|\varepsilon_1|\,|\varepsilon_2| \ldots |\varepsilon_{i-1}|)\,|\varepsilon_1 \wedge \varepsilon_2 \wedge \ldots \wedge \varepsilon_{i-1} \wedge \alpha_i| \quad 6.287$$

- Let us now consider the magnitude of the element $E \wedge \alpha_i$. Again from the formula [6.200] we can write

$$(E \wedge \alpha_i) \ominus (E \wedge \alpha_i) == (E \ominus E)\,(\alpha_i \ominus \alpha_i) - (E \ominus \alpha_i) \ominus (E \ominus \alpha_i)$$

This enables us to write the alternative expression for the inner square of ε_i and hence its magnitude.

$$\varepsilon_i \ominus \varepsilon_i == (E \ominus E)\,((E \ominus E)\,(\alpha_i \ominus \alpha_i) - (E \ominus \alpha_i) \ominus (E \ominus \alpha_i))$$

$$|\varepsilon_i|^2 == |E|^2 \left(|E|^2\,|\alpha_i|^2 - |E \ominus \alpha_i|^2\right) \quad 6.288$$

◆ Example

We can use the results of the simple example in the previous section to confirm formulae [6.287] and [6.288] for the magnitude of the orthogonal factors. The first formula is

$$|\varepsilon_i| == (|\varepsilon_1|\,|\varepsilon_2| \ldots |\varepsilon_{i-1}|)\,|\varepsilon_1 \wedge \varepsilon_2 \wedge \ldots \wedge \varepsilon_{i-1} \wedge \alpha_i|$$

```
F[ε3 ⊖ ε3] == F[ε1 ⊖ ε1] F[ε2 ⊖ ε2] F[(ε1 ∧ ε2 ∧ α3) ⊖ (ε1 ∧ ε2 ∧ α3)]
True
```

The second formula is

$$|\varepsilon_i|^2 == |E|^2 \left(|E|^2\,|\alpha_i|^2 - |E \ominus \alpha_i|^2\right)$$

```
F[ε3 ⊖ ε3] == F[(ε1 ∧ ε2) ⊖ (ε1 ∧ ε2)]
  (F[(ε1 ∧ ε2) ⊖ (ε1 ∧ ε2)] F[α3 ⊖ α3] - F[F[(ε1 ∧ ε2) ⊖ α3] ⊖ F[(ε1 ∧ ε2) ⊖ α3]])
True
```

Explicit orthogonalization

So far our discussion has involved processes of obtaining an orthogonalization by a sequence of operations: each new orthogonalized factor depending on previously determined orthogonalized factors. In this section we explore explicit formulae which give orthogonalized factors in terms only of the original (non-orthogonal) ones. The formulae enable us to compute a factor orthogonal to each factor in a non-orthogonal set without having to orthogonalize the factors in that set.

One way of exploring these formulae fairly directly is by using the triangle components [6.195]. In what follows, we find it instructive to begin from first principles.

Consider again our simple *m*-element A, and write it in the equivalent form where each succeeding factor contains an extra scalar multiple of the previous original factors. By the properties of the exterior product, this will not change A. However it does give us the opportunity to adjust its factors to be the orthogonal factors we seek.

$$A == \alpha_1 \wedge \alpha_2 \wedge \ldots \wedge \alpha_m$$
$$== \alpha_1 \wedge (\alpha_2 + a_{2,1}\,\alpha_1) \wedge (\alpha_3 + a_{3,1}\,\alpha_1 + a_{3,2}\,\alpha_2) \wedge \ldots \wedge (\alpha_m + a_{m,1}\,\alpha_1 + \ldots + a_{m,m-1}\,\alpha_{m-1})$$

Beginning with the first two factors, we can choose the scalar $a_{2,1}$ to make the factors orthogonal.

$$\alpha_1 \ominus (\alpha_2 + a_{2,1}\,\alpha_1) == 0 \quad \Rightarrow \quad a_{2,1} == -\frac{\alpha_1 \ominus \alpha_2}{\alpha_1 \ominus \alpha_1}$$

For the first three factors, we can choose the scalars to make all three factors mutually orthogonal. Let

$$f_{13} = \alpha_1 \ominus (\alpha_3 + a_{3,1}\,\alpha_1 + a_{3,2}\,\alpha_2) == \alpha_1 \ominus \alpha_3 + a_{3,1}\,\alpha_1 \ominus \alpha_1 + a_{3,2}\,\alpha_1 \ominus \alpha_2 == 0$$

$$f_{23} = (\alpha_2 + a_{2,1}\,\alpha_1) \ominus (\alpha_3 + a_{3,1}\,\alpha_1 + a_{3,2}\,\alpha_2)$$
$$== \alpha_2 \ominus (\alpha_3 + a_{3,1}\,\alpha_1 + a_{3,2}\,\alpha_2) + a_{2,1}\,f_{13} == 0$$

Hence we have the two equations for $a_{3,1}$ and $a_{3,2}$ in terms of the scalar products $\alpha_i \ominus \alpha_j$.

$$\alpha_1 \ominus (\alpha_3 + a_{3,1}\,\alpha_1 + a_{3,2}\,\alpha_2) == \alpha_1 \ominus \alpha_3 + a_{3,1}\,\alpha_1 \ominus \alpha_1 + a_{3,2}\,\alpha_1 \ominus \alpha_2 == 0$$
$$\alpha_2 \ominus (\alpha_3 + a_{3,1}\,\alpha_1 + a_{3,2}\,\alpha_2) == \alpha_2 \ominus \alpha_3 + a_{3,1}\,\alpha_2 \ominus \alpha_1 + a_{3,2}\,\alpha_2 \ominus \alpha_2 == 0$$

We can write these as the matrix equation

$$\begin{pmatrix} \alpha_1 \ominus \alpha_1 & \alpha_1 \ominus \alpha_2 \\ \alpha_2 \ominus \alpha_1 & \alpha_2 \ominus \alpha_2 \end{pmatrix} \begin{pmatrix} a_{3,1} \\ a_{3,2} \end{pmatrix} == \begin{pmatrix} -\alpha_1 \ominus \alpha_3 \\ -\alpha_2 \ominus \alpha_3 \end{pmatrix}$$

and its solution for the scalar coefficients as

$$\begin{pmatrix} a_{3,1} \\ a_{3,2} \end{pmatrix} == \frac{1}{\Delta} \begin{pmatrix} \alpha_2 \ominus \alpha_2 & -\alpha_1 \ominus \alpha_2 \\ -\alpha_2 \ominus \alpha_1 & \alpha_1 \ominus \alpha_1 \end{pmatrix} \begin{pmatrix} -\alpha_1 \ominus \alpha_3 \\ -\alpha_2 \ominus \alpha_3 \end{pmatrix}$$
$$== \frac{1}{\Delta} \begin{pmatrix} -(\alpha_2 \ominus \alpha_2)\,(\alpha_1 \ominus \alpha_3) + (\alpha_1 \ominus \alpha_2)\,(\alpha_2 \ominus \alpha_3) \\ (\alpha_2 \ominus \alpha_1)\,(\alpha_1 \ominus \alpha_3) - (\alpha_1 \ominus \alpha_1)\,(\alpha_2 \ominus \alpha_3) \end{pmatrix} == \frac{1}{\Delta} \begin{pmatrix} (\alpha_1 \wedge \alpha_2) \ominus (\alpha_2 \wedge \alpha_3) \\ (\alpha_1 \wedge \alpha_2) \ominus (-\alpha_1 \wedge \alpha_3) \end{pmatrix}$$

Since the determinant Δ may also be written as the inner square of $\alpha_1 \wedge \alpha_2$ we now have the required values for the coefficients.

$$a_{3,1} == \frac{(\alpha_1 \wedge \alpha_2) \ominus (\alpha_2 \wedge \alpha_3)}{(\alpha_1 \wedge \alpha_2) \ominus (\alpha_1 \wedge \alpha_2)} \qquad a_{3,2} == \frac{(\alpha_1 \wedge \alpha_2) \ominus (-\alpha_1 \wedge \alpha_3)}{(\alpha_1 \wedge \alpha_2) \ominus (\alpha_1 \wedge \alpha_2)}$$

If we multiply the first equation of the pair we used to solve for these coefficients by α_2 and the second by α_1 we can combine them into a single equation. We can do this because α_1 and α_2 are, by hypothesis, independent.

$$(\alpha_1 \ominus (\alpha_3 + a_{3,1}\,\alpha_1 + a_{3,2}\,\alpha_2))\,\alpha_2 - (\alpha_2 \ominus (\alpha_3 + a_{3,1}\,\alpha_1 + a_{3,2}\,\alpha_2))\,\alpha_1$$
$$== (\alpha_1 \wedge \alpha_2) \ominus (\alpha_3 + a_{3,1}\,\alpha_1 + a_{3,2}\,\alpha_2) == 0$$

This enables us to write our orthogonality relationship in either of the two following forms

$$(\alpha_1 \wedge \alpha_2) \ominus \left(\alpha_3 + \frac{(\alpha_1 \wedge \alpha_2) \ominus (\alpha_2 \wedge \alpha_3)}{(\alpha_1 \wedge \alpha_2) \ominus (\alpha_1 \wedge \alpha_2)}\,\alpha_1 + \frac{(\alpha_1 \wedge \alpha_2) \ominus (-\alpha_1 \wedge \alpha_3)}{(\alpha_1 \wedge \alpha_2) \ominus (\alpha_1 \wedge \alpha_2)}\,\alpha_2\right) == 0$$

$$(\alpha_1 \wedge \alpha_2) \ominus ((\alpha_1 \wedge \alpha_2) \ominus (\alpha_1 \wedge \alpha_2)\,\alpha_3 + (\alpha_1 \wedge \alpha_2) \ominus (\alpha_2 \wedge \alpha_3)\,\alpha_1 + (\alpha_1 \wedge \alpha_2) \ominus (-\alpha_1 \wedge \alpha_3)\,\alpha_2) == 0$$

And this in turn can be written more simply as

$$(\alpha_1 \wedge \alpha_2) \ominus ((\alpha_1 \wedge \alpha_2 \wedge \alpha_3) \ominus (\alpha_1 \wedge \alpha_2)) == 0$$

Thus we have shown that we can orthogonalize A by replacing

$$\alpha_1 \longrightarrow \alpha_1$$

$$\alpha_2 \longrightarrow \frac{(\alpha_1 \wedge \alpha_2) \ominus \alpha_1}{\alpha_1 \ominus \alpha_1}$$

$$\alpha_3 \longrightarrow \frac{(\alpha_1 \wedge \alpha_2 \wedge \alpha_3) \ominus (\alpha_1 \wedge \alpha_2)}{(\alpha_1 \wedge \alpha_2) \ominus (\alpha_1 \wedge \alpha_2)}$$

$$\alpha_i \longrightarrow \frac{(\alpha_1 \wedge \alpha_2 \wedge \ldots \wedge \alpha_{i-1} \wedge \alpha_i) \ominus (\alpha_1 \wedge \alpha_2 \wedge \ldots \wedge \alpha_{i-1})}{(\alpha_1 \wedge \alpha_2 \wedge \ldots \wedge \alpha_{i-1}) \ominus (\alpha_1 \wedge \alpha_2 \wedge \ldots \wedge \alpha_{i-1})} \qquad 6.289$$

Let A_i be equal to $\alpha_1 \wedge \alpha_2 \wedge \ldots \wedge \alpha_i$, and $\hat{A}_i$ be the corresponding unit element, then we can also write this more succinctly as

$$\alpha_i \longrightarrow \left(\hat{A}_{i-1} \wedge \alpha_i\right) \ominus \hat{A}_{i-1} \qquad 6.290$$

- But we have also shown that

$$\alpha_1 \longrightarrow \alpha_1$$

$$\alpha_2 \longrightarrow \alpha_2 - \frac{\alpha_1 \ominus \alpha_2}{\alpha_1 \ominus \alpha_1} \alpha_1$$

$$\alpha_3 \longrightarrow \alpha_3 + \frac{(\alpha_1 \wedge \alpha_2) \ominus (\alpha_2 \wedge \alpha_3)}{(\alpha_1 \wedge \alpha_2) \ominus (\alpha_1 \wedge \alpha_2)} \alpha_1 - \frac{(\alpha_1 \wedge \alpha_2) \ominus (\alpha_1 \wedge \alpha_3)}{(\alpha_1 \wedge \alpha_2) \ominus (\alpha_1 \wedge \alpha_2)} \alpha_2$$

We can generalize this using the special case of the interior common factor theorem [6.54].

$$(\alpha_1 \wedge \alpha_2 \wedge \ldots \wedge \alpha_m) \ominus \underset{m-1}{\beta} == \sum_{i==1}^{m} (-1)^{m-i} \left((\alpha_1 \wedge \ldots \wedge \Box_i \wedge \ldots \wedge \alpha_m) \ominus \underset{m-1}{\beta}\right) \alpha_i$$

In the notation we have used, this becomes

$$(\alpha_1 \wedge \alpha_2 \wedge \ldots \wedge \alpha_{i-1} \wedge \alpha_i) \ominus (\alpha_1 \wedge \alpha_2 \wedge \ldots \wedge \alpha_{i-1})$$

$$== \sum_{j==1}^{i} (-1)^{i-j} \left((\alpha_1 \wedge \alpha_2 \wedge \ldots \wedge \alpha_{i-1}) \ominus \left(\alpha_1 \wedge \ldots \wedge \Box_j \wedge \ldots \wedge \alpha_i\right)\right) \alpha_j$$

If we extract the term where j is equal to i from the summation and then divide both sides by $A_{i-1} \ominus A_{i-1}$ we get that

$$\alpha_i \longrightarrow \alpha_i + \sum_{j==1}^{i-1} (-1)^{i-j} \frac{(\alpha_1 \wedge \alpha_2 \wedge \ldots \wedge \alpha_{i-1}) \ominus \left(\alpha_1 \wedge \ldots \wedge \Box_j \wedge \ldots \wedge \alpha_i\right)}{(\alpha_1 \wedge \alpha_2 \wedge \ldots \wedge \alpha_{i-1}) \ominus (\alpha_1 \wedge \alpha_2 \wedge \ldots \wedge \alpha_{i-1})} \alpha_j \qquad 6.291$$

The coefficients $a_{i,j}$ with which we began may then be given explicitly in terms of the α_i as

$$\mathbf{A} == \alpha_1 \wedge \alpha_2 \wedge \ldots \wedge \alpha_m$$
$$== \alpha_1 \wedge (\alpha_2 + a_{2,1}\,\alpha_1) \wedge \ldots \wedge (\alpha_m + a_{m,1}\,\alpha_1 + \ldots + a_{m,m-1}\,\alpha_{m-1})$$
$$a_{i,j} == (-1)^{i-j}\,\frac{(\alpha_1 \wedge \alpha_2 \wedge \ldots \wedge \alpha_{i-1}) \ominus \left(\alpha_1 \wedge \ldots \wedge \Box_j \wedge \ldots \wedge \alpha_i\right)}{(\alpha_1 \wedge \alpha_2 \wedge \ldots \wedge \alpha_{i-1}) \ominus (\alpha_1 \wedge \alpha_2 \wedge \ldots \wedge \alpha_{i-1})} \qquad 6.292$$

The Zero Interior Sum Theorem

There is an interesting theorem that Grassmann proves in Section 183 of his *Ausdehnungslehre* of 1862. Roughly translated into the terminology and notation of this book, this theorem states:

> If, from a series of *m* 1-elements, one forms the exterior products of any given grade, and conjoins each of them in an interior product with its supplementary combination, then the sum of these products is zero, that is $S_1 \ominus C_1 + S_2 \ominus C_2 + \ldots == 0$. [Here] $S_1, S_2, \ldots$ are the exterior products of the *m* 1-elements $\alpha_1, \alpha_2, \ldots, \alpha_m$ of any given (say the *k*th) grade, and $C_1, C_2, \ldots$ are the supplementary combinations.

By the *supplementary combination* to an exterior product of *k* 1-elements out of a series of *m* 1-elements, Grassmann is referring to the exterior product of the remaining *m–k* 1-elements ordered consistently with the original series. In our terminology the S_i and C_i are respectively the *k*-span and *k*-cospan of the *m*-element $A = \alpha_1 \wedge \alpha_2 \wedge \ldots \wedge \alpha_m$.

In the terminology and notation of this book, an interior sum is a particular case of an m:k-form (discussed in Chapter 2, section 2.12). In this case we are interested in the simplest one in which the product is the interior product.

$$\star\Sigma\left[(\star\mathbb{S}_k[A]) \odot \left(\star\mathbb{S}^k[A]\right)\right] \quad \Longrightarrow \quad \star\Sigma\left[(\star\mathbb{S}_k[A]) \ominus \left(\star\mathbb{S}^k[A]\right)\right]$$

We assume that A is not zero. If this were the case *all* sums of the above form would be zero, for *any* type of linear product - not just for the interior product. This has been shown in section 2.12.

Remark also that we must exclude the case where *k* is equal to *m* as this would result simply in the expression $A \ominus 1$ which is equal to A, and by assumption, not zero.

This theorem is a useful source of identities involving sums of interior products. In particular, in Volume 2, we will see that it has an important role to play in the development of the generalized Grassmann product - a product which incorporates both the interior and exterior products.

As we will see in the following sections, the Zero Interior Sum Theorem turns out to be another result associated with the invariance of an m:k-form to a factorization of the *m*-element A.

◆ Proving the Zero Interior Sum Theorem

Proving the Interior Sum Theorem is now straightforward and relies on two results:

1. That an m:k-form is invariant to factorizations of its simple *m*-element.
2. That a simple *m*-element may be refactorized into mutually orthogonal factors.

For the first result we rely on the discussion of multilinear forms in section 2.12. For the second result we rely on the results developed earlier in this section, equivalent to the Gram-Schmidt process.

Consider an arbitrary simple m-element A.

$$A = \alpha_1 \wedge \alpha_2 \wedge \alpha_3 \wedge \ldots \wedge \alpha_m$$

Any m:k-form which is the sum of interior products of k-span and k-cospan elements is given by the expression:

$$\star\Sigma\left[(\star\mathbb{S}_k[A]) \ominus \left(\star\mathbb{S}^k[A]\right)\right]$$

This expression is invariant to factorizations of A.

By the Gram-Schmidt process A may be expressed as the exterior product of mutually orthogonal factors.

The form based on this refactorization is composed of sums of terms, each of which involves the interior product of totally orthogonal factors, and hence each of which is zero.

Hence the theorem is proved.

Composing interior sums

In composing interior sums we will find it convenient to use bracketing bars |□| around lists to display them as column matrices. We do this by entering

```
|X_List| := MatrixForm[X]
```

(This should not be confounded with the use of bracketing bars to denote the measure of an element, since that usage has not been defined for lists.)

The *list* components of the m:k-form can then be displayed more comprehensibly as columns. For example entering

$$\left|\star\mathbb{S}_2\left[\underset{3}{\alpha}\right]\right| \ominus \left|\star\mathbb{S}^2\left[\underset{3}{\alpha}\right]\right| == \left|\star\mathbb{S}_2\left[\underset{3}{\alpha}\right] \ominus \star\mathbb{S}^2\left[\underset{3}{\alpha}\right]\right|$$

gives the list (*not matrix*) equation

$$\begin{pmatrix} \alpha_1 \wedge \alpha_2 \\ \alpha_1 \wedge \alpha_3 \\ \alpha_2 \wedge \alpha_3 \end{pmatrix} \ominus \begin{pmatrix} \alpha_3 \\ -\alpha_2 \\ \alpha_1 \end{pmatrix} == \begin{pmatrix} \alpha_1 \wedge \alpha_2 \ominus \alpha_3 \\ \alpha_1 \wedge \alpha_3 \ominus -\alpha_2 \\ \alpha_2 \wedge \alpha_3 \ominus \alpha_1 \end{pmatrix}$$

And entering this expression will return True.

To generate the interior m:k-sum we simply sum the elements of this list. We can do this with *GrassmannAlgebra's* inbuilt shorthand ⋆Σ for Total[Flatten[□]].

$$\star P;\ \star\Sigma\left[\begin{pmatrix} \alpha_1 \wedge \alpha_2 \\ \alpha_1 \wedge \alpha_3 \\ \alpha_2 \wedge \alpha_3 \end{pmatrix} \ominus \begin{pmatrix} \alpha_3 \\ -\alpha_2 \\ \alpha_1 \end{pmatrix}\right]$$

$$(\alpha_1 \wedge \alpha_2) \ominus \alpha_3 + (\alpha_1 \wedge \alpha_3) \ominus (-\alpha_2) + (\alpha_2 \wedge \alpha_3) \ominus \alpha_1$$

We can sum the elements of a column and then simplify them by using ⋆𝒮Σ which is shorthand for GrassmannSimplify[Total[Flatten[□]]]. To show that the sum is zero, we will need to convert the interior products to scalar products as well.

$$\textbf{ToScalarProducts}\left[\star\mathcal{S}\Sigma\left[\begin{pmatrix}\alpha_1\wedge\alpha_2\\ \alpha_1\wedge\alpha_3\\ \alpha_2\wedge\alpha_3\end{pmatrix}\ominus\begin{pmatrix}\alpha_3\\ -\alpha_2\\ \alpha_1\end{pmatrix}\right]\right]$$

0

Below we consider some interior sums, their structure, and their reduction to zero. We will only need to consider m:k-form sums where k is less than m, since the theorem does not apply for k equal to m; and sums where k is greater than or equal to m-k, since otherwise every term is trivially zero by definition of the interior product.

◇ Interior sums for a 2-element

$$\mathbf{A_2} = \star\mathbb{S}_1\left[\underset{2}{\alpha}\right]\ominus\star\mathbb{S}^1\left[\underset{2}{\alpha}\right];\ \left|\star\mathbb{S}_1\left[\underset{2}{\alpha}\right]\right|\ominus\left|\star\mathbb{S}^1\left[\underset{2}{\alpha}\right]\right| == |\mathbf{A_2}|$$

$$\begin{pmatrix}\alpha_1\\ \alpha_2\end{pmatrix}\ominus\begin{pmatrix}\alpha_2\\ -\alpha_1\end{pmatrix} == \begin{pmatrix}\alpha_1\ominus\alpha_2\\ \alpha_2\ominus-\alpha_1\end{pmatrix}$$

$$\star\Sigma[\mathbf{A_2}] == \star\mathcal{S}\Sigma[\mathbf{A_2}]$$

$$\alpha_1\ominus\alpha_2+\alpha_2\ominus-\alpha_1 == 0$$

◇ Interior sums for a 3-element

$$\mathbf{A_3} = \star\mathbb{S}_1\left[\underset{3}{\alpha}\right]\ominus\star\mathbb{S}^1\left[\underset{3}{\alpha}\right];\ \left|\star\mathbb{S}_1\left[\underset{3}{\alpha}\right]\right|\ominus\left|\star\mathbb{S}^1\left[\underset{3}{\alpha}\right]\right| == |\mathbf{A_3}|$$

$$\begin{pmatrix}\alpha_1\\ \alpha_2\\ \alpha_3\end{pmatrix}\ominus\begin{pmatrix}\alpha_2\wedge\alpha_3\\ -(\alpha_1\wedge\alpha_3)\\ \alpha_1\wedge\alpha_2\end{pmatrix} == \begin{pmatrix}\alpha_1\ominus\alpha_2\wedge\alpha_3\\ \alpha_2\ominus-(\alpha_1\wedge\alpha_3)\\ \alpha_3\ominus\alpha_1\wedge\alpha_2\end{pmatrix}$$

$$\star\Sigma[\mathbf{A_3}] == \star\mathcal{S}\Sigma[\mathbf{A_3}]$$

$$\alpha_1\ominus(\alpha_2\wedge\alpha_3)+\alpha_2\ominus(-(\alpha_1\wedge\alpha_3))+\alpha_3\ominus(\alpha_1\wedge\alpha_2) == 0$$

◇ Interior sums for a 4-element

- For k equal to 2:

$$\mathbf{A_{42}} = \star\mathbb{S}_2\left[\underset{4}{\alpha}\right]\ominus\star\mathbb{S}^2\left[\underset{4}{\alpha}\right];\ \left|\star\mathbb{S}_2\left[\underset{4}{\alpha}\right]\right|\ominus\left|\star\mathbb{S}^2\left[\underset{4}{\alpha}\right]\right| == |\mathbf{A_{42}}|$$

$$\begin{pmatrix}\alpha_1\wedge\alpha_2\\ \alpha_1\wedge\alpha_3\\ \alpha_1\wedge\alpha_4\\ \alpha_2\wedge\alpha_3\\ \alpha_2\wedge\alpha_4\\ \alpha_3\wedge\alpha_4\end{pmatrix}\ominus\begin{pmatrix}\alpha_3\wedge\alpha_4\\ -(\alpha_2\wedge\alpha_4)\\ \alpha_2\wedge\alpha_3\\ \alpha_1\wedge\alpha_4\\ -(\alpha_1\wedge\alpha_3)\\ \alpha_1\wedge\alpha_2\end{pmatrix} == \begin{pmatrix}\alpha_1\wedge\alpha_2\ominus\alpha_3\wedge\alpha_4\\ \alpha_1\wedge\alpha_3\ominus-(\alpha_2\wedge\alpha_4)\\ \alpha_1\wedge\alpha_4\ominus\alpha_2\wedge\alpha_3\\ \alpha_2\wedge\alpha_3\ominus\alpha_1\wedge\alpha_4\\ \alpha_2\wedge\alpha_4\ominus-(\alpha_1\wedge\alpha_3)\\ \alpha_3\wedge\alpha_4\ominus\alpha_1\wedge\alpha_2\end{pmatrix}$$

$$\star\Sigma[\mathbf{A_{42}}] == \star\mathcal{S}[\textbf{ToScalarProducts}[\star\mathcal{S}\Sigma[\mathbf{A_{42}}]]]$$

$$(\alpha_1\wedge\alpha_2)\ominus(\alpha_3\wedge\alpha_4)+(\alpha_1\wedge\alpha_3)\ominus(-(\alpha_2\wedge\alpha_4))+(\alpha_1\wedge\alpha_4)\ominus(\alpha_2\wedge\alpha_3)+$$
$$(\alpha_2\wedge\alpha_3)\ominus(\alpha_1\wedge\alpha_4)+(\alpha_2\wedge\alpha_4)\ominus(-(\alpha_1\wedge\alpha_3))+(\alpha_3\wedge\alpha_4)\ominus(\alpha_1\wedge\alpha_2) == 0$$

- For k equal to 3:

$$\mathbf{A_{43}} = \star\mathbb{S}_3\left[\underset{4}{\alpha}\right] \ominus \star\mathbb{S}^3\left[\underset{4}{\alpha}\right]; \quad \left|\star\mathbb{S}_3\left[\underset{4}{\alpha}\right]\right| \ominus \left|\star\mathbb{S}^3\left[\underset{4}{\alpha}\right]\right| == |\mathbf{A_{43}}|$$

$$\begin{pmatrix} \alpha_1 \wedge \alpha_2 \wedge \alpha_3 \\ \alpha_1 \wedge \alpha_2 \wedge \alpha_4 \\ \alpha_1 \wedge \alpha_3 \wedge \alpha_4 \\ \alpha_2 \wedge \alpha_3 \wedge \alpha_4 \end{pmatrix} \ominus \begin{pmatrix} \alpha_4 \\ -\alpha_3 \\ \alpha_2 \\ -\alpha_1 \end{pmatrix} == \begin{pmatrix} \alpha_1 \wedge \alpha_2 \wedge \alpha_3 \ominus \alpha_4 \\ \alpha_1 \wedge \alpha_2 \wedge \alpha_4 \ominus -\alpha_3 \\ \alpha_1 \wedge \alpha_3 \wedge \alpha_4 \ominus \alpha_2 \\ \alpha_2 \wedge \alpha_3 \wedge \alpha_4 \ominus -\alpha_1 \end{pmatrix}$$

```
★Σ[A43] == ★S[ToScalarProducts[★SΣ[A43]]]
```

$$(\alpha_1 \wedge \alpha_2 \wedge \alpha_3) \ominus \alpha_4 + (\alpha_1 \wedge \alpha_2 \wedge \alpha_4) \ominus (-\alpha_3) + (\alpha_1 \wedge \alpha_3 \wedge \alpha_4) \ominus \alpha_2 + (\alpha_2 \wedge \alpha_3 \wedge \alpha_4) \ominus (-\alpha_1) == 0$$

◆ **Verifying the theorem**

Here is a verification for m from 2 to 8 and k from 1 to $m-1$ for each value of m.

```
Table[★𝓑m; ★S[ToScalarProducts[★SΣ[★Sk[α m] ⊖ ★S^k[α m]]]], {m, 2, 8},
  {k, 1, m - 1}]
```

```
{{0}, {0, 0}, {0, 0, 0}, {0, 0, 0, 0},
 {0, 0, 0, 0, 0}, {0, 0, 0, 0, 0, 0}, {0, 0, 0, 0, 0, 0, 0}}
```

6.15 Orthogonal Projection

Orthogonal projection of a 1-element

From the triangle formulae discussed in section 6.11 we have that the orthogonal projection $x^{\#}$ of a 1-element x onto a simple m-element $\underset{m}{A}$ is given by

$$\left|\underset{m}{A}\right|^2 x^{\#} == (-1)^{m-1} \underset{m}{A} \ominus \left(\underset{m}{A} \ominus x\right) \qquad 6.293$$

In section 4.12 we explored how to project an element onto a hyperplane in a projective space. We found that the (projective-space) projection x_p of a simple 1-element x in a direction *parallel* to a simple element B onto a simple element A could be given by the formula $(B \wedge x) \vee A$.

To effect an *orthogonal* projection of x with this formula, all we need to do is to make B orthogonal to A. Or, what is equivalent in this case, equate the complement of B with A. If we do this, we can take our projective-space projection formula, and transform it into an orthogonal projection formula as follows (but note that we are using the congruence equality ≡ (equal up to a scalar multiple), so that we do not have to consider signs or other scalar factors at this stage).

$$x_p \equiv (B \wedge x) \vee A \equiv A \vee (B \wedge x) \equiv A \vee \left(\overline{\overline{B \wedge x}}\right) \equiv A \ominus \left(\overline{B \wedge x}\right)$$
$$\equiv A \ominus \left(\overline{B} \ominus x\right) \equiv A \ominus (A \ominus x) \equiv x^{\#}$$

In section 4.12 we also discovered that if we have several 1-elements we could determine the projection of their exterior product by either taking the exterior product of their projections, or projecting their exterior product. In the next section, we will take the projective-space formula relating these two transformations, and, following the process sketched above, turn it in to the analogous orthogonal projection formula.

Orthogonal projection from projective-space projection

The projective-space formula with which we want to begin is [4.20].

$$((\mathbf{P} \wedge \mathbf{Q}_1) \vee \Pi) \wedge \ldots \wedge ((\mathbf{P} \wedge \mathbf{Q}_k) \vee \Pi) == (\mathbf{P} \wedge \Pi) \vee \ldots \vee (\mathbf{P} \wedge \Pi) \vee (\mathbf{P} \wedge \mathbf{Q}_1 \wedge \ldots \wedge \mathbf{Q}_k) \vee \Pi$$

In the more general notation we want to use for discussing orthogonal projection we rewrite this as

$$((\mathbf{B} \wedge \mathbf{x}_1) \vee \mathbf{A}) \wedge \ldots \wedge ((\mathbf{B} \wedge \mathbf{x}_k) \vee \mathbf{A}) == (\mathbf{B} \wedge \mathbf{A}) \vee \ldots \vee (\mathbf{B} \wedge \mathbf{A}) \vee (\mathbf{B} \wedge \mathbf{x}_1 \wedge \ldots \wedge \mathbf{x}_k) \vee \mathbf{A}$$

Here, the x_i are 1-elements, A is the *m*-element onto which the x_i are being projected, and B is the element *parallel* to which the projection is being made. To effect *orthogonal* projection of the x_i onto A, we will put the complement of B equal to A. There are *k*-1 factors of the form $B \wedge A$ on the right hand side.

The steps below are of the same form as those in the outline process of the previous section. But in each step we will keep track of any sign changes. For simplicity we denote these signs symbolically by s_i, and collect and simplify them at the end. A is of grade *m*. B is of grade *n-m*. We start with the projective-space formula above.

$$((\mathbf{B} \wedge \mathbf{x}_1) \vee \mathbf{A}) \wedge \ldots \wedge ((\mathbf{B} \wedge \mathbf{x}_k) \vee \mathbf{A}) == (\mathbf{B} \wedge \mathbf{A}) \vee \ldots \vee (\mathbf{B} \wedge \mathbf{A}) \vee (\mathbf{B} \wedge \mathbf{x}_1 \wedge \ldots \wedge \mathbf{x}_k) \vee \mathbf{A}$$

Exchange the order of the regressive products with A and the position of the $B \wedge A$ (which causes no change of sign since it is an *n*-element).

$$(-1)^{s_1} (\mathbf{A} \vee (\mathbf{B} \wedge \mathbf{x}_1)) \wedge \ldots \wedge (\mathbf{A} \vee (\mathbf{B} \wedge \mathbf{x}_k))$$
$$== (-1)^{s_2} \mathbf{A} \vee (\mathbf{B} \wedge \mathbf{x}_1 \wedge \ldots \wedge \mathbf{x}_k) \vee (\mathbf{B} \wedge \mathbf{A}) \vee \ldots \vee (\mathbf{B} \wedge \mathbf{A})$$

Doubly complement the exterior products involving B using [5.7].

$$(-1)^{s_1+s_3} \left(\mathbf{A} \vee (\overline{\overline{\mathbf{B} \wedge \mathbf{x}_1}})\right) \wedge \ldots \wedge \left(\mathbf{A} \vee (\overline{\overline{\mathbf{B} \wedge \mathbf{x}_k}})\right)$$
$$== (-1)^{s_2+s_4} \mathbf{A} \vee (\overline{\overline{\mathbf{B} \wedge \mathbf{x}_1 \wedge \ldots \wedge \mathbf{x}_k}}) \vee (\overline{\overline{\mathbf{B} \wedge \mathbf{A}}}) \vee \ldots \vee (\overline{\overline{\mathbf{B} \wedge \mathbf{A}}})$$

Apply the complement axiom [5.3] to the first (lower) complement operation. This involves no change of sign.

$$(-1)^{s_1+s_3} \left(\mathbf{A} \vee (\overline{\overline{\mathbf{B}} \vee \overline{\mathbf{x}_1}})\right) \wedge \ldots \wedge \left(\mathbf{A} \vee (\overline{\overline{\mathbf{B}} \wedge \overline{\mathbf{x}_k}})\right)$$
$$== (-1)^{s_2+s_4} \mathbf{A} \vee (\overline{\overline{\mathbf{B}} \vee \overline{\mathbf{x}_1 \wedge \ldots \wedge \mathbf{x}_k}}) \vee (\overline{\overline{\mathbf{B}} \vee \overline{\mathbf{A}}}) \vee \ldots \vee (\overline{\overline{\mathbf{B}} \vee \overline{\mathbf{A}}})$$

Put the complement of B equal to A, rewriting $\overline{B} \vee \overline{\square}$ as $A \ominus \square$.

$$(-1)^{s_1+s_3} \left(\mathbf{A} \vee (\overline{\mathbf{A} \ominus \mathbf{x}_1})\right) \wedge \ldots \wedge \left(\mathbf{A} \vee (\overline{\mathbf{A} \ominus \mathbf{x}_k})\right)$$
$$== (-1)^{s_2+s_4} \mathbf{A} \vee (\overline{\mathbf{A} \ominus (\mathbf{x}_1 \wedge \ldots \wedge \mathbf{x}_k)}) \vee (\overline{\mathbf{A} \ominus \mathbf{A}}) \vee \ldots \vee (\overline{\mathbf{A} \ominus \mathbf{A}})$$

Do a second conversion of regressive products to interior products.

$$(-1)^{s_1+s_3} (\mathbf{A} \ominus (\mathbf{A} \ominus \mathbf{x}_1)) \wedge \ldots \wedge (\mathbf{A} \ominus (\mathbf{A} \ominus \mathbf{x}_k))$$
$$== (-1)^{s_2+s_4} (((\mathbf{A} \ominus (\mathbf{A} \ominus (\mathbf{x}_1 \wedge \ldots \wedge \mathbf{x}_k))) \ominus (\mathbf{A} \ominus \mathbf{A})) \ominus \ldots) \ominus (\mathbf{A} \ominus \mathbf{A})$$

Since $A \ominus A$ is a scalar, we can take it outside the main interior product.

$$(-1)^{s_1+s_3} (\mathbf{A} \ominus (\mathbf{A} \ominus \mathbf{x}_1)) \wedge \ldots \wedge (\mathbf{A} \ominus (\mathbf{A} \ominus \mathbf{x}_k))$$
$$== (-1)^{s_2+s_4} (\mathbf{A} \ominus \mathbf{A})^{k-1} \mathbf{A} \ominus (\mathbf{A} \ominus (\mathbf{x}_1 \wedge \ldots \wedge \mathbf{x}_k))$$

Now simplify the power signs to get the simplest expression with the same parity.

```
s1 = k (n - m) (n - (n - m + 1));
s2 = (n - m) (n - (n - m + k));
s3 = k (n - m + 1) (n - (n - m + 1));
s4 = (n - m + k) (n - (n - m + k));
SimplifyPowerSigns[(-1)^(s1+s2+s3+s4)]
```

1

Thus finally we can write our transformed formula.

$$(A \ominus (A \ominus x_1)) \wedge \ldots \wedge (A \ominus (A \ominus x_k)) == (A \ominus A)^{k-1} A \ominus (A \ominus (x_1 \wedge \ldots \wedge x_k)) \qquad 6.294$$

Orthogonal projection of a *k*-element

As we have seen, the orthogonal projection $x^\#$ of a 1-element x onto a simple *m*-element A is given by

$$|A|^2 x^\# == (-1)^{m-1} A \ominus (A \ominus x)$$

Now take *k* 1-elements x_i. Their orthogonal projections $x_i{}^\#$ onto A are given by

$$x_i{}^\# == (-1)^{m-1} |A|^{-2} A \ominus (A \ominus x_i)$$

The exterior product of these projections is

$$x^\# == x_1{}^\# \wedge \ldots \wedge x_k{}^\# == (-1)^{k (m-1)} |A|^{-2 k} A \ominus (A \ominus x_1) \wedge \ldots \wedge A \ominus (A \ominus x_k)$$

Substituting from formula [6.294] gives

$$x_1{}^\# \wedge \ldots \wedge x_k{}^\# == (-1)^{k (m-1)} |A|^{-2} A \ominus (A \ominus (x_1 \wedge \ldots \wedge x_k)) \qquad 6.295$$

Thus we have shown that the orthogonal projection formula for a simple *k*-element X onto a simple *m*-element ($k \leq m$) is of the same form as that for a 1-element (see [6.195]).

$$X^\# == (-1)^{k (m-1)} \hat{A} \ominus (\hat{A} \ominus X) \qquad X == x_1 \wedge \ldots \wedge x_k \qquad 6.296$$

Orthogonal projection onto a bivector

In this section we develop the formula for the orthogonal projection of a bivector onto a bivector from first principles. This gives us a specific example to compare with the general derivation above.

The orthogonal projection of a vector x onto a simple unit bivector $\hat{B}$ (that is, the component $x^\#$ of the vector *in* the bivector) has been given in section 6.11 as

$$x^\# == - \hat{B} \ominus (\hat{B} \ominus x)$$

Suppose now, we project another vector y onto $\hat{B}$.

$$y^{\#} == -\hat{B}\ominus(\hat{B}\ominus y)$$

If the vector z is a linear combination of x and y, then its projection will be the same linear combination of $x^{\#}$ and $y^{\#}$.

$$z^{\#} == -\hat{B}\ominus(\hat{B}\ominus z) == -\hat{B}\ominus(\hat{B}\ominus(a\,x+b\,y))$$
$$== a\left(-\hat{B}\ominus(\hat{B}\ominus x)\right) + b\left(-\hat{B}\ominus(\hat{B}\ominus y)\right) == a\,x^{\#} + b\,y^{\#}$$

Since each of the vectors $x^{\#}$ and $y^{\#}$ is in $\hat{B}$, the bivector $x^{\#}\wedge y^{\#}$ is also in $\hat{B}$.

$$x^{\#}\wedge y^{\#} == \left(\hat{B}\ominus(\hat{B}\ominus x)\right)\wedge\left(\hat{B}\ominus(\hat{B}\ominus y)\right)$$

For simplicity, suppose $\hat{B}$ is expressed as the exterior product of two *orthonormal* vectors $\hat{u}$ and $\hat{v}$, then their orthogonal projections can be written more simply.

$$x^{\#} == -\hat{B}\ominus(\hat{B}\ominus x) == -(\hat{u}\wedge\hat{v})\ominus((\hat{u}\wedge\hat{v})\ominus x) == (x\ominus\hat{u})\,\hat{u} + (x\ominus\hat{v})\,\hat{v}$$
$$y^{\#} == -\hat{B}\ominus(\hat{B}\ominus y) == -(\hat{u}\wedge\hat{v})\ominus((\hat{u}\wedge\hat{v})\ominus y) == (y\ominus\hat{u})\,\hat{u} + (y\ominus\hat{v})\,\hat{v}$$
$$x^{\#}\wedge y^{\#} == \left((x\ominus\hat{u})\,\hat{u} + (x\ominus\hat{v})\,\hat{v}\right)\wedge\left((y\ominus\hat{u})\,\hat{u} + (y\ominus\hat{v})\,\hat{v}\right)$$
$$== \left((x\ominus\hat{u})\,(y\ominus\hat{v}) - (x\ominus\hat{v})\,(y\ominus\hat{u})\right)\hat{u}\wedge\hat{v}$$
$$== \left((x\wedge y)\ominus(\hat{u}\wedge\hat{v})\right)\hat{u}\wedge\hat{v}$$

$$x^{\#}\wedge y^{\#} == \hat{B}\ominus(\hat{B}\ominus(x\wedge y)) == \hat{B}\,(\hat{B}\ominus(x\wedge y)) \qquad 6.297$$

Hence the orthogonal projection of a simple bivector X onto a unit bivector $\hat{B}$ can now be written in an analogous form to that for the orthogonal projection of a vector onto a unit vector.

$$X^{\#} == \hat{B}\,(\hat{B}\ominus X) \qquad 6.298$$

◆ Example

We can confirm that this result does not depend on assuming the bivector is expressed as the exterior product of two orthonormal vectors by using the more general formulae in which $\hat{u}$ and $\hat{v}$ are not orthogonal and θ is the angle between them.

$$\hat{B} == \mathrm{Csc}[\theta]\,\hat{u}\wedge\hat{v}$$

$$x^{\#} = \left(\frac{x\ominus\hat{u} - (x\ominus\hat{v})\,\mathrm{Cos}[\theta]}{\mathrm{Sin}[\theta]^2}\right)\hat{u} + \left(\frac{x\ominus\hat{v} - (x\ominus\hat{u})\,\mathrm{Cos}[\theta]}{\mathrm{Sin}[\theta]^2}\right)\hat{v};$$

$$y^{\#} = \left(\frac{y\ominus\hat{u} - (y\ominus\hat{v})\,\mathrm{Cos}[\theta]}{\mathrm{Sin}[\theta]^2}\right)\hat{u} + \left(\frac{y\ominus\hat{v} - (y\ominus\hat{u})\,\mathrm{Cos}[\theta]}{\mathrm{Sin}[\theta]^2}\right)\hat{v};$$

```
⋆⋆S[Cos[θ], Sin[θ], Csc[θ], Cot[θ]];
⋆G[ToScalarProducts[x# ∧ y#]] /.
  ((x⊖û) (y⊖v̂) - (x⊖v̂) (y⊖û)) → (x ∧ y) ⊖ (û ∧ v̂)
```

$(x\wedge y\ominus\hat{u}\wedge\hat{v})\,\mathrm{Csc}[\theta]^2\,\hat{u}\wedge\hat{v}$

Replacing $\hat{u}\wedge\hat{v}$ by $\hat{B}\,\mathrm{Sin}[\theta]$ confirms the result [6.298].

Orthogonal projection onto a trivector

As a second specific example, consider the orthogonal projection of a vector x onto a simple unit trivector $\hat{T}$. This can be written in a manner analogous to that for a bivector. Note that because the grade of the trivector is odd, there is no initial negative sign.

$$x^{\#} == \hat{T} \ominus (\hat{T} \ominus x)$$

To see what this looks like when unwrapped, let the unit trivector be defined as the exterior product of three orthonormal vector $\hat{u}$, $\hat{v}$, and $\hat{w}$.

$$\hat{T} = \hat{u} \wedge \hat{v} \wedge \hat{w};$$

Then a projection $x^{\#}$ can be expanded and expressed in terms of $\hat{u}$, $\hat{v}$ and $\hat{w}$ as expected.

```
x# = ToScalarProducts[T̂ ⊖ (T̂ ⊖ x)] /. {ŵ_ ⊖ ŵ_ :→ 1, û_ ⊖ v̂_ :→ 0}
```

$$(x \ominus \hat{u})\, \hat{u} + (x \ominus \hat{v})\, \hat{v} + (x \ominus \hat{w})\, \hat{w}$$

$$x^{\#} == \hat{T} \ominus (\hat{T} \ominus x) == (x \ominus \hat{u})\, \hat{u} + (x \ominus \hat{v})\, \hat{v} + (x \ominus \hat{w})\, \hat{w} \qquad \hat{T} == \hat{u} \wedge \hat{v} \wedge \hat{w} \qquad 6.299$$

Projecting a second vector y gives a result of the same form.

```
y# = ToScalarProducts[T̂ ⊖ (T̂ ⊖ y)] /. {ŵ_ ⊖ ŵ_ :→ 1, û_ ⊖ v̂_ :→ 0}
```

$$(y \ominus \hat{u})\, \hat{u} + (y \ominus \hat{v})\, \hat{v} + (y \ominus \hat{w})\, \hat{w}$$

We wish now to compare the exterior product of their projections with the projection of their exterior product. The exterior product of their projections is

```
★𝒢[x# ∧ y#]
```

$$(-(x \ominus \hat{v})\, (y \ominus \hat{u}) + (x \ominus \hat{u})\, (y \ominus \hat{v}))\, \hat{u} \wedge \hat{v} +$$
$$(-(x \ominus \hat{w})\, (y \ominus \hat{u}) + (x \ominus \hat{u})\, (y \ominus \hat{w}))\, \hat{u} \wedge \hat{w} +$$
$$(-(x \ominus \hat{w})\, (y \ominus \hat{v}) + (x \ominus \hat{v})\, (y \ominus \hat{w}))\, \hat{v} \wedge \hat{w}$$

The projection of their exterior product is identical:

```
X = ToScalarProducts[T̂ ⊖ (T̂ ⊖ (x ∧ y))] /. {ŵ_ ⊖ ŵ_ :→ 1, û_ ⊖ v̂_ :→ 0}
```

$$(-(x \ominus \hat{v})\, (y \ominus \hat{u}) + (x \ominus \hat{u})\, (y \ominus \hat{v}))\, \hat{u} \wedge \hat{v} +$$
$$(-(x \ominus \hat{w})\, (y \ominus \hat{u}) + (x \ominus \hat{u})\, (y \ominus \hat{w}))\, \hat{u} \wedge \hat{w} +$$
$$(-(x \ominus \hat{w})\, (y \ominus \hat{v}) + (x \ominus \hat{v})\, (y \ominus \hat{w}))\, \hat{v} \wedge \hat{w}$$

Hence we have confirmed that the exterior product of the projections $x^{\#}$ and $y^{\#}$ is equal to the projection of their exterior product $x^{\#} \wedge y^{\#}$.

However, it might be observed that the coefficients of the resulting component bivectors are inner products which can be simplified as in the bivector case.

```
X /. ((x ⊖ û_) (y ⊖ v̂_) - (x ⊖ v̂_) (y ⊖ û_)) :→ (x ∧ y) ⊖ (û ∧ v̂)
```

$$(x \wedge y \ominus \hat{u} \wedge \hat{v})\, \hat{u} \wedge \hat{v} + (x \wedge y \ominus \hat{u} \wedge \hat{w})\, \hat{u} \wedge \hat{w} + (x \wedge y \ominus \hat{v} \wedge \hat{w})\, \hat{v} \wedge \hat{w}$$

Thus, for orthonormal $\hat{u}$, $\hat{v}$, and $\hat{w}$ we have that the projection $X^{\#}$ of a simple *bivector* X onto a unit trivector $\hat{T}$ can be written

$$\mathtt{x}^{\#} == \hat{\mathtt{T}}\ominus\left(\hat{\mathtt{T}}\ominus\mathtt{x}\right) == \left(\hat{u}\wedge\hat{v}\wedge\hat{w}\right)\ominus\left(\left(\hat{u}\wedge\hat{v}\wedge\hat{w}\right)\ominus\mathtt{x}\right)$$
$$== \left(\mathtt{x}\ominus\left(\hat{u}\wedge\hat{v}\right)\right)\hat{u}\wedge\hat{v} + \left(\mathtt{x}\ominus\left(\hat{u}\wedge\hat{w}\right)\right)\hat{u}\wedge\hat{w} + \left(\mathtt{x}\ominus\left(\hat{v}\wedge\hat{w}\right)\right)\hat{v}\wedge\hat{w}$$

If $\hat{B}_1$, $\hat{B}_2$ and $\hat{B}_3$ are a set of orthonormal bivectors in $\hat{T}$, we see that the resulting formula for the projection of a bivector onto a trivector is quite analogous to those obtained for the projection of a bivector onto a bivector, and the projection of a vector onto another vector, a bivector and a trivector.

$$\mathtt{x}^{\#} == \hat{\mathtt{T}}\ominus\left(\hat{\mathtt{T}}\ominus\mathtt{x}\right) == \left(\mathtt{x}\ominus\hat{B}_1\right)\hat{B}_1 + \left(\mathtt{x}\ominus\hat{B}_2\right)\hat{B}_2 + \left(\mathtt{x}\ominus\hat{B}_3\right)\hat{B}_3 \qquad 6.300$$
$$\hat{T} == \hat{u}\wedge\hat{v}\wedge\hat{w} \qquad \hat{B}_1 == \hat{u}\wedge\hat{v} \qquad \hat{B}_2 == \hat{u}\wedge\hat{w} \qquad \hat{B}_3 == \hat{v}\wedge\hat{w}$$

The projection of a bound m-vector onto an m-vector

We have shown above that the orthogonal projection formula for a simple *k*-element X onto a simple unit *m*-element $\hat{A}$ ($k \leq m$) is of the same form as that for a 1-element.

$$\mathtt{X}^{\#} == (-1)^{k\,(m-1)}\,\hat{A}\ominus\left(\hat{A}\ominus\mathtt{X}\right) \qquad \mathtt{X} == \mathtt{x}_1\wedge\ldots\wedge\mathtt{x}_k$$

Suppose now that X is a bound *k*-vector.

$$\mathtt{X} == \mathtt{P}\wedge\underset{k}{\mathtt{x}} == (\star\mathbb{O}+\mathtt{p})\wedge\underset{k}{\mathtt{x}}$$

If we substitute X into the formula for its projection we obtain

$$\mathtt{X}^{\#} == (-1)^{(k+1)\,(m-1)}\left(\hat{A}\ominus\left(\hat{A}\ominus\left(\star\mathbb{O}\wedge\underset{k}{\mathtt{x}}\right)\right) + \hat{A}\ominus\left(\hat{A}\ominus\left(\mathtt{p}\wedge\underset{k}{\mathtt{x}}\right)\right)\right)$$

The first term involving the origin is zero since the origin is taken to be orthogonal to all vectors.

$$\hat{A}\ominus\left(\star\mathbb{O}\wedge\underset{k}{\mathtt{x}}\right) == \left(\hat{A}\ominus\star\mathbb{O}\right)\ominus\underset{k}{\mathtt{x}} == 0$$

Hence we conclude that the projection of a bound *k*-vector onto an *m*-vector is a (*k*+1)-vector.

$$\left(\mathtt{P}\wedge\underset{k}{\alpha}\right)^{\#} == (-1)^{(k+1)\,(m-1)}\,\hat{A}\ominus\left(\hat{A}\ominus\left(\mathtt{p}\wedge\underset{k}{\mathtt{x}}\right)\right) \qquad \mathtt{P} == \star\mathbb{O}+\mathtt{p} \qquad 6.301$$

◆ Example: The projection of a point onto a bivector

The projection of a point onto a unit bivector is the projection of its position vector onto the bivector.

$$\mathtt{P}^{\#} == -\hat{B}\ominus\left(\hat{B}\ominus\mathtt{p}\right) \qquad \mathtt{P} == \star\mathbb{O}+\mathtt{p}$$

◆ Example: The projection of a line onto a bivector

The projection of a line (bound vector) onto a unit bivector $\hat{B}$ can be represented by the projection of a bound-vector $\mathtt{P}\wedge\mathtt{x}$ onto the bivector.

$$(\mathtt{P}\wedge\alpha)^{\#} == \left(\hat{B}\ominus(\mathtt{p}\wedge\alpha)\right)\hat{B} \qquad \mathtt{P} == \star\mathbb{O}+\mathtt{p}$$

The result is a scalar multiple of the unit bivector. If the line is orthogonal to $\hat{B}$, the scalar is zero, and the projection is zero. The line can be orthogonal to $\hat{B}$ if either the vector x of the line or the position vector p of any point on it (or a linear combination of x and p) is orthogonal to $\hat{B}$.

6.16 The Closest Approach of Multiplanes

The shortest distance between two multiplanes

In this section we will explore how to determine the shortest distance between two multiplanes, and the points on the multiplanes between which the shortest distance occurs.

Consider an *m*-plane defined by the bound *m*-vector $P_\alpha \wedge \underset{m}{\alpha}$ and the *k*-plane defined by the bound *k*-vector $P_\beta \wedge \underset{k}{\beta}$. We suppose for the moment that the multiplanes are neither parallel nor intersect - special cases which we address further below. Our initial goal is to find the point $P_\alpha^{\perp}$ on $P_\alpha \wedge \underset{m}{\alpha}$ and the point $P_\beta^{\perp}$ on $P_\beta \wedge \underset{k}{\beta}$ such that their (vector) difference $P_\alpha^{\perp} - P_\beta^{\perp}$ is orthogonal to both multiplanes. Later, we will show that the distance $|P_\alpha^{\perp} - P_\beta^{\perp}|$ is also the minimum distance between (any two points on) the hyperplanes.

Let v be the difference between any two points on the multiplanes. Without loss of generality we can take the points to be P_α and P_β. Thus v is equal to $P_\alpha - P_\beta$. A vector $v^{\perp}$ equal to $P_\alpha^{\perp} - P_\beta^{\perp}$ which is orthogonal to both $\underset{m}{\alpha}$ and $\underset{k}{\beta}$, (and hence to $\underset{m}{\alpha} \wedge \underset{k}{\beta}$) can be obtained from the first term of the triangle (or decomposition) formula (see section 6.11).

$$v == P_\alpha - P_\beta \qquad v^{\perp} == P_\alpha^{\perp} - P_\beta^{\perp} \qquad |v^{\perp}| == |P_\alpha^{\perp} - P_\beta^{\perp}| \qquad 6.302$$

The triangle or decomposition formula for a 1-element is given by formula [6.194].

$$x == \left(\underset{m}{\hat{\alpha}} \wedge x\right) \ominus \underset{m}{\hat{\alpha}} + (-1)^{m-1}\, \underset{m}{\hat{\alpha}} \ominus \left(\underset{m}{\hat{\alpha}} \ominus x\right)$$

The component of x orthogonal to $\underset{m}{\hat{\alpha}}$ is given denoted by $x^{\perp}$, and given by the first term of the formula.

$$x^{\perp} == \left(\underset{m}{\hat{\alpha}} \wedge x\right) \ominus \underset{m}{\hat{\alpha}}$$

Let η be equal to $\underset{m}{\alpha} \wedge \underset{k}{\beta}$ and $\hat{\eta}$ be the associated unit (*m*+*k*)-element.

$$\eta == \underset{m}{\alpha} \wedge \underset{k}{\beta} \qquad |\eta| == \left|\underset{m}{\alpha} \wedge \underset{k}{\beta}\right| \qquad \hat{\eta} == \frac{\eta}{|\eta|} \qquad 6.303$$

This enables us to write the formula for $v^{\perp}$ as:

$$v^{\perp} == (\hat{\eta} \wedge v) \ominus \hat{\eta}$$

From our discussion on the triangle formulae, the magnitude of $v^{\perp}$ can also be written

$$|v^{\perp}| == |\hat{\eta} \wedge v|$$

$$v^{\perp} == \left(\hat{\eta} \wedge v\right) \ominus \hat{\eta} \qquad \left|v^{\perp}\right| == \left|\hat{\eta} \wedge v\right| \qquad 6.304$$

Reverting back to our original notation, the formula for the shortest distance between the multiplanes may now be expressed as:

$$\left|P_\alpha{}^{\perp} - P_\beta{}^{\perp}\right| == \frac{\left|\underset{m}{\alpha} \wedge \underset{k}{\beta} \wedge (P_\alpha - P_\beta)\right|}{\left|\underset{m}{\alpha} \wedge \underset{k}{\beta}\right|} \qquad 6.305$$

Decomposition of the parallel component

In the previous section we have determined the component of v orthogonal to both multivectors. In this section we look at the component $v^\#$ of v parallel to both multivectors (see [6.194]).

$$v^\# == (-1)^{m+k-1} \frac{\left(\underset{m}{\alpha} \wedge \underset{k}{\beta}\right) \ominus \left(\left(\underset{m}{\alpha} \wedge \underset{k}{\beta}\right) \ominus v\right)}{\left(\underset{m}{\alpha} \wedge \underset{k}{\beta}\right) \ominus \left(\underset{m}{\alpha} \wedge \underset{k}{\beta}\right)}$$

This component is a vector - a linear combination of the vector factors α_i and β_i of $\underset{m}{\alpha}$ and $\underset{k}{\beta}$. Let us collect the terms involving α_i as $v_\alpha{}^\#$ and those involving β_i as $v_\beta{}^\#$.

$$v_\alpha{}^\# == \sum_{i=1}^{m} a_i \, \alpha_i \qquad v_\beta{}^\# == \sum_{i=1}^{k} b_i \, \beta_i \qquad v^\# == v_\alpha{}^\# + v_\beta{}^\#$$

Here, the a_i and b_i are scalars which we would like to determine. In section 6.11 on the Triangle Formulae we developed the parallel triangle component into two equivalent forms, the first of which showed its expansion into explicit 1-elements.

$$\left(\underset{m}{\alpha} \ominus \underset{m}{\alpha}\right) x^\# == \sum_{i=1}^{m} (-1)^{m-i} \left((\alpha_1 \wedge \cdots \wedge \square_i \wedge \cdots \wedge \alpha_m \wedge x) \ominus \underset{m}{\alpha}\right) \alpha_i$$

This formula can also be written without the sign by moving x into the blank space.

$$\left(\underset{m}{\alpha} \ominus \underset{m}{\alpha}\right) x^\# == \sum_{i=1}^{m} \left((\alpha_1 \wedge \cdots \wedge \alpha_{i-1} \wedge x \wedge \alpha_{i+1} \wedge \cdots \wedge \alpha_m) \ominus \underset{m}{\alpha}\right) \alpha_i$$

Replacing $\underset{m}{\alpha}$ by η and x by v gives

$$(\eta \ominus \eta) \, v^\# == \sum_{i=1}^{m+k} (\eta \ominus (\eta_1 \wedge \cdots \wedge \eta_{i-1} \wedge v \wedge \eta_{i+1} \wedge \cdots \wedge \eta_{m+k})) \, \eta_i$$

$$\eta == \eta_1 \wedge \cdots \wedge \eta_m \wedge \eta_{m+1} \wedge \cdots \wedge \eta_{m+k}$$

This is a sum of vectors which we can break into two sums.

$$(\eta \ominus \eta)\, v^{\#} == \sum_{i==1}^{m} (\eta \ominus (\eta_1 \wedge \cdots \wedge \eta_{i-1} \wedge v \wedge \eta_{i+1} \wedge \cdots \wedge \eta_{m+k}))\, \eta_i +$$

$$\sum_{i==m+1}^{m+k} (\eta \ominus (\eta_1 \wedge \cdots \wedge \eta_{i-1} \wedge v \wedge \eta_{i+1} \wedge \cdots \wedge \eta_{m+k}))\, \eta_i$$

To obtain the components of v (or $v^{\#}$) parallel to each of the multivectors $\underset{m}{\alpha}$ and $\underset{k}{\beta}$ of our multiplanes we replace η in this formula by $\underset{m}{\alpha} \wedge \underset{k}{\beta}$ and split the decomposition into the vector in $\underset{m}{\alpha}$ and the vector in $\underset{k}{\beta}$.

$$\left(\underset{m}{\alpha} \wedge \underset{k}{\beta}\right) \ominus \left(\underset{m}{\alpha} \wedge \underset{k}{\beta}\right) v^{\#} ==$$
$$\sum_{i==1}^{m} \left(\underset{m}{\alpha} \wedge \underset{k}{\beta}\right) \ominus \left((\alpha_1 \wedge \cdots \wedge \alpha_{i-1} \wedge v \wedge \alpha_{i+1} \wedge \cdots \wedge \alpha_m) \wedge \underset{k}{\beta}\right) \alpha_i +$$
$$\sum_{j==1}^{k} \left(\underset{m}{\alpha} \wedge \underset{k}{\beta}\right) \ominus \left(\underset{m}{\alpha} \wedge (\beta_1 \wedge \cdots \wedge \beta_{j-1} \wedge v \wedge \beta_{j+1} \wedge \cdots \wedge \beta_k)\right) \beta_i \qquad 6.306$$

For further discussion of orthogonal decomposition, see section 6.13.

◆ **Example**

Suppose $\underset{m}{\alpha}$ and $\underset{k}{\beta}$ are vectors α and β as might be the case in determining the common normal to two lines. Then $v^{\#}$ can be written as a linear combination of them.

$$v^{\#} == \frac{(\alpha \wedge \beta) \ominus (v \wedge \beta)}{(\alpha \wedge \beta) \ominus (\alpha \wedge \beta)}\, \alpha + \frac{(\alpha \wedge \beta) \ominus (\alpha \wedge v)}{(\alpha \wedge \beta) \ominus (\alpha \wedge \beta)}\, \beta \qquad 6.307$$

Determining the common normal

We now turn our attention to determining the points $P_{\alpha}{}^{\perp}$ and $P_{\beta}{}^{\perp}$ on the common normal to the two multiplanes. Since $P_{\alpha}{}^{\perp}$ and $P_{\beta}{}^{\perp}$ lie in their respective multiplanes, there will be vectors $v_{\alpha}{}^{\#}$ and $v_{\beta}{}^{\#}$ in these multiplanes joining $P_{\alpha}{}^{\perp}$ and $P_{\beta}{}^{\perp}$ to the original points P_{α} and P_{β}. If we make the following definitions for the vector v and its components

$$v == v^{\perp} + v^{\#} == v^{\perp} + v_{\alpha}{}^{\#} + v_{\beta}{}^{\#}$$

then substituting our original definitions for v and $v^{\perp}$ gives

$$P_{\alpha} - P_{\beta} == P_{\alpha}{}^{\perp} - P_{\beta}{}^{\perp} + v_{\alpha}{}^{\#} + v_{\beta}{}^{\#}$$

Hence we define the orthogonal points $P_{\alpha}{}^{\perp}$ and $P_{\beta}{}^{\perp}$ by

$$P_{\alpha}{}^{\perp} == P_{\alpha} - v_{\alpha}{}^{\#} \qquad P_{\beta}{}^{\perp} == P_{\beta} + v_{\beta}{}^{\#} \qquad 6.308$$

Note carefully the negative sign for the definition of $P_{\alpha}{}^{\perp}$. Had we instead defined v as the difference between P_{β} and P_{α}, the negative sign would have appeared in the formula for $P_{\beta}{}^{\perp}$.

Thus we can write

$$\nu_\alpha{}^{\#} == \sum_{i==1}^{m} \frac{\left(\underset{m}{\alpha} \wedge \underset{k}{\beta}\right) \ominus \left((\alpha_1 \wedge \cdots \wedge \alpha_{i-1} \wedge \nu \wedge \alpha_{i+1} \wedge \cdots \wedge \alpha_m) \wedge \underset{k}{\beta}\right)}{\left(\underset{m}{\alpha} \wedge \underset{k}{\beta}\right) \ominus \left(\underset{m}{\alpha} \wedge \underset{k}{\beta}\right)} \alpha_i \qquad 6.309$$

$$\nu_\beta{}^{\#} == \sum_{j==1}^{k} \frac{\left(\underset{m}{\alpha} \wedge \underset{k}{\beta}\right) \ominus \left(\underset{m}{\alpha} \wedge (\beta_1 \wedge \cdots \wedge \beta_{j-1} \wedge \nu \wedge \beta_{j+1} \wedge \cdots \wedge \beta_k)\right)}{\left(\underset{m}{\alpha} \wedge \underset{k}{\beta}\right) \ominus \left(\underset{m}{\alpha} \wedge \underset{k}{\beta}\right)} \beta_j \qquad 6.310$$

The common normal to the two multiplanes $P_\alpha \wedge \underset{m}{\alpha}$ and $P_\beta \wedge \underset{k}{\beta}$ is now simply the exterior product of these two orthogonal points.

$$\mathbb{N}_{\alpha\beta} == P_\alpha{}^{\llcorner} \wedge P_\beta{}^{\llcorner} == \left(P_\alpha - \nu_\alpha{}^{\#}\right) \wedge \left(P_\beta + \nu_\beta{}^{\#}\right) \qquad 6.311$$

◆ **Example**

The points $P_\alpha{}^{\llcorner}$ and $P_\beta{}^{\llcorner}$ of closest approach on two lines defined by $P_\alpha \wedge \alpha$ and $P_\beta \wedge \beta$ are

$$P_\alpha{}^{\llcorner} == P_\alpha - \frac{(\alpha \wedge \beta) \ominus (\nu \wedge \beta)}{(\alpha \wedge \beta) \ominus (\alpha \wedge \beta)} \alpha \qquad 6.312$$

$$P_\beta{}^{\llcorner} == P_\beta + \frac{(\alpha \wedge \beta) \ominus (\alpha \wedge \nu)}{(\alpha \wedge \beta) \ominus (\alpha \wedge \beta)} \beta \qquad 6.313$$

The closest approach of a point to a multiplane

A point is just a 0-plane, so we can use the formulae above to determine the closest approach of a point P to a multiplane $P_\beta \wedge \underset{k}{\beta}$. To do this we set $\underset{m}{\alpha}$ equal to 1 in the formulae above.

$$\nu == P - P_\beta == \nu^{\llcorner} + \nu_\beta{}^{\#}$$

$$P_\beta{}^{\llcorner} == P_\beta + \sum_{j==1}^{k} \frac{\underset{k}{\beta} \ominus \left(\beta_1 \wedge \cdots \wedge \beta_{j-1} \wedge \nu \wedge \beta_{j+1} \wedge \cdots \wedge \beta_k\right)}{\underset{k}{\beta} \ominus \underset{k}{\beta}} \beta_j \qquad 6.314$$

$$\mathbb{N}_\beta == P \wedge P_\beta{}^{\llcorner} \qquad \left|P - P_\beta{}^{\llcorner}\right| == \left|\hat{\underset{k}{\beta}} \wedge (P - P_\beta)\right| \qquad 6.315$$

Note carefully that in this formulation the vector v is *from* the plane *to* the point. The formulation from the point to the plane would incur a change of sign for v.

◆ **Example: The closest approach of two points**

In the trivial case that the multiplanes are just two points P and Q, the corresponding orthogonal points $P^{\perp}$ and $Q^{\perp}$ are also just P and Q, and the formulae reduce to show, as might be expected, that the *normal* to them both is their vector difference.

◆ **Example: The closest approach of a point to a line**

The point $P_\beta{}^{\perp}$ on a line $P_\beta \wedge \beta$ which is of closest approach to a point P is

$$P_\beta{}^{\perp} == P_\beta + \frac{\beta \ominus v}{\beta \ominus \beta}\,\beta == P_\beta + \left(\hat{\beta} \ominus v\right)\hat{\beta} == P_\beta + \left(\hat{\beta} \ominus (P - P_\beta)\right)\hat{\beta}$$

$$P_\beta{}^{\perp} == P_\beta + \left(\hat{\beta} \ominus (P - P_\beta)\right)\hat{\beta}$$

◆ **Example: The closest approach of a point to a plane**

The closest approach of a point P to a plane defined by $P_{\beta\gamma} \wedge \beta \wedge \gamma$ is

$$v == P - P_{\beta\gamma}$$

$$P_{\beta\gamma}{}^{\perp} == P_{\beta\gamma} + \frac{(\beta \wedge \gamma) \ominus (v \wedge \gamma)}{(\beta \wedge \gamma) \ominus (\beta \wedge \gamma)}\,\beta + \frac{(\beta \wedge \gamma) \ominus (\beta \wedge v)}{(\beta \wedge \gamma) \ominus (\beta \wedge \gamma)}\,\gamma$$

The closest approach of a line to a plane

Suppose we have a line defined by $P_\alpha \wedge \alpha$ and a plane defined by $P_{\beta\gamma} \wedge \beta \wedge \gamma$. We suppose them to be non-parallel and in a space large enough so that they do not intersect. Then

$$P_\alpha{}^{\perp} == P_\alpha - \frac{(\alpha \wedge \beta \wedge \gamma) \ominus (v \wedge \beta \wedge \gamma)}{(\alpha \wedge \beta \wedge \gamma) \ominus (\alpha \wedge \beta \wedge \gamma)}\,\alpha \qquad 6.316$$

$$P_{\beta\gamma}{}^{\perp} == P_{\beta\gamma} + \frac{(\alpha \wedge \beta \wedge \gamma) \ominus (\alpha \wedge v \wedge \gamma)}{(\alpha \wedge \beta \wedge \gamma) \ominus (\alpha \wedge \beta \wedge \gamma)}\,\beta + \frac{(\alpha \wedge \beta \wedge \gamma) \ominus (\alpha \wedge \beta \wedge v)}{(\alpha \wedge \beta \wedge \gamma) \ominus (\alpha \wedge \beta \wedge \gamma)}\,\gamma \qquad 6.317$$

The common normal of two lines

Because of its importance in understanding the general results derived above, and also in its application to geometry and mechanics we will pay special attention in this section to the common normal of two lines. The orthogonal points $P_\alpha{}^{\perp}$ and $P_\beta{}^{\perp}$ of closest approach on two lines defined by $P_\alpha \wedge \alpha$ and $P_\beta \wedge \beta$ have already been determined in a previous example in formulae [6.312] and [6.313].

$$P_\alpha{}^{\perp} == P_\alpha - v_\alpha{}^{\#} == P_\alpha - \frac{(\alpha \wedge \beta) \ominus (v \wedge \beta)}{(\alpha \wedge \beta) \ominus (\alpha \wedge \beta)}\,\alpha$$

$$\mathbf{P}_{\beta}{}^{\llcorner} == \mathbf{P}_{\beta} + \mathrm{v}_{\beta}{}^{\#} == \mathbf{P}_{\beta} + \frac{(\alpha \wedge \beta) \ominus (\alpha \wedge \mathrm{v})}{(\alpha \wedge \beta) \ominus (\alpha \wedge \beta)} \beta$$

Consider the vector $\mathrm{v}_{\alpha}{}^{\#}$.

$$\mathrm{v}_{\alpha}{}^{\#} == \frac{(\alpha \wedge \beta) \ominus (\mathrm{v} \wedge \beta)}{(\alpha \wedge \beta) \ominus (\alpha \wedge \beta)} \alpha$$

This vector determines how far we need to move along the line $\mathrm{P}_{\alpha} \wedge \alpha$ in order to reach the point on it closest to $\mathrm{P}_{\beta} \wedge \beta$. Note that it takes in to account all the important information about both lines: both their directions (α and β) and the vector *difference* v of the points P_{α} and P_{β}. It is only the difference that is important since a translation of both lines in space by the same vector added to the points P_{α} and P_{β} does not change the vector.

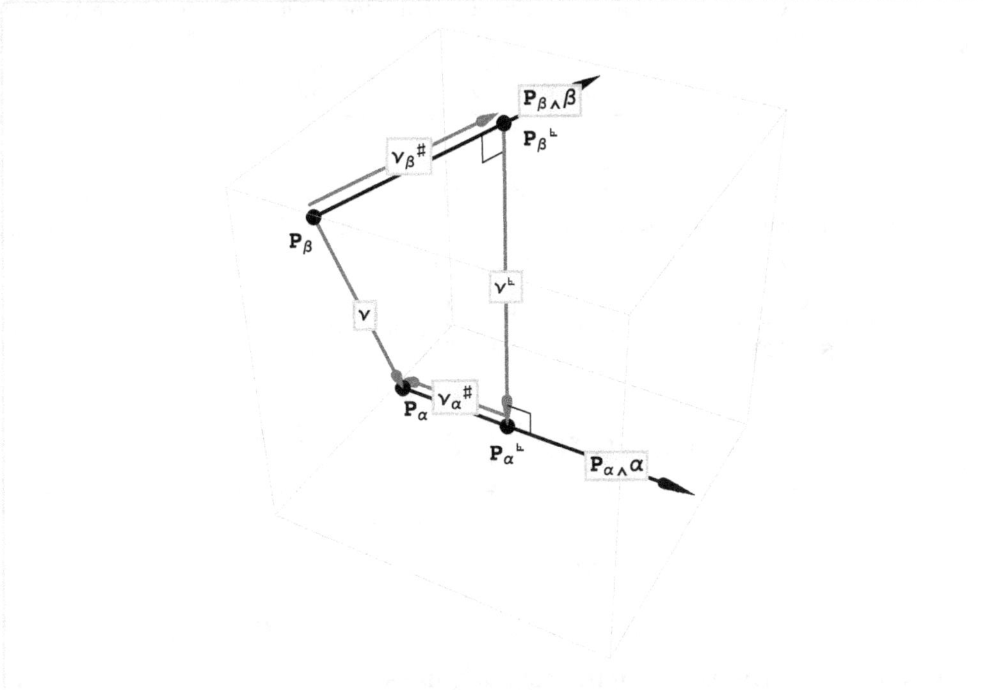

The common normal of two lines

It is useful to see what other forms multiples of this type may take. First, we expand the inner product.

$$\mathrm{v}_{\alpha}{}^{\#} == \frac{(\alpha \wedge \beta) \ominus (\mathrm{v} \wedge \beta)}{(\alpha \wedge \beta) \ominus (\alpha \wedge \beta)} \alpha == \frac{(\alpha \ominus \mathrm{v})\ (\beta \ominus \beta) - (\alpha \ominus \beta)\ (\beta \ominus \mathrm{v})}{(\alpha \ominus \alpha)\ (\beta \ominus \beta) - (\alpha \ominus \beta)\ (\beta \ominus \alpha)} \alpha$$

We can simplify this by forming unit vectors from α and β.

$$\mathrm{v}_{\alpha}{}^{\#} == \frac{\left(\hat{\alpha} \ominus \mathrm{v}\right) - \left(\hat{\alpha} \ominus \hat{\beta}\right) \left(\hat{\beta} \ominus \mathrm{v}\right)}{1 - \left(\hat{\alpha} \ominus \hat{\beta}\right)^{2}} \hat{\alpha} == \frac{\left(\hat{\alpha} - \left(\hat{\alpha} \ominus \hat{\beta}\right) \hat{\beta}\right) \ominus \mathrm{v}}{1 - \left(\hat{\alpha} \ominus \hat{\beta}\right)^{2}} \hat{\alpha}$$

Suppose θ is the angle between the vectors of the lines. Then we can write:

$$\mathrm{v}_{\alpha}{}^{\#} == \frac{\left(\hat{\alpha} - \mathbf{Cos}[\theta]\ \hat{\beta}\right) \ominus \mathrm{v}}{\mathbf{Sin}[\theta]^{2}} \hat{\alpha}$$

Similarly $v_\beta^\#$ may be written:

$$v_\beta^\# == \frac{(\hat{\beta} - \text{Cos}[\theta]\ \hat{\alpha}) \ominus v}{\text{Sin}[\theta]^2}\ \hat{\beta}$$

$$P_\alpha^{\llcorner} == P_\alpha - \frac{(\hat{\alpha} - \text{Cos}[\theta]\ \hat{\beta}) \ominus v}{\text{Sin}[\theta]^2}\ \hat{\alpha} \qquad P_\beta^{\llcorner} == P_\beta + \frac{(\hat{\beta} - \text{Cos}[\theta]\ \hat{\alpha}) \ominus v}{\text{Sin}[\theta]^2}\ \hat{\beta} \qquad 6.318$$

The common normals of two parallel lines

In the case the two lines are parallel, the above formula for the normal vector $v^{\llcorner}$ between them *appears* indeterminate.

$$v^{\llcorner} == P_\alpha^{\llcorner} - P_\beta^{\llcorner} == v - \frac{(\hat{\alpha} - \text{Cos}[\theta]\ \hat{\beta}) \ominus v}{\text{Sin}[\theta]^2}\ \hat{\alpha} - \frac{(\hat{\beta} - \text{Cos}[\theta]\ \hat{\alpha}) \ominus v}{\text{Sin}[\theta]^2}\ \hat{\beta}$$

Interestingly though, if we put $\hat{\alpha}$ equal to $\hat{\beta}$ and take the limit of the quotient involving θ we get the expected normal vector and a normal line which bisects the line joining P_α and P_β - a neat position for an apparently indeterminate normal!

$$\text{Limit}\left[\frac{1 - \text{Cos}[\theta]}{\text{Sin}[\theta]^2}\ (\hat{\alpha} \ominus v)\ \hat{\alpha},\ \theta \to 0\right] == \frac{1}{2}\ (\hat{\alpha} \ominus v)\ \hat{\alpha}$$

$$P_\alpha^{\llcorner} == P_\alpha - \frac{1}{2}\ (\hat{\alpha} \ominus v)\ \hat{\alpha} \qquad P_\beta^{\llcorner} == P_\beta + \frac{1}{2}\ (\hat{\alpha} \ominus v)\ \hat{\alpha} \qquad 6.319$$

$$v^{\llcorner} == P_\alpha^{\llcorner} - P_\beta^{\llcorner} == v - (\hat{\alpha} \ominus v)\ \hat{\alpha} \qquad 6.320$$

The common normals of two intersecting lines

If the two lines intersect, we would expect $P_\alpha^{\llcorner}$ and $P_\beta^{\llcorner}$ to be the same point. In this case we would expect the general formulae to yield no common normal and it is straightforward to confirm that this is indeed the case. That is, *the common normal to a pair of intersecting lines in a space of arbitrary dimension does not exist.*

On the other hand, it is clear that there are usually normals to the plane in which two intersecting lines reside. And we can construct a normal through the point of intersection. An exception is when this plane is the complete space under consideration, since there are no extra dimensions in which the normal can reside.

Given a space of any number of vector dimensions *n*, any vector in the vector-space complement of $\alpha \wedge \beta$ is orthogonal to $\alpha \wedge \beta$. To see this, first form the complement of $\alpha \wedge \beta$. This is a simple element of grade *n*–2, and defines a space of *n*–2 dimensions in which a normal to $\alpha \wedge \beta$ may live. We define a vector v to be in this orthogonal space by asserting that its exterior product with the complement of $\alpha \wedge \beta$ is zero. The complement of this formula asserts that v is orthogonal to $\alpha \wedge \beta$.

$$\overrightarrow{\alpha \wedge \beta \wedge \mathbf{v}} = 0 \quad \Longrightarrow \quad \overrightarrow{\overrightarrow{\alpha \wedge \beta \wedge \mathbf{v}}} = \overrightarrow{\alpha \wedge \beta} \vee \overrightarrow{\mathbf{v}} = (\alpha \wedge \beta) \ominus \mathbf{v} = 0$$

Hence common *normals* are of the form $\mathrm{R} \wedge \mathrm{v}$ where R is the point of intersection of the lines, and v is *any* vector orthogonal to them.

Independent of the dimension of the space, two lines intersect if either of the exterior products $\mathrm{P}_\alpha \wedge \alpha \wedge \mathrm{P}_\beta \wedge \beta$ or $(\mathrm{P}_\alpha - \mathrm{P}_\beta) \wedge \alpha \wedge \beta$ is zero but $\alpha \wedge \beta$ is not zero.

$$\{(\mathbf{P}_\alpha - \mathbf{P}_\beta) \wedge \alpha \wedge \beta = 0,\ \alpha \wedge \beta \neq 0\} \longrightarrow \{\mathbb{N} = \mathbf{R} \wedge \mathbf{v},\ (\alpha \wedge \beta) \ominus \mathbf{v} = 0\} \qquad 6.321$$

The common normal of two lines in three dimensions

All the results of the previous sections apply also to a bound vector 3-space. However, there is one special feature of three vectorial dimensions: there is a unique vector normal to any given two independent vectors. And that this vector may be computed from them using the cross product notation. We shall first review some correspondences between the exterior-interior formalism and the cross-scalar formalism (see section 6.10), and then revisit the results obtained above.

$$\mathbf{u} = \mathbf{w} \times \mathbf{x} = \overrightarrow{\mathbf{w} \wedge \mathbf{x}} \qquad \mathbf{v} = \mathbf{y} \times \mathbf{z} = \overrightarrow{\mathbf{y} \wedge \mathbf{z}}$$

$$\mathbf{u} \ominus \mathbf{v} = (\mathbf{w} \times \mathbf{x}) \ominus (\mathbf{y} \times \mathbf{z}) = \overrightarrow{\overrightarrow{\mathbf{w} \wedge \mathbf{x}} \vee \overrightarrow{\mathbf{y} \wedge \mathbf{z}}} = (\mathbf{w} \wedge \mathbf{x}) \ominus (\mathbf{y} \wedge \mathbf{z})$$

$$\mathbf{u} \ominus \mathbf{v} = (\mathbf{w} \times \mathbf{x}) \ominus \mathbf{v} = \overrightarrow{\mathbf{w} \wedge \mathbf{x}} \vee \overrightarrow{\mathbf{v}} = \overrightarrow{\mathbf{w} \wedge \mathbf{x} \wedge \mathbf{v}} = [\mathbf{w}\ \mathbf{x}\ \mathbf{v}]$$

In the usual three-dimensional vector algebra, the expression denoted by $[\mathrm{w\ x\ v}]$ is called the 'box' product after Grassmann's initial original notation for the exterior product of three vectors (which he assumed to be a scalar). It is clear to see from the complementary trivector form that the vector symbols in the form $(\mathrm{w} \times \mathrm{x}) \ominus \mathrm{v}$ behave anti-symmetrically upon interchange. For example

$$(\mathbf{w} \times \mathbf{x}) \ominus \mathbf{v} = (\mathbf{x} \times \mathbf{v}) \ominus \mathbf{w} = -(\mathbf{w} \times \mathbf{v}) \ominus \mathbf{x}$$

With these correspondences in mind, let us now look at the formulae [6.312] and [6.313] for the points of closest approach of two lines.

$$\mathbf{P}_\alpha{}^{\llcorner} = \mathbf{P}_\alpha - \frac{(\alpha \wedge \beta) \ominus (\mathrm{v} \wedge \beta)}{(\alpha \wedge \beta) \ominus (\alpha \wedge \beta)}\, \alpha \qquad \mathbf{P}_\beta{}^{\llcorner} = \mathbf{P}_\beta + \frac{(\alpha \wedge \beta) \ominus (\alpha \wedge \mathrm{v})}{(\alpha \wedge \beta) \ominus (\alpha \wedge \beta)}\, \beta$$

We can replace the inner products by scalar products of cross products.

$$\mathbf{P}_\alpha{}^{\llcorner} = \mathbf{P}_\alpha - \frac{(\alpha \times \beta) \ominus (\mathrm{v} \times \beta)}{(\alpha \times \beta) \ominus (\alpha \times \beta)}\, \alpha \qquad \mathbf{P}_\beta{}^{\llcorner} = \mathbf{P}_\beta + \frac{(\alpha \times \beta) \ominus (\alpha \times \mathrm{v})}{(\alpha \times \beta) \ominus (\alpha \times \beta)}\, \beta$$

Let the η be the vector $\alpha \times \beta$ normal to α and β. Then these formulae can be written

$$\mathbf{P}_\alpha{}^{\llcorner} = \mathbf{P}_\alpha - \frac{\eta \ominus (\mathrm{v} \times \beta)}{\eta \ominus \eta}\, \alpha \qquad \mathbf{P}_\beta{}^{\llcorner} = \mathbf{P}_\beta + \frac{\eta \ominus (\alpha \times \mathrm{v})}{\eta \ominus \eta}\, \beta$$

Finally, we can write the triple products in the numerators as box products.

$$P_\alpha^{\perp} == P_\alpha - \frac{[\eta \vee \beta]}{\eta \ominus \eta}\alpha \qquad P_\beta^{\perp} == P_\beta + \frac{[\eta\, \alpha \vee]}{\eta \ominus \eta}\beta \tag{6.322}$$

Should the lines intersect in a point R, there are only two common normals $\pm\alpha\times\beta$, so we can write

$$\{(P_\alpha - P_\beta)\wedge\alpha\wedge\beta == 0\ ,\ \alpha\wedge\beta \neq 0\} \longrightarrow \mathbb{N} == \pm R\wedge(\alpha\times\beta) \tag{6.323}$$

The common normal of two lines from first principles

In this final section, by way of consolidating ideas, we revisit the determination of the common normal to two lines from first principles. In order to encourage a fresh reading of the development, we have taken the liberty of using a slightly different notation.

Consider two non-intersecting, non-parallel lines represented by $P\wedge x$ and $Q\wedge y$ with z the vector from P to Q. The line normal to these two lines ($\mathbb{N}$, say) will pass through points $P^\dagger$ and $Q^\dagger$ on them such that the vector from $P^\dagger$ to $Q^\dagger$ ($z^\dagger$, say) is orthogonal to both x and y, or what is equivalent, to their exterior product.

$$(x\wedge y)\ominus z^\dagger == 0$$

Let P, Q, $P^\dagger$, and $Q^\dagger$ have position vectors p, q, $p^\dagger$, and $q^\dagger$ respectively.

$$z == Q - P == q - p \qquad z^\dagger == Q^\dagger - P^\dagger == q^\dagger - p^\dagger$$

Substituting for $z^\dagger$ in the orthogonality relation shows immediately that the components of $P^\dagger$ and $Q^\dagger$ in $x\wedge y$ are equal. And since the origin is orthogonal to all vectors, this also holds for the position vectors $p^\dagger$ and $q^\dagger$.

$$(x\wedge y)\ominus P^\dagger == (x\wedge y)\ominus Q^\dagger == (x\wedge y)\ominus p^\dagger == (x\wedge y)\ominus q^\dagger$$

We suppose that we know the lines $P\wedge x$ and $Q\wedge y$ and the vectors x and y (which we can obtain as the factors of the origin in the expanded expression for the lines). We begin with P and Q as any points on the respective lines and imagine translating P and Q along the lines by vectors $a\,x$ and $b\,y$ respectively until the vector joining them is orthogonal to $x\wedge y$. The translations may then be written in terms of the scalar parameters a and b.

$$P^\dagger == P + a\,x \qquad Q^\dagger == Q + b\,y \qquad z^\dagger == z - a\,x + b\,y$$

To find the minimum distance between the lines we can minimize the expression for the squared magnitude of $z^\dagger$ with respect to the scalar parameters a and b.

$$\partial_a\left(z^\dagger\ominus z^\dagger\right) == -2\,z^\dagger\ominus x == 0$$
$$\partial_b\left(z^\dagger\ominus z^\dagger\right) == 2\,z^\dagger\ominus y == 0$$

These equations lead us to the same equations expressing the orthogonality of $z^\dagger$ to x and y.

$$z^\dagger\ominus x == (z - a\,x + b\,y)\ominus x == z\ominus x - a\,x\ominus x + b\,y\ominus x == 0$$
$$z^\dagger\ominus y == (z - a\,x + b\,y)\ominus y == z\ominus y - a\,x\ominus y + b\,y\ominus y == 0$$

Solving for a and b gives

$$a == \frac{(z\ominus x)\,(y\ominus y) - (z\ominus y)\,(x\ominus y)}{(x\ominus x)\,(y\ominus y) - (x\ominus y)\,(x\ominus y)}$$

$$\mathbf{b} == \frac{(\mathbf{z}\ominus\mathbf{x})\ (\mathbf{x}\ominus\mathbf{y}) - (\mathbf{z}\ominus\mathbf{y})\ (\mathbf{x}\ominus\mathbf{x})}{(\mathbf{x}\ominus\mathbf{x})\ (\mathbf{y}\ominus\mathbf{y}) - (\mathbf{x}\ominus\mathbf{y})\ (\mathbf{x}\ominus\mathbf{y})}$$

These scalar parameters may be expressed a little more compactly in terms of inner products of bivectors.

$$\mathbf{a} == \frac{(\mathbf{z}\wedge\mathbf{y})\ominus(\mathbf{x}\wedge\mathbf{y})}{(\mathbf{x}\wedge\mathbf{y})\ominus(\mathbf{x}\wedge\mathbf{y})} \qquad \mathbf{b} == \frac{(\mathbf{z}\wedge\mathbf{x})\ominus(\mathbf{x}\wedge\mathbf{y})}{(\mathbf{x}\wedge\mathbf{y})\ominus(\mathbf{x}\wedge\mathbf{y})}$$

We can now express the two points $P^\dagger$ and $Q^\dagger$ in terms of the parameters of the original lines.

$$\mathbf{P}^\dagger == \mathbf{P} + \frac{(\mathbf{z}\wedge\mathbf{y})\ominus(\mathbf{x}\wedge\mathbf{y})}{(\mathbf{x}\wedge\mathbf{y})\ominus(\mathbf{x}\wedge\mathbf{y})}\,\mathbf{x} \qquad \mathbf{Q}^\dagger == \mathbf{Q} + \frac{(\mathbf{z}\wedge\mathbf{x})\ominus(\mathbf{x}\wedge\mathbf{y})}{(\mathbf{x}\wedge\mathbf{y})\ominus(\mathbf{x}\wedge\mathbf{y})}\,\mathbf{y}$$
$$\mathbf{z} == \mathbf{Q}-\mathbf{P} == \mathbf{q}-\mathbf{p} \qquad \mathbf{z}^\dagger == \mathbf{Q}^\dagger - \mathbf{P}^\dagger == \mathbf{q}^\dagger - \mathbf{p}^\dagger \qquad 6.324$$

Knowing these points means we can immediately write down the normal line $P^\dagger \wedge Q^\dagger$, the normal vector $z^\dagger$ and the square $z^\dagger \ominus z^\dagger$ of the minimum distance between the lines.

$$\mathbb{N} == \mathbf{P}^\dagger\wedge\mathbf{Q}^\dagger == \mathbf{P}^\dagger\wedge\mathbf{z}^\dagger \qquad 6.325$$

Since x and y are of unspecified magnitude it may be expected that the formulae are also expressible in terms of unit vectors $\hat{x}$ and $\hat{y}$.

$$\mathbf{P}^\dagger == \mathbf{P} + \frac{(\mathbf{z}\wedge\hat{\mathbf{y}})\ominus(\hat{\mathbf{x}}\wedge\hat{\mathbf{y}})}{(\hat{\mathbf{x}}\wedge\hat{\mathbf{y}})\ominus(\hat{\mathbf{x}}\wedge\hat{\mathbf{y}})}\,\hat{\mathbf{x}} \qquad \mathbf{Q}^\dagger == \mathbf{Q} + \frac{(\mathbf{z}\wedge\hat{\mathbf{x}})\ominus(\hat{\mathbf{x}}\wedge\hat{\mathbf{y}})}{(\hat{\mathbf{x}}\wedge\hat{\mathbf{y}})\ominus(\hat{\mathbf{x}}\wedge\hat{\mathbf{y}})}\,\hat{\mathbf{y}} \qquad 6.326$$

The bivector $\hat{x}\wedge\hat{y}$ may be expressed as a scalar multiple of a unit bivector, where θ is the angle between the vectors x and y (and hence the lines), and $\hat{B}$ is a unit bivector in their 2-direction.

$$\hat{\mathbf{x}}\wedge\hat{\mathbf{y}} == \mathbf{Sin}[\theta]\,\hat{\mathbf{B}} \qquad \hat{\mathbf{B}}\ominus\hat{\mathbf{B}} == 1$$

$$\mathbf{P}^\dagger == \mathbf{P} + \frac{(\mathbf{z}\wedge\hat{\mathbf{y}})\ominus\hat{\mathbf{B}}}{\mathbf{Sin}[\theta]}\,\hat{\mathbf{x}} \qquad \mathbf{Q}^\dagger == \mathbf{Q} + \frac{(\mathbf{z}\wedge\hat{\mathbf{x}})\ominus\hat{\mathbf{B}}}{\mathbf{Sin}[\theta]}\,\hat{\mathbf{y}} \qquad 6.327$$

Finally, we can expand the inner products of bivectors following the formula

$$(\mathbf{u}\wedge\mathbf{v})\ominus(\mathbf{x}\wedge\mathbf{y}) == (\mathbf{u}\ominus\mathbf{x})\ (\mathbf{v}\ominus\mathbf{y}) - (\mathbf{u}\ominus\mathbf{y})\ (\mathbf{v}\ominus\mathbf{x})$$

The fractions then become, for example

$$\frac{(\mathbf{z}\wedge\hat{\mathbf{y}})\ominus(\hat{\mathbf{x}}\wedge\hat{\mathbf{y}})}{(\hat{\mathbf{x}}\wedge\hat{\mathbf{y}})\ominus(\hat{\mathbf{x}}\wedge\hat{\mathbf{y}})} == \frac{(\mathbf{z}\ominus\hat{\mathbf{x}}) - (\mathbf{z}\ominus\hat{\mathbf{y}})\ (\hat{\mathbf{x}}\ominus\hat{\mathbf{y}})}{1-(\hat{\mathbf{x}}\ominus\hat{\mathbf{y}})^2}$$

$$== \frac{(\hat{\mathbf{x}} - (\hat{\mathbf{x}}\ominus\hat{\mathbf{y}})\ \hat{\mathbf{y}})\ominus\mathbf{z}}{1-(\hat{\mathbf{x}}\ominus\hat{\mathbf{y}})^2} == \frac{(\hat{\mathbf{x}} - \mathbf{Cos}[\theta]\ \hat{\mathbf{y}})\ominus\mathbf{z}}{\mathbf{Sin}[\theta]^2}$$

$$P^{\dagger} == P + \frac{(\hat{x} - \mathrm{Cos}[\theta]\,\hat{y}) \ominus z}{\mathrm{Sin}[\theta]^2}\,\hat{x} \qquad Q^{\dagger} == Q - \frac{(\hat{y} - \mathrm{Cos}[\theta]\,\hat{x}) \ominus z}{\mathrm{Sin}[\theta]^2}\,\hat{y} \tag{6.328}$$

If we note that the definition of z in these formulae is opposite in sign to the definition of its equivalent $\vee$ in formulae [6.318], it can be seen that both developments give identical results.

It should be remarked that provided the restrictions that the lines be non-intersecting and non-parallel are observed, *these formulae are generally valid in a space of arbitrary dimension*, since nowhere have we invoked the dimension of the space. On the other hand, if the lines do intersect, then expressing z as a linear combination of x and y in [6.324] shows straightforwardly that the formulae give $P^{\dagger}$ equal to $Q^{\dagger}$, and hence both are equal to the point of intersection of the lines.

6.17 Summary

The interior product is the fundamental operation we need for 'measuring' things in the Grassmann algebra. The grade of the interior product $A \ominus B$ of two elements A and B is the difference of their grades and does not depend on the dimension of the space. It may thus be viewed as the 'grade-lowering' operation twin to the 'grade-increasing' operation of the exterior product.

If the grade of A is less than the grade of B, $A \ominus B$ is zero. If the grades of A and B are equal, $A \ominus B$ is of grade zero, and hence a scalar. In this case it may also be called an *inner product*. In the special case that the grades of A and B are both 1, $A \ominus B$ may also be called a *scalar product*. If B is a scalar, $A \ominus B$ is simply that scalar multiple of A.

The interior product was introduced by Grassmann and extensively explored in his works. Because his notation conflated the exterior and regressive products, he was able to denote an interior product with the same symbol (the vertical stroke |) that he used for the complement of an element.

Grassmann's complement operation was fundamentally Euclidean, that is, he did not specifically consider general metric notions in which his basis elements were not mutually orthogonal. Nevertheless his notion of interior product is, with minimal adjustment to account for the difference between scalars and pseudoscalars, eminently transferable to apply to general symmetric metrics.

In this book we have addressed these issues by defining the interior product $A \ominus B$ as the *regressive* product of A with the complement of B. The most immediate formula resulting from this definition is that if B is expressed as an exterior product of factors $B_1 \wedge \ldots \wedge B_k$, $A \ominus B$ may be written as the extended interior product $((A \ominus B_1) \ominus \ldots) \ominus B_k$.

The notion of orthogonality depends on the interior product for its definition. In normal usage we define two elements of grade 1 as orthogonal if their scalar product is zero. If there exists *just one* element of grade 1 in B which is orthogonal to every element of grade 1 in A, then by the extended interior product formula above, $A \ominus B$ is zero, and we say B is *orthogonal* to A.

If the element A is interpreted vectorially, its *measure* is defined as the square root of its inner square $A \ominus A$. If A is a vector, this corresponds to the normally accepted notion of *magnitude*. If A is simple and of grade 2, its measure corresponds to the 'area' of the parallelogram formed from its vector factors. If A is simple and of grade 3, its measure corresponds to the 'volume' of the parallelepiped formed from its vector factors. This measure is invariant to the way in which A is expressed.

To determine the element $A \ominus B$ in terms of the factors of A and B we use the Interior Common Factor Theorem. This theorem gives $A \ominus B$ as a linear combination of all the essentially different exterior products of factors of A of the same grade as $A \ominus B$. A coefficient of a term in this sum is the inner product of B with the term's unused factors of A. This theorem is central to the metric Grassmann algebra.

We can use the Interior Common Factor Theorem to derive formulae for expanding the various types of products of three elements, for example $(A \ominus B) \wedge X$ and $(A \wedge B) \ominus X$. These are called Interior Product Formulae and are useful throughout the algebra.

One of the geometrically enticing results of the usual vector algebra is that the scalar product can be used to 'project' a vector x orthogonally onto another, thus decomposing x into two components, one parallel to the other vector, and one orthogonal to it. The interior product allows us to extend this to the orthogonal projection of a 1-element x onto a simple unit element of any grade, decomposing it into two components, one parallel to the unit element, and one orthogonal to it. Since the two components thus formed are themselves orthogonal, they form a right-angled triangle with x, and the resulting collection of formulae is called the Triangle Formulae.

In this book (Volume 1) we initially called the objects of the Grassmann algebra *elements* to emphasize that they are inherently algebraic without a necessary geometric interpretation. To discuss applications of the algebra to geometry we interpreted the 1-elements geometrically and distinguished two types: points and vectors. A point was defined as the sum of the origin and a vector - its position vector. Previous chapters have explored the power of this interpretation for geometric applications.

In the literature, the point interpretation seemed to have been discarded in favour of the purely vectorial approach, and indeed many geometrical results can be derived without it. Nevertheless, as Chapter 4 has shown, allowing a point-and-vector interpretation to the algebra is conceptually powerful. In this chapter, we have also explored the behaviour of these interpreted elements under the interior product and found that whereas not all formulae give us apparently useful results, there are others which do. For example, following the hybrid metric introduced in Chapter 5, the interior product of a multiplane (represented by a bound unit *m*-vector) with its unit *m*-vector gives (apart from a possible sign) the point on the multiplane closest to the origin.

In Volume 2 we will explore *inter alia* how the exterior and interior products together allow us to define generalized products, and from them, hypercomplex and Clifford algebras.

Coda

Biography

Biographical sources

The principle biographical sources (as of August 2012) to the life of Hermann Grassmann are Michael Crowe's *A History of Vector Analysis* (1967) and more recently Hans-Joachim Petsche's *Hermann Grassmann* (2009). Michael Crowe discusses the lives of the major contributors to the search for 'an algebra of space', and discusses Grassmann's contribution in this wider context. Hans-Joachim Petsche's book, on the other hand, is a definitive biography of Hermann Grassmann. Other major biographical sources are also listed below.

Crowe, Michael J

A History of Vector Analysis
Notre Dame 1967. Republished by Dover 1985.

Petsche, Hans-Joachim

Hermann Graßmann, Biography
Birkhäuser 2009.

Petsche H-J, Kannenberg L, Keßler G, Liskowacka J, editors

Hermann Graßmann, Roots and Traces
Birkhäuser 2009.

Petsche H-J, Lewis A C, Liesen J, and Russ S (eds)

Hermann Graßmann, From Past to Future
Birkhäuser 2011.

Schubring, Gert

Hermann Günther Grassmann (1809–1877)
Kluwer 1996.

A Brief Biography of Hermann Grassmann

Following is a very brief biography of Hermann Grassmann, which relies heavily on Michael Crowe's seminal work *A History of Vector Analysis*. For a more recent and complete evaluation of Grassmann's life, the reader is referred to the excellent works of Hans-Joachim Petsche.

Hermann Günther Grassmann was born in 1809 in Stettin, a town in Pomerania a short distance inland from the Baltic. His father Justus Günther Grassmann taught mathematics and physical sciences at the Stettin Gymnasium. Hermann was no child prodigy. His father used to say that he would be happy if Hermann became a craftsman or a gardener.

In 1827 Grassmann entered the University of Berlin with the intention of studying theology. As his studies progressed he became more and more interested in studying philosophy. At no time whilst a student in Berlin was he known to attend a mathematics lecture.

Grassmann was however only 23 when he made his first important geometric discovery: a method of adding and multiplying lines. This method was to become the foundation of his *Ausdehnungslehre* (extension theory). His own account of this discovery is given below.

Grassmann was interested ultimately in a university post. In order to improve his academic standing in science and mathematics he composed in 1839 a work (over 200 pages) on the study of tides entitled *Theorie der Ebbe und Flut*. This work contained the first presentation of a system of spatial analysis based on vectors including vector addition and subtraction, vector differentiation, and the elements of the linear vector functions, all developed for the first time. His examiners failed to see its importance.

Around Easter of 1842 Grassmann began to turn his full energies to the composition of his first '*Ausdehnungslehre*', and by the autumn of 1843 he had finished writing it. The following is an excerpt from the foreword in which he describes how he made his seminal discovery. The translation is by Lloyd Kannenberg [Grassmann 1844].

> The initial incentive was provided by the consideration of negatives in geometry; I was used to regarding the displacements AB and BA as opposite magnitudes. From this it follows that if A, B, C are points of a straight line, then $AB + BC = AC$ is always true, whether AB and BC are directed similarly or *oppositely*, that is even if C lies between A and B. In the latter case AB and BC are not interpreted merely as lengths, but rather their directions are simultaneously retained as well, according to which they are precisely oppositely oriented. Thus the distinction was drawn between the sum of lengths and the sum of such displacements in which the directions were taken into account. From this there followed the demand to establish this latter concept of a sum, not only for the case that the displacements were similarly or oppositely directed, but also for all other cases. This can most easily be accomplished if the law $AB + BC = AC$ is imposed even when A, B, C do not lie on a single straight line.
>
> Thus the first step was taken toward an analysis that subsequently led to the new branch of mathematics presented here. However, I did not then recognize the rich and fruitful domain I had reached; rather, that result seemed scarcely worthy of note until it was combined with a related idea.
>
> While I was pursuing the concept of product in geometry as it had been established by my father, I concluded that not only rectangles but also parallelograms in general may be regarded as products of an adjacent pair of their sides, provided one again interprets the product, not as the product of their lengths, but as that of the two displacements with their directions taken into account. When I combined this concept of the product with that previously established for the sum, the most striking harmony resulted; thus whether I multiplied the sum (in the sense just given) of two displacements by a third displacement lying in the same plane, or the individual terms by the same displacement and added the products with due regard for their positive and negative values, the same result obtained, and must always obtain.
>
> This harmony did indeed enable me to perceive that a completely new domain had thus been disclosed, one that could lead to important results. Yet this idea remained dormant for some time since the demands of my job led me to other tasks; also, I was initially perplexed by the remarkable result that, although the laws of ordinary multiplication, including the relation of multiplication to addition, remained valid for this new type of product, one could only interchange factors if one simultaneously changed the sign (i.e. changed + into – and vice versa).

As with his earlier work on tides, the importance of this work was ignored. Since few copies were sold, most ended by being used as waste paper by the publisher. The failure to find acceptance for Grassmann's ideas was probably due to two main reasons. The first was that Grassmann was just a simple schoolteacher, and had none of the academic charisma that other contemporaries, like Hamilton for example, had. History seems to suggest that the acceptance of radical discoveries often depends more on the discoverer than the discovery.

The second reason is that Grassmann adopted the format and the approach of the modern mathematician. He introduced and developed his mathematical structure axiomatically and abstractly. The abstract nature of the work, initially devoid of geometric or physical significance, was just too new and formal for the mathematicians of the day and they all seemed to find it too difficult. More fully than any earlier mathematician, Grassmann seems to have understood the associative, commutative and distributive laws; yet still, great mathematicians like Möbius found it unreadable, and Hamilton was led to write to De Morgan that to be able to read Grassmann he 'would have to learn to smoke'.

In the year of publication of the *Ausdehnungslehre* (1844) the Jablonowski Society of Leipzig offered a prize for the creation of a mathematical system fulfilling the idea that Leibniz had sketched in 1679. Grassmann entered with '*Die Geometrische Analyse geknüpft und die von Leibniz Characteristik*', and was awarded the prize. Yet as with the *Ausdehnungslehre* it was subsequently received with almost total silence.

However, in the few years following, three of Grassmann's contemporaries were forced to take notice of his work because of priority questions. In 1845 Saint-Venant published a paper in which he developed vector sums and products essentially identical to those already occurring in Grassmann's earlier works. In 1853 Cauchy published his method of 'algebraic keys' for solving sets of linear equations. Algebraic keys behaved identically to Grassmann's units under the exterior product. In the same year Saint-Venant published an interpretation of the algebraic keys geometrically and in terms of determinants. Since such were fundamental to Grassmann's already published work he wrote a reply for Crelle's Journal in 1855 entitled '*Sur les différentes genres de multiplication*' in which he claimed priority over Cauchy and Saint-Venant and published some new results [Grassmann 1855].

It was not until 1853 that Hamilton heard of the *Ausdehnungslehre*. He set to reading it and soon after wrote to De Morgan.

> I have recently been reading … more than a hundred pages of Grassmann's *Ausdehnungslehre*, with great admiration and interest …. If I could hope to be put in rivalship with Des Cartes on the one hand and with Grassmann on the other, my scientific ambition would be fulfilled.

During the period 1844 to 1862 Grassmann published seventeen scientific papers, including important papers in physics, and a number of mathematics and language textbooks. He edited a political paper for a time and published materials on the evangelization of China. This, on top of a heavy teaching load and the raising of a large family. However, this same period saw only few mathematicians – Hamilton, Cauchy, Möbius, Saint-Venant, Bellavitis and Cremona – having any acquaintance with, or appreciation of, his work.

In 1862 Grassmann published a completely rewritten *Ausdehnungslehre*: *Die Ausdehnungslehre: Vollständing und in strenger Form*. In the foreword Grassmann discussed the poor reception accorded his earlier work and stated that the content of the new book was presented in 'the strongest mathematical form that is actually known to us; that is Euclidean …'. It was a book of theorems and proofs largely unsupported by physical example.

This apparently was a mistake, for the reception accorded this new work was as quiet as that accorded the first, although it contained many new results including a solution to Pfaff's problem. Friedrich Engel (the editor of Grassmann's collected works) comments: 'As in the first *Ausdehnungslehre* so in the second: matters which Grassmann had published in it were later independently rediscovered by others, and only much later was it realized that Grassmann had published them earlier' [Engel in Grassmann 1896].

Thus Grassmann's works were almost totally neglected for forty-five years after his first discovery. In the second half of the 1860s recognition slowly started to dawn on his contemporaries, among them Hankel, Clebsch, Schlegel, Klein, Noth, Sylvester, Clifford and Gibbs. Gibbs discovered Grassmann's works in 1877 (the year of Grassmann's death), and Clifford discovered them in depth about the same time. Both became quite enthusiastic about Grassmann's new mathematics.

Grassmann's activities after 1862 continued to be many and diverse. His contribution to philology rivals his contribution to mathematics. In 1849 he had begun a study of Sanskrit and in 1870 published his *Wörtebuch zum Rig-Veda*, a work of 1784 pages, and his translation of the Rig-Veda, a work of 1123 pages, both still in use today. In addition he published on mathematics, languages, botany, music and religion. In 1876 he was made a member of the American Oriental Society, and received an honorary doctorate from the University of Tübingen.

On 26 September 1877 Hermann Grassmann died, departing from a world only just beginning to recognize the brilliance of the mathematical creations of one of its most outstanding eclectics.

Notation

Symbol types

There are two *types* of symbol used in the book:
symbols which the *GrassmannAlgebra* application understands; and
symbols used in the text for theoretical derivations and textual explanation.

◆ Symbols understood by *GrassmannAlgebra*

Readers interested in using the *GrassmannAlgebra* application are referred to the application's documentation. In brief, *GrassmannAlgebra* allows you to *declare* the symbols you want it to interpret as scalars, vectors, basis elements, or metric elements. It will also recognize any *underscripted* symbol as denoting an element with a grade specified by its underscript. A symbol underscripted with a list of grades will be recognized as denoting a multigraded element. Collectively, these are called *Grassmann symbols*.

For example, the default values for the basis, scalar symbols, vector symbols and metric are:

```
★A
```

```
{{e1, e2, e3}, {a, b, c, d, e, f, g, h},
 {p, q, r, s, t, u, v, w, x, y, z}, {{1, 0, 0}, {0, 1, 0}, {0, 0, 1}}}
```

Any of these symbols or any underscripted symbol is therefore a Grassmann symbol.

$$\star\mathbf{A};\ \mathbf{GrassmannSymbolQ}\left[\left\{\left\{a,\ x,\ e_1,\ \underset{3}{A},\ \underset{\{m,k\}}{\alpha},\ \underset{m}{f_1},\ \underset{m_i}{e}\right\},\ \left\{\alpha,\ j,\ f_1,\ A,\ \underset{m}{e}_{i}\right\}\right\}\right]$$

```
{{True, True, True, True, True, True, True},
 {False, False, False, False, False}}
```

Note carefully that a subscripted symbol with a subscripted underscript is a Grassmann symbol, but a subscripted underscripted symbol is not. The difference can be seen more clearly from their `FullForm`.

$$\left\{\underset{m_i}{e} \longrightarrow \mathbf{FullForm}\left[\underset{m_i}{e}\right],\ \underset{m}{e}_{i} \longrightarrow \mathbf{FullForm}\left[\underset{m}{e}_{i}\right]\right\}$$

$$\left\{\underset{m_i}{e} \longrightarrow \mathtt{Underscript[e,\ Subscript[m,\ i]]},\ \underset{m}{e}_{i} \longrightarrow \mathtt{Subscript[Underscript[e,\ m],\ i]}\right\}$$

◆ Symbols used in the text

Symbols used in the text for theoretical derivations and textual explanation may be of any form, and always include the Grassmann symbols, though they usually conform to one of the following five *forms*.

□	single symbols
$\underset{\square}{\square}$	underscripted symbols
$\underset{\square_\square}{\square}$	underscripted symbols with subscripted underscripts
$\square_\square$	subscripted symbols

$\square_{\square}$ underscripted subscripted symbols
$\underset{\square}{}$

The underscripts may be *lists* of different grades.

If the symbol is a default Grassmann symbol, then its grade may not be specifically stated in the text. Otherwise it will either be clear from the context, or defined.

Symbol forms

◆ Single symbols

a, b, c, d, e, f, g, h	scalars (*GrassmannAlgebra* default)
p, q, r, s, t, u, v, w, x, y, z	1-elements or vectors (*GrassmannAlgebra* default)
P, Q, R, S, T	points or expressions
A, B, C, D, E, F, G, H, U, V, W, X, Y, Z	elements or expressions - may be multigraded
$\mathbb{A}, \mathbb{B}, \mathbb{C}, \mathbb{D}, \mathbb{E}, \mathbb{F}, \mathbb{G}, \mathbb{H}$ $\mathbb{J}, \mathbb{K}, \mathbb{L}, \mathbb{M}, \mathbb{N}, \mathbb{P}, \mathbb{Q}, \mathbb{R}$ $\mathbb{S}, \mathbb{T}, \mathbb{U}, \mathbb{V}, \mathbb{W}, \mathbb{X}, \mathbb{Y}, \mathbb{Z}$	elements representing algebraic, geometric or physical entities: bound vectors, bivectors, lines, planes, screws, forces, moments

Greek symbols may be used to represent elements of varying types depending on context and accepted usage.

◆ Underscripted symbols

An underscript on a symbol denotes its grade. Multigraded elements may be denoted with an underscripted list of grades.

$\underset{0}{a}, a, \underset{0}{x}, \underset{0}{A}, \underset{0}{\mathbb{a}}, \underset{0}{\alpha}, \ldots$	scalars
$\underset{1}{a}, x, \underset{1}{x}, \underset{1}{A}, \underset{1}{\mathbb{a}}, \underset{1}{\alpha}, \ldots$	1-elements or vectors
$\underset{2}{a}, \underset{2}{x}, \underset{2}{A}, \underset{2}{\mathbb{a}}, \underset{2}{\alpha}, \ldots$	2-elements
$\underset{m}{a}, \underset{m}{x}, \underset{m}{A}, \underset{m}{\mathbb{a}}, \underset{m}{\alpha}, \ldots$	*m*-elements
$\underset{\{0,1,2,3\}}{a}, \underset{\{m,k\}}{x}, \ldots$	multigraded elements

An underscripted symbol is considered to represent a different element from the symbol alone.

◆ Underscripted symbols with subscripted underscripts

It may be convenient to create an indexed sequence of underscripts. For example if Z is a multigraded element with grades m_1, m_2, m_3, and m_4, we might find it convenient to compose Z as a sum:

$$Z == \underset{m_1}{Z} + \underset{m_2}{Z} + \underset{m_3}{Z} + \underset{m_4}{Z}$$

◆ Subscripted symbols

- A subscripted symbol is considered to represent a different element from the symbol alone. The subscript indicates a different version of the same type of element.
- A subscripted symbol does not necessarily denote an element of the same grade as the symbol denotes. For example *GrassmannAlgebra*'s composition functions may use the input symbol to compose scalar coefficients or vector factors as appropriate.

$$\star\mathcal{B}_4;\ \{\texttt{ComposeBasisForm[x]},\ \texttt{ComposeSimpleForm}[\underset{3}{z},\ 1]\}$$

$$\{e_1\, x_1 + e_2\, x_2 + e_3\, x_3 + e_4\, x_4,\ z_1 \wedge z_2 \wedge z_3\}$$

◆ Underscripted subscripted symbols

It may be convenient to create an indexed sequence of kernel symbols of the same grade, hence with the same underscript. For example if $\underset{m}{A}$ is a non-simple element of grade m, we might find it convenient to compose it as a sum its simple components

$$\underset{m}{A} == \underset{m}{A_1} + \underset{m}{A_2} + \underset{m}{A_3} + \underset{m}{A_4}$$

Operations

- Symbols for representing mathematical operations

$\wedge$	exterior product operation
$\vee$	regressive product operation
$\ominus$	interior product operation
$\times$	cross product operation
$\underline{\square}$	cobasis element of □
$\overline{\square}$	complement of □
$\vec{\square}$	vector subspace complement of □
$\lvert\square\rvert$	measure of □
$\hat{\square}$	unit element congruent to □
$\langle\square\rangle$	coefficient of the *n*-element □ in the current basis
$\square^{\perp}$	component of □ orthogonal to a given element
$\square^{\#}$	component of □ parallel to a given element
$=$	computationally equal to
$==$	algebraically equal to
$\equiv$	congruent to
$\oplus, \otimes, \odot$	undefined multilinear product operations

Special objects

- Symbols for representing specific objects of the algebra

$\mathbf{1}$ unit scalar

$\underset{n}{\mathbf{1}}$ dual of $\mathbf{1}$ (equal to the unit n-element)

$\overline{\mathbf{1}}$ unit n-element

$\underline{\mathbf{1}}$ n-element in current basis

$|\underline{\mathbf{1}}|$ measure of n-element in current basis

★ five-star symbol used in aliases

★**0** grade of **0**

★𝕆 origin point

★**D** dimension of the currently declared space

★**n** symbolic dimension of a space

★**c** symbolic congruence factor

g_{ij} metric tensor elements

★**g** symbolic determinant of the metric tensor

G matrix of metric tensor elements

★t_i symbolic scalar parameter

Declarations

- Common shorthand aliases for commands to declare the working environment

★**A** declare all default symbols

★$\mathcal{B}_n$ declare a linear or vector space of n dimensions

★$\mathcal{P}_n$ declare a bound n-space comprising an origin point and a vector space of n dimensions

★★**S**[□] declare □ as extra scalar symbols

★★**V**[□] declare □ as extra vector symbols

Spaces

- Symbols for representing the various linear spaces of the algebra

$\underset{0}{\Lambda}$ linear space of scalars or field of scalars

$\underset{1}{\Lambda}$ linear space of 1-elements, underlying linear space

$\underset{m}{\Lambda}$ linear space of m-elements

$\underset{n}{\Lambda}$ linear space of n-elements

Λ linear space of 2^n-elements - the Grassmann algebra

Basis elements

- Symbols for representing types of basis element induced by the default basis of the underlying linear space

e_i basis 1-element (*GrassmannAlgebra* default)

$\underline{e_i}$ cobasis element of e_i

$\underset{m}{e_i}$ basis m-element

$\underline{\underset{m}{e_i}}$ cobasis element of $\underset{m}{e_i}$

Compositors

- Symbols (operations) for composing different types of entity in terms of basis elements. Scalars will be subscripted versions of the symbol entered in place of □.

★$\mathbb{V}_\square$ compose a vector

★$\mathbb{B}_\square$ compose a bivector

★$\mathbb{W}_\square$ compose a simple bivector

★$\mathbb{P}_\square$ compose a point

★$\mathbb{L}_\square$ compose a line element (bound vector)

★$\Pi_\square$ compose a plane element (bound simple bivector)

Commands

- Aliases for commonly used commands

★$\mathcal{E}$ expand an expression (`GrassmannExpand`)

★$\mathcal{S}$ simplify an expression (`GrassmannSimplify`)

★$\mathcal{G}$ expand and simplify an expression (`GrassmannExpandAndSimplify`)

★$\mathcal{F}$ expand and simplify a multilinear form (`ExpandAndSimplifyForm`)

★$\mathcal{C}$ obtains the simplest element congruent to an expression (`CongruenceSimplify`)

★**P** parenthesizes factors in products (`ShowPrecedence`)

★★**P** deactivates `ShowPrecedence` (`HidePrecedence`)

Terminology

◆ Introduction

Grassmann algebra is really quite a simple structure, straightforwardly generated by introducing a product operation onto the elements of a linear space to generate a suite of new linear spaces. If we only wish to describe this algebra and its elements, the terminology can be quite compact and consistent with accepted practice.

However in *applications* of the algebra, for example to geometry and physics, we may want to work in an *interpreted* version of the algebra. This new interpretation may be as simple as viewing one of the basis elements of the underlying linear space as an *origin point*, and the rest as *vectors*, but the interpreted algebraic, and therefore terminological, distinctions are now significantly increased. Whereas the exterior product on an algebraically uninterpreted basis multiplies the basis elements into a single suite of higher grade elements, the exterior product on such an interpreted basis multiplies the basis elements into two suites of interpreted higher grade elements - those containing the origin as a factor, and those which do not.

This is of course more complex, and requires more terminology to make all the distinctions. Deciding on the terminology to use however poses a challenge, because many of the distinctions while new to some, are known to others by different names. For example, E. A. Milne in his *Vectorial Mechanics*, although not mentioning Grassmann or the exterior product uses the term 'line vector' for the entity we model by the exterior product of a point with a vector. Robert Stawell Ball in his *A Treatise on the Theory of Screws*, only briefly mentioning Grassmann, uses the word 'screw' for the entity we model by the sum of a 'line vector' and what is currently known as a bivector.

It is not within my capabilities to extract complete consistency from historically precedent terminology in this highly applicable yet largely unexplored area of mathematics. I have had to adopt some compromises. These compromises have been directed by the desire that the terminology of the book be consistent, historically cognizant, modern, and intuitive. It is certainly not perfect.

One compromise, adopted to avoid tedious extra-locution, is to extend the term *space* in this book to also mean a *Grassmann algebra*. I have done this because it seems to me that when we visualize space, it is not only inhabited by points and vectors, but also by lines, planes, bivectors, trivectors, and other multi-dimensional entities.

◆ Linear space

In this book a *linear space* is defined in the usual way as consisting of an abelian group under addition, a field, and a (scalar) multiplication operation between their elements. A *basis* for a linear space is a set of linearly independent elements which spans the linear space. The *dimension* of a linear space is the number of elements in a basis of the space.

◆ Underlying linear space

The *underlying linear space* of a Grassmann algebra is denoted $\underset{1}{\Lambda}$ and is the linear space which, together with the exterior product operation, generates the algebra. The field of $\underset{1}{\Lambda}$ is denoted $\underset{0}{\Lambda}$, and the dimension of $\underset{1}{\Lambda}$ by n (or, when using *GrassmannAlgebra*, ⋆n). For concreteness in this book, we suppose the field to be that of the real numbers.

◆ Exterior linear space

An *exterior linear space* $\underset{m}{\Lambda}$ is the linear space whose basis consists of all the essentially different exterior products of m basis elements of $\underset{1}{\Lambda}$. The dimension of $\underset{m}{\Lambda}$ is therefore $\binom{n}{m}$. Its elements are called m-elements. The integer m is called the *grade* of $\underset{m}{\Lambda}$ and its elements.

Exterior linear spaces and their elements

Space	Grade	Basis elements	Element
$\underset{0}{\Lambda}$	0	$\{1\}$	0–element
$\underset{1}{\Lambda}$	1	$\{e_1, e_2, \ldots, e_n\}$	1–element
$\underset{2}{\Lambda}$	2	$\{e_1 \wedge e_2, e_1 \wedge e_3, \ldots, e_{n-1} \wedge e_n\}$	2–element
$\underset{m}{\Lambda}$	m	$\{e_1 \wedge e_2 \wedge \ldots \wedge e_m, \ldots, e_{n-m+1} \wedge \ldots \wedge e_{n-1} \wedge e_n\}$	m–element
$\underset{n}{\Lambda}$	n	$\{e_1 \wedge e_2 \wedge \ldots \wedge e_n\}$	n–element

◆ Grassmann algebra

A *Grassmann algebra* Λ is the direct sum $\underset{0}{\Lambda} \oplus \underset{1}{\Lambda} \oplus \underset{2}{\Lambda} \oplus \ldots \oplus \underset{m}{\Lambda} \oplus \ldots \oplus \underset{n}{\Lambda}$ of an underlying linear space $\underset{1}{\Lambda}$, its field $\underset{0}{\Lambda}$, and the exterior linear spaces $\underset{m}{\Lambda}$ $(2 \le m \le n)$. As well as being an algebra, Λ is a linear space of dimension 2^n, whose basis consists of all the basis elements of its exterior linear spaces. Its elements of grade m are called m-elements, and its multigraded elements are called Λ-*elements*. An m-element is called *simple* if it can be expressed as the exterior product of 1-element factors.

◆ Space

In this book, the term *space* is another term for a Grassmann algebra. The term *n-space* is another term for a Grassmann algebra whose underlying linear space is of dimension n. By abuse of terminology we will refer to the dimension and basis of a space as that of its underlying linear space. In *GrassmannAlgebra* you can declare that you want to work in an n-space by entering $\star\mathcal{B}_n$.

The *space of a simple m-element* is the m-space whose underlying linear space consists of all the 1-elements in the m-element. A 1-element is said to be *in* a simple m-element if and only their exterior product is zero. A basis of the underlying linear space of a simple m-element may be taken as any set of m independent factors of the m-element. An n-space and the space of its n-element are identical.

◆ Vector space

A *vector space* is a space in which the elements of the underlying linear space have been *interpreted* as *vectors*. This interpretation views a vector as an (unlocated) *direction* and graphically depicts it as an arrow. An m-element in a vector space is called an *m-vector* or *multivector*. A 2-vector is also called a *bivector*. A 3-vector is also called a *trivector*. A simple m-vector is viewed as a multi-dimensional direction, or *m-direction*.

An element of a vector space may also be called a *geometric entity* (or simply, *entity*) to emphasize that it has a geometric interpretation. An m-vector is an entity of grade m. Vectors are 1-entities, bivectors are 2-entities.

A *vector n-space* is a vector space whose underlying linear space is of dimension n. A vector 3-space is thus richer than the usual 3-dimensional vector algebra, since it also contains bivectors and trivectors, as well as the exterior product operation.

A vector n–space and its entities

Space	Grade	Basis elements	Element
$\underset{0}{\Lambda}$	0	$\{1\}$	scalar
$\underset{1}{\Lambda}$	1	$\{e_1, e_2, \ldots, e_n\}$	vector
$\underset{2}{\Lambda}$	2	$\{e_1 \wedge e_2, e_1 \wedge e_3, \ldots, e_{n-1} \wedge e_n\}$	bivector
$\underset{m}{\Lambda}$	m	$\{e_1 \wedge e_2 \wedge \ldots \wedge e_m, \ldots, e_{n-m+1} \wedge \ldots \wedge e_{n-1} \wedge e_n\}$	m–vector
$\underset{n}{\Lambda}$	n	$\{e_1 \wedge e_2 \wedge \ldots \wedge e_n\}$	n–vector

The *space of a simple m-vector* is the m-space whose underlying linear space consists of all the vectors in the m-vector.

◆ Bound space

A *bound vector space*, or simply a *bound space*, is a vector space to whose basis has been added an *origin*. The origin is an element with the geometric interpretation of a *point*. A *bound vector n-space* is a vector n-space to whose underlying basis has been added an origin. The symbol n refers to *the number of vectors in the basis of the bound space*, thus making the dimension of a bound n-space n+1. In *GrassmannAlgebra* you can declare that you want to work in a bound n-space by entering $\star\mathcal{P}_n$. You can determine the dimension of a space by entering ★D, which in this case would return n+1. Thus the underlying linear space of a bound n-space is of dimension n+1.

An element of a bound space may also be called a *geometric entity* (or simply, *entity*) to emphasize it has a geometric interpretation. An m-entity is an entity of grade m. Vectors and points are 1-entities. Bivectors and bound vectors are 2-entities.

A bound n–space and its entities

Space	Grade	Basis elements	Free Entity	Bound Entity
$\underset{0}{\Lambda}$	0	$\{1\}$	scalar	
$\underset{1}{\Lambda}$	1	$\{\star O, e_1, e_2, \ldots, e_n\}$	vector	weighted point
$\underset{2}{\Lambda}$	2	$\{\star O \wedge e_1, \star O \wedge e_2, \ldots, e_{n-1} \wedge e_n\}$	bivector	bound vector
$\underset{m}{\Lambda}$	m	$\{\star O \wedge e_1 \wedge e_2 \wedge \ldots \wedge e_{m-1}, \ldots, e_{n-m+1} \wedge \ldots \wedge e_{n-1} \wedge e_n\}$	m–vector	bound (m–1)–vector
$\underset{n}{\Lambda}$	n	$\{\star O \wedge e_1 \wedge e_2 \wedge \ldots \wedge e_{n-1}, \ldots, e_2 \wedge \ldots \wedge e_2 \wedge e_n\}$	n–vector	bound (n–1)–vector
$\underset{n+1}{\Lambda}$	n+1	$\{\star O \wedge e_1 \wedge e_2 \wedge \ldots \wedge e_n\}$		bound n–vector

A bound n-space is thus richer than a vector n-space, since as well as containing vectorial (unlocated) entities (vectors, bivectors, ...), it also contains bound (located) entities (points, bound vectors, ...) and sums of these. In contradistinction to the term bound space, a vector space may sometimes be called a *free space*. A bound 3-space is a closer algebraic model to physical 3-space than a vector 3-space, even though its underlying linear space is of four dimensions.

The *space of a bound simple m-vector* is the (m+1)-space whose underlying linear space consists of all the points and vectors in the bound simple m-vector.

◆ Geometric objects

The *geometric object of a bound simple entity* is the *set of all points in the entity*, that is, all points whose exterior product with the bound simple entity is zero.

The geometric object of a bound scalar or weighted point is the point itself. The geometric object of a bound vector is a line (an infinite set of points). The geometric object of a bound simple bivector is a plane (a doubly infinite set of points).

Since the properties of geometric objects are well modelled algebraically by their corresponding entities, we find it convenient to *compute* with geometric objects by computing with their corresponding bound simple entities; thus for example computing the intersection of two lines in the plane by computing the regressive product of two bound vectors.

Geometric objects and their corresponding entities

Space	Grade	Geometric Object	Entity
$\underset{1}{\Lambda}$	1	point	bound scalar
$\underset{2}{\Lambda}$	2	line	bound vector
$\underset{3}{\Lambda}$	3	plane	bound simple bivector
$\underset{m}{\Lambda}$	m	$(m-1)$–plane	bound simple $(m-1)$–vector
$\underset{n}{\Lambda}$	n	hyperplane	bound $(n-1)$–vector
$\underset{n+1}{\Lambda}$	$n+1$	n–plane	bound n–vector

By abuse of terminology where the context is clear, we may refer to bound simple entities by the names of their corresponding geometric objects, for example referring to a bound vector as a line, or a bound bivector as a plane.

◆ Congruence, orientation and sense

Two elements are said to be *congruent* if one is a scalar multiple (not zero) of the other.

The scalar multiple associated with two congruent elements is called the *congruence factor*.

Orientation is a relative concept. Two congruent elements are said to be of *opposite orientation* if their congruence factor is negative.

The term *sense* is equivalent to the term orientation, but is commonly used only for vectors.

Two congruent vectors are of opposite sense if their congruence factor is negative.

◆ Points, vectors and carriers

The *origin of a bound space* is the basis element designated as the origin.

A *vector in a bound space* is a 1-element that does not involve the origin.

A *point* is the sum of the origin with a vector. This vector is called the *position vector of the point*.

A *weighted point* is a scalar multiple of a point. The scalar multiple is called the *weight of the point*.

The *sum of a point and a vector* is another point. The vector may be said to *carry* the first point into the second.

The sum of a bound simple m-vector and a simple (m+1)-vector containing the m-vector is another bound simple m-vector. The simple (m+1)-vector may be said to *carry* the first bound simple m-vector into the second.

◆ Position and direction

Congruent weighted points are said to define the same *position*.

Congruent vectors are said to define the same *direction*.

Congruent simple m-vectors are said to define the same *m-direction*.

Congruent bound simple m-vectors are said to define the same *position and direction*.

A bound simple m-vector that has been carried to a new bound simple m-vector by the addition of an (m+1)-vector may be said to have been *carried to a new position*. The m-direction of the bound simple m-vector is unchanged.

The geometric object of an entity may be said to have the *position and direction of its entity*.

The notions of position and direction are *geometric interpretations* on the algebraic elements.

Glossary

Ausdehnungslehre

The term *Ausdehnungslehre* is variously translated as 'extension theory', 'theory of extension', or 'calculus of extension'. Refers to Grassmann's original work and other early work in the same notational and conceptual tradition.

Basis

A *basis* of a linear space is any set of elements which spans the space. A set of elements spans a linear space if any member of the space can be expressed as a linear combination of the elements.

Bivector

A *bivector* $\mathbb{B}$ is a sum of simple bivectors. A *simple bivector* B is the exterior product of two vectors: $B = x \wedge y$.

Bound vector

A *bound vector* is the exterior product of a point P and a vector x: $P \wedge x$. It may also always be expressed as the exterior product of two points.

Bound vector space

A *bound vector space* is a Grassmann algebra whose underlying linear space has a basis whose elements are interpreted as an origin point $\star\mathbb{O}$ and vectors.

Bound vector *n*-space

A *bound vector n-space* is a Grassmann algebra whose underlying linear space is a vector space of *n* dimensions whose basis has been augmented by an origin point $\star\mathbb{O}$.

The term 'bound vector *n*-space' should be read as bound 'vector *n*-space' rather than 'bound vector' *n*-space. Its dimension is $n+1$.

Bound bivector

A *bound bivector* is the exterior product of a point P and a bivector $\mathbb{B}$: $P \wedge \mathbb{B}$.

Bound simple bivector

A *bound simple bivector* is the exterior product of a point P and a simple bivector B: $P \wedge B$. It may also always be expressed as the exterior product of two points and a vector, or the exterior product of three points.

Cobasis element

The *cobasis element* $\underline{e}$ of a basis element e of a space of dimension n is the $(n-1)$-element formed from the basis elements such that $e \wedge \underline{e}$ is equal to the basis n-element.

Cobasis

The *cobasis* of a space is the list of cobasis elements in order corresponding to the order of the basis elements.

Cofactor

The *cofactor* of a minor M of a matrix A is the signed determinant formed from the rows and columns of A which are not in M.

The sign may be determined from $(-1)^{\Sigma (r_i + c_i)}$, where r_i and c_i are the row and column numbers.

Cometric

The *cometric* of the metric G_m on $\underset{m}{\Lambda}$ is the metric G_{n-m} on $\underset{n-m}{\Lambda}$.

Common factor

The *common factor* of two simple elements is the element of maximal grade they have in common. Common factors can only be determined up to congruence.

Complement

The *complement* of an *m*-element A is the special (*n*–*m*)-element $\overline{A}$ determinable from a formula involving the metric on the space.

Complete span

The *complete span* of a simple *m*-element A is the ordered list of the *k*-spans of A, with *k* ranging from 0 to *m*.

Congruence

Two elements are *congruent* if one is a scalar multiple of the other.

Cospan

A *cospan* of a simple *m*-element is a list of elements determined from a span of the element in the same way that a cobasis may be determined from a basis.

Cross product

The (generalized) *cross product* of two elements is the complement of their exterior product.

Determinant of the metric tensor

The determinant of the metric tensor is equal to the square of the measure of the basis *n*-element.

Dimension of a linear space

The *dimension of a linear space* is the maximum number of independent elements in it.

Dimension of an exterior linear space

The *dimension of an exterior linear space* $\underset{m}{\Lambda}$ is $\binom{n}{m}$ where *n* is the dimension of the underlying linear space $\underset{1}{\Lambda}$.

The dimension $\binom{n}{m}$ is equal to the number of combinations of *n* elements taken *m* at a time.

Dimension of a Grassmann algebra

The *dimension of a Grassmann algebra* is the sum of the dimensions of its component exterior linear spaces.
The dimension of a Grassmann algebra is then given by 2^n, where *n* is the dimension of the underlying linear space.

Direction

A *direction* is the space of a vector and is therefore the set of all vectors parallel to a given vector.

Displacement

A *displacement* is a physical interpretation of a vector. It may also be viewed as the difference of two points.

Duality

A formula B is the *dual* of a formula A involving only exterior and regressive products if B can be obtained from A by the duality transformation algorithm. The dual of the dual of A is A. This duality is based on the axioms for the regressive product being duals of the axioms for the exterior product.

Entity

The term *entity* is used for an element to emphasize that it is being interpreted geometrically or physically.

Essentially different

The term *essentially different* is equivalent to the term *linearly independent*, but is usually applied to *m*-elements rather than 1-elements. Two elements are essentially different if they are not congruent.

Exterior linear space

An *exterior linear space* of grade *m*, denoted $\underset{m}{\Lambda}$, is the linear space generated by *m*-elements.

Exterior product

The exterior product is the fundamental anti-symmetric product operation between the elements of a linear space from which a Grassmann algebra can be constructed.

Force

A *force* is a physical entity represented by a bound vector. This differs from common usage in which a force is represented by a vector. For reasons discussed in the text, common use does not provide a satisfactory model.

Force vector

A *force vector* is the vector of the bound vector representing the force.

Geometrically interpreted 2-element

A *geometrically interpreted 2-element* is the sum of a bound vector $P \wedge x$ and a bivector $\mathbb{B}$: $P \wedge x + \mathbb{B}$

Geometric entities

Points, lines, planes, ... are *geometric entities*. Each is defined as the space of a geometrically interpreted element.

A point is a geometric 1-entity.

A line is a geometric 2-entity.

A plane is a geometric 3-entity.

Geometric interpretations

Points, weighted points, vectors, bound vectors, bivectors, ... are geometric *interpretations* of m-elements.

Geometric representations

Points, weighted points, vectors, bound vectors, bivectors, ... are geometric *representations* of physical entities such as point masses, displacements, velocities, forces, moments.
Physical entities are *represented* by geometrically interpreted elements.

Geometrically interpreted algebra

A *geometrically interpreted algebra* is a Grassmann algebra with a geometrically interpreted underlying linear space.

Grade

The *grade* of an m-element is m.
The grade of a simple m-element is the number of 1-element factors in it.
The grade of the exterior product of an m-element and a k-element is $m+k$.
The grade of the regressive product of an m-element and a k-element is $m+k-n$.
The grade of the complement of an m-element is $n-m$.
The grade of the interior product of an m-element and a k-element is $m-k$.
The grade of a scalar is zero.
The grade of a point or vector is 1.
The grade of a bound vector or bivector is 2.
(The dimension n is the dimension of the underlying linear space.)

GrassmannAlgebra

The concatenated italicized term *GrassmannAlgebra* refers to the *Mathematica* software package available from the *GrassmannAlgebra* website. See the Preface.

A Grassmann algebra

A *Grassmann algebra* is the direct sum of an underlying linear space $\underset{1}{\Lambda}$, its field $\underset{0}{\Lambda}$, and the exterior linear spaces $\underset{m}{\Lambda}$ $(2 \leq m \leq n)$.

$$\underset{0}{\Lambda} \oplus \underset{1}{\Lambda} \oplus \underset{2}{\Lambda} \oplus \cdots \oplus \underset{m}{\Lambda} \oplus \cdots \oplus \underset{n}{\Lambda}$$

A direct sum of linear spaces is the linear space of sums of elements from the component spaces.

The Grassmann algebra

The Grassmann algebra is used to describe that body of algebraic theory and results based on the *Ausdehnungslehre*, but extended to include more recent results and viewpoints.

Grassmann expression

A *Grassmann expression* is a *Mathematica* expression which *GrassmannAlgebra* can understand as a valid expression involving the currently declared scalar, vector, basis and metric symbols.

Hyperplane

A *hyperplane* is a multiplane whose grade is 1 less than the dimension of the space.

Inner product

The *inner product* of two elements is the term given to their interior product when they are of the same grade.

Interior product

The *interior product* of two elements A and B is denoted $A \ominus B$ and defined as the regressive product of A with the complement of B. If the grade of A is less than the grade of B, their interior product is zero.

Intersection of elements

An *intersection* of two simple elements is any of the congruent elements defined by the intersection of their spaces. The common factor of two simple elements is an intersection.

k-span

A k-span of a simple element is any set of $\binom{m}{k}$ essentially different *k*-elements formed from the *m*-element.

Laplace expansion theorem

The *Laplace expansion theorem* states: If any *r* rows are fixed in a determinant, then the value of the determinant may be obtained as the sum of the products of the minors of *r*th order (corresponding to the fixed rows) by their cofactors.

Line

A *line* is the space of a bound vector. Thus a line consists of all the (perhaps weighted) points on it and all the vectors parallel to it.

Linear space

A *linear space* is a mathematical structure defined by a standard set of axioms. It is often referred to simply as a 'space'.

Line at infinity

The *line at infinity* of the projective geometry of the plane is equivalent to any of the congruent bivectors of the plane.

Magnitude

The *magnitude* of an element is its geometrically interpreted measure. The magnitude of vectors, bivectors, and trivectors are interpreted respectively as lengths, areas and volumes.

Measure

The *measure* of an element A is denoted $|A|$, and is defined as the positive square root of the inner product of the element with itself.

Measure of the basis *n*-element

The *measure of the basis n-element* $\underline{1}$ is denoted $|\underline{1}|$. It is also equal to the positive square root of the determinant of the metric tensor G of the space.

Metric

The *metric* of a space is the correspondence set up between the elements of $\underset{1}{\Lambda}$ and $\underset{n-1}{\Lambda}$, that is between basis elements and their complements. The metric tensor (in a given basis) is the array of the regressive products of all the basis elements with all their complements; or equivalently the array of the scalar products of all the basis elements.

Minor

A *minor* of a matrix A is the determinant (or sometimes matrix) of degree (or order) k formed from A by selecting the elements at the intersection of k distinct columns and k distinct rows.

***m*-direction**

An m-direction is the space of a simple m-vector. It is also therefore the set of all vectors parallel to a given simple m-vector.

***m*-element**

An m-element is a sum of simple m-elements.

m:k-form

An *m:k-form* of a simple element A is a multilinear form based on a factorization of A which is invariant to the factorization.

***m*-plane**

An m-plane is the space of a bound simple m-vector. Thus a plane consists of all the (perhaps weighted) points on it and all the vectors parallel to it.

***m*-vector**

An m-vector is a sum of simple m-vectors.

Multilinear form

A multilinear form is a function of 1-elements which is linear in each of them.

Multiplane

A *multiplane* is a geometric interpretation of a bound simple multivector. It is the generalization of the plane to higher dimensions.

Multivector

The term *multivector* is a generic synonym for an m-vector.

***n*-space**

The term n-space is an alias for the phrase *Grassmann algebra with an underlying linear space of n dimensions*.

A bound n-space is an n-space whose basis has been augmented with an origin point, and hence whose underlying linear space is of n+1 dimensions.

Origin

The *origin* ★O is the geometric interpretation of a specific 1-element as a reference point.

Orientation

Orientation is a relative concept. Two congruent elements are said to be of *opposite orientation* if their congruence factor is negative.

Orthogonal

Two elements $\underset{m}{\alpha}$ and $\underset{k}{\beta}$ ($m \geq k$) are *orthogonal* when their interior product $\underset{m}{\alpha} \ominus \underset{k}{\beta}$ is zero.

Two elements $\underset{m}{\alpha}$ and $\underset{k}{\beta}$ ($m \geq k$) are *totally orthogonal* when the interior product $\underset{m}{\alpha} \ominus \beta$ is zero for all β belonging to $\underset{k}{\beta}$.

Parallel

Two entities are said to be *parallel* if the vector subspace of one is a subspace of the other.
Two vectors are parallel if they are congruent.

Plane

A *plane* is the space of a bound simple bivector. Thus a plane consists of all the (perhaps weighted) points on it and all the vectors parallel to it.

Plane at infinity

The *plane at infinity* of the projective geometry of a bound 3-space is equivalent to any of the congruent trivectors of the vector subspace.

Point

A *point* P is the sum of the origin $\star\mathbb{O}$ and a vector x: $P = \star\mathbb{O} + x$.

Point at infinity

The *point at infinity* of the projective geometry of the line is equivalent to any of the congruent vectors of the line.

Point mass

A *point mass* is a physical interpretation of a weighted point.

Physical entities

Point masses, displacements, velocities, forces, moments, angular velocities, ... are *physical entities*. Each is represented by a geometrically interpreted element.
A point mass, displacement or velocity is a physical 1-entity.
A force, moment or angular velocity is a physical 2-entity.

Projective geometry

Projective geometry is geometry without a metric. It is equivalent to a geometrically interpreted Grassmann algebra without a metric.

Regressive product

The *regressive product* operation is the dual operation to the exterior product operation. Its axioms are the duals of the exterior product axioms.

Scalar

A *scalar* is an element of the field $\underset{0}{\Lambda}$ of the underlying linear space $\underset{1}{\Lambda}$.
A scalar is of grade zero.

Scalar product

The *scalar product* of two elements is the term given to their inner product when they are both 1-elements.

Screw

A *screw* is a geometrically interpreted 2-element in a bound 3-space in which the bivector is orthogonal to the vector of the bound vector.
The bivector is necessarily simple since the vector subspace is three-dimensional.

Sense

The term *sense* is equivalent to the term orientation, but is commonly used only for vectors.
Two congruent vectors are of opposite sense if their congruence factor is negative.

Simple element

A *simple element* is an element which may be expressed as the exterior product of 1-elements.

Simple bivector

A *simple bivector* is the exterior product of two vectors: $x \wedge y$.

Simple *m*-element

A *simple m-element* is the exterior product of *m* 1-elements.

Simple *m*-vector

A *simple m-vector* is the exterior product of *m* vectors.

Space

The term *space* is used in this book with several meanings according to context: 'linear space', 'the space of an element', '*n*-space'. A Grassmann algebra based on an underlying linear space of *n* dimensions is also a linear space of 2^n dimensions, but by abuse of terminology may also in this book be referred to as an *n*-space.

Space of a simple *m*-element

The *space of a simple m-element* A is the set of all 1-elements x such that $A \wedge x = 0$.
The space of a simple *m*-element is a linear space of dimension *m*.

Space of a non-simple *m*-element

The *space of a non-simple m-element* is the union of the spaces of its component simple *m*-elements.

Span

A *span* of a simple *m*-element is any set of *m* independent 1-elements whose exterior product is congruent to it.

Trivector

A *trivector* T is a sum of simple trivectors. A *simple trivector* T is the exterior product of three vectors: $T = x \wedge y \wedge z$.

2-direction

A *2-direction* is the space of a simple bivector. It is therefore the set of all vectors parallel to a given simple bivector.

Underlying linear space

The *underlying linear space* of a Grassmann algebra is the linear space $\underset{1}{\Lambda}$ of 1-elements, which together with the exterior product operation, generates the algebra.

Underlying bound vector space

An *underlying bound vector space* is an underlying linear space whose basis elements are interpreted as an origin point $\star\mathbb{O}$ and vectors. It can be shown that from this basis a second basis can be constructed, all of whose basis elements are points.

The dimension of a bound vector space exceeds the dimension of its vector subspace by 1, due to the inclusion of the origin as an extra basis element.

Union of elements

A *union* of two simple elements is any of the congruent elements which define the union of their spaces.

Unit *n*-element

The *unit n-element* is the identity element for the regressive product. Its measure is also unity for any metric.

Vector

A *vector* is a geometric interpretation of a 1-element.

Vector space

A *vector space* is a linear space whose elements are interpreted as vectors.

Weighted point

A *weighted point* is a scalar multiple a of a point P: a P.

Bibliography

A note on sources to Grassmann's work

The best source for Grassmann's contributions to science is his *Collected Works* [Grassmann 1896] which contain in volume 1 both *Die Ausdehnungslehre von 1844* and *Die Ausdehnungslehre von 1862*, as well as *Geometrische Analyse*, his prizewinning essay fulfilling Leibniz's search for an algebra of geometry. Volume 2 contains papers on geometry, analysis, mechanics and physics, while volume 3 contains *Theorie der Ebbe und Flut*.

Die Ausdehnungslehre von 1862, fully titled: *Die Ausdehnungslehre. Vollständig und in strenger Form* is perhaps Grassmann's most important mathematical work. It comprises two main parts: the first devoted basically to the *Ausdehnungslehre* (212 pages) and the second to the theory of functions (155 pages). The Collected Works edition contains 98 pages of notes and comments. The discussion on the *Ausdehnungslehre* includes chapters on addition and subtraction, products in general, progressive and regressive products, interior products, and applications to geometry. A Euclidean metric is assumed.

Both Grassmann's *Ausdehnungslehre* have been translated into English by Lloyd C Kannenberg. The 1844 version is published as *A New Branch of Mathematics: The Ausdehnungslehre of 1844 and Other Works,* Open Court 1995. The translation contains *Die Ausdehnungslehre von 1844*, *Geometrische Analyse*, selected papers on mathematics and physics, a bibliography of Grassmann's principal works, and extensive editorial notes. The 1862 version is published as *Extension Theory*. It contains work on both the theory of extension and the theory of functions. Particularly useful are the editorial and supplementary notes.

Kannenberg has also translated Giuseppe Peano's *Calcolo geometrico secondo l'Ausdehnungslehre di H. Grassmann* [Peano, 1888] as *Geometric Calculus According to the Ausdehnungslehre of H. Grassmann*.

Apart from these translations, probably the best and most complete exposition on the *Ausdehnungslehre* in English is in Alfred North Whitehead's *A Treatise on Universal Algebra* [Whitehead 1898]. Whitehead saw Grassmann's work as one of the foundation stones on which he hoped to build an algebraic theory which united the several new mathematical systems which emerged during the nineteenth century — the algebra of symbolic logic, Grassmann's theory of extension, quaternions, matrices and the general theory of linear algebras.

The second most complete exposition of the *Ausdehnungslehre* is Henry George Forder's *The Theory of Extension* [Forder 1941]. Forder's interest is mainly in the geometric applications of the theory of extension.

The only other books on Grassmann's algebra written in English during the nineteenth and twentieth centuries are those by Edward Wyllys Hyde, *The Directional Calculus* [Hyde 1890] and *Grassmann's Space Analysis* [Hyde 1906]. They treat the theory of extension in two and three-dimensional geometric contexts and include some applications to statics. Several topics such as Hyde's treatment of screws are original contributions.

Seminal papers on the *Ausdehnungslehre* (and Clifford algebra) can be found in William Kingdon Clifford's collected works *Mathematical Papers* [Clifford 1882], republished in a facsimile edition by Chelsea.

Fortunately for those interested in the evolution of the emerging 'geometric' algebras, *The International Association for Promoting the Study of Quaternions and Allied Systems of Mathe-*

matics published a bibliography [Macfarlane 1913] which, together with supplements to 1913, contains about 2500 articles. This therefore most likely contains all the works on the *Ausdehnungslehre* and related subjects up to 1913.

The only other recent text devoted specifically to Grassmann algebra (to the author's knowledge as of 2001) is Arno Zaddach's *Grassmanns Algebra in der Geometrie*, [Zaddach 1994].

Hermann Grassmann was born in 1809. In 2009 a conference on Grassmann was held in Potsdam, Germany to commemorate his diverse and seminal contributions to science, philology and mathematics. Three volumes emerged from the commemoration - see [Petsche 2009], [Petsche *et al* 2009] and [Petsche *et al* 2011]. This set is an invaluable resource for students of Grassmann's work, particularly the biography [Petsche 2009].

Bibliography

The scope of this bibliography is limited. Grassmann algebra has just too many references in recent times, particularly since the expansion of the web, to entertain any sort of completeness. Nevertheless, up until the end of the twentieth century there were still only a few dozen important references *in English* to the *foundational* aspects of the algebra. Since this book (Volume 1) is limited to discussing these foundations, the bibliography will be limited accordingly: to relevant publications appearing before the year 2000. Other applications of the algebra, to calculus and mechanics, and to Clifford and hypercomplex algebras will be discussed in Volume 2. This bibliography only covers these other applications sparsely to provide context on the related research activity during the period.

Since the advent of the web, it is now possible to find much by a simple search. On the other hand a simple search does not return what it has not found. And this becomes more critical for older references. The bibliography therefore lays more emphasis on earlier works.

Although mechanics is covered in Volume 2 rather than in this volume, the bibliography also includes some obscure works on mechanics using line geometry, screws and motor algebra. These are included here because the results may also be obtained using the bound vectors and bivectors discussed in this volume.

Armstrong H L 1959

'On an alternative definition of the vector product in *n*-dimensional vector analysis'
Matrix and Tensor Quarterly, **IX** no 4, pp 107-110.

The author proposes a definition equivalent to the complement of the exterior product of $n-1$ vectors.

Ball R S 1876

The Theory of Screws: A Study in the Dynamics of a Rigid Body
Dublin

See the note on 'The Directional Theory of Screws' [Hyde 1888].

Ball R S 1900

A Treatise on the Theory of Screws
Cambridge University Press (1900). Reprinted 1998.

A classical work on the theory of screws, containing an annotated bibliography in which Ball refers to the *Ausdehnungslehre* of 1862: "This remarkable work … contains much that is of instruction and interest in connection with the present theory … . Here we have a very general theory which includes screw coordinates as a special case." Ball does not use Grassmann's methods in his treatise.

Barnabei M, Brini A and Rota G-C 1985

'On the Exterior Calculus of Invariant Theory'
Journal of Algebra, **96**, pp 120-160.

Contents are: Introduction, Peano spaces, The join, The meet, Cap-products, The Hodge star operator, Alternative laws, Geometric calculus.

Barton H 1927

'A Modern Presentation of Grassmann's Tensor Analysis'
American Journal of Mathematics, **XLIX**, pp 598-614.

This paper covers similar ground to that of Moore (1926).

Birss R R 1980

'Multivector Analysis I: A Comparison with Tensor Algebra'
Physics Letters, **78A**, No 3, pp 223-226.

'Multivector Analysis II: A Comparison with Vector Algebra'
Physics Letters, **78A**, No 3, pp 227-230.

The comparisons are with reference to crystal physics.

Bourbaki N 1948

Algèbra
Actualités Scientifiques et Industrielles No.1044, Paris

Chapter III treats multilinear algebra.

Bowen R M and Wang C-C 1976

Introduction to Vectors and Tensors
Plenum Press. Two volumes.

This is one of the few contemporary texts on vectors and tensors which relates points and vectors via the explicit introduction of the origin into the calculus (p 254).

Brand L 1947

Vector and Tensor Analysis
Wiley, New York.

Contains a chapter on motor algebra which, according to the author in his preface "… is apparently destined to play an important role in mechanics as well as in line geometry". There is also a chapter on quaternions.

Buchheim A 1884-1886

'On the Theory of Screws in Elliptic Space'
Proceedings of the London Mathematical Society
xiv (1884) pp 83-98, **xvi** (1885) pp 15-27, **xvii** (1886) pp 240-254, **xvii** p 88.

The author writes "My special object is to show that the *Ausdehnungslehre* supplies all the necessary materials for a calculus of screws in elliptic space. Clifford was apparently led to construct his theory of biquaternions by the want of such a calculus, but Grassmann's method seems to afford a simpler and more natural means of expression than biquaternions." (*xiv*, p 90) Later he extends this theory to "... all kinds of space." (*xvi*, p 15)

Burali-Forti C 1897

Introduction à la Géométrie Différentielle suivant la Méthode de H. Grassmann
Gauthier-Villars, Paris

This work covers both algebra and differential geometry in the tradition of the Peano approach to the *Ausdehnungslehre*.

Burali-Forti C and Marcolongo R 1910

Éléments de Calcul Vectoriel
Hermann, Paris

Mostly a treatise on standard vector analysis, but it does contain an appendix (pp 176-198) on the methods of the *Ausdehnungslehre* and some interesting historical notes on the vector calculi and their notations. The authors use the wedge $\wedge$ to denote Gibbs' cross product $\times$ and use the $\times$, initially introduced by Grassmann to denote the scalar or inner product.

Available as a print-on-demand volume.

Cartan É 1922

Leçons sur les Invariants Intégraux
Hermann, Paris

In this work Cartan develops the theory of exterior differential forms, remarking that he has called them '*extérieures*' since they obey Grassmann's rules of '*multiplication extérieures*'.

Cartan É 1938

Leçons sur la Théorie des Spineures
Hermann, Paris

In the introduction Cartan writes "One of the principal aims of this work is to develop the theory of spinors systematically by giving a purely geometrical definition of these mathematical entities: because of this geometrical origin, the matrices used by physicists in quantum mechanics appear of their own accord, and we can grasp the profound origin of the property, possessed by Clifford algebras, of representing rotations in space having any number of dimensions". Contains a short section on multivectors and complements (the French term is *supplément*).

An English translation was published by Hermann of Paris in 1966 as *The Theory of Spinors*, and republished in 1981 by Dover.

Cartan É 1946

Leçons sur la Géométrie des Espaces de Riemann
Gautier-Villars, Paris

This is a classic text (originating in 1925) on the application of the exterior calculus to differential geometry. Cartan begins with a chapter on exterior algebra and its expression in tensor index notation. He uses square brackets to denote the exterior product (as did Grassmann), and a wedge to denote the cross product.

Carvallo M E 1892

'La Méthode de Grassmann'
Nouvelles Annales de Mathématiques, serie 3 **XI**, pp 8-37.

An exposition of some of Grassmann's methods applied to three dimensional geometry following the approach of Peano. It does not treat the interior product.

Chevalley C 1955

The Construction and Study of Certain Important Algebras
Mathematical Society of Japan, Tokyo.

Lectures given at the University of Tokyo on graded, tensor, Clifford and exterior algebras.

Clifford W K 1873

'Preliminary Sketch of Biquaternions'
Proceedings of the London Mathematical Society, **IV**, nos 64 and 65, pp 381-395.

This paper includes an interesting discussion of the geometric nature of mechanical quantities. Clifford adopts the term 'rotor' for the bound vector, and 'motor' for the general sum of rotors. By analogy with the quaternion as a quotient of vectors he defines the biquaternion as a quotient of motors.

It is in this paper also that Clifford introduces the symbol ω with the *formal* property that $\omega^2 = 0$, thus enabling a whole new spectrum of quantities to be defined for representing physical phenomena.

Clifford W K 1878

'Applications of Grassmann's Extensive Algebra'
American Journal of Mathematics Pure and Applied, **I**, pp 350-358.

In this paper Clifford lays the foundations for general Clifford algebras.

Clifford W K 1882

Mathematical Papers
Reprinted by Chelsea Publishing Co, New York (1968).

Of particular interest in addition to his two published papers above are the otherwise unpublished notes:
'Notes on Biquaternions' (~1873)
'Further Note on Biquaternions' (1876)
'On the Classification of Geometric Algebras' (1876)
'On the Theory of Screws in a Space of Constant Positive Curvature' (1876).

Coffin J G 1909

Vector Analysis
Wiley, New York.

This is the second English text in the Gibbs-Heaviside tradition. It contains an appendix comparing the various notations in use at the time, including his view of the Grassmannian notation.

Collins J V 1899-1900

'An elementary Exposition of Grassmann's *Ausdehnungslehre* or Theory of Extension'
American Mathematical Monthly,
6 (1899) pp 193-198, 261-266, 297-301; **7** (1900) pp 31-35, 163-166, 181-187, 207-214, 253-258.

This work follows in summary form the *Ausdehnungslehre* of 1862 as regards general theory but differs in its discussion of applications. It includes applications to geometry and brief applications to linear equations, mechanics and logic.

These have been collected into a slim print-on-demand volume.

Coolidge J L 1940

'Grassmann's Calculus of Extension' in
A History of Geometrical Methods
Oxford University Press, pp 252-257.

This brief treatment of Grassmann's work is characterized by its lack of clarity. The author variously describes an exterior product as "essentially a matrix" and as "a vector perpendicular to the factors" (p 254). And confusion arises between Grassmann's matrix and the division of two exterior products (p 256).

Cox H 1882

'On the Application of Quaternions and Grassmann's *Ausdehnungslehre* to different kinds of Uniform Space'
Cambridge Philosophical Transactions, **XIII** part II, pp 69-143.

The author shows that the exterior product is the multiplication required to describe non-metric geometry, for "it involves no ideas of distance" (p 115). He then discusses exterior, regressive and interior products, applying them to geometry, systems of forces, and linear complexes – using the notation of 1844. In other papers Cox applies the *Ausdehnungslehre* to non-Euclidean geometry (1873) and to the properties of circles (1890).

Crowe M J 1967, 1985

A History of Vector Analysis
Notre Dame 1967. Republished by Dover 1985.

This is the most informative work available on the history of vector analysis from the discovery of the geometric representation of complex numbers to the development of the Gibbs-Heaviside system. Crowe's thesis is that the Gibbs-Heaviside system grew mostly out of quaternions rather than from the *Ausdehnungslehre*. His explanation of Grassmannian concepts is particularly accurate in contradistinction to many who supply a more casual reference.

Dibag I 1974

'Factorization in Exterior Algebras'
Journal of Algebra, **30**, pp 259-262

The author develops necessary and sufficient conditions for an m-element to have a certain number of 1-element factors. He also shows that an $(n-2)$-element in an odd dimensional space always has a 1-element factor.

Dimentberg F M 1965

The Screw Calculus and its Applications in Mechanics
Translated by the Foreign Technology Division of the National Technical Information Service, U.S. Department of Commerce. Document number: FTD-HT-23-1632-67.

Dimentberg uses Clifford's dual number approach to develop screw theory applied to mechanics.

Drew T B 1961

Handbook of Vector and Polyadic Analysis
Reinhold, New York

Tensor analysis in invariant notation. Of particular interest here is a section (p 57) on 'polycross products' – an extension of the (three-dimensional) cross product to polyads.

Efimov N V and Rozendorn E R 1975

Linear Algebra and Multi-Dimensional Geometry
MIR, Moscow

Contains a chapter on multivectors and exterior forms.

Fehr H 1899

Application de la Méthode Vectorielle de Grassmann à la Géométrie Infinitésimale
Georges Carré, Paris

Thesis comprising an initial chapter on exterior algebra as well as standard differential geometry. Available as a print-on-demand volume.

Fleming W H 1965

Functions of Several Variables
Addison-Wesley

Contains a chapter on exterior algebra.

Forder H G 1941

The Calculus of Extension
Cambridge (also reprinted by Chelsea)

This text is one of the most recent of the few devoted to an exposition of Grassmann's methods. It is an extensive work (490 pages) largely using Grassmann's notations and relying primarily on the *Ausdehnungslehre* of 1862. Its application is particularly to geometry including many examples well illustrating the power of the methods, a chapter on forces, screws and linear complexes, and a treatment of the algebra of circles.

Gibbs J W 1886

'On multiple algebra'
Address to the American Association for the Advancement of Science
In *Collected Works*, Gibbs 1928, vol 2.

This paper is probably the most authoritative historical comparison of the different 'vectorial' algebras of the time. Gibbs was obviously very enthusiastic about the *Ausdehnungslehre*, and shows himself here to be one of Grassmann's greatest proponents.

Gibbs J W 1891

'Quaternions and the Ausdehnungslehre'
Nature, **44**, pp 79–82. Also in *Collected Works*, Gibbs 1928.

Gibbs compares Hamilton's Quaternions with Grassmann's *Ausdehnungslehre* and concludes that "... Grassmann's system is of indefinitely greater extension ...". Here he also concludes that to Grassmann must be attributed the discovery of matrices. Gibbs published a further three papers in *Nature* (also in *Collected Works*, Gibbs 1928) on the relationship between quaternions and vector analysis, providing an enlightening insight into the quaternion–vector analysis controversy of the time.

Gibbs J W 1928

The Collected Works of J. Willard Gibbs Ph.D. LL.D.
Two volumes. Longmans, New York.

In part 2 of Volume 2 is reprinted Gibbs' only personal work on vector analysis: *Elements of Vector Analysis, Arranged for the Use of Students of Physics* (1881–1884). This was not published elsewhere.

Grassmann H G 1844

Die lineale Ausdehnungslehre: ein neuer Zweig der Mathematik
Leipzig.

The full title is *Die lineale Ausdehnungslehre: ein neuer Zweig der Mathematik dargestellt und durch Andwendungen auf die übrigen Zweige der Mathematik, wie auch auf die Statik, Mechanik, die Lehre vom Magnetismus und die Krystallonomie erläutert*. This first book on Grassmann's new mathematics is known shortly as *Die Ausdehnungslehre von 1844*. It develops the theory of exterior multiplication and division and regressive exterior multiplication. It does not treat complements or interior products in the way of the *Ausdehnungslehre* of 1862. The original *Die Ausdehnungslehre von 1844* was republished in 1878. The best source to this work is Volume 1 of Grassmann's collected works (1896), of which an English translation has been made by Lloyd C. Kannenberg (1995).

Grassmann H G 1845

'Kurze Übersicht über das Wesen der Ausdehnungslehre'
Archiv der Mathematik und Physik (Grunert's Archiv) **VI**

This is a review written by Grassmann, requested by J A Grunert, of his new book published in 1844.

The review was translated by W W Beman and published as 'A Brief Account of the Essential Features of Grassmann's Extensive Algebra' in the The Analyst, VIII, 1881, pp 96-97, 114-124.

Grassmann H G 1855

'Sur les différents genres de multiplication'
Crelle's Journal, **49**, pp 123-141.

This paper was written to claim of Cauchy priority for a method of solving linear equations.

Grassmann H G 1862

Die Ausdehnungslehre. Vollständig und in strenger Form
Berlin.

This is Grassmann's second attempt to publish his new discoveries in book form. It adopts a substantially different approach to the *Ausdehnungslehre* of 1844, relying more on the theorem–proof approach. The work comprises two main parts: the first on the exterior algebra and the second on the theory of functions. The first part includes chapters on addition and subtraction, products in general, combinatorial products (exterior and regressive), the interior product, and applications to geometry. This is probably Grassmann's most important work. The best source is Volume 1 of Grassmann's collected works (1896), of which an English translation has been made by Lloyd C. Kannenberg (2000). In the collected works edition, the editor Friedrich Engel has appended extensive notes and comments, which Kannenberg has also translated.

Short excerpts were translated by M Kormes and published in *A Source Book in Mathematics* (ed D E Smith), McGraw-Hill, 1929, pp 684-696.

Grassmann H G 1878

'Verwendung der Ausdehnungslehre für die allgemeine Theorie der Polaren und den Zusammenhang algebraischer Gebilde'
Crelle's Journal, **84**, pp 273–283.

This is Grassmann's last paper. It contains, among other material, his most complete discussion on the notion of 'simplicity'.

Grassmann H G 1896–1911

Hermann Grassmanns Gesammelte Mathematische und Physikalische Werke
Teubner, Leipzig. Volume 1 (1896), Volume 2 (1902, 1904), Volume 3 (1911).

Grassmann's complete collected works appeared between 1896 and 1911 under the editorship of Friedrich Engel and with the collaboration of Jakob Lüroth, Eduard Study, Justus Grassmann, Hermann Grassmann jr. and Georg Scheffers. The following is a summary of their contents.

Volume 1
Die lineale Ausdehnungslehre: ein neuer Zweig der Mathematik (1844)
Geometrische Analyse: geknüpft an die von Leibniz erfundene geometrische Charakteristik (1847)
Die Ausdehnungslehre. Vollständig und in strenger Form (1862)

Volume 2
Papers on geometry, analysis, mechanics and physics

Volume 3
Theorie der Ebbe und Flut (1840)
Further papers on mathematical physics.

Parts of Volume 1 (*Die lineale Ausdehnungslehre* (1844) and *Geometrische Analyse* (1847)) together with selected papers on mathematics and physics have been translated into English by Lloyd C Kannenberg and published as *A New Branch of Mathematics* (1995). *Geometrische Analyse* is Grassmann's prize-winning essay fulfilling Leibniz' search for an algebra of geometry. The remainder of Volume 1 (*Die Ausdehnungslehre* (1862)) has been translated into

English by Lloyd C Kannenberg and published as *Extension Theory* (2000).

Volume 2 comprises papers on geometry, analysis, analytical mechanics and mathematical physics plus two texts, one on arithmetic and the other on trigonometry. Available as a print-on-demand volume.

Volume 3 comprises Grassmann's earliest major work (*Theorie der Ebbe und Flut* (1840)) and further papers, particularly on wave theory. The Theory of Tides begins to apply Grassmann's new approach to vector analysis.

Grassmann H der Jüngere 1909, 1913, 1927

Projektive Geometrie der Ebene
Teubner, Leipzig.

A comprehensive treatment in three books using the methods of the elder H. Grassmann.

Grassmann R 1895

Die Ausdehnungslehre oder die Wissenschaft von den extensiven Grösen in strenger Formelentwicklung
Stettin.

This is one of the volumes in a series by Hermann Grassmann's younger brother Robert, who published over 40 books on diverse subjects. See the paper by Ivor Grattan-Guinness in Petsche (2011) for an analysis of Robert's work.

Greenberg M J 1976

'Element of area via exterior algebra'
American Mathematical Monthly, **83**, pp 274-275.

The author suggests that the treatment of elements of area by using the exterior product would be a more satisfactory treatment than that normally given in calculus texts.

Greub W H 1967

Multilinear Algebra
Springer-Verlag, Berlin.

Contains chapters on exterior algebra.

Gurevich G B 1964

Foundations of the Theory of Algebraic Invariants
Noordhoff, Groningen

Contains a chapter on *m*-vectors (called here polyvectors) with an extensive treatment of the conditions for divisibility of an *m*-vector by one or more vectors (pp 354-395).

Hamilton W R 1853

Lectures on Quaternions
Dublin

The first English text on quaternions. The introduction briefly mentions Grassmann.

Hamilton W R 1866

Elements of Quaternions
Dublin

The editor Charles Joly in his preface remarks in relation to the quaternion's associativity that "For example, Grassmann's multiplication is sometimes associative, but sometimes it is not". The exterior product is of course associative. However, here it seems Joly may be suffering from a confusion caused by Grassmann's notation for expressions in which both exterior and regressive products appear.
The second edition of 1899 has been reprinted in 1969 by Chelsea Publishing Company.

Hardy A S 1895

Elements of Quaternions
Ginn and Company, Boston

The author claims this to be an introduction to quaternions at an elementary level.

Heath A E 1917

'Hermann Grassmann 1809-1877'
The Monist, **27**, pp 1-21.

'The Neglect of the Work of H. Grassmann'
The Monist, **27**, pp 22-35.

'The Geometric Analysis of Hermann Grassmann and its connection with Leibniz's characteristic'
The Monist, **27**, pp 36-56.

Hestenes D 1966

Space–Time Algebra
Gordon and Breach

This work is a seminal exposition of Clifford algebra emphasizing the geometric nature of the quantities and operations involved. The author writes "... ideas of Grassmann are used to motivate the construction of Clifford algebra and to provide a geometric interpretation of Clifford numbers. This is to be contrasted with other treatments of Clifford algebra which are for the most part formal algebra. By insisting on Grassmann's geometric viewpoint, we are led to look upon the Dirac algebra with new eyes." (p 2).

Hestenes D 1968

'Multivector Calculus'
Journal of Mathematical Analysis and Applications, **24**, pp 313-325.

In the words of the author: "The object of this paper is to show how differential and integral calculus in many dimensions can be greatly simplified using Clifford algebra."

Hestenes D 1999

New Foundations for Classical Mechanics (Second Edition)
Kluwer

In the words of the author: "This is a textbook on classical mechanics at the intermediate level, but its main purpose is to serve as an introduction to a new mathematical language for physics called *geometric algebra*." Geometric algebra and Grassmann algebra are intimately related via the Clifford product.

Hodge W V D 1952

Theory and Applications of Harmonic Integrals
Cambridge

The 'star' operator defined by Hodge has a precursor of similar nature in Grassmann's complement operator.

Hunt K H 1970

Screw systems in Spatial Kinematics (Screw Systems Surveyed and Applied to Jointed Rigid Bodies)
Report MMERS 3, Department of Mechanical Engineering, Monash University, Clayton, Australia

Although in this report the author has intentionally confined himself to well known methods of pure and analytical geometry, he is a strong proponent of the screw as being the natural language for investigating spacial mechanisms.

Hyde E W 1884

'Calculus of Direction and Position'
American Journal of Mathematics, **VI**, pp 1-13.

In this paper the author compares quaternions to the methods of the *Ausdehnungslehre* and concludes that Grassmann's system is far preferable as a system of directed quantities.

Hyde E W 1888

'Geometric Division of Non-congruent Quantities'
Annals of Mathematics, **4**, pp 9-18.

This paper deals with the concept of exterior division more extensively than did Grassmann.

Hyde E W 1888

'The Directional Theory of Screws'
Annals of Mathematics, **4**, pp 137-155.

This paper is an account of the theory of screws using the *Ausdehnungslehre*. Hyde claims that "A screw evidently belongs thoroughly to the realm of the Directional Calculus and will not be easily or naturally treated by Cartesian methods; and Ball's treatment is throughout essentially Cartesian in its nature." Here he is referring to *The Theory of Screws: A Study in the Dynamics of a Rigid Body* (1876). In *A Treatise on the Theory of Screws* (1900) Ball comments: "Prof. Hyde proves by his [sic] calculus many of the fundamental theorems in the present theory in a very concise manner." (p 531).

Hyde E W 1890

The Directional Calculus based upon the Methods of Hermann Grassmann
Ginn and Company, Boston

The author discusses geometric applications in 2 and 3 dimensions including screws and complements of bound elements (for example, points, lines and planes).

Hyde E W 1906

Grassmann's Space Analysis
Wiley, New York

In the words of the author: "This little book is an attempt to present simply and concisely the principles of the 'Extensive Analysis' as fully developed in the comprehensive treatises of Hermann Grassmann, restricting the treatment however to the geometry of two and three dimensional space."

Jahnke E 1905

Vorlesungen über die Vektorenrechnung mit Anwendungen auf Geometrie, Mechanik und Mathematische Physik.
Teubner, Leipzig

This work is full of examples of application of the *Ausdehnungslehre* (notation of 1862) to geometry, mechanics and physics. Only two and three dimensional problems are considered.

Kálnay A J 1976

'A Note on Grassmann Algebras'
Reports on Mathematical Physics, **9**, pp 9-13.

A report on two applications to physics.

Klein F 1908

Elementary Mathematics from an Advanced Standpoint (Volume 2) Geometry
Translated from the third German edition by E R Hedrick and C A Noble.
Dover 1939

Klein begins his discussion of geometric manifolds using Grassmannian concepts. But since he found the style of Grassmann's earlier work (1844) "extraordinarily obscure", he introduces the reader to them using determinant notation.

Lasker E 1896

'An Essay on the Geometrical Calculus'
Proceedings of the London Mathematical Society, **XXVIII**, pp 217-260.

This work differs from most of the other papers of this era on the geometrical applications of the *Ausdehnungslehre* by concentrating on a space of arbitrary dimension rather than two or three.

Lewis G N 1910

'On four-dimensional vector calculus and its application in electrical theory'
Proceedings of the American Academy of Arts and Sciences, **XLVI**, pp 165-181.

A specialization of the *Ausdehnungslehre* to four dimensions and its applications to electromagnetism in Minkowskian terms. The author introduces the new concepts with the minimum of explanation: for example, the anti-symmetric properties of bivectors and trivectors are justified as conventions! (p 167).

Lotze A 1922

Die Grundgleichungen der Mechanik insbesondere Starrer Körper

Teubner, Leipzig

This short monograph is one of the rare works addressing mechanics using the methods of the *Ausdehnungslehre*. It treats the dynamics of systems of particles and the kinematics and dynamics of rigid bodies.

Lounesto P 1997

Clifford Algebras and Spinors
Cambridge

Contains a preparatory chapter on bivectors and exterior algebra.

Macfarlane A 1904

Bibliography of Quaternions and Allied Systems of Mathematics
Dublin

Published for the *International Association for Promoting the Study of Quaternions and Allied Systems of Mathematics*, this bibliography together with supplements to 1913 contains about 2500 articles including many on the *Ausdehnungslehre* and vector analysis.

Marcus M 1966

'The Cauchy-Schwarz inequality in the exterior algebra'
Quarterly Journal of Mathematics, **17**, pp 61-63.

The author shows that a classical inequality for positive definite hermitian matrices is a special case of the Cauchy-Schwarz inequality in the appropriate exterior algebra.

Marcus M 1975

Finite Dimensional Multilinear Algebra
Marcel Dekker, New York

Part II contains chapters on Grassmann and Clifford algebras.

Marcus M and Robinson H 1975

'A Note on the Hodge Star Operator'
Linear Algebra and its Applications, **10**, pp 85-87.

In the author's words: "The usual proof that the Hodge star operator on the Grassmann algebra is independent of the orthonormal basis used to define it requires a decomposition of the operator into a product of three maps, each of which is independent of the basis. The present note contains a very short proof of the result which depends on simple properties of the multiplication and induced inner product in the Grassmann algebra."

Marcus M 1978

'An Inequality for Non-decomposable Elements in the Grassmann Algebra'
Houston Journal of Mathematics, **4**, No 3, pp 417-422.

Here the term 'decomposable' is equivalent to the term 'simple' used in this book. (See also Soule, 1979)

Massey W S 1983

'Cross Products of Vectors in Higher Dimensional Euclidean Spaces'

The American Mathematical Monthly, **90**, No 10, pp 697-701.

The author shows that "... a cross product of vectors exists only in 3-dimensional and 7-dimensional space."

Mehmke R 1880

Andwendung Der Grassmann'schen Ausdehnungslehre Auf Die Geometrie Der Kreise In Ebene
Stuttgart

This is probably the first application of Grassmannian methods to the geometry of circles. Delivered as his inaugural dissertation at the University of Tübingen. Available as a print-on-demand publication.

Mehmke R 1913

Vorlesungen über Punkt- und Vektorenrechnung (2 volumes)
Teubner, Leipzig

Volume 1 (394 pages) deals with the analysis of bound elements (points, lines and planes) and projective geometry.

Milne E A 1948

Vectorial Mechanics
Methuen, London

An exposition of three-dimensional vectorial (and tensorial) mechanics using 'invariant' notation. Milne is one of the rare authors who realizes that physical forces and the linear momenta of particles are better modelled by line vectors (bound vectors) rather than by the usual vector algebra. He treats systems of line vectors (bound vectors) as vector pairs.

Moore C L E 1926

'Grassmannian Geometry in Riemannian Space'
Journal of Mathematics and Physics, **5**, pp 191-200.

This paper treats the complement, and the exterior, regressive and interior products in a Riemannian space in tensor index notation using the alternating tensors and the generalized Kronecker symbol. This classic use of tensor notation does not enhance the readability of the exposition.

Murnaghan F D 1925

'The Generalised Kronecker Symbol and its Application to the Theory of Determinants'
American Mathematical Monthly, **32**, pp 233-241.

The generalized Kronecker symbol is essentially the generalization to an exterior product space of the usual Kronecker symbol.

Murnaghan F D 1925

'The Tensor Character of the Generalised Kronecker Symbol'
Bulletin of the American Mathematical Society, **31**, pp 323-329.

The author states "It will be readily recognized that there is an intimate connection here with Grassmann's *Ausdehnungslehre*, and we believe, in fact, that a systematic exposition of this theory with the aid of the generalized Kronecker symbol would help to make it more widely understood."

Park D 2010

'A precedence operator for the *Mathematica* implementation of *GrassmannAlgebra*'
'Documentation of the *GrassmannAlgebra* application using *WolframWorkbench*'
Private communications.

The results of this work have been incorporated into the *GrassmannAlgebra* application. David Park has provided a number of resources for the *Mathematica* community. They may be found at http://home.comcast.net/~djmpark/Mathematica.html

Peano G 1888

Calcolo geometrico secondo l'Ausdehnungslehre di H. Grassmann
Fratelli Bocca, Torino

This work has been translated into English by Lloyd C Kannenberg and published in 2000 by Birkhauser under the title *Geometric Calculus According to the Ausdehnungslehre of H, Grassmann*.

Peano G 1895

'Essay on Geometrical Calculus'
in *Selected Works of Guiseppe Peano* Chapter XV, p 169-188.
Allen and Unwin, London

The essay is translated from 'Saggio di calcolo geometrico' *Atti, Accad. Sci. Torino*, **31**, (1895-6) pp 952-975. Peano claims to have understood Grassmann's ideas by reconstructing them himself. His geometric ideas come through with clarity, substantiating his claim. Peano's principle exposition of Grassmann's work was in *Calcolo geometrico secondo l'Ausdehnunglehre di H. Grassmann*, Turin (1888) of which a translation of selected passages appears on pp 90-100 of the above *Selected Works*.

Peano G 1901

Formulaire de Mathématiques
Gauthier-Villars, Paris

A compendium of axioms and results. The last part (pp 192-209) is devoted to point and vector spaces and includes some interesting historical comments.

Pedoe D 1967

'On a geometrical theorem in exterior algebra'
Canadian Journal of Mathematics, **19**, pp 1187-1191.

The author remarks "This paper owes its inspiration to the remarkable book by H. G. Forder *The Calculus of Extension* ... Forder introduces many concepts which I find difficult to bring down to earth. But the methods developed in his book are powerful ones, and it is evident that much work can usefully be done in simplifying and interpreting some of the concepts he uses." He does not mention Grassmann.

Petsche H-J 2009

Hermann Graßmann Biography
Birkhäuser, Berlin

Translated by Mark Minnes. The first volume of a three volume set commemorating the 200th anniversary of Grassmann's birth in 1809.

Petsche H-J, Kanneberg L, Keßler G, and Liskowacka J (eds) 2009

Hermann Graßmann Roots and Traces
Birkhäuser, Berlin

Subtitled: "Autographs and Unknown Documents". The second volume of a three volume set commemorating the 200th anniversary of Grassmann's birth in 1809.

Petsche H-J, Lewis A C, Liesen J, and Russ S (eds) 2011

Hermann Graßmann From Past to Future: Graßmann's Work in Context
Graßmann Bicentennial Conference, September 2009
Birkhäuser, Berlin

The third volume of a three volume set commemorating the 200th anniversary of Grassmann's birth in 1809. These are the papers from the conference.

Saddler W 1927

'Apolar triads on a cubic curve'
Proceedings of the London Mathematical Society, Series 2, **26**, pp 249-256.

'Apolar tetrads on the Grassmann quartic surface'
Journal of the London Mathematical Society, **2**, pp 185-189.

Exterior algebra applied to geometric construction.

Sain M 1976

'The Growing Algebraic Presence in Systems Engineering: An Introduction'
Proceedings of the IEEE, **64**, p 96.

A modern algebraic discussion culminating in the definition of the exterior algebra and its relationship to the theory of determinants and some systems theoretical applications.

Schlegel V 1872, 1875

System der Raumlehre (2 volumes)
Leipzig

Geometry using Grassmann's methods. Available as print-on-demand volumes.

Schouten J A 1951

Tensor Analysis for Physicists
Oxford

Contains a chapter on *m*-vectors from a tensor-analytic viewpoint.

Schubring G (ed) 1996

Hermann Günther Graßmann (1809-1877): Visionary Mathematician, Scientist and Neohumanist Scholar
Kluwer

Papers from a sesquicentennial conference on 150 years of Grassmann's work, held in 1994 in Lieschow near his birthplace Stettin (Szczecin).

Schweitzer A R 1950

Bulletin of the American Mathematical Society, **56**
'Grassmann's extensive algebra and modern number theory' (Part I, p 355; Part II, p 458)
'On the place of the algebraic equation in Grassmann's extensive algebra' (p 459)
'On the derivation of the regressive product in Grassmann's geometrical calculus' (p 463)
'A metric generalisation of Grassmann's geometric calculus' (p 464)

Resumés only of these papers are printed.

Scott R F 1880

A Treatise on the Theory of Determinants
Cambridge, London

The author states in the preface that "The principal novelty in the treatise lies in its systematic use of Grassmann's alternate units, by means of which the study of determinants is, I believe, much simplified."

Shepard G C 1966

Vector Spaces of Finite Dimension
Oliver and Boyd, London

Chapter IV contains an introduction to exterior products via tensors and multilinear algebra.

Soule G W 1979

'An Inequality in the Grassmann Algebra'
Houston Journal of Mathematics, **5**, No 2, pp 269-275.

Here the term 'decomposable' is equivalent to the term 'simple' used in this book. (See also Marcus, 1978)

Thomas J M 1962

Systems and Roots
W Byrd Press

The author uses exterior algebraic concepts in some of his network analysis.

Tonti E 1972

Accademia Nazionale die Lincei, Serie VIII, Volume L II
'On the Mathematical Structure of a Large Class of Physical Theories' (Fasc.1, p 48)
'A Mathematical Model for Physical Theories' (Fasc.2-3, p 176, 351)

These papers begin the author's investigations into the structure of physical theories, in which he considers the geometrical calculus to play an important part.

Tonti E 1975

On the Formal Structure of Physical Theories
Report of the *Instituto di Matematica del Politecnico di Milano*

In this report the author constructs a classification scheme for physical quantities and the equations of physical theories. The mathematical structures needed for this, and which are reviewed in this report are algebraic topology, exterior algebra, exterior differential forms, and Clifford algebra. The author shows that the underlying structure of physical theories is basically capable of a geometric interpretation.

von Mises R 1924

Motor Calculus A new theoretical device for mechanics
Translated by E J Baker and K Wohlhart 1996
Institute for Mechanics, University of Technology Graz

A translation of two long articles. The term 'motor' was introduced by Clifford to designate a screw with an associated magnitude. But von Mises eschews the use of Clifford's duality unit and instead uses the concept of ordered line pairs. He does not use Grassmannian methods.

Whitehead A N 1898

A Treatise on Universal Algebra (Volume 1)
Cambridge

No further volumes appeared. This is probably the best and most complete exposition of Grassmann's works in English (586 pages). The author recreates many of Grassmann's results, in many cases extending and clarifying them with original contributions. Whitehead considers non-Euclidean metrics and spaces of arbitrary dimension. However, like Grassmann, he does not distinguish between n-elements and scalars.

Willmore T J 1959

Introduction to Differential Geometry
Oxford

Includes a brief discussion of exterior algebra and its application to differential geometry (p 189).

Wilson E B 1901

Vector Analysis
New York

The first formally published book entirely devoted to presenting the Gibbs-Heaviside system of vector analysis based on Gibbs' lectures and Heaviside's papers in the *Electrician* in 1893. (Wilson uses the term 'bivector', but by it means a vector with real and imaginary parts.)

Wilson E B and Lewis G N 1912

'The space-time manifold of relativity. The non-Euclidean geometry of mechanics and electromagnetics'
Proceedings of the American Academy of Arts and Sciences, **XLVIII**, pp 387-507.

This treatise uses a four-dimensional vector calculus developed by Lewis (see Lewis G N) by specializing the exterior calculus to four dimensions (with the scalar products of time-like vectors negative). This is a good example of the power of the exterior calculus.

Woo L and Freudenstein F 1969

Application of Line Geometry to Theoretical Kinematics and the Kinematic Analysis of Mechanical Systms
New York Scientific Center Technical Report 320-2982

This approach to mechanics uses line geometry. It does not use Grassmannian methods.

Zaddach A 1994

Grassmanns Algebra in der Geometrie
BI-Wissenschaftsverlag

In German only as of 2001.

Ziwet A 1885-6

'A Brief Account of H. Grassmann's Geometrical Theories'
Annals of Mathematics, **2**, (1885 pp 1-11; 1886 pp 25-34).

In the words of the author: "It is the object of the present paper to give in the simplest form possible, a succinct account of Grassmann's mathematical theories and methods in their application to plane geometry." (Follows in the main Schlegel's *System der Raumlehre*.)

Index

The ellipsis … means 'and most pages in between'.
For very common terms, only a few beginning references are given.

A

B

C

⊕ CirclePlus	73									
⊗ CircleTimes	73									
Clifford	1	3	7	11	48	433	519	539	540	543
Clifford algebra	539	549	556							
closest approach of multiplanes			504 … 513							
cobasis	37	52 …	60	123 … 125	294	298	304	370	522	
CobasisPalette	54	299	308							
cofactor	58	59	300	449	530					
cometric	307 … 309		530							
common factor	15	17	22	88	106	110	120	138	174	196
Common Factor Axiom	15	16	27	103	104 … 114	123	262	444		
Common Factor Theorem	16	110 … 126	141	156	227	263	371	388		
complement	23 … 35	105	289 … 301	315 … 325	386	432	442			
complement of a complement		26	292	362	371	381				
complement operation	24 … 27	291	323	377	378	385	443	514		
ComplementPalette	316	317	331	332 … 335	342	344	350 … 353	357		
complete span	78 … 81	147	149	151	414	531				
compose	41	50	51	54	70	71	164	171	182	305
ComposeBasisForm	118 … 120	122	522							
★$\mathbb{B}_\Box$ ComposeBivector64	109	171								
ComposeBoundVector	171									
★$\mathbb{L}_\Box$ ComposeLineElement	191									
★$\mathbb{P}_\Box$ ComposePoint	190	193								
★$\mathbb{W}_\Box$ ComposeSimpleBivector	171									
★$\mathbb{V}_\Box$ ComposeVector	164									
compound matrix	305									
computable formula	149	151	416	418						
congruence	14	21	23	42	85	97	118	159	160	168
★C CongruenceSimplify	230	236	258	261	263	264	270	273		
conic	254 … 279									
construction	254 … 287									
ConvertComplements	325	333 … 346	356	431						
coordinate	183	190 … 206	231	265	303	344	359			
coordinate-free	193	196								
cospan	38	70 … 80	88 … 90	111 … 114	149	189	532			
1-cospan	112									
2-cospan	70	90								
k-cospan	70 … 74	79	495	496						
Cramer's rule	61									
cross product	6	9	33	34	440 … 444	466	511	531		
cubic	255	269 … 271								

D

★$\mathcal{B}_\Box$ DeclareBasis	49 … 54	62	163	164	326	332	451		
★★v DeclareExtraScalarSymbols	71	196	208						
★★s DeclareExtraVectorSymbols	167	261	264						
DeclareMetric	299	304 … 310	326 … 342	345	350	353	407	408	

E

F

Flatten	50	184	297	311	314	315	496	
force	5	169 ... 175		532	536			
force vector	532							
Forder	1	39	183	290	539	545	554	
m:k-form	74 ... 78	81	86	487	495 ... 497	535		
free	5	11	157	189	347	403	404	435	437	527

G

General Product Formula	143	149 ... 154	423	476	477			
geometric algebra	5	550						
geometric dependence	184							
geometric duality	184							
geometric interpretation	10	11	17	30	158	181 ... 183	204	527
Gibbs	1	3	157	287	347	440	519	
grade 11 12	26 ... 33	41	45	48	291	372 ... 378	514	533
Grassmann	1	4	24	39	57	100	105	495	516 ... 519	539
Grassmann algebra	1 ... 8	42	46	50	289	524 ... 526	530 ... 537
GrassmannAlgebra	1 ... 3	38 ... 43	47 ... 52	520 ... 521	525 ... 527
GrassmannComplement	325	330	331	335	336			
GrassmannBases	79	334	335					
★ℰ GrassmannExpand	43	144						
★𝒢 GrassmannExpandAndSimplify	43	46	109	150	177	194	417	479
Grassmann expression	47	48	74	118	330	336	356	533
★𝒮 GrassmannSimplify	43	47	165	208	279	337	345	496

H

half-plane	238	244						
Hamilton	19	163	287	518				
Hestenes	549							
hexagon	279 ... 281	286	287					
HidePrecedence	144	384	385					
Hodge	290	541	550	552				
homogeneous coordinates	219	220						
hybrid metric	189	347 ... 349	361	362	401 ... 403	433	434	454
Hyde	1	183	290	539	540	550	551	
hypercomplex	3	7	48	73	74	373	433	515	540
hyperplane	185	226 ... 233	253	255	256	498	533	

I

independence	2	6	41	...	43	49	61	110	139	193	525
infix	40	73						
inner product	29	289	323	372 ... 374	393 ... 396	406	533	534
Interior Common Factor Theorem	32	379	387 ... 390	394	402	409	488

K

L

M

N

O

P

Q

R

S

W

www.ingramcontent.com/pod-product-compliance
Lightning Source LLC
Chambersburg PA
CBHW061009200526
45171CB00009B/402
* 9 7 8 1 4 7 9 1 9 7 6 3 7 *